Chemical Dynamics in Condensed Phases

OXFORD
UNIVERSITY PRESS

Great Clarendon Street, Oxford OX2 6DP,
United Kingdom

Oxford University Press is a department of the University of Oxford.
It furthers the Universitys objective of excellence in research, scholarship,
and education by publishing worldwide. Oxford is a registered trade mark of
Oxford University Press in the UK and in certain other countries

First published 2006
First published in paperback 2013

Impression: 1

Published in the United States of America by Oxford University
Press 198 Madison Avenue, New York, NY 10016,
United States of America

British Library Cataloguing in Publication Data
Data available

ISBN 978–0–19–852979–8 (hbk.)
ISBN 978–0–19–968668–1 (pbk.)

Printed and bound by CPI Group (UK) Ltd, Croydon, CR0 4YY

To Raya, for her love, patience and understanding

PREFACE

In the past half century we have seen an explosive growth in the study of chemical reaction dynamics, spurred by advances in both experimental and theoretical techniques. Chemical processes are now measured on timescales as long as many years and as short as several femtoseconds, and in environments ranging from high vacuum isolated encounters to condensed phases at elevated pressures. This large variety of conditions has lead to the evolution of two branches of theoretical studies. On one hand, "bare" chemical reactions involving isolated molecular species are studied with regard to the effect of initial conditions and of molecular parameters associated with the relevant potential surface(s). On the other, the study of chemical reactions in high-pressure gases and in condensed phases is strongly associated with the issue of environmental effects. Here the bare chemical process is assumed to be well understood, and the focus is on the way it is modified by the interaction with the environment.

It is important to realize that not only does the solvent environment modify the equilibrium properties and the dynamics of the chemical process, it often changes the nature of the process and therefore the questions we ask about it. The principal object in a bimolecular gas phase reaction is the collision process between the molecules involved. In studying such processes we focus on the relation between the final states of the products and the initial states of the reactants, averaging over the latter when needed. Questions of interest include energy flow between different degrees of freedom, mode selectivity, and yields of different channels. Such questions could be asked also in condensed phase reactions, however, in most circumstances the associated observable cannot be directly monitored. Instead questions concerning the effect of solvent dynamics on the reaction process and the inter-relations between reaction dynamics and solvation, diffusion and heat transport become central.

As a particular example consider photodissociation of iodine $I_2 \leftrightarrow I + I$ that was studied by many authors in the past 70 years.[1] In the gas phase, following optical excitation at wavelength $\sim$500 nm the I_2 molecule dissociates and this is the end of the story as far as we are concerned. In solutions the process is much more complex. The molecular absorption at $\sim$500 nm is first bleached (evidence of depletion of ground state molecules) but recovers after 100–200 ps. Also some transient state

[1] For a review see A. L. Harris, J. K. Brown, and C. B. Harris, Ann. Rev. Phys. Chem. **39**, 341 (1988).

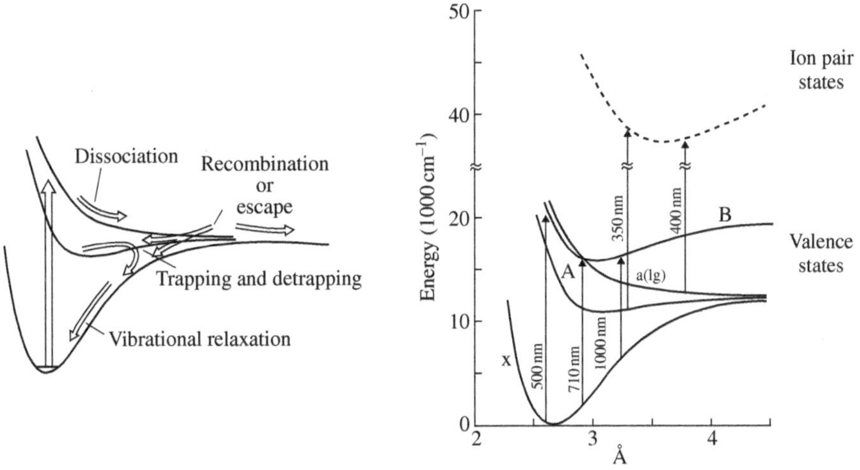

FIG. 0.1 A simplified energy level diagram for I_2 (right), with the processes discussed in the text (left). (Based on Harris et al. (see footnote 1).)

which absorbs at ~350 nm seems to be formed. Its lifetime strongly depends on the solvent (60 ps in alkane solvents, 2700 ps (=2.7 ns) in CCl_4). Transient IR absorption is also observed and can be assigned to two intermediate species. These observations can be interpreted in terms of the schematic potential energy diagram shown in Fig. 0.1 which depicts several electronic states: The ground state X, bound excited states A and B and a repulsive state that correlates with the ground state of the dissociated species. A highly excited state corresponding to the ionic configuration $\mathbf{I^+I^-}$ is also shown. Note that the energy of the latter will be very sensitive to the solvent polarity. Also note that these are just a few representative electronic states of the I_2 molecule. The ground state absorption, which peaks at 500 nm, corresponds to the X→B transition, which in the low-pressure gas phase leads to molecular dissociation after crossing to the repulsive state. In solution the dissociated pair finds itself in a solvent cage, with a finite probability to recombine. This recombination yields an iodine molecule in the excited A state or in the higher vibrational levels of the ground X states. These are the intermediates that give rise to the transient absorption signals.

Several solvent induced relaxation processes are involved in this process: Diffusion, trapping, geminate recombination, and vibrational relaxation. In addition, the A→X transition represents the important class of nonadiabatic reactions, here induced by the solute–solvent interaction. Furthermore, the interaction between the molecular species and the radiation field, used to initiate and to monitor the process, is modified by the solvent environment. Other important solvent induced processes: Diffusion controlled reactions, charge (electron, proton) transfer, solvation dynamics, barrier crossing and more, play important roles in other condensed phase chemical dynamics phenomena.

In modeling such processes our general strategy is to include, to the largest extent possible, the influence of the environment in the dynamical description of the system, while avoiding, as much as possible, a specific description of the environment itself. On the most elementary level this strategy results in the appearance of phenomenological coefficients, for example dielectric constants, in the forces that enter the equations of motion. In other cases the equations of motions are modified more drastically, for example, replacing the fundamental Newton equations by the phenomenological diffusion law. On more microscopic levels we use tools such as coarse graining, projections, and stochastic equations of motion.

How much about the environment do we need to know? The answer to this question depends on the process under study and on the nature of the knowledge required about this process. A student can go through a full course of chemical kinetics without ever bringing out the solvent as a participant in the game—all that is needed is a set of rate coefficients (sometimes called "constants"). When we start asking questions about the origin of these coefficients and investigate their dependence on the nature of the solvent and on external parameters such as temperature and pressure, then some knowledge of the environment becomes essential.

Timescales are a principle issue in deciding this matter. In fact, the need for more microscopic theories arises from our ability to follow processes on shorter timescales. To see how time becomes of essence consider the example shown in Fig. 0.2 that depicts a dog trying to engage a hamburger. In order to do so it has to go across a barrier that is made of steps of the following property: When you stand on a step for more than 1 s the following step drops to the level on which you stand. The (hungry) dog moves at constant speed but if it runs too fast he will spend less than one second on each step and will have to work hard to climb the barrier. On the other hand, moving slowly enough it will find itself walking effortlessly through a plane.

In this example, the 1 second timescale represents the characteristic relaxation time of the environment—here the barrier. The dog experiences, when it moves

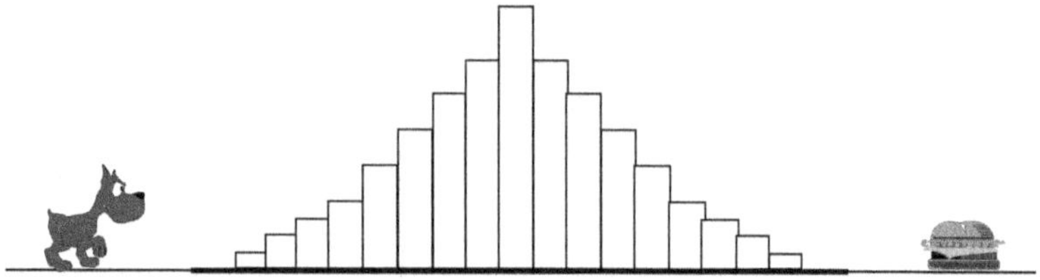

FIG. 0.2 The hamburger–dog dilemma as a lesson in the importance of timescales.

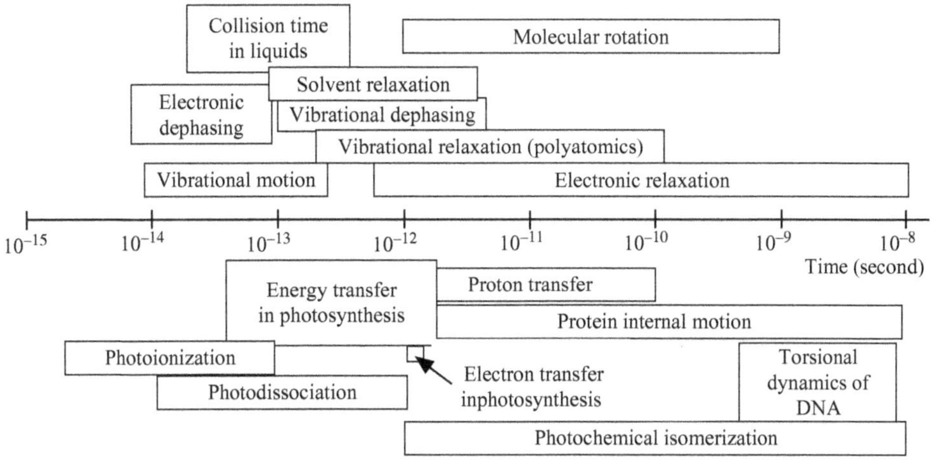

FIG. 0.3 Typical condensed phase molecular timescales in chemistry and biology. (Adapted from G. R. Fleming and P. G. Wolynes, *Physics Today*, p. 36, May 1990).

slowly or quickly relative to this timescale, very different interactions with this environment. A major theme in the study of molecular processes in condensed phases is to gauge the characteristic molecular times with characteristic times of the environment. Some important molecular processes and their characteristic times are shown in Fig. 0.3.

The study of chemical dynamics in condensed phases therefore requires the understanding of solids, liquids, high-pressure gases, and interfaces between them, as well as of radiation–matter interaction, relaxation and transport processes in these environments. Obviously such a broad range of subjects cannot be treated comprehensively in any single text. Instead, I have undertaken to present several selected prototype processes in depth, together with enough coverage of the necessary background to make this book self contained. The reader will be directed to other available texts for more thorough coverage of background subjects.

The subjects covered by this text fall into three categories. The first five chapters provide background material in quantum dynamics, radiation–matter interaction, solids and liquids. Many readers will already have this background, but it is my experience that many others will find at least part of it useful. Chapters 6–12 cover mainly methodologies although some applications are brought in as examples. In terms of methodologies this is an intermediate level text, covering needed subjects from nonequilibrium statistical mechanics in the classical and quantum regime as well as needed elements from the theory of stochastic processes, however, without going into advanced subjects such as path integrals, Liouville-space Green functions or Keldysh nonequilibrium Green functions.

The third part of this text focuses on several important dynamical processes in condensed phase molecular systems. These are vibrational relaxation (Chapter 13), Chemical reactions in the barrier controlled and diffusion controlled regimes (Chapter 14), solvation dynamics in dielectric environments (Chapter 15), electron transfer in bulk (Chapter 16), and interfacial (Chapter 17) systems and spectroscopy (Chapter 18). These subjects pertain to theoretical and experimental developments of the last half century; some such as single molecule spectroscopy and molecular conduction—of the last decade.

I have used this material in graduate teaching in several ways. Chapters 2 and 9 are parts of my core course in quantum dynamics. Chapters 6–12 constitute the bulk of my course on nonequilibrium statistical mechanics and its applications. Increasingly over the last 15 years I have been using selected parts of Chapters 6–12 with parts from Chapters 13 to 18 in the course "Chemical Dynamics in Condensed Phases" that I taught at Tel Aviv and Northwestern Universities.

A text of this nature is characterized not only by what it includes but also by what it does not, and many important phenomena belonging to this vast field were left out in order to make this book-project finite in length and time. Proton transfer, diffusion in restricted geometries and electromagnetic interactions involving molecules at interfaces are a few examples. The subject of numerical simulations, an important tool in the arsenal of methodologies, is not covered as an independent topic, however, a few specific applications are discussed in the different chapters of Part 3.

ACKNOWLEDGEMENTS

In the course of writing the manuscript I have sought and received information, advice and suggestions from many colleagues. I wish to thank Yoel Calev, Graham Fleming, Michael Galperin, Eliezer Gileadi, Robin Hochstrasser, Joshua Jortner, Rafi Levine, Mark Maroncelli, Eli Pollak, Mark Ratner, Joerg Schröder, Zeev Schuss, Ohad Silbert, Jim Skinner and Alessandro Troisi for their advice and help. In particular I am obliged to Misha Galperin, Mark Ratner and Ohad Silbert for their enormous help in going through the whole manuscript and making many important observations and suggestions, and to Joan Shwachman for her help in editing the text. No less important is the accumulated help that I got from a substantially larger group of students and colleagues with whom I collaborated, discussed and exchanged ideas on many of the subjects covered in this book. This include a wonderful group of colleagues who have been meeting every two years in Telluride to discuss issues pertaining to chemical dynamics in condensed phases. My interactions with all these colleagues and friends have shaped my views and understanding of this subject. Thanks are due also to members of the publishing office at OUP, whose gentle prodding has helped keep this project on track.

Finally, deep gratidude is due to my family, for providing me with loving and supportive environment without which this work could not have been accomplished.

On the occasion of the second printing (paperback edition) I would like to thank all readers, in particular Michael Galperin and Joe Subotnik, who pointed out typographical errors (now corrected) in the original book.

CONTENTS

PART I

BACKGROUND

1

REVIEW OF SOME MATHEMATICAL AND PHYSICAL SUBJECTS

> The lawyers plead in court or draw up briefs,
> The generals wage wars, the mariners
> Fight with their ancient enemy the wind,
> And I keep doing what I am doing here:
> Try to learn about the way things are
> And set my findings down in Latin verse . . .
>
> Such things as this require a basic course
> In fundamentals, and a long approach
> By various devious ways, so, all the more,
> I need your full attention . . .
>
> *Lucretius (c.99–c.55* BCE*) "The way things are"*
> *translated by Rolfe Humphries, Indiana University Press, 1968.*

This chapter reviews some subjects in mathematics and physics that are used in different contexts throughout this book. The selection of subjects and the level of their coverage reflect the author's perception of what potential users of this text were exposed to in their earlier studies. Therefore, only brief overview is given of some subjects while somewhat more comprehensive discussion is given of others. In neither case can the coverage provided substitute for the actual learning of these subjects that are covered in detail by many textbooks.

1.1 Mathematical background

1.1.1 Random variables and probability distributions

A random variable is an observable whose repeated determination yields a series of numerical values ("realizations" of the random variable) that vary from trial to trial in a way characteristic of the observable. The outcomes of tossing a coin or throwing a die are familiar examples of discrete random variables. The position of a dust particle in air and the lifetime of a light bulb are continuous random variables. Discrete random variables are characterized by probability distributions; P_n denotes the probability that a realization of the given random variable is n. Continuous random variables are associated with probability density functions $P(x)$: $P(x_1)dx$

denotes the probability that the realization of the variable x will be in the interval $x_1 \ldots x_1+dx$. By their nature, probability distributions have to be normalized, that is,

$$\sum_n P_n = 1; \qquad \int dx P(x) = 1 \tag{1.1}$$

The jth moments of these distributions are

$$M_j = \langle x^j \rangle \equiv \int dx\, x^j P(x) \quad \text{or} \quad \langle n^j \rangle \equiv \sum_n n^j P_n \tag{1.2}$$

Obviously, $M_0 = 1$ and M_1 is the average value of the corresponding random variable. In what follows we will focus on the continuous case. The second moment is usually expressed by the variance,

$$\langle \delta x^2 \rangle \equiv \langle (x - \langle x \rangle)^2 \rangle = M_2 - M_1^2 \tag{1.3}$$

The standard deviation

$$\langle \delta x^2 \rangle^{1/2} = \sqrt{M_2 - M_1^2} \tag{1.4}$$

is a measure of the spread of the fluctuations about the average M_1. The *generating function*[1] for the moments of the distribution $P(x)$ is defined as the average

$$g(\alpha) = \langle e^{\alpha x} \rangle = \int dx P(x) e^{\alpha x} \tag{1.5}$$

the name generating function stems from the identity (obtained by expanding $e^{\alpha x}$ inside the integral)

$$g(\alpha) = 1 + \alpha \langle x \rangle + (1/2)\alpha^2 \langle x^2 \rangle + \cdots + (1/n!)\alpha^n \langle x^n \rangle + \cdots \tag{1.6}$$

which implies that all moments $\langle x^n \rangle$ of $P(x)$ can be obtained from $g(\alpha)$ according to

$$\langle x^n \rangle = \left[\frac{\partial^n}{\partial \alpha^n} g(\alpha) \right]_{\alpha=0} \tag{1.7}$$

Following are some examples of frequently encountered probability distributions: *Poisson distribution*. This is the discrete distribution

$$P(n) = \frac{a^n e^{-a}}{n!} \qquad n = 0, 1, \ldots \tag{1.8}$$

[1] Sometimes referred to also as the *characteristic function* of the given distribution.

which is normalized because $\sum_n a^n/n! = e^a$. It can be easily verified that

$$\langle n \rangle = \langle \delta n^2 \rangle = a \tag{1.9}$$

Binomial distribution. This is a discrete distribution in finite space: The probability that the random variable n takes any integer value between 0 and N is given by

$$P(n) = \frac{N! p^n q^{N-n}}{n!(N-n)!}; \qquad p+q=1; \quad n=0,1,\ldots,N \tag{1.10}$$

The normalization condition is satisfied by the binomial theorem since $\sum_{n=0}^{N} P(n) = (p+q)^N$. We discuss properties of this distribution in Section 7.3.3.

Gaussian distribution. The probability density associated with this continuous distribution is

$$P(x) = \frac{1}{\sqrt{2\pi\sigma^2}} \exp(-[(x-\bar{x})^2/2\sigma^2]); \qquad -\infty < x < \infty \tag{1.11}$$

with average and variance

$$\langle x \rangle = \bar{x}, \qquad \langle \Delta x^2 \rangle = \sigma^2 \tag{1.12}$$

In the limit of zero variance this function approaches a δ function (see Section 1.1.5)

$$P(x) \underset{\sigma \to 0}{\longrightarrow} \delta(x-\bar{x}) \tag{1.13}$$

Lorentzian distribution. This continuous distribution is defined by

$$P(x) = \frac{\gamma/\pi}{(x-\bar{x})^2 + \gamma^2}; \qquad -\infty < x < \infty \tag{1.14}$$

The average is $\langle x \rangle = \bar{x}$ and a δ function, $\delta(x-\bar{x})$, is approached as $\gamma \to 0$, however higher moments of this distribution diverge. This appears to suggest that such a distribution cannot reasonably describe physical observables, but we will see that, on the contrary it is, along with the Gaussian distribution quite pervasive in our discussions, though indeed as a common *physical approximation to observations made near the peak of the distribution.* Note that even though the second moment diverges, γ measures the width at half height of this distribution.

A general phenomenon associated with sums of many random variables has far reaching implications on the random nature of many physical observables. Its mathematical expression is known as the *central limit theorem*. Let $x_1, x_2, \ldots, x_n$

($n \gg 1$) be independent random variables with $\langle x_j \rangle = 0$ and $\langle x_j^2 \rangle = \sigma_j^2$. Consider the sum

$$X_n = x_1 + x_2 + \cdots + x_n \tag{1.15}$$

Under certain conditions that may be qualitatively stated by (1) all variables are alike, that is, there are no few variables that dominate the others, and (2) certain convergence criteria (see below) are satisfied, the probability distribution function $F(X_n)$ of X_n is given by[2]

$$F(X_n) \cong \frac{1}{S_n\sqrt{2\pi}} \exp\left(-\frac{X_n^2}{2S_n^2}\right) \tag{1.16}$$

where

$$S_n^2 = \sigma_1^2 + \sigma_2^2 + \cdots + \sigma_n^2 \tag{1.17}$$

This result is independent of the forms of the probability distributions $f_j(x_j)$ of the variables x_j provided they satisfy, as stated above, some convergence criteria. A sufficient (but not absolutely necessary) condition is that all moments $\int dx_j x_j^n f(x_j)$ of these distributions exist and are of the same order of magnitude.

In applications of these concepts to many particle systems, for example in statistical mechanics, we encounter the need to approximate discrete distributions such as in Eq. (1.10), in the limit of large values of their arguments by continuous functions. The Stirling Approximation

$$N! \sim e^{N \ln N - N} \qquad \text{when } N \to \infty \tag{1.18}$$

is a very useful tool in such cases.

1.1.2 Constrained extrema

In many applications we need to find the maxima or minima of a given function $f(x_1, x_2, \ldots, x_n)$ subject to some constraints. These constraints are expressed as given relationships between the variables that we express by

$$g_k(x_1, x_2 \ldots x_n) = 0; \qquad k = 1, 2, \ldots, m \quad (\text{for } m \text{ constraints}) \tag{1.19}$$

[2] If $\langle x_j \rangle = a_j \neq 0$ and $A_n = a_1 + a_2 + \cdots + a_n$ then Eq. (1.16) is replaced by $F(X_n) \cong (S_n\sqrt{2\pi})^{-1} \exp(-(X_n - A_n)^2/2S_n^2)$.

Such constrained extrema can be found by the Lagrange multipliers method: One form the "Lagrangian"

$$L(x_1,\ldots,x_n) = f(x_1,\ldots,x_n) - \sum_{k=1}^{m} \lambda_k g_k(x_1,\ldots,x_n) \tag{1.20}$$

with m unknown constants $\{\lambda_k\}$. The set of $n+m$ equations

$$\left.\begin{array}{l} \dfrac{\partial L}{\partial x_1} = \cdots = \dfrac{\partial L}{\partial x_n} = 0 \\ g_1 = 0, \ldots, g_m = 0 \end{array}\right\} \tag{1.21}$$

then yield the extremum points $(x_1,\ldots,x_n)$ and the associated Lagrange multipliers $\{\lambda_k\}$.

1.1.3 Vector and fields

1.1.3.1 Vectors

Our discussion here refers to vectors in three-dimensional Euclidean space, so vectors are written in one of the equivalent forms $\mathbf{a} = (a_1, a_2, a_3)$ or (a_x, a_y, a_z). Two products involving such vectors often appear in our text. The scalar (or dot) product is

$$\mathbf{a}\cdot\mathbf{b} = \sum_{n=1}^{3} a_n b_n \tag{1.22}$$

and the vector product is

$$\mathbf{a}\times\mathbf{b} = -\mathbf{b}\times\mathbf{a} = \begin{vmatrix} \mathbf{u}_1 & \mathbf{u}_2 & \mathbf{u}_3 \\ a_1 & a_2 & a_3 \\ b_1 & b_2 & b_2 \end{vmatrix} \tag{1.23}$$

where $\mathbf{u}_j (j = 1, 2, 3)$ are unit vectors in the three cartesian directions and where $|\ |$ denotes a determinant. Useful identities involving scalar and vector products are

$$\mathbf{a}\cdot(\mathbf{b}\times\mathbf{c}) = (\mathbf{a}\times\mathbf{b})\cdot\mathbf{c} = \mathbf{b}\cdot(\mathbf{c}\times\mathbf{a}) \tag{1.24}$$

$$\mathbf{a}\times(\mathbf{b}\times\mathbf{c}) = \mathbf{b}(\mathbf{a}\cdot\mathbf{c}) - \mathbf{c}(\mathbf{a}\cdot\mathbf{b}) \tag{1.25}$$

1.1.3.2 Fields

A field is a quantity that depends on one or more continuous variables. We will usually think of the coordinates that define a position of space as these variables, and the rest of our discussion is done in this language. A scalar field is a map that assigns a scalar to any position in space. Similarly, a vector field is a map that assigns a vector to any such position, that is, is a vector function of position. Here we summarize some definitions and properties of scalar and vector fields.

The gradient, ∇S, of a scalar function $S(\mathbf{r})$, and the divergence, $\nabla \cdot \mathbf{F}$, and rotor (curl), $\nabla \times \mathbf{F}$, of a vector field $\mathbf{F}(\mathbf{r})$ are given in cartesian coordinates by

$$\nabla S = \frac{\partial S}{\partial x}\mathbf{u}_x + \frac{\partial S}{\partial y}\mathbf{u}_y + \frac{\partial S}{\partial z}\mathbf{u}_z \tag{1.26}$$

$$\nabla \cdot \mathbf{F} = \frac{\partial F_x}{\partial x} + \frac{\partial F_y}{\partial y} + \frac{\partial F_z}{\partial z} \tag{1.27}$$

$$\nabla \times \mathbf{F} = \begin{vmatrix} \mathbf{u}_x & \mathbf{u}_y & \mathbf{u}_z \\ \partial/\partial x & \partial/\partial y & \partial/\partial z \\ F_x & F_y & F_z \end{vmatrix} = \mathbf{u}_x\left(\frac{\partial F_z}{\partial y} - \frac{\partial F_y}{\partial z}\right) - \mathbf{u}_y\left(\frac{\partial F_z}{\partial x} - \frac{\partial F_x}{\partial z}\right) + \mathbf{u}_z\left(\frac{\partial F_y}{\partial x} - \frac{\partial F_x}{\partial y}\right) \tag{1.28}$$

where $\mathbf{u}_x$, $\mathbf{u}_y$, $\mathbf{u}_z$ are again unit vectors in the x, y, z directions. Some identities involving these objects are

$$\nabla \times (\nabla \times \mathbf{F}) = \nabla(\nabla \cdot \mathbf{F}) - \nabla^2 \mathbf{F} \tag{1.29}$$

$$\nabla \cdot (S\mathbf{F}) = \mathbf{F} \cdot \nabla S + S\nabla \cdot \mathbf{F} \tag{1.30}$$

$$\nabla \cdot (\nabla \times \mathbf{F}) = 0 \tag{1.31}$$

$$\nabla \times (\nabla S) = 0 \tag{1.32}$$

The Helmholtz theorem states that any vector field $\mathbf{F}$ can be written as a sum of its transverse $\mathbf{F}^{\perp}$ and longitudinal $\mathbf{F}^{\parallel}$ components

$$\mathbf{F} = \mathbf{F}^{\perp} + \mathbf{F}^{\parallel} \tag{1.33}$$

which have the properties

$$\nabla \cdot \mathbf{F}^{\perp} = 0 \tag{1.34a}$$

$$\nabla \times \mathbf{F}^{\parallel} = 0 \tag{1.34b}$$

that is, the transverse component has zero divergence while the longitudinal component has zero curl. Explicit expressions for these components are

$$\mathbf{F}^{\perp}(\mathbf{r}) = \frac{1}{4\pi}\nabla \times \int d^3r' \frac{\nabla' \times \mathbf{F}(\mathbf{r}')}{|\mathbf{r}-\mathbf{r}'|} \tag{1.35a}$$

$$\mathbf{F}^{\parallel}(\mathbf{r}) = \frac{1}{4\pi}\nabla \int d^3r' \frac{\nabla' \cdot \mathbf{F}(\mathbf{r}')}{|\mathbf{r}-\mathbf{r}'|} \tag{1.35b}$$

1.1.3.3 *Integral relations*

Let $V(S)$ be a volume bounded by a closed surface S. Denote a three-dimensional volume element by d^3r and a surface vector element by $d\mathbf{S}$. $d\mathbf{S}$ has the magnitude of the corresponding surface area and direction along its normal, facing outward. We sometimes write $d\mathbf{S} = \hat{\mathbf{n}}d^2x$ where $\hat{\mathbf{n}}$ is an outward normal unit vector. Then for any vector and scalar functions of position, $\mathbf{F}(\mathbf{r})$ and $\phi(\mathbf{r})$, respectively

$$\int_V d^3r(\nabla \cdot \mathbf{F}) = \oint_S d\mathbf{S} \cdot \mathbf{F} \qquad \text{(Gauss's divergence theorem)} \tag{1.36}$$

$$\int_V d^3r(\nabla\phi) = \oint_S d\mathbf{S}\phi \tag{1.37}$$

$$\int_V d^3r(\nabla \times \mathbf{F}) = \oint_S d\mathbf{S} \times \mathbf{F} \tag{1.38}$$

In these equations $\oint_S$ denotes an integral over the surface S.

Finally, the following theorem concerning the integral in (1.37) is of interest: Let $\phi(\mathbf{r})$ be a periodic function in three dimensions, so that $\phi(\mathbf{r}) = \phi(\mathbf{r}+\mathbf{R})$ with $\mathbf{R} = n_1\mathbf{a}_1 + n_2\mathbf{a}_2 + n_3\mathbf{a}_3$ with $\mathbf{a}_j (j = 1, 2, 3)$ being three vectors that characterize the three-dimensional periodicity and n_j any integers (see Section 4.1). The function is therefore characterized by its values in one unit cell defined by the three **a** vectors. Then the integral (1.37) vanishes if the volume of integration is exactly one unit cell.

To prove this statement consider the integral over a unit cell

$$I(\mathbf{r}') = \int_V d^3r\phi(\mathbf{r}+\mathbf{r}') \tag{1.39}$$

Since ϕ is periodic and the integral is over one period, the result should not depend on $\mathbf{r}'$. Therefore,

$$0 = \nabla_{\mathbf{r}'} I(\mathbf{r}') = \int_V d^3 r \nabla_{\mathbf{r}'} \phi(\mathbf{r} + \mathbf{r}') = \int_V d^3 r \nabla_{\mathbf{r}} \phi(\mathbf{r} + \mathbf{r}') = \int_V d^3 r \nabla \phi(\mathbf{r}) \tag{1.40}$$

which concludes the proof.

1.1.4 Continuity equation for the flow of conserved entities

We will repeatedly encounter in this book processes that involve the flow of conserved quantities. An easily visualized example is the diffusion of nonreactive particles, but it should be emphasized at the outset that the motion involved can be of any type and the moving object(s) do not have to be particles. The essential ingredient in the following discussion is a conserved entity Q whose distribution in space is described by some time-dependent density function $\rho_Q(\mathbf{r}, t)$ so that its amount within some finite volume V is given by

$$Q(t) = \int_V d^3 r \rho_Q(\mathbf{r}, t) \tag{1.41}$$

The conservation of Q implies that any change in $Q(t)$ can result only from flow of Q through the boundary of volume V. Let S be the surface that encloses the volume V, and $d\mathbf{S}$—a vector surface element whose direction is normal to the surface in the outward direction. Denote by $\mathbf{J}_Q(\mathbf{r}, t)$ the flux of Q, that is, the amount of Q moving in the direction of $\mathbf{J}_Q$ per unit time and per unit area of the surface perpendicular to $\mathbf{J}_Q$. The Q conservation law can then be written in the following mathematical form

$$\frac{dQ}{dt} = \int_V d^3 r \frac{\partial \rho_Q(\mathbf{r}, t)}{\partial t} = -\int_S d\mathbf{S} \cdot \mathbf{J}_Q(\mathbf{r}, t) \tag{1.42}$$

Using Eq. (1.36) this can be recast as

$$\int_V d^3 r \frac{\partial \rho_Q(\mathbf{r}, t)}{\partial t} = -\int_V d^3 r \nabla \cdot \mathbf{J}_Q \tag{1.43}$$

which implies, since the volume V is arbitrary

$$\frac{\partial \rho_Q(\mathbf{r}, t)}{\partial t} = -\nabla \cdot \mathbf{J}_Q \tag{1.44}$$

Equation (1.44), the local form of Eq. (1.42), is the continuity equation for the conserved Q. Note that in terms of the velocity field $\mathbf{v}(\mathbf{r},t) = \dot{\mathbf{r}}(\mathbf{r},t)$ associated with the Q motion we have

$$\mathbf{J}_Q(\mathbf{r},t) = \mathbf{v}(\mathbf{r},t)\rho_Q(\mathbf{r},t) \tag{1.45}$$

It is important to realize that the derivation above does not involve any physics. It is a mathematical expression of conservation of entities that that can change their position but are not created or destroyed in time. Also, it is not limited to entities distributed in space and could be applied to objects moving in other dimensions. For example, let the function $\rho_1(\mathbf{r},\mathbf{v},t)$ be the particles density in position and velocity space (i.e. $\rho_1(\mathbf{r},\mathbf{v},t)d^3rd^3v$ is the number of particles whose position and velocity are respectively within the volume element d^3r about $\mathbf{r}$ and the velocity element d^3v about $\mathbf{v}$). The total number, $N = \int d^3r \int d^3v\rho(\mathbf{r},\mathbf{v},t)$, is fixed. The change of ρ_1 in time can then be described by Eq. (1.44) in the form

$$\frac{\partial\rho_1(\mathbf{r},\mathbf{v},t)}{\partial t} = -\nabla_{\mathbf{r}}\cdot(\mathbf{v}\rho_1) - \nabla_{\mathbf{v}}(\dot{\mathbf{v}}\rho_1) \tag{1.46}$$

where $\nabla_{\mathbf{r}} = (\partial/\partial x, \partial/\partial y, \partial/\partial z)$ is the gradient in position space and $\nabla_{\mathbf{v}} = (\partial/\partial v_x, \partial/\partial v_y, \partial/\partial v_z)$ is the gradient in velocity space. Note that $\mathbf{v}\rho$ is a flux in position space, while $\dot{\mathbf{v}}\rho$ is a flux in velocity space.

1.1.5 Delta functions

The delta function[3] (or: Dirac's delta function) is a generalized function that is obtained as a limit when a normalized function $\int_{-\infty}^{\infty} dxf(x) = 1$ becomes zero everywhere except at one point. For example,

$$\delta(x) = \lim_{a\to\infty}\sqrt{\frac{a}{\pi}}e^{-ax^2} \quad \text{or} \quad \delta(x) = \lim_{a\to 0}\frac{a/\pi}{x^2+a^2} \quad \text{or}$$

$$\delta(x) = \lim_{a\to\infty}\frac{\sin(ax)}{\pi x} \tag{1.47}$$

Another way to view this function is as the derivative of the Heaviside step function ($\eta(x) = 0$ for $x < 0$ and $\eta(x) = 1$ for $x \geq 0$):

$$\delta(x) = \frac{d}{dx}\eta(x) \tag{1.48}$$

[3] Further reading: http://mathworld.wolfram.com/DeltaFunction.html and references therein.

In what follows we list a few properties of this function

$$\int_a^b dx f(x)\,\delta(x-x_0) = \begin{cases} f(x_0) & \text{if } a < x_0 < b \\ 0 & \text{if } x_0 < a \text{ or } x_0 > b \\ (1/2)f(x_0) & \text{if } x_0 = a \text{ or } x_0 = b \end{cases} \tag{1.49}$$

$$\delta[g(x)] = \sum_j \frac{\delta(x-x_j)}{|g'(x_j)|} \tag{1.50}$$

where x_j are the roots of $g(x) = 0$, for example,

$$\delta(ax) = \frac{1}{|a|}\delta(x) \tag{1.51}$$

and

$$\delta(x^2 - a^2) = \frac{1}{2|a|}(\delta(x+a) + \delta(x-a)) \tag{1.52}$$

The derivative of the δ function is also a useful concept. It satisfies (from integration by parts)

$$\int dx f(x)\,\delta'(x-x_0) = -f'(x_0) \tag{1.53}$$

and more generally

$$\int dx f(x)\delta^{(n)}(x) = -\int dx \frac{\partial f}{\partial x}\delta^{(n-1)}(x) \tag{1.54}$$

Also (from checking integrals involving the two sides)

$$x\delta'(x) = -\delta(x) \tag{1.55}$$

$$x^2\delta'(x) = 0 \tag{1.56}$$

Since for $-\pi < a < \pi$ we have

$$\int_{-\pi}^{\pi} dx \cos(nx)\delta(x-a) = \cos(na) \quad \text{and} \quad \int_{-\pi}^{\pi} dx \sin(nx)\delta(x-a) = \sin(na) \tag{1.57}$$

it follows that the Fourier series expansion of the δ function is

$$\begin{aligned}\delta(x-a) &= \frac{1}{2\pi} + \frac{1}{\pi}\sum_{n=1}^{\infty}[\cos(na)\cos(nx) + \sin(na)\sin(nx)] \\ &= \frac{1}{2\pi} + \frac{1}{\pi}\sum_{n=1}^{\infty}\cos(n(x-a)) \end{aligned} \tag{1.58}$$

Also since

$$\int_{-\infty}^{\infty} dx e^{ikx}\delta(x-a) = e^{ika} \tag{1.59}$$

it follows that

$$\delta(x-a) = \frac{1}{2\pi}\int_{-\infty}^{\infty} dk e^{-ik(x-a)} = \frac{1}{2\pi}\int_{-\infty}^{\infty} dk e^{ik(x-a)} \tag{1.60}$$

Extending δ functions to two and three dimensions is simple in cartesian coordinates

$$\begin{aligned}\delta^2(\mathbf{r}) &= \delta(x)\delta(y) \\ \delta^3(\mathbf{r}) &= \delta(x)\delta(y)\delta(z)\end{aligned} \tag{1.61}$$

In spherical coordinates care has to be taken of the integration element. The result is

$$\delta^2(\mathbf{r}) = \frac{\delta(r)}{\pi r} \tag{1.62}$$

$$\delta^3(\mathbf{r}) = \frac{\delta(r)}{2\pi r^2} \tag{1.63}$$

1.1.6 Complex integration

Integration in the complex plane is a powerful technique for evaluating a certain class of integrals, including those encountered in the solution of the time-dependent Schrödinger equation (or other linear initial value problems) by the Laplace transform method (see next subsection). At the core of this technique are the Cauchy theorem which states that the integral along a closed contour of a function $g(z)$, which is analytic on the contour and the region enclosed by it, is zero:

$$\oint dz g(z) = 0 \tag{1.64}$$

and the Cauchy integral formula—valid for a function $g(z)$ with the properties defined above,

$$\oint dz \frac{g(z)}{z-\alpha} = 2\pi i g(\alpha) \tag{1.65}$$

where the integration contour surrounds the point $z = \alpha$ at which the integrand has a simple singularity, and the integration is done in the counter-clockwise direction (reversing the direction yields the result with opposite sign).

Both the theorem and the integral formula are very useful for many applications: the Cauchy theorem implies that any integration path can be distorted in the complex plane as long as the area enclosed between the original and modified path does not contain any singularities of the integrand. This makes it often possible to modify integration paths in order to make evaluation easier. The Cauchy integral formula is often used to evaluate integrals over unclosed path—if the contour can be closed along a line on which the integral is either zero or easily evaluated. An example is shown below, where the integral (1.78) is evaluated by this method.

Problem 1.1. Use complex integration to obtain the identity for $\varepsilon > 0$

$$\frac{1}{2\pi}\int_{-\infty}^{\infty} d\omega e^{-i\omega t}\frac{1}{\omega-\omega_0+i\varepsilon} = \begin{cases} 0 & \text{for } t<0 \\ -ie^{-i\omega_0 t-\varepsilon t} & \text{for } t>0 \end{cases} \tag{1.66}$$

In quantum dynamics applications we often encounter this identity in the limit $\varepsilon \to 0$. We can rewrite it in the form

$$\frac{1}{2\pi}\int_{-\infty}^{\infty} d\omega e^{-i\omega t}\frac{1}{\omega-\omega_0+i\varepsilon} \xrightarrow{\varepsilon\to 0+} -i\eta(t)e^{-i\omega_0 t} \tag{1.67}$$

where $\eta(t)$ is the step function defined above Eq. (1.48).

Another useful identity is associated with integrals involving the function $(\omega - \omega_0 + i\varepsilon)^{-1}$ and a real function $f(x)$. Consider

$$\int_a^b d\omega \frac{f(\omega)}{\omega-\omega_0+i\varepsilon} = \int_a^b d\omega \frac{(\omega-\omega_0)f(\omega)}{(\omega-\omega_0)^2+\varepsilon^2} - i\int_a^b d\omega \frac{\varepsilon}{(\omega-\omega_0)^2+\varepsilon^2} f(\omega) \tag{1.68}$$

where the integration is on the real ω axis and where $a < \omega_0 < b$. Again we are interested in the limit $\varepsilon \to 0$. The imaginary term in (1.68) is easily evaluated to be $-i\pi f(\omega_0)$ by noting that the limiting form of the Lorentzian function that multiplies $f(\omega)$ is a delta-function (see Eq. (1.47)). The real part is identified as

$$\int_a^b d\omega \frac{(\omega-\omega_0)f(\omega)}{(\omega-\omega_0)^2+\varepsilon^2} \xrightarrow{\varepsilon\to 0} \mathrm{PP}\int_a^b d\omega \frac{f(\omega)}{\omega-\omega_0} \tag{1.69}$$

Here PP stands for the so-called Cauchy principal part (or principal value) of the integral about the singular point ω_0. In general, the Cauchy principal value of a finite integral of a function $f(x)$ about a point x_0 with $a < x_0 < b$ is given by

$$\mathrm{PP}\int_a^b dx f(x) = \lim_{\varepsilon\to 0+}\left\{\int_a^{x_0-\varepsilon} dx f(x) + \int_{x_0+\varepsilon}^b dx f(x)\right\} \tag{1.70}$$

We sometimes express the information contained in Eqs (1.68)–(1.70) and (1.47) in the concise form

$$\frac{1}{\omega-\omega_0+i\varepsilon} \xrightarrow{\varepsilon\to 0+} \mathrm{PP}\frac{1}{\omega-\omega_0} - i\pi\delta(\omega-\omega_0) \tag{1.71}$$

1.1.7 Laplace transform

A function $f(t)$ and its Laplace transform $\tilde{f}(z)$ are related by

$$\tilde{f}(z) = \int_0^\infty dt e^{-zt} f(t) \tag{1.72}$$

$$f(t) = \frac{1}{2\pi i}\int_{-\infty i+\varepsilon}^{\infty i+\varepsilon} dz e^{zt}\tilde{f}(z) \tag{1.73}$$

where ε is chosen so that the integration path is to the right of all singularities of $\tilde{f}(z)$. In particular, if the singularities of $\tilde{f}(z)$ are all on the imaginary axis, ε can be taken arbitrarily small, that is, the limit $\varepsilon \to 0_+$ may be considered. We will see below that this is in fact the situation encountered in solving the time-dependent Schrödinger equation.

Laplace transforms are useful for initial value problems because of identities such as

$$\int_0^\infty dt e^{-zt}\frac{df}{dt} = [e^{-zt}f]_0^\infty + z\int_0^\infty dt e^{-zt}f(t) = z\tilde{f}(z) - f(t=0) \tag{1.74}$$

and

$$\int_0^\infty dte^{-zt}\frac{d^2f}{dt^2} = z^2\tilde{f}(z) - zf(t=0) - f'(t=0) \tag{1.75}$$

which are easily verified using integration by parts. As an example consider the equation

$$\frac{df}{dt} = -\alpha f \tag{1.76}$$

Taking Laplace transform we get

$$z\tilde{f}(z) - f(0) = -\alpha\tilde{f}(z) \tag{1.77}$$

that is,

$$\tilde{f}(z) = (z+\alpha)^{-1}f(0) \qquad \text{and} \qquad f(t) = (2\pi i)^{-1}\int_{\varepsilon-i\infty}^{\varepsilon+i\infty} dze^{zt}(z+\alpha)^{-1}f(0). \tag{1.78}$$

If α is real and positive ε can be taken as 0, and evaluating the integral by closing a counter-clockwise contour on the negative-real half z plane[4] leads to

$$f(t) = e^{-\alpha t}f(0) \tag{1.79}$$

1.1.8 The Schwarz inequality

In its simplest form the Schwarz inequality expressed an obvious relation between the products of magnitudes of two real vectors $\mathbf{c}_1$ and $\mathbf{c}_2$ and their scalar product

$$c_1c_2 \geq \mathbf{c}_1 \cdot \mathbf{c}_2 \tag{1.80}$$

It is less obvious to show that this inequality holds also for complex vectors, provided that the scalar product of two complex vectors $\mathbf{e}$ and $\mathbf{f}$ is defined by $\mathbf{e}^* \cdot \mathbf{f}$. The inequality is of the form

$$|\mathbf{e}||\mathbf{f}| \geq |\mathbf{e}^* \cdot \mathbf{f}| \tag{1.81}$$

[4] The contour is closed at $z \to -\infty$ where the integrand is zero. In fact the integrand has to vanish faster than z^{-2} as $z \to -\infty$ because the length of the added path diverges like z^2 in that limit.

Note that in the scalar product $\mathbf{e}^* \cdot \mathbf{f}$ the order is important, that is, $\mathbf{e}^* \cdot \mathbf{f} = (\mathbf{f}^* \cdot \mathbf{e})^*$. To prove the inequality (1.81) we start from

$$(a^*\mathbf{e}^* - b^*\mathbf{f}^*) \cdot (a\mathbf{e} - b\mathbf{f}) \geq 0 \tag{1.82}$$

which holds for any scalars a and b. Using the choice

$$a = \sqrt{(\mathbf{f}^* \cdot \mathbf{f})(\mathbf{e}^* \cdot \mathbf{f})} \qquad \text{and} \qquad b = \sqrt{(\mathbf{e}^* \cdot \mathbf{e})(\mathbf{f}^* \cdot \mathbf{e})} \tag{1.83}$$

in Eq. (1.82) leads after some algebra to $\sqrt{(\mathbf{e}^* \cdot \mathbf{e})(\mathbf{f}^* \cdot \mathbf{f})} \geq |\mathbf{e}^* \cdot \mathbf{f}|$, which implies (1.81).

Equation (1.80) can be applied also to real functions that can be viewed as vectors with a continuous ordering index. We can make the identification $c_k = (\mathbf{c}_k \cdot \mathbf{c}_k)^{1/2} \Rightarrow (\int dx c_k^2(x))^{1/2}$; $k = 1, 2$ and $\mathbf{c}_1 \cdot \mathbf{c}_2 \Rightarrow \int dx c_1(x) c_2(x)$ to get

$$\left(\int dx c_1^2(x) \right) \left(\int dx c_2^2(x) \right) \geq \left(\int dx c_1(x) c_2(x) \right)^2 \tag{1.84}$$

The same development can be done for Hilbert space vectors. The result is

$$\langle \psi | \psi \rangle \langle \phi | \phi \rangle \geq |\langle \psi | \phi \rangle|^2 \tag{1.85}$$

where ψ and ϕ are complex functions so that $\langle \psi | \phi \rangle = \int d\mathbf{r} \psi^*(\mathbf{r}) \phi(\mathbf{r})$. To prove Eq. (1.85) define a function $y(\mathbf{r}) = \psi(\mathbf{r}) + \lambda \phi(\mathbf{r})$ where λ is a complex constant. The following inequality is obviously satisfied

$$\int d\mathbf{r} \, y^*(\mathbf{r}) y(\mathbf{r}) \geq 0$$

This leads to

$$\int d\mathbf{r} \psi^*(\mathbf{r}) \psi(\mathbf{r}) + \lambda \int d\mathbf{r} \psi^*(\mathbf{r}) \phi(\mathbf{r}) + \lambda^* \int d\mathbf{r} \phi^*(\mathbf{r}) \psi(\mathbf{r}) + \lambda^* \lambda \int d\mathbf{r} \phi^*(\mathbf{r}) \phi(\mathbf{r}) \geq 0 \tag{1.86}$$

or

$$\langle \psi | \psi \rangle + \lambda \langle \psi | \phi \rangle + \lambda^* \langle \phi | \psi \rangle + |\lambda|^2 \langle \phi | \phi \rangle \geq 0 \tag{1.87}$$

Now choose

$$\lambda = -\frac{\langle \phi | \psi \rangle}{\langle \phi | \phi \rangle}; \qquad \lambda^* = -\frac{\langle \psi | \phi \rangle}{\langle \phi | \phi \rangle} \tag{1.88}$$

and also multiply (1.87) by $\langle\phi|\phi\rangle$ to get

$$\langle\psi|\psi\rangle\langle\phi|\phi\rangle - |\langle\psi|\phi\rangle|^2 - |\langle\phi|\psi\rangle|^2 + |\langle\psi|\phi\rangle|^2 \geq 0 \tag{1.89}$$

which leads to (1.85).

An interesting implication of the Schwarz inequality appears in the relationship between averages and correlations involving two observables $\mathcal{A}$ and $\mathcal{B}$. Let P_n be the probability that the system is in state n and let A_n and B_n be the values of these observables in this state. Then $\langle A^2\rangle = \sum_n P_n A_n^2$, $\langle B^2\rangle = \sum_n P_n B_n^2$, and $\langle AB\rangle = \sum_n P_n A_n B_n$. The Schwarz inequality now implies

$$\langle A^2\rangle\langle B^2\rangle \geq \langle AB\rangle^2 \tag{1.90}$$

Indeed, Eq. (1.90) is identical to Eq. (1.80) written in the form $(\mathbf{a}\cdot\mathbf{a})(\mathbf{b}\cdot\mathbf{b}) \geq (\mathbf{a}\cdot\mathbf{b})^2$ where $\mathbf{a}$ and $\mathbf{b}$ are the vectors $a_n = \sqrt{P_n}A_n$; $b_n = \sqrt{P_n}B_n$.

1.2 Classical mechanics

1.2.1 Classical equations of motion

Time evolution in classical mechanics is described by the Newton equations

$$\begin{aligned}\dot{\mathbf{r}}_i &= \frac{1}{m_i}\mathbf{p}_i \\ \dot{\mathbf{p}}_i &= \mathbf{F}_i = -\nabla_i U\end{aligned} \tag{1.91}$$

$\mathbf{r}_i$ and $\mathbf{p}_i$ are the position and momentum vectors of particle i of mass m_i, $\mathbf{F}_i$ is the force acting on the particle, U is the potential, and ∇_i is the gradient with respect to the position of this particle. These equations of motion can be obtained from the Lagrangian

$$L = K(\{\dot{x}\}) - U(\{x\}) \tag{1.92}$$

where K and U are, respectively, the total kinetic and potential energies and $\{x\}$, $\{\dot{x}\}$ stands for all the position and velocity coordinates. The Lagrange equations of motion are

$$\frac{d}{dt}\frac{\partial L}{\partial \dot{x}_i} = \frac{\partial L}{\partial x_i} \text{ (and same for } y, \text{z)} \tag{1.93}$$

The significance of this form of the Newton equations is its invariance to coordinate transformation.

Another useful way to express the Newton equations of motion is in the Hamiltonian representation. One starts with the generalized momenta

$$p_j = \frac{\partial L}{\partial \dot{x}_j} \tag{1.94}$$

and define the Hamiltonian according to

$$H = -\left[L(\{x\}, \{\dot{x}\}) - \sum_j p_j \dot{x}_j \right] \tag{1.95}$$

The mathematical operation done in (1.95) transforms the function L of variables $\{x\}$, $\{\dot{x}\}$ to a new function H of the variables $\{x\}$, $\{p\}$.[5] The resulting function, $H(\{x\}, \{p\})$, is the Hamiltonian, which is readily shown to satisfy

$$H = U + K \tag{1.96}$$

that is, it is the total energy of the system, and

$$\dot{x}_j = \frac{\partial H}{\partial p_j}; \qquad \dot{p}_j = -\frac{\partial H}{\partial q_j} \tag{1.97}$$

which is the Hamiltonian form of the Newton equations. In a many-particle system the index j goes over all generalized positions and momenta of all particles.

The specification of all positions and momenta of all particles in the system defines the dynamical state of the system. Any *dynamical variable*, that is, a function of these positions and momenta, can be computed given this state. Dynamical variables are precursors of macroscopic observables that are defined as suitable averages over such variables and calculated using the machinery of statistical mechanics.

1.2.2 Phase space, the classical distribution function, and the Liouville equation

In what follows we will consider an N particle system in Euclidian space. The classical equations of motion are written in the form

$$\dot{\mathbf{r}}^N = \frac{\partial H(\mathbf{r}^N, \mathbf{p}^N)}{\partial \mathbf{p}^N}; \qquad \dot{\mathbf{p}}^N = -\frac{\partial H(\mathbf{r}^N, \mathbf{p}^N)}{\partial \mathbf{r}^N} \tag{1.98}$$

---

[5] This type of transformation is called a *Legendre transform*.

which describe the time evolution of all coordinates and momenta in the system. In these equations $\mathbf{r}^N$ and $\mathbf{p}^N$ are the $3N$-dimensional vectors of coordinates and momenta of the N particles. The $6N$-dimensional space whose axes are these coordinates and momenta is refereed to as the *phase space* of the system. A *phase point* $(\mathbf{r}^N, \mathbf{p}^N)$ in this space describes the instantaneous state of the system. The probability distribution function $f(\mathbf{r}^N, \mathbf{p}^N; t)$ is defined such that $f(\mathbf{r}^N, \mathbf{p}^N; t)d\mathbf{r}^N d\mathbf{p}^N$ is the probability at time t that the phase point will be inside the volume $d\mathbf{r}^N d\mathbf{p}^N$ in phase space. This implies that in an ensemble containing a large number $\mathcal{N}$ of identical systems the number of those characterized by positions and momenta within the $d\mathbf{r}^N d\mathbf{p}^N$ neighborhood is $\mathcal{N}f(\mathbf{r}^N, \mathbf{p}^N; t)d\mathbf{r}^N d\mathbf{p}^N$. Here and below we use a shorthand notation in which, for example, $d\mathbf{r}^N = d\mathbf{r}_1 d\mathbf{r}_2 \ldots d\mathbf{r}_N = dx_1 dx_2 dx_3 dx_4 dx_5 dx_6, \ldots, dx_{3N-2} dx_{3N-1} dx_{3N}$ and, $(\partial F/\partial \mathbf{r}^N)(\partial G/\partial \mathbf{p}^N) = \sum_{j=1}^{3N} (\partial F/\partial x_j)(\partial G/\partial p_j)$.

As the system evolves in time according to Eq. (1.98) the distribution function evolves accordingly. We want to derive an equation of motion for this distribution. To this end consider first any dynamical variable $\mathcal{A}(\mathbf{r}^N, \mathbf{p}^N)$. Its time evolution is given by

$$\frac{d\mathcal{A}}{dt} = \frac{\partial \mathcal{A}}{\partial \mathbf{r}^N}\dot{\mathbf{r}}^N + \frac{\partial \mathcal{A}}{\partial \mathbf{p}^N}\dot{\mathbf{p}}^N = \frac{\partial \mathcal{A}}{\partial \mathbf{r}^N}\frac{\partial H}{\partial \mathbf{p}^N} - \frac{\partial \mathcal{A}}{\partial \mathbf{p}^N}\frac{\partial H}{\partial \mathbf{r}^N} = \{H, \mathcal{A}\} \equiv i\mathcal{L}\mathcal{A}$$
$$\mathcal{L} = -i\{H, \} \qquad (1.99)$$

The second equality in Eq. (1.99) defines the *Poisson brackets* and $\mathcal{L}$ is called the (classical) Liouville operator. Consider next the ensemble average $A(t) = \langle \mathcal{A} \rangle_t$ of the dynamical variable $\mathcal{A}$. This average, a time-dependent observable, can be expressed in two ways that bring out two different, though equivalent, roles played by the function $\mathcal{A}(\mathbf{r}^N, \mathbf{p}^N)$. First, it is a function in phase space that gets a distinct numerical value at each phase point. Its average at time t is therefore given by

$$A(t) = \int d\mathbf{r}^N \int d\mathbf{p}^N f(\mathbf{r}^N, \mathbf{p}^N; t)\mathcal{A}(\mathbf{r}^N, \mathbf{p}^N) \qquad (1.100)$$

At the same time the value of $\mathcal{A}$ at time t is given by $\mathcal{A}(\mathbf{r}^N(t), \mathbf{p}^N(t))$ and is determined uniquely by the initial conditions $(\mathbf{r}^N(0), \mathbf{p}^N(0))$. Therefore,

$$A(t) = \int d\mathbf{r}^N \int d\mathbf{p}^N f(\mathbf{r}^N, \mathbf{p}^N; 0)\mathcal{A}(\mathbf{r}^N(t), \mathbf{p}^N(t)) \qquad (1.101)$$

An equation of motion for f can be obtained by equating the time derivatives of Eqs (1.100) and (1.101):

$$\int d\mathbf{r}^N \int d\mathbf{p}^N \frac{\partial f(\mathbf{r}^N, \mathbf{p}^N; t)}{\partial t} \mathcal{A}(\mathbf{r}^N, \mathbf{p}^N) = \int d\mathbf{r}^N \int d\mathbf{p}^N f(\mathbf{r}^N, \mathbf{p}^N; 0) \frac{d\mathcal{A}}{dt}$$
$$= \int d\mathbf{r}^N \int d\mathbf{p}^N f(\mathbf{r}^N, \mathbf{p}^N; 0) \left(\frac{\partial \mathcal{A}}{\partial \mathbf{r}^N} \frac{\partial H}{\partial \mathbf{p}^N} - \frac{\partial \mathcal{A}}{\partial \mathbf{p}^N} \frac{\partial H}{\partial \mathbf{r}^N} \right) \tag{1.102}$$

Using integration by parts while assuming that f vanishes at the boundary of phase space, the right-hand side of (1.102) may be transformed according to

$$\int d\mathbf{r}^N \int d\mathbf{p}^N f(\mathbf{r}^N, \mathbf{p}^N; 0) \left(\frac{\partial \mathcal{A}}{\partial \mathbf{r}^N} \frac{\partial H}{\partial \mathbf{p}^N} - \frac{\partial \mathcal{A}}{\partial \mathbf{p}^N} \frac{\partial H}{\partial \mathbf{r}^N} \right)$$
$$= \int d\mathbf{r}^N \int d\mathbf{p}^N \left(\frac{\partial H}{\partial \mathbf{r}^N} \frac{\partial f}{\partial \mathbf{p}^N} - \frac{\partial H}{\partial \mathbf{p}^N} \frac{\partial f}{\partial \mathbf{r}^N} \right) \mathcal{A}(\mathbf{r}^N, \mathbf{p}^N) \tag{1.103}$$
$$= \int d\mathbf{r}^N \int d\mathbf{p}^N (-i\mathcal{L}f) \mathcal{A}(\mathbf{r}^N, \mathbf{p}^N)$$

Comparing this to the left-hand side of Eq. (1.102) we get

$$\frac{\partial f(\mathbf{r}^N, \mathbf{p}^N; t)}{\partial t} = -i\mathcal{L}f = -\sum_{j=1}^{3N} \left\{ \frac{\partial f}{\partial x_j} \frac{\partial H}{\partial p_j} - \frac{\partial f}{\partial p_j} \frac{\partial H}{\partial x_j} \right\}$$
$$= -\sum_{j=1}^{3N} \left\{ \frac{\partial f}{\partial x_j} \dot{x}_j + \frac{\partial f}{\partial p_j} \dot{p}_j \right\} \tag{1.104}$$

This is the classical *Liouville equation.* An alternative derivation of this equation that sheds additional light on the nature of the phase space distribution function $f(\mathbf{r}^N, \mathbf{p}^N; t)$ is given in Appendix 1A.

An important attribute of the phase space distribution function f is that it is globally constant. Let us see first what this statement means mathematically. Using

$$\frac{d}{dt} = \frac{\partial}{\partial t} + \dot{\mathbf{r}}^N \frac{\partial}{\partial \mathbf{r}^N} + \dot{\mathbf{p}}^N \frac{\partial}{\partial \mathbf{p}^N} = \frac{\partial}{\partial t} + \frac{\partial H}{\partial \mathbf{p}^N} \frac{\partial}{\partial \mathbf{r}^N} - \frac{\partial H}{\partial \mathbf{r}^N} \frac{\partial}{\partial \mathbf{p}^N} = \frac{\partial}{\partial t} + i\mathcal{L} \tag{1.105}$$

Equation (1.104) implies that

$$\frac{df}{dt} = 0 \tag{1.106}$$

that is, f has to satisfy the following identity:

$$f(\mathbf{r}^N(0), \mathbf{p}^N(0); t = 0) = f(\mathbf{r}^N(t), \mathbf{p}^N(t); t) \tag{1.107}$$

As the ensemble of systems evolves in time, each phase point moves along the trajectory $(\mathbf{r}^N(t), \mathbf{p}^N(t))$. Equation (1.107) states that the density of phase points appears constant when observed *along the trajectory*.

Another important outcome of these considerations is the following. The uniqueness of solutions of the Newton equations of motion implies that phase point trajectories do not cross. If we follow the motions of phase points that started at a given volume element in phase space we will therefore see all these points evolving in time into an equivalent volume element, not necessarily of the same geometrical shape. The number of points in this new volume is the same as the original one, and Eq. (1.107) implies that also their density is the same. Therefore, the new volume (again, not necessarily the shape) is the same as the original one. If we think of this set of points as molecules of some multidimensional fluid, the nature of the time evolution implies that this fluid is totally incompressible. Equation (1.107) is the mathematical expression of this incompressibility property.

1.3 Quantum mechanics

In quantum mechanics the state of a many-particle system is represented by a wavefunction $\Psi(\mathbf{r}^N, t)$, observables correspond to hermitian operators[6] and results of measurements are represented by expectation values of these operators,

$$\begin{aligned}\langle A\rangle(t) &= \langle\Psi(\mathbf{r}^N, t)|\hat{A}|\Psi(\mathbf{r}^N, t)\rangle \\ &= \int d\mathbf{r}^N \Psi^*(\mathbf{r}^N, t)\hat{A}\Psi(\mathbf{r}^N, t)\end{aligned} \tag{1.108}$$

When $\hat{A}$ is substituted with the unity operator, Eq. (1.108) shows that acceptable wavefunctions should be normalized to 1, that is, $\langle\psi|\psi\rangle = 1$. A central problem is the calculation of the wavefunction, $\Psi(\mathbf{r}^N, t)$, that describes the time-dependent state of the system. This wavefunction is the solution of the time-dependent

[6] The hermitian conjugatre of an operator $\hat{A}$ is the operator $\hat{A}^\dagger$ that satisfies

$$\int d\mathbf{r}^N \phi^*(\mathbf{r}^N)\hat{A}\psi(\mathbf{r}^N) = \int d\mathbf{r}^N (\hat{A}^\dagger\phi(\mathbf{r}^N))^*\psi(\mathbf{r}^N)$$

for all ϕ and ψ in the hilbert state of the system.

Schrödinger equation

$$\frac{\partial \Psi}{\partial t} = -\frac{i}{\hbar}\hat{H}\Psi \tag{1.109}$$

where the Hamiltonian $\hat{H}$ is the operator that corresponds to the energy observable, and in analogy to Eq. (1.196) is given by

$$\hat{H} = \hat{K} + \hat{U}(\mathbf{r}^N) \tag{1.110}$$

In the so-called *coordinate representation* the potential energy operator $\hat{U}$ amounts to a simple product, that is, $\hat{U}(\mathbf{r}^N)\Psi(\mathbf{r}^N, t) = U(\mathbf{r}^N)\Psi(\mathbf{r}^N, t)$ where $U(\mathbf{r}^N)$ is the classical potential energy. The kinetic energy operator is given in cartesian coordinates by

$$\hat{K} = -\hbar^2 \sum_{j=1}^{N} \frac{1}{2m_j}\nabla_j^2 \tag{1.111}$$

where the Laplacian operator is defined by

$$\nabla_j^2 = \frac{\partial^2}{\partial x_j^2} + \frac{\partial^2}{\partial y_j^2} + \frac{\partial^2}{\partial z_j^2} \tag{1.112}$$

Alternatively Eq. (1.111) can be written in the form

$$\hat{K} = \sum_{j=1}^{N} \frac{\hat{\mathbf{p}}_j^2}{2m_j} = \sum_{j=1}^{N} \frac{\hat{\mathbf{p}}_j \cdot \hat{\mathbf{p}}_j}{2m_j} \tag{1.113}$$

where the momentum vector operator is

$$\hat{\mathbf{p}}_j = \frac{\hbar}{i}\left(\frac{\partial}{\partial x_j}, \frac{\partial}{\partial y_j}, \frac{\partial}{\partial z_j}\right); \qquad i = \sqrt{-1} \tag{1.114}$$

The solution of Eq. (1.109) can be written in the form

$$\Psi(\mathbf{r}^N, t) = \sum_n \psi_n(\mathbf{r}^N) e^{-iE_n t/\hbar} \tag{1.115}$$

where ψ_n and E_n are solutions of the time-independent Schrödinger equation

$$\hat{H}\psi_n = E_n\psi_n \tag{1.116}$$

Equation (1.116) is an eigenvalue equation, and ψ_n and E_n are eigenfunctions and corresponding eigenvalues of the Hamiltonian. If at time $t = 0$ the system is in a state which is one of these eigenfunctions, that is,

$$\Psi(\mathbf{r}^N, t = 0) = \psi_n(\mathbf{r}^n) \tag{1.117}$$

then its future (and past) evolution is obtained from (1.109) to be

$$\Psi(\mathbf{r}^N, t) = \psi_n(\mathbf{r}^N)e^{-iE_nt/\hbar} \tag{1.118}$$

Equation (1.108) then implies that all observables are constant in time, and the eigenstates of $\hat{H}$ thus constitute *stationary states* of the system.

The set of energy eigenvalues of a given Hamiltonian, that is, the energies that characterize the stationary states of the corresponding system is called the *spectrum* of the Hamiltonian and plays a critical role in both equilibrium and dynamical properties of the system. Some elementary examples of single particle Hamiltonian spectra are:

A particle of mass m in a one-dimensional box of infinite depth and width a,

$$E_n = \frac{(2\pi\hbar)^2 n^2}{8ma^2}; \qquad n = 1, 2 \ldots \tag{1.119}$$

A particle of mass m moving in a one-dimensional harmonic potential $U(x) = (1/2)kx^2$,

$$E_n = \hbar\omega\left(n + \frac{1}{2}\right) \qquad n = 0, 1 \ldots; \qquad \omega = \sqrt{k/m} \tag{1.120}$$

A rigid rotator with moment of inertia I,

$$E_n = \frac{n(n+1)\hbar^2}{2I}; \qquad w_n = 2n + 1 \tag{1.121}$$

where w_n is the degeneracy of level n. Degeneracy is the number of different eigenfunctions that have the same eigenenergy.

An important difference between quantum and classical mechanics is that in classical mechanics stationary states exist at all energies while in quantum mechanics of finite systems the spectrum is discrete as shown in the examples above. This difference disappears when the system becomes large. Even for a single particle system, the spacings between allowed energy levels become increasingly smaller as the size of accessible spatial extent of the system increases, as seen, for example, in Eq. (1.119) in the limit of large a. This effect is tremendously amplified when the number of degrees of freedom increases. For example the three-dimensional analog of (1.119), that is, the spectrum of the Hamiltonian describing a particle in a three-dimensional infinitely deep rectangular box of side lengths a, b, c is

$$E(n_x, n_y, n_z) = \frac{(2\pi\hbar)^2}{8m}\left(\frac{n_x^2}{a^2} + \frac{n_y^2}{b^2} + \frac{n_z^2}{c^2}\right); \qquad n_x, n_y, n_z = 1, 2, \ldots \tag{1.122}$$

showing that in any energy interval the number of possible states is much larger because of the various possibilities to divide the energy among the three degrees of freedom.

For a many particle system this argument is compounded many times and the spectrum becomes essentially continuous. In this limit the details of the energy levels are no longer important. Instead, the density of states $\rho_E(E)$ becomes the important characteristic of the system spectral properties. $\rho_E(E)$ is defined such that $\rho_E(E)\Delta E$ is the number of system eigenstates with energy in the interval $E, \ldots, E + \Delta E$. For an example of application of this function see, for example, Section 2.8.2. Note that the density of states function can be defined also for a system with a dense but discrete spectrum, see Eqs (1.181) and (1.182) below.

1.4 Thermodynamics and statistical mechanics

1.4.1 Thermodynamics

The first law of thermodynamics is a statement of the law of energy conservation. The change in the system energy when its state changes from A to B is written as the sum of the work W done on the system, and the heat flow Q into the system, during the process. The mathematical statement of the first law is then

$$\Delta E = E_B - E_A = Q + W \tag{1.123}$$

The differential form of this statement is

$$dE = TdS - Pd\Omega \tag{1.124}$$

where S is the system entropy, T is the temperature, P is the pressure, and Ω is the system volume, respectively, and where we have assumed that all the mechanical work is an expansion against some pressure, that is, $dW = -Pd\Omega$. If the material composition in the system changes during the process a corresponding contribution to the energy appears and Eq. (1.124) becomes

$$dE = TdS - Pd\Omega + \sum_j \mu_j dN_j \tag{1.125}$$

where N_j is the number of molecules of species j and μ_j is the chemical potential of this species. An important observation is that while the energy is a function of the state of the system, the components of its change, W and Q are not—they depend on the path taken to reach that state. The entropy S is also a function of state; its difference between two equilibrium states of the system is

$$\Delta S = \int_A^B \left(\frac{dQ}{T}\right)_{\text{rev}} \tag{1.126}$$

where $()_{\text{rev}}$ denotes a reversible process—a change that is slow relative to the timescale of molecular relaxation processes, so that at each point along the way the system can be assumed to be at equilibrium.

When conditions for reversibility are not satisfied, that is, when the transition from A to B is not much slower than the internal system relaxation, the system cannot be assumed in equilibrium and in particular its temperature may not be well defined during the process. Still $\Delta S = S_B - S_A$ is well defined as the difference between entropies of two equilibrium states of the system. The second law of thermodynamics states that for a nonreversible path between states A and B

$$\Delta S > \int_A^B \frac{dQ}{T} \tag{1.127}$$

where T is the temperature of the surroundings (that of the system is not well defined in such an irreversible process).

Finally, the third law of thermodynamics states that the entropy of perfect crystalline substances vanishes at the absolute zero temperature.

The presentation so far describes an equilibrium system in terms of the extensive variables (i.e. variables proportional to the size of the system) $E, \Omega, S, \{N_j; j = 1, \ldots, n\}$. The intensive (size-independent) variables $P, T, \{\mu_j; j = 1, \ldots, N\}$ can be defined according to Eq. (1.125)

$$T = \left(\frac{\partial E}{\partial S}\right)_{\Omega,\{N\}}; \qquad P = -\left(\frac{\partial E}{\partial \Omega}\right)_{S,\{N\}}; \qquad \mu_j = \left(\frac{\partial E}{\partial N_j}\right)_{S,\Omega,\{N\}\neq N_j} \tag{1.128}$$

however, the independent variables in this representation are E (or S), Ω and $\{N_j\}$ that characterize a closed system.

Other representations are possible. The enthalpy

$$H = E + P\Omega \tag{1.129}$$

is a function of the independent variables S, P, and $\{N_j\}$, as can be seen by using Eq. (1.125) in $dH = dE + \Omega dP + Pd\Omega$ to get

$$dH = TdS + \Omega dP + \sum_j \mu_j dN_j \tag{1.130}$$

The Helmholtz free energy

$$F = E - TS \tag{1.131}$$

similarly satisfies

$$dF = -SdT - Pd\Omega + \sum_j \mu_j dN_j \tag{1.132}$$

This characterizes it as a function of the variables T, Ω, and $\{N_j\}$. The Gibbs free energy

$$G = E + P\Omega - TS \tag{1.133}$$

is then a function of T, P, and $\{N_j\}$. Indeed

$$dG = -SdT + VdP + \sum_j \mu_j dN_j \tag{1.134}$$

These thermodynamic functions can be shown to satisfy important extremum principles. The entropy of a closed system (characterized by the variables E, Ω, and $\{N_j\}$) at equilibrium is maximum in the sense that it is greater than the entropy of any other closed system characterized by the same extensive variables but with more internal restrictions. (A restriction can be, for example, a wall dividing the system and forcing molecules to stay on either one or the other side of it.) The energy of a closed equilibrium system with given entropy, volume, and particle numbers, is smaller than that of any similar system that is subjected to additional internal restrictions. The most useful statements are however those concerning the free energies. The Helmholtz free energy assumes a minimum value for an equilibrium system characterized by a given volume, given particle numbers, and a *given temperature*, again compared to similar systems with more imposed restrictions. Finally, the Gibbs free energy is minimal (in the same sense) for systems with given temperature, pressure, and particle numbers.

The study of thermodynamics involves the need to navigate in a space of many-variables, to transform between these variables, and to identify physically meaningful subspaces. Some mathematical theorems are useful in this respect. The

Euler theorem concerns the so-called homogeneous functions of order n, defined by the property

$$f(\lambda x_1 \ldots \lambda x_N) = \lambda^n f(x_1, \ldots, x_N) \tag{1.135}$$

It states that such functions satisfy

$$\sum_{j=1}^{N} x_j \frac{\partial f(x_1, \ldots, x_N)}{\partial x_j} = nf(x_1, \ldots, x_N) \tag{1.136}$$

We can use this theorem to address extensive functions of extensive variables, which are obviously homogeneous functions of order 1 in these variables, for example, the expression

$$E(\lambda S, \lambda \Omega, \{\lambda N_j\}) = \lambda E(S, \Omega, \{N_j\}) \tag{1.137}$$

just says that all quantities here are proportional to the system size. Using (1.136) with $n = 1$ then yields

$$E = S\frac{\partial E}{\partial S} + \Omega \frac{\partial E}{\partial \Omega} + \sum_j N_j \frac{\partial E}{\partial N_j} \tag{1.138}$$

Using (1.128) then leads to

$$E = TS - P\Omega + \sum_j N_j \mu_j \tag{1.139}$$

and, from (1.33)

$$G = \sum_j N_j \mu_j \tag{1.140}$$

Furthermore, since at constant T and P (from (1.134))

$$(dG)_{T,P} = \sum_j \mu_j dN_j \tag{1.141}$$

it follows, using (1.140) and (1.141) that

$$\sum_j N_j (d\mu_j)_{T,P} = 0 \tag{1.142}$$

The result (1.142) is the Gibbs–Duhem equation, the starting point in the derivation of the equations of chemical equilibria.

1.4.2 Statistical mechanics

Statistical mechanics is the branch of physical science that studies properties of macroscopic systems from the microscopic starting point. For definiteness we focus on the dynamics of an N-particle system as our underlying microscopic description. In classical mechanics the set of coordinates and momenta, $(\mathbf{r}^N, \mathbf{p}^N)$ represents a state of the system, and the microscopic representation of observables is provided by the dynamical variables, $A(\mathbf{r}^N, \mathbf{p}^N, t)$. The equivalent quantum mechanical objects are the quantum state $|j\rangle$ of the system and the associated expectation value $A_j = \langle j|\hat{A}|j\rangle$ of the operator $\hat{A}$ that corresponds to the classical variable A. The corresponding observables can be thought of as time averages

$$\langle A \rangle_t = \lim_{t\to\infty} \frac{1}{2t} \int_{-t}^{t} dt' A(t') \tag{1.143}$$

or as ensemble averages: if we consider an ensemble of $\mathcal{N}$ macroscopically identical systems, the ensemble average is

$$\langle A \rangle_e = \lim_{\mathcal{N}\to\infty} \frac{1}{\mathcal{N}} \sum_{j=1}^{\mathcal{N}} A_j \tag{1.144}$$

Obviously the time average (1.143) is useful only for stationary systems, that is, systems that do not macroscopically evolve in time. The *ergodic hypothesis* (sometimes called ergodic theorem) assumes that for large stationary systems the two averages, (1.143) and (1.144) are the same. In what follows we discuss equilibrium systems, but still focus on ensemble averages that lead to more tractable theoretical descriptions. Time averages are very useful in analyzing results of computer simulations.

The formulation of statistical mechanics from ensemble averages can take different routes depending on the ensemble used. Our intuition tells us that if we focus attention on a small (but still macroscopic) part of a large system, say a glass of water from the Atlantic ocean, its thermodynamic properties will be the same when open to the rest of the ocean, that is, exchanging energy and matter with the outside world, as when closed to it. Three theoretical constructs correspond to these scenarios. The microcanonical ensemble is a collection of microscopically identical closed systems characterized by energy, E, volume Ω, and number of particles N. The canonical ensemble is a collection of systems characterized by their volume and number of particles, and by their temperature; the latter is determined by keeping open the possibility to exchange energy with a thermal bath of temperature T. The grand canonical ensemble is a collection of systems that are in equilibrium and can exchange both energy and matter with a bath characterized by a given temperature T and a chemical potential of the material system, μ.

For each of these ensembles of $\mathcal{N}$ systems let $f_j(\mathcal{N})$ be the fraction of systems occupying a given microscopic state j. The ensemble probability P_j is defined by $P_j = \lim_{\mathcal{N}\to\infty} f_j(\mathcal{N})$. The macroscopic observable that corresponds to the dynamical variable $\mathcal{A}$ is then

$$\langle A \rangle_e = \sum_j P_j \mathcal{A}_j \tag{1.145}$$

In the grand canonical formulation the sum over j should be taken to include also a sum over number of particles.

1.4.2.1 *Microcanonical ensemble*

The probability that a system is found in a state of energy E_j is given by

$$P_j = \frac{1}{\rho_E(E,\Omega,N)} \delta(E - E_j) \tag{1.146}$$

where $\rho_E(E,\Omega,N)$ is the density of energy states, the same function that was discussed at the end of Section 1.3. Its formal definition

$$\rho_E(E,\Omega,N) = \sum_j \delta(E - E_j(\Omega,N)) \tag{1.147}$$

insures that P_j is normalized and makes it clear that the integral $\int_E^{E+\Delta E} dE\, \rho_E(E,\Omega,N)$ gives the number of energy states in the interval between E and $E + \Delta E$. Equation (1.146) expresses a basic postulate of statistical mechanics, that all microscopic states of the same energy have the same probability.

One thing that should be appreciated about the density of states of a macroscopic system is how huge it is. For a system of N structureless (i.e. no internal states) particles of mass m confined to a volume Ω but otherwise moving freely it is given by[7]

$$\rho_E(E,\Omega,N) = \frac{1}{\Gamma(N+1)\Gamma(3N/2)} \left(\frac{m}{2\pi\hbar^2}\right)^{3N/2} \Omega^N E^{(3N/2)-1} \tag{1.148}$$

where Γ are gamma-functions ($\Gamma(N) = (N-1)!$ for an integer N) that for large N can be evaluated from the Stirling formula (1.18). For a system of linear size

[7] D. A. McQuarrie, *Statistical Mechanics* (Harper and Row, New York, 1976), Chapter 1.

$l = 1$ cm ($\Omega = l^3$) and energy $E = 3Nk_BT/2$ with $m = 10^{-22}$ g, $N = 10^{23}$, and $T = 300$ K, the dimensionless parameter $\xi = ml^2E/(2\pi\hbar^2)$ is $\sim 10^{41}$ so that $\rho_E = E^{-1}(10^{41})^{3N/2}/e^{(5/2)N \ln N} \sim 10^{3N}E^{-1}$.

In the present context $\rho_E(E, \Omega, N)$ is the *microcanonical partition function*—a sum over the un-normalized probabilities. This function is in turn directly related to the system entropy

$$S(E, \Omega, N) = k_B \ln \rho_E(E, \Omega, N) \tag{1.149}$$

where k_B is the Boltzmann constant. From (1.125), written in the form

$$dS = \frac{1}{T}dE + \frac{P}{T}d\Omega - \frac{\mu}{T}dN \tag{1.150}$$

it immediately follows that

$$\frac{1}{k_BT} = \left(\frac{\partial \ln \rho_E}{\partial E}\right)_{N,\Omega} \tag{1.151}$$

$$\frac{P}{k_BT} = \left(\frac{\partial \ln \rho_E}{\partial \Omega}\right)_{N,E} \tag{1.152}$$

$$\frac{\mu}{k_BT} = -\left(\frac{\partial \ln \rho_E}{\partial N}\right)_{\Omega,E} \tag{1.153}$$

1.4.2.2 *Canonical ensemble*

For an ensemble of systems that are in equilibrium with an external heat bath of temperature T, the probability to find a system in state j of energy E_j is given by

$$P_j = e^{-\beta E_j}/Q \tag{1.154}$$

where

$$Q(T, \Omega, N) = \sum_j e^{-\beta E_j(\Omega, N)}; \qquad \beta = (k_BT)^{-1} \tag{1.155}$$

is the *canonical partition function*. For a macroscopic system the energy spectrum is continuous and Eq. (1.155) can be rewritten as (setting the energy scale so that the ground state energy is nonnegative)

$$Q(T, \Omega, N) = \int_0^\infty dE\, \rho_E(E, \Omega, N) e^{-\beta E(\Omega, N)} \tag{1.156}$$

The canonical partition function is found to be most simply related to the Helmholtz free energy

$$F = -k_B T \ln Q \tag{1.157}$$

Using (1.132) it follows that

$$S = k_B T \left(\frac{\partial \ln Q}{\partial T} \right)_{N,\Omega} + k_B \ln Q \tag{1.158}$$

$$P = k_B T \left(\frac{\partial \ln Q}{\partial \Omega} \right)_{N,T} \tag{1.159}$$

$$\mu = -k_B T \left(\frac{\partial \ln Q}{\partial N} \right)_{\Omega,T} \tag{1.160}$$

In addition, it follows from Eqs (1.154) and (1.155) that the average energy in the system is

$$E = \sum_j E_j P_j = k_B T^2 \left(\frac{\partial \ln Q}{\partial T} \right)_{N,\Omega} \tag{1.161}$$

It is easily verified that that the analog expression for the pressure

$$P = -\sum_j \frac{\partial E_j}{\partial \Omega} P_j \tag{1.162}$$

is consistent with Eq. (1.159).

It is important to understand the conceptual difference between the quantities E and S in Eqs (1.161) and (1.158), and the corresponding quantities in Eq. (1.149). In the microcanonical case E, S, and the other derived quantities (P, T, μ) are unique numbers. In the canonical case these, except for T which is defined by the external bath, are ensemble averages. Even T as defined by Eq. (1.151) is not the same as T in the canonical ensemble. Equation (1.151) defines a temperature for a closed equilibrium system of a given total energy while as just said, in the canonical ensemble T is determined by the external bath. For macroscopic observations we often disregard the difference between average quantities that characterize a system open to its environment and the deterministic values of these parameters in the equivalent closed system. Note however that fluctuations from the average are themselves often related to physical observables and should be discussed within their proper ensemble.

An interesting microscopic view of the first law of thermodynamics is obtained from using (1.161) to write

$$dE = d\sum_j E_j P_j = \underbrace{\sum_j E_j dP_j}_{\substack{\text{reversible}\\ \text{heat}}} + \underbrace{\sum_j P_j dE_j}_{\substack{\text{reversible}\\ \text{work}}} \tag{1.163}$$

The use of the word "reversible" here is natural: any infinitesimal process is by definition reversible. The change in the average energy of a system is seen to be made of a contribution associated with the change in the occupation probability of different energy states—which is what we associate with changing temperature, that is, reversible heat exchange with the surrounding, and another contribution in which these occupation probabilities are fixed but the energies of the state themselves change—as will be the case if the volume of the system changed as a result of mechanical work.

1.4.2.3 Grand-canonical ensemble

For an ensemble of systems that are in contact equilibrium with both heat and matter reservoirs characterized by a temperature T and a chemical potential μ, respectively, the probability to find a system with N particles and in the energy level $E_{jN}(\Omega)$ is given by

$$P_{jN} = \frac{e^{-\beta(E_{jN}(\Omega)-\mu N)}}{\Xi} \tag{1.164}$$

where the grand-canonical partition function is

$$\begin{aligned} \Xi = \Xi(T,\Omega,\mu) &= \sum_N \sum_j e^{-\beta E_{jN}(\Omega)} e^{\beta\mu N} \\ &= \sum_{N=0}^{\infty} Q(\Omega,T,N)\lambda^N; \qquad \lambda = e^{\beta\mu} \end{aligned} \tag{1.165}$$

Its connection to average thermodynamic observables can be obtained from the fundamental relationship

$$P\Omega = k_B T \ln \Xi \tag{1.166}$$

and the identity

$$d(P\Omega) = SdT + Pd\Omega + Nd\mu \tag{1.167}$$

Together (1.166) and (1.167) imply

$$S = k_B \ln \Xi + k_B T \left(\frac{\partial (\ln \Xi)}{\partial T} \right)_{\Omega,\mu} \tag{1.168}$$

$$\bar{P} = k_B T \left(\frac{\partial \ln \Xi}{\partial \Omega} \right)_{T,\mu} \tag{1.169}$$

and

$$\bar{N} = k_B T \left(\frac{\partial \ln \Xi}{\partial \mu} \right)_{T,\Omega} \tag{1.170}$$

1.4.3 Quantum distributions

The quantum analogs of the phase space distribution function and the Liouville equation discussed in Section 1.2.2 are the density operator and the quantum Liouville equation discussed in Chapter 10. Here we mention for future reference the particularly simple results obtained for equilibrium systems of identical noninteracting particles. If the particles are distinguishable, for example, atoms attached to their lattice sites, then the canonical partitions function is, for a system of N particles

$$Q(T,\Omega,N) = q^N; \qquad q = \sum_i e^{-\beta\varepsilon_i} \tag{1.171}$$

where ε_i is the energy of the single particle state i and q is the single particle partition function. If the particles are non-distinguishable, we need to account for the fact that interchanging between them does not produce a new state. In the high-temperature limit, where the number of energetically accessible states greatly exceeds the number of particles this leads to

$$Q(T,\Omega,N) = \frac{q^N}{N!} \tag{1.172}$$

Using Eq. (1.161), both Eqs (1.171) and (1.172) lead to the same expression for the average system energy

$$\bar{E} = N \sum_i \varepsilon_i f_i \tag{1.173}$$

where f_i, the probability that a molecule occupies the state i, is

$$f_i = e^{-\beta\varepsilon_i}/q \tag{1.174}$$

At low temperature the situation is complicated by the fact that the difference between distinguishable and indistinguishable particles enters only when they occupy different states. This leads to different statistics between fermions and bosons and to the generalization of (1.174) to

$$f_i = \frac{1}{e^{\beta(\varepsilon_i - \mu)} \pm 1} \tag{1.175}$$

where μ is the chemical potential and where the (+) sign is for fermions while the (−) is for bosons. μ is determined from the condition $\sum_i f_i = 1$. This condition also implies that when $T \to \infty$ individual occupation probabilities should approach zero, which means that $\mu \to -\infty$ so that

$$f_i \xrightarrow{T\to\infty} e^{-\beta(\varepsilon_i - \mu)} \quad \text{and} \quad \mu \xrightarrow{T\to\infty} -k_B T \ln\left(\sum_i e^{-\beta\varepsilon_i}\right) \tag{1.176}$$

1.4.4 Coarse graining

Consider the local density, $\rho(\mathbf{r}, t)$, of particles distributed and moving in space. We explicitly indicate the position and time dependence of this quantity in order to express the fact the system may be non-homogeneous and out of equilibrium. To define the local density we count the number of particles $n(\mathbf{r}, t)$ in a volume $\Delta\Omega$ about position $\mathbf{r}$ at time t (for definiteness we may think of a spherical volume centered about $\mathbf{r}$). The density

$$\rho^{\Delta\Omega}(\mathbf{r}, t) = n(\mathbf{r}, t)/\Delta\Omega \tag{1.177}$$

is obviously a fluctuating variable that depends on the size of $\Delta\Omega$. To define a meaningful local density $\Delta\Omega$ should be large relative to the interparticle spacing and small relative to the scale of inhomogeneities in the local density that we wish to describe. Alternatively, we can get a meaningful density by averaging the instantaneous density over predefined time intervals.

We can make these statements more quantitative by defining the dynamical density variable (see Section 1.2.1) according to

$$\rho(\mathbf{r}, t) = \rho(\mathbf{r}, \{\mathbf{r}_i(t)\}) \equiv \sum_i \delta(\mathbf{r} - \mathbf{r}_i(t)) \tag{1.178}$$

where δ is the three-dimensional Dirac delta function, $\mathbf{r}_i(t)$ is the position of particle i at time t and the sum is over all particles. This dynamical variable depends on the positions of all particles in the system and does not depend on their momenta.

The local density defined in Eq. (1.177) is then given by

$$\rho^{\Delta\Omega}(\mathbf{r},t) = \frac{1}{\Delta\Omega}\int_{\Delta\Omega} d\mathbf{r}'\rho(\mathbf{r}',\{\mathbf{r}_i(t)\}) \tag{1.179}$$

where the integral is over a volume $\Delta\Omega$ about the point $\mathbf{r}$. Furthermore, for an equilibrium system we could also perform a local time average

$$\rho^{\Delta\Omega,\Delta t}(\mathbf{r},t) = \frac{1}{\Delta t}\int_{t-\Delta t/2}^{t+\Delta t/2} dt'\rho^{\Delta\Omega}(\mathbf{r},t') \tag{1.180}$$

The processes (1.179) and (1.180) by which we transformed the dynamical variable $\rho(\mathbf{r},t)$ to its "smoother" counterpart $\rho^{\Delta\Omega}(\mathbf{r},t)$ is an example of *coarse graining*.[8]

What was achieved by this coarse-graining process? Consider the spatial coarse graining (1.179). As a function of $\mathbf{r}$, ρ of Eq. (1.178) varies strongly on a length scale of the order of a particle size—showing a spike at the position of each particle,[9] however variations on these length scales are rarely of interest. Instead we are often interested in more systematic inhomogeneities that are observed in hydrodynamics or in electrochemistry, or those that can be probed by light scattering (with typical length-scale determined by the radiation wavelength). Such variations, without the irrelevant spiky structure, are fully contained in $\rho^{\Delta\Omega}$ provided that the volume elements $\Delta\Omega$ are taken large relative to the inter-particle distance and small relative to the inhomogeneous features of interest. Clearly, $\rho^{\Delta\Omega}(\mathbf{r})$ cannot describe the system structure on a length scale smaller than $l \sim (\Delta\Omega)^{1/3}$, but it provides a simpler description of those system properties that depend on longer length scales.

Coarse graining in time is similarly useful. It converts a function that is spiky (or has other irregularities) in time to a function that is smooth on timescales shorter than Δt, but reproduces the relevant slower variations of this function. This serves to achieve a mathematically simpler description of a physical system on the timescale of interest. The attribute "of interest" may be determined by the experiment—it is

---

[8] Another systematic way to coarse grain a function $f(\mathbf{r})$ is to express it as a truncated version of its Fourier transform

$$f^{cg}(\mathbf{r}) = \int_{|\mathbf{k}|<k_c} d\mathbf{k}\hat{f}(\mathbf{k})e^{i\mathbf{k}\cdot\mathbf{r}} \qquad \text{where } \hat{f}(\mathbf{k}) = (1/(2\pi)^3)\int d\mathbf{r}f(\mathbf{r})e^{-i\mathbf{k}\cdot\mathbf{r}}$$

where k_c is some cutoff that filters high k components out of the coarse-grained function $f^{cg}(\mathbf{r})$.

[9] In fact, if we were interested on variations on such length scales, we should have replaced the delta function in Eq. (1.178) by a function that reflects this size, for example, zero outside the volume $\Delta\Omega$ occupied by the particle, and $(\Delta\Omega)^{-1}$ inside this volume.

often useless to describe a system on a timescale shorter than what is measurable. Our brain performs such coarse graining when we watch motion pictures, so we sense a continuously changing picture rather than jumping frames.

It should be noticed that coarse graining is a reduction process: We effectively reduce the number of random variables used to describe a system. This statement may appear contradictory at first glance. In (1.179) we convert the function $\rho(\mathbf{r}, t)$, Eq. (1.178), which is completely specified by $3N$ position variables (of N particles) to the function $\rho^{\Delta\Omega}(\mathbf{r}, t)$ that appears to depend on any point in continuous space. However, spatial variations in the latter exists only over length scales larger than $(\Delta\Omega)^{1/3}$, so the actual number of independent variables in the coarse-grained system is of order $\Omega/\Delta\Omega$ where Ω is the system volume. This number is much smaller than N if (as we usually take) $(\Delta\Omega)^{1/3}$ is much larger than both molecular size and intermolecular spacing.

Finally, consider the eigenvalues $\{E_j\}$ of some Hamiltonian $\hat{H}$ of interest. We can define the density of states function

$$\rho(E) = \sum_j \delta(E - E_j) \tag{1.181}$$

that has the property that the integral $\int_{E_a}^{E_b} dE\rho(E)$ gives the number of energy eigenvalues in the interval $E_a, \ldots, E_b$. When the spectrum of $\hat{H}$ becomes very dense it is useful to define a continuous coarse-grained analog of (1.181)

$$\rho(E) \to \frac{1}{\Delta E} \int\limits_{E-(1/2)\Delta E}^{E+(1/2)\Delta E} \rho(E) dE \tag{1.182}$$

where ΔE is large relative to the spacing between consecutive E_js. This coarse-grained density of states is useful in applications where the spectrum is dense enough so that ΔE can be taken small relative to any experimentally meaningful energy interval. In such applications $\rho(E)$ in (1.181) and (1.182) can be used interchangeably, and we will use the same notation for both.

The central limit theorem of probability theory (Section 1.1.1) finds its most useful application in statistical mechanics through applications of the coarse-graining idea. The coarse-graining procedure essentially amounts to generating a new "coarse grained" random variable by summing up many random variables in a certain interval. Indeed the reduction (1.179) amounts to replacing a group of random variables ($\rho(\mathbf{r})$ for all $\mathbf{r}$ in the interval $\Delta\Omega$) by their sum. If this interval is large relative to the correlation distance[10] between these variables, and if the probability distribution that governs these variables satisfies the required convergence conditions, then the probability distribution for the coarse-grained variable is Gaussian. An example is given in Chapter 7 by the derivation of Eq. (7.37).

1.5 Physical observables as random variables

1.5.1 Origin of randomness in physical systems

Classical mechanics is a deterministic theory, in which the time evolution is uniquely determined for any given initial condition by the Newton equations (1.98). In quantum mechanics, the physical information associated with a given wavefunction has an inherent probabilistic character, however the wavefunction itself is uniquely determined, again from any given initial wavefunction, by the Schrödinger equation (1.109). Nevertheless, many processes in nature appear to involve a random component in addition to their *systematic* evolution. What is the origin of this random character? There are two answers to this question, both related to the way we observe physical systems:

1. *The initial conditions are not well characterized.* This is the usual starting point of statistical mechanics. While it is true that given the time evolution of a physical system is uniquely defined by the initial state, a full specification of this state includes all positions and momenta of all N particles of a classical system or the full N-particle wavefunction of the quantum system. Realistic initial conditions are never specified in this way—only a few averaged system properties (e.g. temperature, volume) are given. Even studies of microscopic phenomena start with a specification of a few coordinates that are judged to be interesting, while the effect of all others is again specified in terms of macroscopic averages. Repeating the experiment (or the calculation) under such ill-defined initial conditions amounts to working with an ensemble of systems characterized by these conditions. The observables are now random variables that should be averaged over this ensemble.

2. *We use a reduced description of the system (or process) of interest.* In many cases, we seek simplified descriptions of physical processes by focusing on a small subsystem or on a few observables that characterize the process of interest. These observables can be macroscopic, for example, the energy, pressure, temperature, etc., or microscopic, for example, the center of mass position, a particular bond length, or the internal energy of a single molecule. In the reduced space of these "important" observables, the microscopic influence of the other $\sim 10^{23}$ degrees of

[10] The correlation distance r_{cor} for $\rho(\mathbf{r})$ is defined as the distance above which $\rho(\mathbf{r})$ and $\rho(\mathbf{r}+\hat{\mathbf{n}}r_{\text{cor}})$ are statistically independent ($\hat{\mathbf{n}}$ is a unit vector in any direction).

freedom appears as random fluctuations that give these observables an apparently random character. For example, the energy of an individual molecule behaves as a random function of time (i.e. a stochastic process) even in a closed system whose total energy is strictly constant.

1.5.2 Joint probabilities, conditional probabilities, and reduced descriptions

Most readers of this text have been exposed to probability theory concepts (Section 1.1.1) in an elementary course in statistical thermodynamics. As outlined in Section 1.2.2, a state of a classical N-particle system is fully characterized by the $6N$-dimensional vector $(r^N, p^N) \equiv (\mathbf{r}_1, \mathbf{r}_2, \ldots, \mathbf{r}_N, \mathbf{p}_1, \mathbf{p}_2, \ldots \mathbf{p}_N)$ (a point in the $6N$-dimensional phase space). A probability density function $f(r^N, p^N)$ characterizes the equilibrium state of the system, so that $f(r^N, p^N)dr^N dp^N$ is the probability to find the system in the neighborhood $dr^N dp^N = d\mathbf{r}_1, \ldots, d\mathbf{p}_N$ of the corresponding phase point. In a canonical ensemble of equilibrium systems characterized by a temperature T the function $f(r^N, p^N)$ is given by

$$f(r^N, p^N) = \frac{e^{-\beta H(r^N, p^N)}}{\int dr^N \int dp^N e^{-\beta H(r^N, p^N)}}; \qquad \beta = (k_B T)^{-1} \tag{1.183}$$

where k_B is the Boltzmann constant and H is the system Hamiltonian

$$H(r^N, p^N) = \sum_{i=1}^{N} \frac{\mathbf{p}_i^2}{2m_i} + U(r^N) \tag{1.184}$$

Here $p_i^2 = p_{ix}^2 + p_{iy}^2 + p_{iz}^2$ and U is the potential associated with the inter-particle interaction. The function $f(r^N, p^N)$ is an example of a *joint probability density* function (see below). The structure of the Hamiltonian (1.184) implies that f can be factorized into a term that depends only on the particles' positions and terms that depend only on their momenta. This implies, as explained below, that at equilibrium positions and momenta are statistically independent. In fact, Eqs (1.183) and (1.184) imply that individual particle momenta are also statistically independent and so are the different cartesian components of the momenta of each particle.

Let us consider these issues more explicitly. Consider two random variables x and y. The *joint probability density* $P_2(x, y)$ is defined so that $P_2(x, y)dxdy$ is the probability of finding the variable x at $x, \ldots, x + dx$ *and* y at $y, \ldots, y + dy$. We refer by the name *reduced description* to a description of the system in terms of partial specification of its state. For example, the probability that the variable x is at the interval $x, \ldots, x + dx$ *irrespective of the value of* y is $P_1^{(x)}(x) = \int dy P_2(x, y)$. Similarly, $P_1^{(y)}(y) = \int dx P_2(x, y)$. Note that the functional forms of $P_1^{(x)}$ and $P_1^{(y)}$ are not necessarily the same. Also note that all these functions satisfy the normalization conditions

$$\int dxdyP_2(x,y) = \int dxP_1^{(x)}(x) = \int dyP_1^{(y)}(y) = 1 \qquad (1.185)$$

The two random variables x and y are called independent or uncorrelated if

$$P_2(x,y) = P_1^{(x)}(x)P_1^{(y)}(y) \qquad (1.186)$$

The *conditional probability distribution* $P(x|y)$ is defined so that $P(x|y)dx$ is the probability that the value of x is in the interval $x,\ldots,x+dx$ given that the variable y takes the value y. From this definition it follows that

$$P(x|y) = \frac{P_2(x,y)}{P_1^{(y)}(y)}; \qquad P(y|x) = \frac{P_2(x,y)}{P_1^{(x)}(x)} \qquad (1.187)$$

or rather

$$P(x|y)dx \cdot P_1^{(y)}(y)dy = P(x,y)dxdy \qquad (1.188)$$

The term $P(x|y)dx$ is the probability that if y has a given value then x is in the range $x,\ldots,x+dx$. The term $P_1^{(y)}(y)dy$ is the probability that y is in the range $y,\ldots,y+dy$. Their product is the joint probability that x and y have particular values within their respective intervals.

Problem 1.2. Show that if x and y are independent random variables then $P(x|y)$ does not depend on y.

Reduced descriptions are not necessarily obtained in terms of the original random variables. For example, given the probability density $P(x,y)$ we may want the probability of the random variable

$$z = x + y \qquad (1.189)$$

This is given by

$$P_1^{(z)}(z) = \int dxP_2(x,z-x) \qquad (1.190)$$

More generally, if $z = f(x,y)$ then

$$P_1^{(z)}(z) = \int dx \int dy\delta(z-f(x,y))P_2(x,y) \qquad (1.191)$$

1.5.3 Random functions

Consider a set of random variables, $x_1, x_2, \ldots, x_N$ and the associated probability density $P_N(x_1, \ldots, x_N)$. Here, the ordering indices $n = 1, \ldots, N$ are integers. Alternatively, the ordering index may be continuous so that, for example, $x(\nu)$ is a random variable for each real ν. We say that $x(\nu)$ is a random function of ν: a random function assigns a random variable to each value of its argument(s). The corresponding joint probability density $P[x(\nu)]$ is a functional of this function.

The most common continuous ordering parameters in physics and chemistry are position and time. For example, the water height in a lake on a windy day is a random function $h(x, y, t)$ of the positions x and y in the two-dimensional lake plane and of the time. For any particular choice, say x_1, y_1, t_1 of position and time $h(x_1, y_1, t_1)$ is a random variable in the usual sense that its repeated measurements (over an ensemble of lakes or in different days with the same wind characteristics) will yield different results, predictable only in a probabilistic sense.

1.5.4 Correlations

When Eq. (1.186) does not hold, the variables x and y are said to be correlated. In this case the probability to realize a certain value of x depends on the value realized by y, as expressed by the conditional probability density $P(x|y)$. When x and y are not correlated Eqs (1.186) and (1.187) imply that $P(x|y) = P_1^{(x)}(x)$.

The moments of $P_2(x,y)$ are defined by integrals such as

$$
\begin{aligned}
\langle x^k \rangle &= \int dxdyx^k P_2(x,y) = \int dxx^k P_1^{(x)}(x) \\
\langle y^k \rangle &= \int dxdyy^k P_2(x,y) = \int dyy^k P_1^{(y)}(y) \\
\langle x^k y^l \rangle &= \int dxdyx^k y^l P_2(x,y)
\end{aligned}
\tag{1.192}
$$

Again, if x and y are uncorrelated then

$$
\langle x^k y^l \rangle = \int dxx^k P_1^{(x)}(x) \int dyy^l P_1^{(y)}(y) = \langle x^k \rangle \langle y^l \rangle
$$

The difference

$$
\langle x^k y^l \rangle - \langle x^k \rangle \langle y^l \rangle \tag{1.193}
$$

therefore measures the correlation between the random variables x and y. In particular, if x and y are random functions of some variable z, then

$C_{xy}(z_1, z_2) = \langle x(z_1)y(z_2)\rangle - \langle x(z_1)\rangle\langle y(z_2)\rangle$ is referred to as the *correlation function* of these variables.

Two types of correlation functions are particularly important in the description of physical and chemical systems:

1. *Spatial correlation functions*. Consider, for example, the density of liquid molecules as a function of position, $\rho(\mathbf{r})$. In macroscopic thermodynamics $\rho(\mathbf{r})$ is an ensemble average. However, if we actually count the number of molecules $n(\mathbf{r})$ in a given volume ΔV about $\mathbf{r}$, then

$$\rho^{\Delta V}(\mathbf{r}) = n(\mathbf{r})/\Delta V \tag{1.194}$$

is a random variable, and, taken as a function of $\mathbf{r}$, is a random function of position. It should be emphasized that the random variable defined in this way depends on the coarse graining (see Section 1.4.3) volume ΔV, however for the rest of this section we will suppress the superscript denoting this fact.

In a homogeneous equilibrium system the ensemble average $\langle\rho(\mathbf{r})\rangle = \rho$ is independent of $\mathbf{r}$, and the difference $\delta\rho(\mathbf{r}) \equiv \rho(\mathbf{r}) - \rho$ is a random function of position that measures local fluctuations from the average density. Obviously $\langle\delta\rho(\mathbf{r})\rangle = 0$, while a measure of the magnitude of *density fluctuations* is given by $\langle\delta\rho^2\rangle = \langle\rho^2\rangle - \langle\rho\rangle^2$. The density–density spatial correlation function measures the correlation between the random variables $\delta\rho(\mathbf{r}')$ and $\delta\rho(\mathbf{r}'')$, that is, $C(\mathbf{r}', \mathbf{r}'') = \langle\delta\rho(\mathbf{r}')\delta\rho(\mathbf{r}'')\rangle$. In a homogeneous system it depends only on the distance $\mathbf{r}' - \mathbf{r}''$, that is,

$$C(\mathbf{r}', \mathbf{r}'') = C(\mathbf{r}) = \langle\delta\rho(\mathbf{r})\delta\rho(0)\rangle = \langle\delta\rho(0)\delta\rho(\mathbf{r})\rangle; \qquad \mathbf{r} = \mathbf{r}' - \mathbf{r}'' \tag{1.195}$$

and in an isotropic system—only on its absolute value $r = |\mathbf{r}|$. Both $\langle\delta\rho^2\rangle$ and $C(\mathbf{r})$ are measurable and contain important information on the equilibrium system.

Problem 1.3. Show that in a homogeneous system

$$\langle\delta\rho(\mathbf{r}')\delta\rho(\mathbf{r}'')\rangle = \langle\rho(\mathbf{r}')\rho(\mathbf{r}'')\rangle - \rho^2$$

2. *Time correlation functions*. If we look at $\delta\rho(\mathbf{r}, t)$ at a given $\mathbf{r}$ as a function of time, its time evolution is an example of a stochastic process (see Chapter 7). In a given time t_1 the variables $\rho(\mathbf{r}, t_1)$ and $\delta\rho(\mathbf{r}, t_1) = \rho(\mathbf{r}, t_1) - \rho$ are random variables in the sense that repeated measurements done on different identical systems will give different realizations for these variables about the average ρ. Again, for the random variables $x = \delta\rho(\mathbf{r}, t')$ and $y = \delta\rho(\mathbf{r}, t'')$ we can look at the correlation

function $\langle xy \rangle$. This function

$$C(\mathbf{r}, t', t'') = \langle \delta\rho(\mathbf{r}, t')\delta\rho(\mathbf{r}, t'') \rangle = \langle \rho(\mathbf{r}, t')\rho(\mathbf{r}, t'') \rangle - \rho^2 \tag{1.196}$$

is the time correlation function. In a stationary system, for example, at equilibrium, this function depends only on the time difference

$$C(\mathbf{r}, t', t'') = C(\mathbf{r}, t); \qquad t = t' - t'' \tag{1.197}$$

Many time correlation functions are observables and contain important information on dynamical system properties. We can also study time and space correlation functions

$$C(\mathbf{r} - \mathbf{r}', t - t') = \langle \rho(\mathbf{r}, t)\rho(\mathbf{r}', t') \rangle - \rho^2 \tag{1.198}$$

that contain information on the time evolution of the system's structure.

1.5.5 Diffusion

As a demonstration of the use of the concepts introduced above consider the well known process of diffusion. Consider a system of diffusing particles and let $P(\mathbf{r}, t)$ be the probability density to find a particle in position r at time t, that is, $P(\mathbf{r}, t)d^3r$ is the probability that a particle is in the neighborhood d^3r of $\mathbf{r}$ at this time. $P(\mathbf{r}, t)$ is related to the concentration $c(\mathbf{r}, t)$ by a change of normalization

$$c(\mathbf{r}, t) = NP(\mathbf{r}, t) \tag{1.199}$$

where N is the total number of particles. The way by which $c(\mathbf{r}, t)$ and $P(\mathbf{r}, t)$ evolve with time is known from experimental observation to be given by the *diffusion equation.* In one dimension

$$\frac{\partial P(x, t)}{\partial t} = D\frac{\partial^2}{\partial x^2}P(x, t) \tag{1.200}$$

This evolution equation demonstrates the way in which a reduced description (see Section 1.5.1) yields dynamics that is qualitatively different than the fundamental one: A complete description of the assumed classical system involves the solution of a huge number of coupled Newton equations for all particles. Focusing on the position of one particle and realizing that the ensuing description has to be probabilistic, we find (in the present case experimentally) that the evolution is fundamentally different. For example, in the absence of external forces the particle position changes linearly with time, $x = vt$, while (see below) Eq. (1.200) implies that the mean square displacement $\langle x^2 \rangle$ changes linearly with time. Clearly the reason for this is that microscopically, the evolution of the particle position involves multiple collisions with many other particles whose detailed motions do not appear in (1.200). Consequently, Eq. (1.200) is valid only on timescales long relative to the time between collisions and on length scales long relative to the mean free path, that is, it is a coarse-grained description of the particle's motion.

If the distribution depends on time, so do its moments. Suppose the particle starts at the origin, $x = 0$. Its average position at time t is given by

$$\langle x \rangle_t = \int_{-\infty}^{\infty} dx x P(x,t) \tag{1.201}$$

Therefore,

$$\frac{\partial \langle x \rangle}{\partial t} = D \int_{-\infty}^{\infty} dx x \frac{\partial^2}{\partial x^2} P(x,t) \tag{1.202}$$

Integrating on the right-hand side by parts, using the fact that P and its derivatives have to vanish at $|x| \to \infty$, leads to[11]

$$\frac{\partial \langle x \rangle}{\partial t} = 0, \qquad \text{that is, } \langle x \rangle = 0 \text{ at all times} \tag{1.203}$$

Consider now the second moment

$$\langle x^2 \rangle_t = \int_{-\infty}^{\infty} dx x^2 P(x,t) \tag{1.204}$$

whose time evolution is given by

$$\frac{\partial \langle x^2 \rangle}{\partial t} = D \int_{-\infty}^{\infty} dx x^2 \frac{\partial^2 P}{\partial x^2} \tag{1.205}$$

[11] To obtain Eqs (1.203) and (1.206) we need to assume that P vanishes as $x \to \infty$ faster than x^{-2}. Physically this must be so because a particle that starts at $x = 0$ cannot reach beyond some finite distance at any finite time if only because its speed cannot exceed the speed of light. Of course, the diffusion equation does not know the restrictions imposed by the Einstein relativity theory (similarly, the Maxwell–Boltzmann distribution assigns finite probabilities to find particles with speeds that exceed the speed of light). The real mathematical reason why P has to vanish faster than x^{-2} is that in the equivalent three-dimensional formulation $P(\mathbf{r})$ has to vanish faster than r^{-2} as $r \to \infty$ in order to be normalizable.

Again, integration by parts of the right-hand side and using the boundary conditions at infinity, that is,

$$\int_{-\infty}^{\infty} dx x^2 \frac{\partial^2 P}{\partial x^2} = \left[x^2 \frac{\partial P}{\partial x} \right]_{-\infty}^{\infty} - \int_{-\infty}^{\infty} dx \cdot 2x \frac{\partial P}{\partial x} = -2[xP]_{-\infty}^{\infty} + 2 \int_{-\infty}^{\infty} dx P = 2 \tag{1.206}$$

leads to $\partial \langle x^2 \rangle / \partial t = 2D$, therefore, since $\langle x^2 \rangle_0 = 0$,

$$\langle x^2 \rangle_t = 2Dt \tag{1.207}$$

For three-dimensional diffusion in an isotropic system the motions in the x, y, and z directions are independent (the equation $\partial P(\mathbf{r}, t)/\partial t = D(\partial^2/\partial x^2 + \partial^2/\partial y^2 + \partial^2/\partial z^2) P(\mathbf{r}, t)$ is separable), so

$$\langle \mathbf{r}^2 \rangle_t = \langle x^2 \rangle_t + \langle y^2 \rangle_t + \langle z^2 \rangle_t = 6Dt \tag{1.208}$$

This exact solution of the diffusion equation is valid only at long times because the diffusion equation itself holds for such times. The diffusion coefficient may therefore be calculated from

$$D = \lim_{t \to \infty} \frac{1}{6t} \langle (\mathbf{r}(t) - \mathbf{r}(0))^2 \rangle \tag{1.209}$$

1.6 Electrostatics

1.6.1 Fundamental equations of electrostatics

Unless otherwise stated, we follow here and elsewhere the electrostatic system of units. The electric field at position $\mathbf{r}$ associated with a distribution of point charges q_i at positions $\boldsymbol{r}_i$ in vacuum is given by the Coulomb law

$$\mathcal{E}(\mathbf{r}) = \sum_{i=1}^{n} q_i \frac{(\mathbf{r} - \mathbf{r}_i)}{|\mathbf{r} - \mathbf{r}_i|^3} \tag{1.210}$$

For a continuous charge distribution $\rho(\mathbf{r})$ the equivalent expression is

$$\mathcal{E}(\mathbf{r}) = \int d\mathbf{r}' \, \rho(\mathbf{r}') \frac{(\mathbf{r} - \mathbf{r}')}{|\mathbf{r} - \mathbf{r}'|^3} \tag{1.211}$$

Note that taking ρ to be a distribution of point charges, $\rho(\mathbf{r}) = \sum_i q_i \delta(\mathbf{r} - \mathbf{r}_i)$, leads to Eq. (1.210).

Another expression of the Coulomb law is the Gauss law, which states that the electric field associated with a charge distribution $\rho(\mathbf{r})$ satisfies the relationship

$$\oint_S ds\, \boldsymbol{\mathcal{E}} \cdot \mathbf{n} = 4\pi \int_\Omega d\mathbf{r}\rho(\mathbf{r}) \tag{1.212}$$

In (1.212) Ω denotes a volume that is enclosed by the surface S, $\mathbf{n}$ is a unit vector at a surface element *ds* of S in the outward direction and $\oint_S$ is an integral over the surface S. The Gauss law (1.212) relates the surface-integrated field on the boundary of a volume to the total net charge *inside* the volume. Using the divergence theorem $\oint_S \mathbf{B} \cdot \mathbf{n} da = \int_\Omega \nabla \cdot \mathbf{B}\, d\mathbf{r}$ for any vector field $\mathbf{B}$ leads to the differential form of the Gauss theorem

$$\nabla \cdot \boldsymbol{\mathcal{E}} = 4\pi\rho \tag{1.213}$$

The electrostatic potential Φ is related to the electrostatic field by

$$\boldsymbol{\mathcal{E}} = -\nabla\Phi \tag{1.214}$$

This and Eq. (1.211) imply that

$$\Phi(\mathbf{r}) = \int d\mathbf{r}' \frac{\rho(\mathbf{r}')}{|\mathbf{r} - \mathbf{r}'|} \tag{1.215}$$

Equations (1.213) and (1.214) together yield

$$\nabla^2\Phi = -4\pi\rho \tag{1.216}$$

which is the Poisson equation. In regions of space in which $\rho = 0$ this becomes the Laplace equation, $\nabla^2\Phi = 0$.

The energy needed to bring a charge q from a position where $\Phi = 0$ to a position with an electrostatic potential Φ is $q\Phi$. This can be used to obtain the energy needed to assemble a charge distribution $\rho(\mathbf{r})$:

$$W = \frac{1}{2}\int d\mathbf{r} \int d\mathbf{r}' \frac{\rho(\mathbf{r})\rho(\mathbf{r}')}{|\mathbf{r} - \mathbf{r}'|} = \frac{1}{2}\int d\mathbf{r}\rho(\mathbf{r})\Phi(\mathbf{r}) \tag{1.217}$$

Using (1.216) we get

$$W = -\frac{1}{8\pi}\int d\mathbf{r}\, \Phi(\mathbf{r})\nabla^2\Phi(\mathbf{r}) \tag{1.218}$$

and, upon integrating by parts while assuming that $\Phi = 0$ on the boundary of the system (e.g. at infinity) this leads to

$$W = \frac{1}{8\pi} \int d\mathbf{r} |\nabla \Phi|^2 = \frac{1}{8\pi} \int d\mathbf{r} |\mathcal{E}(\mathbf{r})|^2 \tag{1.219}$$

we can thus identify the energy density w in an electrostatic field:

$$w(\mathbf{r}) = \frac{1}{8\pi} |\mathcal{E}(\mathbf{r})|^2 \tag{1.220}$$

Consider now the electrostatic potential, Eq. (1.215), whose source is a charge distribution $\rho(\mathbf{r}')$. Assume that $\rho(\mathbf{r}')$ is localized within some small volume whose center is at $\mathbf{r}_0$, and that $\mathbf{r}$ is outside this volume and far from $\mathbf{r}_0$. In this case we can expand

$$\frac{1}{|\mathbf{r} - \mathbf{r}'|} = \frac{1}{|\mathbf{r} - \mathbf{r}_0 - (\mathbf{r}' - \mathbf{r}_0)|} = \frac{1}{|\mathbf{r} - \mathbf{r}_0|} - (\mathbf{r}' - \mathbf{r}_0) \cdot \nabla_r \frac{1}{|\mathbf{r} - \mathbf{r}_0|} + \cdots \tag{1.221}$$

Disregarding the higher-order terms and inserting into (1.215) leads to

$$\begin{aligned} \Phi(\mathbf{r}) &= \left[\int d\mathbf{r}' \rho(\mathbf{r}') \right] \frac{1}{|\mathbf{r} - \mathbf{r}_0|} - \left[\int d\mathbf{r}' \rho(\mathbf{r}')(\mathbf{r}' - \mathbf{r}_0) \right] \cdot \nabla_r \frac{1}{|\mathbf{r} - \mathbf{r}_0|} \\ &= \frac{q(\mathbf{r}_0)}{|\mathbf{r} - \mathbf{r}_0|} + \frac{\mathbf{d}(\mathbf{r}_0) \cdot (\mathbf{r} - \mathbf{r}_0)}{|\mathbf{r} - \mathbf{r}_0|^3} \end{aligned} \tag{1.222}$$

where $q(\mathbf{r}_0) = \int d\mathbf{r}' \rho(\mathbf{r}')$ is the net charge about $\mathbf{r}_0$ and $\mathbf{d}(\mathbf{r}_0) = \int d\mathbf{r}' \rho(\mathbf{r}')(\mathbf{r}' - \mathbf{r}_0)$ is the net dipole about that point. Higher-order terms will involve higher moments (multipoles) of the charge distribution $\rho(\mathbf{r})$, and the resulting expansion is referred to as the *multipole expansion*. In the next section this expansion is used as a starting point of a brief overview of dielectric continua.

1.6.2 Electrostatics in continuous dielectric media

The description of electrostatic phenomena in condensed molecular environments rests on the observation that charges appear in two kinds. First, molecular electrons are confined to the molecular volume so that molecules move as neutral polarizable bodies. Second, free mobile charges (e.g. ions) may exist. In a continuum description the effect of the polarizable background is expressed by the dielectric response of such environments.

Consider such an infinite environment (in real systems we assume that the effects of the system boundary can be disregarded). Divide it into small nonoverlapping volumes $\Delta^3 r$ that are large relative to molecular size and consider the electrostatic potential at point $\mathbf{r}$, taken to be far from the center of all these volumes.[12] Using Eq. (1.222) we can write

$$\Phi(\mathbf{r}) = \sum_j \left(\int_{\mathbf{r}'_j} d\mathbf{r}'_j \rho(\mathbf{r}'_j) \frac{1}{|\mathbf{r}-\mathbf{r}_j|} + \left[\int d\mathbf{r}'_j \rho(\mathbf{r}'_j)(\mathbf{r}'_j - \mathbf{r}_j) \right] \cdot \nabla_r \frac{1}{|\mathbf{r}-\mathbf{r}_j|} \right)$$
$$= \sum_j \left(\frac{q(\mathbf{r}_j)}{|\mathbf{r}-\mathbf{r}_j|} + \mathbf{d}(\mathbf{r}_j) \cdot \nabla_r \frac{1}{|\mathbf{r}-\mathbf{r}_j|} \right) \tag{1.223}$$

where $\int d\mathbf{r}'_j$ is an integral over the small volume j whose center is at $\mathbf{r}_j$, and where the sum is over all such volumes. For what follows it is convenient to write $q(\mathbf{r}_j) = \rho(\mathbf{r}_j)\Delta^3 r$ and $\mathbf{d}(\mathbf{r}_j) = \mathbf{P}(\mathbf{r}_j)\Delta^3 r$ where $\rho(\mathbf{r}_j)$ and $\mathbf{P}(\mathbf{r}_j)$ are coarse-grained charge and dipole density. The later is also called *polarization*. In the continuum limit the sum over j is replaced by an integral over the system volume.

$$\Phi(\mathbf{r}) = \int d\mathbf{r}' \left(\frac{\rho(\mathbf{r}')}{|\mathbf{r}-\mathbf{r}'|} + \mathbf{P}(\mathbf{r}') \cdot \nabla_r \frac{1}{|\mathbf{r}-\mathbf{r}'|} \right) = \int d\mathbf{r}' \left(\frac{\rho(\mathbf{r}')}{|\mathbf{r}-\mathbf{r}'|} - \frac{\nabla_{\mathbf{r}'} \cdot \mathbf{P}(\mathbf{r}')}{|\mathbf{r}-\mathbf{r}'|} \right) \tag{1.224}$$

To obtain the second equality we have integrated by parts using Eq. (1.30). According to (1.224) the electrostatic potential field is seen to arise from two charge densities: the "regular" $\rho(\mathbf{r})$ and an additional contribution associated with the dipole density $\rho_P(\mathbf{r}) \equiv -\nabla_\mathbf{r} \cdot \mathbf{P}(\mathbf{r})$. We will refer to $\rho(\mathbf{r})$ as the *external charge density*. This reflects a picture of a dielectric solvent with added ionic charges.

Equation (1.224) together with (1.214) imply that the Poisson equation (1.216) now takes the form

$$\nabla \cdot \mathcal{E} = 4\pi(\rho + \rho_P) = 4\pi(\rho - \nabla \cdot \mathbf{P}) \tag{1.225}$$

that is, the electric field originates not only from the external charges but also from the polarization. It is convenient to define an additional field, the *electric displacement*, which is associated with the external charges only:

$$\nabla \cdot \mathcal{D} = 4\pi\rho, \qquad \text{that is, } \mathcal{D} = \mathcal{E} + 4\pi\mathbf{P} \tag{1.226}$$

[12] We disregard the fact that there is at least one volume, that surrounding $\mathbf{r}$, for which this assumption cannot be made.

The electric field $\mathcal{E}$ and the electrostatic potential Φ continue to have the meaning taught in elementary courses: $\mathcal{E}(\mathbf{r})\delta q$ is the force experienced by an infinitesimal charge δq added at position $\mathbf{r}$, and $\Phi(\mathbf{r})\delta q$ is the work to add this charge. (The reason this statement is formulated with an infinitesimal charge δq is that in a dielectric medium a finite charge q can cause a change in $\mathcal{E}$ and Φ.) $\mathcal{E}$, however, has a contribution that arises from the polarization $\mathbf{P}$. The latter, an expression of microscopic separation of bound positive and negative charges within the molecules, may be considered as the response of the dielectric system to the external field $\mathcal{D}$. A fundamental ingredient of *linear* dielectric theory is the assumption that this response $\mathbf{P}$ depends linearly on its cause Δ, that is,

$$\mathbf{P}(\mathbf{r},t) = \int d\mathbf{r}' \int dt' \boldsymbol{\alpha}(\mathbf{r},\mathbf{r}';t,t')\mathcal{D}(\mathbf{r}',t') \tag{1.227}$$

$\boldsymbol{\alpha}$ is the *polarizability* tensor. The tensor character of $\boldsymbol{\alpha}$ expresses the fact that the direction of $\mathbf{P}$ can be different from that of $\mathcal{D}$. In an isotropic system the response is the same in all directions, so $\mathbf{P}$ and $\mathcal{D}$ are parallel and $\boldsymbol{\alpha} = \alpha\mathbf{I}$ where α is a scalar and $\mathbf{I}$ is the unit tensor. In a homogeneous (all positions equivalent) and stationary (all times equivalent) system, $\boldsymbol{\alpha}(\mathbf{r},\mathbf{r}';t,t') = \boldsymbol{\alpha}(\mathbf{r}-\mathbf{r}';t-t')$. The time dependence of $\boldsymbol{\alpha}(\mathbf{r},t)$ reflects the fact that an external field at some position at some time can cause a response at other positions and times (e.g. a sudden switch-on of a field in position $\mathbf{r}$ can cause a molecular dipole at that position to rotate, thereby affecting the field seen at a later time at a different place). In many experimental situations we can approximate $\boldsymbol{\alpha}(\mathbf{r}-\mathbf{r}';t-t')$ by $\boldsymbol{\alpha}\delta(\mathbf{r}-\mathbf{r}')\delta(t-t')$, that is, take $\mathbf{P}(\mathbf{r},t) = \boldsymbol{\alpha}\mathcal{D}(\mathbf{r},t)$. This is the case when the time and length scales of interest are large relative to those that characterize $\boldsymbol{\alpha}(\mathbf{r}-\mathbf{r}';t-t')$. We refer to the response in such cases as *local* in time and place. A common approximation used for molecular system is to take $\boldsymbol{\alpha}$ to be local in space but not in time,

$$\boldsymbol{\alpha}(\mathbf{r}-\mathbf{r}';t-t') = \boldsymbol{\alpha}(t-t')\delta(\mathbf{r}-\mathbf{r}') \tag{1.228}$$

Proceeding for simplicity with a homogeneous and isotropic system and with local and isotropic response, $\mathbf{P} = \alpha\mathcal{D}$, and defining the *dielectric constant* ε from $\varepsilon^{-1} = 1 - 4\pi\alpha$, we get from (1.226)

$$\mathcal{E} = \frac{1}{\varepsilon}\mathcal{D} \tag{1.229}$$

From (1.226) it also follows that

$$\mathbf{P} = \frac{\varepsilon - 1}{4\pi}\mathcal{E} \equiv \chi\mathcal{E} \tag{1.230}$$

The linear response coefficient χ is called the dielectric *susceptibility*.

Equivalent expressions can be obtained when describing systems that are homogeneous and isotropic but whose response is not local in space and time. This is done by taking the Fourier transform ($\mathbf{r} \to \mathbf{k}, t \to \omega$) of

$$\mathbf{P}(\mathbf{r}, t) = \int d\mathbf{r}' \int dt' \boldsymbol{\alpha}(\mathbf{r} - \mathbf{r}'; t - t')\boldsymbol{\mathcal{D}}(\mathbf{r}', t') \tag{1.231}$$

to get

$$\begin{aligned} \mathbf{P}(\mathbf{k}, \omega) &= \alpha(\mathbf{k}, \omega)\boldsymbol{\mathcal{D}}(\mathbf{k}, \omega) \\ \boldsymbol{\mathcal{E}}(\mathbf{k}, \omega) &= \varepsilon^{-1}(\mathbf{k}, \omega)\boldsymbol{\mathcal{D}}(\mathbf{k}, \omega) \\ \mathbf{P}(\mathbf{k}, \omega) &= \chi(\mathbf{k}, \omega)\boldsymbol{\mathcal{E}}(\mathbf{k}, \omega) \end{aligned} \tag{1.232}$$

with

$$\begin{aligned} \varepsilon^{-1}(\mathbf{k}, \omega) &= 1 - 4\pi\alpha(\mathbf{k}, \omega) \\ \chi(\mathbf{k}, \omega) &= (\varepsilon(\mathbf{k}, \omega) - 1)/4\pi \end{aligned} \tag{1.233}$$

Problem 1.4. Show that if the response is local in space but not in time the equivalent expressions for homogeneous stationary systems are

$$\mathbf{P}(\mathbf{r}, t) = \int dt' \alpha(\mathbf{r}; t - t')\boldsymbol{\mathcal{D}}(\mathbf{r}, t') \tag{1.234}$$

$$\begin{aligned} \mathbf{P}(\mathbf{r}, \omega) &= \alpha(\omega)\boldsymbol{\mathcal{D}}(\mathbf{r}, \omega) \\ \boldsymbol{\mathcal{E}}(\mathbf{r}, \omega) &= \varepsilon^{-1}(\omega)\boldsymbol{\mathcal{D}}(\mathbf{r}, \omega) \\ \mathbf{P}(\mathbf{r}, \omega) &= \chi(\omega)\boldsymbol{\mathcal{E}}(\mathbf{r}, \omega) \end{aligned} \tag{1.235}$$

$$\begin{aligned} \varepsilon^{-1}(\omega) &= 1 - 4\pi\alpha(\omega) \\ \chi(\omega) &= (\varepsilon(\omega) - 1)/4\pi \end{aligned} \tag{1.236}$$

Explain the equality $\alpha(\omega) = \lim_{\mathbf{k}\to 0} \alpha(\mathbf{k}, \omega)$ (and similarly for $\varepsilon(\omega)$ and $\chi(\omega)$).

In molecular systems the polarization $\mathbf{P}$ results from the individual molecular dipoles and has two main contributions. One is associated with the average orientation induced by an external field in the distribution of permanent molecular dipoles. The other results from the dipoles induced in each individual molecule by the local electrostatic field. The characteristic timescale associated with the first effect is that of nuclear orientational relaxation, τ_n, typically 10^{-11} s for small molecule fluids at room temperature. The other effect arises mostly from the distortion of the molecular electronic charge distribution by the external field, and its typical response time τ_e is of the order 10^{-16} s. Accordingly, we can define three dielectric response constants:

$$\mathbf{P} = (\alpha_e + \alpha_n)\mathcal{D} \equiv \alpha_s \mathcal{D} \tag{1.237}$$

α_e expresses the electronic response (induced dipoles), α_n is associated with the average orientation induced in the distribution of permanent molecular dipoles, and α_s denotes the total response. These contributions can in principle be monitored experimentally: Immediately following a sudden switch-on of an external field $\mathcal{D}$, the instantaneous locally averaged induced dipole is zero, however after a time large relative to τ_e but small with respect to τ_n the polarization becomes $\mathbf{P}_e = \alpha_e \mathcal{D}$. Equation (1.237) is satisfied only after a time long relative to τ_n. Similarly we can define two dielectric constants, ε_e and ε_s such that $\mathcal{E} = \varepsilon_e^{-1}\mathcal{D}$ and $\mathbf{P}_e = [(\varepsilon_e - 1)/4\pi]\mathcal{E}$ are satisfied for $t_e \ll t \ll t_n$ while $\mathcal{E} = \varepsilon_s^{-1}\mathcal{D}$ and $\mathbf{P} = [(\varepsilon_s - 1)/4\pi]\mathcal{E}$ hold for $t \gg t_n$.

Problem 1.5. Show that for $t \gg t_n$ the contribution to the polarization of a dielectric solvent that arises from the orientation of permanent dipoles is given by

$$\mathbf{P}_n = \mathbf{P} - \mathbf{P}_e = \frac{1}{4\pi}\left(\frac{1}{\varepsilon_e} - \frac{1}{\varepsilon_s}\right)\mathcal{D} \tag{1.238}$$

Note: The factor $C_{\mathrm{Pekar}} = (1/\varepsilon_e) - (1/\varepsilon_s)$ is often referred to as the Pekar factor.

1.6.2.1 Electrostatic energy

Equation (1.219) was an expression for the energy in an electrostatic field in vacuum. How is it modified in a dielectric environment?

Starting from a system with given (position-dependent) electric, electric displacement, and polarization fields, the change in energy upon adding an infinitesimal charge distribution $\delta\rho(\mathbf{r})$ is

$$\delta W = \int d\mathbf{r}\, \delta\rho(\mathbf{r})\Phi(\mathbf{r}) \tag{1.239}$$

The corresponding change in the electric displacement $\delta\mathcal{D}$ satisfies the Poisson equation $\nabla \cdot \delta\mathcal{D} = 4\pi\delta\rho$. Therefore, $\delta W = (4\pi)^{-1}\int d\mathbf{r}\,\Phi(\mathbf{r})\nabla \cdot \delta\mathcal{D}$. Integrating by parts, assuming that $\delta\rho$ is local so $\delta\mathcal{D} \to 0$ at infinity and using (1.214), yields

$$\delta W = \frac{1}{4\pi}\int d\mathbf{r}\mathcal{E} \cdot \delta\mathcal{D} = \frac{1}{8\pi}\int d\mathbf{r}\delta(\mathcal{E} \cdot \mathcal{D}) \tag{1.240}$$

To obtain the second equality we have made the assumption that the dielectric response is linear and local, that is, $\mathcal{E}(\mathbf{r})\delta\mathcal{D}(\mathbf{r}) = \varepsilon(\mathbf{r})\mathcal{E}(\mathbf{r})\delta\mathcal{E}(\mathbf{r}) = (1/2)\varepsilon(\mathbf{r})\delta(\mathcal{E}(\mathbf{r}) \cdot \mathcal{E}(\mathbf{r}))$. Now assume that all the buildup of the $\mathcal{E}$ and $\mathcal{D}$ fields in the system results from the added charge. This means that integrating over the added charge will give the total energy

$$W = \frac{1}{8\pi} \int d\mathbf{r}\, \mathcal{E}(\mathbf{r}) \cdot \mathcal{D}(\mathbf{r}) \tag{1.241}$$

Accordingly, the energy density is $w(\mathbf{r}) = (\mathcal{D}(\mathbf{r}))^2/(8\pi\varepsilon(\mathbf{r}))$.

As an application of these results consider the work needed to charge a conducting sphere of radius a in a dielectric environment characterized by a dielectric constant ε. Taking the center of the sphere to be at the origin and to carry a charge q, the electric displacement outside the sphere is q/r^2 and the electric field is $q/(\varepsilon r^2)$. Equation (1.241) then yields

$$\frac{q^2}{8\pi\varepsilon} \int_a^\infty d\mathbf{r} \frac{1}{r^4} = \frac{q^2}{2\varepsilon} \int_a^\infty dr \frac{1}{r^2} = \frac{q^2}{2\varepsilon a} \tag{1.242}$$

The energy needed to move a charged sphere from vacuum ($\varepsilon = 1$) to the interior of a dielectric medium is therefore,

$$W_B = -\frac{q^2}{2a}\left(1 - \frac{1}{\varepsilon}\right) \tag{1.243}$$

This is the Born expression for the *dielectric solvation energy*.

1.6.3 Screening by mobile charges

Next consider the implications of the existence of mobile charge carriers in the system. These can be ions in an electrolyte solution or in molten salts, electrons in metals and semiconductors, and electrons and ions in plasmas. For specificity we consider an ionic solution characterized by bulk densities n_+^B and n_-^B of positive and negative ions. The ionic charges are

$$q_+ = z_+ e \qquad \text{and} \qquad q_- = -z_- e \tag{1.244}$$

where e is the absolute value of the electron charge. On a coarse-grained level of description in which we consider quantities averaged over length scales that contain many such ions the system is locally electroneutral

$$\rho_q = n_+^B q_+ + n_-^B q_- = 0 \tag{1.245}$$

Consider now such a semi-infinite system, confined on one side by an infinite planar surface, and assume that a given potential Φ_S is imposed on this surface. The interior bulk potential is denoted Φ_B. Having $\Phi_S \neq \Phi_B$ implies that the mobile charges move under the resulting electric field until drift and diffusion balance each other. The resulting equilibrium densities n_+ and n_- are different from their bulk values and may depend on the distance from the surface. At issue is the question how do the electrostatic potential and these ionic densities approach their bulk values as we go from the surface into the interior of this solution.

In what follows we take that the direction perpendicular to the surface and pointing into the solution as the positive x direction. At any point the potential $\delta\Phi(x) = \Phi(x) - \Phi_B$ may be found as the solution of the Poisson equation (1.226), written in the form

$$\frac{\partial^2 \delta\Phi(x)}{\partial x^2} = -\frac{4\pi}{\varepsilon}\delta\rho_q(x) \tag{1.246}$$

where ε is the dielectric constant and where the excess charge $\delta\rho_q$ is given by

$$\delta\rho_q(x) = (n_+(x) - n_+^B)q_+ + (n_-(x) - n_-^B)q_- = n_+(x)q_+ + n_-(x)q_- \tag{1.247}$$

In the second equality we have used Eq. (1.245). The densities $n_{+/-}(x)$ are related to their bulk value by the Boltzmann equilibrium relations

$$\begin{aligned} n_+ &= n_+^B e^{-\beta q_+ \delta\Phi} \xrightarrow{q_+\delta\Phi \ll k_BT} n_+^B(1 - \beta q_+ \delta\Phi) \\ n_- &= n_-^B e^{-\beta q_- \delta\Phi} \xrightarrow{q_-\delta\Phi \ll k_BT} n_-^B(1 - \beta q_- \delta\Phi) \end{aligned} \tag{1.248}$$

We continue with the assumption that the conditions for expanding the exponential Boltzmann factors to linear order as in (1.248) hold, and that the expansion to first order is valid. Using this together with (1.245) in (1.247) leads to

$$\begin{aligned} \delta\rho_q(x) &= -\beta\delta\Phi(x)(n_+^B q_+^2 + n_-^B q_-^2) \\ &= -\beta\delta\Phi(x) n_+^B q_+(q_+ - q_-) = -\beta\delta\Phi(x) n_+^B z_+ e^2(z_+ + z_-) \end{aligned} \tag{1.249}$$

We can improve the appearance of this result by symmetrizing it, using (cf. Eqs (1.244) and (1.245)) $n_+^B z_+ = (1/2)(n_+^B z_+ + n_-^B z_-)$. We finally get

$$\delta\rho_q(x) = -\frac{1}{2}\beta(n_+^B z_+ + n_-^B z_-)e^2(z_+ + z_-)\delta\Phi(x) \tag{1.250}$$

Using this in (1.246) leads to

$$\frac{\partial^2 \delta\Phi}{\partial x^2} = \kappa^2 \delta\Phi, \tag{1.251}$$

where

$$\kappa^2 = \frac{2\pi e^2}{k_B T \varepsilon}(z_+ + z_-)(z_+ n_+^B + z_- n_-^B) \tag{1.252}$$

The solution of (1.251) that satisfies the boundary condition $\delta\Phi(x = 0) = \Phi_S - \Phi_B$ and $\delta\Phi(x \to \infty) = 0$ is $\delta\Phi = (\Phi_S - \Phi_B)^{-\kappa x}$, that is,

$$\Phi(x) = \Phi_B + (\Phi_S - \Phi_B)e^{-\kappa x} \tag{1.253}$$

We have found that in an electrolyte solution the potential on the surface approaches its bulk value on a length scale κ^{-1}, known as the *Debye screening length.*

The theory outlined above is a takeoff on the Debye Huckel theory of ionic solvation. In the electrochemistry literature it is known as the Gouy–Chapman theory. The Debye screening length is seen to depend linearly on $\sqrt{T}$ and to decrease as $(z_+ n_+^B + z_- n_-^B)^{-1/2}$ with increasing ionic densities. For a solution of monovalent salt, where $z_+ = z_- = 1$ and $n_+^B = n_-^B \equiv n^B$, this length is given by

$$\kappa^{-1} = \left(\frac{k_B T \varepsilon}{8\pi e^2 n^B}\right)^{1/2} \tag{1.254}$$

Typical screening lengths in aqueous ionic solutions are in the range of 10–100 Å. At $T = 300$ K, and using $\varepsilon = 80$ and salt concentration 0.01 M, that is, $n \sim 6 \times 10^{18}$ cm^{-3}, yields a length of the order ~30 Å.

Appendix 1A Derivation of the classical Liouville equation as a conservation law

Here we describe an alternative derivation of the Liouville equation (1.104) for the time evolution of the phase space distribution function $f(\mathbf{r}^N, \mathbf{p}^N; t)$. The derivation below is based on two observations: First, a change in f reflects only the change in positions and momenta of particles in the system, that is, of motion of phase points in phase space, and second, that phase points are *conserved*, neither created nor destroyed.

Consider an ensemble of $\mathcal{N}$ macroscopically identical systems that are represented by $\mathcal{N}$ points moving in phase space. Consider a given volume υ in this space. The number of systems (phase points) within this volume at time t is

$$n(t) = \mathcal{N} \int_{\upsilon} d\mathbf{p}^N d\mathbf{r}^N f(\mathbf{r}^N(t), \mathbf{p}^N(t); t) \tag{1.255}$$

and the rate at which it changes is given by

$$\frac{dn}{dt} = \mathcal{N} \int_{\upsilon} d\mathbf{p}^N d\mathbf{r}^N \frac{\partial f}{\partial t} \tag{1.256}$$

Since phase points are neither created nor destroyed, this rate should be equal to the rate at which phase points flow into the volume υ (negative rate means flowing out of the volume). The velocity of a phase point, i.e., the rate at which its "position" in the $6N$-dimensional phase space is changing, is represented by the $6N$ velocity vector $\mathbf{u} = (\dot{\mathbf{r}}^N, \dot{\mathbf{p}}^N)$. The flux of phase points at phase-space "position" is $\mathcal{N} f \mathbf{u}$.[13] Therefore,

$$\frac{dn}{dt} = -\mathcal{N} \int_{S} f \mathbf{u} \cdot d\mathbf{S} \tag{1.257}$$

where the integral is over the phase-space surface surrounding the volume υ, and where $d\mathbf{S}$ is a surface element vector whose direction is normal to the surface in the outward direction. Using Gauss theorem to transform the surface integral into a volume integral we get

$$\frac{dn}{dt} = -\mathcal{N} \int_{\upsilon} \nabla \cdot (f\mathbf{u}) d\mathbf{r}^N d\mathbf{p}^N \tag{1.258}$$

(Note that Eq. (1.258)) is the multidimensional analog of Eq. (1.36)). Comparing to (1.256) and noting that the volume υ is arbitrary, we find that

$$\begin{aligned} \frac{\partial f}{\partial t} &= -\nabla \cdot (f\mathbf{u})) = -\sum_{j=1}^{3N} \left\{ \frac{\partial}{\partial x_j}(f\,\dot{x}_j) + \frac{\partial}{\partial p_j}(f\,\dot{p}_j) \right\} \\ &= -\sum_{j=1}^{3N} \left\{ \frac{\partial f}{\partial x_j}\dot{x}_j + \frac{\partial f}{\partial p_j}\dot{p}_j \right\} - \sum_{j=1}^{3N} \left\{ \frac{\partial \dot{x}_j}{\partial x_j} + \frac{\partial \dot{p}_j}{\partial p_j} \right\} f \end{aligned} \tag{1.259}$$

Note that the first line of (1.259) and the way it was derived are analogous to the derivation of the continuity equation in Section 1.1.4. Equation (1.259) expresses

[13] "Position" in phase space is the $6N$-dimensional point $\mathbf{q} = (\mathbf{r}^N, \mathbf{p}^N)$. Phase point velocity is $\mathbf{u} = \dot{\mathbf{q}} = (\dot{\mathbf{r}}^N, \dot{\mathbf{p}}^N)$. The flux of moving particles (number of particles going through a unit area normal to the flux vector per unit time) is given by the product of particle velocity and particle density, in the present case of $\mathbf{u}$ and $\mathcal{N} f$.

the fact that the number of points in phase space, that is, systems in our ensemble, is conserved. In the present case we obtain an additional simplification, noting that the Hamilton equations (1.98) imply that $\partial\dot{x}_j/\partial x_j + \partial\dot{p}_j/\partial p_j = 0$. Equation (1.259) then becomes

$$\frac{\partial f}{\partial t} = -\sum_{j=1}^{3N}\left\{\frac{\partial f}{\partial x_j}\frac{\partial H}{\partial p_j} - \frac{\partial f}{\partial p_j}\frac{\partial H}{\partial x_j}\right\} \tag{1.260}$$

which is again the Liouville equation. This additional step from (1.259) to (1.260) expresses the incompressibility property of the "Liouville space fluid" discussed at the end of Section 1.2.2.

2

QUANTUM DYNAMICS USING THE TIME-DEPENDENT SCHRÖDINGER EQUATION

I have taught how everything begins,
The nature of those first particles, their shape,
Their differences, their voluntary course,
Their everlasting motion and the way
Things are created by them...

Lucretius (c.99–c.55 BCE) "The way things are" translated by Rolfe Humphries, Indiana University Press, 1968.

This chapter focuses on the time-dependent Schrödinger equation and its solutions for several prototype systems. It provides the basis for discussing and understanding quantum dynamics in condensed phases, however, a full picture can be obtained only by including also dynamical processes that destroy the quantum mechanical phase. Such a full description of quantum dynamics cannot be handled by the Schrödinger equation alone; a more general approach based on the quantum Liouville equation is needed. This important part of the theory of quantum dynamics is discussed in Chapter 10.

2.1 Formal solutions

Given a system characterized by a Hamiltonian $\hat{H}$, the time-dependent Schrödinger equation is

$$\frac{\partial \Psi}{\partial t} = -\frac{i}{\hbar}\hat{H}\Psi \tag{2.1}$$

For a closed, isolated system $\hat{H}$ is time independent; time dependence in the Hamiltonian enters via effect of time-dependent external forces. Here we focus on the earlier case. Equation (1) is a first-order linear differential equation that can be solved as an initial value problem. If $\Psi(t_0)$ is known, a formal solution to Eq. (1) is given by

$$\Psi(t) = \hat{U}(t, t_0)\Psi(t_0) \tag{2.2}$$

where the *time evolution operator* $\hat{U}$ is[1]

$$\hat{U}(t, t_0) = e^{-(i/\hbar)\hat{H}(t-t_0)} \tag{2.3}$$

A more useful solution for $\Psi(t)$ may be obtained by expanding it in the complete orthonormal basis of eigenstates of $\hat{H}$, $\{\psi_n\}$, which satisfy

$$\hat{H}\psi_n = E_n\psi_n \tag{2.4}$$

Writing

$$\Psi(t_0) = \sum_n c_n(t_0)\psi_n \qquad \text{with } c_n(t_0) = \langle\psi_n|\Psi(t_0)\rangle \tag{2.5}$$

we get either from (2.1) or from (2.2) the result

$$\Psi(t) = \sum_n c_n(t)\psi_n \qquad \text{with } c_n(t) = e^{-(i/\hbar)E_n(t-t_0)}c_n(t_0) \tag{2.6}$$

Problem 2.1. Show how Eq. (2.6) is obtained from Eq. (2.1) and how it is obtained from Eq. (2.2)

A solution of the time-dependent Schrödinger equation may be obtained also in terms of any complete orthonormal basis $\{\phi_n\}$, not necessarily the one that diagonalizes the Hamiltonian $\hat{H}$. In this basis the Hamiltonian is represented as a matrix $H_{nm} = \langle\phi_n|H|\phi_m\rangle$ and the wavefunction $\Psi(t)$ is written as

$$\Psi(t) = \sum_n b_n(t)\phi_n \tag{2.7}$$

Inserting (2.7) into (2.1) and using the orthonormality conditions $\langle\phi_n|\phi_m\rangle = \delta_{nm}$ leads to a set of equations for the b coefficients

$$\frac{db_n}{dt} = -\frac{i}{\hbar}\sum_m H_{nm}b_m \tag{2.8}$$

or in obvious vector-matrix notation

$$\frac{d}{dt}\mathbf{b} = -\frac{i}{\hbar}\mathbf{Hb} \tag{2.9}$$

[1] If $F(x)$ is an analytical function of x in a given domain that contains the point $x = 0$, the function $\hat{F}(\hat{A})$ of an operator $\hat{A}$ is defined in the same domain by the Taylor expansion $\hat{F}(\hat{A}) = \sum_n (1/n!)F^{(n)}(x=0)\hat{A}^n$ where $F^{(n)}$ is the nth derivative of F.

Problem 2.2.

1. Derive Eqs (2.8) and (2.9)
2. Show that the time evolution defined by Eqs (2.2) and (2.3) corresponds to solving Eqs (2.8) (or equivalently (2.9) with the initial conditions

$$b_n(t_0) = \langle \phi_n | \Psi(t_0) \rangle \tag{2.10}$$

Problem 2.3. Let $\Psi(\mathbf{r},t)$ be the solution of a 1-particle Schrödinger equation with the Hamiltonian $\hat{H} = -(\hbar^2/2m)\nabla^2 + V(\mathbf{r})$, and let $\rho(\mathbf{r},t) = |\Psi(\mathbf{r},t)|^2$ be the corresponding probability density. Prove the identity

$$\frac{\partial \rho}{\partial t} = -\nabla \cdot \mathbf{J}; \qquad \mathbf{J} \equiv \frac{\hbar}{2im}(\Psi^* \nabla \Psi - \Psi \nabla \Psi^*) = \frac{\hbar}{m}\mathrm{Im}(\Psi^* \nabla \Psi) \tag{2.11}$$

The first equality in (2.11) has the form of a continuity equation (see Section 1.1.4) that establishes $\mathbf{J}$ as a flux (henceforth referred to as a probability flux).

2.2 An example: The two-level system

Consider a two-level system whose Hamiltonian is a sum of a "simple" part, $\hat{H}_0$, and a "perturbation" $\hat{V}$.

$$\hat{H} = \hat{H}_0 + \hat{V} \tag{2.12}$$

The eigenfunctions of $\hat{H}_0$ are $|\phi_a\rangle, |\phi_b\rangle$, with the corresponding eigenvalues E_a, E_b. We will interchangeably use the notation $\langle i|\hat{O}|j\rangle = \langle \phi_i|\hat{O}|\phi_j\rangle = O_{i,j}$ for any operator $\hat{O}$. Without loss of generality we may assume that $V_{a,a} = V_{b,b} = 0$ (otherwise we may include the diagonal part of V in H_0). In the basis of the functions $|\phi_a\rangle$ and $|\phi_b\rangle$ $\hat{H}$ is then represented by the matrix

$$\hat{H} = \begin{pmatrix} E_a & V_{a,b} \\ V_{b,a} & E_b \end{pmatrix}; \qquad V_{a,b} = V_{b,a}^* = \langle \phi_a | V | \phi_b \rangle \tag{2.13}$$

The coupling elements $V_{i,j}$ are in principle complex, and we express them as

$$V_{a,b} = Ve^{-i\eta}; \qquad V_{b,a} = Ve^{i\eta} \tag{2.14}$$

with V taken real and positive. It should be emphasized that the two-level problem represented by $\hat{H}$ is not more difficult to solve than that given by $\hat{H}_0$, however, there are situations where it helps to discuss the problem in terms of both Hamiltonians. For example, the system may be represented by the Hamiltonian $\hat{H}_0$ and exists in the stationary state ϕ_a, then at some time taken to be $t = 0$ the perturbation $\hat{V}$ is switched on. A typical question is what is the probability $P_b(t)$ to find the system in state b following this switch-on of the perturbation that couples between the two eigenstates of $\hat{H}_0$.

The simplest approach to solving this problem is to diagonalize the Hamiltonian $\hat{H}$, that is, to find its eigenstates, denoted ψ_+ and ψ_-, and eigenvalues E_+ and E_-, respectively.

Problem 2.4. Show that given $P_a(t = 0) = 1; P_b(t = 0) = 1 - P_a(t = 0) = 0$, then the probability to find the system in state b at time t is given in terms of the eigenstates and eigenvalues of $\hat{H}$ by the form

$$P_b(t) = \left| \langle \phi_b | \psi_+ \rangle \langle \psi_+ | \phi_a \rangle e^{-(i/\hbar)E_+ t} + \langle \phi_b | \psi_- \rangle \langle \psi_- | \phi_a \rangle e^{-(i/\hbar)E_- t} \right|^2 \tag{2.15}$$

Diagonalization of $\hat{H}$. The functions ψ_+ and ψ_- and eigenvalues E_+ and E_- are solutions of the time-independent Schrödinger equation $\hat{H}\psi = E\psi$. They are found by writing a general solution in the form

$$|\psi\rangle = c_a |\phi_a\rangle + c_b |\phi_b\rangle \tag{2.16}$$

which in the basis of $|\phi_a\rangle$ and $|\phi_b\rangle$ is represented by the vector $\begin{pmatrix} c_a \\ c_b \end{pmatrix}$. The Schrödinger equation in this representation is

$$\begin{pmatrix} E_a & V_{a,b} \\ V_{b,a} & E_b \end{pmatrix} \begin{pmatrix} c_a \\ c_b \end{pmatrix} = E \begin{pmatrix} c_a \\ c_b \end{pmatrix} \tag{2.17}$$

The requirement of existence of a nontrivial solution leads to the secular equation

$$(E_a - E)(E_b - E) = V^2 \tag{2.18}$$

which yields two solutions for E, given by

$$E_{\pm} = \frac{E_a + E_b \pm \sqrt{(E_a - E_b)^2 + 4V^2}}{2} \tag{2.19}$$

The following identities

$$\frac{E_+ - E_a}{V} = \frac{V}{E_+ - E_b} = \frac{V}{E_a - E_-} \equiv X \tag{2.20}$$

where X is real and positive, are also easily verified. The coefficients c_a and c_b satisfy

$$c_b = -\frac{E_a - E}{V} e^{i\eta} c_a \tag{2.21}$$

For the absolute values we have

$$\frac{|c_b|}{|c_a|} = \frac{|E_a - E|}{V}; \qquad |c_a|^2 + |c_b|^2 = 1 \tag{2.22}$$

Consider first $E = E_+$

$$|c_a|^2 = \frac{1}{1+X^2}; \qquad |c_b|^2 = \frac{X^2}{1+X^2} \tag{2.23}$$

The phase factor $e^{i\eta}$ in (2.21) may be distributed at will between c_a and c_b and a particular choice is expressed by

$$|\psi_+\rangle = \cos\theta e^{-i\eta/2}|\phi_a\rangle + \sin\theta e^{i\eta/2}|\phi_b\rangle \tag{2.24a}$$

$$|\psi_-\rangle = -\sin\theta e^{-i\eta/2}|\phi_a\rangle + \cos\theta e^{i\eta/2}|\phi_b\rangle \tag{2.24b}$$

or

$$\begin{pmatrix} |\psi_+\rangle \\ |\psi_-\rangle \end{pmatrix} = \begin{pmatrix} \cos\theta & \sin\theta \\ -\sin\theta & \cos\theta \end{pmatrix} \begin{pmatrix} |\phi_a\rangle\, e^{-i\eta/2} \\ |\phi_b\rangle\, e^{i\eta/2} \end{pmatrix} \tag{2.25}$$

where

$$\theta \equiv \arctan X; \qquad 0 < \theta < \pi/2 \tag{2.26}$$

or

$$\sin\theta = \frac{X}{(1+X^2)^{1/2}}; \qquad \cos\theta = \frac{1}{(1+X^2)^{1/2}} \tag{2.27}$$

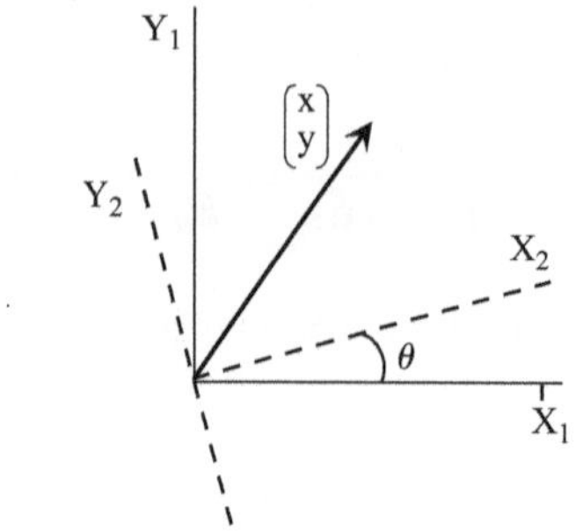

Figure 2.1 The components $\begin{pmatrix} x_1 \\ y_1 \end{pmatrix}$ and $\begin{pmatrix} x_2 \\ y_2 \end{pmatrix}$ of a vector expressed in the two systems of coordinates 1 and 2 shown are related to each other by the transformation $\begin{pmatrix} x_2 \\ y_2 \end{pmatrix} = \begin{pmatrix} \cos\theta & \sin\theta \\ -\sin\theta & \cos\theta \end{pmatrix}\begin{pmatrix} x_1 \\ y_1 \end{pmatrix}$.

The inverse transformation is

$$|\phi_a\rangle = (\cos\theta|\psi_+\rangle - \sin\theta|\psi_-\rangle)e^{i\eta/2} \tag{2.28a}$$

$$|\phi_b\rangle = (\sin\theta|\psi_+\rangle + \cos\theta|\psi_-\rangle)e^{-i\eta/2} \tag{2.28b}$$

Some readers may note that that Eqs (2.25) and (2.28) constitute a rotation in two-dimensional space as seen in Fig. 2.1.

Calculating the time evolution. The required time evolution is now obtained from Eq. (2.15). Using (cf. (2.24) and (2.28))

$$\begin{aligned} \langle\phi_b|\psi_+\rangle &= -\langle\psi_-|\phi_a\rangle = \sin(\theta)e^{i\eta/2} \\ \langle\phi_b|\psi_-\rangle &= \langle\psi_+|\phi_a\rangle = \cos(\theta)e^{i\eta/2} \end{aligned} \tag{2.29}$$

we get

$$P_b(t) = |\langle\phi_b|\Psi(t)\rangle|^2 = \left|\sin\theta\cos\theta(e^{-iE_+t/\hbar} - e^{-iE_-t/\hbar})\right|^2 \tag{2.30}$$

and

$$P_a(t) = |\langle\phi_a|\Psi(t)\rangle|^2 = \left|\cos^2\theta e^{-iE_+t/\hbar} + \sin^2\theta e^{-iE_-t/\hbar}\right|^2 = 1 - P_b \tag{2.31}$$

Using Eqs (2.19), (2.20), and (2.26), Eq. (2.30) can be recast in the form

$$P_b(t) = \frac{4|V_{ab}|^2}{(E_a - E_b)^2 + 4|V_{ab}|^2}\sin^2\left[\frac{\Omega_R}{2}t\right] \tag{2.32}$$

where

$$\Omega_R = \frac{1}{\hbar}\sqrt{(E_a - E_b)^2 + 4|V_{ab}|^2} \tag{2.33}$$

is known as the *Rabi frequency*.

2.3 Time-dependent Hamiltonians

Much of the formal discussion below, as well as the two-level example discussed above deal with systems and processes characterized by time-independent Hamiltonians. Many processes of interest, however, are described using time-dependent Hamiltonians; a familiar example is the semi-classical description of a system interacting with the radiation field. In this latter case the system-field interaction can be described by a time-dependent potential, for example,

$$\hat{H} = \hat{H}_0 + \hat{\boldsymbol{\mu}} \cdot \boldsymbol{\mathcal{E}}(t) \tag{2.34}$$

where $\hat{\boldsymbol{\mu}}$ is the dipole moment operator of the system and $\boldsymbol{\mathcal{E}}$ is the electric field associated with the *external* time-dependent classical radiation field. The emphasis here is on the word "external": The electromagnetic field is taken as an entity outside the system, affecting the system but not affected by it, and its physical character is assumed to be described by the classical Maxwell theory. This is of course an approximation, even if intuitively appealing and quite useful (see Sections 10.5.2 and 18.7). An exact description can be obtained (see Chapter 3) only by taking the field to be part of the system.

We can formalize this type of approximation in the following way. Consider a Hamiltonian that describes two interacting systems, 1 and 2. In what follows we use M_1, R_1 and M_2, R_2 as shorthand notations for the masses and coordinates of systems 1 and 2, which are generally many-body systems

$$\begin{aligned} \hat{H} &= \hat{H}_1 + \hat{H}_2 + \hat{V}_{12}(\hat{R}_1, \hat{R}_2) \\ \hat{H}_k &= -\frac{\hbar^2}{2M_k}\nabla_k^2 + \hat{V}_k(\hat{R}_k); \qquad k = 1, 2 \end{aligned} \tag{2.35}$$

In writing (2.35) we have deviated from our standard loose notation that does not usually mark the difference between a coordinate and the corresponding operator. For reasons that become clear below we emphasize that $\hat{R}_1, \hat{R}_2$ are operators, on equal footings with other operators such as $\hat{H}$ or ∇.

Next we assume that the solution of the time-dependent Schrödinger equation can be written as a simple product of normalized wavefunctions that describe the individual systems, that is,

$$\Psi(t) = \Psi_1(t)\Psi_2(t) \tag{2.36}$$

where, for non-interacting subsystems, each function $\Psi_k(t); k = 1, 2$ satisfies the corresponding Schrödinger equation

$$i\hbar\frac{\partial \Psi_k}{\partial t} = \hat{H}_k \Psi_k \tag{2.37}$$

The time-dependent Schrödinger equation for the overall system is

$$i\hbar\left(\Psi_2\frac{\partial \Psi_1}{\partial t} + \Psi_1\frac{\partial \Psi_2}{\partial t}\right) = \Psi_2\hat{H}_1\Psi_1 + \Psi_1\hat{H}_2\Psi_2 + \hat{V}_{12}\Psi_1\Psi_2 \tag{2.38}$$

Multiplying (2.38) by Ψ_2^* and integrating over the coordinates of system 2 yields

$$i\hbar\left(\frac{\partial \Psi_1}{\partial t} + \Psi_1\left\langle \Psi_2 \middle| \frac{\partial \Psi_2}{\partial t}\right\rangle_2\right) = \hat{H}_1\Psi_1 + \Psi_1\left\langle \Psi_2\right| \hat{H}_2 \left|\Psi_2\right\rangle_2 + \langle \Psi_2|\, \hat{V}_{12}\, |\Psi_2\rangle_2\, \Psi_1 \tag{2.39}$$

where the subscript k in $\langle\ \rangle_k$ indicates that the integration is taken over the subspace of system k ($k = 1, 2$). Using (2.37) with $k = 2$ this yields[2]

$$i\hbar\frac{\partial \Psi_1}{\partial t} = \left(-\frac{\hbar^2}{2M_1}\nabla_1^2 + \hat{V}_{1,\text{eff}}(\hat{R}_1)\right)\Psi_1 \tag{2.40a}$$

with an effective potential for system 1 given by

$$\hat{V}_{1,\text{eff}}(\hat{R}_1) = \hat{V}_1(R_1) + \left\langle \Psi_2 \middle| \hat{V}_{12} \middle| \Psi_2 \right\rangle_2 \tag{2.41a}$$

similarly, for system 2 we get

$$i\hbar\frac{\partial \Psi_2}{\partial t} = \left(-\frac{\hbar^2}{2M_2}\nabla_2^2 + \hat{V}_{2,\text{eff}}(R_2)\right)\Psi_2 \tag{2.40b}$$

$$\hat{V}_{2,\text{eff}}(R_2) = \hat{V}_2(R_2) + \left\langle \Psi_1 \middle| \hat{V}_{12} \middle| \Psi_1 \right\rangle_1 \tag{2.41b}$$

[2] This is a gross oversimplification. A proper derivation of Eqs. (2.40) relies on time dependent variational theory (see, e.g., A. D. McLachlan, Mol. Phys. **7**, 139 (1964))

The result, Eq. (2.40) is known as the *time-dependent mean field* or *time-dependent Hartree* approximation. In this approximation each system is moving in the average field of the other system.

At this point the two systems are treated on the quantum level, however if there is reason to believe that classical mechanics provides a good approximation for the dynamics of system 2, say, we may replace Eq. (2.40b) by its classical counterpart

$$\ddot{R}_2 = -\frac{1}{M_2}\nabla V_{2,\mathrm{eff}}(R_2) \tag{2.42}$$

and at the same time replace $\langle\Psi_2|V_{12}|\Psi_2\rangle_2$ in Eq. (2.41a) by $\hat{V}_{12}(\hat{R}_1; R_2(t))$, that is, the interaction potential is taken to be parametrically dependent on the instantaneous configuration R_2 of the classical system. In $\hat{V}_{12}(\hat{R}_1; R_2(t))$ $\hat{R}_1$ is an operator while $R_2(t)$ is a classical quantity. The equation describing the dynamics of system 1

$$\begin{aligned} i\hbar\frac{\partial\Psi_1}{\partial t} &= \left(-\frac{\hbar^2}{2M_1}\nabla_1^2 + \hat{V}_{1,\mathrm{eff}}(\hat{R}_1, R_2(t))\right)\Psi_1 \\ \hat{V}_{1,\mathrm{eff}}(\hat{R}_1, R_2(t)) &= \hat{V}_1(\hat{R}_1) + \hat{V}_{12}(\hat{R}_1; R_2(t)) \end{aligned} \tag{2.43}$$

together with Eq. (2.42) now describe a set of coupled quantum and classical equations of motion that can be solved self-consistently: The quantum system 1 moves in a potential that depends on the configuration of the classical system 2, while the latter moves in an effective potential that is the expectation value of $\hat{V}_{12}$ with the instantaneous wavefunction $\Psi_1(t)$.

The validity of this mixed quantum-classical scheme is far from obvious, and important questions regarding its applicability may be raised. For example, does this coupled system of quantum and classical degrees of freedom conserve the total energy as it should (the answer is a qualified yes: it may be shown that the sum of the energy of the classical system and the instantaneous expectation value of the Hamiltonian of the quantum system is a constant of motion of this dynamics). Experience shows that in many cases this scheme provides a good approximation, at least for the short-time dynamics.

A further approximation is possible: There are situations in which we may have reason to believe that while system 1 is strongly affected by system 2, the opposite is not so, and the classical dynamics of system 2 may be treated while disregarding system 1. In this case, $R_2(t)$ is a classical function of time obtained as the solution of the independent equation $\ddot{R}_2 = -(1/M_2)\nabla V_2(R_2)$ and Eq. (2.43) is a Schrödinger equation for system 1 with a time-dependent Hamiltonian, similar in spirit to Eq. (2.34). System 2 now became "external" and the issue of energy conservation is not relevant anymore: we are interested only in the energy of system 1 which is obviously not conserved.

We can intuitively identify possible conditions under which such an approximation may be useful. The interaction of a system with a radiation field is a familiar example. If we are interested in the probability that a molecule absorbs a photon from the radiation field, it feels "right" to assume that the field is not affected by the loss of a photon, so its dynamics might be described with no regard to the molecule. Similarly, when a heavy and a light particles exchange a given amount of energy upon collision, the trajectory of the light particle is changed considerably while the heavy particle is hardly affected, again a reason to disregard the light particle when considering the motion of the heavy one. It should be kept in mind, however, that the success of any approximation may depend on the observable under study. For example, Eq. (2.34) can be useful for describing absorption or induced emission by a molecule interacting with the radiation field, but it cannot describe the phenomenon of spontaneous emission. Indeed, the latter process can add a photon to an empty (no photons or "vacuum") field, and an approximation that disregards the effect of the molecule on the field can hardly be expected to describe such a change in the field state. On the other hand, when favorable conditions exist, this approximation is very successful in describing short-time phenomena such as the outcome of single collision events.

2.4 A two-level system in a time-dependent field

As a specific example, consider again the two-level model, but now with a time-dependent Hamiltonian affected by some external force. There are three frequently encountered problems of this kind:

1. A two-level system, Eq. (2.13), where $\hat{V}$ is a periodic function of time, for example,

$$\hat{V}(t) = \hat{\boldsymbol{\mu}} \cdot \mathbf{E}_0 \cos(\omega t) \tag{2.44}$$

   so that in Eq. (2.13) is $V_{ab} = \hat{\boldsymbol{\mu}}_{ab} \cdot \mathbf{E}_0 \cos(\omega t)$. This is a standard semiclassical model for describing atoms and molecules in a radiation field that will be further discussed in Chapter 18.
2. A two-level system, Eq. (2.13), with a coupling that simulates a collision process, that is, $\hat{V}(t) = \hat{V}_0 f(t)$ where $f(t)$ is a function that has a maximum

at (say) $t = 0$ and satisfies $f(t) \to 0$ as $|t| \to \infty$. Writing $\Psi(t) = c_a(t)\phi_a + c_b(t)\phi_b$ (see Eq. (2.16)), a typical concern is the values of $|c_a(t)|^2$ and $|c_b(t)|^2$ at $t \to \infty$ given that, say, $c_a(t \to -\infty) = 1$ (hence $c_b(t \to -\infty) = 0$).[3]

3. The Landau Zener (LZ) problem:[4] The two-level Hamiltonian as well as the basis used to describe it are taken to depend on a parameter R in the form

$$\begin{aligned}\hat{H} &= \hat{H}_0 + \hat{V}\\ \hat{H}_0 &= E_a(R)\,|\phi_a\rangle\,\langle\phi_a| + E_b(R)\,|\phi_b\rangle\,\langle\phi_b| \\ \hat{V} &= V_{ab}\,|\phi_a\rangle\,\langle\phi_b| + V_{ba}\,|\phi_b\rangle\,\langle\phi_a|\end{aligned} \tag{2.45}$$

and the parameter R is a known function of time, for example it may correspond to the distance between two molecules colliding with each other. In this respect this problem is similar to the previous one, however, the following detail characterizes the LZ problem: The time dependence is such that at $t = 0$ (say), where $R(t = 0) = R^*$, the zero order energies are equal, $E_a = E_b$, while at $t \to \pm\infty$ $|E_a - E_b|$ is much larger than $|V_{ab}|$. In reality the basis functions ϕ_a, ϕ_b as well as the coupling elements V_{ab} can also depend on R, but this dependence is assumed weak, and is disregarded in what follows. The question posed is as before: given that at $t \to -\infty$ the system starts at state ϕ_a, what is the probability that it will cross into state ϕ_b at $t \to \infty$.

We dwell briefly on the last problem that will be relevant to later discussions. The picture described above constitutes a semiclassical model for nonadiabatic transitions between two electronic states. In this model R may represent the coordinate(s) of the nuclear system, while a and b denote two electronic states obtained for each nuclear configuration by disregarding the nuclear kinetic energy as well as other residual interactions V (e.g. spin–orbit coupling). The resulting electronic energies $E_a(R)$ and $E_b(R)$ constitute potential surfaces for the nuclear motions in these electronic states. (The reader may consult the following section for further discussion of potential energy surfaces.) These surfaces cross as $R = R^*$,[5] see Fig. 2.2. The time dependence of R is depicted in this figure in a way that represents a collision process. The motion starts at $t \to -\infty, R \to -\infty$ and proceeds to $t \to \infty, R \to \infty$ after going through a configuration $R = R^*$ (at

[3] For an example of using such an approach to model collisional transitions in the semiclassical approximation see F. E. Heidrich, K. R. Wilson, and D. Rapp 1971, J. Chem. Phys., **54**, 3885.

[4] L. Landau, 1932, Phyz. Z. Sowjetunion **1**, 89; 1932, **2**, 46; C. Zener, 1933, Proc. Roy. Soc. **A137**, 696; 1933, **A140**, 660; E. C. G. Stueckelberg 1932, Hel. Phys. Acta **5**, 369. For an update of recent development of this subject see H. Nakamura, *Nonadiabatic Transitions* (World Scientific, Singapore, 2002).

[5] The subspace defined by $R = R^*$ is not necessarily a point, but a lower dimensionality surface.

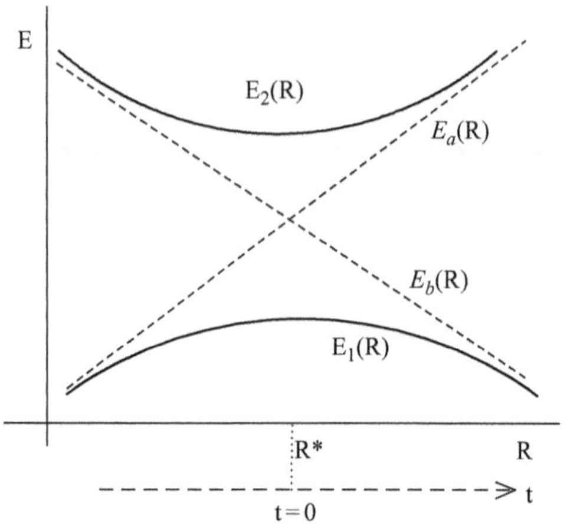

FIG. 2.2 A schematic description of the LZ problem: Two quantum states and the coupling between them depend parametrically on a classical variable R. The energies of the zero-order states a and b cross at $R = R^*$. The energies obtained by diagonalizing the Hamiltonian at any point R (adiabatic states) are $E_1(R)$ and $E_2(R)$.

time set to be $t = 0$) in which interaction between the two colliding particles has caused E_a and E_b to become equal. The corresponding states ϕ_a and ϕ_b are sometime referred to as *diabatic*. The exact adiabatic energies $E_1(R)$, $E_2(R)$ and wavefunctions ϕ_1, ϕ_2 are obtained by diagonalizing the Hamiltonian at each nuclear configuration R. In the case of a one-dimensional motion the diagonalized surfaces do not cross (the *non-crossing rule*); instead, the approach of the surfaces $E_1(R)$ and $E_2(R)$ into close proximity is the *avoided crossing* seen in Fig. 2.2. In a higher dimensional system crossing may occur, but only on a lower dimensionality surface.

We already know that when $|V_{ab}| \ll |E_a - E_b|$ the transition between states a and b can be disregarded (see Eq. (2.32)). Thus, the transition probability effectively vanishes at $t \to \pm\infty$. In particular, if the slopes $|(dE_a(R)/dR)_{R^*}|$ and $|(dE_b(R)/dR)_{R^*}|$ are sufficiently different from each other and if $|V_{ab}|$ is small enough the transition will be limited to a small neighborhood of R^*. (This is an argument for disregarding the R dependence of $V_{ab}(R)$, $\phi_a(R)$, and $\phi_b(R)$ by setting $R = R^*$ in these functions.) Outside this neighborhood the effect of the coupling V_{ab} is negligible; the adiabatic and diabatic representations are essentially identical. Thus, a system that starts at $t \to -\infty$ in state a (or equivalently state 1) moves initially on the potential surface $E_a(R)$ (equivalently $E_1(R)$).

The description (not the actual physics) of the subsequent time evolution depends on the representation used. Because of the coupling V_{ab} and/or the time dependence of R, transitions between states can take place, so that at each point R (or equivalently time t) the state of the system is a linear combination

$$\psi(R) = C_1(R)\psi_1(R) + C_2(R)\psi_2(R) = C_a(R)\phi_a(R) + C_b(R)\phi_b(R) \tag{2.46}$$

with $|C_1(R)|^2 + |C_2(R)|^2 = 1$, $|C_a(R)|^2 + |C_b(R)|^2 = 1$, and $|C_1(R \to -\infty)|^2 = |C_a(R \to -\infty)|^2 = 1$. At $t \to \infty$ ($R \to \infty$) the two states are again effectively noninteracting, the transition process has ended and the transition probabilities can be determined. As seen in Fig. 2.2, the diabatic state a is in this limit the same as the adiabatic state 2, so that $|C_a(R \to \infty)|^2 = |C_2(R \to \infty)|^2$ represents the probability $P_{a\leftarrow a}$ to stay in the same diabatic state a but also the probability $P_{2\leftarrow 1}$ to cross into the other adiabatic state. (We sometimes say that hopping between the two adiabatic potential surfaces occurred with probability $P_{2\leftarrow 1}$.) Similarly $|C_b(R \to \infty)|^2 = |C_1(R \to \infty)|^2$ is the probability $P_{b\leftarrow a}$ to change diabatic state but also the probability $P_{1\leftarrow 1}$ to stay on the original adiabatic surface. The LZ approximate solution to this problem is

$$P_{1\leftarrow 1} = P_{b\leftarrow a} = 1 - \exp\left\{ -\frac{2\pi |V_{ab}|^2}{\hbar |(d/dt)(E_a(R) - E_b(R))|} \right\}_{R=R^*} \tag{2.47}$$

where the time dependence of the energy spacing between states a and b stems from their dependence on R. Consequently,

$$\frac{d}{dt}(E_a(R) - E_b(R)) = \dot{R}|F_b - F_a| \tag{2.48}$$

where $\dot{R}$ is the nuclear velocity and $F_i = -\partial E_i/\partial R$ is the force on the system when it moves on the potential surface E_i. All quantities are to be evaluated at the crossing point $R = R^*$.

Two limits of the result (2.47) are particularly simple. In the weak coupling/high speed limit, $2\pi|V_{ab}|^2 \ll \hbar\dot{R}|F_b - F_a|$ we get[6]

$$P_{1\leftarrow 1} = P_{b\leftarrow a} = \left\{ \frac{2\pi |V_{ab}|^2}{\hbar\dot{R}\,|F_b - F_a|} \right\}_{R=R^*} \tag{2.49}$$

The probability to remain on the adiabatic surface 1 is very small in this limit, and it is more appealing to think of the process as a low probability non-adiabatic transition between the diabatic states a and b. This case is often referred to as the *non-adiabatic limit* of the LZ problem. In the opposite limit (large coupling, slow motion—adiabatic limit) we get $P_{1\leftarrow 1} = P_{b\leftarrow a} = 1$, namely, the system moves adiabatically on a single potential surface.

---

[6] In many dimensions $\dot{R}|F_b - F_a|$ stands for a scalar product of the vectors $\dot{\mathbf{R}}$ and $|\mathbf{F}_b - \mathbf{F}_a|$.

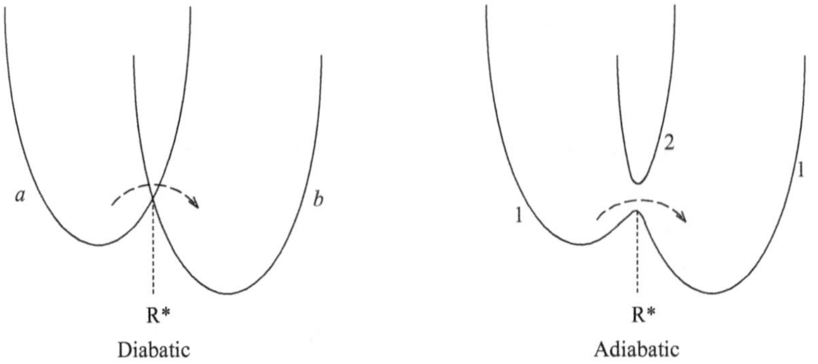

FIG. 2.3 Motion through the potential-crossing region depicted in the adiabatic representation (right) and in the diabatic representation (left).

Even though the above discussion of curve crossing dynamics was presented in terms of a collision process, the same phenomenology applies in other situations encountered in molecular physics. Figure 2.3 depicts the potential energy surfaces for the nuclear motion associated with two stable electronic states. Such states are usually obtained from a level of description (e.g. the Born–Oppenheimer approximation) that neglects some small coupling terms in the molecular Hamiltonian. The smallness of the terms neglected is gauged against the interstate energy spacing, and when this spacing vanishes (at $R = R^*$) the coupling becomes important. The left and right panels show respectively the diabatic and adiabatic picture of this situation. Both pictures show potential energy surfaces that are obtained from a Born–Oppenheimer approximation—neglecting the effect of nuclear *motion* (not to be confused with nuclear *position*) on the electronic states and energies, however, the "diabatic" picture is obtained by further neglecting terms in the electronic Hamiltonian that couple the states on the left and the right (for a more detailed discussion of this point see Section 2.5). The arrow indicates the reaction under discussion. When the interaction is small, the diabatic picture is more convenient; the reaction is regarded as a non-adiabatic transition $a \rightarrow b$. In the opposite limit the adiabatic picture may be more convenient. Indeed, if the interaction is large enough the splitting between the adiabatic surfaces 1 and 2 is large and transitions between them may be disregarded. At low temperature the presence of state 2 may be disregarded altogether, and the reaction may be regarded as a barrier crossing process taking place on the adiabatic potential surface 1.

Reaction rates. In studying processes of this kind it is often the case that the relevant observable is not the transition probability (2.47) but the *reaction rate*. How are the two connected?

In the non-adiabatic limit, where the $a \to b$ transition probability is small, the system may oscillate in the reactant well a for a relatively long time, occasionally passing through the transition point (or lower dimensionality subspace) R^*, that is, through a configuration in which the $a \to b$ transition probability may be significant. We refer to such as passage as an "attempt" to cross over to the product state b. If we assume that successive crossing attempts are independent of each other, and if the number of such attempts per unit time is ν, then the rate is roughly given by

$$k_{b\leftarrow a} = \nu P_{b\leftarrow a}$$

This is a crude description. As we have seen, the probability $P_{b\leftarrow a}$ depends on the crossing speed, and a proper thermal averaging must be taken. We will come back to these issues in Sections 14.3.5 and 16.2.

2.5 A digression on nuclear potential surfaces

The basis for separating the electronic and nuclear dynamics of, say, a molecular system is the Born–Oppenheimer (BO) approximation. A system of electrons and nuclei is described by the Hamiltonian

$$\begin{aligned} \hat{H} &= \hat{H}_{\text{el}}(\mathbf{r}) + \hat{H}_{\text{N}}(\mathbf{R}) + \hat{V}_{\text{el–N}}(\mathbf{r},\mathbf{R}) \\ \hat{H}_{\text{el}} &= \hat{T}_{\text{el}} + \hat{V}_{\text{el}}(\mathbf{r}); \qquad \hat{H}_{\text{N}} = \hat{T}_{\text{N}} + \hat{V}_{\text{N}}(\mathbf{R}) \end{aligned} \tag{2.50}$$

Here $\hat{H}_{\text{el}}$ is the Hamiltonian of the electronic subsystem, $\hat{H}_{\text{N}}$—that of the nuclear subsystem (each a sum of kinetic energy and potential energy operators) and $V_{\text{el–N}}(\mathbf{r},\mathbf{R})$ is the electrons–nuclei (electrostatic) interaction that depends on the electronic coordinates $\mathbf{r}$ and the nuclear coordinates $\mathbf{R}$. The BO approximation relies on the large mass difference between electron and nuclei that in turn implies that electrons move on a much faster timescale than nuclei. Exploring this viewpoint leads one to look for solutions for eigenstates of $\hat{H}$ of the form $\psi_\nu^{(n)}(\mathbf{r},\mathbf{R}) = \phi_n(\mathbf{r},\mathbf{R})\chi_\nu^{(n)}(\mathbf{R})$, or a linear combination of such products. Here $\phi(\mathbf{r},\mathbf{R})$ are solutions of the electronic Schrödinger equation in which the nuclear configuration $\mathbf{R}$is taken constant

$$(\hat{H}_{\text{el}} + \hat{V}_{\text{el–N}}(\mathbf{r},\mathbf{R}))\phi_n(\mathbf{r},\mathbf{R}) = E_{\text{el}}^{(n)}(\mathbf{R})\phi_n(\mathbf{r},\mathbf{R}) \tag{2.51}$$

while $\chi_\nu^{(n)}(\mathbf{R})$ are solutions of the nuclear Schrödinger equation with $E_{\text{el}}^{(n)}(\mathbf{R})$ as a potential

$$(\hat{T}_{\text{N}} + V_{\text{N}}(\mathbf{R}) + E_{\text{el}}^{(n)}(\mathbf{R}))\chi_\nu^{(n)}(\mathbf{R}) = E_{n\nu}\chi_\nu^{(n)}(\mathbf{R}) \tag{2.52}$$

The function $E_n(\mathbf{R}) = V_{\mathrm{N}}(\mathbf{R}) + E_{\mathrm{el}}^{(n)}(\mathbf{R})$ is the *adiabatic* potential surface for the nuclear motion when the system is in electronic state n. The wavefunctions $\phi_n(\mathbf{r}, \mathbf{R})$ are referred to as the adiabatic electronic wavefunctions and the electronic-nuclear wavefunctions $\psi_\nu^{(n)}(\mathbf{r}, \mathbf{R}) = \phi_n(\mathbf{r}, \mathbf{R})\chi_\nu^{(n)}(\mathbf{R})$ represent *vibronic states* in the adiabatic representation. The *non-adiabatic coupling* between vibronic states stems from the parametric dependence of the electronic wavefunctions on $\mathbf{R}$. Specifying for simplicity to a single nuclear coordinate it is given by (for $n \neq n'$)

$$\left\langle \psi_\nu^{(n)}(\mathbf{r}, \mathbf{R}) \middle| \hat{H} \middle| \psi_{\nu'}^{(n')}(\mathbf{r}, \mathbf{R}) \right\rangle = \left\langle \chi_\nu^{(n)}(R) \middle| \langle \phi_n(\mathbf{r}, R) | \hat{T}_{\mathrm{N}} | \phi_{n'}(\mathbf{r}, R) \rangle_{\mathbf{r}} \middle| \chi_{\nu'}^{(n')}(R) \right\rangle_{\mathbf{R}}$$

$$= -\frac{\hbar^2}{2M} \int dR \chi_\nu^{(n)*}(R) \chi_{\nu'}^{(n')}(R) \left\langle \phi_n(\mathbf{r}, R) \middle| \frac{\partial^2}{\partial R^2} \middle| \phi_{n'}(\mathbf{r}, R) \right\rangle_{\mathbf{r}}$$

$$- \frac{\hbar^2}{M} \int dR \chi_\nu^{(n)*}(R) \frac{\partial}{\partial R} \chi_{\nu'}^{(n')}(R) \left\langle \phi_n(\mathbf{r}, R) \middle| \frac{\partial}{\partial R} \middle| \phi_{n'}(\mathbf{r}, R) \right\rangle_{\mathbf{r}} \qquad (2.53)$$

where the subscripts $\mathbf{R}$ and $\mathbf{r}$ denote integrations in the nuclear or electronic spaces, respectively.

Diabatic states are obtained from a similar approach, except that additional term (or terms) in the Hamiltonian are disregarded in order to adopt a specific physical picture. For example, suppose we want to describe a process where an electron e is transferred between two centers of attraction, A and B, of a molecular systems. We may choose to work in a basis of vibronic states obtained for the e-A system in the absence of e-B attraction, and for the e-B system in the absence of the e-A attraction. To get these vibronic states we again use a Born–Oppenheimer procedure as described above. The potential surfaces for the nuclear motion obtained in this approximation are the corresponding diabatic potentials. By the nature of the approximation made, these potentials will correspond to electronic states that describe an electron localized on A or on B, and electron transfer between centers A and B implies that the system has crossed from one diabatic potential surface to the other.

To clarify these general statements lets consider a simple example (Fig. 2.4). A single electron can move between two identical atoms X, fixed in space. A single nuclear coordinate is exemplified by the angle θ of the orientation of a dipole that represents a solvent molecule.

Consider first the two-center system without the "solvent" dipole. Denote the ground state of the electron about the isolated left center by $\phi_{\mathrm{L}}(\mathbf{r})$ and the equivalent ground state about the isolated right center by $\phi_{\mathrm{R}}(\mathbf{r})$. $\phi_{\mathrm{L}}(\mathbf{r})$ is the electronic ground state of a Hamiltonian in which the interaction of the electron with the right center was neglected. Similarly, $\phi_{\mathrm{R}}(\mathbf{r})$ corresponds to the Hamiltonian that

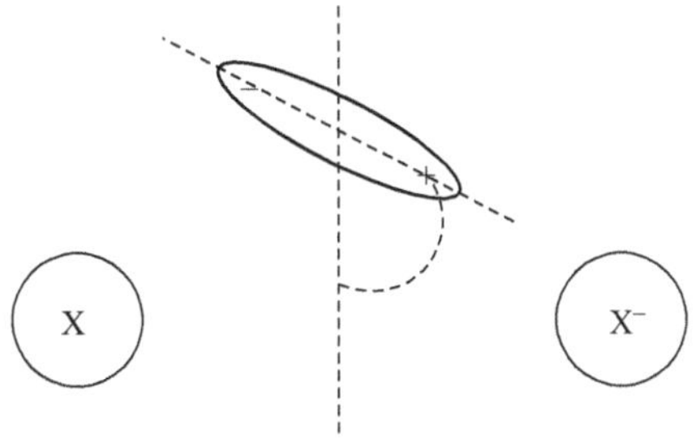

FIG. 2.4 An example that demonstrates the origin of diabatic potential surfaces and the difference between them and the corresponding adiabatic surfaces. The process is an electron transfer between the two centers X, witnessed by a nuclear coordinates represented by the orientation θ of a "solvent dipole".

does not contain the electron interaction with the left center. These functions describe an electron localized about the left center and the right center, respectively. Suppose also that other electronic states are far higher in energy and can be disregarded. When both centers are present the true ground state is a linear combination $2^{-1/2}(\phi_L(\mathbf{r}) + \phi_R(\mathbf{r}))$ which describes an electron with equal amplitudes on these centers.

Next consider the two-center system together with the "solvent." Let us construct a Born–Oppenheimer description of the electronic ground state. When only the right center is present the electronic function $\phi_R(\mathbf{r}, \theta)$ still represents an electron localized about the right center. The corresponding ground state energy $E_R(\theta)$ constitutes a potential surface for the orientational motion of the dipole in the presence of this center. This surface has a single minimum, attained when the angle θ is such that the dipole is oriented toward the negative charge, as shown in the figure. Similarly $\phi_L(\mathbf{r}, \theta)$ and $E_L(\theta)$ are the ground electronic state and the corresponding orientational potential surface when the electron is localized about the left center. $E_R(\theta)$ and $E_L(\theta)$ are the two diabatic potential surfaces. In contrast the adiabatic ground electronic state is the exact ground state of the full Hamiltonian, that is, $2^{-1/2}(\phi_L(\mathbf{r}, \theta) + \phi_R(\mathbf{r}, \theta))$. The corresponding ground state adiabatic potential, $E(\theta)$, will have two symmetric minima as a function of θ, reflecting the fact that the dipole equally prefers to be oriented toward either one of the partial charges on the two centers.

The first picture above yields two diabatic potential surfaces, each with a minimum reflecting the tendency of the dipole to point toward the localized electron. The second yields the lower adiabatic potential surface—a double minimum potential. The relationship between these surfaces can be understood by noting that the most important effect of the residual interaction (i.e. the originally neglected interaction of the electron on the left center with the right center and vice versa) on the electronic energy is to split the degeneracy at the point θ^* when the two diabatic surfaces cross. This leads to the picture seen in Figs 2.2 and 2.3.

Which representation is "better"? The answer depends on our objective. If this objective is to find accurate energy levels of a molecular system in order to predict a molecular spectrum, say, then the adiabatic representation gives us a starting point closer to our ultimate goal. On the other end, it often happens that a system is initially prepared in a state which is more closely represented by a diabatic basis as in the example discussed above, and the ensuing transfer process is investigated. In this case the diabatic picture provides a more physical description of the transfer process, though the adiabatic representation remains useful in the strong coupling limit.

2.6 Expressing the time evolution in terms of the Green's operator

We now return to time-independent Hamiltonians and describe another method for solving the time-dependent Schrödinger equation. Linear initial value problems described by time-independent operators are conveniently solved using Laplace transforms (Section 1.1.7). In Section 1.1.7 we have seen an example where the equation

$$\frac{df}{dt} = -\alpha f; \qquad \alpha \text{ real and positive} \tag{2.54}$$

was solved by such a procedure. The solution could be expressed as the inverse Laplace transform, Eq. (1.78), which could be evaluated for α real and positive by closing a counter-clockwise contour on the negative-real half plane of z, leading to the result (1.79). To make the procedure more similar to that used below we repeat that development in a slightly different form: Define $z = -i\omega$, so that $dz = -id\omega$. In terms of the new integration variable ω, Eq. (1.78) becomes

$$f(t) = -\frac{1}{2\pi i} \int_{-\infty+i\varepsilon}^{\infty+i\varepsilon} d\omega e^{-i\omega t} \frac{1}{\omega + i\alpha} f(0) \tag{2.55}$$

Closing a clockwise contour in the lower half complex plane along $\text{Im}(z) \to -\infty$, leads again to the result (1.79).

Consider now the initial value problem represented by Eq. (2.1) with a given $\Psi(t_0 = 0)$. The Laplace transform of Eq. (2.1) is

$$z\tilde{\Psi}(z) - \Psi(0) = \frac{-i}{\hbar} \hat{H} \tilde{\Psi}(z) \tag{2.56}$$

which leads to[7]

$$\tilde{\Psi}(z) = \frac{1}{z + (i/\hbar)\hat{H}}\Psi(0) \tag{2.57}$$

The time-dependent solution $\Psi(t)$ of Eq. (2.1) is obtained from the inverse Laplace transform of (2.57)

$$\Psi(t) = \frac{1}{2\pi i}\int_{-\infty i+\varepsilon}^{\infty i+\varepsilon} dz e^{zt}\frac{1}{z + (i/\hbar)\hat{H}}\Psi(0) \tag{2.58}$$

with $\varepsilon > 0$. Here ε may be taken as small as we wish because the eigenvalues of $\hat{H}$ are all real and consequently all singularities of the integrand in (2.58) are on the imaginary z axis. It is again convenient to use the substitution $z = -i\omega$, $dz = -id\omega$, which transforms Eq. (2.58) to

$$\Psi(t) = -\frac{1}{2\pi i}\int_{-\infty+i\varepsilon}^{\infty+i\varepsilon} d\omega e^{-i\omega t}\frac{1}{\omega - \hat{H}/\hbar}\Psi(0)$$

or, changing integration variable according to $E = \hbar(\omega - i\varepsilon)$

$$\Psi(t) = -\frac{1}{2\pi i}\int_{-\infty}^{\infty} dE e^{-i(E+i\varepsilon)t/\hbar}\frac{1}{E - \hat{H} + i\varepsilon}\Psi(0); \qquad \varepsilon \to 0 \tag{2.59}$$

where ε was redefined with an additional factor $\hbar$. The $i\varepsilon$ term in the exponent can be disregarded in the $\varepsilon \to 0$ limit, however, the corresponding term in the denominator has to be handled more carefully since the spectrum of $\hat{H}$ is real. The time-dependent wavefunction is seen to be essentially a Fourier transform of the function $\hat{G}(E)\Psi(t = 0)$, where $\hat{G}(E) \equiv (E - \hat{H} + i\varepsilon)^{-1}$ is the *retarded Green's function* (or, rather, Green's operator). In particular, the probability amplitude for the system to remain in state $\Psi(0)$ at time t is given by the Fourier transform of a

[7] Throughout this text we use operator expressions such as $(z - \hat{H})^{-1}$ with a scalar z to denote $(z\hat{I} - \hat{H})^{-1}$ where $\hat{I}$ is the unit operator.

diagonal matrix element of this operator

$$\langle\Psi(0)|\Psi(t)\rangle = -\frac{1}{2\pi i}\int_{-\infty}^{\infty} dE e^{-iEt/\hbar} G_{00}^{+}(E) \tag{2.60}$$

$$G_{00}^{+}(E) = \left\langle \Psi(0) \left| \frac{1}{E-\hat{H}+i\varepsilon} \right| \Psi(0) \right\rangle; \qquad \varepsilon \to 0 \tag{2.61}$$

Equations (2.59)–(2.61) constitute a formal solution of the time-dependent Schrödinger equation, expressed in terms of the Green operator. We will later see how this formal solution can be applied.

2.7 Representations

2.7.1 The Schrödinger and Heisenberg representations

Consider again the time evolution equations (2.1)–(2.3). If $\hat{A}$ is an operator representing a physical observable, the expectation value of this observable at time t is $\langle A\rangle_t = \langle\Psi(t)|\hat{A}|\Psi(t)\rangle$. We can express this same quantity differently. Define

$$\Psi_{\mathrm{H}} \equiv \hat{U}^{\dagger}(t)\Psi(t) \tag{2.62a}$$

$$\hat{A}_{\mathrm{H}}(t) = \hat{U}^{\dagger}(t)\hat{A}\hat{U}(t) \tag{2.62b}$$

where

$$\hat{U}(t) \equiv \hat{U}(t,0) = e^{-(i/\hbar)\hat{H}t} \tag{2.63}$$

Obviously, Ψ_{H} is simply $\Psi(t = 0)$ and is by definition time independent. Equation (2.62) is a unitary transformation on the wavefunctions and the operators at time t. The original *representation* in which the wavefunctions are time dependent while the operators are not, is transformed to another representation in which the operators depend on time while the wavefunctions do not. The original formulation is referred to as the *Schrödinger representation*, while the one obtained using (2.62) is called the *Heisenberg representation*. We sometimes use the subscript S to emphasize the Schrödinger representation nature of a wavefunction or an operator, that is,

$$\Psi_{\mathrm{S}}(t) = \Psi(t); \qquad \hat{A}_{\mathrm{S}}(t) = \hat{A}_{\mathrm{H}}(t=0) = \hat{A} \tag{2.64}$$

Either representation can be used to describe the time evolution of any observable quantity. Indeed

$$\langle\hat{A}\rangle_t = \langle\Psi(t)|\hat{A}|\Psi(t)\rangle = \langle\Psi_{\mathrm{H}}|\hat{A}_{\mathrm{H}}(t)|\Psi_{\mathrm{H}}\rangle \tag{2.65}$$

Problem 2.5. Prove this identity.

Note that the invariance of quantum observables under unitary transformations has enabled us to represent quantum time evolutions either as an evolution of the wavefunction with the operator fixed, or as an evolution of the operator with constant wavefunctions. Equation (2.1) describes the time evolution of wavefunctions in the Schrödinger picture. In the Heisenberg picture the wavefunctions do not evolve in time. Instead we have a time evolution equation for the Heisenberg operators:

$$\frac{d}{dt}\hat{A}_{\mathrm{H}}(t) = \frac{i}{\hbar}\left[\hat{H}, \hat{A}_{\mathrm{H}}(t)\right] \tag{2.66}$$

Problem 2.6. Use Eqs (2.62)–(2.63) to prove Eq. (2.66)

Equation (2.66) is referred to as the *Heisenberg equation of motion*. Note that it should be solved as an initial value problem, given that $\hat{A}_{\mathrm{H}}(t = 0) = \hat{A}$. In fact, Eq. (2.62b) can be regarded as the formal solution of the Heisenberg equation (2.66) in the same way that the expression $\Psi(t) = e^{-(i/\hbar)\hat{H}t}\Psi(t = 0)$ is a formal solution to the Schrödinger equation (2.1).

To end this section we note that the entire time evolution referred to in the above discussion arises from the Schrödinger equation. In general the operator $\hat{A}$ may have an explicit dependence on time, in which case the transformation to the Heisenberg representation may again be carried out, however, the resulting Heisenberg equation is

$$\frac{d}{dt}\hat{A}_{\mathrm{H}}(t) = \frac{i}{\hbar}\left[\hat{H}, \hat{A}_{\mathrm{H}}(t)\right] + \frac{\partial \hat{A}_{\mathrm{H}}(t)}{\partial t} \tag{2.67}$$

Problem 2.7. Use the definition $\hat{A}_{\mathrm{H}}(t) = \exp(i\hat{H}t/\hbar)\hat{A}_{\mathrm{S}}(t)\exp(-i\hat{H}t/\hbar)$ to verify (2.67).

2.7.2 The interaction representation

Obviously, any unitary transformation can be applied to the wavefunctions and operators and used to our advantage. In particular, for any Hamiltonian that is written as

$$\hat{H} = \hat{H}_0 + \hat{V} \tag{2.68}$$

the interaction representation is defined by the transformation

$$\hat{A}_{\mathrm{I}}(t) = e^{(i/\hbar)\hat{H}_0 t}\hat{A}e^{-(i/\hbar)\hat{H}_0 t} \tag{2.69}$$

$$\Psi_{\mathrm{I}}(t) = e^{(i/\hbar)\hat{H}_0 t}\Psi_{\mathrm{S}}(t) = e^{(i/\hbar)\hat{H}_0 t}e^{-(i/\hbar)\hat{H}t}\Psi(0) \tag{2.70}$$

Problem 2.8. Show that

$$\langle\Psi_{\mathrm{S}}(t)|\hat{A}_{\mathrm{S}}|\Psi_{\mathrm{S}}(t)\rangle = \langle\Psi_{\mathrm{H}}|\hat{A}_{\mathrm{H}}(t)|\Psi_{\mathrm{H}}\rangle = \langle\Psi_{\mathrm{I}}(t)|\hat{A}_{\mathrm{I}}(t)|\Psi_{\mathrm{I}}(t)\rangle \tag{2.71}$$

The time evolution equations in the interaction representation are easily derived from these definitions

$$\frac{d\hat{A}_{\mathrm{I}}}{dt} = \frac{i}{\hbar}\left[\hat{H}_0, \hat{A}_{\mathrm{I}}\right] \tag{2.72}$$

and

$$\begin{aligned}\frac{d\Psi_{\mathrm{I}}}{dt} &= \frac{i}{\hbar}e^{(i/\hbar)\hat{H}_0 t}(\hat{H}_0 - \hat{H})e^{-(i/\hbar)\hat{H}t}\Psi(0) \\ &= -\frac{i}{\hbar}e^{(i/\hbar)\hat{H}_0 t}Ve^{-(i/\hbar)\hat{H}_0 t}e^{(i/\hbar)\hat{H}_0 t}e^{-(i/\hbar)\hat{H}t}\Psi(0) \\ &= -\frac{i}{\hbar}\hat{V}_{\mathrm{I}}(t)\Psi_{\mathrm{I}}(t)\end{aligned} \tag{2.73}$$

Equations (2.72) and (2.73) indicate that in the interaction representation the time evolution of the operators is carried by $\hat{H}_0$, while that of the wavefunctions is determined by the interaction $\hat{V}$, or rather by its interaction representation that is itself a time-dependent operator.

2.7.3 Time-dependent perturbation theory

Equation (2.73) is particularly useful in cases where the time evolution carried by $\hat{H}_0$ can be easily evaluated, and the effect of $\hat{V}$ is to be determined perturbatively. Equation (2.73) is a direct route to such a perturbation expansion. We start by integrating it to get

$$\Psi_{\mathrm{I}}(t) = \Psi_{\mathrm{I}}(0) - \frac{i}{\hbar}\int_0^t dt_1\,\hat{V}_{\mathrm{I}}(t_1)\Psi_{\mathrm{I}}(t_1) \tag{2.74}$$

and continue by substituting the same expression (2.74) for $\Psi_I(t_1)$ on the right. This yields

$$\Psi_I(t) = \Psi_I(0) + \left(-\frac{i}{\hbar}\right) \int_0^t dt_1 \hat{V}_I(t_1)\Psi_I(0) + \left(-\frac{i}{\hbar}\right)^2 \int_0^t dt_1 \int_0^{t_1} dt_2 \hat{V}_I(t_1)\hat{V}_I(t_2)\Psi_I(t_2) \tag{2.75}$$

$$\Psi_I(t) = \left(1 + \sum_{n=1}^{\infty} \left(-\frac{i}{\hbar}\right)^n \int_0^t dt_1 \int_0^{t_1} dt_2 \cdots \int_0^{t_{n-1}} dt_n \hat{V}_I(t_1)\hat{V}_I(t_2) \cdots \hat{V}_I(t_n)\right) \Psi_I(0) \tag{2.76}$$

Note that the order of the operators $V_I(t)$ inside the integrand is important: These operators do not in general commute with each other because they are associated with different times. It is seen from Eq. (2.76) that the order is such that operators associated with later times appear more to the left.

Problem 2.9. Show that Eq. (2.74) is equivalent to the operator identity

$$e^{-(i/\hbar)\hat{H}t} = e^{-(i/\hbar)\hat{H}_0 t} - \frac{i}{\hbar} \int_0^t dt' e^{-(i/\hbar)\hat{H}_0(t-t')} \hat{V} e^{-(i/\hbar)\hat{H}t'} \tag{2.77}$$

Problem 2.10. Confirm the following operator identity

$$\exp[\beta(\hat{S} + \hat{R})] = \exp(\beta\hat{S}) \left(1 + \int_0^\beta d\lambda e^{-\lambda\hat{S}} \hat{R} e^{[\lambda(\hat{S}+\hat{R})]}\right) \tag{2.78}$$

by multiplying both sides by $\exp(-\beta\hat{S})$ and taking derivative with respect to β. Verify that Eqs (2.78) and (2.77) are equivalent.

2.8 Quantum dynamics of the free particles

2.8.1 Free particle eigenfunctions

The Hamiltonian of a free particle of mass m is

$$\hat{H} = -\frac{\hbar^2}{2m}\nabla^2 \tag{2.79}$$

and the corresponding time-independent Schrödinger equation is

$$\nabla^2\psi = -k^2\psi; \qquad k^2 = \frac{2mE}{\hbar^2} \tag{2.80}$$

It is convenient to use the set of eigenfunctions normalized in a box of dimensions (L_x, L_y, L_z) with periodic boundary conditions

$$\Psi(x, y, z) = \Psi(x + n_x L_x, y + n_y L_y, z + n_z L_z); \qquad n = 0, \pm 1, \pm 2, \ldots \tag{2.81}$$

A set of such functions is

$$\begin{aligned} |\mathbf{k}\rangle \equiv \psi_{\mathbf{k}}(\mathbf{r}) &= \frac{1}{\sqrt{\Omega}} e^{i\mathbf{k}\cdot\mathbf{r}}; \qquad \mathbf{k} = (k_x, k_y, k_z) \\ k_j &= \frac{2\pi}{L_j} n_j; \qquad n_j = 0, \pm 1, \ldots; \qquad j = x, y, z \\ \Omega &= L_x L_y L_z \end{aligned} \tag{2.82}$$

with the eigenvalues

$$E_{\mathbf{k}} = \frac{\hbar^2 k^2}{2m} \tag{2.83}$$

These functions constitute an orthonormal set

$$\langle \mathbf{k} | \mathbf{k}' \rangle \equiv \int_{\Omega} d^3\mathbf{r}\, \psi_{\mathbf{k}}^*(\mathbf{r}) \psi_{\mathbf{k}'}(\mathbf{r}) = \delta_{\mathbf{k},\mathbf{k}'} \tag{2.84}$$

and can be used to express the time evolution of any function $\Psi(\mathbf{r}, t = 0)$ that satisfies the periodic boundary conditions (2.81). Following Eqs (2.5) and (2.6) we get

$$\Psi(\mathbf{r}, t) = \sum_{\mathbf{k}} \langle \psi_{\mathbf{k}} | \Psi(\mathbf{r}, 0) \rangle e^{-(i/\hbar)[\hbar^2 k^2/(2m)]t} | \psi_{\mathbf{k}} \rangle \tag{2.85}$$

where

$$\langle \psi_{\mathbf{k}} \mid \Psi(\mathbf{r}, 0) \rangle = \frac{1}{\sqrt{\Omega}} \int_{\Omega} d^3 r e^{-i\mathbf{k}\cdot\mathbf{r}} \Psi(\mathbf{r}, 0) \tag{2.86}$$

We have seen that the normalization condition $\int_{\Omega} dx |\psi(x)|^2 = 1$ implies that free particle wavefunctions vanish everywhere like $\psi(x) \sim \Omega^{-1/2}$ as $\Omega \to \infty$. The probability $|\psi(x)|^2 dx$ to find the particle at position $x \ldots x+dx$ vanishes like Ω^{-1} in this limit. As such, these functions are *by themselves* meaningless. We will see that meaningful physics can be obtained in two scenarios: First, if we think of the process as undergone by a distribution of N identical independent particles where the number N is proportional to the volume Ω so that the density $\rho(x) = N|\psi(x)|^2$ is finite. We may thus work with the single particle wavefunctions, keeping in mind that (1) such functions are normalized by $\Omega^{-1/2}$ and (2) that physically meaningful quantities are obtained by multiplying observables by the total number of particles to get the single particle density factored in.

Second, several observables are obtained as products of matrix elements that scale like $\psi(x)^2$ (therefore like Ω^{-1}) and the density of states that scales like Ω (Eq. (2.95) or (2.97) below). A well-known example is the golden rule formula (9.25) for the inverse lifetime of a local state interacting with a continuum. Such products remain finite and physically meaningful even when $\Omega \to \infty$.

Anticipating such scenarios, we use in many applications periodic boundary conditions as a trick to represent infinite systems by taking the periodic box dimensions to infinity at the end of the calculation. We will see several examples below.

Problem 2.11. Show that if the wavefunction $\Psi(\mathbf{r}, t = 0)$ is square-integrable (i.e. $\int d^3 r |\Psi(\mathbf{r}, t = 0)|^2$ (integral over all space) is finite, so is $\Psi(\mathbf{r}, t)$ at any later time.

Problem 2.12. For the function

$$\psi(x) = \frac{1}{(2\pi D^2)^{1/4}} \exp\left[-\frac{x^2}{4D^2}\right] \tag{2.87}$$

Consider the expansion in terms of the one-dimensional free particle eigenstates

$$\psi(x) = \sum_k c_k \psi_k(x) \tag{2.88}$$

$$\psi_k(x) = L^{-1/2} e^{ikx}; \qquad k = (2\pi/L)n; \quad n = 0, \pm 1, \ldots \tag{2.89}$$

Explain why this expansion is meaningful only in the limit $L \gg D$. Show that in this limit

$$c_k = \left(\frac{8\pi D^2}{L^2}\right)^{1/4} e^{-D^2k^2} \tag{2.90}$$

2.8.2 Free particle density of states

The density of states concept was introduced in Section 1.3. For any operator $\hat{A}$ characterized by the eigenvalue spectrum $\{a_j\}$ we can define a density function $\rho_A(a)$ such that the number of eigenvalues that satisfy $a \le a_j \le a+\Delta a$ is $\rho_A(a)\Delta a$. Such a density can be introduced for a discrete spectrum

$$\rho_A(a) = \sum_j \delta(a - a_j) \tag{2.91}$$

but it is most useful when the spectrum becomes so dense that either our measurement cannot resolve the individual eigenvalues or we are not interested in these high resolution details. This is obviously the case for the momentum operator whose eigenfunctions are the free particle wavefunctions (2.82), in the limit $L_x, L_y, L_z \to \infty$.

In what follows we describe one possible way to obtain the density of momentum states for this problem. Consider Eqs (2.87)–(2.90). Obviously the identity

$$\sum_k |c_k|^2 = 1 \tag{2.92}$$

has to be satisfied. Using (2.90) in (2.92) we get

$$\frac{\sqrt{8\pi}D}{L}\sum_k \exp[-2D^2k^2] = \frac{\sqrt{8\pi}D\rho}{L}\int_{-\infty}^{\infty} dk \exp[-2D^2k^2] = 1 \tag{2.93}$$

where the conversion to an integral is suggested by the fact that when $\Omega \to \infty$ the allowed values of the wavevectors k constitute a dense set (cf. Eq. (2.82)). ρ is the desired density of states in this set, defined such that $\rho\Delta k$ is the number of allowed states, ($k_j = (2\pi/L)n; n = 0, \pm 1, \ldots$) in the interval $k \ldots k + \Delta k$. In principle ρ could depend on k, but it does not in the present case: this is seen from the fact that the spacing $2\pi/L$ between allowed values of k is a constant independent of k. On evaluating the Gaussian integral in (2.93) we can find the explicit expression for ρ,

$$\rho = \frac{L}{2\pi} \tag{2.94}$$

The same reasoning in three dimensions will yield

$$\rho = \frac{L_x L_y L_z}{(2\pi)^3} = \frac{\Omega}{(2\pi)^3} \tag{2.95}$$

This number, multiplied by $d^3k = dk_x dk_y dk_z$ is the number of quantum states in the k-space volume between k_x and $k_x + dk_x$, k_y and $k_y + dk_y$, and k_z and $k_z + dk_z$. The fact that this result does not depend on $\mathbf{k} = (k_x, k_y, k_z)$ implies that the distribution of free particle eigenstates in k-space is homogeneous.

It is useful also to cast the density of states in other representations, most notably the energy. We may thus seek a density ρ_E so that the number of states with energy between E and $E + \Delta E$ is $\rho_E \Delta E$. In one dimension the indicated energy interval corresponds to the k-axis interval $2\hbar^{-1}(\sqrt{2m(E + dE)} - \sqrt{2mE}) = \hbar^{-1}(2m/E)^{1/2} dE$ (the factor 2 comes from the fact that a given interval in E corresponds to two intervals in k, for positive and negative values) so that

$$\rho_E = \frac{L}{2\pi\hbar}\sqrt{\frac{2m}{E}}; \qquad (d = 1) \tag{2.96}$$

We can get the same result from the formal connection $\rho_k dk = \rho_E dE$, which implies that $\rho_E = \rho_k (dE/dk)^{-1}$. In three dimensions the interval between E and $E + \Delta E$ corresponds in k-space to a spherical shell whose surface area is $4\pi k^2$ and width is Δk, where $k^2 = 2mE/\hbar^2$ and $\Delta k = (dE/dk)^{-1}\Delta E = \hbar^{-1}(m/(2E))^{1/2}\Delta E$. This yields

$$\rho_E = \frac{\Omega}{(2\pi)^3} \times 4\pi \frac{2mE}{\hbar^2} \times \frac{1}{\hbar}\left(\frac{m}{2E}\right)^{1/2} = \frac{\Omega}{2\pi^2}\frac{m}{\hbar^3}\sqrt{2mE}; \quad (d = 3) \tag{2.97}$$

Note that ρ_E is a function of the energy E.

2.8.3 Time evolution of a one-dimensional free particle wavepacket

Consider now the time evolution of a free particle moving in one dimension that starts at $t = 0$ in the normalized state

$$\Psi(x, t = 0) = \frac{1}{(2\pi D^2)^{1/4}} \exp\left[-\frac{(x - x_0)^2}{4D^2} + \frac{ip_0 x}{\hbar}\right] \tag{2.98}$$

We refer to the wavefunction (2.98) in the present context as a wavepacket: It is a local function in position space that can obviously be expanded in the complete set of free particle waves (2.82).[8] Below we see that this function is also localized in momentum space.

Problem 2.15. Show that for the wavefunction (2.98) the expectation values of the position and momentum are

$$\langle x \rangle_{t=0} = \int_{-\infty}^{\infty} dx \Psi^*(x, t=0)\hat{x}\Psi(x, t=0) = x_0 \tag{2.99}$$

and

$$\langle p \rangle_{t=0} = \int_{-\infty}^{\infty} dx \Psi^*(x, t=0)\left(-i\hbar\frac{\partial}{\partial x}\right)\Psi(x, t=0) = p_0 \tag{2.100}$$

Also show that the position variance associated with this wavefunction is

$$\langle (x - \langle x \rangle)^2 \rangle_{t=0} = \langle x^2 \rangle - x_0^2 = D^2 \tag{2.101}$$

and that the momentum variance is

$$\langle (p - \langle p \rangle)^2 \rangle_{t=0} = \langle p^2 \rangle - p_0^2 = \frac{\hbar^2}{4D^2} \tag{2.102}$$

Note that $\Delta x_0 \equiv [\langle (x - \langle x \rangle)^2 \rangle_{t=0}]^{1/2}$ and $\Delta p_0 \equiv [\langle (p - \langle p \rangle)^2 \rangle_{t=0}]^{1/2}$ satisfy the Heisenberg uncertainty rule as an equality: $\Delta x_0 \Delta p_0 = \hbar/2$. For this reason we refer to (2.98) as a minimum uncertainty wavepacket.

The expansion of (2.98) in the set of eigenstates $\psi_k(x) = L^{-(1/2)}e^{ik\cdot x}$ yields

$$\Psi(x, t=0) = \frac{1}{\sqrt{L}}\sum_k c_k e^{ikx} \tag{2.103}$$

[8] The periodic boundary conditions are inconsequential here provided that the range in which this wavefunction is significantly different from zero is far smaller than L.

where

$$c_k = \langle \psi_k \mid \Psi(x,0)\rangle = \frac{1}{\left(2\pi L^2 D^2\right)^{1/4}} \int_{-\infty}^{\infty} dx \exp\left[-\frac{(x-x_0)^2}{4D^2} - i(k-k_0)x\right] \tag{2.104}$$

where $k_0 = p_0/\hbar$. Evaluating the Fourier transform in (2.104) we get

$$c_k = \left(\frac{8\pi D^2}{L^2}\right)^{1/4} \exp[-D^2(k-k_0)^2 - ix_0(k-k_0)] \tag{2.105}$$

This implies that for a particle whose quantum state is (2.98), the probability to find it with momentum $\hbar k$ is

$$|c_k|^2 = \frac{\sqrt{8\pi}D}{L} \exp[-2D^2(k-k_0)^2] \tag{2.106}$$

Note that Eqs (2.87)–(2.90) represent a special case of these results.

The time evolution that follows from Eq. (2.98) may now be found by using Eq. (2.85). In one dimension it becomes

$$\Psi(x,t) = \frac{1}{\sqrt{L}} \sum_k c_k e^{ikx-(i/\hbar)[\hbar^2k^2/(2m)]t} \tag{2.107}$$

The probability that the system described initially by $\Psi(x, t=0)$ stays in this initial state is given by

$$P(t) = |\langle \Psi(x,t=0) \mid \Psi(x,t)\rangle|^2 \tag{2.108}$$

Using Eqs (2.103) and (2.107) as well as the one-dimensional version of (2.84) yields

$$P(t) = \left|\sum_k |c_k|^2 \, e^{-(i/\hbar)[\hbar^2k^2/(2m)]t}\right|^2 \tag{2.109}$$

Inserting Eq. (2.106) and converting the sum over k to an integral, $\sum_k \to (L/(2\pi)) \int dk$ finally leads to

$$P(t) = \left|\sqrt{\frac{2}{\pi}}D \int_{-\infty}^{\infty} dk \exp\left[-\frac{i\hbar}{2m}tk^2 - 2D^2(k-k_0)^2\right]\right|^2 \tag{2.110}$$

Note that the dependence on L does not appear in any observable calculated above.

We can also find the time-dependent wavefunction explicitly. Combining Eqs (2.107), (2.105), and (2.94), and converting again the sum over k to an integral leads to

$$\Psi(x,t) = \left(\frac{D^2}{2\pi^3}\right)^{1/4} e^{ik_0x_0} \int_{-\infty}^{\infty} dk \exp\left[ik(x-x_0) - D^2(k-k_0)^2 - \frac{i\hbar k^2}{2m}t\right] \tag{2.111}$$

This Gaussian integral can be evaluated in a straightforward way. To get some physical insight consider the result obtained from the initial wavepacket with $x_0 = p_0 = k_0 = 0$. In this case (2.111) yields

$$\Psi(x,t) = \left(\frac{1}{2\pi}\right)^{1/4} \left(D + \frac{i\hbar t}{2mD}\right)^{-1/2} \exp\left[-\frac{x^2}{4D^2 + 2i\hbar t/m}\right] \tag{2.112}$$

that leads to

$$|\Psi(x,t)|^2 = \left\{2\pi\left[D^2 + \frac{\hbar^2 t^2}{4m^2D^2}\right]\right\}^{-1/2} \exp\left[-\frac{x^2}{2[D^2 + \hbar^2t^2/(4m^2D^2)]}\right] \tag{2.113}$$

This wavepacket remains at the peak position $x = 0$, with its width increases with time according to

$$\langle(\Delta x)^2\rangle^{1/2} = [D^2 + \hbar^2 t^2/(4m^2D^2)]^{1/2} \xrightarrow{t\to\infty} \frac{\hbar t}{2mD} \tag{2.114}$$

2.8.4 The quantum mechanical flux

In a classical system of moving particles the magnitude of the flux vector is the number of particles going per unit time through a unit area perpendicular to that vector. If $\rho(\mathbf{r})$ and $\mathbf{v}(\mathbf{r})$ are the density and average speed of particles at point $\mathbf{r}$, the flux is given by

$$\mathbf{J}(\mathbf{r}) = \mathbf{v}(\mathbf{r})\rho(\mathbf{r}) \tag{2.115a}$$

It can be written as a classical dynamical variable, that is, a function of positions and momenta of all particles in the system, in the form

$$\mathbf{J}(\mathbf{r}) = \sum_j (\mathbf{p}_j/m_j)\,\delta(\mathbf{r} - \mathbf{r}_j) \tag{2.115b}$$

where the sum is over all particles.

The quantum analog of this observable should be an operator. To find it we start for simplicity in one dimension and consider the time-dependent Schrödinger equation and its complex conjugate

$$\frac{\partial \Psi}{\partial t} = -\frac{i}{\hbar}\left[-\frac{\hbar^2}{2m}\frac{\partial^2}{\partial x^2}\Psi + V(x)\Psi\right] \tag{2.116}$$

$$\frac{\partial \Psi^*}{\partial t} = \frac{i}{\hbar}\left[-\frac{\hbar^2}{2m}\frac{\partial^2}{\partial x^2}\Psi^* + V(x)\Psi^*\right] \tag{2.117}$$

Multiply (2.116) by Ψ^* and (2.117) by Ψ and add the resulting equations to get

$$\frac{\partial(\Psi^*\Psi)}{\partial t} = \frac{i\hbar}{2m}\left(\Psi^*\frac{\partial^2}{\partial x^2}\Psi - \Psi\frac{\partial^2}{\partial x^2}\Psi^*\right) \tag{2.118}$$

We next integrate this equation between two points, x_1 and x_2. $\Psi^*(x,t)\Psi(x,t)dx$ is the probability at time t to find the particle in $x \ldots x + dx$, hence the integral on the left yields the rate of change of the probability P_{1-2} to find the particle in the range between x_1 and x_2. We therefore get

$$\begin{aligned}\frac{dP_{1-2}(t)}{dt} &= \frac{i\hbar}{2m}\int_{x_1}^{x_2} dx\left(\Psi^*\frac{\partial^2\Psi}{\partial x^2} - \Psi\frac{\partial^2\Psi^*}{\partial x^2}\right) = \frac{i\hbar}{2m}\int_{x_1}^{x_2} dx\frac{\partial}{\partial x}\left(\Psi^*\frac{\partial\Psi}{\partial x} - \Psi\frac{\partial\Psi^*}{\partial x}\right)\\ &= -\frac{\hbar}{m}\int_{x_1}^{x_2} dx\frac{\partial}{\partial x}\left(\mathrm{Im}\left[\Psi^*(x,t)\frac{\partial\Psi(x,t)}{\partial x}\right]\right)\end{aligned} \tag{2.119}$$

This can be integrated to give

$$\frac{dP_{1-2}(t)}{dt} = J(x_1,t) - J(x_2,t) \tag{2.120}$$

where

$$J(x,t) \equiv \frac{\hbar}{m}\mathrm{Im}\left[\Psi^*(x,t)\frac{\partial\Psi(x,t)}{\partial x}\right] = \frac{\hbar}{2mi}\left(\Psi^*\frac{\partial\Psi}{\partial x} - \Psi\frac{\partial\Psi^*}{\partial x}\right) \tag{2.121}$$

is defined as the *probability flux* at point x. Note that while $\mathbf{J}$ in Eq. (2.115) is a particle flux, Eq. (2.121) is the probability flux, and should be multiplied by N (in a system of N noninteracting particles) to give the particle flux.

Equation (2.120) may be recognized as a conservation law: NP_{1-2} is the number of particles in the (x_1,x_2) interval and $NJ(x)$ is the number of particles moving per unit time through point x. Equation (2.120) tells us that a change in number of particles in the $x_1, \ldots, x_2$ interval is caused by particles entering or leaving at the boundaries x_1 and x_2, that is, particles cannot be created or destroyed.

Problem 2.14. Use the results above to prove that at steady state of a one-dimensional system the flux has to be independent of both position and time.

Note that in our one-dimensional formulation the dimensionality of Ψ is $[\text{length}]^{-1/2}$ and the flux J has dimensionality of t^{-1}. The generalization of (2.121) to more than one dimension is found by repeating the procedure of Eqs (2.116)–(2.121), with the gradient operator ∇ replacing $\partial/\partial x$ everywhere. Equations (2.119) and (2.120) become

$$\begin{aligned}\frac{dP_\Omega(t)}{dt} &= -\frac{\hbar}{m}\int_\Omega d^3r\nabla\cdot\left(\mathrm{Im}\left[\Psi^*(\mathbf{r},t)\,\nabla\Psi(\mathbf{r},t)\right]\right)\\ &= -\frac{\hbar}{m}\int_{S_\Omega} d\mathbf{s}\mathbf{n}_s\cdot\mathrm{Im}\left[\Psi^*(\mathbf{r},t)\,\nabla\Psi(\mathbf{r},t)\right]\end{aligned}\tag{2.122}$$

where Ω here denotes a finite volume whose boundary is the surface S_Ω, and where $\mathbf{n}_s$ is a unit vector normal to the surface element $\mathbf{ds}$ in the outward direction. In getting the second line of Eq. (2.122) we have used the divergence theorem (Eq. (1.36)). In fact, the mathematical structure of Eq. (2.122) reflects the fact that in a closed system the quantum mechanical probability is a globally conserved quantity (See Section 1.1.4). It also enables us to identify the probability flux: The second line of (2.122) is the analog of the right-hand side of Eq. (2.120), where the flux is now defined by the analog of Eq. (2.121).

$$\mathbf{J}(\mathbf{r},t) = \frac{\hbar}{2mi}\left[\Psi^*(\mathbf{r},t)\,(\nabla\Psi(\mathbf{r},t)) - \Psi(\mathbf{r},t)\left(\nabla\Psi^*(\mathbf{r},t)\right)\right]\tag{2.123}$$

In three dimensions Ψ has the dimensionality $[\text{length}]^{-3/2}$ and the dimension of flux is $[tl^2]^{-1}$. When multiplied by the total number of particles N, the flux vector gives the number of particles that cross a unit area normal to its direction per unit time.

As an example consider the free particle wavefunctions $\psi_1(\mathbf{r}) = A\exp(i\mathbf{k}\cdot\mathbf{r})$ and $\psi_2(\mathbf{r}) = A\cos(\mathbf{k}\cdot\mathbf{r})$. From Eq. (2.123) it is clear that the flux associated with ψ_2 is zero. This is true for any wavefunction that is real or can be made real by multiplication by a position independent phase factor. On the other hand, using

$\Psi=\psi_1$ in (2.123) yields

$$\mathbf{J}(\mathbf{r}) = \frac{\hbar\mathbf{k}}{m}|A|^2 = \mathbf{v}|A|^2 \tag{2.124}$$

The flux is defined up to the constant A. Taking $|A|^2 = \Omega^{-1}$ (a single particle wavefunction normalized in the volume Ω) implies that the relevant observable is $N\mathbf{J}(\mathbf{r})$, that is, is the particle flux for a system with a total of N particles with $N \sim \Omega$. Sometimes it is convenient to normalize the wavefunction to unit flux, $J = 1$ by choosing $A = \sqrt{m/(\hbar k)}$.

2.9 Quantum dynamics of the harmonic oscillator

2.9.1 Elementary considerations

The classical Hamiltonian for a one-dimensional harmonic oscillator of mass m centered about $x = 0$,

$$H = \frac{p^2}{2m} + \frac{1}{2}kx^2 \tag{2.125}$$

implies the classical equations of motion

$$\dot{x} = p/m; \quad \dot{p} = -m\omega^2 x \qquad \text{with } \omega = \sqrt{\frac{k}{m}} \tag{2.126}$$

It is convenient for future reference to define the dimensionless position ξ and momentum ϕ

$$\xi = \alpha x \qquad \text{where } \alpha = \sqrt{\frac{m\omega}{\hbar}} \tag{2.127}$$

$$\phi = p/\sqrt{\hbar m\omega} \tag{2.128}$$

In terms of these variables the Hamiltonian (2.125) takes the form

$$H = \frac{\hbar\omega}{2}(\xi^2 + \phi^2) \tag{2.129}$$

and the classical equations of motion become

$$\dot{\xi} = \omega\phi; \qquad \dot{\phi} = -\omega\xi \tag{2.130}$$

In quantum mechanics the momentum corresponds to the operator $\hat{p} = -i\hbar\partial/\partial x$, or

$$\hat{\phi} = -i\partial/\partial\xi \tag{2.131}$$

the position and momentum operator satisfy the familiar commutation relationship

$$[\hat{x}, \hat{p}] = i\hbar \rightarrow [\hat{\xi}, \hat{\phi}] = i \tag{2.132}$$

and the quantum Hamiltonian is

$$\hat{H} = -\frac{\hbar^2}{2m}\frac{d^2}{dx^2} + \frac{1}{2}m\omega^2\hat{x}^2 \tag{2.133}$$

or in dimensionless form

$$\frac{\hat{H}}{\hbar\omega} = \frac{1}{2}\left(-\frac{\partial^2}{\partial\xi^2} + \hat{\xi}^2\right) \tag{2.134}$$

The solutions of the time-independent Schrödinger equation $\hat{H}\psi = E\psi$ are the (orthonormal) eigenfunctions and the corresponding eigenvalues

$$\psi_n(x) = N_n H_n(\alpha x)\, e^{-(1/2)(\alpha x)^2} \qquad E_n = (n + 1/2)\,\hbar\omega \tag{2.135}$$

where $H_n(\xi)$ are the Hermit polynomials that can be obtained from

$$H_{n+1}(\xi) = 2\xi H_n(\xi) - 2nH_{n-1}(\xi) \tag{2.136}$$

$$H_0(\xi) = 1; \qquad H_1(\xi) = 2\xi \tag{2.137}$$

and N_n is the normalization factor

$$N_n = \sqrt{\frac{\alpha}{\pi^{1/2}2^n n!}} \tag{2.138}$$

chosen so that $\int_{-\infty}^{\infty} dx\psi_n^2(x) = 1$. In particular the ground state wavefunction and energy are

$$\psi_0(x) = \sqrt{\frac{\alpha}{\sqrt{\pi}}} e^{-(1/2)\alpha^2 x^2}; \qquad E_0 = (1/2)\hbar\omega \tag{2.139}$$

The Hermite polynomials also satisfy the identity

$$\frac{d}{d\xi} H_n(\xi) = 2nH_{n-1}(\xi) \tag{2.140}$$

and the eigenfunctions satisfy

$$\langle \psi_n | \hat{x} | \psi_m \rangle = \int_{-\infty}^{\infty} dx x \psi_n(x) \psi_m(x) = \begin{cases} \alpha^{-1}\sqrt{(n+1)/2}; & m = n+1 \\ \alpha^{-1}\sqrt{n/2}; & m = n-1 \\ 0 & \text{otherwise} \end{cases} \tag{2.141}$$

Consider now solutions of the time-dependent Schrödinger equation

$$\frac{\partial \Psi(x,t)}{\partial t} = -\frac{i}{\hbar}\left(-\frac{\hbar^2}{2m}\frac{d^2}{dx^2} + \frac{1}{2} m\omega^2 x^2 \right) \Psi(x,t) \tag{2.142}$$

Knowing the eigenfunctions and eigenvalues implies that any solution to this equation can be written in the form (2.6), with the coefficients determined from the initial condition according to $c_n(t_0) = \langle \psi_n(x) | \Psi(x, t_0) \rangle$. The following problem demonstrates an important property of properly chosen wavepackets of harmonic oscillator wavefunctions.

Problem 2.15.

1. Show by direct substitution that the solution of (2.142) with the initial condition

$$\Psi(x, t = 0) = \sqrt{\frac{\alpha}{\sqrt{\pi}}} e^{-(1/2)\alpha^2 (x - x_0)^2} = \psi_0(x - x_0) \tag{2.143}$$

is given by

$$\Psi(x,t) = \sqrt{\frac{\alpha}{\sqrt{\pi}}} e^{-(1/2)i\omega t} e^{-(1/2)\alpha^2 [x - \bar{x}(t)]^2 + (i/\hbar)\bar{p}(t)[x - (1/2)\bar{x}(t)]} \tag{2.144}$$

where

$$\bar{x}(t) = x_0 \cos(\omega t)\,; \qquad \bar{p}(t) = -m\omega x_0 \sin(\omega t) \tag{2.145}$$

satisfy the classical equation of motions (2.126), that is, $\dot{\bar{x}} = \bar{p}/m$ and $\dot{\bar{p}} = -m\omega^2\bar{x}$. (Note that in terms of the reduced position $\bar{\xi}(t)$ and momentum $\bar{\phi}(t)$, Eq. (2.144) takes the form

$$\Psi(x,t) = \sqrt{\frac{\alpha}{\sqrt{\pi}}} e^{-(1/2)i\omega t} e^{-(1/2)[\xi-\bar{\xi}(t)]^2 + i\bar{\phi}(t)[\xi-(1/2)\bar{\xi}(t)]} \tag{2.146}$$

where $\bar{\xi}(t) = \xi_0 \cos(\omega t)$ and $\bar{\phi}(t) = -\xi_0 \sin(\omega t)$ (with $\xi_0 = \alpha x_0$) satisfy Eq. (2.130).)

2. Show that the $t = 0$ wavepacket (2.143) is just the ground state wavefunction of the harmonic oscillator (Eq. (2.135) with $n = 0$) with the equilibrium position shifted from 0 to x_0.
3. Show that at time t the average position and momentum associated with the wavefunction (2.144) satisfy

$$\langle x\rangle(t) \equiv \int_{-\infty}^{\infty} dx \Psi^*(x,t)\hat{x}\Psi(x,t) = \bar{x}(t) \tag{2.147}$$

$$\langle p\rangle(t) \equiv \int_{-\infty}^{\infty} dx \Psi^*(x,t)\hat{p}\Psi(x,t) = \bar{p}(t) \tag{2.148}$$

while the variances

$$(\Delta x(t))^2 \equiv \int_{-\infty}^{\infty} dx \Psi^*(x,t)(\hat{x} - \langle x\rangle)^2\Psi(x,t) \tag{2.149}$$

$$(\Delta p(t))^2 \equiv \int_{-\infty}^{\infty} dx \Psi^*(x,t)(\hat{p} - \langle p\rangle)^2\Psi(x,t) \tag{2.150}$$

do not depend on time and satisfy Eqs (2.101) and (2.102), respectively, with $\alpha^2 = (2D^2)^{-1}$. (Note that consequently, the uncertainty $\Delta x \cdot \Delta p = (1/2)\hbar$ is also time independent.)

Thus, the solution (2.144) oscillates with frequency ω in a way that resembles the classical motion: First, the expectation values of the position and momentum oscillate, as implied by Eqs (2.145), according to the corresponding classical equations of motion. Second, the wavepacket as a whole executes such oscillations, as can be most clearly seen from the probability distribution

$$|\Psi(x,t)|^2 = \frac{\alpha}{\sqrt{\pi}} e^{-\alpha^2[x-\bar{x}(t)]^2} \tag{2.151}$$

that is, the wavepacket oscillates just by shifting the position of its center in a way that satisfies the classical Newton equations. In particular, unlike in the free particle case (see Eq. (2.114)), the width of this wavepacket does not change with time.

2.9.2 The raising/lowering operators formalism

Focusing again on Eqs (2.125)–(2.134) it is convenient to define a pair of operators, linear combinations of the position and momentum, according to

$$\begin{aligned} \hat{a} &= \sqrt{\frac{m\omega}{2\hbar}}\hat{x} + \frac{i}{\sqrt{2\hbar m\omega}}\hat{p} = \frac{1}{\sqrt{2}}(\hat{\xi} + i\hat{\phi}) \\ \hat{a}^\dagger &= \sqrt{\frac{m\omega}{2\hbar}}\hat{x} - \frac{i}{\sqrt{2\hbar m\omega}}\hat{p} = \frac{1}{\sqrt{2}}(\hat{\xi} - i\hat{\phi}) \end{aligned} \tag{2.152}$$

so that

$$\hat{x} = \sqrt{\frac{\hbar}{2m\omega}}(\hat{a}^\dagger + \hat{a}); \qquad \hat{p} = i\sqrt{\frac{m\hbar\omega}{2}}(\hat{a}^\dagger - \hat{a}) \tag{2.153}$$

Using Eq. (2.132) we find that $\hat{a}$ and $\hat{a}^\dagger$ satisfy

$$[\hat{a}, \hat{a}^\dagger] = 1 \tag{2.154}$$

and can be used to rewrite the Hamiltonian in the form

$$\hat{H} = \hbar\omega\left(\hat{a}^\dagger\hat{a} + \frac{1}{2}\right) = \hbar\omega\left(\hat{N} + \frac{1}{2}\right) \tag{2.155}$$

Here $\hat{N} = \hat{a}^\dagger\hat{a}$ is called the "number operator" for reasons given below. This operator satisfies the commutation relations

$$\begin{aligned} [\hat{N}, \hat{a}] &= -\hat{a} \\ [\hat{N}, \hat{a}^\dagger] &= \hat{a}^\dagger \end{aligned} \tag{2.156}$$

To see the significance of these operators we use Eqs (2.131), (2.135), (2.136) and (2.140) to derive the following identities:

$$\hat{\xi}\psi_n = \frac{1}{\sqrt{2}}(\sqrt{n}\psi_{n-1} + \sqrt{n+1}\psi_{n+1})$$
$$\hat{\phi}\psi_n = \frac{1}{i\sqrt{2}}(\sqrt{n}\psi_{n-1} - \sqrt{n+1}\psi_{n+1}) \tag{2.157}$$

Using Eqs (2.152) this implies

$$\hat{a}\,|n\rangle = \sqrt{n}\,|n-1\rangle\,; \qquad \hat{a}^{\dagger}\,|n\rangle = \sqrt{n+1}\,|n+1\rangle \tag{2.158}$$

where we have used $|n\rangle$ to denote ψ_n. The operators $\hat{a}^{\dagger}$ and $\hat{a}$ are seen to have the property that when operating on an eigenfunction of the Harmonic oscillator Hamiltonian they yield the eigenfunction just above or below it, respectively. $\hat{a}^{\dagger}$ and $\hat{a}$ will therefore be referred to as the harmonic oscillator raising (or creation) and lowering (or annihilation) operators, respectively.[9]

Equation (2.152) also leads to

$$\hat{N}\,|n\rangle = n\,|n\rangle \tag{2.159}$$

(hence the name "number operator") and to the representation of the nth eigenstate in the form

$$|n\rangle = \frac{1}{\sqrt{n!}}(\hat{a}^{\dagger})^{n}\,|0\rangle \tag{2.160}$$

Furthermore it is easy to derive the following useful relations:

$$\langle n|\,\hat{a} = \sqrt{n+1}\,\langle n+1|$$
$$\langle n|\,\hat{a}^{\dagger} = \sqrt{n}\,\langle n-1| \tag{2.161}$$

$$\langle n'|\hat{a}|n\rangle = \sqrt{n}\delta_{n',n-1}$$
$$\langle n'|\hat{a}^{\dagger}|n\rangle = \sqrt{n+1}\delta_{n',n+1} \tag{2.162}$$

[9] The terms "creation" and "annihilation" arise in applications where the system of interest is a group of harmonic oscillators with a given distribution of frequencies. Photons in the radiation field and phonons in an elastic field (see Chapters 3 and 4 respectively) correspond to excitations of such oscillators. $\hat{a}_{\omega}^{\dagger}$ is then said to create a phonon (or a photon) of frequency ω and $\hat{a}_{\omega}$ destroys such a "particle."

Problem 2.16. Use Eqs (2.153), (2.154), and (2.161) to prove that

$$\langle n|x|n'\rangle = \sqrt{\frac{\hbar}{2m\omega}}(\sqrt{n+1}\delta_{n',n+1} + \sqrt{n}\delta_{n',n-1}) \tag{2.163}$$

2.9.3 The Heisenberg equations of motion

An important advantage of formulating harmonic oscillators problems in terms of raising and lowering operators is that these operators evolve very simply in time. Using the Heisenberg equations of motion (2.66), the expression (2.155) and the commutation relations for $\hat{a}$ and $\hat{a}^\dagger$ leads to

$$\dot{\hat{a}}(t) = -i\omega_0\hat{a}(t); \qquad \dot{\hat{a}}^\dagger(t) = i\omega_0\hat{a}^\dagger(t) \tag{2.164}$$

where now $\hat{a}(t)$ and $\hat{a}^\dagger(t)$ are in the Heisenberg representation. To simplify notation we will often omit the subscript H that denotes this representation (see Eq. (2.66)) when the identity of operators as Heisenberg representation operators is clear from the text. Eq. (2.164) yields the explicit time dependence for these operators

$$\hat{a}(t) = \hat{a}e^{-i\omega t}; \qquad \hat{a}^\dagger(t) = \hat{a}^\dagger e^{i\omega t} \tag{2.165}$$

Consequently, the Heisenberg representations of the position and momentum operators are

$$\hat{x}(t) = \sqrt{\frac{\hbar}{2m\omega}}(\hat{a}^\dagger e^{i\omega t} + \hat{a}e^{-i\omega t}); \qquad \hat{p}(t) = i\sqrt{\frac{m\hbar\omega}{2}}(\hat{a}^\dagger e^{i\omega t} - \hat{a}e^{-i\omega t}) \tag{2.166}$$

As an example for the use of this formulation let us calculate the (in-principle time-dependent) variance, $\langle \Delta x(t)^2 \rangle$, defined by Eq. (2.149) for a Harmonic oscillator in its ground state. Using the expression for position operator in the Heisenberg representation from Eq. (2.166) and the fact that $\langle 0|\Delta x(t)^2|0\rangle = \langle 0|x(t)^2|0\rangle$ for an oscillator centered at the origin, this can be written in the from

$$\begin{aligned}\langle 0|\Delta x(t)^2\,|0\rangle &= \frac{\hbar}{2m\omega}\langle 0|\,(\hat{a}^\dagger e^{i\omega t} + \hat{a}e^{-i\omega t})^2\,|0\rangle = \frac{\hbar}{2m\omega}\langle 0|\,\hat{a}^\dagger\hat{a} + \hat{a}\hat{a}^\dagger\,|0\rangle \\ &= \frac{\hbar}{2m\omega}\langle 0|\,2\hat{a}^\dagger\hat{a} + 1\,|0\rangle = \frac{\hbar}{2m\omega}\end{aligned} \tag{2.167}$$

where we have also used the commutation relation (2.154). A reader that evaluates Eq. (2.149) using the explicit wavefunction (2.139) can appreciate the great simplification offered by this formulation.

2.9.4 The shifted harmonic oscillator

Problems involving harmonic oscillators that are shifted in their equilibrium positions relative to some preset origin are ubiquitous in simple models of quantum dynamical processes. We consider a few examples in this section.

2.9.4.1 Harmonic oscillator under an additional constant force

Consider a particle of charge q moving in one dimension (along the x-axis, say) in a harmonic potential. The Hamiltonian describing its motion is

$$\hat{H} = \frac{\hat{p}^2}{2m} + \frac{1}{2}m\omega^2\hat{x}^2 \tag{2.168}$$

Let us switch on an external uniform electrostatic field $\mathcal{E}$ along the same direction. The Hamiltonian becomes

$$\hat{H}_{\text{S}} = \frac{\hat{p}^2}{2m} + \frac{1}{2}m\omega^2\hat{x}^2 - q\mathcal{E}\hat{x} \tag{2.169}$$

It is easy to find the eigenfunctions and eigenvalues of the Hamiltonian (2.169) given the corresponding eigenfunctions and eigenvalues of (2.168). Making the transformation

$$\bar{x} = x - \frac{q\mathcal{E}}{m\omega^2} \tag{2.170}$$

the Hamiltonian (2.169) becomes

$$\hat{H}_{\text{S}} = \frac{\hat{p}^2}{2m} + \frac{1}{2}m\omega^2\hat{\bar{x}}^2 - \frac{q^2\mathcal{E}^2}{2m\omega^2} \tag{2.171}$$

In Eqs (2.169) and (2.171) $\hat{p} = -i\hbar\partial/\partial x = -i\hbar\partial/\partial\bar{x}$. The Hamiltonian (2.169) is thus shown to represent a harmonic oscillator in the absence of external field with an energy spectrum that is shifted uniformly by the last term on the right of (2.171), and whose equilibrium position is shifted according to Eq. (2.170). The new eigenstates are therefore shifted harmonic oscillator wavefunctions:

$$\psi_{\text{S}}(x;\varepsilon) = \psi\left(x - \frac{q\mathcal{E}}{m\omega^2};0\right) \tag{2.172}$$

The position shift operator. Consider the operator

$$\hat{U}(\lambda) \equiv e^{-\lambda(\partial/\partial x)} \tag{2.173}$$

Since the operator $\partial/\partial x$ is anti-hermitian (i.e. $(\partial/\partial x)^\dagger = -\partial/\partial x$) $\hat{U}(\lambda)$ is unitary ($\hat{U}^\dagger = \hat{U}^{-1}$) for real λ. The identity

$$e^{-\lambda(\partial/\partial x)}\psi(x) = \left(1 - \lambda\frac{\partial}{\partial x} + \frac{1}{2}\lambda^2\frac{\partial^2}{\partial x^2} + \cdots \frac{(-1)^n}{n!}\lambda^n\frac{\partial^n}{\partial x^n} + \cdots\right)\psi(x) = \psi(x-\lambda) \tag{2.174}$$

identifies this unitary operator as the position shift operator. In terms of the operators a and $a^\dagger$ this operator takes the form

$$\hat{U}(\lambda) = e^{\bar{\lambda}(\hat{a}^\dagger - \hat{a})} \tag{2.175}$$

with

$$\bar{\lambda} = \lambda\sqrt{\frac{m\omega}{2\hbar}} \tag{2.176}$$

Under the unitary transformation defined by $\hat{U}$ the position and momentum operators transform in a simple way. For example, since $\hat{U}$ is unitary, the following identity must hold for all functions $\psi(x)$ and $\phi(x)$

$$\langle\psi(x)|\hat{x}|\phi(x)\rangle = \langle\hat{U}\psi(x)|\hat{U}\hat{x}\hat{U}^\dagger|\hat{U}\phi(x)\rangle = \langle\psi(x-\lambda)|\hat{U}\hat{x}\hat{U}^\dagger|\phi(x-\lambda)\rangle \tag{2.177}$$

For this identity to be true we must have

$$\hat{U}(\lambda)\hat{x}\hat{U}^\dagger(\lambda) = \hat{x} - \lambda \tag{2.178a}$$

Also, since $\hat{p}$ and $\hat{U}$ commute it follows that

$$\hat{U}(\lambda)\hat{p}\hat{U}^\dagger(\lambda) = \hat{p} \tag{2.178b}$$

Using Eqs (2.178) and (2.152) it is easy to show also that

$$\begin{aligned}\hat{U}(\lambda)\hat{a}\hat{U}^\dagger(\lambda) &= \hat{a} - \bar{\lambda}\\ \hat{U}(\lambda)\hat{a}^\dagger\hat{U}^+(\lambda) &= \hat{a}^\dagger - \bar{\lambda}\end{aligned} \tag{2.179}$$

Appendix 2A (see entry 6) presents a more direct proof of these equalities using operator algebra relationships obtained there.

Franck–Condon factors. As an application of the raising and lowering operator formalism we next calculate the *Franck–Condon factor* in a model of shifted harmonic potential surfaces. Franck–Condon factors are absolute square overlap integrals between nuclear wavefunctions associated with different electronic potential surfaces. Such overlap integrals appear in calculations of transition rates between molecular electronic states, for example they determine the relative intensities of vibrational transitions that dress electronic spectral lineshapes: such intensities are determined by matrix elements of the dipole moment operator between two molecular vibronic states $\mu_{nv,n'v'} = \langle \phi_n(\mathbf{r},\mathbf{R})\chi_v^{(n)}(\mathbf{R})|\hat{\mu}(\hat{\mathbf{r}})|\phi_{n'}(\mathbf{r},\mathbf{R})\chi_{v'}^{(n')}(\mathbf{R})\rangle_{\mathbf{r},\mathbf{R}}$ where $\phi_n(\mathbf{r},\mathbf{R})$ and $\chi_v^{(n)}(\mathbf{R})$ are electronic and nuclear wavefunctions, respectively, obtained in the Born–Oppenheimer approximation (see Section 2.5), $\mathbf{r}$ and $\mathbf{R}$ are electronic and nuclear coordinates, respectively, and $\langle\ \rangle_{\mathbf{r},\mathbf{R}}$ indicates that integration is both in the nuclear and the electronic subspaces. In the so called Condon approximation one assumes that the dependence of the electronic integral $\mu_{n,n'}(\mathbf{R}) = \langle \phi_n(\mathbf{r},\mathbf{R})|\hat{\mu}(\hat{\mathbf{r}})|\phi_{n'}(\mathbf{r},\mathbf{R})\rangle_{\mathbf{r}}$ on the nuclear coordinate $\mathbf{R}$ is small and removes this term from the integral over $\mathbf{R}$, leading to

$$\mu_{nv,n'v'} = \mu_{n,n'}\langle \chi_v^{(n)}(\mathbf{R})|\chi_{v'}^{(n')}(\mathbf{R})\rangle_{\mathbf{R}} \Rightarrow |\mu_{nv,n'v'}|^2 = |\mu_{n,n'}|^2(\mathrm{FC})_{v,v'}^{(n,n')}$$

We will calculate the Franck–Condon factor in a model where the nuclear potential surfaces are identical one-dimensional harmonic potentials that are horizontally shifted with respect to each other, that is,

$$V_1(x) = (1/2)m\omega^2x^2; \qquad V_2(x) = (1/2)m\omega^2\,(x-\lambda)^2 \tag{2.180}$$

The FC factors arising from the overlap integral between vth excited state on the harmonic potential 1, say, and v'th excited state on the harmonic potential 2 is

$$(\mathrm{FC})_{v,v'}^{(1,2)} = \left|\int_{-\infty}^{\infty} dx\chi_v^{(1)*}(x)\chi_{v'}^{(2)}(x)\right|^2 \tag{2.181}$$

For simplicity we will consider the case where $v' = 0$, that is, where $\chi_{v'}^{(2)}(x)$ is the ground vibrational state on the harmonic surface 2. Now, from Eq. (2.180) it follows that $\chi_{v'}^{(2)}(x) = \chi_{v'}^{(1)}(x-\lambda)$. The desired FC factor is therefore

$$(\mathrm{FC})_{v,0}^{(1,2)} = (\mathrm{FC})_{v,0}\,(\lambda) = \left|\int_{-\infty}^{\infty} dx\chi_v^*(x)\chi_0(x-\lambda)\right|^2 \tag{2.182}$$

where both wavefunctions are defined on the same potential surface 1 whose explicit designation is now omitted. Note that the only relevant information concerning the electronic states 1 and 2 is the relative shift of their corresponding potential surfaces.

Now, from Eqs (2.174)–(2.175) we have $\chi_0(x-\lambda) = e^{\bar{\lambda}(\hat{a}^\dagger - \hat{a})}\chi_0(x)$, therefore

$$I \equiv \int_{-\infty}^{\infty} dx \chi_\nu^*(x)\chi_0(x-\lambda) = \langle \nu | \, e^{\bar{\lambda}(\hat{a}^\dagger - \hat{a})} \, |0\rangle \tag{2.183}$$

Note $|\nu\rangle$ and $|0\rangle$ are states defined on the same harmonic potential and are not shifted with respect to each other. Using Eq. (2.225) to replace $\exp(\bar{\lambda}(\hat{a}^\dagger - \hat{a}))$ by $\exp(-(1/2)\bar{\lambda}^2)\exp(\bar{\lambda}\hat{a}^\dagger)\exp(-\bar{\lambda}\hat{a})$, and using the Taylor expansion to verify that $\exp(-\bar{\lambda}\hat{a})|0\rangle = |0\rangle$ this leads to

$$I = e^{-(1/2)\bar{\lambda}^2}\langle \nu | e^{\bar{\lambda}\hat{a}^\dagger} |0\rangle \tag{2.184}$$

Again making a Taylor expansion, now of the operator $\exp(\bar{\lambda}\hat{a}^\dagger)$, it is easily seen that the only term that contributes is $(\bar{\lambda}^\nu/\nu!)(\hat{a}^\dagger)^\nu$. Using also Eq. (2.160) leads to

$$I = e^{-(1/2)\bar{\lambda}^2}\frac{\bar{\lambda}^\nu}{\sqrt{\nu!}} \tag{2.185}$$

Using also Eq. (2.176) finally yields the result

$$(\mathrm{FC})_{\nu,0}(\lambda) = |I|^2 = \exp\left(-\frac{m\omega\lambda^2}{2\hbar}\right)\frac{\left(m\omega\lambda^2/2\hbar\right)^\nu}{\nu!} \tag{2.186}$$

Time evolution of a shifted oscillator. We have already considered (see Problem 2.15) the time evolution of a state obtained by shifting the equilibrium position of the ground state $|0\rangle$ of a harmonic oscillator, that is,

$$\Psi(x, t=0) = \sqrt{\frac{\alpha}{\sqrt{\pi}}}e^{-(1/2)\alpha^2(x-\lambda)^2} = \psi_0(x-\lambda) \tag{2.187}$$

Let us repeat it using the shift operator (2.175). The initial shifted state takes the form

$$\Psi(t=0) = e^{\bar{\lambda}(\hat{a}^\dagger - \hat{a})}\,|0\rangle \equiv |\bar{\lambda}\rangle \tag{2.188}$$

which can be rewritten, using Eq. (2.225), in the form

$$|\bar{\lambda}\rangle = e^{-(1/2)|\bar{\lambda}|^2}e^{\bar{\lambda}\hat{a}^\dagger}\,|0\rangle; \qquad \bar{\lambda} = \lambda\sqrt{\frac{m\omega}{2\hbar}} \tag{2.189}$$

Such a state is sometimes referred to as a *coherent state* of the Harmonic oscillator.

Problem 2.17.

1. Show that a coherent state (2.189) is an eigenstate of the lowering operator. Specifically

$$\hat{a}\,|\bar{\lambda}\rangle = \bar{\lambda}\,|\bar{\lambda}\rangle \tag{2.190}$$

2. Show that the coherent state (2.189) is normalized

$$\langle\bar{\lambda}|\bar{\lambda}\rangle = 1 \tag{2.191}$$

The time evolution of this state can be now calculated in a straightforward way

$$\Psi(t) = e^{-(i/\hbar)\hat{H}t}e^{-(1/2)|\bar{\lambda}|^2}e^{\bar{\lambda}\hat{a}^\dagger}|0\rangle = e^{-(1/2)|\bar{\lambda}|^2}e^{-(i/\hbar)\hat{H}t}e^{\bar{\lambda}\hat{a}^\dagger}e^{(i/\hbar)\hat{H}t}e^{-(i/\hbar)\hat{H}t}|0\rangle \tag{2.192}$$

Using $e^{-(i/\hbar)\hat{H}t}|0\rangle = e^{-(i/2)\omega t}|0\rangle$ and $e^{-(i/\hbar)\hat{H}t}e^{\bar{\lambda}\hat{a}^\dagger}e^{(i/\hbar)\hat{H}t} = e^{\bar{\lambda}e^{-i\omega t}\hat{a}^\dagger}$ this leads to

$$\Psi(t) = e^{-(1/2)i\omega t}e^{-(1/2)|\bar{\lambda}|^2}e^{\bar{\lambda}e^{-i\omega t}\hat{a}^\dagger}|0\rangle = e^{-(1/2)i\omega t}|\bar{\lambda}e^{-i\omega t}\rangle \tag{2.193}$$

Except for a phase factor, the time evolution is given by an oscillating position shift, $\bar{\lambda} \to \bar{\lambda}e^{-i\omega t}$. Using this and (2.176) in (2.187) yields the result

$$\Psi(x,t) = \sqrt{\frac{\alpha}{\sqrt{\pi}}}e^{-(1/2)i\omega t}e^{-(1/2)\alpha^2(x-\lambda(t))^2}; \qquad \lambda(t) = \lambda e^{-i\omega t} \tag{2.194}$$

Problem 2.18. Show that (2.194) is identical to (2.144)–(2.145).

2.9.5 Harmonic oscillator at thermal equilibrium

Harmonic oscillators are often used as approximate models for realistic systems. A common application is their use as convenient models for the thermal environments of systems of interest (see Section 6.5). Such models are mathematically simple, yet able to account for the important physical attributes of a thermal bath: temperature, coupling distribution over the bath normal modes, and characteristic timescales. Their prominence in such applications is one reason why we study them in such detail in this chapter.

The treatment of quantum systems in thermal equilibrium, and of systems interacting with their thermal environments is expanded on in Chapter 10. For now it is enough to recall the statistical mechanics result for the average energy of a harmonic oscillator of frequency ω at thermal equilibrium

$$E = \hbar\omega\left(\langle n\rangle_T + \frac{1}{2}\right) \tag{2.195}$$

where $\langle n\rangle_T$ is the average excitation, that is, the average number of quanta $\hbar\omega$ in the oscillator, given by

$$\langle n\rangle_T = \langle a^\dagger a\rangle_T = \frac{\sum_n e^{-\beta n\hbar\omega}\langle n|a^\dagger a|n\rangle}{\sum_n e^{-\beta n\hbar\omega}} = \frac{1}{e^{\beta\hbar\omega} - 1} \tag{2.196}$$

In addition we may write

$$\langle\hat{a}\rangle_T = \langle\hat{a}^\dagger\rangle_T = \langle\hat{a}\hat{a}\rangle_T = \langle\hat{a}^\dagger\hat{a}^\dagger\rangle_T = 0 \tag{2.197}$$

because the diagonal elements of the operators involved are zero.

Problem 2.19.

1. Show that $\langle aa^\dagger\rangle_T = 1/(1 - e^{-\beta\hbar\omega})$.
2. Use these results to find the thermal averages $\langle\hat{x}^2\rangle_T$ and $\langle\hat{p}^2\rangle_T$, of the squared position and momentum operators.

2.10 Tunneling

In classical mechanics a particle with total energy E cannot penetrate a spatial regions $\mathbf{r}$ with potential energy $V(\mathbf{r}) > E$. Such a region therefore constitutes an impenetrable barrier for this particle. In quantum mechanics this is not so, and the possibility of the quantum wavefunction to penetrate into classically forbidden regions leads to the phenomenon of tunneling, whereupon a particle located at one side of a classically impenetrable barrier may, with a finite probability, appear on the other side.

2.10.1 Tunneling through a square barrier

Figure 2.5 depicts a simple example. A particle with energy E collides with a rectangular potential barrier of height $U_B > E$. In classical mechanics it will be simply reflected back. In reality there is a finite probability that it will tunnel to the

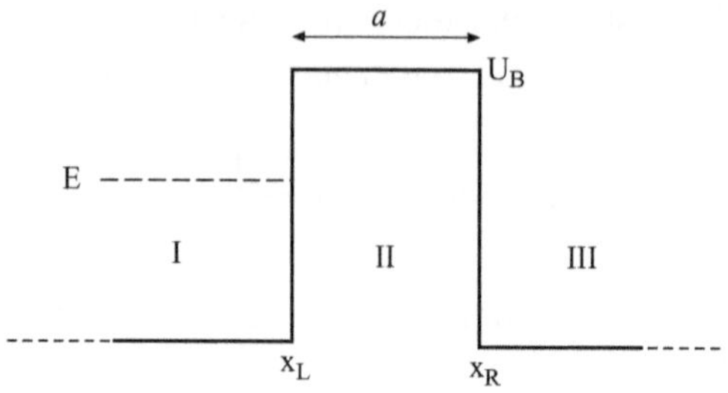

FIG. 2.5 Tunneling through a rectangular potential barrier characterized by a width a and a height U_B. E is the energy of the tunneling particle relative to the bottom of the potential shown.

other side of the barrier. This probability is expressed in terms of a *transmission coefficient*, a property of the barrier/particle system that is defined below.

Our problem is defined by the Hamiltonian

$$\hat{H} = -\frac{\hbar^2}{2m}\frac{\partial^2}{\partial x^2} + \hat{V}(x) \tag{2.198}$$

with

$$V(x) = \begin{cases} 0; & x < x_L; \qquad x > x_R \qquad \text{(region I)} \\ U_B; & x_L \le x \le x_R = x_L + a \qquad \text{(regions II, III)} \end{cases} \tag{2.199}$$

Consider the solution of the time-independent Schrödinger equation $\hat{H}\psi = E\psi$ for a given energy E. In regions I and III, where $U_B = 0$, it is the free particle equations whose solutions are

$$\psi_I(x) = Ae^{ikx} + Be^{-ikx}; \qquad (x < x_L) \tag{2.200}$$

$$\psi_{III}(x) = Ce^{ikx} + De^{-ikx}; \qquad (x > x_R) \tag{2.201}$$

In both regions, k corresponds to the given energy

$$k = \frac{1}{\hbar}\sqrt{2mE} \tag{2.202}$$

In the barrier region II the wavefunction is a solution of the equation

$$-\frac{\hbar^2}{2m}\frac{\partial^2}{\partial x^2}\psi = (E - U_B)\psi; \qquad (x_L \ge x \le x_R) \tag{2.203}$$

Denoting (since we are interested primarily in the case $E < U_B$)

$$\kappa = \frac{1}{\hbar}\sqrt{2m(U_B - E)} \tag{2.204}$$

this becomes $d^2\psi/dx^2 = \kappa^2\psi$, which yields

$$\psi_{\mathrm{II}}(x) = Fe^{\kappa x} + Ge^{-\kappa x} \tag{2.205}$$

In Eqs (2.200), (2.201), and (2.205), the coefficients A, B, C, D, F, and G are constants that should be determined from the boundary conditions. Such conditions stem from the requirement that the wavefunction and its derivative should be continuous everywhere, in particular at the boundaries $x = x_{\mathrm{L}}$ and $x = x_{\mathrm{R}}$, that is,

$$\psi_{\mathrm{I}}(x_{\mathrm{L}}) = \psi_{\mathrm{II}}(x_{\mathrm{L}}); \qquad [d\psi_{\mathrm{I}}(x)/dx]_{x=x_{\mathrm{L}}} = [d\psi_{\mathrm{II}}(x)/dx]_{x=x_{\mathrm{L}}} \tag{2.206}$$

and the same with x_{R} and ψ_{III} replacing x_{L} and ψ_{I}. This leads to the four equations

$$\begin{aligned}
Ae^{ikx_{\mathrm{L}}} + Be^{-ikx_{\mathrm{L}}} &= Fe^{\kappa x_{\mathrm{L}}} + Ge^{-\kappa x_{\mathrm{L}}} \\
ikAe^{ikx_{\mathrm{L}}} - ikBe^{-ikx_{\mathrm{L}}} &= \kappa Fe^{\kappa x_{\mathrm{L}}} - \kappa Ge^{-\kappa x_{\mathrm{L}}} \\
Ce^{ikx_{\mathrm{R}}} + De^{-ikx_{\mathrm{R}}} &= Fe^{\kappa x_{\mathrm{R}}} + Ge^{-\kappa x_{\mathrm{R}}} \\
ikCe^{ikx_{\mathrm{R}}} - ikDe^{-ikx_{\mathrm{R}}} &= \kappa Fe^{\kappa x_{\mathrm{R}}} - \kappa Ge^{-\kappa x_{\mathrm{R}}}
\end{aligned} \tag{2.207}$$

Note that we have only four conditions but six coefficients. The other two coefficients should be determined from the physical nature of the problem, for example, the boundary conditions at $\pm\infty$.[10] In the present case we may choose, for example, $D = 0$ to describe a process in which an incident particle comes from the left. The wavefunction then has an incident ($\exp(ikx)$) and reflected ($\exp(-ikx)$) components in region I, and a transmitted component ($\exp(ikx)$) in region III. Dividing the four equations (2.207) by A, we see that we have just enough equations to determine the four quantities B/A, C/A, F/A, and G/A. As discussed below, the first two are physically significant. We obtain for this case

$$\frac{B}{A} = \frac{(k^2+\kappa^2)\left(1-e^{-2\kappa a}\right)e^{2ikx_{\mathrm{L}}}}{(k+i\kappa)^2-(k-i\kappa)^2e^{-2\kappa a}}; \qquad \frac{C}{A} = \frac{4ik\kappa e^{-ika-\kappa a}}{(k+i\kappa)^2-(k-i\kappa)^2e^{-2\kappa a}} \tag{2.208}$$

whence

$$\mathcal{R}(E) \equiv \left|\frac{B}{A}\right|^2 = \frac{1}{1+\left(4E\,(U_{\mathrm{B}}-E)/U_{\mathrm{B}}^2\sinh^2(\kappa a)\right)} \tag{2.209}$$

[10] For example, the free particle wavefunction (2.200), a solution of a differential equation of the second-order, is also characterized by two coefficients, and we may choose $B = 0$ to describe a particle going in the positive x direction or $A = 0$ to describe a particle going in the opposite direction. The other coefficient can be chosen to express normalization as was done in Eq. (2.82).

and

$$\mathcal{T}(E) \equiv \left|\frac{C}{A}\right|^2 = \frac{1}{1 + (U_B^2 \sinh^2(\kappa a)/4E(U_B - E))} \tag{2.210}$$

so that

$$\mathcal{R}(E) + \mathcal{T}(E) = 1 \tag{2.211}$$

Obviously, another solution of the same Schrödinger equation with $A = 0$ corresponds to a similar process, where the incident particle comes onto the barrier from the right, and would yield results similar to (2.209) and (2.210) for $|D/C|^2$ and $|B/C|^2$, respectively.

The ratios $\mathcal{R}$ and $\mathcal{T}$ are called reflection and transmission coefficients, respectively. In the deep tunneling limit, $\kappa a \gg 1$, these coefficients take the forms

$$\mathcal{T} = \frac{16E(U_B - E)}{U_B^2} e^{-2\kappa a}; \qquad \mathcal{R} = 1 - \mathcal{T} \tag{2.212}$$

Tunneling, a classically forbidden process, is seen to be a very low probability process when the barrier is substantial, that is, wide and high, and when the particle is more classical-like, that is, heavier. For a typical molecular distance, $a = 3$ Å, and barrier height $U_B - E = 0.1$eV we find for the exponential factor $\exp(-2\kappa a) = \exp[-(2a/\hbar)\sqrt{2m(U_B - E)}]$ the values ~ 0.38 for an electron ($m = 9.11 \times 10^{-28}$ g), $\sim 8.4 \times 10^{-19}$ for a hydrogen atom ($m = 1.67 \cdot 10^{-24}$ g) and $\sim 2.4 \times 10^{-63}$ for a carbon atom ($m = 2 \cdot 10^{-23}$ g). Tunneling is seen to be potentially important for electron dynamics and sometimes (for shorter distances and/or lower barriers) also for proton or hydrogen atom dynamics, but it rarely appears as a factor of importance in processes involving other atomic species.

Very few potential barrier models, including the rectangular barrier model discussed above, yield exact results for the tunneling problem. In general one needs to resort to numerical calculations or approximations. A very useful approximation is the WKB formula,[11] which generalizes the solution $\exp(\pm ikx)$ of the free particle Schrödinger equation to the form

$$\psi(x) \sim \frac{1}{\sqrt{k(x)}} e^{\pm i \int dx k(x)}; \quad k(x) = \begin{cases} \hbar^{-1}\sqrt{2m[E - U(x)]}; & E \ge U(x) \\ -i\hbar^{-1}\sqrt{2m[U(x) - E]}; & E \le U(x) \end{cases} \tag{2.213}$$

[11] Named after G. Wentzel [Zeits. f. Phys, *38*, 518 (1926)], H. A. Kramers [Zeits. f. Phys, *39*, 828 (1926)] and L. Brillouin [Comptes Rendus *183*, 24 (1926)] who independently applied this method to problems involving the Schrödinger equation in the early days of quantum mechanics.

in the presence of a potential $U(x)$, provided that the potential varies smoothly so that $dk(x)/dx \ll k^2(x)$. These WKB wavefunctions, constructed for parts I, II, and III of the one-dimensional space as in Eqs (2.200)–(2.205) can be used again to construct the full tunneling wavefunction. The resulting transmission coefficient in the WKB approximation is

$$\mathcal{T} \sim \exp\left[-2\int_{-x_L}^{x_r} dx\,|k(x)|\right] \tag{2.214}$$

2.10.2 Some observations

2.10.2.1 Normalization

The problem solved above is an example of a scattering process, treated here within a one-dimensional model. Unlike bound state systems such as the harmonic oscillator of Section 2.9, in a scattering process all energies are possible and we seek a solution at a given energy E, so we do not solve an eigenvalue problem. The wavefunction does not vanish at infinity, therefore normalization as a requirement that $\int_{-\infty}^{\infty} dx|\psi(x)|^2 = 1$ is meaningless.

Still, as discussed in Section 2.8.1, normalization is in some sense still a useful concept even for such processes. As we saw in Section 2.8.1, we may think of an infinite system as a $\Omega \to \infty$ limit of a finite system of volume Ω. Intuition suggests that a scattering process characterized by a short range potential should not depend on system size. On the other hand the normalization condition $\int_\Omega dx|\psi(x)|^2 = 1$ implies that scattering wavefunctions will vanish everywhere like $\psi(x) \sim \Omega^{-1/2}$ as $\Omega \to \infty$. We have noted (Section 2.8) that physically meaningful results are associated either with products such as $N|\psi(x)|^2$ or $\rho|\psi(x)|^2$, where N, the total number of particles, and ρ, the density of states, are both proportional to Ω. Thus, for physical observables the volume factor cancels.

2.10.2.2 Steady states

The process discussed above has an intuitively clear history: A particle incident on the barrier from the left emerges later as a reflected particle on the left or a transmitted particle on the right. This sounds as a problem that should be (and indeed can be) described in a time-dependent framework. However, the theoretical treatment above does not explicitly depend on time. How can a time-independent wavefunction $\psi = [\psi_{\text{I}}$ (in region I), ψ_{II} (in II), ψ_{III} (in III)] describe a process that appears to have a past and a future as described above?

The answer lies in the realization that the time-independent Schrödinger equation can describe stationary states of two kinds. The first are states characterized by zero flux, where not only the wavefunction is constant except for a phase factor

$\exp(-iEt/\hbar)$, but also all currents are zero. (See Section 2.8.4 for a discussion of quantum currents.) The Eigenfunctions of a system Hamiltonian that describe bound states are always of this kind. States of the other kind are also constant in time, but they describe systems with constant finite fluxes. Such states are designated as *steady states*. Time-independent scattering theory, including the procedure described by Eqs (2.198)–(2.210), is in fact a theory for steady-state processes (see also Section 9.5).

To be specific, Eq. (2.208) may be understood as the answer to the following question: What is the steady state in a system in which a constant flux of particles, described by the incident wavefunction $\psi_{\mathrm{I}}(x) = Ae^{ikx}$, impinges on the barrier from the left in region I? This solution is given not by specifying quantum states and their energies (which is what is usually required for zero flux problems), but rather by finding the way in which the incident flux is distributed between different *channels*, in the present case the transmission and reflection channels.

Consider now the steady-state solution of our tunneling problem. For the solution $\psi_{\mathrm{I}}(x) = Ae^{ikx} + Be^{-ikx}$, and $\psi_{\mathrm{III}}(x) = Ce^{ikx}$ associated with the case of a particle incident from the left, we find from Eq. (2.121) the fluxes in regions I and III to be

$$J_{\mathrm{I}} = \frac{\hbar k}{m}\left(|A|^2 - |B|^2\right) \tag{2.215}$$

and

$$J_{\mathrm{III}} = \frac{\hbar k}{m}|C|^2 \tag{2.216}$$

At steady state the current has to be the same everywhere (See Problem 2.15), implying the identity

$$|A|^2 - |B|^2 = |C|^2 \tag{2.217}$$

which is indeed satisfied by our solution (2.208). In the form $|A|^2 = |B|^2 + |C|^2$ this identity implies that the incident flux, whose intensity is proportional to $|A|^2$, is split during the scattering process into two components: The reflected flux, proportional to $|B|^2$ and the transmitted flux given by $|C|^2$. The designation of the ratios $\mathcal{R} = |B|^2/|A|^2$ and $\mathcal{T} = |C|^2/|A|^2$ as the corresponding reflection and transmission coefficients, respectively, thus become clear as ratios between fluxes. The identity (2.217) is again an expression of particle (or probability) conservation.

2.10.2.3 *Tunneling observables*

Consider the tunneling problems represented by the three potentials depicted in Figure 2.6. Figure 2.6a represents a scattering problem similar to that solved above. For a general potential surface it can be solved numerically or, for a smooth barrier

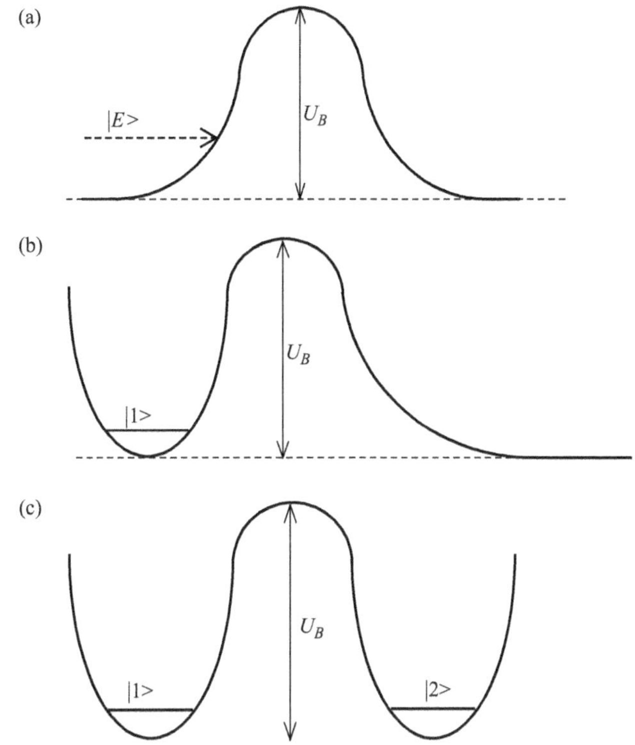

FIG. 2.6 Three different tunneling processes. (a) Tunneling through a simple barrier in a scattering event. (b) Tunneling induced escape from a single well (c) Tunneling in a double well structure.

in the WKB approximation, to give the transmission probability $\mathcal{T}(E)$ that depends on the mass of the tunneling particle and on the barrier height and width as discussed above.

Figure 2.6b corresponds to a problem of a different kind. Here a particle is initially in the well on the left, and can tunnel outside through the potential barrier. Such a problem is encountered, for example, in autoionization of excited atoms and in radioactive decay of unstable nuclei. The relevant observable is not a transmission coefficient but the *decay rate*, that is, the rate at which the probability to find the particle in the well decreases.

Figure 2.6c describes yet a another problem, where a particle initially localized in the left well can appear on the right due to tunneling through the separating barrier. This is a bound state problem and the dynamics can be evaluated by solving for the eigenstates of the corresponding Hamiltonian (such eigenstates have amplitudes in both wells), expanding the initial states in these eigenstates and employing Eq. (5). In particular, in the symmetric double well problem, if the wells are deep (i.e. the barrier between them is high) and if the particle starts in the lowest energy state

$|1\rangle$ supported by the left well, it is reasonable to expect that the only relevant state in the right well is the lowest energy state $|2\rangle$. These local states are conveniently described by the ground states of the corresponding wells when they decouple from each other, for example, when the barrier width becomes infinite. In the actual finite barrier case these two zero order states are coupled to each other and the problem becomes identical to the two-state problem of Section 2.2. The resulting dynamics shows the particle oscillating between the two wells (cf. Eq. (2.32)) with a frequency proportional to the coupling (cf. Eq. (2.33) with $E_a = E_b$). The explicit magnitude of this coupling is not immediately obvious, however, as shown in Section 2.2, this oscillation frequency corresponds to the energy splitting between the two exact eigenstates of the double-barrier problem. Experimentally this *tunneling splitting* frequency can be measured either by monitoring the dynamics, or spectroscopically if the two states can be resolved energetically. An important observation is that *this frequency is essentially a measure of the tunneling coupling between states localized in the individual wells.*

It should be appreciated that the three phenomena described above correspond to very different physical processes: scattering, decay of an initially localized state and dynamics in a bound state system that can be often approximated as a two state system. The relevant observables are different as well: Transmission coefficient, lifetime, or decay rate and tunnel-splitting. Common to these processes is the fact that they are all associated with tunneling through a potential barrier and will therefore show a characteristic dependence on the mass of the tunneling particle (an attribute usually explored experimentally in processes involving tunneling by hydrogen and its isotopes) and on the barrier height and width.

An interesting observation can be made without further computation. Assuming that the same "tunneling coupling" V_{tun} controls the three processes described above, we already saw (cf. Eq. (2.19) with $E_a = E_b$) that the tunnel splitting between the eigenstates in Fig. 2.6c is proportional to V_{tun}. On the other hand the decay rate of a particle tunneling out of a well, Fig. 2.6b, is a problem of a discrete state interacting with a continuum of states (see Section 9.1) where the "golden rule formula", (Eq. 9.25), implies that the decay rate should be proportional to V_{tun}^2. The same holds for the transmission coefficient of Fig. 2.6a (see Section 9.5). From the WKB theory we expect that

$$V_{\text{tun}}(E) \sim \exp\left\{-\frac{1}{\hbar}\int_{x_1}^{x_2} dx\sqrt{2m\left[U(x)-E\right]}\right\} \tag{2.218}$$

so the tunneling splitting in Fig. 2.6c is proportional to this factor while the transmission coefficient in Fig. 2.6a and the decay rate in Fig. 2.6b are proportional to its square.

Appendix 2A: Some operator identities

Here we derive some operator identities involving the raising and lowering operators of the harmonic oscillators, which are used in Section 2.9 and in many applications discussed in this book.

1.

$$[\hat{a}, (\hat{a}^\dagger)^n] = [\hat{a}, \hat{a}^\dagger] n(\hat{a}^\dagger)^{n-1} = n(\hat{a}^\dagger)^{n-1} \tag{2.219a}$$

$$[\hat{a}^\dagger, \hat{a}^n] = [\hat{a}^\dagger, \hat{a}] n\hat{a}^{n-1} = -n\hat{a}^{n-1} \tag{2.219b}$$

(note that (2.219b) is the Hermitian conjugate of (2.219a)). The proof can be carried by induction. Assume that Eq. (2.219a) holds and show that

$$[\hat{a}, (\hat{a}^\dagger)^{n+1}] = (n+1)(\hat{a}^\dagger)^n \tag{2.220}$$

follows. The left-hand side of (2.220) can be manipulated as follows:

$$[\hat{a}, (\hat{a}^\dagger)^{n+1}] = \underbrace{\hat{a}(\hat{a}^\dagger)^{n+1}} - (\hat{a}^\dagger)^{n+1}\hat{a}$$

$$\downarrow$$

$$\begin{aligned} \hat{a}(\hat{a}^\dagger)^{n+1} &= \hat{a}(\hat{a}^\dagger)^n \hat{a}^\dagger = [(\hat{a}^\dagger)^n \hat{a} + n(\hat{a}^\dagger)^{n-1}]\hat{a}^\dagger \\ &= \underbrace{(\hat{a}^\dagger)^n \underbrace{\hat{a}\hat{a}^\dagger}_{\hat{a}^\dagger\hat{a}+1}}_{(\hat{a}^\dagger)^{n+1}\hat{a}+(\hat{a}^\dagger)^n} + n(\hat{a}^\dagger)^n = (\hat{a}^\dagger)^{n+1}\hat{a} + (n+1)(\hat{a}^\dagger)^n \end{aligned}$$

which yields the right-hand side of (2.219a).

2. A corollary follows after observing that (2.219a) (say) can be written as $[\hat{a}, (\hat{a}^\dagger)^n] = [(d/dx)x^n]_{x=\hat{a}^\dagger}$. Since a function $f(a^\dagger)$ is defined by its Taylor series, we have, for every analytic function f

$$[\hat{a}, f(\hat{a}^\dagger)] = \left[\frac{d}{dx} f(x)\right]_{x=\hat{a}^+} \tag{2.221a}$$

and similarly

$$[\hat{a}^\dagger, f(\hat{a})] = \left[-\frac{d}{dx} f(x)\right]_{x=\hat{a}} \tag{2.221b}$$

3. The identities (2.221) are special cases of the following theorem: If the operators $\hat{A}$ and $\hat{B}$ commute with their commutator $[\hat{A}, \hat{B}]$, that is, $[\hat{A}, [\hat{A}, \hat{B}]] = [\hat{B}, [\hat{A}, \hat{B}]] = 0$, then

$$[\hat{A}, F(\hat{B})] = [\hat{A}, \hat{B}] F'(\hat{B}) \tag{2.222}$$

To prove this identity we note that since $\hat{F}(\hat{B})$ can be expanded in powers of $\hat{B}$ it suffices, as in proving (2.221), to show that $[\hat{A}, \hat{B}^n] = [\hat{A}, \hat{B}]n\hat{B}^{n-1}$. This is shown by repeated use of the commutation relation to get $\hat{A}\hat{B}^n = \hat{B}\hat{A}\hat{B}^{n-1} + [\hat{A}, \hat{B}]\hat{B}^{n-1} = \cdots = \hat{B}^n\hat{A} + n[\hat{A}, \hat{B}]\hat{B}^{n-1}$.

4. We can use induction as above to prove the following identity

$$(\hat{a}^\dagger a)^n \hat{a}^\dagger = \hat{a}^\dagger (\hat{a}^\dagger \hat{a} + 1)^n \tag{2.223}$$

and consequently also for an analytical function $f(x)$

$$\begin{aligned} f(\hat{a}^\dagger \hat{a})\hat{a}^\dagger &= \hat{a}^\dagger f(\hat{a}^\dagger \hat{a} + 1) \\ \hat{a} f(\hat{a}^\dagger \hat{a}) &= f(\hat{a}^\dagger \hat{a} + 1)\hat{a} \end{aligned} \tag{2.224}$$

5. The following important identity holds for any two operators $\hat{A}$ and $\hat{B}$ under the condition that, as in 3 above, both commute with their commutator:

$$e^{\hat{A}+\hat{B}} = e^{\hat{A}} e^{\hat{B}} e^{-\frac{1}{2}[\hat{A}, \hat{B}]} \tag{2.225}$$

In particular $\hat{A}$ and $\hat{B}$ can be any linear combination of $\hat{x}$, $\hat{p}$, $\hat{a}$, and $\hat{a}^\dagger$.

To prove (2.225) consider the operator $\hat{F}(t) = e^{\hat{A}t} e^{\hat{B}t}$ defined in terms of two operators $\hat{A}$ and $\hat{B}$, and a parameter t. Take its derivative with respect to t

$$\frac{d\hat{F}}{dt} = \hat{A} e^{\hat{A}t} e^{\hat{B}t} + e^{\hat{A}t} e^{\hat{B}t} \hat{B} = (\hat{A} + e^{\hat{A}t} B e^{-\hat{A}t})\hat{F}(t) \tag{2.226}$$

Next, use the identity $[\hat{B}, e^{-\hat{A}t}] = [\hat{B}, \hat{A}](-t)e^{-\hat{A}t}$ that follows from (2.222). From this and the fact that $[\hat{B}, \hat{A}]$ commutes with $\hat{A}$ it follows that $\hat{B}e^{-\hat{A}t} = e^{-\hat{A}t}\hat{B} - te^{-\hat{A}t}[\hat{B}, \hat{A}]$. Using the last identity in (2.226) leads to

$$\frac{d\hat{F}}{dt} = (\hat{A} + \hat{B} + t[\hat{A}, \hat{B}])\hat{F}(t) \tag{2.227}$$

The two operators, $\hat{A} + \hat{B}$ and $[\hat{A}, \hat{B}]$ commute with each other, and can be viewed as scalars when integrating this equation. We get

$$\hat{F}(t) = \hat{F}(0) e^{(\hat{A}+\hat{B})t + \frac{1}{2}[\hat{A}, \hat{B}]t^2} = e^{(\hat{A}+\hat{B})t} e^{\frac{1}{2}[\hat{A}, \hat{B}]t^2} \tag{2.228}$$

It remains to substitute $t = 1$ in Eq. (2.228) to obtain (2.225).

6. The identities (2.179) can now be verified directly. For example

$$
\begin{aligned}
&\hat{U}(\lambda)\hat{a}\hat{U}^{\dagger}(\lambda) = e^{+\bar{\lambda}(\hat{a}^{\dagger}-\hat{a})}\hat{a}e^{-\bar{\lambda}(\hat{a}^{\dagger}-\hat{a})} \\
&\xrightarrow{1} e^{+\bar{\lambda}(\hat{a}^{\dagger}-\hat{a})}\hat{a}e^{+\bar{\lambda}\hat{a}}e^{-\bar{\lambda}\hat{a}^{\dagger}}e^{(1/2)\bar{\lambda}^2[\hat{a},\hat{a}^{\dagger}]} = e^{(1/2)\bar{\lambda}^2}e^{+\bar{\lambda}(\hat{a}^{\dagger}-\hat{a})}e^{+\bar{\lambda}\hat{a}}\hat{a}e^{-\bar{\lambda}\hat{a}^{\dagger}} \\
&\xrightarrow{2} e^{(1/2)\bar{\lambda}^2}e^{+\bar{\lambda}(\hat{a}^{\dagger}-\hat{a})}e^{+\bar{\lambda}\hat{a}}e^{-\bar{\lambda}\hat{a}^{\dagger}}(\hat{a}-\bar{\lambda}) \xrightarrow{3} e^{+\bar{\lambda}(\hat{a}^{\dagger}-\hat{a})}e^{-\bar{\lambda}(\hat{a}^{\dagger}-\hat{a})}(\hat{a}-\lambda) \\
&= \hat{a}-\lambda
\end{aligned}
\tag{2.229}
$$

where, in the steps marked 1 and 3 we used (2.228) and in the step marked 2 we used (2.221a).

Further Reading

C. Cohen-Tannoudji, B. Diu and F. Laloe, Quantum Mechanics, (Wiley, New York, 1977).
A. S. Davydov, Quantum Mechanics, translated and edited by D. Ter Haar (Pergamon, Oxford, 1965).
G. C. Schatz and M. A. Ratner, Quantum Mechanics in Chemistry (Prentice Hall, Englewood Cliffs, 1993).

3

AN OVERVIEW OF QUANTUM ELECTRODYNAMICS AND MATTER–RADIATION FIELD INTERACTION

For light is much more mobile, is composed
Of finer particles, yet has more power,
And once it clears the roadways of the eyes,
Removing the dark barriers and blocks,
At once the images of things begin
To move in our direction, driving on
Out of the light to help us see. From light
We can not see into darkness, for that air
Is slower moving, thicker, bound to fill
All opening, so no image can move
Across the solid massiveness of dark...

Lucretius (c.99–c.55 BCE*) "The way things are"*
translated by Rolfe Humphries, Indiana University Press, 1968

Many dynamical processes of interest are either initiated or probed by light, and their understanding requires some knowledge of this subject. This chapter is included in order to make this text self contained by providing an overview of subjects that are used in various applications later in the text. In particular, it aims to supplement the elementary view of radiation–matter interaction as a time-dependent perturbation in the Hamiltonian, by describing some aspects of the quantum nature of the radiation field. This is done on two levels: The main body of this chapter is an essentially qualitative overview that ends with a treatment of spontaneous emission as an example. The Appendix gives some more details on the mathematical structure of the theory.

3.1 Introduction

In elementary treatments of the interaction of atoms and molecules with light, the radiation field is taken as a classical phenomenon. Its interaction with a molecule is often expressed by

$$\hat{H}_{\mathrm{MR}} = -\hat{\boldsymbol{\mu}} \cdot \boldsymbol{\mathcal{E}}(t), \tag{3.1}$$

where $\hat{\boldsymbol{\mu}}$ is the molecular dipole operator while $\boldsymbol{\mathcal{E}}(t)$ is the time-dependent electric field associated with the local electromagnetic field at the position of the molecule. In fact, much can be accomplished with this approach including most applications discussed in this text. One reason to go beyond this simple description of radiation-field–matter interaction is that, as will be seen, the formalism of quantum electrodynamics is sometimes simpler to handle. However, more important is the fact that the quantum description provides a picture of the radiation–matter interaction which is conceptually different from the classical one, including the possibility to describe the state of the field in terms of particles (photons).

An important conceptual issue already appears in the classical description. According to Eq. (3.1) the interaction between a material system and the electromagnetic field vanishes when the field $\mathcal{E}$ is zero. We know that this is not so, or else spontaneous emission of radiation from an excited atom, or from a classical oscillating dipole, would not occur. The fact that it does occur implies not only that the presence of a field can change the state of the system but also that the presence of a system can change the state of the radiation field, creating radiation where it did not exist before. One needs to reconcile Eq. (3.1) with this observation. In fact, all we need is to realize that one *should distinguish between the presence of a field and the state of this field* in much the same way that this is done for material systems, and that the magnitude of $\mathcal{E}$ is a designation of the state, not existence, of the field.

As an example consider two particles, 1 and 2 with coordinates x_1 and x_2, and suppose that the interaction between them has the form $\alpha x_1 x_2$. The statement $x_2 = 0$ refers not to the existence of particle 2, only to its state in position space. When particle 1 has a finite energy it can transfer some of it to particle 2 even if initially the state of the latter is $x_2 = 0$. In a similar way, the entity called "electromagnetic field" always exists and $\boldsymbol{\mathcal{E}}$ in Eq. (3.1) plays the role of a coordinate that may be zero in some state. In the lowest energy (ground) state of this entity the amplitudes of both the electric field $\boldsymbol{\mathcal{E}}$ and the magnetic field $\boldsymbol{\mathcal{H}}$ are zero, while excited states correspond to nonzero values of these amplitudes. Indeed, classical electromagnetic theory yields the following expression for the energy associated with the electromagnetic field in homogeneous space with local dielectric constant ε and magnetic permeability μ[1]

$$E = \frac{1}{8\pi} \int d\mathbf{r}(\varepsilon|\boldsymbol{\mathcal{E}}(\mathbf{r})|^2 + \mu|\boldsymbol{\mathcal{H}}(\mathbf{r})|^2) \tag{3.2}$$

[1] While we usually attempt not to use overlapping notations, because magnetic susceptibility does not appear in this text beyond this chapter we denote it by μ, the same symbol used for the dipole moment. The distinction between these variables should be clear from the text.

This picture is developed to a high level of sophistication within the classical theory of the electromagnetic field, where dynamics is described by the Maxwell equations. Some basics of this theory are described in Appendix 3A. Here we briefly outline some of the important results of this theory that are needed to understand the nature of the interaction between a radiation field and a molecular system.

3.2 The quantum radiation field

3.2.1 Classical electrodynamics

When we study processes that involve interaction between two systems, it is almost always a prerequisite to understand each system separately. For definiteness we consider one molecule in the radiation field, and assume that the molecular problem has been solved in the sense that we know the eigenfunctions and eigenvalues of the molecular Hamiltonian. We require similar knowledge of the radiation field, that is, we need to solve the Maxwell equations, Eqs (3.32(a–d)) of Appendix 3A, for some given boundary conditions. The way this solution is obtained is described in Appendix 3A. There are many representations ("gauges") in which this can be done and in a particular one, the Coulomb gauge, one can represent the solutions of the Maxwell equations in terms of one transverse (see Appendix) vector function $\mathbf{A}(\mathbf{r}, t)$, called the vector potential. $\mathbf{A}(\mathbf{r}, t)$, itself a solution of a vector differential equation (Appendix 3A, Eq. (3.46)), yields the physical electric and magnetic fields via the relationships (in gaussian units; cf. Appendix 3A, Eq. (3.47))

$$\mathcal{B} = \nabla \times \mathbf{A}; \qquad \mathcal{E} = -\frac{1}{c}\frac{\partial \mathbf{A}}{\partial t} \tag{3.3}$$

where c is the speed of light. For linear (paramagnetic or diamagnetic) media the magnetic induction $\mathcal{B}$ is related to the magnetic field $\mathcal{H}$ by $\mathcal{B} = \mu\mathcal{H}$. Furthermore, $\mathbf{A}$ is found to be conveniently represented as a superposition of contributions from independent degrees of freedom (modes) in the form

$$\mathbf{A}(\mathbf{r}, t) = \sum_{\mathbf{k},\sigma_\mathbf{k}} \mathbf{A}_{\mathbf{k},\sigma_\mathbf{k}}(\mathbf{r}, t) \tag{3.4}$$

where $\mathbf{k}$ and $\boldsymbol{\sigma}_\mathbf{k}$ are wave vectors and polarization vectors, respectively (see below). The functional form of terms $\mathbf{A}_{\mathbf{k},\sigma_\mathbf{k}}(\mathbf{r}, t)$ depends on the boundary conditions. For an infinite homogeneous system it is convenient to use periodic boundary conditions within a rectangular box of volume $\Omega = L_x L_y L_z$, and to set $\Omega \to \infty$ at the end of a calculation. In this case we find

$$\mathbf{A}_{\mathbf{k},\sigma_\mathbf{k}}(\mathbf{r}, t) = c\sqrt{\frac{2\pi\hbar}{\varepsilon\Omega\omega_k}}\boldsymbol{\sigma}_\mathbf{k}(a_{\mathbf{k},\sigma_\mathbf{k}}(t)e^{i\mathbf{k}\cdot\mathbf{r}} + a^*_{\mathbf{k},\sigma_\mathbf{k}}(t)e^{-i\mathbf{k}\cdot\mathbf{r}}) \tag{3.5}$$

where

$$a_{\mathbf{k},\sigma_\mathbf{k}}(t) = a_{\mathbf{k},\sigma_\mathbf{k}} e^{-i\omega_k t}; \qquad a^*_{\mathbf{k},\sigma_\mathbf{k}}(t) = a^*_{\mathbf{k},\sigma_\mathbf{k}} e^{-i\omega_k t} \tag{3.6}$$

with $\omega_k = kc/\sqrt{\mu\varepsilon}$ and $k = |\mathbf{k}|$. Here the amplitudes $a_{\mathbf{k},\sigma}$ are scalar constants whose magnitudes reflect the degree of excitation of the corresponding modes. The components $k_j (j = x, y, z)$ of the vector $\mathbf{k}$ have to be integer multiples of the corresponding factors $2\pi/L_j$ in order to satisfy the periodic boundary conditions.

Equation (3.5) implies that different modes are distinguished by their time and space dependence, characterized by two vectors: A wave vector $\mathbf{k}$ that points to the direction of spatial modulation of $\mathbf{A}_{\mathbf{k},\sigma_\mathbf{k}}(\mathbf{r}, t)$ and a polarization unit-vector $\boldsymbol{\sigma}_k$, that specifies the direction of $\mathbf{A}$ itself. The transversality of $\mathbf{A}$ expresses the fact that $\mathbf{k}$ and $\mathbf{A}$ are perpendicular to each other, that is,

$$\boldsymbol{\sigma}_\mathbf{k} \cdot \mathbf{k} = 0 \tag{3.7}$$

Thus, for every wave vector there are two possible polarization directions perpendicular to it.

Given $\mathbf{A}$, that is given $a_{\mathbf{k},\sigma_\mathbf{k}}$ for every $(\mathbf{k}, \boldsymbol{\sigma}_\mathbf{k})$, the electric and magnetic fields can be found as sums over modes using Eqs (3.4) and (3.3). For example, this leads to

$$\mathcal{E}(\mathbf{r}, t) = i \sum_\mathbf{k} \sum_{\sigma_\mathbf{k}} \sqrt{\frac{2\pi\hbar\omega_k}{\varepsilon\Omega}} \boldsymbol{\sigma}_\mathbf{k} \left(a_{\mathbf{k},\sigma_\mathbf{k}} e^{-i\omega_k t + i\mathbf{k}\cdot\mathbf{r}} - a^*_{\mathbf{k},\sigma_\mathbf{k}} e^{i\omega_k t - i\mathbf{k}\cdot\mathbf{r}} \right) \tag{3.8}$$

Using (3.8), and the similar equation derived from (3.3) for $\mathcal{H} = \mathcal{B}/\mu$, in Eq. (3.2) leads to

$$E = \sum_{\mathbf{k},\sigma_\mathbf{k}} \hbar\omega_\mathbf{k} |a_{\mathbf{k},\sigma_\mathbf{k}}|^2 = \frac{1}{2} \sum_{\mathbf{k},\sigma_\mathbf{k}} \hbar\omega_\mathbf{k} (a^*_{\mathbf{k},\sigma_\mathbf{k}} a_{\mathbf{k},\sigma_\mathbf{k}} + a_{\mathbf{k},\sigma_\mathbf{k}} a^*_{\mathbf{k},\sigma_\mathbf{k}}) \tag{3.9}$$

The second form of this result is written in anticipation of the analogous quantum result.

3.2.2 Quantum electrodynamics

A crucial step motivated by experimental observations and theoretical considerations is the quantization of the radiation field, whereupon the electric and magnetic fields assume operator character. We first notice that the functions $q(t)$ and $p(t)$ defined for each $(\mathbf{k}, \boldsymbol{\sigma}_\mathbf{k})$ by

$$q(t) = \sqrt{\frac{\hbar}{2\omega}} (a^*(t) + a(t)); \qquad p(t) = i\sqrt{\frac{\hbar\omega}{2}} (a^*(t) - a(t)) \tag{3.10}$$

satisfy (as seen from (3.6)) the time evolution equations

$$\dot{q} = p; \qquad \dot{p} = -\omega^2 q \tag{3.11}$$

These equations have the form of harmonic oscillator equations of motion for a "coordinate" q and "momentum" p. Indeed, Eqs (3.11) can be derived from the Hamiltonian

$$h = (1/2)(p^2 + \omega^2 q^2) \tag{3.12}$$

using the Hamilton equations $\dot{q} = \partial h/\partial p$; $\dot{p} = -\partial h/\partial q$. It turns out that the correct quantization of the radiation field is achieved by replacing these coordinates and momenta by the corresponding quantum operators that obey the commutation relations

$$[\hat{q}_{\mathbf{k},\sigma_\mathbf{k}}, \hat{p}_{\mathbf{k},\sigma_\mathbf{k}}] = i\hbar \tag{3.13}$$

with operators associated with different modes taken to commute with each other. The classical functions $a(t)$ and $a^*(t)$ also become operators, $\hat{a}(t)$ and $\hat{a}^\dagger(t)$ (see Eqs (3.62)) that satisfy equations similar to (2.154)

$$[\hat{a}_{\mathbf{k},\sigma_\mathbf{k}}, \hat{a}^\dagger_{\mathbf{k}',\sigma'_{\mathbf{k}'}}] = \delta_{\mathbf{k},\mathbf{k}'}\delta_{\sigma_\mathbf{k},\sigma'_{\mathbf{k}'}} \tag{3.14}$$

$$[\hat{a}_{\mathbf{k},\sigma_\mathbf{k}}, \hat{a}_{\mathbf{k}',\sigma'_{\mathbf{k}'}}] = [\hat{a}^\dagger_{\mathbf{k},\sigma_\mathbf{k}}, \hat{a}^\dagger_{\mathbf{k}',\sigma'_{\mathbf{k}'}}] = 0 \tag{3.15}$$

This identifies $\hat{a}$ and $\hat{a}^\dagger$ as lowering and raising operators of the corresponding harmonic modes. Equation (3.6) is recognized as the Heisenberg representation of these operators

$$\hat{a}_{\mathbf{k},\sigma_\mathbf{k}}(t) = \hat{a}_{\mathbf{k},\sigma_\mathbf{k}} e^{-i\omega_k t}; \qquad \hat{a}^\dagger_{\mathbf{k},\sigma_\mathbf{k}}(t) = \hat{a}^\dagger_{\mathbf{k},\sigma_\mathbf{k}} e^{i\omega_k t} \tag{3.16}$$

and the energy in the radiation field, Eq. (3.9) becomes the Hamiltonian

$$\hat{H}_\mathrm{R} = \sum_{\mathbf{k},\sigma_\mathbf{k}} \hbar\omega_\mathbf{k}(\hat{a}^\dagger_{\mathbf{k},\sigma_\mathbf{k}}\hat{a}_{\mathbf{k},\sigma_\mathbf{k}} + 1/2) \tag{3.17}$$

which describes a system of independent harmonic modes. A mode $(\mathbf{k}, \boldsymbol{\sigma}_\mathbf{k})$ of frequency ω_k can therefore be in any one of an infinite number of discrete states of energies $\hbar\omega_\mathbf{k} n_{\mathbf{k},\sigma_\mathbf{k}}$. The degree of excitation, $n_{\mathbf{k},\sigma_\mathbf{k}}$, is referred to as the occupation number or number of photons in the corresponding mode. Note that $(\mathbf{k}, \boldsymbol{\sigma}_\mathbf{k})$ is collection of five numbers characterizing the wavevector and polarization associated with the particular mode. The vector potential $\mathbf{A}$ and the fields derived from it by

Eq. (3.3) become operators whose Schrödinger representations are

$$\hat{\mathbf{A}} = c\sum_{\mathbf{k}}\sum_{\sigma_{\mathbf{k}}}\sqrt{\frac{2\pi\hbar}{\varepsilon\Omega\omega_k}}\boldsymbol{\sigma}_{\mathbf{k}}(\hat{a}_{\mathbf{k},\sigma_{\mathbf{k}}}e^{i\mathbf{k}\cdot\mathbf{r}} + \hat{a}^{\dagger}_{\mathbf{k},\sigma_{\mathbf{k}}}e^{-i\mathbf{k}\cdot\mathbf{r}}) \tag{3.18}$$

and

$$\hat{\boldsymbol{\mathcal{E}}} = i\sum_{\mathbf{k}}\sum_{\sigma_{\mathbf{k}}}\sqrt{\frac{2\pi\hbar\omega_k}{\varepsilon\Omega}}\boldsymbol{\sigma}_{\mathbf{k}}(\hat{a}_{\mathbf{k},\sigma_{\mathbf{k}}}e^{i\mathbf{k}\cdot\mathbf{r}} - a^{\dagger}_{\mathbf{k},\sigma_{\mathbf{k}}}e^{-i\mathbf{k}\cdot\mathbf{r}}) \tag{3.19}$$

In many applications we encounter such sums of contributions from different modes, and because in the limit $\Omega \to \infty$ the spectrum of modes is continuous, such sums are converted to integrals where the density of modes enters as a weight function. An important attribute of the radiation field is therefore the density of modes per unit volume in **k**-space, $\rho_{\mathbf{k}}$, per unit frequency range, ρ_ω, or per unit energy, ρ_E $(E = \hbar\omega)$. We find (see Appendix 3A)

$$\rho_{\mathbf{k}}(\mathbf{k}) = 2\frac{\Omega}{(2\pi)^3} \tag{3.20a}$$

$$\rho_\omega(\omega) = \hbar\rho_E(E) = \frac{(\mu\varepsilon)^{3/2}}{\pi^2}\frac{\omega^2}{c^3}\Omega \tag{3.20b}$$

Note that expression (3.20a) is the same result as Eq. (2.95), obtained for the density of states of a free quantum particle except for the additional factor 2 in (3.20a) that reflects the existence of two polarization modes for a given **k** vector. Eq (3.20b) is obtained from (3.20a) by using $\omega = |k|\bar{c}$ where $\bar{c} = c/\sqrt{\varepsilon\mu}$ to get $\rho_\omega(\omega) = [4\pi k^2 dk \times \rho_{\mathbf{k}} d\mathbf{k}]_{k=\omega/\bar{c}}$ (compare to the derivation of Eq (2.97)).

To see the physical significance of these results consider the Hamiltonian that describes the radiation field, a single two-level molecule located at the origin, and the interaction between them, using for the latter the fully quantum analog of Eq. (3.1) in the Schrödinger representation

$$\hat{H} = \hat{H}_{\mathrm{M}} + \hat{H}_{\mathrm{R}} + \hat{H}_{\mathrm{MR}} \tag{3.21}$$

$$\hat{H}_{\mathrm{M}} = E_1|1\rangle\langle 1| + E_2|2\rangle\langle 2| \tag{3.22}$$

$$\hat{H}_{\mathrm{R}} = \sum_{\mathbf{k},\sigma_{\mathbf{k}}}\hbar\omega_k\hat{a}^{\dagger}_{\mathbf{k},\sigma_{\mathbf{k}}}\hat{a}_{\mathbf{k},\sigma_{\mathbf{k}}} \tag{3.23}$$

$$\hat{H}_{\mathrm{MR}} = -\hat{\boldsymbol{\mu}}\cdot\hat{\boldsymbol{\mathcal{E}}}(\mathbf{r}=0) \tag{3.24}$$

Taking $\mathbf{r} = 0$ in (3.24) implies that the variation of $\hat{\boldsymbol{\mathcal{E}}}(\mathbf{r} = 0)$ over the molecule is neglected. This approximation is valid if the mode wavelength $\lambda = 2\pi/k$ is much larger than the molecular size for all relevant modes. This holds for most cases encountered in molecular spectroscopy, and for this reason the factors $\exp(i\mathbf{k}\cdot\mathbf{r})$ can be approximated by unities.[2] In the basis of molecular eigenstates the interaction (3.24) is

$$\hat{H}_{\text{MR}} = -\hat{\boldsymbol{\mathcal{E}}}(0)\sum_{j=1}^{2}\sum_{l=1}^{2}\langle j|\hat{\boldsymbol{\mu}}|l\rangle|j\rangle\langle l| \tag{3.25}$$

In this approximation we get using (3.19)

$$\hat{\boldsymbol{\mathcal{E}}}(0)\cdot\langle j|\hat{\boldsymbol{\mu}}|l\rangle = i\sum_{\mathbf{k}}\sum_{\sigma_{\mathbf{k}}}\sqrt{\frac{2\pi\hbar\omega_k}{\varepsilon\Omega}}(\boldsymbol{\mu}_{jl}\cdot\boldsymbol{\sigma}_{\mathbf{k}})(\hat{a}_{\mathbf{k},\sigma_{\mathbf{k}}} - \hat{a}^{\dagger}_{\mathbf{k},\sigma_{\mathbf{k}}}) \tag{3.26}$$

Next, assuming that the molecule has no permanent dipole in either of its two states, the dipole operator $\hat{\boldsymbol{\mu}}$ can have only non-diagonal matrix elements in the molecular basis representation. Equation (3.25) then becomes

$$\hat{H}_{\text{MR}} = -i\sum_{\mathbf{k}}\sum_{\sigma_{\mathbf{k}}}\sqrt{\frac{2\pi\hbar\omega_k}{\varepsilon\Omega}}[(\boldsymbol{\mu}_{12}\cdot\boldsymbol{\sigma}_{\mathbf{k}})|1\rangle\langle 2| + (\boldsymbol{\mu}_{21}\cdot\boldsymbol{\sigma}_{\mathbf{k}})|2\rangle\langle 1|](\hat{a}_{\mathbf{k},\sigma_{\mathbf{k}}} - \hat{a}^{\dagger}_{\mathbf{k},\sigma_{\mathbf{k}}}) \tag{3.27}$$

The molecule–radiation-field interaction is seen to be a sum, over all the field modes, of products of two terms, one that changes the molecular state and another that changes the photon number in different modes.

Problem 3.1. Write, under the same approximation, the interaction equivalent to (3.27) for the case of a multilevel molecule.

The interaction (3.27) couples between eigenstates of $\hat{H}_0 = \hat{H}_{\text{M}} + \hat{H}_{\text{R}}$. Such states are direct products of eigenstates of $\hat{H}_{\text{M}}$ and of $\hat{H}_{\text{R}}$ and may be written as $|j,\{n\}\rangle$ where the index j (in our model $j = 1, 2$) denotes the molecular state and $\{n\}$ is the set of photon occupation numbers. From (3.27) we see that $\hat{V}_{\text{MR}}$ is a sum of terms that couple between states of this kind that differ both in their molecular-state character *and* in the occupation number of one mode.

Suppose now that in (3.22) $E_2 > E_1$. Equation (3.27) displays two kinds of terms: $|2\rangle\langle 1|\hat{a}_{\mathbf{k},\sigma_{\mathbf{k}}}$ and $|1\rangle\langle 2|\hat{a}^{\dagger}_{\mathbf{k},\sigma_{\mathbf{k}}}$ describe physical processes that are "acceptable" in the

[2] Note that this approximation does not hold when the states j and l belong to different molecules unless the distance between these molecules is much smaller than the radiation wavelength.

sense that they may conserve energy: They describe a molecule going up while absorbing a photon or down while emitting one. The other two terms $|2\rangle\langle 1|\hat{a}_{\mathbf{k},\sigma_{\mathbf{k}}}$ and $|1\rangle\langle 2|\hat{a}^{\dagger}_{\mathbf{k},\sigma_{\mathbf{k}}}$ are in this sense "unphysical": They describe the molecule going up while emitting a photon or down while absorbing one. It should be emphasized that these designations are much too simplistic. In terms of perturbation theory the apparently unphysical interaction terms can contribute to physical processes when considered in higher than first-order. On the other hand, if we expect a process to be well described within low-order perturbation theory, we may disregard terms in the interaction that cannot conserve energy on this low-order level. This approximation is known as the *rotating wave approximation* (RWA). It leads to an approximate interaction operator of the form (for $E_2 > E_1$)

$$\hat{H}_{\mathrm{MR}}^{(\mathrm{RWA})} = -i \sum_{\mathbf{k}} \sum_{\sigma_{\mathbf{k}}} \sqrt{\frac{2\pi\hbar\omega_k}{\varepsilon\Omega}} [(\boldsymbol{\mu}_{21} \cdot \boldsymbol{\sigma}_{\mathbf{k}})|2\rangle\langle 1|\hat{a}_{\mathbf{k},\sigma_{\mathbf{k}}} - (\boldsymbol{\mu}_{12} \cdot \boldsymbol{\sigma}_{\mathbf{k}})|1\rangle\langle 2|\hat{a}^{\dagger}_{\mathbf{k},\sigma_{\mathbf{k}}}] \tag{3.28}$$

3.2.3 Spontaneous emission

As an application of the results obtained above we consider the spontaneous emission rate from our molecule after it is prepared in the excited state $|2\rangle$. In terms of zero-order states of the Hamiltonian $\hat{H}_0 = \hat{H}_{\mathrm{M}} + \hat{H}_{\mathrm{R}}$ the initial state is $|2, \{0\}\rangle$ and it is coupled by the interaction (3.28) to a continuum of 1-photon states $|1, \{0 \ldots 0, 1, 0, \ldots, 0\}\rangle$. The decay rate is given by the golden rule formula

$$k_{\mathrm{R}} = \frac{2\pi}{\hbar} |V|^2 \rho \tag{3.29}$$

where V is the coupling matrix elements calculated for final states with photon frequency $\omega_{21} = E_{21}/\hbar = (E_2 - E_1)/\hbar$ and where $\rho = \rho(E_{21})$ is the density (number per unit energy) of such 1-photon states. More details on the origin of the golden rule formula in the present context are given in Chapter 9 and in Section 9.2.3. From Eq. (3.28) we find[3]

$$|V|^2 = \frac{2\pi\hbar\omega_{12}}{\varepsilon\Omega} |\mu_{21}|^2 \tag{3.30}$$

[3] Care need to be taken in order to accommodate the vector nature of μ and of the incident field. For spherically symmetric molecules, each of the two directions perpendicular to the wavevector $\mathbf{k}$ of a given mode contributes equally: We can use for $\hat{\mu}$ any component, say μ_x, of the transition dipole, and (as already noted below Eq. (3.20)), the density of states used below takes an extra factor of 2 for the two possible directions of the polarization.

Now consider the density ρ of 1-photon states. Because each of these states is characterized by one mode being in the first excited state while all others are in the ground state, the number of states is the same as the number of modes and the required density of states per unit energy is given by ρ_E of Eq. (3.20). Using this, together with (3.30) in (3.29) leads to

$$k_R = \frac{2\varepsilon^{1/2}\mu^{3/2}}{\hbar}\left(\frac{\omega_{21}}{c}\right)^3 |\mu_{12}|^2 \tag{3.31}$$

As an order of magnitude estimate, take typical values for electronic transitions, for example, $\omega_{21} = 20\,000\ \text{cm}^{-1} \approx 4 \times 10^{15}\ \text{s}^{-1}$, $|\mu_{12}| = 10^{-17}$ esu cm and $\varepsilon = \mu = 1$ to find $k_R \simeq 4 \times 10^8\ \text{s}^{-1}$.

Several observations should be made regarding this result. First, while regular chemical solvents are characterized by $\mu \simeq 1$, different solvents can differ considerably in their dielectric constants ε, and Eq. (3.31) predicts the way in which the radiative lifetime changes with ε. Note that a dependence on ε may appear also in ω_{12} because different molecular states may respond differently to solvation, so a test of this prediction should be made by monitoring both k_R and ω_{21} as functions of the dielectric constant in different embedding solvents.

Second, the dependence on ω^3 is a very significant property of the radiative decay rates. Assuming similar transition dipoles for allowed transitions, Eq. (3.31) predicts that lifetimes of electronically excited states (ω_{21} of order 10^4cm^{-1}) are shorter by a factor of $\sim 10^3$ than those of vibrational excitations (ω_{21} of order 10^3cm^{-1}), while the latter are $\sim 10^3$ shorter than those of rotational excitations (ω_{21} of order 10^2cm^{-1}), as indeed observed.

Finally, we have obtained the result (3.31) by using expression (3.1) for the molecule–radiation field interaction. This form, written as an extension of an electrostatic energy term to the case of time varying field, is an approximation, that is discussed further in the appendix.

Appendix 3A: The radiation field and its interaction with matter

We start with the Maxwell equations, themselves a concise summary of many experimental observations. In gaussian units these are

$$\nabla \cdot \mathcal{D} = 4\pi\rho \tag{3.32a}$$

$$\nabla \cdot \mathcal{B} = 0 \tag{3.32b}$$

$$\nabla \times \mathcal{E} + \frac{1}{c}\frac{\partial \mathcal{B}}{\partial t} = 0 \tag{3.32c}$$

$$\nabla \times \mathcal{H} - \frac{1}{c}\frac{\partial \mathcal{D}}{\partial t} = \frac{4\pi}{c}\mathbf{J} \tag{3.32d}$$

where ρ and $\mathbf{J}$ are the charge density and current density associated with *free charges* in the system and where the electric fields $\mathcal{E}$ and displacement $\mathcal{D}$, and the magnetic field $\mathcal{H}$ and induction $\mathcal{B}$ are related through the polarization $\mathbf{P}$ (electric dipole density) and the magnetization $\mathbf{M}$ (magnetic dipole density) in the medium according to

$$\mathcal{E} = \mathcal{D} - 4\pi\mathbf{P} \tag{3.33a}$$

$$\mathcal{H} = \mathcal{B} - 4\pi\mathbf{M} \tag{3.33b}$$

Equation (3.32a) is a differential form of Coulomb's law. Equation (3.32b) is an equivalent equation for the magnetic case, except that, since magnetic monopoles do not exist, the "magnetic charge" density is zero. Equation (3.32c) expresses Faraday's law (varying magnetic flux induces a circular electric field) in a differential form and, finally, Eq. (3.32d) is Maxwell's generalization of Ampere's law (induction of a magnetic field by a current). We usually assume a linear relationship between the dipole densities and the corresponding local fields. For example, for simple homogeneous systems we take

$$\mathbf{P} = \chi_e \mathcal{E}; \qquad \mathbf{M} = \chi_h \mathcal{H} \tag{3.34}$$

so that

$$\mathcal{D} = \varepsilon \mathcal{E}; \qquad \mathcal{B} = \mu \mathcal{H} \tag{3.35}$$

where

$$\varepsilon = 1 + 4\pi\chi_e; \qquad \mu = 1 + 4\pi\chi_h \tag{3.36}$$

are constants. The energy in the field may be shown to be given by

$$E = \frac{1}{8\pi} \int d\mathbf{r} (\varepsilon |\mathcal{E}(\mathbf{r})|^2 + \mu |\mathcal{H}(\mathbf{r})|^2) \tag{3.37}$$

It is important to realize that Eqs (3.32) are macroscopic equations, where bound atomic and molecular charges have been coarse grained to yield the macroscopic electric and magnetic dipole densities $\mathbf{P}$ and $\mathbf{M}$.[4] Such coarse-graining operation (see Section 1.4.4) involves averaging over a length scale l that is assumed to be (1) small relative to distances encountered in the applications of the resulting macroscopic equations and (2) large relative to atomic dimensions over which these

[4] See J. D. Jackson, *Classical Electrodynamics*, 2nd Edition (Wiley, New York, 1975, section 6.7), for details of this averaging procedure.

bound charges are moving. This implies that l should be larger than, say, 10 Å, which makes questionable the use of such equations for applications involving individual molecules. The same question arises with respect to the use of macroscopic electrostatic models to describe molecular phenomena and constitutes a continuing enigma in many models constructed to treat chemical energetics and dynamics in dielectric media. We will confront this issue again in later chapters.

In what follows we use some results from the calculus of vector fields that are summarized in Section 1.1.3. The solution of Eqs (3.32) is facilitated by introducing the so called scalar potential $\Phi(\mathbf{r}, t)$ and vector potential $\mathbf{A}(\mathbf{r}, t)$, in terms of which

$$\mathcal{B} = \nabla \times \mathbf{A} \tag{3.38}$$

$$\mathcal{E} = -\nabla\Phi - \frac{1}{c}\frac{\partial \mathbf{A}}{\partial t} \tag{3.39}$$

The forms (3.38) and (3.39) automatically satisfy Eqs (3.32b) and (3.32c). It is important to remember that the physical fields are $\mathcal{E}$ and $\mathcal{B}$, while $\mathbf{A}$ and Φ are mathematical constructs defined for convenience. In fact, infinitely many choices of these fields give the same $\mathcal{B}$ and $\mathcal{E}$: Any scalar function of space and time $S(\mathbf{r}, t)$ can be used to transform between these choices as follows:

$$\mathbf{A}(\mathbf{r}, t) \rightarrow \mathbf{A}(\mathbf{r}, t) + \nabla S(\mathbf{r}, t) \tag{3.40a}$$

$$\Phi(\mathbf{r}, t) \rightarrow \Phi(\mathbf{r}, t) - \frac{1}{c}\frac{\partial S(\mathbf{r}, t)}{\partial t} \tag{3.40b}$$

The transformation (3.40) is called gauge transformation, and a solution obtained with a particular choice of S is referred to as the solution in the corresponding gauge.

Problem 3.2. Show, using the identity (1.32), that Eqs (3.38) and (3.39) are indeed invariant to this transformation.

For the discussion of free radiation fields and their quantization a particular choice of gauge, called the Coulomb (or transverse) gauge, is useful. It is defined by the requirement (which can always be satisfied with a proper choice of S) that

$$\nabla \cdot \mathbf{A} = 0 \tag{3.41}$$

Note that Eq. (3.38) than implies (using Eqs (1.31)–(1.34)) that in this gauge $\mathcal{B} = \mathcal{B}^{\perp}$ is a transversal field, while the two contributions to $\mathcal{E}$ in (3.39) are its transversal and longitudinal components

$$\mathcal{E}^{\perp} = -\frac{1}{c}\frac{\partial \mathbf{A}}{\partial t}; \qquad \mathcal{E}^{\parallel} = -\nabla\Phi \tag{3.42}$$

Limiting ourselves to homogeneous systems, for which Eqs (3.35) are valid with constant ε and μ, Eqs (3.32a) and (3.39) then imply

$$\nabla^2 \Phi = -\frac{4\pi\rho}{\varepsilon} \tag{3.43}$$

This is the Poisson equation, known from electrostatics as the differential form of Coulomb's law. Also, Eqs (3.32d), (3.38), (3.39), (3.41), and (1.29) lead to

$$\nabla^2 \mathbf{A} - \frac{\varepsilon\mu}{c^2}\frac{\partial^2 \mathbf{A}}{\partial t^2} = -\frac{4\pi\mu}{c}\mathbf{J} + \frac{\varepsilon\mu}{c}\nabla\frac{\partial\Phi}{\partial t} \tag{3.44}$$

Equation (3.43) identifies the scalar potential *in this gauge* as the (instantaneous) Coulomb potential associated with free charges in the system. Its solution is the familiar Coulomb-law expression

$$\Phi(\mathbf{r}, t) = \frac{1}{\varepsilon}\int d^3r' \frac{\rho(\mathbf{r}', t)}{|\mathbf{r} - \mathbf{r}'|} \tag{3.45}$$

In (3.44), the terms on the right-hand side can be viewed as the sources of the radiation field. Two sources are seen: A current (moving charge) and a time variation in the magnitude of the charge. If $\mathbf{A} = 0$ (ground state of the radiation field) and such sources are absent, the field will remain in this ground state. Obviously there exist other states of the free radiation field, solutions of Eq. (3.44) in the absence of sources,

$$\nabla^2 \mathbf{A} = \frac{1}{\bar{c}^2}\frac{\partial^2 \mathbf{A}}{\partial t^2}; \qquad \bar{c} = c/\sqrt{\varepsilon\mu} \tag{3.46}$$

Before considering the solutions of this equation we note that given these solutions, the physical fields are obtained from (3.38) and (3.39). In particular, in the absence of free charges ($\rho = 0$, hence $\Phi =$ constant)

$$\mathcal{B} = \nabla \times \mathbf{A}; \qquad \mathcal{E} = -\frac{1}{c}\frac{\partial \mathbf{A}}{\partial t} \tag{3.47}$$

and the energy in the field, Eq. (3.37), is given by

$$E = \frac{1}{8\pi\mu}\int d\mathbf{r}\left(\frac{1}{\bar{c}^2}\left(\frac{\partial \mathbf{A}}{\partial t}\right)^2 + (\nabla \times \mathbf{A})^2\right) \tag{3.48}$$

Consider now the solutions of (3.46), which is a wave equation. In order to get a feeling for its properties lets consider a one-dimensional version,

$$\frac{\partial^2 A}{\partial x^2} = \frac{1}{\bar{c}^2}\frac{\partial^2 A}{\partial t^2} \tag{3.49}$$

The obvious existence of solutions of the form $A(x,t) = A(x \pm \bar{c}t)$ shows that $\bar{c}$ plays the role of a propagation speed. Explicit solutions may be found by separating variables. Assuming solutions of the form $A(x,t) = \alpha b(x)q(t)$, α being any constant, we get after inserting into (3.49)

$$\frac{1}{b(x)}\frac{d^2b(x)}{dx^2} = \frac{1}{\bar{c}^2}\frac{1}{q(t)}\frac{d^2q(t)}{dt^2} \tag{3.50}$$

This implies that each side of this equation is a constant. A negative constant, to be denoted $-k^2$ with real k, will yield wave-like solutions. A is in general a linear combination of such solutions, that is,

$$A(x,t) = \alpha \sum_l q_l(t)b_l(x) \tag{3.51}$$

$$\frac{d^2q_l}{dt^2} + \omega_l^2 q_l = 0; \qquad \omega_l = \bar{c}k_l \tag{3.52}$$

$$\frac{d^2b_l}{dx^2} + k_l^2 b_l = 0 \tag{3.53}$$

Equation (3.51) expresses the general solution for $A(x,t)$ as a sum over independent "normal modes." $q_l(t)$, obtained from Eq. (3.52), determines the time evolution of a mode, while $b_l(x)$, the solution to Eq. (3.53), determines its spatial structure in much the same way as the time-independent Schrödinger equation determine the intrinsic eigenfunctions of a given system. In fact, Eq. (3.53) has the same structure as the time-independent Schrödinger equation for a free particle, Eq. (2.80). It admits similar solutions that depend on the imposed boundary conditions. If we use periodic boundary conditions with period L we find, in analogy to (2.82),

$$b_l(x) = \frac{1}{\sqrt{L}}e^{ik_l x}; \qquad k_l = \frac{2\pi}{L}l; \quad (l = 0, \pm 1, \pm 2 \ldots) \tag{3.54}$$

Furthermore, as in the free particle case, Eq. (3.54) implies that the density of modes (i.e. the number, per unit interval along the k-axis, of possible values of k) is $L/2\pi$. Obviously, different solution are orthogonal to each other, $\int_L b_l^*(x)b_{l'}(x)dx = \delta_{l,l'}$, and the choice of normalization can be arbitrary because the amplitude of the mode is determined by the solution of Eq. (3.52).

In the general case, where $\mathbf{A}$ is a vector and (3.49) is a three-dimensional vector equation, a generalization of (3.51)[5]

$$\mathbf{A}(\mathbf{r},t) = \sqrt{\frac{4\pi}{\varepsilon}}c\sum_l q_l(t)\mathbf{b}_l(\mathbf{r}) \tag{3.55}$$

yields again Eq. (3.52) and a generalization of Eq. (3.53)

$$\frac{d^2q_l}{dt^2} + \omega_l^2 q_l = 0; \qquad \omega_l = \bar{c}k_l \tag{3.56}$$

$$\nabla^2\mathbf{b}_l + \frac{\omega_l^2}{\bar{c}^2}\mathbf{b}_l = 0 \tag{3.57}$$

If $\mathbf{b}_l(\mathbf{r})$ was a scalar function $b_l(\mathbf{r})$, (3.57) would be equivalent to the Schrödinger equation for a three-dimensional free particle, yielding, for periodic boundary conditions, solutions of the form

$$b_l(\mathbf{r}) = \Omega^{-1/2}e^{i\mathbf{k}_l\cdot\mathbf{r}} \qquad (\Omega = L_xL_yL_z) \tag{3.58}$$

characterized by a wavevector $\mathbf{k}$ that satisfies $|\mathbf{k}_l| = \omega_l\bar{c}$ with components of the form (3.54). This remains true also when $\mathbf{b}$ is a vector, however, in addition to the three numbers comprising the wavevector $\mathbf{k}_l$, the mode is also characterized by the direction of $\mathbf{b}_l$. This extra information, called *polarization*, can be conveyed by a unit vector, $\boldsymbol{\sigma}_l$, in the direction of $\mathbf{b}_l$, that is, $\mathbf{b}_l = \Omega^{-1/2}\boldsymbol{\sigma}_l e^{i\mathbf{k}_l\cdot\mathbf{r}}$. This form, together with the Coulomb gauge property (3.41), implies that $\boldsymbol{\sigma}_l$ has to satisfy the transversality condition $\boldsymbol{\sigma}_l \cdot \mathbf{k}_l = 0$. (This results from Eq. (1.22) and the identity $\nabla e^{i\mathbf{k}\cdot\mathbf{r}} = i\mathbf{k}e^{i\mathbf{k}\cdot\mathbf{r}}$.)

Additional important insight is obtained by using Eqs (3.55), (3.47), and (3.37) to find the energy contained in the field. Because different $\mathbf{b}_l(\mathbf{r})$ functions constitute an orthogonal set, different modes contribute independently. This calculation is rather technical even if conceptually straightforward. The result is

$$E = \sum_l E_l; \qquad E_l = (1/2)(\dot{q}_l^2 + \omega_l^2 q_l^2) \tag{3.59}$$

We recall that $q(t)$ is just the time-dependent amplitude of the vector potential, and by (3.47) $\dot{q}(t)$ is related to the electric field. On the other hand, Eq. (3.56) has

[5] In (3.55) we chose a particular value for the constant α in order to simplify the form of subsequent expressions. This choice is in effect a scaling of q that yields the simple form of Eq. (3.59).

the form of the Newton equation for a harmonic oscillator of frequency ω_l with coordinate q_l and momentum $p_l = \dot{q}_l$ derived from the Hamiltonian

$$h_l = (1/2)(p_l^2 + \omega_l^2 q_l^2) \tag{3.60}$$

in which q_l plays the role of coordinate while p_l is the conjugated momentum. Note that these coordinate and momentum are related to the standard quantities by mass renormalization so that their dimensionalities are $m^{1/2}l$ and $m^{1/2}l/t$, respectively.

If we take this seriously, the radiation field appears to have the character of a harmonic medium described by a set of normal modes $\{q_l\}$ and by the (still classical) Hamiltonian

$$H = \sum_l (1/2)(p_l^2 + \omega_l^2 q_l^2) \tag{3.61}$$

When such a mode is excited a time-dependent oscillation is set up in the system as determined by Eq. (3.56). The oscillating object is an electromagnetic field whose spatial variation is determined by Eq. (3.57). This spatial dependence is characterized by a wavevector $\mathbf{k}_l$ and a polarization $\boldsymbol{\sigma}_l$ that satisfy the conditions $\omega_l = \bar{c}|\mathbf{k}_l|$ and $\boldsymbol{\sigma}_l \cdot \mathbf{k}_l = 0$. The mode index l represents the five numbers that determine $\mathbf{k}_l$ and $\boldsymbol{\sigma}_l$. The energy associated with the mode is determined by Eq. (3.59).

Should we take it seriously? Experiments such as studies of blackbody radiation not only answer in the affirmative, but tell us that we should go one step further and assign a quantum character to the field, where each normal mode is taken to represent a quantum oscillator characterized by operator analog of Eq. (3.60) in which q and p become operators, $\hat{q}$ and $\hat{p}$, that satisfy $[\hat{q}, \hat{p}] = i\hbar$. This follows from the observation that the thermal properties of blackbody radiation can be understood only if we assume that a mode of frequency ω can be only in states of energies $n\hbar\omega$, with interger n, above the ground state. We refer to a mode in such state as being occupied by n photons. The state of the overall field is then characterized by the set $\{n_l\}$ of occupation numbers of all modes, where the ground state corresponds to $n_l = 0$ for all l. Note that the vector potential $\mathbf{A}$ then becomes an operator (since $q_l(t)$ in (3.55) are now operators; the time dependence should be interpreted as the corresponding Heisenberg representation), and the derived fields $\mathcal{E}$ and $\mathcal{B}$ in Eq. (3.47) are operators as well.

It is convenient to use raising and lowering operators in this context. Define

$$\hat{a}_l = \frac{1}{\sqrt{2\hbar\omega_l}}(\omega_l \hat{q}_l + i\hat{p}_l); \qquad \hat{a}_l^\dagger = \frac{1}{\sqrt{2\hbar\omega_l}}(\omega_l \hat{q}_l - i\hat{p}_l) \tag{3.62}$$

that obey the commutation relation $[\hat{a}_l, \hat{a}_l^\dagger] = 1$, with the inverse transformation

$$\hat{q}_l = \sqrt{\frac{\hbar}{2\omega_l}}(\hat{a}_l^\dagger + \hat{a}_l); \qquad \hat{p}_l = i\sqrt{\frac{\hbar\omega_l}{2}}(\hat{a}_l^\dagger - \hat{a}_l) \tag{3.63}$$

The Hamiltonian (3.61) than becomes

$$\hat{H} = \sum_l \hbar\omega_l(\hat{a}_l^\dagger\hat{a}_l + (1/2)) \tag{3.64}$$

the corresponding time-dependent (Heisenberg) operators are

$$\hat{a}_l(t) = \hat{a}_l e^{-i\omega_l t}; \qquad \hat{a}_l^\dagger(t) = \hat{a}_l^\dagger e^{i\omega_l t} \tag{3.65}$$

and the vector potential operator takes the form (cf. Eq. (3.55))

$$\begin{aligned}\hat{\mathbf{A}}(\mathbf{r}, t) &= c\sqrt{\frac{4\pi}{\varepsilon}}\sum_l \boldsymbol{\sigma}_l b_l(\mathbf{r})\hat{q}_l(t) \\ &= c\sum_l \boldsymbol{\sigma}_l\sqrt{\frac{2\pi\hbar}{\varepsilon\omega_l}} b_l(\mathbf{r})(\hat{a}_l^\dagger e^{i\omega_l t} + \hat{a}_l e^{-i\omega_l t})\end{aligned} \tag{3.66}$$

When periodic boundary conditions are used, the spatial functions $b_l(\mathbf{r})$ are given by Eq. (3.58). Because we look for real solutions of the Maxwell equation $\hat{\mathbf{A}}$ takes the form analogous to the corresponding real classical solution

$$\hat{\mathbf{A}}(\mathbf{r}, t) = c\sum_l \sqrt{\frac{2\pi\hbar}{\varepsilon\Omega\omega_l}}\boldsymbol{\sigma}_l(\hat{a}_l e^{-i\omega_l t + i\mathbf{k}_l\cdot\mathbf{r}} + \hat{a}_l^\dagger e^{i\omega_l t - i\mathbf{k}_l\cdot\mathbf{r}}) \tag{3.67}$$

(The assignment of $-i\mathbf{k}_l$ to $+i\omega_l$ is arbitrary because the sum over l implies summation over all positive and negative $\mathbf{k}$ components). Since l encompasses the five numbers $(\mathbf{k}, \boldsymbol{\sigma})$ we can write (3.67) in the alternative form

$$\hat{\mathbf{A}}(\mathbf{r}, t) = c\sum_{\mathbf{k}}\sum_{\sigma_{\mathbf{k}}}\sqrt{\frac{2\pi\hbar}{\varepsilon\Omega\omega_k}}\boldsymbol{\sigma}_{\mathbf{k}}(\hat{a}_{\mathbf{k},\sigma_{\mathbf{k}}} e^{-i\omega_k t + i\mathbf{k}\cdot\mathbf{r}} + \hat{a}_{\mathbf{k},\sigma_{\mathbf{k}}}^\dagger e^{i\omega_k t - i\mathbf{k}\cdot\mathbf{r}}) \tag{3.68}$$

with $\omega_k = k\bar{c}$ and $k = |\mathbf{k}|$. Consequently the electric field operator is, from Eq. (3.47)

$$\hat{\boldsymbol{\mathcal{E}}}(\mathbf{r}, t) = i\sum_{\mathbf{k}}\sum_{\sigma_{\mathbf{k}}}\sqrt{\frac{2\pi\hbar\omega_k}{\varepsilon\Omega}}\boldsymbol{\sigma}_{\mathbf{k}}(\hat{a}_{\mathbf{k},\sigma_{\mathbf{k}}} e^{-i\omega_k t + i\mathbf{k}\cdot\mathbf{r}} - \hat{a}_{\mathbf{k},\sigma_{\mathbf{k}}}^\dagger e^{i\omega_k t - i\mathbf{k}\cdot\mathbf{r}}) \tag{3.69}$$

As noted above this is the Heisenberg representation. The corresponding Schrödinger form is

$$\hat{\boldsymbol{\mathcal{E}}}(\mathbf{r}) = i\sum_{\mathbf{k}}\sum_{\sigma_{\mathbf{k}}}\sqrt{\frac{2\pi\hbar\omega_k}{\varepsilon\Omega}}\boldsymbol{\sigma}_{\mathbf{k}}(\hat{a}_{\mathbf{k},\sigma_{\mathbf{k}}}e^{i\mathbf{k}\cdot\mathbf{r}} - \hat{a}^{\dagger}_{\mathbf{k},\sigma_{\mathbf{k}}}e^{-i\mathbf{k}\cdot\mathbf{r}}) \tag{3.70}$$

Finally consider the interaction between a molecule (or any system of particles) and the radiation field. A simple expression for this interaction is provided by Eq. (3.1) or, when applied to a single molecule, by the simpler version (3.24). From (3.24) and (3.70) we finally get Eq. (3.26) for the desired interaction operator.

We will be using this form of the molecule–field interaction repeatedly in this text, however, it should be kept in mind that it is an approximation on several counts. Already Eq. (3.1), an electrostatic energy expression used with a time varying field, is an approximation. Even in this electrostatic limit, Eq. (3.1) is just the first term in an infinite multipole expansion in which the higher-order terms depend on higher spatial derivatives of the electric field.

$$\hat{H}_{\mathrm{MR}} = q\Phi - \boldsymbol{\mu}\cdot\boldsymbol{\mathcal{E}}(0) - \frac{1}{6}\sum_i\sum_j Q_{ij}\frac{\partial\mathcal{E}_i}{\partial x_j}(0) + \cdots \tag{3.71}$$

were Q_{ij} is the molecular quadrupole tensor and (0) denotes the molecular center. Other contributions to the interaction are associated with the motion of charged particles in a magnetic field. Another approximation is associated with the fact that the radiation field equations and the field operators were constructed from the macroscopic forms of the Maxwell equations, where the phenomenological constants ε and $\boldsymbol{\mu}$ already contain elements of field–matter interaction. This corresponds to a picture in which both the atomic system of interest and the radiation field exist in an ambient medium characterized by these dielectric and magnetic response constants. A fully microscopic theory would start with the microscopic Maxwell equations and from a fundamental form for matter–radiation field interaction, and derive these results in a systematic way.

Such a general theory of interaction of radiation with matter has been formulated. It yields the following expression for the Hamiltonian of a system that comprises matter (say a molecule) and radiation

$$\hat{H} = \sum_j \frac{(\hat{\mathbf{p}}_j - (q_j/c)\hat{\mathbf{A}}(\mathbf{r}_j))^2}{2m_j} + \hat{U}_{\mathrm{M}}(\{\mathbf{r}_j\}) + \hat{H}_{\mathrm{R}} \tag{3.72}$$

Here $\hat{U}_{\mathrm{M}}(\{\mathbf{r}_j\})$ is the molecular potential operator that depend on all electronic and nuclear coordinates, and $\hat{\mathbf{p}}_j$, $\hat{\mathbf{r}}_j$, and q_j are respectively the momentum and coordinate operators and the charge associated with molecular particle j. Nothing reminiscent of the form (3.1) appears in Eq. (3.72), still it may be shown that the interaction $-\sum_j (q_j/m_j c)\hat{\mathbf{p}}_j \cdot \hat{\mathbf{A}}(\mathbf{r}_j)$ implied by Eq. (3.72) yields the same interaction matrix elements needed in the calculation of optical transitions, provided that the conditions that lead to Eq. (3.26) are satisfied, that is, $\lambda = 2\pi/k \gg$ molecular dimension.[6]

More generally, it can be shown that if magnetic interactions are disregarded then, in a semiclassical approximation in which the electromagnetic field is treated classically while the material degrees of freedom retain their quantum nature, Eq. (3.72) yields the following Hamiltonian for the material system

$$\hat{H} = \hat{H}_{\mathrm{M}} - \int d\mathbf{r}\hat{P}(\mathbf{r}) \cdot \mathcal{E}^{\perp}(\mathbf{r}, t) \tag{3.73}$$

where $\mathcal{E}^{\perp}(\mathbf{r}, t)$ is the transverse part of the electric field and $\hat{\mathbf{P}}(\mathbf{r})$ is the dipole density operator. In the limit of molecular point dipoles (i.e. when (3.24) applies for any molecule taken at the origin), this operator is given by

$$\hat{\mathbf{P}}(\mathbf{r}) = \sum_m \hat{\boldsymbol{\mu}}_m \delta(\mathbf{r}_m - \mathbf{r}) \tag{3.74}$$

where the sum is over all molecules. For a single molecule Eqs (3.73) and (3.74) yield Eq. (3.24). In the many molecules case the molecular part, $\hat{H}_{\mathrm{M}}$, must include also the dipole–dipole interaction operators between the molecules.

The Hamiltonian (3.73) is a time-dependent operator for the molecular system, where the electromagnetic field appears through the time-dependent electric field. A useful starting point for analyzing nonlinear optical processes in molecular systems is obtained by supplementing (3.73) by an equation of motion for this time-dependent electric field. Such an equation can be derived from the Maxwell equations (3.32) and (3.33). Limiting ourselves to systems without free charges, so that ρ and $\mathbf{J}$ are zero, and to non-magnetic materials so that $\mathbf{M} = 0$ and $\mathcal{H} = \mathcal{B}$, Eq. (3.32d) with (3.33a) and (3.32c) yield

$$\nabla \times \nabla \times \mathcal{E}(\mathbf{r}, t) + \frac{1}{c^2}\frac{\partial^2 \mathcal{E}(\mathbf{r}, t)}{\partial t^2} = -\frac{4\pi}{c^2}\frac{\partial^2 \langle \mathbf{P}(\mathbf{r}, t)\rangle}{\partial t^2} \tag{3.75}$$

Note that if $\mathcal{E}$ is transverse, that is, $\nabla \cdot \mathcal{E} = 0$, then $\nabla \times \nabla \times \mathcal{E}(\mathbf{r}, t) = -\nabla^2 \mathcal{E}(\mathbf{r}, t)$. Equations (3.73) and (3.75) together with the definition $\langle \mathbf{P}(\mathbf{r}, t)\rangle = \mathrm{Tr}[\hat{\rho}\hat{\mathbf{P}}]$ where $\hat{\rho}$

[6] An interesting difference is that while in Eq. (3.27) we find the photon frequency ω_k as a multiplying factor, in the calculation based on the interaction $-\sum_j (q_j/m_j c)\hat{\mathbf{p}}_j \cdot \mathbf{A}(\mathbf{r}_j)$ we get instead a factor of $\omega_{ss'}$—the transition frequency between the molecular states involved. For physical processes that conserve energy the two are equal.

is the density operator (see Chapter 10) constitute the desired closed set of equation for the molecular system and for the classical radiation field, which should now be solved self consistently.

Further reading

C. Cohen-Tannoudji, J. Dupont-Roc, and G. Grynberg, *Atom–Photon Interactions: Basic Processes and Applications* (Wiley, New York, 1992).

J. D. Jackson, *Classical Electrodynamics* 2nd edn, (Wiley, NYC, 1975, chapter 6).

W. H. *Louisell, Quantum Statistical Properties of Radiation* (Wiley, New York, 1973).

S. Mukamel, *Principles of Nonlinear Optical Spectroscopy* (Oxford University Press, Oxford, 1995).

4

INTRODUCTION TO SOLIDS AND THEIR INTERFACES

Tight-knit, must have more barbs and hooks to hold them,
Must be more interwoven, like thorny branches
In a closed hedgerow; in this class of things
We find, say, adamant, flint, iron, bronze
That shrieks in protest if you try to force
The stout oak door against the holding bars...

Lucretius (c.99–c.55 BCE*) "The way things are"*
translated by Rolfe Humphries, Indiana University Press, 1968

The study of dynamics of molecular processes in condensed phases necessarily involves properties of the condensed environment that surrounds the system under consideration. This chapter provides some essential background on the properties of solids while the next chapter does the same for liquids. No attempt is made to provide a comprehensive discussion of these subjects. Rather, this chapter only aims to provide enough background as needed in later chapters in order to take into consideration two essential attributes of the solid environment: Its interaction with the molecular system of interest and the relevant timescales associated with this interaction. This would entail the need to have some familiarity with the relevant degrees of freedom, the nature of their interaction with a guest molecule, the corresponding densities of states or modes, and the associated characteristic timescales. Focusing on the solid crystal environment we thus need to have some understanding of its electronic and nuclear dynamics.

4.1 Lattice periodicity

The geometry of a crystal is defined with respect to a given lattice by picturing the crystal as made of periodically repeating unit cells. The atomic structure within the cell is a property of the particular structure (e.g. each cell can contain one or more molecules, or several atoms arranged within the cell volume in some given way), however, the cells themselves are assigned to lattice points that determine the periodicity. This periodicity is characterized by three *lattice vectors*, $\mathbf{a}_i$, $i = 1, 2, 3$, that determine the *primitive lattice cell*—a parallelepiped defined by these three vectors. The lattice itself is then the collection of all points (or all vectors) defined by

$$\mathbf{R} = n_1\mathbf{a}_1 + n_2\mathbf{a}_2 + n_3\mathbf{a}_3 \tag{4.1}$$

where (here and below) n_1, n_2, n_3 are all integers. It will prove useful to define also the *reciprocal lattice*: The collection of all vectors $\mathbf{G}$ that satisfy

$$\mathbf{R} \cdot \mathbf{G} = 2\pi m, \qquad m \text{ integer} \tag{4.2}$$

It can be shown that these vectors $\mathbf{G}$ are of the form

$$\mathbf{G} = n_1\mathbf{b}_1 + n_2\mathbf{b}_2 + n_3\mathbf{b}_3 \tag{4.3}$$

with the primitive vectors of the reciprocal lattice given by

$$\mathbf{b}_1 = 2\pi \frac{\mathbf{a}_2 \times \mathbf{a}_3}{\mathbf{a}_1 (\mathbf{a}_2 \times \mathbf{a}_3)}; \qquad \mathbf{b}_2 = 2\pi \frac{\mathbf{a}_3 \times \mathbf{a}_1}{\mathbf{a}_1 \cdot \mathbf{a}_2 \times \mathbf{a}_3}; \qquad \mathbf{b}_3 = 2\pi \frac{\mathbf{a}_1 \times \mathbf{a}_2}{\mathbf{a}_1 \cdot \mathbf{a}_2 \times \mathbf{a}_3} \tag{4.4}$$

For example, in one-dimension the direct lattice is na and the reciprocal lattice is $(2\pi/a)n$ $(n = 0, \pm 1, \pm 2, \ldots)$. The *First Brillouin zone* is a cell in the reciprocal lattice that encloses points closer to the origin $(n_1, n_2, n_3 = 0)$ than to any other lattice point.[1] Obviously, for a one-dimensional lattice the first Brilloin zone is $-(\pi/a) \ldots (\pi/a)$.

4.2 Lattice vibrations

Periodicity is an important attribute of crystals with significant implications for their properties. Another important property of these systems is the fact that the amplitudes of atomic motions about their equilibrium positions are small enough to allow a harmonic approximation of the interatomic potential. The resulting theory of atomic motion in harmonic crystals constitutes the simplest example for many-body dynamics, which is discussed in this section.

4.2.1 Normal modes of harmonic systems

As in molecules, the starting point of a study of atomic motions in solid is the potential surface on which the atoms move. This potential is obtained in principle from the Born–Oppenheimer approximation (see Section 2.5). Once given, the

[1] Such a cell is also called a Wigner–Seitz cell.

many-body atomic motion in a system of N atoms is described by a Hamiltonian of the form

$$H = \sum_{j=1}^{N} \frac{p_j^2}{2m_j} + V(x_1, x_2, \ldots, x_j, \ldots, x_N) \tag{4.5}$$

A harmonic approximation is obtained by expanding the potential about the minimum energy configuration and neglecting terms above second order. This leads to

$$V(x^N) = V(x_0^N) + \frac{1}{2} \sum_{i,j} k_{i,j}(x_i - x_{i0})(x_j - x_{j0}) \tag{4.6}$$

where we use the notation $x^N = (x_1, x_2, \ldots, x_N)$ and where $k_{i,j} = (\partial^2 V / \partial x_i \partial x_j)_{x_0^N}$. The resulting Hamiltonian corresponds to a set of particles of mass m_j, attached to each other by harmonic springs characterized by a force constants $k_{j,l}$. The classical equations of motion are

$$\ddot{x}_j = -\frac{1}{m_j} \sum_l k_{j,l}(x_l - x_{l,0}) \tag{4.7}$$

In (4.6) and (4.7) x_0 are the equilibrium positions. For simplicity we will redefine $x_j \equiv x_j - x_{j0}$. So

$$\ddot{x}_j = -\frac{1}{m_j} \sum_l k_{j,l} x_l \tag{4.8}$$

In terms of the renormalized positions and force constants

$$y_j = \sqrt{m_j} x_j; \qquad K_{j,l} = \frac{k_{j,l}}{\sqrt{m_j m_l}} \tag{4.9}$$

we get $\ddot{y}_j = -\sum_l K_{j,l} y_l$ or

$$\ddot{\mathbf{y}} = -\mathbf{K}\mathbf{y} \tag{4.10}$$

$\mathbf{K}$ is a real symmetric matrix, hence its eigenvalues are real. Stability requires that these eigenvalues are positive; otherwise small deviations from equilibrium will spontaneously grow in time. We will denote these eigenvalues by ω_j^2, that is,

$$\mathbf{TKT}^{-1} = \begin{pmatrix} \omega_1^2 & & & 0 \\ & \omega_2^2 & & \vdots \\ & & \ddots & \vdots \\ 0 & \cdots & \cdots & \omega_N^2 \end{pmatrix}; \qquad \mathbf{u} = \mathbf{Ty} \tag{4.11}$$

where $\mathbf{T}$ is the unitary transformation that diagonalizes $\mathbf{K}$. The components of $\mathbf{u}$ are the amplitudes of the normal modes of the system defined by Eqs. (4.5) and (4.6). Their equations of motion are those of independent harmonic oscillators

$$\ddot{u}_j = -\omega_j^2 u_j \tag{4.12}$$

The individual atomic motions are now obtained from the inverse transformation

$$y_j = \sum_k (\mathbf{T}^{-1})_{jk} u_k; \qquad x_j = (m_j)^{-1/2} y_j \tag{4.13}$$

This linear problem is thus exactly soluble. On the practical level, however, one cannot carry out the diagonalization (4.11) for macroscopic systems without additional considerations, for example, by invoking the lattice periodicity as shown below. The important physical message at this point is that *atomic motions in solids can be described, in the harmonic approximation, as motion of independent harmonic oscillators*. It is important to note that even though we used a classical mechanics language above, what was actually done is to replace the interatomic potential by its expansion to quadratic order. Therefore, an identical independent harmonic oscillator picture holds also in the quantum regime.

4.2.2 Simple harmonic crystal in one dimension

As a simple example we consider a monatomic one-dimensional solid with identical atoms, one per unit cell, characterized by the Hamiltonian

$$H = \sum_{n=1}^{N} \frac{m}{2}\dot{x}_n^2 + \sum_{n=1}^{N} \frac{1}{2}\kappa(x_n - x_{n-1})^2 \tag{4.14}$$

where x_n is the deviation of the nth atom from its equilibrium position. It is convenient to use periodic boundary conditions by imposing

$$x_{n+N} = x_n \tag{4.15}$$

and to take the limit $N \to \infty$ at the end of the calculation. This makes our system a ring of N elastically bound atoms. The equations of motion for x_n

$$m\ddot{x}_n = \kappa(x_{n+1} + x_{n-1} - 2x_n); \qquad n = 1, \ldots, N \tag{4.16}$$

are solved by using the ansatz

$$x_n(t) = u_\phi(t)\, e^{in\phi} \tag{4.17}$$

in (4.16). This leads to

$$m\ddot{u}_\phi = \kappa u_\phi(e^{i\phi} + e^{-i\phi} - 2) = -\left(4\kappa \sin^2 \frac{\phi}{2}\right) u_\phi \tag{4.18}$$

which is an equation of motion for a harmonic oscillator, $\ddot{u}_\phi = -\omega^2(\phi)u_\phi$, characterized by the frequency

$$\omega(\phi) = 2\omega_0 \sin \frac{\phi}{2} \tag{4.19}$$

where

$$\omega_0 = \sqrt{\frac{\kappa}{m}} \tag{4.20}$$

It is convenient to define a wave-vector **k** in the direction of the particle's motion, whose magnitude is ϕ/a, where a is the lattice spacing

$$e^{in\phi} = e^{i(\phi/a)na} = e^{ikna}; \qquad k = \frac{\phi}{a} \tag{4.21}$$

na is the characteristic position of an atom in the chain. Thus, for each value of k we got an independent equation of motion

$$\ddot{u}_k = -\omega^2(k)u_k \tag{4.22}$$

whose solution can be written in terms of initial conditions for u and $\dot{u}$

$$u_k(t) = u_k(0)\cos(\omega_k t) + \frac{\dot{u}_k(0)}{\omega_k}\sin(\omega_k t) \tag{4.23}$$

These are the normal modes of this harmonic system. A motion of this type is a collective motion of all atoms according to (from (4.17))

$$x_n(t) = u_k(t)\, e^{ikna} \tag{4.24}$$

Each such oscillation constitutes a wave of wavelength $\lambda = 2\pi/|k|$ and a corresponding frequency[2]

$$\omega_k = 2\omega_0|\sin[ka/2]| \tag{4.25}$$

These modes of motion with wavelengths and frequencies determined by k are called phonons. A relationship such as (4.25) between k and ω is called a *dispersion*

[2] From Eq. (4.23) we see that it is enough to consider positive frequencies.

relation. The allowed values of k are determined from the periodicity of the model and the imposed boundary conditions. First note that e^{ikna} remains the same if k is shifted according to $k \rightarrow k+(2\pi/a)j$, where j is any integer. Therefore, independent solutions are obtained only for values of k within an interval $k_0 \dots k_0 + 2\pi/a$. If we choose $k_0 = -\pi/a$, then

$$-\frac{\pi}{a} \le k \le \frac{\pi}{a} \tag{4.26}$$

namely, all physically different values of k are represented within the first Brillouin zone of the reciprocal lattice. Second, Eq. (4.15) implies that $e^{ikNa} = 1$. To satisfy this k must be of the form $(2\pi/Na)l$, with integer l. Together with (4.26) this implies

$$k = \frac{2\pi}{Na} l; \qquad l = 0, \pm 1, \pm 2, \dots, \pm \frac{N-1}{2}, (+ \text{ or } -)\frac{N}{2} \tag{4.27}$$

When the lattice becomes infinitely long, $N \rightarrow \infty$, k becomes a continuous parameter. In the long wavelength (small k) limit these phonons should become the familiar sound waves. In this limit, $k \rightarrow 0$, we can expand Eq. (4.25)

$$\omega_k = 2\omega_0 \sin \frac{|k|a}{2} \rightarrow \omega_0 a|k| \tag{4.28}$$

This is indeed a dispersion relation for a sound wave of speed

$$c = \omega_0 a \tag{4.29}$$

Typically $\omega_0 = 10^{13}$ s^{-1} and $a = 10^{-8}$ cm, therefore $c \sim 10^5$cm/s^{-1}. This is indeed the order of magnitude of sound velocity in solids. When k increases $\omega(k)$ becomes nonlinear in k, that is, the "velocity"[3] $d\omega/dk$ depends on k.

What was achieved above is an implementation of the general solution of Section 4.2.1 for a system of harmonically connected atoms whose equilibrium positions lie on a one-dimensional periodic lattice. Indeed, Eq. (4.24) connects (up to a normalization constant) between the amplitude of each atomic motion and that of each normal mode. Consequently, the transformation (4.11) has the explicit form

$$\mathbf{T}_{nk} = \frac{1}{\sqrt{N}} e^{ikna}; \qquad (\mathbf{T}^{-1})_{nk} = \frac{1}{\sqrt{N}} e^{-ikna} \tag{4.30}$$

The normalization constant $(\sqrt{N})^{-1}$ is needed to satisfy the unitarity requirement $(\mathbf{TT}^{+})_{mn} = \delta_{nm}$ (see also Problem 4.1 below).

[3] $\partial\omega/\partial k$ is known as the *group velocity* of the wave motion.

Problem 4.1. Consider the transformation (cf. Eq. (4.24)), written for the displacement x_n of the atom at site n when many phonon modes are excited in a one-dimensional lattice of N sites

$$x_n = \sum_k u_k e^{ikna} \tag{4.31}$$

Prove the identity $\sum_n e^{ikna} = N\delta_{k,0}$ (the sum is over all lattice points) and use it to show that (4.31) implies $u_k = N^{-1}\sum_n x_n e^{-ikna}$. (Note that k takes only the values (4.27)). It is convenient to redefine the normal-mode coordinates according to $\sqrt{N}u_k \to u_k$ so that the transformation takes the more symmetric form $x_n = (\sqrt{N})^{-1}\sum_k u_k e^{ikna}$ and $u_k = (\sqrt{N})^{-1}\sum_n x_n e^{-ikna}$ as implied by (4.30).

4.2.3 Density of modes

In problems addressed in this text, solids appear not as the system of principal interest but as an environment, a host, of our system. We therefore focus on those properties of solids that are associated with their effect on a molecular guest. One such property is the density of modes, a function $g(\omega)$ defined such that the number of modes in any frequency interval $\omega_1 \le \omega \le \omega_2$ is $\int_{\omega_1}^{\omega_2} d\omega\, g(\omega)$. As a formal definition we may write

$$g(\omega) \equiv \sum_j \delta(\omega - \omega_j) \tag{4.32}$$

In fact, this function dominates also thermal and optical properties of the solids themselves because experimental probes do not address individual normal modes but rather collective mode motions that manifest themselves through the mode density. For example, the vibrational energy of a harmonic solid is given by

$$E = \sum_j \hbar\omega_j(n_j + \frac{1}{2}) = \int d\omega g(\omega)\hbar\omega\left(\langle n(\omega)\rangle + \frac{1}{2}\right) \tag{4.33}$$

where n_j is the occupation of the mode j. Note that the density of modes $g(\omega)$ is an analog of the density of states, Eq. (1.181), and its use in (4.83) results from a coarse-graining process equivalent to (1.182). The second equality in (4.33) becomes exact in the infinite system limit where the spectrum of normal mode frequencies is continuous.

At thermal equilibrium

$$\langle n(\omega)\rangle = \frac{1}{e^{\beta\hbar\omega} - 1} \tag{4.34}$$

The heat capacity is the derivative of E with respect to T. We get

$$C_V = \left(\frac{\partial E}{\partial T}\right)_{V,N} = k_B \int d\omega g(\omega) \left(\frac{\hbar\omega}{k_B T}\right)^2 \frac{e^{\beta\hbar\omega}}{(e^{\beta\hbar\omega} - 1)^2} \tag{4.35}$$

The density of modes is seen to be the only solid property needed for a complete evaluation of these thermodynamic quantities. In what follows we consider this function within the one-dimensional model of Section 4.2.2.

Consider the one-dimensional solid analyzed in Section 4.2.2. From the expression for the allowed value of $k = (2\pi/Na)l, l = 0, \pm 1, \ldots$ we find that the number of possible k values in the interval $k, \ldots, k + \Delta k$ is $(Na/2\pi)\Delta k$, so the density of modes per unit interval in k is

$$g(k) = \frac{Na}{2\pi}, \qquad \text{that is } g(|k|) = \frac{Na}{\pi} \tag{4.36}$$

The difference between these two forms stems from the fact that in one dimension there are two values of k for a given $|k|$. The density of modes in frequency space is obtained from the requirement that the number of modes in a given interval of $|k|$ is the same as in the corresponding interval of ω,

$$g(|k|)d|k| = g(\omega)d\omega \tag{4.37}$$

so that

$$g(\omega) = \frac{Na}{\pi}\left(\frac{d\omega}{d|k|}\right)^{-1} \tag{4.38}$$

and using the dispersion relation, Eq (4.28)

$$g(\omega) = \frac{N}{\pi\omega_0 \cos(|k|a/2)} = \frac{N}{\pi\omega_0\sqrt{1 - (\omega/2\omega_0)^2}} \tag{4.39}$$

In the long wavelength limit $\omega = ck$; $c = \omega_0 a$, $g(\omega) = N/\pi\omega_0$. For larger k, that is, larger ω, $g(\omega)$ depends on ω and becomes singular at the Brillouin zone boundary $k = \pm\pi/a$, $|\omega| = 2\omega_0$.

While a one-dimensional model is not very realistic, the analytical result (4.39) shows an important feature of a general nature—the fact that the phonon spectrum is bound: There are no modes of frequency larger than $2\omega_0$. Note that this upper bound is associated with wavevectors at the Brillouin zone boundary, that is, wavelengths comparable to the interatomic distance.

Next, consider the three-dimensional case, focusing on a simple cubic lattice. Rewriting Eq. (4.36) in the form $g(k) = L/(2\pi)$ where $L = Na$ is the lattice length, the three-dimensional analog is clearly $L_x L_y L_z/(2\pi)^3 = \Omega/(2\pi)^3$ where Ω is the volume. In terms of the absolute value of the wavevector **k** the number of modes in the interval between $|k|$ and $|k| + d|k|$ is

$$g(|k|)d|k| = 4\pi k^2 \frac{\Omega}{(2\pi)^3} d|k| \tag{4.40}$$

Using again Eq. (4.37) we now get

$$g(\omega) = 4\pi k^2 \frac{\Omega}{(2\pi)^3} \left(\frac{d\omega}{d|k|} \right)^{-1} \tag{4.41}$$

To proceed, we need the dispersion relation $\omega = \omega(|k|)$ in three dimensions. At this point one can either resort to numerical evaluation of this function, or to a simple model constructed according to available data and physical insight. In the next section we take the second route.

4.2.4 Phonons in higher dimensions and the heat capacity of solids

The analysis that leads to Eq. (4.39) can be repeated for three-dimensional systems and for solids with more than one atom per unit cell, however analytical results can be obtained only for simple models. Here we discuss two such models and their implications with regard to thermal properties of solids. We will focus on the heat capacity, Eq. (4.35), keeping in mind that the integral in this expression is actually bound by the maximum frequency. Additional information on this maximum frequency is available via the obvious sum rule

$$\int_0^{\omega_{\max}} d\omega g(\omega) = 3N - 6 \simeq 3N \tag{4.42}$$

where $3N$-6 is the number of vibrational degrees of freedom atoms in the N-atom crystal. In what follows we consider two simple models for $g(\omega)$ and their implications for the heat capacity.

4.2.4.1 The Einstein model

This model assumes that all the normal mode frequencies are the same. Taking Eq. (4.42) into account the density of modes then takes the form

$$g(\omega) = 3N\delta(\omega - \omega_e) \tag{4.43}$$

Using this in (4.35) yields

$$C_V = 3Nk_B \left(\frac{\hbar\omega_e}{k_B T} \right)^2 \frac{e^{\beta\hbar\omega_e}}{(e^{\beta\hbar\omega_e} - 1)^2} \tag{4.44}$$

For $T \to \infty$ this gives $C_V = 3Nk_B$. This result is known as the Dulong–Petit law that is approximately obeyed for many solids at high temperature. This law reflects the thermodynamic result that in a system of *classical* oscillators, each vibrational degree of freedom contributes an amount k_B ($(1/2)k_B$ for each kinetic and each positional mode of motion) to the overall heat capacity.

In the low temperature limit Eq. (4.44) predicts that the heat capacity vanishes like

$$C_V \xrightarrow{T\to 0} e^{-\hbar\omega/k_B T} \tag{4.45}$$

This is in qualitative agreement with experimental results. The heat capacity indeed goes to zero at low T—reflecting the fact that a quantum oscillator of frequency ω cannot accept energy from its thermal environment if $k_B T \ll \hbar\omega$. However, the observed low temperature behavior of the heat capacity of nonconducting solids is $C_V \sim T^3$.

4.2.4.2 *The Debye model*

The fact that a quantum oscillator of frequency ω does not interact effectively with a bath of temperature smaller than $\hbar\omega/k_B$ implies that if the low temperature behavior of the solid heat capacity is associated with vibrational motions, it must be related to the low frequency phonon modes. The Debye model combines this observation with two additional physical ideas: One is the fact that the low frequency (long wavelength) limit of the dispersion relation must be

$$\omega = c|k| \tag{4.46}$$

with c being the speed of sound, and the other is the existence of the sum rule (4.42). Using (4.41) with (4.46) leads to

$$\frac{1}{\Omega} g(\omega) d\omega = \frac{\omega^2}{2\pi^2 c^3} d\omega \tag{4.47}$$

More rigorously, there are three branches of modes associated with each $|k|$: Two transverse, with polarization perpendicular to $\mathbf{k}$, and one longitudinal, with polarization along the $\mathbf{k}$ direction. The speed associated with the transverse modes, c_t is somewhat different from that of the longitudinal mode, c_l. For our purpose this distinction is immaterial, and we take

$$\frac{1}{\Omega} g(\omega) d\omega = \left(\frac{2}{c_t^3} + \frac{1}{c_\ell^3}\right) \frac{\omega^2}{2\pi^2} d\omega \equiv \frac{3\omega^2}{2\pi^2 c^3} d\omega \tag{4.48}$$

The last equality defines the average speed of sound c.

We know that Eq. (4.48) describes correctly the low frequency limit. We also know that the total number of modes is $3N$ and that there is an upper bound to the frequency spectrum. The Debye model determines this upper bound by fitting Eq. (4.48) to the sum rule (4.42). Denoting the maximum frequency ω_{D} (Debye frequency) this implies

$$\frac{3}{2\pi^2 c^3}\int_0^{\omega_{\mathrm{D}}} d\omega\omega^2 = \frac{3N}{\Omega} \tag{4.49}$$

whence

$$\omega_{\mathrm{D}} = \left(\frac{6\pi^2 N}{\Omega}\right)^{1/3} c \tag{4.50}$$

and

$$\frac{1}{\Omega}g(\omega) = \frac{9N}{\Omega}\frac{\omega^2}{\omega_{\mathrm{D}}^3} \tag{4.51}$$

To reiterate the statements made above, this model shares two important features with reality: First $g(\omega) \sim \omega^2$ as $\omega \to 0$, and second, the existence of a characteristic cutoff associated with the total number of normal modes. The fact that the model accounts for the low-frequency spectrum of lattice vibrations enables it to describe correctly the low-temperature behavior of the phonon contribution to the heat capacity. Indeed, using (4.51) in (4.35) leads to

$$C_{\mathrm{V}} = \frac{9k_B N}{\omega_{\mathrm{D}}^3}\int_0^{\omega_{\mathrm{D}}} d\omega\,\omega^2 \left(\frac{\hbar\omega}{k_B T}\right)^2 \frac{e^{\hbar\omega/(k_B T)}}{(e^{\hbar\omega/(k_B T)}-1)^2} \tag{4.52}$$

Denoting $\hbar\omega/(k_B T) = x$ and defining the *Debye temperature*

$$\Theta_{\mathrm{D}} \equiv \frac{\hbar\omega_{\mathrm{D}}}{k_B} \tag{4.53}$$

(4.52) becomes

$$C_{\mathrm{V}} = 9k_B N\left(\frac{T}{\Theta_{\mathrm{D}}}\right)^3 \int_0^{(\Theta_{\mathrm{D}}/T)} dx\,x^4 \frac{e^x}{(e^x-1)^2} \tag{4.54}$$

Note that all the properties of the particular crystal enter only through Θ_{D} and that C_{V} in this model is a universal function of T/Θ_{D}. In the high T limit, $\Theta_{\mathrm{D}}/T \to 0$, the relevant x in the integrand of (4.54) is small

$$\frac{x^4 e^x}{(e^x-1)^2} \sim x^2 \tag{4.55}$$

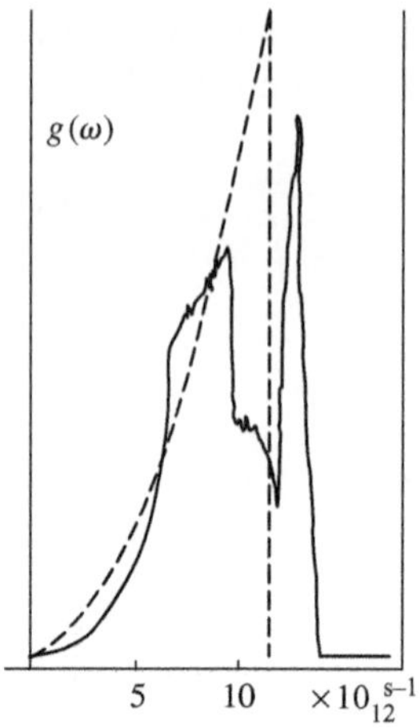

FIG. 4.1 Density of modes in lead. Full line – numerical calculation based on lattice geometry and interatomic potential of lead. Dashed line: The Debye model fitted to the density and speed of sound of lead. (From the Cornell Solid State Simulation site, R.H. Silbee, http://www.physics.cornell.edu/sss/.)

leading to the Dulong–Petit law

$$C_V = 9k_BN\left(\frac{T}{\Theta_D}\right)^3 \int_0^{(\Theta_D/T)} dx\, x^2 = 3k_BN \tag{4.56}$$

In the opposite limit the integral in (4.54) becomes

$$\int_0^\infty dx\, x^4 \frac{e^x}{(e^x - 1)^2} = \frac{4\pi^4}{15} \tag{4.57}$$

and does not depend on T, so that Eq. (4.56) shows the expected T^3 dependence of C_V, in agreement with experiment.

It should be emphasized that although this success of the Debye model has made it a standard starting point for qualitative discussions of solid properties associated with lattice vibrations, it is only a qualitative model with little resemblance to real normal-mode spectra of solids. Figure 4.1 shows the numerically calculated density of modes of lead in comparison with the Debye model for this metal as obtained from the experimental speed of sound. Table 4.1 list the Debye temperature for a few selected solids.

TABLE 4.1 Debye temperatures of several solids.

Solid	Debye temperature (K)
Na	158
Ag	215
Cu	315
Al	398
Ar (solid)	82

4.3 Electronic structure of solids

In addition to the thermal bath of nuclear motions, important groups of solids—metals and semiconductors provide continua of electronic states that can dominate the dynamical behavior of adsorbed molecules. For example, the primary relaxation route of an electronically excited molecule positioned near a metal surface is electron and/or energy transfer involving the electronic degrees of freedom in the metal. In this section we briefly outline some concepts from the electronic structure of solids that are needed to understand the interactions of molecules with such environments.

4.3.1 The free electron theory of metals: Energetics

The simplest electronic theory of metals regards a metallic object as a box filled with noninteracting electrons. (A slightly more elaborate picture is the *jellium model* in which the free electrons are moving on the background of a continuous positive uniform charge distribution that represents the nuclei.) The Drude model, built on this picture, is characterized by two parameters: The density of electrons n (number per unit volume) and the relaxation time τ. The density n is sometimes expressed in terms of the radius r_s of a sphere whose volume is the volume per electron in the metal

$$r_s = \left(\frac{3}{4\pi n}\right)^{1/3} \tag{4.58}$$

The density of states of a free particle as a function of its energy E was obtained in Section 2.8.2. It is given by

$$\rho(E) = \frac{\Omega}{\pi^2}\frac{m}{\hbar^3}\sqrt{2mE} \quad (d = 3) \tag{4.59}$$

where Ω is the volume and m is the particle mass. The additional multiplicative factor of 2 added to the result (2.97) accounts for the electronic spin states.

Let N be the total number of free electrons and n their density, so that $N = n\Omega$. Being Fermions, we can have at most one electron per state. This implies that at

$T = 0$ the highest occupied energy, the Fermi energy E_F, has to satisfy

$$\int_0^{E_F} dE\rho(E) = N \tag{4.60}$$

which implies

$$E_F = \left(\frac{9\pi^4}{8}\right)^{1/3} \frac{\hbar^2}{m} n^{2/3} \tag{4.61}$$

Problem 4.2. Show that the ground state energy of this N electron system is given by

$$E_0 = \frac{3}{5} N E_F \tag{4.62}$$

At finite temperature the picture described above changes slightly. The probability that a single electron level of energy E is occupied is given by the Fermi–Dirac distribution

$$f(E) = \frac{1}{e^{\beta(E-\mu)} + 1}; \quad \beta = (k_B T)^{-1} \tag{4.63}$$

where μ is the chemical potential of the free electron system, which is in principle obtained from the equation

$$N = \sum_j f(E_j) = \int_0^{\infty} dE\rho(E) f(E) \tag{4.64}$$

It is important to get a notion of the energy scales involved. Taking sodium metal as an example, using the mass density 0.97 g cm^{-3} and assuming that each sodium atom contributes one free electron to the system, we get using Eq. (4.61) $E_F = 3.1$ eV. For $T = 300$ K we find that $E_F/(k_B T) \sim 118.6$. Noting that according to Eq. (4.63) $f(E)$ falls from 1 to 0 in an energy interval of the order $k_B T$, it follows that at room temperature the Fermi–Dirac distribution still carries many characteristics of the zero-temperature step function. In particular, the electronic chemical potential μ is approximated well by the Fermi energy. It may indeed be shown that

$$\mu = E_F \left(1 - \mathrm{O}\left(\frac{k_B T}{E_F}\right)^2\right) \tag{4.65}$$

The quantum distribution of electrons in metals has a profound effect on many of their properties. As an example consider their contribution to a metal heat capacity.

The molar heat capacity of a gas of classical structureless noninteracting particles, that is, a classical ideal gas, is $(3/2)R$ where R is the gas constant, $R = k_B\mathcal{A}$, where $\mathcal{A}$ is the Avogadro's number. This is because each degree of freedom that can accept energy contributes $(1/2)k_B$ to the heat capacity. In a quantum ideal low temperature ($k_BT \ll E_F$) Fermi gas most particles cannot accept energy since their state cannot change to that of an occupied level. The only electrons that can accept energy are those in a range $\sim k_BT$ about the Fermi level, that is, only a fraction $\sim k_BT/E_F$ of the total number.

This observation has two consequences that can be confirmed by a rigorous calculation:

1. The molar heat capacity of electrons in metal is about a factor $\sim k_BT/E_F$ smaller than that of a classical ideal gas.
2. This electronic contribution to the heat capacity is linear in the temperature T. This should be contrasted with the cubic form of the low temperature dependence of the phonon contribution, Eqs (4.54) and (4.57).

4.3.2 The free electron theory of metals: Motion

Next consider the motion of these electrons. It was already mentioned that in addition to their density, metallic electrons are characterized, at this level of theory, by a relaxation time τ. In the Drude theory this enters via a simple friction force by assuming that under a given force $\mathbf{f}(t)$ the electron moves according to

$$\dot{\mathbf{r}} = \mathbf{v}$$
$$\frac{d\mathbf{v}}{dt} = \frac{1}{m}\mathbf{f}(t) - \frac{1}{\tau}\mathbf{v}(t) \tag{4.66}$$

This implies that at steady state under a constant force the electron moves with a constant speed $\mathbf{v} = m^{-1}\tau\mathbf{f}$. Using $\mathbf{f} = -e\mathcal{E}$ where $\mathcal{E}$ is the electric field and $-e$ is the electron charge, and the expression for the electric current density in terms of the electron density n, charge $-e$ and speed $\mathbf{v}$,

$$\mathbf{j} = -ne\mathbf{v} \tag{4.67}$$

we find

$$\mathbf{j} = \frac{ne^2\tau}{m}\mathcal{E} \tag{4.68}$$

The coefficient that relates the current density to the electric field is the conductivity[4] σ. We found

$$\sigma = \frac{ne^2\tau}{m} \tag{4.69}$$

The conductivity obtained from the Drude model is seen to be proportional to the electron density and to the relaxation time, and inversely proportional to the electron mass.

Note that the conductivity σ has the dimensionality of inverse time. The Drude model is characterized by two time parameters: τ, that can be thought of as the time between collision suffered by the electron, and σ. Typical values of metallic resistivities are in the range of $10^{-6}\Omega$ cm, that is, $\sigma = 10^6(\Omega\,\text{cm})^{-1} = 10^{18}\,\text{s}^{-1}$. Using this in (4.69) together $n \sim 10^{23}\,\text{cm}^{-3}$, $e \sim 4.8 \times 10^{-10}$ esu and $m \sim 9 \times 10^{-28}$ g leads to τ of the order $\sim 10^{-14}$ s. Several points should be made:

1. The arguments used to get Eq. (4.69) and consequently the estimate $\tau \sim 10^{-14}$ s are classical, and their validity for metallic electrons cannot be taken for granted without further justification.
2. The conductivity (4.69) depends on the carrier charge as e^2, therefore a measurement of electric conductivity cannot identify the sign of this charge.

Information about the sign of the mobile charge may be obtained from another observable, the *Hall coefficient*. The Hall effect is observed when a current carrying conductor (current in direction x, say) is placed in a magnetic field $\mathcal{H}$ perpendicular to the current direction, the z direction say. An electric field $\mathcal{E}$ is formed in the direction y perpendicular to both the current and to the applied magnetic field, and the ratio $R_{\text{H}} = \mathcal{E}_y/j_x\mathcal{H}_z$ is the Hall coefficient. A theory done on the same classical level as used above leads to

$$R_{\text{H}} = -\frac{1}{nec} \tag{4.70}$$

where c is the speed of light. Here the sign of the charge carriers is seen to matter, and Hall effect measurements gave the first indications that charge carriers in metals (e.g. Al) can be effectively positive. Such an observation cannot be explained in the framework of the classical theory described above. Understanding this, as well as many other electronic properties of crystalline solids, requires a more detailed electronic structure theory of solids that takes into account their periodic structure.

[4] The conductivity σ is the inverse of the resistivity—the resistance per unit length of a conductor of unit surface cross-section. Equation (4.68) is a local version of Ohm's law.

4.3.3 Electronic structure of periodic solids: Bloch theory

In order to study the implications of the periodic structure of lattices on the electronic structure of the corresponding solids we consider a single electron Hamiltonian of the form

$$\hat{H} = \hat{T} + \hat{U}(\mathbf{r}) \tag{4.71}$$

where $\hat{T}$ and $\hat{U}$ are respectively kinetic and potential energy operators, and where periodicity enters through

$$U(\mathbf{r}+\mathbf{R}) = U(\mathbf{r}) \tag{4.72}$$

with any lattice vector $\mathbf{R}$, given by Eq. (4.1).

It is convenient to use periodic boundary conditions. For simplicity we consider a cubic lattice, so we take the system to be a rectangular prism with sides $L_1 = N_1a_1$, $L_2 = N_2a_2$, $L_3 = N_3a_3$, that is infinitely reproduced to form infinite space. As in the free particle problem (Section 2.8) this is used just for mathematical convenience, assuming that bulk properties of the system do not depend on the boundary conditions for $L_1, L_2, L_3 \to \infty$.

In the absence of the periodic potential our problem is reduced again to that of a free particle. Eigenfunctions of $\hat{H} = \hat{T}$ that satisfy the periodic boundary conditions are of the form

$$e^{i\mathbf{k}\cdot\mathbf{r}} = e^{ik_1x_1}e^{ik_2x_2}e^{ik_3x_3} \tag{4.73}$$

and the wavevector $\mathbf{k} = (k_1, k_2, k_3)$ needs to satisfy

$$e^{ik_j(x_j+L_j)} = e^{ik_jx_j}, \text{that is } e^{ik_jL_j} = 1; \qquad j = 1,2,3 \tag{4.74}$$

This in turn implies that the allowed values of k_j are[5]

$$k_j = \frac{2\pi}{L_j}n_j = \frac{2\pi}{a_j}\frac{n_j}{N_j}; \quad j = 1,2,3, \quad n_j = 0, \pm 1, \pm 2, \ldots \tag{4.75}$$

These waves satisfy the orthonormality relation:

$$\int_0^{L_1} dx_1 e^{i(2\pi/L_1)n_1x_1} e^{-i(2\pi/L_1)n_1'x_1} = L_1\delta_{n_1,n_1'} \tag{4.76}$$

[5] This is a result for a cubic lattice. The generalization for any lattice is $\mathbf{k} = \sum_{j=1}^{3}(n_j/N_j)\hat{\mathbf{b}}_j$, where $\hat{\mathbf{b}}_j$ ($j = 1,2,3$) are the primitive vectors of the reciprocal lattice.

and in three dimensions

$$\int_{L_1 L_2 L_3} d^3 r e^{i(\mathbf{k}-\mathbf{k}')\cdot\mathbf{r}_i} = \delta_{\mathbf{k},\mathbf{k}'} L_1 L_2 L_3 = \delta_{\mathbf{k},\mathbf{k}'} \Omega \tag{4.77}$$

The presence of the periodic potential $U(\mathbf{r})$ has important consequences with regard to the solutions of the time-independent Schrödinger equation associated with the Hamiltonian (4.71). In particular, a fundamental property of eigenfunctions of such a Hamiltonian is expressed by the *Bloch theorem.*

4.3.3.1 Bloch's theorem

The Bloch theorem states that the eigenfunctions of the Hamiltonian (4.71), (4.72) are products of a wave of the form (4.73) and a function that is periodic on the lattice, that is,

$$\psi_{n\mathbf{k}}(\mathbf{r}) = e^{i\mathbf{k}\cdot\mathbf{r}} u_{n\mathbf{k}}(\mathbf{r}) \tag{4.78}$$

$$u_{n\mathbf{k}}(\mathbf{r}) = u_{n\mathbf{k}}(\mathbf{r} + \mathbf{R}) \tag{4.79}$$

A corollary of Eqs (4.78) and (4.79) is that such functions also satisfy

$$\psi_{n\mathbf{k}}(\mathbf{r} + \mathbf{R}) = e^{i\mathbf{k}\cdot\mathbf{R}} \psi_{n\mathbf{k}}(\mathbf{r}) \tag{4.80}$$

where $\mathbf{R}$ is a lattice vector. For a free particle $u_{n\mathbf{k}}$ is constant and Eqs (4.78)–(4.80) are satisfied for all $\mathbf{R}$. The vector $\mathbf{k}$ and the number(s) n are quantum numbers: $\mathbf{k}$ is associated with the wave property of these functions, while n stands for any quantum number needed to specify the wavefunction beyond the information contained in $\mathbf{k}$.

The proof of Bloch's theorem can be found in any text of solid state physics and will not be reproduced here. In the course of that proof it is shown that the eigenfunctions $\psi_{n\mathbf{k}}$ $(\mathbf{r})$ can be written in the form

$$\psi_{n\mathbf{k}}(\mathbf{r}) = e^{i\mathbf{k}\cdot\mathbf{r}} \sum_{\mathbf{G}} C_{\mathbf{k}-\mathbf{G}}^{(n)} e^{-i\mathbf{G}\cdot\mathbf{r}} \tag{4.81}$$

where $C_{\mathbf{k}}^{(n)}$ are constant coefficients and where the sum is over all vectors $\mathbf{G}$ of the reciprocal lattice. By definition, the function $u_{n\mathbf{k}}(\mathbf{r}) = \sum_{\mathbf{G}} C_{\mathbf{k}-\mathbf{G}}^{(n)} e^{-i\mathbf{G}\cdot\mathbf{r}}$ satisfies $u_{n\mathbf{k}}(\mathbf{r}) = u_{n\mathbf{k}}(\mathbf{r} + \mathbf{R})$, so $\psi_{\mathbf{k}}(\mathbf{r})$ of Eq. (4.81) is indeed a Bloch function.

Several consequences follow immediately:

1. From (4.81) it follows that

$$\psi_{n,\mathbf{k}+\mathbf{G}'}(\mathbf{r}) = e^{i\mathbf{k}\cdot\mathbf{r}} \sum_{\mathbf{G}} C_{\mathbf{k}-(\mathbf{G}-\mathbf{G}')}^{(n)} e^{-i(\mathbf{G}-\mathbf{G}')\cdot\mathbf{r}} = \psi_{n,\mathbf{k}}(\mathbf{r}) \tag{4.82}$$

(The last equality follows from the fact that a sum over $\mathbf{G}$ and over $\mathbf{G} - \mathbf{G}'$ are identical—both cover the whole reciprocal space.) Furthermore, from $\psi_{n,\mathbf{k}+\mathbf{G}} = \psi_{n,\mathbf{k}}$ we find that the eigenenergies also satisfy

$$E_{n,\mathbf{k}+\mathbf{G}} = E_{n,\mathbf{k}} \tag{4.83}$$

that is, both the eigenfunctions ψ and eigenvalues E are periodic in the wavevector $\mathbf{k}$ with the periodicity of the reciprocal lattice.

2. Under the imposed periodic boundary conditions, the wave component $\exp(i\mathbf{kr})$ again has to satisfy $\exp(ik_jL_j) = 1$ ($j = 1, 2, 3$), with the same implications for the possible values of $\mathbf{k}$ as above, that is, Eq. (4.75) for a cubic lattice. Furthermore, Eqs (4.82) and (4.83) imply that for any reciprocal lattice vector $\mathbf{G}$ the wavevectors $\mathbf{k}$ and $\mathbf{k} + \mathbf{G}$ are equivalent. This implies that all different $\mathbf{k}$ vectors can be mapped into a single unit cell, for example the first Brillouin zone, of the reciprocal lattice. In particular, for a one-dimensional lattice, they can be mapped into the range $-(\pi/a)\dots(\pi/a)$. The different values of k are then $k = (2\pi/a)(n/N)$, where $N = L/a$ (L is the length that defines the periodic boundary conditions) is chosen even and where the integer n takes the N different values $n = -(N/2), -(N/2) + 1, \dots, (N/2) - 1$.
3. Even though in similarity to free particle wavefunctions the Bloch wavefunctions are characterized by the wavevector $\mathbf{k}$, and even though Eq. (4.80) is reminiscent of free particle behavior, the functions $\psi_{n\mathbf{k}}(\mathbf{r})$ are not eigenfunctions of the momentum operator. Indeed for the Bloch function (Eqs (4.78) and (4.79)) we have

$$\hat{p}\psi_k = \frac{\hbar}{i}\nabla\psi_k = \frac{\hbar}{i}\nabla(e^{i\mathbf{k}\cdot\mathbf{r}}u_{\mathbf{k}}(\mathbf{r})) = \hbar\mathbf{k}\psi + e^{i\mathbf{k}\cdot\mathbf{r}}\frac{\hbar}{i}\nabla u_{\mathbf{k}}(\mathbf{r}) \tag{4.84}$$

that is, ψ_k is not an eigenfunction of the momentum operator. $\hbar\mathbf{k}$ is sometimes called the *crystal momentum*.

Problem 4.3. Show that in a three-dimensional lattice the number of distinctly different $\mathbf{k}$ vectors is $N_1N_2N_3$. Since these vectors can all be mapped into the first Brillouin zone whose volume is $\mathbf{b}_1 \cdot (\mathbf{b}_2 \times \mathbf{b}_3) = (2\pi)^3/w$ where $w = \mathbf{a}_1 \cdot (\mathbf{a}_2 \times \mathbf{a}_3)$ is the volume of the primitive unit cell of the direct lattice, we can infer that per unit volume of the reciprocal lattice there are $N_1N_2N_3/[(2\pi)^3/w] = wN_1N_2N_3/(2\pi)^3 = \Omega/(2\pi)^3$ states, where $\Omega = L_1L_2L_3$ is the system volume. Show that this implies that the density (in k-space) of allowed k states is $1/(2\pi)^3$ per unit system volume, same result as for free particle.

4.3.4 The one-dimensional tight binding model

This model consists of a row of identical atoms arranged so that their centers lie on a one-dimensional lattice (along the x-axis, say) with lattice-spacing a. Denote the electronic Hamiltonian for atom j by $\hat{h}_j$ and the corresponding atomic orbitals by ϕ_{jn}

$$\hat{h}_j\phi_{jn} = \varepsilon_n^0\phi_{jn} \tag{4.85}$$

where ε_n^0 are the energy levels of an individual atom. By symmetry

$$\phi_{jn}(\mathbf{r}) = \phi_n(x - ja, y, z) \tag{4.86}$$

The Hamiltonian of the full system is $\hat{H} = \Sigma_j\hat{h}_j + \hat{V}$, where $\hat{V}$ is the interatomic interaction. We focus on the low-energy regime where the atomic orbitals are well localized about their corresponding atomic centers, and use this set of electronic states as a basis for the representation of the full problem. When $a \to \infty$, that is, the atoms are infinitely far from each other, $\hat{V} \to 0$ and we have

$$\langle\phi_{jn}|\hat{H}|\phi_{j'n'}\rangle = \varepsilon_n^0\delta_{jj'}\delta_{nn'} \tag{4.87}$$

For finite a both diagonal and non-diagonal elements of $\hat{H}$ change, and in particular $\langle\phi_{jn}|\hat{H}|\phi_{j'n'}\rangle \neq 0$.

Consider then the Hamiltonian matrix in this atomic orbital representation. We denote

$$\langle\phi_{jn}|\hat{H}|\phi_{jn}\rangle = \varepsilon_n \tag{4.88}$$

By symmetry, these diagonal elements do not depend on j. We see that in the sub-matrix of $\hat{H}$ associated with the same atomic level n defined on each atom, all diagonal elements are the same ε_n. Non-diagonal elements of $\hat{H}$ result from the interatomic coupling, and if the atomic centers are not too close to each other these elements will be small relative to the spacings between different ε_n's, that is,

$$\langle\phi_{jn}|\hat{H}|\phi_{j'n'}\rangle \ll |\varepsilon_n - \varepsilon_{n'}|$$

(Note that we take the atomic levels n and n' to be nondegenerate. Degenerate levels have to be included within the same sub-matrix). In this case the existence of the non-diagonal matrix elements of $\hat{H}$ will have an appreciable effect only within the sub-matrices defined above. *Disregarding non-diagonal matrix elements of H outside these blocks constitutes the tight binding model.* Explicitly, we take

$$\langle\phi_{jn}|\hat{H}|\phi_{j'n'}\rangle = 0 \qquad \text{for } j \neq j' \text{ unless } n = n' \tag{4.89}$$

In this case our problem is reduced to diagonalizing each Hamiltonian sub-matrix associated with the same atomic level n (or with a group of degenerate atomic levels) and with the different atomic centers.

Further simplification is achieved by an additional approximation. When the atomic orbitals are closely localized about their corresponding centers it is reasonable to assume that interatomic couplings are appreciable only between nearest neighbors

$$\langle \phi_{jn} | \hat{H} | \phi_{j'n} \rangle = \varepsilon_n \delta_{jj'} + \beta_n \delta_{j,j'\pm 1} \tag{4.90}$$

From now on we drop the index n and denote $\varepsilon_n = \alpha, \beta_n = \beta$. The corresponding Hamiltonian sub-matrix is

$$\hat{H} = \begin{pmatrix} \ddots & \beta & 0 & 0 \\ \beta & \alpha & \beta & 0 \\ 0 & \beta & \alpha & \beta \\ 0 & 0 & \beta & \ddots \end{pmatrix} \tag{4.91}$$

and the Schrödinger equation

$$\begin{pmatrix} \ddots & \beta & 0 & 0 \\ \beta & \alpha - E & \beta & 0 \\ 0 & \beta & \alpha - E & \beta \\ 0 & 0 & \beta & \ddots \end{pmatrix} \begin{pmatrix} \vdots \\ C_j \\ C_{j+1} \\ \vdots \end{pmatrix} = 0 \tag{4.92}$$

will yield the coefficients of the expansion of the eigenfunctions in terms of the atomic orbitals

$$\psi_k(\mathbf{r}) = \sum_j C_{kj} \phi_j(\mathbf{r}) \tag{4.93}$$

and the corresponding eigenvalues E_k. The index k corresponds to the different solutions of (4.92).

Now, Eq. (4.92) is equivalent to the set of coupled equations

$$\beta C_{j-1} + (\alpha - E) C_j + \beta C_{j+1} = 0 \qquad \text{(for all integer } j) \tag{4.94}$$

whose solutions are

$$C_{kj} = (e^{ija})^k = e^{ikx_j} \qquad (x_j = ja \text{ is the position of atomic center } j) \tag{4.95}$$

Inserting (4.95) to (4.94) leads to an equation for the eigenvalue $E(k)$

$$\beta e^{-ika} + (\alpha - E(k)) + \beta e^{ika} = 0 \tag{4.96}$$

which yields

$$E(k) = \alpha + 2\beta \cos ka \tag{4.97}$$

An explicit form for the eigenfunctions is obtained from (4.95), (4.93), and (4.86)

$$\psi_k(\mathbf{r}) = \sum_j e^{ikja}\phi_j(\mathbf{r}) = \sum_j e^{ikja}\phi(x - ja, y, z) \tag{4.98}$$

For any lattice vector $R = la$ (l integer) this function satisfies

$$\begin{aligned}\psi_k(x + R, y, z) &= \sum_j e^{ikja}\phi(x - ja + R, y, z) \\ &= e^{ikR}\sum_j e^{ik(ja-R)}\phi(x - (ja - R), y, z) = e^{ikR}\psi_k(x, y, z)\end{aligned} \tag{4.99}$$

comparing to Eq. (4.80) we see that this is a Bloch function in one dimension. Alternatively, we can rewrite Eq. (4.98) in the form

$$\psi_k(\mathbf{r}) = e^{ikx}u(\mathbf{r}) \tag{4.100}$$

and show that $u(\mathbf{r}) = \sum_i e^{-ik(x-ja)}\phi(x - ja, y, z)$ has the periodicity of the lattice, satisfying the Bloch condition (4.79) on the one-dimensional lattice.

Problem 4.4. Show that $u(\mathbf{r})$ defined above satisfies $u(x + la, y, z) = u(x, y, z)$

Going back to the eigenvalues, Eq. (4.97) three observations can be made. First, when the atoms are far from each other $\beta = 0$ and $E(k) = \alpha$. This is our zero-order solution—all states associated with the same quantum level on the different atomic centers are degenerate. Second, when the coupling β between nearest neighbor atoms is switched on, this degeneracy is lifted. The infinite number of degenerate levels now become a *band* of states spanning a range of energies of width 4β between $\alpha - 2\beta$ and $\alpha + 2\beta$. Finally, as a function of k, $E(k)$ is periodic, with the period $2\pi/a$—a special case of Eq. (4.83).

4.3.5 The nearly free particle model

In the tight binding model we start from electronic states localized on individual atoms and explore the consequence of coupling between these atomic centers. Here our starting point is the free electron, and the periodic lattice potential enters as a small perturbation. Thus, writing

$$H = H_0 + H_1 \qquad \text{with } H_0 = T;\ H_1 = U(\mathbf{r}) = U(\mathbf{r} + \mathbf{R}) \tag{4.101}$$

the free particle model draws its simplicity from the assumption that H_1 is small. This smallness should be measured relative to the energy range considered, that is, if

$$H\psi = E\psi \tag{4.102}$$

we assume that $U \ll E$.

How can we use this to simplify our problem in the present context? Consider one of the eigenfunctions of the unperturbed Hamiltonian H_0

$$\psi_k = e^{i\mathbf{k}\cdot\mathbf{r}}; \qquad E_k = \frac{\hbar^2}{2m}k^2 \tag{4.103}$$

We have found that the perturbation U couples each such eigenfunction to other zero-order wavefunctions according to (cf. Eq. (4.81))

$$e^{i\mathbf{k}\cdot\mathbf{r}} \rightarrow \sum_{G} C_{\mathbf{k}-\mathbf{G}} e^{i(\mathbf{k}-\mathbf{G})\cdot\mathbf{r}} = C_{\mathbf{k}} e^{i\mathbf{k}\cdot\mathbf{r}} + \sum_{\mathbf{G}\neq 0} C_{\mathbf{k}-\mathbf{G}} e^{i(\mathbf{k}-\mathbf{G})\cdot\mathbf{r}} \tag{4.104}$$

Inserting (4.104) into the Schrödinger equation, $\hat{H}\psi = E\psi$ we find that the coefficients $C_{\mathbf{k}}$ are the solutions of

$$\left(\frac{\hbar^2}{2m}k^2 - E\right) C_{\mathbf{k}} + \sum_{\mathbf{G}} U_{\mathbf{G}} C_{\mathbf{k}-\mathbf{G}} = 0 \tag{4.105}$$

where $\mathbf{G}$ belongs to the reciprocal lattice and where

$$U_{\mathbf{G}} = \frac{1}{\Omega_{PC}} \int_{\Omega_{PC}} d^3 r e^{-i\mathbf{G}\cdot\mathbf{r}} U(\mathbf{r}) \qquad \text{(integral over the primitive cell)} \tag{4.106}$$

Here Ω_{PC} is the volume of the primitive cell. Note that we can take $U_0 = 0$ without loss of generality. This just means that we have taken the average lattice potential to be zero, that is, $\int_{\Omega_{PC}} d^3 r U(\mathbf{r}) = 0$. Equation (4.105) represents a set of coupled equations for all the coefficients $C_{\mathbf{k}}$ associated with the original $\mathbf{k}$ and all the $\mathbf{k}'$ derived from it by $\mathbf{k}' = \mathbf{k} - \mathbf{G}$.

Suppose for the moment that only $\mathbf{G} = 0$ and one other reciprocal lattice vector are involved. The coupled equations are

$$\left(\frac{\hbar^2 k^2}{2m} - E\right) C_{\mathbf{k}} + U_{\mathbf{G}} C_{\mathbf{k}-\mathbf{G}} = 0 \tag{4.107}$$

$$\left(\frac{\hbar^2 (\mathbf{k}-\mathbf{G})^2}{2m} - E\right) C_{\mathbf{k}-\mathbf{G}} + U_{\mathbf{G}}^* C_{\mathbf{k}} = 0 \tag{4.108}$$

where we have used $U_{-\mathbf{G}} = U^*_{\mathbf{G}}$. The condition for a nontrivial solution is a secular equation for E that yields

$$E = \frac{1}{2}\left(\varepsilon^0_{\mathbf{k}} + \varepsilon^0_{\mathbf{k}-\mathbf{G}}\right) \pm \left[\left(\frac{\varepsilon^0_{\mathbf{k}} - \varepsilon^0_{\mathbf{k}-\mathbf{G}}}{2}\right)^2 + |U_{\mathbf{G}}|^2\right]^{1/2}; \quad \varepsilon^0_{\mathbf{k}} = \frac{\hbar^2 k^2}{2m} \tag{4.109}$$

If $|U_{\mathbf{G}}|^2$ is much smaller than $(\varepsilon^0_{\mathbf{k}} - \varepsilon^0_{\mathbf{k}-\mathbf{G}})^2$ we can expand in their ratio to get the two solutions

$$E_1 = \varepsilon^0_{\mathbf{k}} + \frac{|U_{\mathbf{G}}|^2}{\varepsilon^0_{\mathbf{k}} - \varepsilon^0_{\mathbf{k}-\mathbf{G}}}; \qquad E_2 = \varepsilon^0_{\mathbf{k}-\mathbf{G}} - \frac{|U_{\mathbf{G}}|^2}{\varepsilon^0_{\mathbf{k}} - \varepsilon^0_{\mathbf{k}-\mathbf{G}}} \tag{4.110}$$

showing small corrections to the free particle energies. In the other extreme limit, if $\varepsilon^0_{\mathbf{k}} = \varepsilon^0_{\mathbf{k}-\mathbf{G}} \equiv \varepsilon^0$ we get

$$E_1 = \varepsilon_0 + U_{\mathbf{G}}; \qquad E_2 = \varepsilon_0 - U_{\mathbf{G}} \tag{4.111}$$

This splitting has the same origin as the splitting that takes place between any two coupled levels that are degenerate in zero order, see, for example, the treatment of Section 2.2. Indeed Eq. (4.109) is the same as Eq. (2.19) with E_a, E_b replaced by $\varepsilon^0_{\mathbf{k}}, \varepsilon^0_{\mathbf{k}-\mathbf{G}}$, and V_{12} replaced by $U_{\mathbf{G}}$. In summary we may say that in the "almost free particle limit" the free particle energies are only slightly modified (Eq. (4.110)) except when $\mathbf{k}$ satisfies for some reciprocal lattice vector $\mathbf{G}$ the equality $\varepsilon^0_{\mathbf{k}} = \varepsilon^0_{\mathbf{k}-\mathbf{G}}$. This condition implies that $k^2 = (\mathbf{k} - \mathbf{G})^2$, that is, $2\mathbf{k} \cdot \mathbf{G} = G^2$ or

$$\mathbf{k} \cdot \hat{\mathbf{G}} = \frac{1}{2}G \tag{4.112}$$

where $\hat{\mathbf{G}} = \mathbf{G}/G$ is a unit vector in the direction of $\mathbf{G}$.

What is the physical meaning of this condition? In one-dimension it implies that $k = \pm(1/2)G$, and since $G = (2\pi/a)n$ (n integer or zero) the smallest k that satisfies this condition is $k = \pm\pi/a$. Since distinct values of k lie in the range $-\pi/a, \ldots, \pi/a$ (the first Brillouin zone) we find that the one-dimensional equivalent to (4.112) is the statement that $\mathbf{k}$ lies at the edge of the Brillouin zone.

Equation (4.112) is a generalization of this statement to three dimensions. The set of equations (4.105) represent, in the weak periodic potential limit, a set of uncoupled waves (i.e. we can practically disregard the second term on the left-hand side of (4.105)) except when (4.112) is satisfied, namely when $\mathbf{k}$ is at the edge of the Brillouin zone. At that point the zero-order energies associated with the waves $\mathbf{k}$ and $\mathbf{k} - \mathbf{G}$ (and just these two waves) are the same, therefore these states are strongly coupled, leading to the energy splitting given by Eq. (4.111).

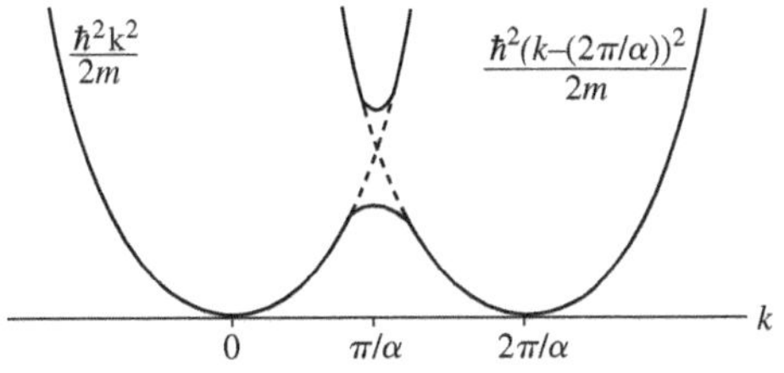

FIG. 4.2 A graphic display of the origin of band structure in the nearly free electron model.

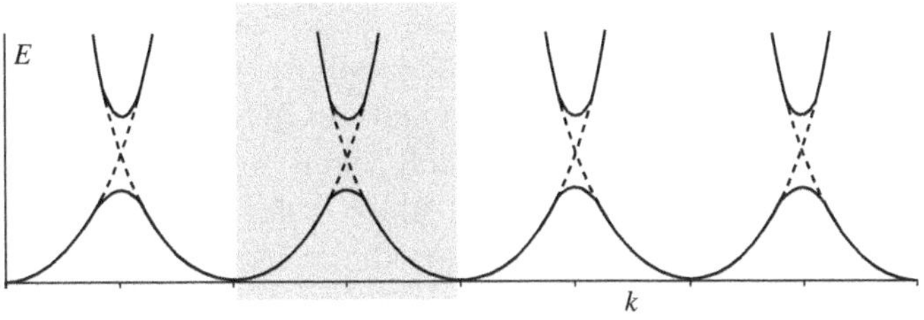

FIG. 4.3 The origin of band structure in the nearly free electron model: An extended picture that shows also the origin of the k-periodicity of $E(k)$.

A graphical interpretation of this situation is shown in Fig. 4.2. The energy as a function of k is shown for two waves whose origins in k-space differ by $G = 2\pi/a$. Each energy curve corresponds to the free particle parabola. Their coupling by the periodic potential does not appreciably change them except in the neighborhood of $k = \pi/a$. The effective strong coupling in this neighborhood leads to the splitting shown.

An extended picture of this situation is depicted in Fig. 4.3. Here we show the parabolas representing the free particle energies associated with each of the k vectors (in one-dimension) that are coupled to each other according to Eq. (4.105), that is, k, $k \pm 2\pi/a$, $k \pm 4\pi/a, \ldots$. At each point where two parabolas cross, the energy spacing between the two zero-order energies is small relative to the amplitude of the periodic potential. This leads to splitting and to the band structure in the energy spectrum. Also, the emergence of the reciprocal lattice periodicity of $E(k)$ is clearly seen. Again we note that a reduced picture may be obtained by focusing on the first Brillouin zone (marked in Fig. 4.3 as the shaded area)—the equivalent zones in k-space represent physically equivalent descriptions.

4.3.6 Intermediate summary: Free electrons versus noninteracting electrons in a periodic potential

Let us summarize the important differences between the free electron model of a metal and models based on the electronic band structure as discussed above. The first disregards, while the second takes into account, the periodic lattice potential on which the electrons move. The electron spin will not be an issue; we just keep in mind that a factor of 2 in the single electron density of state arises for the multiplicity of spin states.

1. In the free electron model the electronic wavefunctions are characterized by a wavevector $\mathbf{k}$ $(-\infty < k_1, k_2, k_3 < \infty)$ and can be chosen to be also eigenfunctions of the momentum operator with eigenvalues $\hbar\mathbf{k}$. The eigenstates of an electron moving in a periodic potential are also characterized by a wavevector $\mathbf{k}$ (the crystal momentum) whose independent values lie within a single primitive cell of the reciprocal lattice. Another quantum number n take discrete integer values and distinguishes between different bands.
2. The eigenfunctions of the free particle Hamiltonian can be written as free waves, $\psi_{\mathbf{k}}(\mathbf{r}) = \Omega^{-1/2}\exp(i\mathbf{k}\cdot\mathbf{r})$. Bloch states have the form $\psi_{n,\mathbf{k}}(\mathbf{r}) = e^{i\mathbf{k}\cdot\mathbf{r}}u_{n,\mathbf{k}}(\mathbf{r})$ where u has the lattice periodicity, that is, $u_{n,\mathbf{k}}(\mathbf{r}+\mathbf{R}) = u_{n,\mathbf{k}}(\mathbf{r})$ where $\mathbf{R}$ is any lattice vector.
3. The energy eigenvalues of the free particle are $E(k) = \hbar^2k^2/(2m)$ where m is the particle mass. As such, the energy is a continuous variable that can take values in the interval $(0, \infty)$. The energy eigenvalues that correspond to Bloch states satisfy $E_n(\mathbf{k}+\mathbf{G}) = E_n(\mathbf{k})$ and, as seen in Sections 4.3.4 and 4.3.5, are arranged in bands separated by forbidden energy gaps.
4. For free particles, $\hbar\mathbf{k}$ is the momentum and the corresponding velocity is $\hbar\mathbf{k}/m = \hbar^{-1}\nabla E(k)$. These momentum and speed change under the operation of an external force $\mathbf{F}_{\text{external}}$. It may be shown that as long as this force does not change too fast in space and time, a classical-like equation of motion

$$\hbar\dot{\mathbf{k}} = \mathbf{F}_{\text{ext}} \tag{4.113}$$

holds.[6] Indeed, we have used equivalent expressions, for example, (4.66), in the analysis of Section 4.3.2. As pointed out above, the "crystal momentum" $\hbar\mathbf{k}$ is not really an eigenvalue of the electron's momentum operator. Still, under certain conditions it is possible to show that the function

$$\mathbf{v}_n(\mathbf{k}) = \hbar^{-1}\nabla E_n(\mathbf{k}) \tag{4.114}$$

still represents the speed of the electron in the state $(n, \mathbf{k})$, that is, in a given band and with a given crystal momentum. Furthermore, Eq. (4.113) for the rate of change of the crystal momentum remains approximately valid under

[6] The validity of Eq. (4.113) is a nontrivial issue that should be examined carefully under any given conditions.

these conditions. Equations (4.113) and (4.114) constitute the basis to what is known as the semiclassical model of electron dynamics.

Problem 4.5. Show that if the external force $\mathbf{F}_{\text{ext}}$ is derived from a time-independent potential, $\mathbf{F}_{\text{ext}} = -\nabla U_{\text{ext}}(\mathbf{r})$, and if the total energy of an electron in Bloch state (n, k) that moves in this external potential is taken as

$$E_{\text{tot}}(\mathbf{r}) = E_n(k) + U_{\text{ext}}(\mathbf{r}) \tag{4.115}$$

than Eq. (4.113) follows from (4.114) and the requirement that energy is conserved.

Solution: Conservation of $E_{\text{tot}}(\mathbf{r})$ during the electron motion implies

$$0 = \frac{dE_{\text{tot}}(\mathbf{r})}{dt} = \dot{\mathbf{k}} \cdot \nabla_{\mathbf{k}} E_n(\mathbf{k}) + \dot{\mathbf{r}} \cdot \nabla_{\mathbf{r}} U_{\text{ext}}(\mathbf{r}) \tag{4.116}$$

We have used subscripts **k** and **r** to distinguish between the corresponding gradient. Using (4.114) in the form $\dot{\mathbf{r}} = \hbar^{-1}\nabla_{\mathbf{k}} E_n(\mathbf{k})$ we get $\hbar\dot{\mathbf{k}} = -\nabla_{\mathbf{r}} U_{\text{ext}}(\mathbf{r}) = \mathbf{F}_{\text{ext}}$.

4.3.7 Further dynamical implications of the electronic band structure of solids

An immediate and most important consequence of the band structure of crystalline solids is the distinction between metals and nonmetals that reflects the position of the Fermi energy vis-à-vis the band energy. Before addressing this issue it is important to consider the energy scales involved. The following points are relevant for this consideration:

1. In atoms and molecules the characteristic electronic energy scale is typically a few electronvolts. This is the order of energy spacing between the lower electronic energies of atoms and molecules. It is also the order of interatomic coupling (e.g. interaction of electrons on one atom with the nucleus of its nearest neighbor, that is, the β parameter in (4.91)) and of the Fermi energy calculated from (4.61). We thus expect the bandwidths and band gaps to be of the same order of up to a few electron volts.
2. These characteristic energies scales are larger by about two orders of magnitude than another important energy scale—the thermal energy. Indeed, at $T = 300$ K we have $1\ \text{eV}/(k_B T) = 38.7$.

These two observations imply that most electrons in a solid cannot contribute to dynamical processes that require energy exchange of magnitude $k_B T$ or less—the argument is similar to that made in Section 4.3.1. One can furthermore assert that filled bands, that is, band for which all states lie below the Fermi energy do not contribute to the electric or thermal conduction of a solid. To see this note first that when a small electric field or a small temperature gradient is applied, the energies involved are not sufficient to add or remove electrons to/from the band. The energy and electrical fluxes associated with a filled band are then given by

$$\mathbf{j} = -\frac{e}{\Omega}\int d\mathbf{k}\rho(\mathbf{k})\mathbf{v}(\mathbf{k}) \tag{4.117}$$

and

$$\mathbf{j}_E = \frac{1}{\Omega}\int d\mathbf{k}\rho(\mathbf{k})E(\mathbf{k})\mathbf{v}(\mathbf{k}) \tag{4.118}$$

where $\rho(\mathbf{k}) = 2 \times \Omega/(2\pi)^3$ is the density of states per unit volume of k-space (the factor 2 comes from the spin multiplicity) and Ω is the system volume. Using Eq. (4.114) for the electron speed, these expressions become

$$\mathbf{j} = -\frac{e}{\hbar}\int \frac{d\mathbf{k}}{4\pi^3}\nabla E(\mathbf{k}) \tag{4.119}$$

and

$$\mathbf{j}_E = \frac{1}{2\hbar}\int \frac{d\mathbf{k}}{4\pi^3}\nabla E^2(\mathbf{k}) \tag{4.120}$$

These integrals are done over the volume of a primitive cell of the reciprocal lattice. Using a theorem (see Section 1.1.3) that states that the integral over a period of the gradient of a periodic function is zero, we find that both $\mathbf{j}$ and $\mathbf{j}_E$ vanish.

Thus, we have found that filled bands do not contribute to the charge and energy transport properties of solids. Empty bands obviously do not contribute either. We may conclude that solids in which all bands are either full or empty are insulators. In this case the Fermi energy, or more generally the electronic chemical potential, is located in the gap, far (relative to $k_B T$) from the nearest bands above and below it, so that all lower bands are fully occupied and all upper ones are empty.

In the other extreme case the Fermi energy is found in the interior of a band and we are dealing with a metal. As long as it is far (relative to $k_B T$) from the band edges, the situation is not much different from that described by the free electron model discussed in Sections 4.3.1 and 4.3.2, and this model provides a reasonable simple approximation for the transport properties.

In the interesting case where the Fermi energy is in the gap but its distance from the nearest band is not very large, this band may be thermally populated. This leads to a characteristic temperature dependence of the density of mobile charge carriers and the associated transport properties. Such solids are called semiconductors, and are discussed next.

4.3.8 Semiconductors

When $T = 0$ semiconductors are insulator in which the gap between the highest filled band (henceforth referred to as *valence band*) and the lowest empty one (referred to as *conduction band*) is relatively small. At elevated temperatures that are still lower than the melting point enough electrons are transferred from the valence to the conduction band and form a population of mobile charges that contributes to electronic transport. Alternatively, the source of electrons in the conduction band and/or their deficiency in the valence band can be the result of electron transfer to/from impurity species as discussed below.

Figure 4.4 displays a schematic electronic structure showing the valence and conduction bands and the gap between them. The two bands are arranged in a way that is reminiscent of what was seen in the nearly free electron model, Fig. 4.3, except that in general the minimum conduction band energy and the maximum valence band energy are not necessarily aligned vertically above each other.[7] Real band structure diagrams are far more complex both because different bands can overlap in energy and because in the three-dimensional k-space $E(\mathbf{k})$ can behave differently in different k directions. Still, this simple picture suffices for conveying some fundamental issues:

1. Semiconductors are low bandgap insulators. "Low" is defined qualitatively, so that an appreciable density of electrons can be thermally excited into the conduction band at temperatures that are technologically relevant. In silicon, a large gap semiconductor ($E_g = 1.12$ eV; $\exp(-E_g/k_BT) \sim 1.6 \times 10^{-19}$ at 300 K), this density is very small at room temperature. Germanium ($E_g = 0.67$) and indium-antimonide (InSb, $E_g = 0.16$ eV; $\exp(-E_g/k_BT) \sim 2 \times 10^{-3}$ at 300 K) are examples of lower gap semiconductors. For comparison, in diamond $E_g = 5.5$ eV.

2. When electrons are excited, thermally or optically to the bottom of the conduction band they behave essentially as free mobile charge carriers. Indeed, we may expand the conduction band energy $E_c(\mathbf{k})$ about the bottom, at $\mathbf{k} = \mathbf{k}_c$, of the

[7] This observation is experimentally significant. It can be shown that photoinduced electronic excitation from the valence to the conduction band obeys a selection rule by which the $\mathbf{k}$ vector remains essentially unchanged. When the minimum valence band energy and the maximum conduction band energy are aligned exactly above each other in this diagram, the minimum absorption energy determines the band gap. Otherwise, when the minima and maxima occur at different points in k-space, the minimum absorption energy is larger than the band gap. In the semiconductor literature these processes are referred to as direct transitions and indirect transitions, respectively.

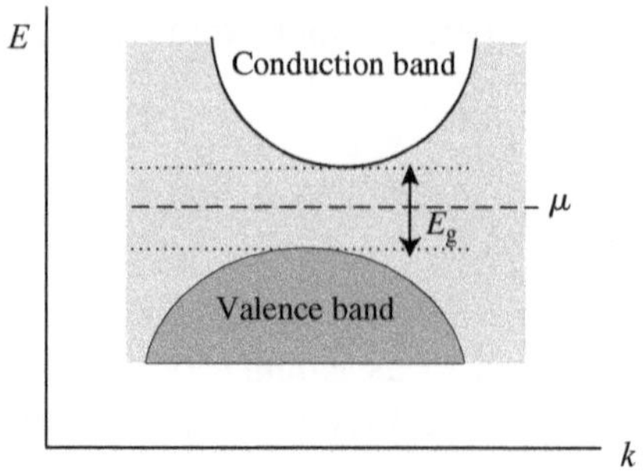

FIG. 4.4 A portion of a schematic band structure diagram showing the energy as a function of k in a particular direction of k-space, for the valence and conduction bands. The minimum energy difference E_g is the band gap. μ is the electron chemical potential.

conduction band in the form

$$E_c(\mathbf{k}) = E_c + \frac{\hbar^2(\mathbf{k} - \mathbf{k}_c)^2}{2m_c} \tag{4.121}$$

m_c is defined from this expansion.[8] The electron near the bottom of the conduction band may be regarded as a free particle of mass m_c. We refer to this mass parameter as the *effective mass* of the conduction band electron.

3. As discussed in Section 4.3.2, mobile charge carriers move as free particles between scattering events. The conductivity σ, Eq. (4.69), depends on their density n and on the relaxation time τ. In metals n does not depend on temperature while τ decreases with increasing T because it is partly determined by electron–phonon scattering that increases at higher temperatures. Therefore, metallic conduction decreases with increasing T. In semiconductors, the strong exponential temperature dependence of the density n of mobile charge carriers dominates the temperature dependence of the conductivity, which therefore increases with temperature.

4. The above discussion pertains to conduction by electrons in the conduction band without addressing their source, and would remain the same also if these electrons are injected into the conduction band from the outside. It should be intuitively clear that if, instead, we remove electrons from the valence band the resulting "electron vacancies" or "holes" contribute to the conduction in a similar way: Electrons

[8] Equation (4.121) is a simplified form. The general expansion takes the form

$$E_c(\mathbf{k}) = E_c + (1/2)\hbar^2 \sum_{i,j=1}^{3} (k_i - k_{ci})(\mathbf{m}_c^{-1})_{i,j}(k_j - k_{cj})$$

and defines the *effective mass tensor* $\mathbf{m}$.

move to fill the vacancies, which amounts to an effective motion of positive holes in the opposite direction. The rigorous formulation of this statement rests on two points:

a) Writing Eq. (4.117) in the form

$$\mathbf{j} = -e \int_{\substack{\text{occupied} \\ \text{band states}}} \frac{d\mathbf{k}}{4\pi^3}\mathbf{v}(\mathbf{k}); \qquad \mathbf{v}(\mathbf{k}) = \hbar^{-1}\nabla E(\mathbf{k}) \tag{4.122}$$

and using the fact that when the integral in (4.122) is carried over all states in the band the result vanishes, imply that the current density $\mathbf{j}$ is also given by

$$\mathbf{j} = e \int_{\substack{\text{unoccupied} \\ \text{band states}}} \frac{d\mathbf{k}}{4\pi^3}\mathbf{v}\,(\mathbf{k}) \tag{4.123}$$

Rather than looking at the occupied states in the almost filled valence band we can focus on the few empty states. The current density is given according to (4.123) as an integral over these unoccupied states, or "states occupied by holes," and its form is the same as (4.122) except with positive, rather than negative, charge.

b) The dispersion relationship for these states, near the top of the valence band, is obtained by expanding the band energy $E_V(\mathbf{k})$ about its maximum at $\mathbf{k} = \mathbf{k}_V$, leading to an equation similar to (4.121)

$$E_V(\mathbf{k}) = E_V - \frac{\hbar^2(\mathbf{k} - \mathbf{k}_V)^2}{2m_V} \tag{4.124}$$

(again a more general expression should usually be used[8]). The important thing to note is that since we now expand near the band maximum, the curvature of the dispersion curve is negative, that is, the particle behaves as if its effective mass is $-m_V$. This means that an external force in a given direction should induce motion in the opposite direction ($\hbar\dot{\mathbf{k}} = \mathbf{F}_{\text{ext}}$ and $\dot{\mathbf{v}} = -(1/m_V)\hbar\dot{\mathbf{k}}$). Equivalently, since the forces relevant to the problem are derived from the interaction of the electrons with electrostatic or electromagnetic fields, they are proportional to the particle charge. The resulting hole acceleration can therefore be taken to correspond to a particle of positive mass m_V but with a charge of opposite, that is, positive, sign. Referring to Eq. (4.123) we may conclude that the motion of holes, that is, the acceleration resulting from the action of an external field, reflect particles carrying positive charge. An additional factor e is needed to get the electric current so this current is proportional to e^2 as was already asserted in discussing Eq. (4.69).

We may conclude that electron and holes contribute additively to the observed conductivity of the semiconductor, $\sigma = \sigma_e + \sigma_h$. The contributions σ_e of the electron density in the conduction band and σ_v of holes in the valence band can be approximately assessed from Eq. (4.69),

$$\sigma_e = \frac{n_c e^2 \tau_c}{m_c}; \qquad \sigma_h = \frac{p_v e^2 \tau_v}{m_v} \tag{4.125}$$

where the densities of electron and holes in conduction and valence bands are denoted n_c and p_v, respectively.

5. In the picture portrayed so far, the existence of electrons in the conduction band must result from thermal excitation of these electrons out of the valence band, hence

$$n_c = p_v \qquad \text{(intrinsic semiconductors)} \tag{4.126}$$

This always holds when the semiconductor is clean, without any added impurities. Such semiconductors are called *intrinsic*. The balance (4.126) can be changed by adding impurities that can selectively ionize to release electrons into the conduction band or holes into the valence band. Consider, for example, an arsenic impurity (with five valence electrons) in germanium (four valence electrons). The arsenic impurity acts as an electron donor and tends to release an electron into the system conduction band. Similarly, a gallium impurity (three valence electrons) acts as an acceptor, and tends to take an electron out of the valence band. The overall system remains neutral, however now $n_c \neq p_v$ and the difference is balanced by the immobile ionized impurity centers that are randomly distributed in the system. We refer to the resulting systems as *doped* or *extrinsic semiconductors* and to the added impurities as *dopants*. Extrinsic semiconductors with excess electrons are called *n-type*. In these systems the negatively charged electrons constitute the *majority carrier*. Semiconductors in which holes are the majority carriers are called *p-type*.

6. The main variables that determine the transport and screening (see below) of both intrinsic and extrinsic semiconductors are the mobile carrier densities n_c and p_v. Given the energetic information, that is, the electronic band structure, and the dopant concentrations, these densities can be evaluated from equilibrium statistical mechanics. For example, the density of electrons in the conduction band is

$$n_c = \frac{1}{\Omega} \int_{E_c}^{\infty} dE \rho_c(E) \frac{1}{e^{\beta(E-\mu)} + 1} \tag{4.127}$$

where Ω is the system volume and $\rho_c(E)$ is the density of single electron states in the conduction band. The determination of the chemical potential μ is discussed below. In what follows we will denote by $\bar{\rho} = \rho/\Omega$ the density of states for unit volume.

In the *effective mass approximation* we assume that the expansion (4.121) is valid in the energy range (near the conduction band edge) for which the integrand in (4.127) is appreciable. In this case we can use for $\rho_c(E)$ the free particle expression (cf. Eq. (4.59))

$$\rho_c(E) = \frac{\Omega}{\pi^2}\frac{m_c}{\hbar^3}\sqrt{2m_c(E-E_c)} \equiv \Omega\bar{\rho}_c(E) \tag{4.128}$$

Similarly

$$p_v = \int_{-\infty}^{E_v} dE\bar{\rho}_v(E)(1-\frac{1}{e^{\beta(E-\mu)}+1}) = \int_{-\infty}^{E_v} dE\bar{\rho}_v(E)\frac{1}{e^{\beta(\mu-E)}+1} \tag{4.129}$$

where, again in the effective mass approximation, the hole density of states is given by an equation like (4.128) with m_v and $|E-E_v|$ replacing m_c and $E-E_c$. Note that the function $f_h(E) = [\exp(\beta(\mu-E))+1]^{-1}$ that appears in (4.129) can be thought of as the average hole occupation of a level at energy E.

We have seen that for most room temperature semiconductors $E_g \gg k_BT$. Simpler expressions may be obtained in the often encountered situation when the inequalities

$$E_c-\mu \gg k_BT; \qquad \mu-E_v \gg k_BT \tag{4.130}$$

are also satisfied. In this case we can simplify the occupation factors according to

$$\frac{1}{e^{\beta(E-\mu)}+1} \approx e^{-\beta(E-\mu)}; \qquad E > E_c \text{ (conduction band)} \tag{4.131a}$$

$$\frac{1}{e^{\beta(\mu-E)}+1} \approx e^{-\beta(\mu-E)}; \qquad E < E_v \text{(valence band)} \tag{4.131b}$$

In this case Eqs (4.127) and (4.128) take the simpler forms

$$n_c(T) = N_c(T)e^{-\beta(E_c-\mu)} \tag{4.132a}$$

$$p_v(T) = P_v(T)e^{-\beta(\mu-E_v)} \tag{4.132b}$$

where

$$N_c(T) = \int_{E_c}^{\infty} dE\bar{\rho}_c(E)e^{-\beta(E-E_c)} \tag{4.133a}$$

$$P_v(T) = \int_{-\infty}^{E_v} dE\bar{\rho}_v(E)e^{-\beta(E_v-E)} \tag{4.133b}$$

Problem 4.6.

1. Show that under the approximation that leads to (4.132) we can write

$$n_c p_v = N_c P_v e^{-\beta E_g} \tag{4.134}$$

2. Show that in the effective mass approximation

$$N_c(T) = \frac{1}{4}\left(\frac{2m_c k_B T}{\pi \hbar^2}\right)^{3/2} \tag{4.135a}$$

$$P_v(T) = \frac{1}{4}\left(\frac{2m_v k_B T}{\pi \hbar^2}\right)^{3/2} \tag{4.135b}$$

Using expressions (4.135) with the free electron mass replacing m_c or m_v yields 2.5×10^{19} cm^{-3} at $T = 300$ K for these parameters.

The only yet unknown quantity in Eqs (4.132) is the chemical potential μ. It can be determined from the condition of local charge neutrality, which for intrinsic semiconductors is simply $n_c = p_v$.

Problem 4.7. Show that for intrinsic semiconductors, assuming the validity of (4.132),

$$\mu = \frac{1}{2}\left[(E_v + E_c) + k_B T \ln\left(\frac{P_v}{N_c}\right)\right] \tag{4.136}$$

In the extrinsic case, the expression of overall charge neutrality should take into account the existence of immobile positive centers of donors that lost electrons and/or negative centers of acceptors that gained electrons. Also, the validity of the approximations (4.131) may some times become questionable. We will not dwell on the details of these calculations but it should be clear that they have now been reduced to merely technical issues.

4.4 The work function

Chapter 17 of this text focuses on the interface between molecular systems and metals or semiconductors and in particular on electron exchange processes at such interfaces. Electron injection or removal processes into/from metals and semiconductors underline many other important phenomena such as contact potentials (the potential gradient formed at the contact between two different metals), thermionic emission (electron ejection out of hot metals), and the photoelectric effect (electron emission induced by photon absorption). Two energetic quantities are central to the understanding of these phenomena: The electron chemical potential and the work function.

Let us start from individual atoms. The minimum energy required to remove an electron from a given atom is the atomic *ionization potential*, IP. The energy released upon inserting an electron to the atom is the *electron affinity*, EA, of that atom. (A negative electron affinity implies that energy is required to insert the electron.) For a given atom IP $\neq$ EA because different electronic energy levels of the atom are involved in the two processes: An electron is removed from the highest occupied atomic orbital and is inserted to the lowest unoccupied one. Obviously, the electron affinity of a given atom is equal to the ionization potential of the corresponding negative ion.

Things are somewhat more complicated already with molecules. While the concepts of ionization potential and electron affinity remain the same, the underlying nuclear motion can affect the observed energies. Two issues are at play: First, the equilibrium nuclear configuration of a molecule is usually different from that of the corresponding molecular ions, and second, that the timescale for nuclear motions is much slower than that which characterizes the electronic process. For this reason, what is usually observed is the sudden, or *vertical*, energy to remove the electron, which is larger than the actual, so called *adiabatic*, ionization potential. Figure 4.5 depicts the difference between these quantities.

A macroscopic solid can be regarded as a very large molecule, and the situation pictured above remains in principle the same. Some differences however should be noted:

1. In metals, the ionization potential and the electron affinity are the same, and are given by the electron chemical potential (or the Fermi energy at $T = 0$) measured with respect to the vacuum energy.[9] To be specific we write, for $T = 0$,

$$(\mathrm{IP})_{\mathrm{metal}} = -E_{\mathrm{F}} \tag{4.137}$$

where the vacuum energy is taken as the energy origin. In a zero-temperature semiconductor the ionization potential is the difference between vacuum energy and the top of the valence band, while the electron affinity is the corresponding difference between vacuum and the bottom of the conduction band. This implies that

$$\mathrm{IP} - \mathrm{EA} = E_g$$

[9] Unless otherwise stated, the term "vacuum energy" is taken to refer to the ground state energy of a single free electron in infinite space.

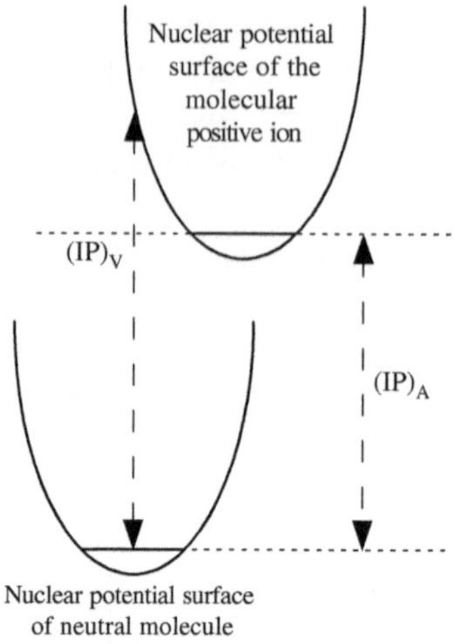

FIG. 4.5 A comparison between the vertical and the adiabatic molecular ionization potentials. The parabolas represent the nuclear potential surfaces of the molecule and the molecular ion. The horizontal shifts correspond to the different equilibrium nuclear configurations of these species. Electronic energies are measured from the corresponding ground vibrational levels. $(IP)_V$ and $(IP)_A$ are the vertical and adiabatic ionization potentials, respectively.

2. The nuclear relaxation energy (the difference between the vertical and adiabatic ionization potentials) is expected to be negligible for metals: The electronic states involved in losing or gaining an electron by the metal are delocalized and the effect on the nuclear configuration of removing or adding a single electron to the system is therefore negligible.

3. The energy needed to remove an electron from the interior of a metal to vacuum at infinity is given by (4.137). However, in practical measurements, the probe that determines electron exit from the molecule (in, say, photoemission or thermionic emission experiments) is located at distances from the metal surface that are small relative to the metal size. At such distances the measured workfunction (as determined, for example, from the photocurrent energy threshold in a photoemission experiment) depends on the excess charge density on the metal surface. Such excess charge results from the fact that the metal surface provides a different local environment for the metal electrons than the bulk, therefore if electrons were distributed homogeneously in all parts of the metal including its surface, the local electron chemical potential at the surface would be different then in the bulk. This leads to a redistribution of the electron density and to excess (positive or negative) surface charge. In this case the workfunction is given by

$$W = -E_F + W_s \tag{4.138}$$

where W_s is the additional work associated with this surface charge. This additional contribution to the electron removal energy can be in the order of 5–10% of the workfunction, and usually depends on the particular metal surface involved in the experiment. In qualitative estimates we often disregard this issue and use quoted experimental values of the workfunction as measures of the true ionization potential (or Fermi energy) of the metal.

4.5 Surface potential and screening

4.5.1 General considerations

Metals, semiconductors, electrolyte solutions, and molten salts have in common the fact that they contain given or variable densities of mobile charge carriers. These carriers move to screen externally imposed or internal electrostatic fields, thus substantially affecting the physics and chemistry of such systems. The Debye–Huckel theory of screening of an ionic charge in an electrolyte solution is an example familiar to many readers.

When two such phases come into contact, charge may be transferred between them, creating a potential difference between the two phases. This is already observed when two different metals come into contact. At equilibrium we should be able to move an electron through this contact without energy cost. However, if the work to extract an electron from one metal is its work function W_1, and the work gained by inserting the electron to the other metal is the second work function W_2, then energy conservation implies that there must be an interfacial electric field that does work $W_1 - W_2$ on the electron, that is, a potential difference between the two metal faces (called *contact potential*) given by

$$-e\Delta\Phi = W_1 - W_2 \tag{4.139}$$

A potential difference may be also imposed externally. One may expect intuitively that far enough from the interface the system exhibits the properties of a pure homogeneous system with no potential gradients (this statement is a rephrasing of the familiar principle that the electrostatic field must vanish in a homogeneous system containing mobile charge carriers). Therefore, the potential change (the terms "potential distribution" or "potential fall" are often used) must take place near the interface. The following example demonstrates the importance of knowing the way the potential is distributed across such interfaces: We consider a molecule seated near a semiconductor surface (Fig. 4.6). The molecule is characterized by its highest occupied molecular orbital (HOMO) and lowest unoccupied molecular orbital (LUMO), and the semiconductor is characterized by its valence and conduction bands, VB and CB, and their edges, E_v and E_c, respectively. Suppose we are interested in the possibility to transfer an electron from the molecule to the semiconductor following an optical excitation that transfers an electron from the HOMO to the LUMO molecular level. When the energy relationships are as shown

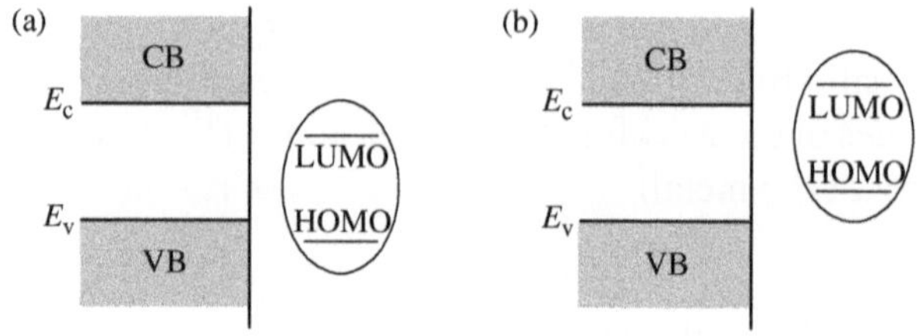

FIG. 4.6 A molecule (represented by its HOMO and LUMO levels) next to a semiconductor surface characterized by its conduction and valence band edges, E_C and E_V. Following excitation that populates the HOMO, electron transfer into the conduction band of the semiconductor can take place when the alignment of molecular and semiconductor levels are as in (b), but not (a) (see also Fig. 4.8).

in panel (a) electron transfer is energetically forbidden because the LUMO is positioned next to the semiconductor band gap with no available levels to accept the electron. Electron transfer could take place after populating the LUMO if E_{lumo} was higher than E_c so the LUMO is energetically degenerate with empty conduction band states in the semiconductor. This could happen if a potential bias is imposed between the molecule and the semiconductor, so that the molecule side is at negative potential bias relative to the semiconductor surface as shown in panel (b).

Now, a potential bias can be practically imposed only between the *interiors* of the semiconductor and the molecular phases. The implications for the process under discussion are related to the way this bias is reflected in the potential fall at the semiconductor-molecule interface. This is the issue under consideration. Before addressing this issue we need to understand how an electrostatic potential is distributed in each phase separately.

4.5.2 The Thomas–Fermi theory of screening by metallic electrons

It should be appreciated that in contrast to the simple free electron models used in much of our discussion of metals and semiconductors, a treatment of screening necessarily involves taking into account, on some level, the interaction between charge carriers. In the Thomas–Fermi theory this is done by combining a semiclassical approximation for the response of the electron density to an external potential with a mean field approximation on the Hartree level—assuming that each electron is moving in the mean electrostatic potential of the other electrons.

Consider a semi-infinite metal represented by the gray area in Fig. 4.7. The homogeneous bulk metal is taken to be locally neutral, the electronic charge is compensated by the positive background and the potential is constant. Near impurities or at the surface this is not necessarily so. Suppose that the potential is given to be Φ_S on the metal surface and Φ_B in its interior and consider the potential

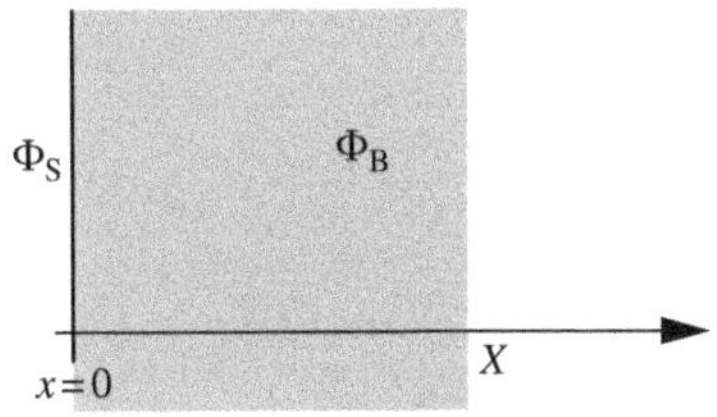

FIG. 4.7 A semi-infinite metal with a surface potential $\Phi(x = 0) = \Phi_S$ and bulk potential $\Phi(x \to \infty) = \Phi_B$: A model for discussing metallic screening.

distribution in between these regions and what is the charge associated with this distribution.

The system is infinite in the y and z directions, so our problem is one-dimensional. In the absence of potential bias we have free electrons that occupy eigenstates of the kinetic energy operator up to the Fermi energy. The density of states per unit volume is (cf. Eq. (4.59))

$$\bar{\rho}(E) = \frac{1}{\pi^2}\frac{\sqrt{2}m^{3/2}}{\hbar^3}\sqrt{E} \tag{4.140}$$

In the presence of an external potential we use a semiclassical argument as in (4.115), by which the electronic states remain the free wave eigenstates of the kinetic energy operator associated with eigenvalues E_K, however the corresponding electronic energies become position-dependent according to

$$E(x) = E_K - e(\Phi(x) - \Phi_B) = E_K - e\delta\Phi(x) \tag{4.141}$$

The Fermi energy is the same everywhere, however (4.141) implies that the ground state energy becomes position-dependent. Equivalently, we may regard the zero energy as uniformly fixed everywhere in the system but the Fermi energy becoming position-dependent

$$E_F \to E_F + e\delta\Phi(x) \tag{4.142}$$

The excess density of electrons at position x is therefore given by

$$\delta n(x) = \int_0^{E_F+e\delta\Phi(x)} dE\bar{\rho}(E) - \int_0^{E_F} dE\bar{\rho}(E) \tag{4.143}$$

which, using (4.140) yields

$$\delta n(x) = \frac{(2m)^{3/2}}{3\pi^2\hbar^3}[(E_F + e\delta\Phi)^{3/2} - E_F^{3/2}]$$
$$\xrightarrow{e\Phi\ll E_F} \frac{(2m)^{3/2}e\sqrt{E_F}}{2\pi^2\hbar^3}\delta\Phi(x) \tag{4.144}$$

The excess charge density associated with the surface potential is therefore

$$\delta\rho_q(x) = -e\delta n(x) = -\frac{(2m)^{3/2}e^2\sqrt{E_F}}{2\pi^2\hbar^3}\delta\Phi(x) \tag{4.145}$$

Using the one-dimensional version of the Poisson equation (1.216), $\partial^2\delta\Phi/\partial x^2 = -4\pi\delta\rho_q$, this yields an equation for the potential distribution

$$\frac{\partial^2\delta\Phi}{\partial x^2} = \kappa_{TF}^2\delta\Phi \tag{4.146}$$

where κ_{TF} is the inverse *Thomas Fermi screening length*

$$k_{TF} = \frac{4\pi e}{(2\pi\hbar)^{3/2}}(2m)^{3/4}E_F^{1/4} \tag{4.147}$$

The general solution of (4.146) is $\delta\Phi(x) = A\exp(k_{TF}x) + B\exp(-k_{TF}x)$ and using $\delta\Phi(x=0) = \Phi_S - \Phi_B$ and $\delta\Phi(x\to\infty) = 0$ leads to the final solution in the form

$$\Phi(x) = \Phi_B + (\Phi_S - \Phi_B)e^{-\kappa_{TF}x} \tag{4.148}$$

The Thomas–Fermi length, k_{TF}^{-1} characterizes screening by metallic electrons: Given the potential on the metal surface, the potential inside the metal approaches its bulk value within this length scale. Using the electron charge and mass together with a value for E_F in the range of, say, 5 eV, yields ~0.6 Å for this length. The metal is seen to screen efficiently any potential imposed on its surface: The interior of the metal does not see the surface potential beyond a "skin depth" of the order of ~1 Å.

4.5.3 Semiconductor interfaces

Contacts between semiconductors on one side, and metals, electrolyte solutions, and other semiconductors are pervasive in today's technology. Contacts between semiconductors and various types of molecular environments are increasingly found in advanced application such as organic light-emitting diodes. Understanding the electrical properties of semiconductor interfaces starts again with the relatively simple example of Fig. 4.7: Taking the gray area in this figure to be a semiconductor of known properties (band structure, doping, dielectric constant, and temperature) and given a difference between the surface electrostatic potential Φ_S and the bulk potential Φ_B, how is the potential distributed at the semiconductor interface (here assumed planar), namely what is the dependence of the potential on the distance x from the surface going inwards toward to interior?

A detailed answer to this question can be found in many texts on semiconductors and semiconductor interfaces. Here we just outline the main points of this theory and make a few observations that will be referred to elsewhere in this text:

1. As in Section 4.5.2, the solution to this problem is obtained from the Poisson equation (1.219), which is again needed in one dimension

$$\frac{\partial^2 \delta\Phi(x)}{\partial x^2} = -\frac{4\pi}{\varepsilon}\delta\rho_q(x) \tag{4.149}$$

where ε is the dielectric constant[10] and $\delta\rho_q$ is the excess charge density.

2. In turn, the excess charge density $\delta\rho_q(x)$ depends on the local potential. To see this consider Eqs (4.127) and (4.129) for the densities of electrons in the conduction band and holes in the valence bands. These equations where written for a system where the potential is uniform everywhere (and can therefore be taken zero). The presence of an additional potential $\delta\Phi(x)$ at position x has the effect of shifting the local electron energy by $-e\delta\Phi(x)$.[11] Under the approximation that yields (4.132) the corresponding local electron and hole densities become

$$n_c(x;T) = N_c(T)e^{-\beta(E_c - e\delta\Phi(x) - \mu)} \tag{4.150a}$$

and

$$p_v(x;T) = P_v(T)e^{-\beta(\mu - E_v + e\delta\Phi(x))} \tag{4.150b}$$

3. For intrinsic semiconductors the net excess charge is

$$\delta\rho_q(x) = n_c(x) + p_v(x) \tag{4.151}$$

This case is completely analogous to the case of ionic solution that was treated in Section 1.6.3. Indeed, Eq. (4.151) is identical to (1.247). For $|e\delta\Phi| \ll k_B T$ we can proceed along the same lines as in that treatment to obtain (cf. Eq. (1.253))

$$\Phi(x) = \Phi_B + (\Phi_S - \Phi_B)e^{-\kappa x} \tag{4.152}$$

---

[10] Note that in the corresponding equation used in the Thomas–Fermi theory, Section 4.5.2, one takes $\varepsilon = 1$: It is assumed that the dielectric response is dominated by the free metallic electrons, which are treated explicitly.

[11] Here we apply the same semiclassical approximation that was used in (4.115) and (4.141).

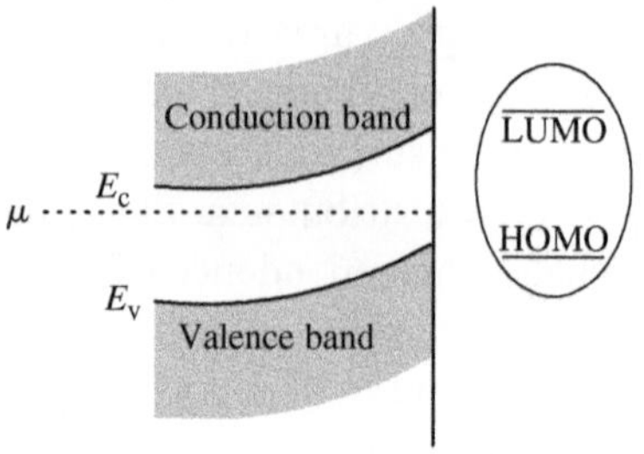

FIG. 4.8 A molecule (represented by its HOMO and LUMO levels next to a semiconductor surface) characterized by its conduction and valence band edges, E_c and E_V—same as Fig. 4.6, except that the common effect of an interfacial electrostatic potential is shown in the semiconductor band bending near its surface. In the case shown the surface potential is lower than in the semiconductor bulk, leading to up-bending of the band edges.

where κ is a screening length given by Eq. (1.252) with $z_+ = z_- = 1$, $n_-^B \to N_c e^{-\beta(E_c-\mu)}$, $n_+^B \to P_v(T)e^{-\beta(\mu-E_v)}$.

4. For extrinsic semiconductors the calculation is somewhat more involved because of the presence of immobile charged centers, but as long as linearization of Eq. (4.150) in $\delta\Phi$ can be implemented the result will again be similar in form to (4.152) with a screening length which is essentially of the same form (1.252). It again depends on the density of mobile carriers densities, which may now be dominated by the doping characteristics of the semiconductor.

5. Equation (4.150) reveals an important characteristic of semiconductor surfaces: The effect of the surface potential can be represented by defining local band edges,

$$E_c, E_v \to E_c - e\delta\Phi(x), E_v - e\delta\Phi(x) \tag{4.153}$$

Since the electronic properties of semiconductors are determined by the relative positioning of the electronic chemical potential and the band edges, this would imply that the surface potential modifies the electronic behavior of semiconductor surfaces relative to their bulk, including the surface charge density and the propensity for accepting or releasing electrons. An example is shown in Fig. (4.8). Note that at equilibrium the electronic chemical potential is a position-independent constant over all the semiconductor volume. While we will not develop this subject further here, it should be evident that understanding electrostatic effects on band structures at semiconductor interfaces is a prerequisite to understanding charge transfer reactions at semiconductor surfaces.[12]

[12] For further reading see A. Many, *Semiconductor Surfaces* (North Holland, New York, 1965) or W. Mönch, *Semiconductor Surfaces and Interfaces* (Springer, Berlin, 1995).

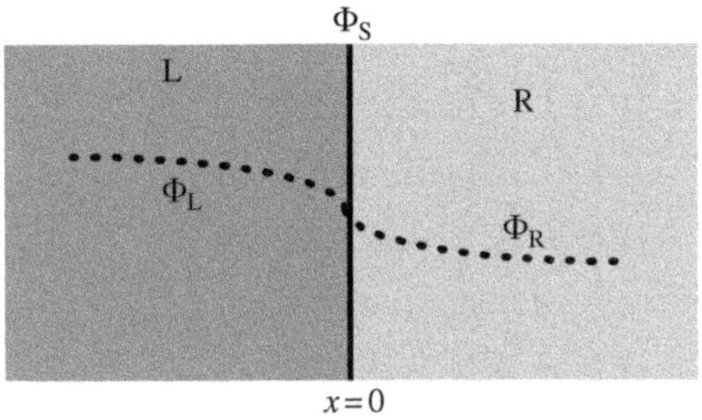

FIG. 4.9 The potential bias distribution in an interface between two systems containing mobile charges.

6. It is of interest to compare screening in a "typical" semiconductor to that found in a "typical" ionic solution, keeping in mind that the screening length is inversely proportional to the square root of the mobile charge density. In a 1M fully ionized monovalent salt solution the total carrier density is of the order $\sim 10^{21}$ ions cm^{-3}. In most intrinsic semiconductors the number is smaller by orders of magnitude, as can be seen from Eqs (4.134) and (4.135) and the estimates underneath which imply $n_c = p_v = \sqrt{N_c P_v} e^{-\beta E_g/2} \approx 2.5 \times 10^{19} e^{-\beta E_g/2}$. For highly doped extrinsic semiconductors the density of majority carriers is approximately determined by the density of the corresponding impurities and can be as high as 10^{19} cm^{-3}, same as in a 0.01 M electrolyte solution. We may conclude that the screening length in a semiconductor is at most comparable to that of $\sim$10 mM electrolyte solution.

4.5.4 Interfacial potential distributions

It is remarkable that the surface potential fall toward its bulk value is a similar exponential function, (4.148) or (4.152), in the semiclassical Thomas–Fermi theory of electronic screening in the Debye–Huckel/Gouy–Chapman theory of ionic screening and at semiconductor interfaces. Here we consider the following issue: When two such phases come into contact as in Fig. 4.9, and a potential bias is set between their interiors, how is the potential drop distributed at the interface?

Denote the two systems by L and R and let their corresponding inverse screening lengths be κ_L and κ_R. The potentials in the interior bulk systems are given as Φ_L and Φ_R, respectively. Φ_S denotes the yet unknown potential at the interface, where $x = 0$. At issue is the magnitudes of the partial contributions $\Phi_L - \Phi_S$ and $\Phi_S - \Phi_R$ to the overall potential bias $\Phi_L - \Phi_R$.

Using (4.152) we can write

$$\Phi_L(x) = \Phi_L + (\Phi_S - \Phi_L)e^{\kappa_L x}; \qquad x \leq 0 \tag{4.154a}$$

$$\Phi_R(x) = \Phi_R + (\Phi_S - \Phi_R)e^{-\kappa_R x}; \qquad x \geq 0 \tag{4.154b}$$

This form contains the information concerning the bulk potentials and also the fact that $\Phi_L(x = 0) = \Phi_R(x = 0) = \Phi_S$. To find Φ_S in terms of Φ_L and Φ_R we use the electrostatic continuity relationship

$$\left(\frac{\partial \Phi_L}{\partial x}\right)_{x=0} = \left(\frac{\partial \Phi_R}{\partial x}\right)_{x=0} \tag{4.155}$$

This leads to

$$\Phi_S = \frac{\kappa_R \Phi_R + \kappa_L \Phi_L}{\kappa_L + \kappa_R} \tag{4.156}$$

and to

$$\begin{aligned} \frac{\Phi_S - \Phi_R}{\Phi_L - \Phi_R} &= \frac{\kappa_L}{\kappa_L + \kappa_R} = \frac{\kappa_R^{-1}}{\kappa_L^{-1} + \kappa_R^{-1}} \\ \frac{\Phi_L - \Phi_S}{\Phi_L - \Phi_R} &= \frac{\kappa_R}{\kappa_L + \kappa_R} = \frac{\kappa_L^{-1}}{\kappa_L^{-1} + \kappa_R^{-1}} \end{aligned} \tag{4.157}$$

Equation (4.157) is the mathematical expression of the intuitive result: *The interfacial potential between two phases in contact is distributed between these two phases in proportion to the corresponding screening lengths.*

Further reading

N. W. Ashcroft and N. D. Mermin, Solid State Physics (Brooke Cole, Philadelphia, 1976).

5

INTRODUCTION TO LIQUIDS[1]

Fluid substances must be composed
Of smooth and rounded particles. Poppy seeds
Might serve as an example, being round
And small and smooth, mercurial as drops
Of water, almost never held together...

Lucretius (c. 99–c. 55 BCE) "The way things are" translated by Rolfe Humphries, Indiana University Press, 1968.

The statistical mechanics of atomic motion in gases and solids have convenient starting points. For gases it is the ideal gas limit where intermolecular interactions are disregarded. In solids, the equilibrium structure is pre-determined, the dynamics at normal temperature is characterized by small amplitude motions about this structure and the starting point for the description of such motions is the harmonic approximation that makes it possible to describe the system in terms of noninteracting normal modes (phonons). Liquids are considerably more difficult to describe on the atomic/molecular level: their densities are of the same order as those of the corresponding solids, however, they lack symmetry and rigidity and, with time, their particles execute large-scale motions. Expansion about a noninteracting particle picture is therefore not an option for liquids. On the other hand, with the exclusion of low molecular mass liquids such as hydrogen and helium, and of liquid metals where some properties are dominated by the conduction electrons, classical mechanics usually provides a reasonable approximation for liquids at and above room temperature.[2] For such systems concepts from probability theory (see Section 1.1.1) will be seen to be quite useful.

[1] This chapter follows closely part of *D. Chandler's Introduction to Modern Statistical Mechanics*, (Oxford University Press, 1987, chapter 7).

[2] An often used criterion for the validity of classical mechanics is that the De Broglie wavelength $\lambda = h/p$ (h is the Planck constant and p—the particle momentum) should be small relative to the intermolecular length scale. If we use $\{p = \langle |p| \rangle_T\}$ (where $\langle\rangle_T$ denotes thermal averaging) this becomes essentially the thermal De Broglie wavelength, $\lambda \approx \lambda_T \equiv \hbar\sqrt{2\pi/(mk_BT)}$. At 300 K and for a molecular weight of nitrogen, say, we get $\lambda = 0.18$ Å, small compared to characteristic distances in liquids—atomic sizes and range of interatomic potentials.

This chapter introduces the reader to basic concepts in the theory of classical liquids. It should be emphasized that the theory itself is general and can be applied to classical solids and gases as well, as exemplified by the derivation of the virial expansion is Section 5.6 below. We shall limit ourselves only to concepts and methods needed for the rest of our discussion of dynamical processes in such environments.

5.1 Statistical mechanics of classical liquids

The microscopic state of a classical system of N atoms is characterized by a point in *phase space*, $(\mathbf{r}^N, \mathbf{p}^N) \equiv (\mathbf{p}_1, \mathbf{p}_2, \ldots, \mathbf{p}_N, \mathbf{r}_1, \mathbf{r}_2, \ldots, \mathbf{r}_N)$. The classical Hamiltonian is

$$H(\mathbf{r}^N, \mathbf{p}^N) = \sum_{i=1}^{N} \frac{\mathbf{p}_i^2}{2m_i} + U(\mathbf{r}^N) \tag{5.1}$$

where U is the potential energy which depends on the positions of all atoms in the system. The probability to find the system in the neighborhood $d\mathbf{r}^N d\mathbf{p}^N$ of the point $(\mathbf{r}^N, \mathbf{p}^N)$ is $f(\mathbf{r}^N, \mathbf{p}^N) d\mathbf{r}^N d\mathbf{p}^N$, where

$$f(\mathbf{r}^N, \mathbf{p}^N) = \frac{e^{-\beta H}}{\int d\mathbf{r}^N \int d\mathbf{p}^N e^{-\beta H}} \tag{5.2}$$

The denominator in Eq. (5.2) is related to the classical canonical partition function.[3]

Using Eq. (5.1) this distribution function can be written as a product of momentum and position parts

$$f(\mathbf{r}^N, \mathbf{p}^N) = \Phi(\mathbf{p}^N) P(\mathbf{r}^N) \tag{5.3}$$

where

$$\Phi(\mathbf{p}^N) = \frac{e^{-\beta \sum_i \mathbf{p}_i^2/2m}}{\int d\mathbf{p}^N e^{-\beta \sum_i \mathbf{p}_i^2/2m}} \tag{5.4}$$

is the probability distribution for the momentum sub-space, itself separable into a product of factors associated with individual degrees of freedom

$$\Phi(\mathbf{p}^N) = \prod_{i=1}^{3N} \phi(p_i); \qquad \phi(p_i) = \frac{e^{-\beta p_i^2/2m}}{\int_{-\infty}^{\infty} dp_i e^{-\beta p_i^2/2m}} = (2\pi m k_B T)^{-1/2} e^{-\beta p_i^2/2m} \tag{5.5}$$

[3] Quantum mechanics implies the uncertainty restriction on the determination of positions and momenta, limiting the number of possible quantum states. This leads to the canonical partition function for a system of N identical particles $Q = (N! h^{3N})^{-1} \int d\mathbf{r}^N d\mathbf{p}^N e^{-\beta H(\mathbf{r}^N, \mathbf{p}^N)}$.

and where

$$P(\mathbf{r}^N) = \frac{e^{-\beta U(\mathbf{r}^N)}}{\int d\mathbf{r}^N e^{-\beta U(\mathbf{r}^N)}} \tag{5.6}$$

is the probability distribution to observe the system at configuration phase point $\mathbf{r}^N$. The denominator in Eq. (5.6), $Z_N \equiv \int d\mathbf{r}^N e^{-\beta U(\mathbf{r}^N)}$ the *Configurational partition function.*

The potential $U(\mathbf{r}^N)$ is a sum over all intra- and intermolecular interactions in the fluid, and is assumed known. In most applications it is approximated as a sum of binary interactions, $U(\mathbf{r}^N) = \sum_{i>j} u(\mathbf{r}_{ij})$ where $\mathbf{r}_{ij}$ is the vector distance from particle i to particle j. Some generic models are often used. For atomic fluids the simplest of these is the hard sphere model, in which $u(\mathbf{r}) = 0$ for $r > a$ and $u(\mathbf{r}) = \infty$ for $r \leq a$, where a is the hard sphere radius. A more sophisticated model is the Lennard Jones potential

$$u(r) = 4\varepsilon \left[\left(\frac{\sigma}{r}\right)^{12} - \left(\frac{\sigma}{r}\right)^{6} \right] \tag{5.7}$$

Here σ is the *collision diameter* and ε is the depth of the potential well at the minimum of $u(r)$. For molecules we often use combinations of atomic pair potentials, adding several body potentials that describe bending or torsion when needed. For dipolar fluids we have to add dipole–dipole interactions (or, in a more sophisticated description, Coulomb interactions between partial charges on the atoms) and for ionic solutions also Coulomb interactions between the ionic charges.

5.2 Time and ensemble average

Consider an equilibrium thermodynamic ensemble, say a set of atomic systems characterized by the macroscopic variables T (temperature), Ω (volume), and N (number of particles). Each system in this ensemble contains N atoms whose positions and momenta are assigned according to the distribution function (5.2) subjected to the volume restriction. At some given time each system in this ensemble is in a particular microscopic state that corresponds to a point $(\mathbf{r}^N, \mathbf{p}^N)$ in phase space. As the system evolves in time such a point moves according to the Newton equations of motion, defining what we call a phase space trajectory (see Section 1.2.2). The ensemble corresponds to a set of such trajectories, defined by their starting point and by the Newton equations. Due to the uniqueness of solutions of the Newton's equations, these trajectories do not intersect with themselves or with each other.

In this microscopic picture, any dynamical property of the system is represented by a dynamical variable—a function of the positions and momenta of all particles,

$$\mathcal{A} = \mathcal{A}(\mathbf{r}^N, \mathbf{p}^N) \tag{5.8}$$

The associated thermodynamic property is the *ensemble average*

$$A = \langle \mathcal{A}(\mathbf{r}^N, \mathbf{p}^N) \rangle = A(\Omega, T, N) \tag{5.9}$$

that is,

$$\begin{aligned} A(\Omega, T, N) &= \int\int d\mathbf{r}^N d\mathbf{p}^N f(\mathbf{r}^N, \mathbf{p}^N) \mathcal{A}(\mathbf{r}^N, \mathbf{p}^N) \\ &= \frac{\int\int d\mathbf{r}^N d\mathbf{p}^N e^{-\beta H(\mathbf{r}^N, \mathbf{p}^N)} \mathcal{A}(\mathbf{r}^N, \mathbf{p}^N)}{\int\int d\mathbf{r}^N d\mathbf{p}^N e^{-\beta H(\mathbf{r}^N, \mathbf{p}^N)}} \end{aligned} \tag{5.10}$$

Equation (5.10) defines an *ensemble average*. Alternatively we could consider another definition of the thermodynamic quantity, using a *time average*

$$A(\Omega, T, N) = \lim_{\tau \to \infty} \left(\frac{1}{\tau} \int_0^\tau \mathcal{A}(\mathbf{r}^N(t)\mathbf{p}^N(t))dt \right) \tag{5.11}$$

The *ergodic* "theorem" of statistical mechanics (see also Section 1.4.2) states that, for "realistic" systems, these two kinds of averaging, Eqs (5.10) and (5.11) yield identical results. As example of an application of this theorem consider the total kinetic energy of the system. The corresponding dynamical variable is

$$\mathcal{A} = \frac{1}{2} \sum_{i=1}^{N} \frac{\mathbf{p}_i^2}{m_i}; \qquad \mathbf{p}_i^2 = p_{ix}^2 + p_{iy}^2 + p_{iz}^2 \tag{5.12}$$

Using Eqs (5.3)–(5.5), Eqs (5.10) and (5.12) yield

$$A(\Omega, T, N) = \frac{3}{2} N k_B T \tag{5.13}$$

Therefore, under the ergodic theorem, Eq. (5.11) implies that

$$T = \frac{1}{3Nk_B} \lim_{\tau \to \infty} \frac{1}{\tau} \int_0^\tau dt \sum_{i=1}^{N} \frac{\mathbf{p}_i(t)^2}{m_i} \tag{5.14}$$

This observation has an important practical consequence: In numerical simulation we usually follow a single-system trajectory *in time*, and the system temperature can be obtained from such an equilibrium trajectory using Eq. (5.14).[4] Note that

[4] In practice, the operation $\lim_{\tau \to \infty}(1/\tau) \int_0^\tau d\tau$ is replaced by an average over a finite number of points sampled along the equilibrium trajectory.

Eq. (5.14) holds separately for each atom, that is,

$$T = \frac{1}{3k_B} \lim_{\tau\to\infty} \frac{1}{\tau} \int_0^\tau \frac{\mathbf{p}_i(t)^2}{m_i}$$

for any atom i.

5.3 Reduced configurational distribution functions

Consider the configuration space distribution function $P(\mathbf{r}^N)$, Eq. (5.6). Mathematically, it is the *joint distribution function* (see Section 1.5.2) to find the N particles of the system in their respective positions in configuration space, that is, $P(\mathbf{r}^N)d\mathbf{r}^N$ is the probability to find particle 1 in the range $d\mathbf{r}_1$ near $\mathbf{r}_1$, *and* particle 2 in the range $d\mathbf{r}_2$ near $\mathbf{r}_2$, *and* particle 3 in the range $d\mathbf{r}_3$ near $\mathbf{r}_3$, and so on.

We may also define a reduced distribution (see Section 1.5.2). The probability to find particle 1, say, in the neighborhood $d\mathbf{r}_1$ of $\mathbf{r}_1$ irrespective of the positions of all other particles is $P^{(1/N)}(\mathbf{r}_1)d\mathbf{r}_1$, where

$$P^{(1/N)}(\mathbf{r}_1) = \int d\mathbf{r}_2 d\mathbf{r}_3 d\mathbf{r}_4, \ldots, d\mathbf{r}_N P(\mathbf{r}^N) \tag{5.15}$$

In a homogeneous system of volume Ω this is obviously $P^{(1/N)}(\mathbf{r}_1) = 1/\Omega$. We may similarly define a reduced joint distribution function to find the two particles 1 and 2 at location $\mathbf{r}_1$, $\mathbf{r}_2$, respectively, irrespective of the positions of all other particles

$$P^{(2/N)}(\mathbf{r}_1, \mathbf{r}_2) = \int d\mathbf{r}_3 d\mathbf{r}_4, \ldots, d\mathbf{r}_N P(r^N) \tag{5.16}$$

Note that $P^{(2/N)}$ is normalized, that is,

$$\int d\mathbf{r}_1 d\mathbf{r}_2 P^{(2/N)}(\mathbf{r}_1, \mathbf{r}_2) = \int d\mathbf{r}^N P(\mathbf{r}^N) = 1 \tag{5.17}$$

If all the particles in the system are identical then $\mathbf{r}_1$ and $\mathbf{r}_2$ can be the coordinates of any two particles in the system. It is sometimes convenient to use a normalization that will express the fact that, if we look at the corresponding neighborhoods of $\mathbf{r}_1$ and $\mathbf{r}_2$, the probability to find these neighborhoods occupied by *any two* particles increases in a statistically determined way with the number of particles in the system. This is achieved by multiplying the joint distribution function $P^{(2/N)}(\mathbf{r}_1, \mathbf{r}_2)$ by the number, $N(N-1)$, of distinct pairs in the system. This yields the *pair distribution function*

$$\rho^{(2/N)}(\mathbf{r}_1, \mathbf{r}_2) = N(N-1)P^{(2/N)}(\mathbf{r}_1, \mathbf{r}_2). \tag{5.18}$$

Noting that $N(N-1)$ is the total number of pairs in the system, $\rho^{(2/N)}$ represents the density of such pairs per unit square volume. This concept can be generalized: the reduced joint distribution function for particles $1, \ldots, n$ is given by

$$P^{(n/N)}(\mathbf{r}_1, \ldots, \mathbf{r}_n) = \int d\mathbf{r}_{n+1}, d\mathbf{r}_{n+2}, \ldots, d\mathbf{r}_N P(r^N) \tag{5.19}$$

and the n particle distribution function is defined by

$$\begin{aligned}\rho^{(n/N)}(\mathbf{r}_1, \ldots, \mathbf{r}_n) &= \frac{N!}{(N-n)!} P^{(n/N)}(\mathbf{r}_1, \ldots, \mathbf{r}_n) \\ &= \frac{N!}{(N-n)!} \int d\mathbf{r}^{N-n} \frac{e^{-\beta U(\mathbf{r}^N)}}{\int d\mathbf{r}^N e^{-\beta U(\mathbf{r}^N)}}\end{aligned} \tag{5.20}$$

where $d\mathbf{r}^{N-n} = d\mathbf{r}_{n+1}, \ldots, d\mathbf{r}_N$. To get a better intuition about these density functions it is useful to note that the relation of $\rho^{(2/N)}(\mathbf{r}_1, \mathbf{r}_2)$ to $P^{(2/N)}(\mathbf{r}_1, \mathbf{r}_2)$is the analog of the relationship (in a homogeneous system) between $\rho^{(1/N)} = N/\Omega$ and $P^{(1/N)} = 1/\Omega$. The distributions $P^{(n/N)}$ are always normalized to 1. On the other hand, $\rho^{(1/N)}$ is normalized to the number of particles N, $\rho^{(2/N)}$ is normalized to the number of pairs, $N(N-1)$, etc. (Note that, for indistinguishable particles, the number of *distinct* pairs is $N(N-1)/2$. The normalization we chose is convenient because it satisfies relations such as Eq. (5.23) below).

As already noted, in a homogeneous fluid $P^{(1/N)}$ does not depend on the particle's position and therefore

$$P^{(1/N)} = \frac{1}{\Omega}; \qquad \rho^{(1/N)} = \frac{N}{\Omega} = \rho \tag{5.21}$$

that is, $\rho^{(1/N)}$ is just the density ρ. In an ideal gas there are no correlations between particles, therefore in an isotropic system

$$P^{(2/N)}(\mathbf{r}_1, \mathbf{r}_2) = P^{(1/N)}(\mathbf{r}_1) P^{(1/N)}(\mathbf{r}_2) = \frac{1}{\Omega^2} \tag{5.22}$$

Hence, the pair distribution function for an isotropic ideal gas is given by

$$\rho^{(2/N)}(\mathbf{r}_1, \mathbf{r}_2) = \frac{N(N-1)}{\Omega^2} \approx \rho^2 \tag{5.23}$$

Correlations in the system caused by deviation from ideality can be measured by the *pair correlation functions*

$$g(\mathbf{r}_1, \mathbf{r}_2) = \rho^{(2/N)}(\mathbf{r}_1, \mathbf{r}_2)/\rho^2 \tag{5.24}$$

or

$$h(\mathbf{r}_1, \mathbf{r}_2) = \frac{\rho^{(2/N)} - \rho^2}{\rho^2} = g(\mathbf{r}_1, \mathbf{r}_2) - 1 \tag{5.25}$$

It should be intuitively clear that the correlation between any two particles vanishes as $|\mathbf{r}_1 - \mathbf{r}_2| \to \infty$. Therefore $g \to 1$ and $h \to 0$ in this limit. For homogeneous fluids all positions are equivalent, and it follows that $g(\mathbf{r}_1, \mathbf{r}_2) = g(\mathbf{r}_1 - \mathbf{r}_2)$. For homogeneous-isotropic fluids $g(\mathbf{r}_1, \mathbf{r}_2) = g(|\mathbf{r}_1 - \mathbf{r}_2|)$, and similarly for h. In this case we refer to these functions as *radial distribution functions*.

The physical meaning of the pair correlation function g can be elucidated by using the conditional probability concept introduced in Section 1.5.2. In analogy with Eq. (1.187), the single particle *conditional* distribution function in a homogeneous system is given by

$$P^{(1/N)}(\mathbf{r}_1 \mid \mathbf{r}_2)d\mathbf{r}_1 = \frac{P^{(2/N)}(\mathbf{r}_1, \mathbf{r}_2)}{P^{(1/N)}(\mathbf{r}_2)}d\mathbf{r}_1 = \Omega P^{(2/N)}(\mathbf{r}_1, \mathbf{r}_2)d\mathbf{r}_1 \tag{5.26}$$

(the second equality follows from Eq. (5.21) that holds for homogeneous systems). $P^{(1/N)}(\mathbf{r}_1 \mid \mathbf{r}_2)d\mathbf{r}_1$ is the conditional probability to find particle 1 in the neighborhood $d\mathbf{r}_1$ of $\mathbf{r}_1$ given that particle 2 (or, if all particles are identical, any particle) is at $\mathbf{r}_2$. Using Eqs (5.18) and (5.24) this can be rewritten in the form

$$\rho g(\mathbf{r}_1, \mathbf{r}_2) = NP^{(1/N)}(\mathbf{r}_1 \mid \mathbf{r}_2) \tag{5.27}$$

The product on the right is the conditional density (number of particles per unit volume) of particles at $\mathbf{r}_1$ given that a particle is at $\mathbf{r}_2$. For a homogeneous system this can be rephrased as follows:

$\rho g(\mathbf{r})$ is the density of particles at $\mathbf{r}$ given that a particle is located at the origin $\mathbf{r} = 0$.

If the system is also isotropic, g depends only on the modulus r of $\mathbf{r}$. In the absence of correlations between particles, $g = 1$ and the conditional density is simply ρ irrespective of whether there is a particle at the origin or not. When correlations exist, g describes their effect on the fluid structure.

Figure 5.1 shows the pair correlation function of a typical Lennard–Jones liquid. Two general features are seen: First, the short range structure that shows that atoms in liquids arrange themselves about a central atom in a way that reflects their atomic diameters (here expressed by the Lennard–Jones parameter σ), and, second, the relative fast decay of this short-range order, expressed by the rapid approach of $g(r)$ to 1.

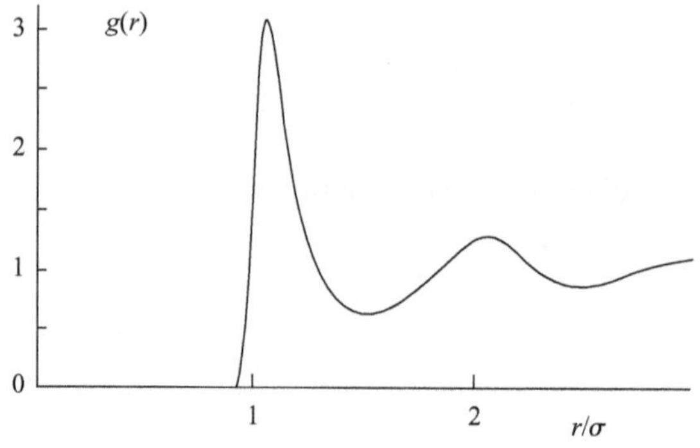

FIG. 5.1 The pair correlation function of a Lennard–Jones fluid.

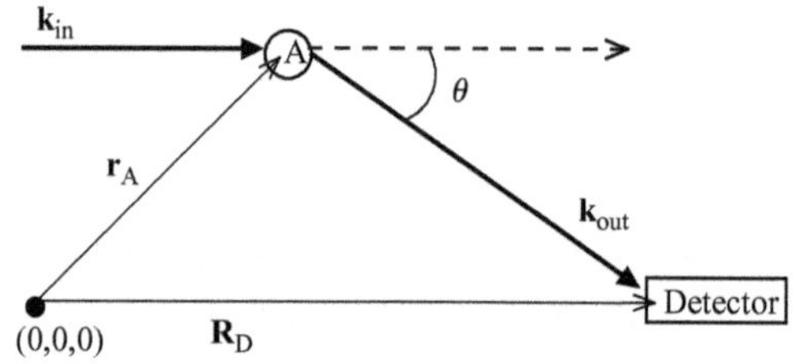

FIG. 5.2 A schematic view of a wave scattering off an atom.

5.4 Observable implications of the pair correlation function

5.4.1 X-ray scattering

For normal liquids the characteristic structural distance is of order ~1 Å. A probe of liquid structure should therefore have a characteristic wavelength λ in this range. This calls for using X-rays or light atomic particles as probes. In liquids we are interested in the short-range structure, on the scale of intermolecular distances. This implies the need to apply short range interactions and therefore the use of particles should be limited to neutral ones, such as neutrons.

To see how g can be measured by X-rays or neutron scattering consider the scheme of a scattering experiment shown in Figure 5.2. An atom A at $\mathbf{r}_A$ scatters an incoming wave with wavevector $\mathbf{k}_{\text{in}}$ and the scattered wave with wavevector $\mathbf{k}_{\text{out}}$ is monitored by the detector at $\mathbf{R}_D$. The scattering angle is θ, as shown. In what follows we consider elastic scattering only, that is, $|\mathbf{k}_{\text{in}}| = |\mathbf{k}_{\text{out}}|$.

The scattered wave at the detector is

$$S = f(k)\frac{1}{|\mathbf{R}_D - \mathbf{r}_A|}e^{i\mathbf{k}_{\text{in}}\cdot\mathbf{r}_A + i\mathbf{k}_{\text{out}}\cdot(\mathbf{R}_D - \mathbf{r}_A)} \approx \frac{f(k)}{|\mathbf{R}_D - \mathbf{R}_C|}e^{i\mathbf{k}_{\text{out}}\cdot\mathbf{R}_D}e^{-i\mathbf{k}\cdot\mathbf{r}_A} \quad (5.28)$$

where $f(k)$ is an atomic scattering factor, $\mathbf{k} = \mathbf{k}_{\text{out}} - \mathbf{k}_{\text{in}}$ and $\mathbf{R}_{\text{C}}$ is the center of the scattering sample. All vectors are related to some origin at (0,0,0). Note the approximation that was made in order to obtain the last part of Eq. (5.28) (see Problem 5.1). The appearance in the denominator of the distance between the atom and the detector results from the fact that the scattered amplitude depends inversely on this distance.

Problem 5.1. Discuss the approximation made in Eq. (5.28). (a) Under what conditions can we replace $\mathbf{r}_{\text{A}}$ by $\mathbf{R}_{\text{C}}$ in the denominator, as done? (b) Why is it impossible to make this substitution in the phase factor $e^{-i\mathbf{k}\cdot\mathbf{r}_{\text{A}}}$?

Because $|\mathbf{k}_{\text{in}}| = |\mathbf{k}_{\text{out}}|$, the scattering angle θ and the modulus of the scattered wavevector are related to each other by

$$k = 2|\mathbf{k}_{\text{in}}| \sin\frac{\theta}{2} = \frac{4\pi}{\lambda}\sin\frac{\theta}{2} \tag{5.29}$$

The total scattered intensity is the absolute-value square of the scattered amplitude, which is in turn a combination of scattered waves like Eq. (5.28) summed over all scattering centers. The signal at the detector is therefore

$$I(\theta) = \left| f(k)\frac{e^{i\mathbf{k}_{\text{out}}\cdot\mathbf{R}_{\text{D}}}}{|\mathbf{R}_{\text{C}} - \mathbf{R}_{\text{D}}|}\sum_{j=1}^{N} e^{-i\mathbf{k}\cdot\mathbf{r}_j} \right|^2 = \frac{|f(k)|^2}{|\mathbf{R}_{\text{C}} - \mathbf{R}_{\text{D}}|^2} NS(k) \tag{5.30}$$

where S is the *structure factor*, the factor in $I(\theta)$ that depends on the fluid structure:

$$S(\mathbf{k}) \equiv \frac{1}{N}\left\langle \sum_{l,j=1}^{N} e^{i\mathbf{k}\cdot(\mathbf{r}_l - \mathbf{r}_j)} \right\rangle \tag{5.31}$$

To find a more useful form for $S(\mathbf{k})$ we first separate it to its diagonal ($l = j$) and non-diagonal parts. The diagonal part yields unity. In a homogeneous isotropic system all $N(N-1)$ non-diagonal terms are identical. We get

$$\begin{aligned} S(\mathbf{k}) &= 1 + \frac{1}{N}N(N-1)\langle e^{i\mathbf{k}\cdot(\mathbf{r}_1-\mathbf{r}_2)}\rangle \\ &= 1 + \frac{N(N-1)}{N}\frac{\int dr^N e^{i\mathbf{k}\cdot(\mathbf{r}_1-\mathbf{r}_2)}e^{-\beta U}}{\int dr^N e^{-\beta U}} \\ &= 1 + \frac{1}{N}\int d\mathbf{r}_1 \int d\mathbf{r}_2 \rho^{(2/N)}(\mathbf{r}_1, \mathbf{r}_2) e^{i\mathbf{k}\cdot(\mathbf{r}_1-\mathbf{r}_2)} \end{aligned} \tag{5.32}$$

The last equality was obtained using Eqs (5.16) and (5.18). Since the system is homogeneous we have $\rho^{(2/N)}(\mathbf{r}_1, \mathbf{r}_2) = \rho^2 g(\mathbf{r}_{12})$ with $\mathbf{r}_{12} = \mathbf{r}_1 - \mathbf{r}_2$. Therefore the integrand is a function of $\mathbf{r}_{12}$ only and $\int d\mathbf{r}_1 \int d\mathbf{r}_2 = \int d\mathbf{r}_{12} \int d\mathbf{r}_1 = \Omega \int d\mathbf{r}_{12}$. This yields

$$S(\mathbf{k}) = 1 + \int d\mathbf{r}_{12} \rho g(\mathbf{r}_{12}) e^{i\mathbf{k}\cdot\mathbf{r}_{12}} \tag{5.33}$$

This identifies $S(\mathbf{k})$ with the Fourier transform of the pair correlation function, so the latter may be obtained from the structure factor by inverting the transform. Finally, denoting $r = |\mathbf{r}_{12}|$ and using, for an isotropic system, $g(\mathbf{r}_{12}) = g(r)$ we get

$$\begin{aligned} S(\mathbf{k}) &= 1 + 2\pi\rho \int_0^\infty dr\, r^2 g(r) \int_0^\pi d\theta \sin\theta e^{ikr\cos\theta} \\ &= 1 + \frac{4\pi\rho}{k} \int_0^\infty dr\, r \sin(kr) g(r) \end{aligned} \tag{5.34}$$

5.4.2 The average energy

Because some of the important interactions used to model atomic and molecular fluids are binary, the corresponding averages that define macroscopic thermodynamic quantities can be expressed in terms of the pair correlation function. The following calculation of the average energy is an example. We consider the average potential energy in a homogeneous atomic fluid of identical particles with an interatomic potential given by

$$U(\mathbf{r}^N) = \sum_i \sum_j u(\mathbf{r}_{ij}) \tag{5.35}$$

We already know that the average kinetic energy is $(3/2)Nk_BT$, so once the average potential energy has been calculated we will have a full microscopic expression for the macroscopic energy of the system. This average potential energy is

$$\begin{aligned} \langle U \rangle &= \sum_{i>j} \langle u(\mathbf{r}_{ij}) \rangle = \frac{N(N-1)}{2} \langle u(\mathbf{r}_{12}) \rangle \\ &= \frac{1}{2} \frac{N(N-1) \int d\mathbf{r}^N u(\mathbf{r}_{12}) e^{-\beta u(\mathbf{r}^N)}}{\int d\mathbf{r}^N e^{-\beta U(\mathbf{r}^N)}} \\ &= \frac{N(N-1)}{2} \int d\mathbf{r}_1 d\mathbf{r}_2 u(\mathbf{r}_{12}) \frac{\int d\mathbf{r}^{N-2} e^{-\beta u(\mathbf{r}^N)}}{\int d\mathbf{r}^N e^{-\beta u(\mathbf{r}^N)}} \end{aligned} \tag{5.36}$$

Using again the definition of the pair correlation function g, this leads to

$$\begin{aligned}\langle U\rangle &= \frac{1}{2}\rho^2\int d\mathbf{r}_1\int d\mathbf{r}_2 g(\mathbf{r}_{12})u(\mathbf{r}_{12})\\ &= \frac{1}{2}\rho N\int d\mathbf{r} g(\mathbf{r})u(\mathbf{r}) = 2\pi\rho N\int_0^\infty drr^2 g(r)u(r)\end{aligned} \tag{5.37}$$

The last equality holds for an isotropic and homogeneous system. This result can be understood intuitively: For each of the N particles in the system (taken to be at the origin), the potential energy is obtained as a volume integral over the density of interaction energy associated with this particle. The latter is $\rho g(\mathbf{r})$ (density of other particles at position $\mathbf{r}$), multiplied by $u(\mathbf{r})$. This will lead to double-counting of all interactions and should therefore be divided by 2. The result is (5.37).

5.4.3 Pressure

Next consider the pressure. It may be obtained from the canonical partition function

$$P = k_B T\left(\frac{\partial \ln Q}{\partial \Omega}\right)_{N,T} = k_B T\left(\frac{\partial \ln Z_N}{\partial \Omega}\right)_{N,T} \tag{5.38}$$

The second equality results form the fact that in the expression for Q

$$Q = \frac{Q_{\text{internal}}}{N!}\left(\frac{2\pi m k_B T}{h^2}\right)^{3N/2} Z_N; \qquad Z_N = \int d\mathbf{r}^N e^{-\beta U(\mathbf{r}^N)} \tag{5.39}$$

Z is the only term that depends on the volume. Since we expect that the macroscopic properties of the system will not depend on the shape of the container, we can consider a cubic box of dimension $\Omega^{1/3}$. To make this dependence explicit we will scale all coordinates by $\Omega^{1/3}$ where V is the volume, so that $x_k = \Omega^{1/3}\bar{x}_k$. In terms of the new coordinates $\bar{x}$, Z_N is

$$Z_N = \Omega^N \int_0^1 \ldots \int_0^1 e^{-\beta\bar{U}(\bar{\mathbf{r}}^N)} d\bar{\mathbf{r}}^N \tag{5.40}$$

where

$$\bar{U}(\bar{\mathbf{r}}^N) = \sum_{i<j} u(\Omega^{1/3}\bar{\mathbf{r}}_{ij}) \tag{5.41}$$

With the dependence of Z on Ω expressed explicitly we can take the derivative

$$\frac{\partial Z_N}{\partial \Omega} = \frac{N}{\Omega} Z_N + \Omega^N \frac{d}{d\Omega}\left(\int_0^1 \dots \int_0^1 d\bar{\mathbf{r}}^N e^{-\beta \sum_{i<j} u(\Omega^{1/3}\bar{\mathbf{r}}_{ij})}\right) \tag{5.42}$$

In the second term on the right the derivative with respect to volume may be found as follows:

$$\frac{d}{d\Omega}\left(\int_0^1 \dots \int_0^1 d\bar{\mathbf{r}}^N e^{-\beta \sum_{i<j} u(\Omega^{1/3}\bar{\mathbf{r}}_{ij})}\right)$$

$$= -\beta \int_0^1 \dots \int_0^1 d\bar{\mathbf{r}}^N \sum_{i<j} (u'(\Omega^{1/3}\bar{\mathbf{r}}_{ij})(1/3)\Omega^{-2/3} \cdot \bar{\mathbf{r}}_{ij}) e^{-\beta \sum_{i<j} u(\Omega^{1/3}\bar{\mathbf{r}}_{ij})}$$

$$\xrightarrow{\text{going back to unscaled coordinates}} -\frac{\beta}{3\Omega^{N+1}} \int d\mathbf{r}^N \sum_{i<j} (u'(r_{ij}) \cdot \mathbf{r}_{ij}) e^{-\beta \sum_{i<j} u(\mathbf{r}_{ij})}$$

$$= -\frac{\beta}{6\Omega^{N+1}} N(N-1) \int d\mathbf{r}^N \mathbf{r}_{12} \cdot u'(\mathbf{r}_{12}) e^{-\beta U} \tag{5.43}$$

Here we have used the notation $u'(\mathbf{r}) = \nabla u(\mathbf{r})$. Using this in (5.42) yields

$$\frac{\partial \ln Z_N}{\partial \Omega} = \frac{1}{Z_N}\frac{\partial Z_N}{\partial \Omega} = \frac{N}{\Omega} - \frac{1}{6k_B T \Omega} N(N-1)\langle \mathbf{r}_{12} u'(\mathbf{r}_{12})\rangle \tag{5.44}$$

Using the same arguments that led to $N(N-1)\langle u(\mathbf{r}_{12})\rangle = \rho N \int d\mathbf{r} g(\mathbf{r}) u(\mathbf{r})$, (see Eqs (5.36) and (5.37)) we now get $N(N-1)\langle \mathbf{r}_{12} u'(\mathbf{r}_{12})\rangle = \rho N \int d\mathbf{r}\, g(\mathbf{r})\, \mathbf{r} \cdot u'(\mathbf{r})$, so

$$\frac{P}{k_B T} = \left(\frac{\partial \ln Z_N}{\partial \Omega}\right)_{N,T} = \rho - \frac{\rho^2}{6k_B T} \int_0^\infty d\mathbf{r} g(\mathbf{r})\, \mathbf{r} \cdot u'(\mathbf{r}) \tag{5.45}$$

To get explicit results for the thermodynamic quantities discussed above (energy, pressure) we need an expression for $g(\mathbf{r})$. This is considered next.

5.5 The potential of mean force and the reversible work theorem

Consider again our equilibrium system of N interacting particles in a given volume V at temperature T. The *average* force exerted on particle 1 by particle 2 is defined by

$$\langle \mathbf{F}_{12} \rangle (\mathbf{r}_1, \mathbf{r}_2) = -\left\langle \frac{\partial}{\partial \mathbf{r}_1} U(r^N) \right\rangle_{\mathbf{r}_1, \mathbf{r}_2} = -\frac{\int d\mathbf{r}_3 \ldots \int d\mathbf{r}_N (\partial U / \partial \mathbf{r}_1) e^{-\beta U}}{\int d\mathbf{r}_3 \ldots \int d\mathbf{r}_N e^{-\beta U}} \tag{5.46}$$

Here the total force acting on particle 1 was averaged over all of the system's configurations in which particles 1 and 2 are fixed in their positions $\mathbf{r}_1$ and $\mathbf{r}_2$, respectively. In an isotropic system this mean force depends only on the vector distance, $\mathbf{r}_{12}$, between particles 1 and 2, and if the particles are structureless only on the magnitude of this distance.

Knowing $\langle \mathbf{F}_{12} \rangle (\mathbf{r}_{12})$ makes it possible to integrate it in order to calculate the *reversible work* needed to bring particles 1 and 2 from infinite mutual distance to a relative distance $\mathbf{r}$ between them, while averaging over the positions of all other particles. The latter are assumed to be at equilibrium for any instantaneous configuration of particles 1 and 2. Note that the word "reversible" enters here as it enters in thermodynamics: A reversible process is one where a system changes slowly enough so that equilibrium prevails throughout the process. This work $W(T, \Omega, N; \mathbf{r})$ is called *the potential of mean force*. It is intuitively clear that its dependence on Ω and N enters only through their ratio, $\rho = N/\Omega$. As defined, W satisfies

$$\frac{dW(\mathbf{r}_{12})}{d\mathbf{r}_1} = -\mathbf{F}_{12} \tag{5.47}$$

and from Eq. (5.46) we get

$$\begin{aligned} \frac{dW}{d\mathbf{r}_1} &= \frac{\int d\mathbf{r}_3 \ldots \int d\mathbf{r}_N (\partial U / \partial \mathbf{r}_1) e^{-\beta U}}{\int d\mathbf{r}_3 \ldots \int d\mathbf{r}_N e^{-\beta U}} = -k_B T \frac{(d/d\mathbf{r}_1) \int d\mathbf{r}_3 \ldots \int d\mathbf{r}_N e^{-\beta U}}{\int d\mathbf{r}_3 \ldots \int d\mathbf{r}_N e^{-\beta U}} \\ &= -k_B T \frac{d}{d\mathbf{r}_1} \left[\ln \int d\mathbf{r}_3 \ldots \int d\mathbf{r}_N e^{-\beta U} \right] \end{aligned} \tag{5.48}$$

Next we change the integral on the RHS of Eq. (5.48) in the following way:

$$\int d\mathbf{r}_3 \ldots \int d\mathbf{r}_N e^{-\beta U} \Rightarrow \frac{N(N-1)}{\rho^2 \int d\mathbf{r}_1 \ldots \int d\mathbf{r}_N e^{-\beta U}} \int d\mathbf{r}_3 \ldots \int d\mathbf{r}_N e^{-\beta U} \tag{5.49}$$

This can be done because the added factors do not depend on $\mathbf{r}_1$. Using the definition of the pair correlation function $g(\mathbf{r})$ we find

$$\frac{dW(\mathbf{r}_{12})}{d\mathbf{r}_1} = -k_B T \frac{d}{d\mathbf{r}_1} \ln g(\mathbf{r}_{12}) \tag{5.50}$$

In particular, for isotropic systems where $g(\mathbf{r}_{12}) = g(|\mathbf{r}_{12}|)$

$$\frac{dW(r)}{dr} = -k_B T \frac{d}{dr} \ln g(r) \tag{5.51}$$

and integrating between some distance r between two particles and infinity (where $W = 0$ and $g = 1$) yields

$$W(r) = -k_B T \ln g(r) \tag{5.52}$$

For an isotropic system at equilibrium at given T, Ω, N we then have

$$g(r) = e^{-\beta W(r)} \tag{5.53}$$

As defined, $W(r)$ is the reversible (minimal) work needed to bring two tagged particles in the system from infinite separation to a distance r between them. For our equilibrium system at a given volume and temperature this is the change in Helmholtz free energy of the system in this process. If the particles repel each other we need to put work into this process, and the change in free energy is positive, that is, $W > 0$. In the opposite case of attraction between the particles $W < 0$. Expression (5.53) reflects the corresponding change in the probability to find two such particles with a distance r between them relative to the completely uncorrelated case where $g = 1$.

In the low density limit the work $W(r)$ is dominated by the two-body potential, so $W(r) \to u(r)$. Therefore the low density limit of $g(r)$ is

$$g(r) \overset{\rho \to 0}{\longrightarrow} \exp(-\beta u(r)). \tag{5.54}$$

5.6 The virial expansion—the second virial coefficient

In the absence of inter-atomic interactions the system is an ideal gas that satisfies $P/(k_B T) = \rho$. The *virial expansion* is an expansion of $P/(k_B T)$ in a power series in ρ:

$$\frac{P}{k_B T} = \rho + B_2 \rho^2 + \dots \tag{5.55}$$

We can get the first correction from Eq. (5.45) by using the fact that the potential of mean force between two atoms in the fluid is the bare interaction potential corrected by a term that vanishes with the fluid density

$$g(r) = e^{-\beta W(r)}; \qquad W(r) = u(r) + \mathrm{O}(\rho) \tag{5.56}$$

Therefore in Eq. (5.45) it is sufficient to use $W = u$ in order to get the first-order correction to the ideal gas law. We find

$$B_2 = -\frac{1}{6k_B T} \int d\mathbf{r} e^{-\beta u(\mathbf{r})} u'(\mathbf{r}) \cdot \mathbf{r} = \frac{1}{6} \int d\mathbf{r} (\nabla e^{-\beta u(\mathbf{r})} \cdot \mathbf{r}) \tag{5.57}$$

and for an isotropic system

$$B_2 = \frac{4\pi}{6} \int_0^\infty dr\, r^3 \frac{d}{dr} e^{-\beta u(r)} \tag{5.58}$$

This can be simplified by performing integration by parts, leading to the final result for the *second virial coefficient*

$$B_2 = -2\pi \int_0^\infty dr r^2 (e^{-\beta u(r)} - 1) \tag{5.59}$$

Further reading

U. Balucani and M. Zoppi, Dynamics of the Liquid State (Clarendon Press, Oxford, 1994).
D. Chandler, Introduction to Modern Statistical Mechanics (Oxford University Press, Oxford, 1987).
J. P. Hansen and I. R. McDonald, Theory of Simple Liquids, 2nd Edition (Elsevier, London, 1986).

PART II

METHODS

6

TIME CORRELATION FUNCTIONS

It is no wonder
That while the atoms are in constant motion
Their total seems to be at total rest,
Save here and there some individual stir.
Their nature lies beyond our range of sense,
Far far beyond. Since you can not get to see
The things themselves, they are bound to hide their moves,
Especially since things we can see, often
Conceal their movement too when at a distance...

Lucretius (c. 99–c. 55 BCE*) "The way things are"*
translated by Rolfe Humphries, Indiana University Press, 1968

In the previous chapter we have seen how spatial correlation functions express useful structural information about our system. This chapter focuses on time correlation functions (see also Section 1.5.4) that, as will be seen, convey important dynamical information. Time correlation functions will repeatedly appear in our future discussions of reduced descriptions of physical systems. A typical task is to derive dynamical equations for the time evolution of an interesting subsystem, in which only relevant information about the surrounding thermal environment (bath) is included. We will see that dynamic aspects of this relevant information usually enter via time correlation functions involving bath variables. Another type of reduction aims to derive equations for the evolution of macroscopic variables by averaging out microscopic information. This leads to kinetic equations that involve rates and transport coefficients, which are also expressed as time correlation functions of microscopic variables. Such functions are therefore instrumental in all discussions that relate macroscopic dynamics to microscopic equations of motion.

6.1 Stationary systems

It is important to keep in mind that dynamical properties are not exclusively relevant only to nonequilibrium system. One may naively think that dynamics is unimportant at equilibrium because in this state there is no evolution on the average. Indeed in such systems all times are equivalent, in analogy to the fact that in spatially homogeneous systems all positions are equivalent. On the other hand, just as in the previous chapter we analyzed equilibrium structures by examining correlations between particles located at different spatial points, also here we can gain dynamical information by looking at the correlations between events that occur at different temporal points.

Time correlation functions are our main tools for conveying this information in stationary systems. These are systems at thermodynamic equilibrium or at steady state with steady fluxes present. In such systems macroscopic observables do not evolve in time and there is no time origin that specifies the "beginning" of a process. However, it is meaningful to consider conditional probabilities such as $P(B, t_2 \mid A, t_1)dB$—the probability that a dynamical variable B will have a value in the range $(B, \ldots, B + dB)$ at time t_2 *if* another dynamical variable A had the value A at time t_1, and the joint probability $P(B, t_2; A, t_1)dBdA$ that A will be in the range $(A, \ldots, A + \mathrm{d}A)$ at time $t = t_1$ *and* B will be in $(B, \ldots, B + \mathrm{d}B)$ at time t_2. These two probabilities are connected by the usual relation (cf. Eq. (1.188))

$$P(B, t_2; A, t_1) = P(B, t_2 \mid A, t_1)P(A, t_1) \tag{6.1}$$

where $P(A, t_1)dA$ is the probability that A has a value in the range $(A, \ldots, A + dA)$ at time t_1. In a stationary system the latter obviously does not depend on time, $P(A, t_1) = P(A)$ and the conditional and joint probabilities depend only on the time difference

$$P(B, t_2; A, t_1) = P(B, t_2 - t_1; A, 0); \quad P(B, t_2 \mid A, t_1) = P(B, t_2 - t_1 \mid A, 0) \tag{6.2}$$

where $t = 0$ is arbitrary.

The time correlation function of two dynamical variables A and B can formally be defined by (see also Eq. (7.42a))

$$C_{AB}(t_1, t_2) = \langle A(t_1)B(t_2)\rangle = \iint dAdB\, AB\, P(B, t_2; A, t_1) \tag{6.3}$$

In a stationary system it is a function of the time difference only

$$\langle A(t_1)B(t_2)\rangle = \langle A(0)B(t)\rangle = \langle A(-t)B(0)\rangle; \quad t = t_2 - t_1 \tag{6.4}$$

Regarding Eq. (6.3), note that we did not say anything about the joint probability function. While it seems intuitively clear that such function exists, its evaluation involves analysis of the time evolution of the system. To see this more clearly let us focus on classical mechanics, and recall that the observables A and B correspond to dynamical variables A and B that are function of positions and momenta of all particles in the system

$$A(t) = \mathcal{A}[\mathbf{r}^N(t), \mathbf{p}^N(t)]; \qquad B(t) = \mathcal{B}[\mathbf{r}^N(t), \mathbf{p}^N(t)] \tag{6.5}$$

The phase space trajectory $\mathbf{r}^N(t)$, $\mathbf{p}^N(t)$ is uniquely determined by the initial conditions $\mathbf{r}^N(t=0) = \mathbf{r}^N$; $\mathbf{p}^N(t=0) = \mathbf{p}^N$. There are therefore no probabilistic issues in the time evolution from $t = 0$ to t. The only uncertainty stems from the fact that our knowledge of the initial condition is probabilistic in nature. The phase space definition of the equilibrium time correlation function is therefore,

$$C_{AB}(t_1, t_2) = \int d\mathbf{r}^N d\mathbf{p}^N f(\mathbf{r}^N, \mathbf{p}^N) A[t_1; \mathbf{r}^N, \mathbf{p}^N, t = 0] B[t_2; \mathbf{r}^N, \mathbf{p}^N, t = 0] \tag{6.6}$$

where, for example, $A[t_1; \mathbf{r}^N, \mathbf{p}^N, t = 0]$ is the value of A at time t_1, that is, $A[\mathbf{r}^N(t_1), \mathbf{p}^N(t_1)]$, given that the state of the system was $(\mathbf{r}^N, \mathbf{p}^N)$ at $t = 0$, and where $f(\mathbf{r}^N, \mathbf{p}^N)$ is the phase space distribution function for this initial state. In stationary system this "initial" state distribution does not depend on time.

How do the definitions (6.3) and (6.6) relate to each other? While a formal connection can be made, it is more important at this stage to understand their range of applicability. The definition (6.6) involves the detailed time evolution of all particles in the system. Equation (6.3) becomes useful in reduced descriptions of the system of interest. In the present case, if we are interested only in the mutual dynamics of the observables A and B we may seek a description in the subspace of these variables and include the effect of the huge number of all other microscopic variables only to the extent that it affects the dynamics of interest. This leads to a reduced space dynamics that is probabilistic in nature, where the functions $P(B, t_2; A, t_1)$ and $P(B, t_2 \mid A, t_1)$ emerge. We will dwell more on these issues in Chapter 7. Common procedures for evaluating time correlation functions are discussed in Section 7.4.1.

6.2 Simple examples

Here we describe two simple examples, one based on classical and the other on quantum mechanics, that demonstrate the power of time correlation functions in addressing important observables.

6.2.1 The diffusion coefficient

The diffusion coefficient describes the coarse-grained dynamics of particles in condensed systems (see Section 1.5.5). To get an explicit expression we start from (cf. Eq. (1.209))

$$D = \lim_{t\to\infty} \frac{1}{6t} \langle (\mathbf{r}(t) - \mathbf{r}(0))^2 \rangle \tag{6.7}$$

and use

$$\mathbf{r}(t) - \mathbf{r}(0) = \int_0^t dt' \mathbf{v}(t') \tag{6.8}$$

to get

$$\langle(\mathbf{r}(t) - \mathbf{r}(0))^2\rangle = \int_0^t dt' \int_0^t dt''\langle\mathbf{v}(t'') \cdot \mathbf{v}(t')\rangle = 2\int_0^t dt' \int_0^{t'} dt''\langle\mathbf{v}(t'') \cdot \mathbf{v}(t')\rangle \tag{6.9}$$

The last equality holds because $C_\mathbf{v}(t) \equiv \langle\mathbf{v}(t_1) \cdot \mathbf{v}(t_1 + t)\rangle = \langle\mathbf{v}(t_1 + t) \cdot \mathbf{v}(t_1)\rangle = C_\mathbf{v}(-t)$. Note that we rely here on the classical identity $\mathbf{v}(t_1) \cdot \mathbf{v}(t_1 + t) = \mathbf{v}(t_1 + t) \cdot \mathbf{v}(t_1)$. Therefore,

$$D = \lim_{t\to\infty} \frac{1}{3t} \int_0^t dt'' \int_0^{t''} dt' C_\mathbf{v}(t'' - t') \tag{6.10}$$

This can be simplified by changing variables: $\theta = t'' - t'$, with θ goes from t'' to 0 and $dt' = -d\theta$. This leads to

$$3D = \lim_{t\to\infty} \frac{1}{t} \int_0^t dt'' \int_0^{t''} d\theta C_\mathbf{v}(\theta) \tag{6.11}$$

The integral is done over the shaded area in Fig. 6.1. Using this picture it is easily seen that the order of integration in (6.11) may be changed so as to give

$$\begin{aligned} 3D &= \lim_{t\to\infty} \frac{1}{t} \int_0^t d\theta \int_\theta^t dt'' C_\mathbf{v}(\theta) \\ &= \lim_{t\to\infty} \frac{1}{t} \int_0^t d\theta (t - \theta) C_\mathbf{v}(\theta) \end{aligned} \tag{6.12}$$

The correlation function $C_\mathbf{v}(t) \equiv \langle\mathbf{v}(0) \cdot \mathbf{v}(t)\rangle$ vanishes at long time because velocities at different times become quickly (on a timescale of a few collisions) uncorrelated which implies $\langle\mathbf{v}(0) \cdot \mathbf{v}(t)\rangle \to \langle\mathbf{v}(0)\rangle\langle\mathbf{v}(t)\rangle = 0$. For this reason the

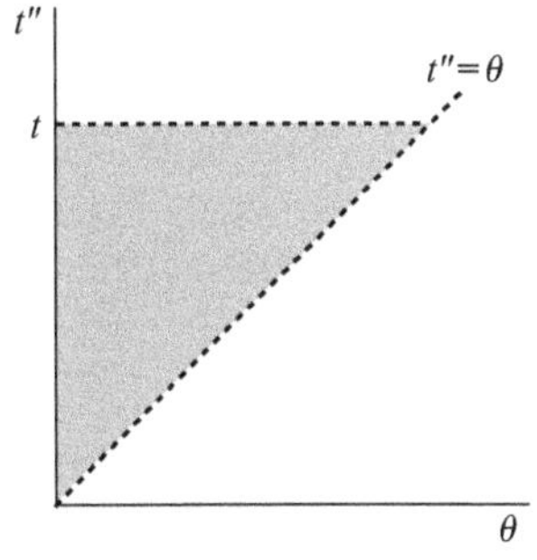

FIG. 6.1 The integration interval for Eq. (6.11).

range of θ that contributes in (6.12) is limited to such rather short times. Therefore, when $t \to \infty$ Eq. (6.12) yields

$$3D = \lim_{t\to\infty} \frac{1}{t} \int_0^t d\theta (t-\theta) C_{\mathbf{v}}(\theta) = \int_0^\infty d\theta C_{\mathbf{v}}(\theta) \tag{6.13}$$

A time correlation function that involves the same observable at two different times is called an autocorrelation function. We have found that the self-diffusion coefficient is the time integral of the velocity auto-correlation function

$$D = \frac{1}{3} \int_0^\infty dt \langle \mathbf{v}(t) \cdot \mathbf{v}(0) \rangle = \frac{1}{6} \int_{-\infty}^\infty dt \langle \mathbf{v}(t) \cdot \mathbf{v}(0) \rangle = \frac{1}{6} \tilde{C}_{\mathbf{v}}(\omega = 0) \tag{6.14}$$

where

$$\tilde{C}_{\mathbf{v}}(\omega) = \int_{-\infty}^{\infty} dt C_{\mathbf{v}}(t) e^{i\omega t} = \int_{-\infty}^{\infty} dt \langle \mathbf{v}(t) \cdot \mathbf{v}(0) \rangle e^{i\omega t} \tag{6.15}$$

Equation (6.14) associates the zero frequency component of the velocity time correlation function with the long-time diffusive dynamics. We will later find (see Section 6.5.4) that the high frequency part of the same Fourier transform, Eq. (6.15), is related to the short-time dynamics of the same system as expressed by its spectrum of *instantaneous normal modes*.

6.2.2 Golden rule rates

The golden rule rate expression is a standard quantum-mechanical result for the relaxation rate of a prepared state $|i\rangle$ interacting with a continuous manifold of states $\{|f\rangle\}$. The result, derived in Section 9.1, for this rate is the Fermi golden rule formula according to which the rate is $k_i = \Gamma_i/\hbar$ where (cf. Eq. (9.28))

$$\Gamma_i = 2\pi\overline{(|V_{if}|^2\rho_f)}_{E_f=E_i} = 2\pi\sum_f |V_{if}|^2\delta(E_i - E_f) \tag{6.16}$$

where $V_{if} = \langle i|\hat{V}|f\rangle$ and $\hat{V}$ is the coupling responsible for the relaxation, and where ρ_f is the density of states in the manifold $\{f\}$. If at $t = 0$ the system is represented by a thermal ensemble of initial states, the decay rate is the average[1]

$$\Gamma = 2\pi\sum_i P_i \sum_f |V_{if}|^2\delta(E_i - E_f) \tag{6.17a}$$

$$P_i = Q^{-1}e^{-\beta E_i}; \qquad Q = \sum_i e^{-\beta E_i}; \qquad \beta = (k_B T)^{-1} \tag{6.17b}$$

From (cf. Eq. (1.60)) $2\pi\delta(E_i - E_f) = \hbar^{-1}\int_{-\infty}^{\infty} dt \exp(i(E_i - E_f)t/\hbar)$ and $E_k|k\rangle = \hat{H}_0|k\rangle$ $(k = i,f)$ we get

$$\begin{aligned} 2\pi\delta(E_i - E_f)e^{-\beta E_i}\langle i|\hat{V}|f\rangle &= \hbar^{-1}\int_{-\infty}^{\infty} dt \langle i|e^{-\beta\hat{H}_0}(e^{i\hat{H}_0 t/\hbar}\hat{V}e^{-i\hat{H}_0 t/\hbar})|f\rangle \\ &= \hbar^{-1}\int_{-\infty}^{\infty} dt\langle i|e^{-\beta\hat{H}_0}\hat{V}_I(t)|f\rangle \end{aligned} \tag{6.18}$$

where $\hat{V}_I(t) = e^{i\hat{H}_0 t/\hbar}\hat{V}e^{-i\hat{H}_0 t/\hbar}$ is the interaction representation of the operator $\hat{V}$. Inserting (6.18) into (6.17) and using[2]

$$\sum_f |f\rangle\langle f| = \hat{I} - \sum_i |i\rangle\langle i|; \qquad \langle i|\hat{V}|i'\rangle = 0 \tag{6.19}$$

where $\hat{I}$ is the unit operator, we get

$$\Gamma = \hbar^{-1}\int_{-\infty}^{\infty} dt\langle\hat{V}_I(t)\hat{V}_I(0)\rangle_T \tag{6.20}$$

[1] In fact what is needed for Eq. (6.17) to be a meaningful transition rate is that thermal relaxation (caused by interaction with the thermal environment) in the manifold of initial states is fast relative to Γ. See Section 12.4 for further discussion.

[2] The equality $\langle i|\hat{V}|i'\rangle = 0$ for all i, i' is a model assumption: The picture is of two manifolds of states, $\{i\}$ and $\{f\}$ and an interaction that couples between, but not within, them.

Here $\langle \ldots \rangle_T$ is the quantum thermal average, $\langle \ldots \rangle_T = \mathrm{Tr}[e^{-\beta \hat{H}_0} \ldots]/\mathrm{Tr}[e^{-\beta \hat{H}_0}]$, and Tr denotes a trace over the initial manifold $\{i\}$. We have thus identified the golden rule rate as an integral over time of a quantum time correlation function associated with the interaction representation of the coupling operator.

6.2.3 Optical absorption lineshapes

A variation on the same theme is the optical absorption lineshape, where the transition from the group of levels $\{i\}$ to the group $\{f\}$ is accompanied by the absorption of a photon of frequency ω. The transition rate monitored as a function of ω is essentially the absorption lineshape $L(\omega)$. Except for the fact that the process is accompanied by the absorption of a photon, the rate is again given by the golden rule expression (6.16) modified to account for the given energetics and with coupling that is proportional to the dipole moment operator (for dipole allowed transitions) $\hat{\mu}$,

$$L(\omega) = A \sum_i P_i \sum_f |\langle i|\hat{\mu}|f\rangle|^2 \delta(E_i - E_f + \hbar\omega) \tag{6.21}$$

where $P_i = e^{-\beta E_i}/\sum_i e^{-\beta E_i}$, $\hat{\mu}$ is the dipole moment operator, and A is a numerical constant. The same procedure that leads to (6.21) now yields

$$L(\omega) \sim \frac{A}{2\pi\hbar} \int_{-\infty}^{\infty} dt e^{i\omega t} \langle \hat{\mu}(t)\hat{\mu}(0)\rangle; \qquad \hat{\mu}(t) = e^{i\hat{H}_0 t/\hbar} \hat{\mu} e^{-i\hat{H}_0 t/\hbar} \tag{6.22}$$

where $\hat{H}_0$ is the system Hamiltonian that does not include the interaction with the radiation field Eq. (6.22) is an expression for the golden rule rate of a process that involves absorption of energy $\hbar\omega$ from an external source.

In many applications we use models that are more explicit about the nature of the initial and final states involved in this transition. A common model (see Chapter 12) is a two-level system that interacts with its thermal environment. The lineshape of interest then corresponds to the photon-induced transition from state 1 to state 2, dressed by states of the thermal environment. The initial and final states are now $|i\rangle = |1, \alpha\rangle$ and $|f\rangle = |2, \alpha'\rangle$ where α and α' are states of the bath. Equation (6.21) can then be rewritten as[3]

$$L(\omega) = A \sum_\alpha P_\alpha \sum_{\alpha'} |\mu_{1\alpha,2\alpha'}|^2 \delta(E_1 + \varepsilon_\alpha - E_2 - \varepsilon_{\alpha'} + \hbar\omega) \tag{6.23}$$

[3] The form (6.23) relies on a weak system-bath coupling, whereupon the energies are written as additive contributions, for example, $E_1 + \varepsilon_\alpha$, of these subsystems.

where we have used $P_i = e^{-\beta(E_1+\varepsilon_\alpha)}/\sum_\alpha e^{-\beta(E_1+\varepsilon_\alpha)} = e^{-\beta\varepsilon_\alpha}/\sum_\alpha e^{-\beta\varepsilon_\alpha} = P_\alpha$ and $\mu_{1\alpha,2\alpha'} = \langle 1\alpha|\hat{\mu}|2\alpha'\rangle$. The operator $\hat{\mu}$ can be expressed in the form

$$\hat{\mu} = \hat{\mu}_{12}|1\rangle\langle 2| + \hat{\mu}_{21}|2\rangle\langle 1| \tag{6.24}$$

in which $\hat{\mu}_{12}$ and $\hat{\mu}_{21}$ are operators in the bath subspace. Denoting $\hbar\omega_{21} = E_2 - E_1$ and repeating the procedure that leads to (6.20) gives

$$\begin{aligned} L(\omega) &= \frac{A}{2\pi\hbar}\int_{-\infty}^{\infty} dte^{-i(\omega-\omega_{21})t}\sum_\alpha P_\alpha \sum_{\alpha'}\langle\alpha|\hat{\mu}_{12}|\alpha'\rangle\langle\alpha'|e^{i\hat{H}_B t/\hbar}\hat{\mu}_{21}e^{-i\hat{H}_B t/\hbar}|\alpha\rangle \\ &= \frac{A}{2\pi\hbar}\int_{-\infty}^{\infty} dte^{-i(\omega-\omega_{21})t}\langle\hat{\mu}_{12}\hat{\mu}_{21}(t)\rangle_B \end{aligned} \tag{6.25}$$

Let us have a closer look at the two forms, (6.22) and (6.25). The form (6.22) was obtained from a picture that looked at a system as a whole, and follows from a golden-rule type expression (Eq. (6.21)) for the transition between two groups of states, $\{i\}$ and $\{f\}$, separated along the energy axis by the photon energy $\hbar\omega$. Equation (6.25) was obtained for a model in which these groups of states are chosen in a particular way that looks at our overall system as a two-level system interacting with a bath. In (6.22) the time evolution and averaging is done in the space of the overall system. In contrast, in (6.25) the operators $\hat{\mu}_{12}$, $\hat{\mu}_{21}$ are bath operators and the correlation function is defined in the bath subspace. If the two-level system is not coupled to its environment then $\hat{\mu}_{21}(t)$ becomes a time-independent scalar and (6.25) gives $L(\omega) \sim \delta(\omega - \omega_{21})$. The result is a δ-function because in the model as defined above no source of broadening except the coupling to the thermal bath was taken into account. The environmental broadening originates from the fact that in the presence of system-bath coupling the correlation function $\langle\hat{\mu}_{12}\hat{\mu}_{21}(t)\rangle_B$ depends on time in a way that reflects the effect of the bath dynamics on the system.

The simplest model for the time-dependent dipole correlation function is an exponentially decaying function, $\langle\hat{\mu}(t)\hat{\mu}(0)\rangle \sim \exp(-\Gamma|t|)$.[4] This form leads to a Lorentzian lineshape

$$L(\omega) \sim \int_{-\infty}^{\infty} dte^{-i(\omega-\omega_{21})t}e^{-\Gamma|t|} = \frac{2\Gamma}{(\omega-\omega_{21})^2+\Gamma^2} \tag{6.26}$$

[4] This is a useful model but we should keep in mind its limitations: (1) It cannot describe the correct dynamics near $t = 0$ because the function $\exp(-\Gamma|t|)$ is not analytic at that point. Also, it does not obey the fundamental identity (6.72), and therefore corresponds to a high temperature approximation.

Such Lorentzian lineshapes will be also obtained from microscopic quantum (Section 9.3; see Eq. (9.49)) and classical (Section 8.2.4; Eq. (8.41)) models.

6.3 Classical time correlation functions

The time correlation function of two observables, A and B was defined as

$$C_{AB}(t_1, t_2) = \langle A(t_1)B(t_2)\rangle \tag{6.27}$$

where, in phase space the average should be understood as an ensemble average over the initial distribution as detailed in Eq. (6.6).

For a stationary (e.g. equilibrium) system the time origin is irrelevant and

$$C_{AB}(t_1, t_2) = C_{AB}(t_1 - t_2) \tag{6.28}$$

In this case the correlation function can also be calculated as the time average

$$C_{AB}(t_1, t_2) = \lim_{\tau\to\infty} \frac{1}{2\tau} \int_{-\tau}^{\tau} A(t_1 + t)B(t_2 + t)dt \tag{6.29}$$

The equality between the averages computed by Eqs (6.27) and (6.29) is implied by the *ergodic "theorem"* of statistical mechanics (see Section 1.4.2).

At $t \to 0$ the correlation function, $C_{AB}(t)$ becomes the static correlation function $C_{AB}(0) = \langle AB\rangle$. In the opposite limit $t \to \infty$ we may assume that correlations vanish so that

$$\lim_{t\to\infty} C_{AB}(t) = \langle A\rangle\langle B\rangle \tag{6.30}$$

We will often measure dynamical variables relative to their average values, that is, use $A - \langle A\rangle$ rather than A. Under such convention the limit in (6.30) vanishes. In what follows we list some other properties of classical time correlation functions that will be useful in our future discussions.

1. The stationarity property (6.28) implies that $\langle A(t + s)B(s)\rangle$ does not depend on s. It follows that

$$\begin{aligned} 0 = \frac{d}{ds}\langle A(t + s)B(s)\rangle &= \langle \dot{A}(t + s)B(s)\rangle + \langle A(t + s)\dot{B}(s)\rangle \\ &= \langle \dot{A}(t)B(0)\rangle + \langle A(t)\dot{B}(0)\rangle \end{aligned} \tag{6.31}$$

Thus, we have found that for a stationary system

$$\langle \dot{A}(t)B(0)\rangle = -\langle A(t)\dot{B}(0)\rangle \tag{6.32}$$

An alternative proof provides insight into the nature of time averaging: From Eq. (6.29) we have

$$\langle \dot{A}(t)B(0)\rangle = \lim_{\tau\to\infty}\frac{1}{\tau}\int_0^{\tau}\dot{A}(t+t')B(t')dt' \tag{6.33}$$

Integrating by parts leads to

$$= \lim_{\tau\to\infty}\frac{1}{\tau}[A(t+t')B(t')]_{t'=0}^{t'=\tau} - \lim_{\tau\to\infty}\frac{1}{\tau}\int_0^{\tau}A(t+t')\dot{B}(t')dt = -\langle A(t)\dot{B}\rangle \tag{6.34}$$

(the first term vanishes because A and B, being physical variables, are bounded). We will henceforth use the notation $B \equiv B(t=0)$.

2. An immediate corollary of (6.32) is that the equal time correlation of a classical dynamical variable with its time derivative is zero

$$\langle A\dot{A}\rangle = 0. \tag{6.35}$$

Problem 6.1. Using similar reasoning as in (6.31) and (6.34) show that

$$\langle \ddot{A}(t)B\rangle = -\langle \dot{A}(t)\dot{B}\rangle \tag{6.36}$$

and more generally

$$\langle A^{(2n)}(t)B\rangle = (-1)^n\langle A^{(n)}(t)B^{(n)}\rangle \tag{6.37}$$

where $A^{(n)}$ denotes the nth time derivative.

Problem 6.2. For a classical harmonic oscillator whose position and momentum are x and p define the complex amplitude

$$a(t) = x(t) + (i/(m\omega))p(t) \tag{6.38}$$

so that $x = (1/2)(a + a^*)$ and $p = -(im\omega/2)(a - a^*)$. a and a^*evolve according to $\dot{a} = -i\omega a$; $\dot{a}^* = i\omega a^*$. Show that for a harmonic oscillator at thermal equilibrium

$$\langle a^2\rangle = \langle (a^*)^2\rangle = 0; \qquad \langle |a|^2\rangle = 2k_BT/(m\omega^2) \tag{6.39}$$

Use it to show that $\langle x(t)x(0)\rangle = (k_BT/m\omega^2)\cos(\omega t)$.

3. An important property of time correlation functions is derived from the time reversal symmetry of the equations of motion. The time reversal operation, that is, inverting simultaneously the sign of time and of all momenta, reverses the direction of the system's trajectory in phase space. At the moment the transformation is made each dynamical variable is therefore transformed according to

$$A \rightarrow \varepsilon_A A \tag{6.40}$$

where $\varepsilon_A = +1$ or -1 depending on whether A is even or odd in the combined power of the momenta. Except for that, the dependence of A on r and p remains the same and, since the time-reversed trajectory is a legitimate system's trajectory, it samples in equilibrium the same system's states as the original trajectory. This yields the same result for Eq. (6.29). It follows that

$$C_{AB}(t) = \langle A(t)B(0)\rangle = \varepsilon_A\varepsilon_B\langle A(-t)B(0)\rangle = \varepsilon_A\varepsilon_B\langle A(0)B(t)\rangle = \varepsilon_A\varepsilon_B C_{BA}(t) \tag{6.41}$$

It also follows that autocorrelation functions ($A = B$) are even functions of time

$$C_{AA}(t) = C_{AA}(-t) \tag{6.42}$$

4. Dynamical variables that correspond to observables are real, however it is sometimes convenient to work with complex quantities such as the variables a and a^* in Eq. (6.38). For a complex dynamical variable the autocorrelation function is conventionally defined as

$$C_{AA}(t) = \langle A(t)A^*\rangle \tag{6.43}$$

This insures that $C_{AA}(t)$ is a real function of t because, by the same argument as in (6.41)

$$C_{AA}(t) = \langle A(t)A^*\rangle = \langle A(-t)A^*\rangle = \langle AA^*(t)\rangle = \langle A^*(t)A\rangle = C_{AA}(t)^* \tag{6.44}$$

Note that this reasoning assumes that A and $A^*(t)$ commute, and therefore holds only in classical mechanics.

5. An important property of all correlation functions follows from the Schwarz's inequality, Eq. (1.87)

$$\langle AB\rangle^2 \leq \langle AA\rangle\langle BB\rangle \tag{6.45}$$

or, if A and B are complex,

$$|\langle AB^*\rangle|^2 \leq \langle AA^*\rangle\langle BB^*\rangle \tag{6.46}$$

From this it follows that

$$|C_{AA}(t)| \leq |C_{AA}(0)| \tag{6.47}$$

that is, the magnitude of an autocorrelation function is never larger than its initial value.

6. Consider the Fourier transform of an autocorrelation function

$$\tilde{C}_{AA}(\omega) = \int_{-\infty}^{\infty} dt e^{i\omega t} C_{AA}(t) \tag{6.48}$$

This function is sometimes called the power spectrum or the spectral function of the observable $A(t)$, and can be related to the absolute square of the Fourier transform of $A(t)$ itself. To show this start from

$$A_T(\omega) = \int_{-T}^{T} e^{i\omega t} A(t) dt \tag{6.49}$$

whence

$$\langle |A_T(\omega)|^2 \rangle = \int_{-T}^{T} dt \int_{-T}^{T} dt' e^{i\omega(t-t')} \langle A(t-t') A^*(0) \rangle \xrightarrow{T\to\infty} 2T \int_{-\infty}^{\infty} d\tau e^{i\omega\tau} C_{AA}(\tau) \tag{6.50}$$

This result is known as the *Wiener–Khintchine theorem* (see also Section 7.5.2). From the obvious inequality $\langle |A_T(\omega)|^2 \rangle \geq 0$ it follows that

$$C_{AA}(\omega) \geq 0 \tag{6.51}$$

that is, the spectrum of an autocorrelation function is nonnegative.

7. Consider again the autocorrelation function

$$C_{AA}(t) = \langle A(t) A^* \rangle$$

In real systems, where the interaction potential is continuous and finite for all relevant interatomic distances, all forces are finite and the time evolution is continuous. $C_{AA}(t)$ is then an analytical function of time, and in particular can be expanded about $t = 0$. From $C_{AA}(t) = C_{AA}(-t)$ it follows that only even powers of t contribute to this expansion:

$$C_{AA}(t) = \sum_{n=0}^{\infty} \frac{t^{2n}}{(2n)!} \langle A^{(2n)} A^* \rangle = \sum_{n=0}^{\infty} (-1)^n \frac{t^{2n}}{(2n)!} \langle |A^{(n)}|^2 \rangle \tag{6.52}$$

where again $A^{(n)}$ is the nth time derivative of A, and where we have used Eq. (6.37). For the velocity autocorrelation function of a given particle this leads to

$$\langle \mathbf{v}(t) \cdot \mathbf{v} \rangle = \frac{3k_B T}{m}\left(1 - \frac{1}{2}\Omega_0^2 t^2 + \cdots\right) \tag{6.53}$$

where

$$\Omega_0^2 = \frac{m}{3k_B T}\langle \dot{\mathbf{v}} \cdot \dot{\mathbf{v}} \rangle = \frac{1}{3mk_B T}\langle |\nabla U|^2 \rangle \tag{6.54}$$

where U is the potential and $-\nabla U$ is a gradient with respect to the position of the particle, that is, the total force on that particle. For a fluid of identical particles Ω_0 is sometimes referred to as the "Einstein frequency" of the fluid. This is the frequency at which a tagged particle would vibrate for small amplitude oscillations in the averaged (over all frozen configurations) potential well created by all other particles. Since the forces in liquids are similar to those in solids, we expect this frequency to be of the same order as the Debye frequency of solids, of the order of 10^{13}–10^{14} s^{-1}.

Problem 6.3. Show that

$$\Omega_0^2 = \frac{1}{3m}\langle \nabla^2 U \rangle \tag{6.55}$$

and use this result to show that if the potential U is a sum of binary interactions, $U = (1/2)\sum_{i,j} u(r_{ij})$ then

$$\Omega_0^2 = \frac{2\pi\rho}{3m}\int_0^\infty \nabla^2 u(r) g(r) r^2 dr \tag{6.56}$$

where $g(r)$ is the (radial) pair correlation function.

Solution: Start from $\langle |\nabla_1 U|^2 \rangle = \int d\mathbf{r}^N e^{-\beta U(\mathbf{r}^N)} \nabla_1 U(\mathbf{r}^N) \cdot \nabla_1 U(\mathbf{r}^N) / \int d\mathbf{r}^N e^{-\beta U(\mathbf{r}^N)}$, where the subscript 1 refers to the tagged particle that serves to specify the force, and rewrite it in the form

$$\langle |\nabla_1 U|^2 \rangle = -\frac{1}{\beta}\frac{\int d\mathbf{r}^N \nabla_1 e^{-\beta U(\mathbf{r}^N)} \cdot \nabla_1 U(\mathbf{r}^N)}{\int d\mathbf{r}^N e^{-\beta U(\mathbf{r}^N)}} \tag{6.57}$$

The expression in the numerator is $\int d\mathbf{r}^N \nabla_1 e^{-\beta U(\mathbf{r}^N)} \cdot \nabla_1 U(\mathbf{r}^N) = \int d\mathbf{r}^N (\nabla_1 \cdot (e^{-\beta U} \nabla_1 U)) - \int d\mathbf{r}^N e^{-\beta U} \nabla_1^2 U$ and the contribution of first term on the right may be shown to be zero. Indeed, from the divergence theorem (Eq. (1.36))

$$\int d\mathbf{r}^N (\nabla_1 \cdot (e^{-\beta U} \nabla_1 U)) \Big/ \int d\mathbf{r}^N e^{-\beta U(\mathbf{r}^N)}$$
$$= \int d\mathbf{r}^{N-1} \int_{S_1} ds_1 e^{-\beta U} (\hat{\mathbf{n}} \cdot \nabla_1 U) \Big/ \int d\mathbf{r}^N e^{-\beta U(\mathbf{r}^N)} \tag{6.58}$$

Where $\hat{\mathbf{n}}$ is a unit vector in the subspace of the tagged particle in the direction normal to the surface S_1 of this subspace. The integrand in (6.58) is seen to be the averaged force exerted on the system's surface by the tagged particle. For a system with short-range forces this quantity will vanish like (system size)$^{-1}$ which is the ratio between the number of particles near the surface to the total number of particles. This implies that $\langle |\nabla U|^2 \rangle = k_B T \langle \nabla^2 U \rangle$ from which follows (6.55). To obtain the expression (6.56) one follows the procedure of Section 5.4.2, except that the relevant sum is not over all pairs but only over pairs that involve the tagged particle.

Problem 6.4. Show that the short time expansion of the position autocorrelation function is

$$\langle \mathbf{r}(t) \cdot \mathbf{r}(0) \rangle = \langle r^2 \rangle - \frac{1}{2} t^2 \langle \mathbf{v}^2 \rangle + \cdots = \langle r^2 \rangle - \frac{3}{2} \frac{k_B T}{m} t^2 + \cdots \tag{6.59}$$

Note that the negative sign in the $O(t^2)$ term is compatible with the inequality (6.47).

6.4 Quantum time correlation functions

The quantum mechanical analog of the equilibrium correlation function (6.6) is

$$C_{AB}(t_1, t_2) = \langle \hat{A}(t_1) \hat{B}(t_2) \rangle = \mathrm{Tr}\left(\frac{e^{-\beta \hat{H}}}{Q} \hat{A}(t_1) \hat{B}(t_2) \right) \tag{6.60}$$

where $Q = \mathrm{Tr}(e^{-\beta \hat{H}})$ and where $\hat{X}(t) = \hat{X}_{\mathrm{H}}(t) = e^{i\hat{H}t/\hbar} \hat{X} e^{-i\hat{H}t/\hbar}$ ($\hat{X}_{\mathrm{H}}$ denotes the Heisenberg representation, see Section 2.7.1, however for notational simplicity we will suppress this subscript in the remainder of this section). The quantum mechanical thermal average and the quantum density operator $\hat{\rho} = Q^{-1}\exp(-\beta\hat{H})$ are discussed in more detail in Section 10.1.5. Several properties follow directly:

1. From the cyclic property of the trace and the fact that the equilibrium density operator $\hat{\rho} = Q^{-1}\exp(-\beta\hat{H})$ commutes with the time evolution operator $\exp(-i\hat{H}t/\hbar)$ it follows that

$$\langle\hat{A}(t_1)\hat{B}(t_2)\rangle = \langle\hat{A}(t_1 - t_2)\hat{B}(0)\rangle \quad \text{or} \quad C_{AB}(t_1, t_2) = C_{AB}(t_1 - t_2) \tag{6.61}$$

and that

$$\langle\hat{A}(t)\hat{B}(0)\rangle = \langle\hat{A}(0)\hat{B}(-t)\rangle \tag{6.62}$$

These identities are identical to their classical counterparts, for example, (6.28).

2. If the Schrödinger representations $\hat{A}$ and $\hat{B}$ are real operators (in the sense that $\hat{A}^*$ is obtained from $\hat{A}$ by replacing all parameters by their complex conjugates)

$$\langle\hat{A}(t)\hat{B}(0)\rangle = \langle\hat{A}(-t)\hat{B}(0)\rangle^* \tag{6.63}$$

3. If $\hat{A}$ and $\hat{B}$ are hermitian operators then the identities

$$C_{BA}(-t) = \langle\hat{B}(-t)\hat{A}(0)\rangle = \langle\hat{B}(0)\hat{A}(t)\rangle = \langle(\hat{B}(0)\hat{A}(t))^\dagger\rangle^* = \langle\hat{A}(t)\hat{B}(0)\rangle^*$$

(where we have used the identity (10.37)) implies that for such operators

$$\langle\hat{A}(t)\hat{B}(0)\rangle = \langle\hat{B}(0)\hat{A}(t)\rangle^*, \qquad \text{that is, } C_{AB}(t) = C^*_{BA}(-t) \tag{6.64}$$

A special case of the identity (6.64) is $\langle\hat{A}(t)\hat{A}(0)\rangle = \langle\hat{A}(-t)\hat{A}(0)\rangle^*$. This shows that (6.63) holds also for the autocorrelation function of a hermitian (not necessarily real) operator.

4. For hermitian $\hat{A}$ and $\hat{B}$, the Fourier transform

$$\tilde{C}_{AB}(\omega) = \int_{-\infty}^{\infty} dt e^{i\omega t} C_{AB}(t) \tag{6.65}$$

satisfies the symmetry property

$$\begin{aligned}\tilde{C}_{AB}(-\omega) &= \int_{-\infty}^{\infty} dt e^{-i\omega t} C_{AB}(t) \overset{t\to -t}{\longrightarrow} \int_{-\infty}^{\infty} dt e^{i\omega t} C_{AB}(-t) \\ &= \int_{-\infty}^{\infty} dt e^{i\omega t} C^*_{BA}(t) = \tilde{C}^*_{BA}(\omega)\end{aligned} \tag{6.66}$$

The reader should notice the difference between $\tilde{C}^*_{BA}(\omega) = \int_{-\infty}^{\infty} dt e^{i\omega t} C^*_{BA}(t)$ and $(\tilde{C}_{BA}(\omega))^* = \int_{-\infty}^{\infty} dt e^{-i\omega t} C^*_{BA}(t)$.

Problem 6.5. For the real and imaginary part of the quantum correlation function

$$\begin{aligned} C^+_{AB}(t) &\equiv C_{AB}(t) + C^*_{AB}(t) = 2\mathrm{Re}C_{AB}(t) \\ C^-_{AB}(t) &\equiv C_{AB}(t) - C^*_{AB}(t) = 2i\mathrm{Im}C_{AB}(t) \end{aligned} \tag{6.67}$$

Show that

$$C^+_{AB}(-t) = C^+_{BA}(t), \qquad C^-_{AB}(-t) = -C^-_{BA}(t) \tag{6.68}$$

and

$$\begin{aligned} \tilde{C}^+_{AB}(\omega) &= \int_{-\infty}^{\infty} dt e^{i\omega t} C^+_{AB}(t) = \tilde{C}_{AB}(\omega) + \tilde{C}_{BA}(-\omega) \\ \tilde{C}^-_{AB}(\omega) &= \tilde{C}_{AB}(\omega) - \tilde{C}_{BA}(-\omega) \end{aligned} \tag{6.69}$$

5. In the representation defined by the complete set of eigenstates of the Hamiltonian Eqs (6.60) and (6.65) yield

$$\begin{aligned} \tilde{C}_{AB}(\omega) &= \int_{-\infty}^{\infty} dt e^{i\omega t} \langle \hat{A}(t)\hat{B}(0)\rangle = \int_{-\infty}^{\infty} dt e^{i\omega t} \mathrm{Tr}\left(\frac{e^{-\beta\hat{H}}}{Q} e^{i\hat{H}t/\hbar}\hat{A}e^{-i\hat{H}t/\hbar}\hat{B}\right) \\ &= \frac{1}{Q}\int_{-\infty}^{\infty} dt e^{i\omega t} \sum_n \sum_m e^{-\beta E_n} e^{i(E_n - E_m)t/\hbar} A_{nm} B_{mn} \end{aligned} \tag{6.70}$$

The integral over time yields $2\pi\hbar\delta(E_n - E_m + \hbar\omega)$, so

$$\tilde{C}_{AB}(\omega) = \frac{2\pi\hbar}{Q} \sum_n e^{-\beta E_n} \sum_m A_{nm} B_{mn} \delta(E_n - E_m + \hbar\omega) \tag{6.71}$$

from this follows the following interesting identities

$$\tilde{C}_{AB}(\omega) = e^{\beta\hbar\omega} \tilde{C}_{BA}(-\omega) \tag{6.72}$$

or

$$\int_{-\infty}^{\infty} dte^{i\omega t}\langle \hat{A}(t)\hat{B}(0)\rangle = e^{\beta\hbar\omega}\int_{-\infty}^{\infty} dte^{-i\omega t}\langle \hat{B}(t)\hat{A}(0)\rangle = e^{\beta\hbar\omega}\int_{-\infty}^{\infty} dte^{i\omega t}\langle \hat{B}(0)\hat{A}(t)\rangle \tag{6.73}$$

Equations (6.72) and (6.73) are obtained by replacing in (6.70) $\exp(-\beta E_n) \to \exp(-\beta(E_m - \hbar\omega))$. The possibility for this substitution is implied by (6.71). We then get

$$\begin{aligned}\int_{-\infty}^{\infty} dte^{i\omega t}\langle \hat{A}(t)\hat{B}(0)\rangle &= e^{\beta\hbar\omega}\int_{-\infty}^{\infty} dte^{i\omega t}\sum_n\sum_m \frac{e^{-\beta E_m}}{Q}\langle m|\hat{B}|n\rangle\langle n|e^{i\hat{H}t/\hbar}\hat{A}e^{-i\hat{H}t/\hbar}|m\rangle \\ &= e^{\beta\hbar\omega}\int_{-\infty}^{\infty} dte^{i\omega t}\mathrm{Tr}\left(\frac{e^{-\beta\hat{H}}}{Q}\hat{B}e^{i\hat{H}t/\hbar}\hat{A}e^{-i\hat{H}t/\hbar}\right)\end{aligned} \tag{6.74}$$

which is the expression on the right-hand side of (6.73). The middle form in (6.73) is obtained by transforming the integration variable $t \to -t$. For $\hat{B} = \hat{A}$ Eq. (6.73) becomes

$$\begin{aligned}\int_{-\infty}^{\infty} dte^{i\omega t}\langle \hat{A}(t)\hat{A}(0)\rangle &= e^{\beta\hbar\omega}\int_{-\infty}^{\infty} dte^{i\omega t}\langle \hat{A}(-t)\hat{A}(0)\rangle \\ &= e^{\beta\hbar\omega}\int_{-\infty}^{\infty} dte^{-i\omega t}\langle \hat{A}(t)\hat{A}(0)\rangle\end{aligned} \tag{6.75}$$

Note that in classical mechanics the mutual commutativity of $A(t)$ and $A(0)$ implies that these identities are satisfied without the factors $e^{\beta\hbar\omega}$. This is consistent with the fact that the classical limit is achieved when $\beta\hbar\omega \ll 1$. We will see that the identities (6.73) and (6.75) imply the existence of detailed balance in thermally equilibrated systems.

6.5 Harmonic reservoir

The Debye model discussed in Section 4.2.4 rests on three physical observations: The fact that an atomic system characterized by small oscillations about the point of minimum energy can be described as a system of independent harmonic oscillators, the observation that the small frequency limit of the dispersion relation stems from the characterization of long wavelength modes as sound waves and the realization that an upper frequency cutoff is suggested by the finite number of atoms and the finite interatomic distance. None of these features rely on some underlying crystal periodicity. Indeed, in describing systems interacting with their condensed thermal environment we often model the latter as a bath of harmonic oscillators even when this environment is not a periodic solid or even a solid at all. This modeling is suggested by the mathematical simplicity of the harmonic oscillator problem on one hand, and by timescale considerations on the other. We will return to the latter issue below. In what follows we assume the thermal environment (referred to as "bath" below) may indeed be described by a set of independent oscillators ("modes") and explore the dynamical properties of such an environment.

6.5.1 Classical bath

We consider a classical equilibrium system of independent harmonic oscillators whose positions and velocities are denoted $x_j, v_j = \dot{x}_j$, respectively. In fact, dealing with normal modes implies that we have gone through the linearization and diagonalization procedure described in Section 4.2.1. In this procedure it is convenient to work in mass-normalized coordinates, in particular when the problem involves different particle masses. This would lead to mass weighted position and velocities, $y_j = \sqrt{m_j}x_j$; $\dot{y}_j = \sqrt{m_j}v_j$ and to the normal modes coordinates $(u_j, \dot{u}_j)$,[5] in terms of which the bath Hamiltonian is

$$H = (1/2)\sum_j (\dot{u}_j^2 + \omega_j^2 u_j^2). \tag{6.76}$$

The phase space probability distribution is

$$\begin{aligned} P(\{\dot{u}_k, u_k\}) &= \prod_k P_k(\dot{u}_k, u_k) \\ P_k(\dot{u}_k, u_k) &= \frac{\beta\omega_k}{2\pi} e^{-(1/2)\beta(\dot{u}_k^2 + \omega_k^2 u_k^2)} \end{aligned} \tag{6.77}$$

[5] The normal modes are derived from the mass weighted coordinates, therefore u has the dimensionality $[l][m]^{1/2}$. In a system of identical atoms it is sometimes convenient to derive the normal modes from the original coordinates so as to keep the conventional dimensionality of u and $\dot{u}$. In this case the Hamiltonian is $H = (m/2)\sum_j (\dot{u}_j^2 + \omega_j^2 u_j^2)$ and Eqs (6.78) take the forms

$$\langle u_k u_{k'} \rangle = k_B T/(m\omega^2)\delta_{kk'}; \qquad \langle \dot{u}_k \dot{u}_{k'} \rangle = (k_B T/m)\delta_{kk'}$$

The coefficient in front of the exponent is a normalization factor. This leads to

$$\langle u_k u_{k'} \rangle = \frac{k_B T}{\omega_k^2} \delta_{kk'}; \qquad \langle \dot{u}_k \dot{u}_{k'} \rangle = k_B T \delta_{kk'}; \qquad \langle u_k \dot{u}_k \rangle = 0 \tag{6.78}$$

This bath is characterized by the density of modes function, $g(\omega)$, defined such that $g(\omega)d\omega$ is the number of modes whose frequency lies in the interval $\omega, \ldots, \omega + d\omega$. Let

$$A = \sum\nolimits_j c_j u_j, \qquad B = \sum\nolimits_j c_j \dot{u}_j = \dot{A} \tag{6.79}$$

and consider the time correlation functions

$$C_{AA}(t) = \langle A(t)A(0) \rangle, \qquad C_{BB}(t) = \langle B(t)B(0) \rangle, \quad \text{and} \quad C_{AB}(t) = \langle A(t)B(0) \rangle \tag{6.80}$$

Such correlation functions are often encountered in treatments of systems coupled to their thermal environment, where the model for the system–bath interaction is taken as a product of A or B with a system variable. In such treatments the coefficients c_j reflect the distribution of the system–bath coupling among the different modes. In classical mechanics these functions can be easily evaluated explicitly from the definition (6.6) by using the general solution of the harmonic oscillator equations of motion

$$u_j(t) = u_j(0) \cos(\omega_j t) + \omega_j^{-1} \dot{u}_j(0) \sin(\omega_j t) \tag{6.81a}$$

$$\dot{u}_j(t) = -\omega_j u_j(0) \sin(\omega_j t) + \dot{u}_j(0) \cos(\omega_j t) \tag{6.81b}$$

Problem 6.6.

(1) Show that if Eq. (6.78) holds for $u_j(t = 0)$ and $\dot{u}_j(t = 0)$ then it holds at any time, for example, $\omega_j^2 \langle u_j(t) u_{j'}(t) \rangle = \langle \dot{u}_j(t) \dot{u}_{j'}(t) \rangle = k_B T \delta_{j,j'}$.

(2) Using Eqs (6.81) show that the velocity time correlation function of a classical harmonic oscillator in thermal equilibrium satisfies

$$\langle \dot{u}_j(0) \dot{u}_{j'}(t) \rangle = k_B T \cos(\omega_j t) \delta_{j,j'} \tag{6.82}$$

(3) Use Eq. (2.166) to show that the quantum analog of (6.82) is

$$\langle \dot{u}_j(0)\dot{u}_j(t)\rangle = \frac{\hbar\omega_j}{e^{\beta\hbar\omega_j}-1}\cos(\omega_j t) + \frac{\hbar\omega_j}{2}e^{i\omega_j t} \tag{6.83}$$

(4) Show that in an isotropic harmonic system of N identical atoms with atomic mass m the density of normal modes is related to the atomic velocity correlation function by

$$\frac{g(\omega)}{N} = \frac{3m}{\pi k_B T}\int_{-\infty}^{\infty} dt\langle \dot{x}(0)\dot{x}(t)\rangle e^{-i\omega t} \tag{6.84}$$

Solution to 6(4). Derivation of Eq. (6.84):
From Eqs (4.13) and (6.81) it follows that

$$\begin{aligned}\langle \dot{y}_j(0)\dot{y}_{j'}(t)\rangle &= \sum_k\sum_{k'}(\mathbf{T}^{-1})_{jk}(\mathbf{T}^{-1})_{j'k'}\langle \dot{u}_k(0)\dot{u}_{k'}(t)\rangle \\ &= k_BT\sum_k(\mathbf{T}^{-1})_{jk}\mathbf{T}_{kj'}\cos(\omega_k t)\end{aligned} \tag{6.85}$$

In the second equality we have used the unitarity of $\mathbf{T}$. Using also $\sum_j(\mathbf{T}^{-1})_{jk}\mathbf{T}_{kj} = 1$ we get from (6.85)

$$\sum_j\langle \dot{y}_j(0)\dot{y}_j(t)\rangle = k_BT\sum_k\cos\omega_k t = k_BT\int_0^{\infty} d\omega g(\omega)\cos(\omega t) \tag{6.86}$$

If all atoms are identical, the left-hand side of (6.86) is $3Nm\langle \dot{x}(0)\dot{x}(t)\rangle$ where $\dot{x}$ is the atomic velocity (see Eq. (4.9)). *Defining* $g(-\omega) = g(\omega)$, we rewrite (6.86) in the form

$$\langle \dot{x}(0)\dot{x}(t)\rangle = \frac{k_BT}{6mN}\int_{-\infty}^{\infty} d\omega g(\omega)e^{i\omega t} \tag{6.87}$$

which leads, by inverting the Fourier transform, to (6.84).

Using Eqs (6.81) and (6.78) in (6.80) yields

$$C_{AA}(t) = k_BT\sum_j \frac{c_j^2}{\omega_j^2}\cos(\omega_j t) = \frac{2k_BT}{\pi}\int_0^\infty d\omega \frac{J(\omega)}{\omega}\cos(\omega t) \tag{6.88a}$$

$$C_{BB}(t) = k_BT\sum_j c_j^2\cos(\omega_j t) = \frac{2k_BT}{\pi}\int_0^\infty d\omega\omega J(\omega)\cos(\omega t) \tag{6.88b}$$

and

$$C_{AB}(t) = k_BT\sum_j \frac{c_j^2}{\omega_j}\sin(\omega_j t)\cos(\omega_j t) = \frac{2k_BT}{\pi}\int_0^\infty d\omega J(\omega)\sin(\omega t)\cos(\omega t) \tag{6.89}$$

where $J(\omega)$ is the bath *spectral density*, defined by

$$J(\omega) \equiv \frac{\pi}{2}\sum_j \frac{c_j^2}{\omega_j}\delta(\omega - \omega_j) \tag{6.90}$$

The function is defined as a sum of delta-functions, however for macroscopic systems this sum can be handled as a continuous function of ω in the same way that the density of modes, $g(\omega) = \sum_j \delta(\omega - \omega_j)$ is.[6] Defining the coupling density by

$$c^2(\omega)g(\omega) \equiv \sum\nolimits_j c_j^2\delta(\omega - \omega_j) \tag{6.91}$$

Equation (6.90) can also be written as

$$J(\omega) = \frac{\pi g(\omega)c^2(\omega)}{2\omega} \tag{6.92}$$

6.5.2 The spectral density

The spectral density, Eqs (6.90) and (6.92) is seen to be a weighted density of modes that includes as weights the coupling strengths $c^2(\omega)$. The harmonic frequencies ω_j

[6] This is a coarse-graining procedure that is valid if the spacing between frequencies ω_j is much smaller than the inverse time resolution of any conceivable observation. See Section 1.4.4 and Eq. (1.182).

are obviously positive, therefore by this narrow definition $J(\omega)$ (and $g(\omega)$) are zero for negative values of the frequency. It is sometimes useful to extend these functions to negative frequencies by defining $g(-\omega)c^2(-\omega) \equiv g(\omega)c^2(\omega)$. From (6.90) or (6.92) it follows that under this definition $J(\omega)$ is an antisymmetric function of ω,

$$J(-\omega) = -J(\omega) \tag{6.93}$$

The spectral density (see also Sections (7-5.2) and (8-2.5)) plays a prominent role in models of thermal relaxation that use harmonic oscillators description of the thermal environment and where the system-bath coupling is taken linear in the bath coordinates and/or momenta. We will see (an explicit example is given in Sections 8.2.5–8.2.6) that $J(\omega)$ characterizes the dynamics of the thermal environment *as seen by the relaxing system*, and consequently determines the relaxation behavior of the system itself. Two simple models for this function are often used:

The *Ohmic spectral density*

$$J(\omega) = \eta\omega \exp(-|\omega|/\omega_c) \tag{6.94}$$

is characterized by a cutoff frequency ω_c, a linear dependence on ω for $\omega \ll \omega_c$ and an exponential drop to zero for $\omega > \omega_c$. We will see in Section 8.2.6 that in the limit where ω_c^{-1} is smaller than all the characteristic system timescales this model bath affects a simple constant friction on the system.

The *Drude* (sometimes called *Debye*) *spectral density*

$$J(\omega) = \frac{\eta\omega}{1 + (\omega/\omega_c)^2} \tag{6.95}$$

also grows linearly with ω for $\omega \ll \omega_c$, however it vanishes as ω^{-1} when $\omega \to \infty$.

Problem 6.7. In Section 8.2.5 we will encounter the "memory function" $Z(t) = (2/\pi m)\int_0^\infty d\omega (J(\omega)/\omega) \cos(\omega t)$. Calculate the memory functions associated with the Ohmic and the Drude models and compare their time evolutions.

6.5.3 Quantum bath

For a system of harmonic oscillators it is easy to derive the quantum equivalents of the results obtained above, as was already exemplified by Problem 6.6(3). By way of demonstration we focus on the time correlation function, $C_{AA}(t) = \sum_j c_j^2 \langle \hat{u}_j(t)\hat{u}_j(0)\rangle$. The normal mode position operator can be written in terms of the raising and lowering operators (cf. Eq. (2.153); note that the mass factor does not appear below because normal modes coordinates are mass weighted variables)

$$\hat{u}_j = \sqrt{\hbar/(2\omega_j)}(\hat{a}_j^\dagger + \hat{a}_j) \tag{6.96}$$

Using Eqs (2.165) and (2.196), (2.197) we get

$$C_{AA}(t) = \sum_j \frac{\hbar c_j^2}{2\omega_j}((n_j + 1)e^{-i\omega_j t} + n_j e^{i\omega_j t})$$
$$= \frac{\hbar}{\pi}\int_0^\infty d\omega J(\omega)[(n(\omega) + 1)e^{-i\omega t} + n(\omega)e^{i\omega t}] \tag{6.97}$$

where $n_j = n(\omega_j) = (e^{\beta\hbar\omega_j} - 1)^{-1}$. In the classical limit where $n(\omega) = (k_B T/\hbar\omega) \gg 1$ this leads again to (6.88a). $C_{BB}(t)$ and $C_{AB}(t)$ are obtained along similar lines, and for example, $C_{AB}(t) = (i\hbar/\pi)\int_0^\infty d\omega\,\omega J(\omega)[(n(\omega) + 1)e^{-i\omega t} - n(\omega)e^{i\omega t}]$. At time $t = 0$ this gives $C_{AB}(0) = (i\hbar/\pi)\int_0^\infty \omega J(\omega)$, demonstrating that the identity (6.35) does not hold for quantum correlation functions.

Problem 6.8.

(1) Show that $C_{AA}(t)$ satisfies the identity (6.64).
(2) Show that $\int_{-\infty}^\infty dt e^{i\omega t} C_{AA}(t) = 2\hbar[J(\omega)(n(\omega) - n(-\omega) + 1)]$.
(3) Show that $\int_{-\infty}^\infty dt e^{i\omega t} C_{AA}(t) = e^{\beta\hbar\omega}\int_{-\infty}^\infty dt e^{i\omega t} C_{AA}(-t) = e^{\beta\hbar\omega}\int_{-\infty}^\infty dt e^{-i\omega t} C_{AA}(t)$ as implied by Eq. (6.73).

6.5.4 Why are harmonic baths models useful?

Consider Eq. (6.84). This result was obtained for a harmonic system of identical and equivalent atoms. We could however reverse our reasoning and *define* a vibrational spectrum for a dense atomic system from the velocity autocorrelation function according to Eq. (6.84). Since this function can be computed for all systems, including liquids and disordered solids, we may use (6.84) as a definition of a spectrum that may be interpreted as density of modes function for such media. We can then use it in expressions such as (4.33), and (6.92). Is this approach to dynamics in condensed phases any good?

We can also take another route. Obviously, we can repeat the development that lead to Eq. (4.12) for any harmonic system. We can define such a system by expanding of the interatomic potential about some configuration x_0^N, not necessarily a minimum point, and neglecting higher-order terms, as in (4.6). This can be done also for liquids, taking x_0^N to be any instantaneous configuration. The matrix **K**, Eq. (4.9), is again diagonalized and the spectrum of eigenvalues in (4.11) is used to construct the density of modes. The resulting $\langle g_{\text{inst}}(\omega)\rangle$, averaged over all relevant configurations (usually a thermal distribution) is the averaged density of *instantaneous normal modes* of the liquid.

The two distributions, $g(\omega)$ defined by (6.84) and $\langle g_{\text{inst}}(\omega)\rangle$, are not identical, and neither should be taken as a representation of a real harmonic system. Indeed, while the eigenvalues defined by Eq. (4.11) are all positive, so that the corresponding frequencies are real, there is no reason to expect that this will be the case for the "instantaneous frequencies." Imaginary instantaneous frequencies (negative eigenvalues in (4.11)) just reflect the fact that configurations other than local minimum points of the potential surface may have negative curvature in some directions. As for the velocity correlation function, while intuition tells us that high-frequency Fourier components in (6.84) indeed reflect a local oscillatory motion, low-frequency ones seem more likely to reflect longer-range motion. Indeed, the zero frequency Fourier transform $\int_{-\infty}^{\infty} dt\langle\dot{x}(0)\dot{x}(t)\rangle$ has been shown (see Eq. (6.14)) to be related to the diffusion coefficient of the corresponding particle.

Still, these concepts are found to be useful and their usefulness stems from timescale considerations. We will repeatedly see that for many chemical processes the relevant timescales for environmental interactions are short. This does not mean that the system sees its environment for just a short time, but that the dynamics is determined by a succession of short time interactions. If subsequent interactions are uncorrelated with each other, each can be treated separately and for this treatment a harmonic bath picture might suffice. Two conditions need to be satisfied for this to be a good approximation:

1. In the instantaneous normal mode picture, the density should be high enough and the temperature low enough so that the solvent stays not too far from its equilibrium configuration and therefore the contribution of modes of imaginary frequencies can be disregarded.
2. The timescale of environmental (solvent) motions that determine the solvent dynamical effect on the process of interest should be shorter than the timescale on which the solvent can be described as a harmonic medium.

Figure 6.2 shows an example where these ideas were applied to water as a solvent. The process under investigation is solvation dynamics (see Chapter 15), in this particular case—solvation of electron in water. Figure 6.2(a) shows the instantaneous normal mode density for water at 300 K obtained from numerical simulations. By

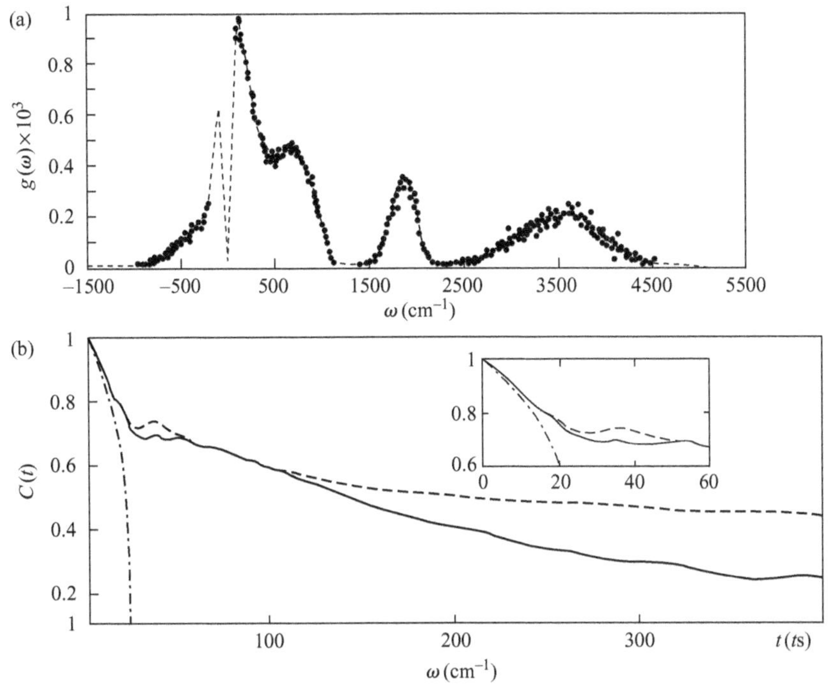

FIG. 6.2 (a) Instantaneous normal modes in room temperature water as obtained from molecular dynamics simulations. The negative frequency axis is used to show the density of imaginary frequencies. (b) The solvation response function (see Chapter 15) for electron solvation in water, calculated from direct classical MD simulations (full line), from the instantaneous normal mode representation of water (dash-dotted line), and from a similar instantaneous normal mode representation in which the imaginary frequency modes were excluded (dashed line). The inset in Fig. 6.2 shows the short time behavior of the same data. (From C.-Y. Yang, K. F. Wong, M. S. Skaf, and P. J. Rossky, J. Chem. Phys. **114**, 3598 (2001).)

convention the modes of imaginary frequency are shown on the negative side of the ω axis. The peaks about 1700 cm^{-1} and 3600 cm^{-1} correspond to the internal vibrational modes of water. Figure 6.2(b) shows the result of a calculation pertaining to solvation dynamics of electron in water, comparing the result of a full calculation (full line) to results obtained by representing water as a normal mode fluid with normal mode density taken from Fig. 6.2(a) (dashed-dotted-line) and from a similar calculation (dashed line) that uses only the real frequency modes. The agreement, for up to $t \approx 100$ fs, between the full calculation and the calculation based on instantaneous normal modes of real frequencies, suggests that at least as a practical tool a harmonic view of water can be useful for describing processes whose dynamics is determined by shorter timescales.

The use of harmonic baths to model the thermal environment of molecular systems does not rest only on such timescale arguments. We have seen in Chapter 3 that the radiation field constitutes an harmonic environment that determines the radiative relaxation properties of all material systems. We will see in Chapters 15 and 16 that dielectric solvents can be modeled as harmonic environments in which the harmonic modes are motions of the polarization field. In the latter case the harmonic environment picture does not rely on a short time approximation, but rather stems from the long wavelength nature of the motions involved that makes it possible to view the solvent as a continuum dielectric.

Further reading

J. P. Hansen and I. R. McDonald, *Theory of Simple Liquids*, 2nd edition (Elsevier, London, 1986) Chapter 7.

D. A. McQuarrie, Statistical Mechanics (Harper and Row, New York, 1976) Chapters 21–22.

7

INTRODUCTION TO STOCHASTIC PROCESSES

Once you make
The count of atoms limited, what follows
Is just this kind of tossing, being tossed,
Flotsam and jetsam through the seas of time,
Never allowed to join in peace, to dwell
In peace, estranged from amity and growth...

Lucretius (c.99–c.55 BCE) "The way things are" translated by Rolfe Humphries, Indiana University Press, 1968

As discussed in Section 1.5, the characterization of observables as random variables is ubiquitous in descriptions of physical phenomena. This is not immediately obvious in view of the fact that the physical equations of motion are deterministic and this issue was discussed in Section 1.5.1. Random functions, ordered sequences of random variable, were discussed in Section 1.5.3. The focus of this chapter is a particular class of random functions, stochastic processes, for which the ordering parameter is time. Time is a continuous ordering parameter, however in many practical situations observations of the random function $z(t)$ are made at discrete time $0 < t_1 < t_2, \ldots, < t_n < T$. In this case the sequence $\{z(t_i)\}$ is a discrete sample of the stochastic process $z(t)$.

7.1 The nature of stochastic processes

Let us start with an example. Consider a stretch of highway between two intersections, and let the variable of interest be the number of cars within this road segment at any given time, $N(t)$. This number is obviously a random function of time whose properties can be deduced from observation and also from experience and intuition. First, this function takes positive integer values but this is of no significance: we could redefine $N \to N - \langle N \rangle$ and the new variable will assume both positive and negative values. Second and more significantly, this function is characterized by several timescales:

1. Let τ_1 is the average time it takes a car to go through this road segment, for example 1 min, and compare $N(t)$ and $N(t + \Delta t)$ for $\Delta t \ll \tau_1$ and $\Delta t \gg \tau_1$.

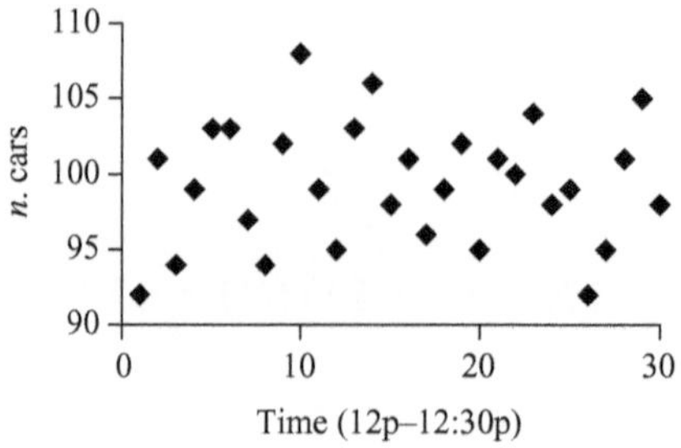

FIG. 7.1 An illustrative example of stochastic processes: The number of cars in a given road segment during a period of 30 min. Sampling is taken every 1 min which is the average time it takes a car to pass through this road segment.

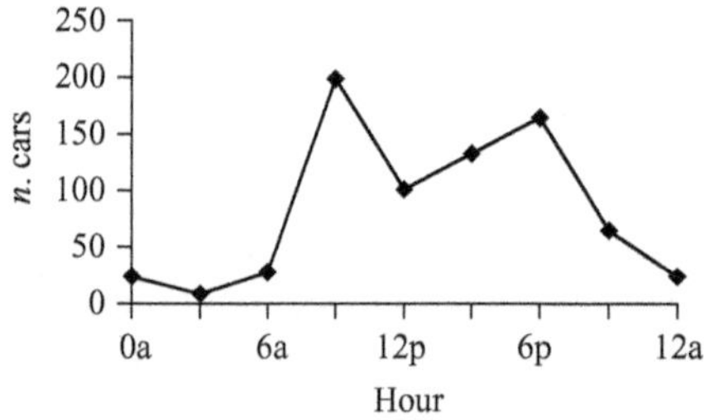

FIG. 7.2 The same process as in Fig. 7.1, observed over a longer timescale. Sampling is taken every 3 h over a period of 24 h.

Obviously $N(t) \approx N(t + \Delta t)$ for $\Delta t \ll \tau_1$ while in the opposite case the random numbers $N(t)$ and $N(t + \Delta t)$ will be almost uncorrelated. Figure 7.1 shows a typical result of one observation of this kind. The apparent lack of correlations between successive points in this data set expresses the fact that numbers sampled at intervals equal to or longer than the time it takes to traverse the given distance are not correlated.

2. The apparent lack of systematic component in the time series displayed here reflects only a relatively short-time behavior. For time exceeding another characteristic time, τ_2, typically of the order $\sim$1 h for this problem, we observe what appears to be a systematic trend as seen in Fig. 7.2.

   Here sampling is made every 3 h over a period of 24 hours and the line connecting the results has been added to aid the eye. Alternatively, we may perform coarse graining in the spirit of Eq. (1.180) using time intervals of, say, $\Delta t = \tau_2 = 1$ h, which will lead to a smoother display. The systematic trend shows the high and low traffic volumes at different times of day.
3. If we extend our observation to longer times we will see other trends that occur on longer timescales. In this example, we may distinguish between

long timescale (multi-year) evolution associated with changing population characteristics, monthly evolution related to seasons, daily evolution associated with work-weekend schedules, hourly evolution (timescale τ_2) related to day/night and working hours, and short time evolution (timescale τ_1) associated with individual driver's timings. All but the last one are systematic phenomena that could be analyzed and predicted. This last stochastic component could be eliminated from our considerations by coarse graining over time period longer than τ_1 and shorter than τ_2. The resulting description will be adequate for many practical applications, for example, planning the highway system.

Note that in the example considered above, the systematic behavior and the stochastic fluctuations arise from different causes. Sometimes the random motion itself gives, in time, a systematic signal. If we put an ink drop at the center of a pool of water the systematic spread of the ink by diffusion is caused by random motions of the individual molecules, each characterized by zero mean displacement.

Consider another example where the variable of interest is the number n of molecules in some small volume ΔV about a point $\mathbf{r}$ in a homogeneous equilibrium fluid. Again, this is a random variable that can be monitored as a function of time, and the result $n(t)$ is again a stochastic process. In the homogeneous fluid all spatial positions are equivalent and we may regard a set of such points $\{\mathbf{r}_j\}$ and volumes ΔV about them, sufficiently far from each other so that these small volumes are statistically independent. This defines an ensemble of identical systems. The corresponding ensemble of stochastic trajectories, $n_j(t); j = 1, \ldots, N$, is a collection of different *realizations* of the stochastic process under consideration. We can also obtain such different realizations if we focus on a single system and generate trajectories $n(t)$ from different starting times. The equivalence of both ways for generating realizations stems from the fact that the stochastic process under discussion is stationary, displaying no average (systematic) time evolution.

The following table represents a data set collected at discrete time points for N such systems

	$t_1 = 0$	$t_2 = t_1 + \Delta t$	$t_3 = t_1 + 2\Delta t$	$\ldots$
n_1	$n_1(t_1)$	$n_1(t_2)$	$n_1(t_3)$	$\ldots$
n_2	$n_2(t_1)$	$n_2(t_2)$	$n_2(t_3)$	$\ldots$
$\vdots$	$\vdots$	$\vdots$	$\vdots$	$\ldots$
n_N	$n_N(t_1)$	$n_N(t_2)$	$n_N(t_3)$	$\ldots$

Each line in this set represents a realization of our stochastic process. In fact each set of numbers starting at any time, for example, $n_1(t_3), n_1(t_4 - t_3), n_1(t_5 - t_3), \ldots,$ represents such a realization. Each column, for example, $n_j(t_1); j = 1, \ldots, N$, contains different realizations of the random variable that represents the number of particles in volume ΔV at the given time.

Two important observations can now be made

1. The data set given above can be used to generate the probability distribution associated with the random variable n. For each column (i.e. at each time t) we can count the number of systems in the ensemble, $S_n(t)$, that contain at time t exactly n molecules. A plot of S_n against n is called *histogram*. The required probability distribution at time t is given by this histogram after normalization

$$P(n, t) = \lim_{N\to\infty} \frac{S_n(t)}{\sum_n S_n(t)} = \lim_{N\to\infty} \frac{S_n(t)}{N} \tag{7.1}$$

In practical applications a finite large N is used. In a stationary system like our equilibrium fluid, where $P(n, t) = P(n)$ does not depend on time, we can reduce statistical errors by further averaging (7.1) over different time data (i.e. different columns).

Problem 7.1. Consider the observation of the number of cars on a road segment as a function of time discussed above. In order to obtain the distribution $P(n, t)$ for this process one needs to obtain a representative set of stochastic trajectories. Using the timescales scenario discussed above for this system, suggest a procedure for obtaining an approximation for $P(n, t)$.

2. The stationary nature of our system and the ergodic theorem (see Section 1.4.2) imply that time and ensemble averaging are equivalent. This by no means implies that the statistical information in a row of the table above is equivalent to that in a column. As defined, the different systems $j = 1, \ldots, N$ are statistically independent, so, for example, $\langle n_1(t_1)n_2(t_1)\rangle = \langle n_1(t_1)\rangle\langle n_2(t_1)\rangle$. In contrast, when two times, t_1 and t_2, are close to each other the numbers $n_1(t_1)$ and $n_1(t_2)$ may not be statistically independent so that $\langle n_1(t_1)n_1(t_2)\rangle \neq \langle n_1(t_1)\rangle\langle n_1(t_2)\rangle$. The time series provide information about time correlations that is absent from a single time ensemble data. The stationary nature of our system does imply, as discussed in Section 6.1, that $\langle n_1(t_1)n_1(t_2)\rangle$ depends only on the time difference $t_2 - t_1$.

More generally, while in a stationary system $P(n, t) = P(n)$ contains no dynamical information, such information is contained in joint distributions like $P(n_2, t_2; n_1, t_1)$ (the probability to observe n_1 molecules at time t_1 and n_2 molecules at time t_2) and conditional distributions such as $P(n_2, t_2 \mid n_1, t_1)$ (the probability to observe n_2 molecules at time t_2 *given that* n_1 molecules were observed at time t_1). We will expand on these issues in Section 7.4.

7.2 Stochastic modeling of physical processes

Given the initial conditions of a classical system of N particles (i.e. all initial $3N$ positions and $3N$ momenta) its time evolution is determined by the Newton equations of motion. For a quantum system, the corresponding N-particle wavefunction is determined by evolving the initial wavefunction according to the Schrödinger equation. In fact these initial conditions are generally not known but can often be characterized by a probability distribution (e.g. the Boltzmann distribution for an equilibrium system). The (completely deterministic) time evolution associated with any given initial state should be averaged over this distribution. This is an ensemble average of deterministic trajectories.

As discussed in Section 1.5.1, we often seek simplified descriptions of physical processes by focusing on a small subsystem or on a few observables that characterize the process of interest, and these variables then assume random character. As a particular example consider the center of mass position $\mathbf{r}_i$ of an isotopically substituted molecule i in a homogeneous equilibrium fluid containing a macroscopic number N of normal molecules. The trajectory $\mathbf{r}_i(t)$ of this molecule shows an erratic behavior, changing direction (and velocity) after each collision. This trajectory is just a projection of a deterministic trajectory in the $6N$-dimensional phase space on the coordinate of interest, however solving this $6N$-dimensional problem may be intractable and, moreover, may constitute a huge waste of effort because it yields the time dependence of $6N$ momenta and positions of all N particles while we are interested only in the position $\mathbf{r}_i(t)$ of a particular particle i. Instead we may look for a reduced description of $\mathbf{r}_i(t)$ only. We may attempt to get it by a systematical reduction of the $6N$-coupled equations of motion. Alternatively, we may construct a phenomenological model for the motion of this coordinate under the influence of all other motions. As we shall see, both ways lead to the characterization of $\mathbf{r}_i(\mathrm{t})$ as a stochastic process.

As another example consider the internal vibrational energy of a diatomic solute molecule, for example, CO, in a simple atomic solvent (e.g. Ar). This energy can be monitored by spectroscopic methods, and we can follow processes such as thermal (or optical) excitation and relaxation, energy transfer, and energy migration. The observable of interest may be the time evolution of the average vibrational energy per molecule, where the average is taken over all molecules of this type in the system (or in the observation zone). At low concentration these molecules do not affect each other and all the information can be obtained by observing (or theorizing on) the energy $E_j(t)$ of a single molecule j. The average molecular energy $\langle E(t)\rangle$ is obtained as an ensemble average of $E_j(t)$ over many such molecules, or over repeated independent observations on a single molecule.

With respect to the average vibrational energy, it is often observed following vibrational excitation that this observable relaxes as an exponential function of time, $\langle E(t)\rangle = E(0)\exp(-\gamma t)$. A single trajectory $E_j(t)$ (also observable in principle by a technique called *single molecule spectroscopy*, see Section 18.6.3) is however much more complicated. As before, to predict its exact course of evolution we need to know the initial positions and velocities of all the particles in the system (in quantum mechanics—the initial many-particle wavefunction), then to solve the Newton or the Schrödinger equation with these initial conditions. Again, the resulting trajectory in phase space is completely deterministic, however $E_j(t)$ appears random. In particular, it will look different in repeated experiments because in setting up such experiments only the initial value of E_j is specified, while the other degrees of freedom are subjected only to a few conditions (such as temperature and density). In this *reduced* description $E_j(t)$ may be viewed as a stochastic variable. The role of the theory is to determine its statistical properties and to investigate their consequences.

Obviously, for a given physical process, different stochastic models can be considered by employing different levels of reduction, that is, different subspaces in which the process is described. For example, the time evolution of the vibrational energy of a single diatomic molecule can be described as a stochastic evolution of just this variable, or by studying the stochastic dynamics in the subspace of the coordinate (the internuclear distance) and momentum of the intramolecular nuclear motion, or by focusing on the atomic coordinates and velocities associated with the molecule and its nearest neighbors, etc. These increasingly detailed reduced descriptions lead to greater accuracy at the cost of bigger calculations. The choice of the level of reduction is guided by the information designated as relevant based on available experiments, and by considerations of accuracy based on physical arguments. In particular, timescale and interaction-range considerations are central to the theory and practice of reduced descriptions.

The relevance of stochastic descriptions brings out the issue of their theoretical and numerical evaluation. Instead of solving the equations of motion for $\sim 6\times 10^{23}$ degrees of freedom we now face the much less demanding, but still challenging need to construct and to solve *stochastic equations of motion* for the few relevant variables. The next section describes a particular example.

7.3 The random walk problem

Random walk is a stochastic process that represents a reduced molecular description of diffusion, in which a particle starts from a given position and executes a stochastic motion, occasionally changing direction in a random way. This random change in direction represents a collision process, after which the particle assumes a new, randomly chosen, direction, moves a certain length Δx (of the order of the mean free path) until the next collision. As a simple example consider a one-dimensional lattice model with lattice spacing Δx, in which the particle moves between nearest neighbor lattice points. During a time interval Δt the particle may move to the right with probability $p_r = k_r \Delta t$ and to the left with probability $p_l = k_l \Delta t$. The probability that it stays in its original position is $1 - p_r - p_l$. k_l and k_r are rate coefficients, measuring the probabilities per unit time that the corresponding events will occur. In an isotropic homogeneous system the rates to move to the right and the left are the same, $k_l = k_r$, and are position-independent. Inequality may reflect the existence of some force that makes these rates different from each other. Obviously p_l and p_r are linear in Δt only for Δt sufficiently small, so that these numbers are substantially less than 1.

7.3.1 Time evolution

Starting from $t = 0$, we want to know the probability $P(n, t) = P(n, N, \Delta t)$ that the particle has made a net number n of steps to the right (a negative n implies that the particle has actually moved to the left) after time $t = N\Delta t$. In other words, for a particle that starts at position $n = 0$ we seek the probability to find it at position n (i.e. at distance $n\Delta l$ from the origin) at time t, after making at total of N steps. An equation for $P(n, t)$ can be found by considering the propagation from time t to $t + \Delta t$:

$$P(n, t+\Delta t) = P(n, t) + k_r \Delta t\,(P(n-1, t) - P(n, t)) + k_l \Delta t\,(P(n+1, t) - P(n, t)) \tag{7.2}$$

In Eq. (7.2) the terms that add to $P(n, t)$ on the right hand side result from the walk. Thus, for example, $k_r \Delta t P(n-1, t)$is the increase in $P(n, t)$ due to the possibility of a jump from position n-1 to position n during a time interval Δt, while $-k_r \Delta t P(n, t)$ is the decrease in $P(n, t)$ because of transition from n to n+1 in the same period. Rearranging Eq. (7.2) and dividing by Δt we get, when $\Delta t \to 0$,

$$\frac{\partial P(n, t)}{\partial t} = k_r\,(P(n-1, t) - P(n, t)) + k_l\,(P(n+1, t) - P(n, t)) \tag{7.3}$$

Note that in (7.3) time is a continuous variable, while the position n is discrete. We may go into a continuous representation also in position space by substituting

$n \to n\Delta x = x,\ n-1 \to x - \Delta x,\ n+1 \to x + \Delta x$, to get

$$\frac{\partial P(x,t)}{\partial t} = k_r\,(P(x-\Delta x,t) - P(x,t)) + k_l\,(P(x+\Delta x,t) - P(x,t)) \tag{7.4}$$

Here $P(x,t)$ may be understood as the probability to find the particle in an interval of length Δx about x. Introducing the density $f(x,t)$ so that $P(x,t) = f(x,t)\Delta x$ and expanding the right hand side of (7.4) up to second order in Δx we obtain

$$\frac{\partial f(x,t)}{\partial t} = -v\frac{\partial f(x,t)}{\partial x} + D\frac{\partial^2 f(x,t)}{\partial x^2} \tag{7.5}$$

where

$$v = (k_r - k_l)\Delta x = (p_r - p_l)\,(\Delta x/\Delta t) \tag{7.6}$$

and where

$$D = (1/2)(k_r + k_l)\Delta x^2 = (p_r + p_l)[\Delta x^2/(2\Delta t)] \tag{7.7}$$

Note that even though in (7.5) we use a continuous representation of position and time, the nature of our physical problem implies that Δx and Δt are finite, of the order of the mean free path and the mean free time, respectively.

To get a feeling for the nature of the solution of Eq. (7.5) consider first the case $D = 0$ (that is also obtained if we truncate the expansion that led to (7.5) after first order). The solutions of the equation $\partial f/\partial t = -v\partial f/\partial x$ have the form $f(x,t) = f(x - vt)$, that is, any structure defined by f moves to the right with speed v (*drift velocity*). This behavior is expected under the influence of a constant force that makes k_r and k_l different. The first term of (7.5) reflects the effect of the systematic motion resulting from this force.

Problem 7.2.

(1) Under what physical circumstances does a constant force lead to motion with a constant speed that depends linearly on the force?
(2) Suppose the force on the particles is derived from a potential $U = -Fx$. Assume that the rates k_r and k_l satisfy

$$k(x \to x+\Delta x) = \begin{cases} A\exp[-\beta(U(x+\Delta x) - U(x))] & \text{for } U(x+\Delta x) > U(x) \\ A & \text{for } U(x+\Delta x) \le U(x) \end{cases}$$

where $\beta = (k_B T)^{-1}$. Assuming that $|F\Delta x| \ll k_B T$ derive an expression for the *mobility* u, a (temperature-dependent) parameter defined by $v = uF$.

Next consider Eq. (7.5) for the case $v = 0$, that is, when $k_r = k_l$. In this case Eq. (7.5) becomes the diffusion equation

$$\frac{\partial f(x,t)}{\partial t} = D\frac{\partial^2 f(x,t)}{\partial x^2} \tag{7.8}$$

The solution of this equation for the initial condition $f(x,0) = \delta(x - x_0)$ is

$$f(x,t \mid x_0, t = 0) = \frac{1}{(4\pi Dt)^{1/2}} \exp\left(-\frac{(x - x_0)^2}{4Dt}\right) \tag{7.9}$$

Note that the left-hand side is written as a conditional probability density. This is the probability density about point x at time t given that the particle was at x_0 at time $t = 0$. Note also that the initial density $f(x,0) = \delta(x - x_0)$ reflects the initial condition that the particle was, with probability 1, at x_0. The diffusion process is the actual manifestation of the random walk that leads to a symmetric spread of the density about the initial position.

Problem 7.3.

(1) Show that the transformation $x - vt \rightarrow x$ transforms Eq. (7.5) into the form (7.8). What is the implication of this observation?
(2) Show that the solution of (7.5) under the initial condition $f(x,0) = \delta(x - x_0)$ is

$$f(x,t \mid x_0, t = 0) = \frac{1}{(4\pi Dt)^{1/2}} \exp\left(-\frac{(x - vt - x_0)^2}{4Dt}\right) \tag{7.10}$$

The solution (7.10) shows both drift and the diffusion spread.

7.3.2 Moments

Further insight into the nature of this drift–diffusion process can be obtained by considering the moments of this probability distribution. Equation (7.3) readily yields equations that describe the time evolution of these moments.

Problem 7.4. Show that both sides of Eq. (7.3) yield zero when summed over all n from $-\infty$ to ∞, while multiplying this equation by n and n^2 then performing the summation lead to

$$\frac{d\langle n\rangle}{dt} = k_r - k_l \tag{7.11}$$

and

$$\frac{d\langle n^2\rangle}{dt} = 2\langle n\rangle (k_r - k_l) + k_r + k_l \tag{7.12}$$

Assuming $\langle n\rangle(t = 0) = \langle n^2\rangle(t = 0) = 0$, that is, that the particle starts its walk from the origin, $n = 0$, Eq. (7.11) results in

$$\langle n\rangle_t = (k_r - k_l)t \tag{7.13}$$

while Eq. (7.12) leads to

$$\langle n^2\rangle_t = (k_r - k_l)^2 t^2 + (k_r + k_l)t \tag{7.14}$$

From Eqs (7.13) and (7.14) it follows that

$$\langle \delta n^2\rangle_t = \langle n^2\rangle_t - \langle n\rangle_t^2 = (k_r + k_l)t = (p_r + p_l)\frac{t}{\Delta t} = (p_r + p_l)N \tag{7.15}$$

for a walker that has executed a total of N steps during time $t = N\Delta t$.

Similar results may be obtained from Eq. (7.5). Suppose the particle starts at the origin, $x = 0$. Its average position at time t is given by

$$\langle x\rangle_t = \int_{-\infty}^{\infty} dx x f(x, t) \tag{7.16}$$

Therefore,

$$\frac{\partial \langle x\rangle}{\partial t} = -v \int_{-\infty}^{\infty} dx\, x \frac{\partial}{\partial x} f(x, t) + D \int_{-\infty}^{\infty} dx\, x \frac{\partial^2}{\partial x^2} f(x, t) \tag{7.17}$$

Integrating on the right-hand side by parts, using the fact that f and its derivatives have to vanish at $|x| \to \infty$, leads to[1]

$$\frac{\partial \langle x \rangle}{\partial t} = v, \qquad \text{that is, } \langle x \rangle = vt \text{ at all times} \tag{7.18}$$

Consider now the second moment

$$\langle x^2 \rangle_t = \int_{-\infty}^{\infty} dx x^2 f(x,t) \tag{7.19}$$

whose time evolution is given by

$$\frac{\partial \langle x^2 \rangle}{\partial t} = -v \int_{-\infty}^{\infty} dx x^2 \frac{\partial f(x,t)}{\partial x} + D \int_{-\infty}^{\infty} dx x^2 \frac{\partial^2 f(x,t)}{\partial x^2} \tag{7.20}$$

Again, integration by parts of the terms on the right-hand side and using the above boundary conditions at infinity, that is,

$$\int_{-\infty}^{\infty} dx x^2 \frac{\partial f}{\partial x} = \left[x^2 f \right]_{-\infty}^{\infty} - \int_{-\infty}^{\infty} dx 2xf = -2 \langle x \rangle_t$$

$$\int_{-\infty}^{\infty} dx x^2 \frac{\partial^2 f}{\partial x^2} = \left[x^2 \frac{\partial f}{\partial x} \right]_{-\infty}^{\infty} - \int_{-\infty}^{\infty} dx 2x \frac{\partial f}{\partial x} = -2 \left[xf \right]_{-\infty}^{\infty} + 2 \int_{-\infty}^{\infty} dx f = 2 \tag{7.21}$$

leads to $\partial \langle x^2 \rangle / \partial t = 2v^2 t + 2D$, therefore, since $\langle x^2 \rangle_0 = 0$,

$$\langle x^2 \rangle_t = v^2 t^2 + 2Dt = \langle x \rangle_t^2 + 2Dt \tag{7.22}$$

or

$$\langle \delta x^2 \rangle_t = 2Dt \tag{7.23}$$

---

[1] To obtain Eqs (7.18) and (7.21) we need to assume that f vanishes as $x \to \infty$ faster than x^{-2}. Physically this must be so because a particle that starts at $x = 0$ cannot reach beyond some finite distance at any finite time if only because its speed cannot exceed the speed of light. Of course, the diffusion equation does not know the restrictions imposed by the Einstein relativity theory (similarly, the Maxwell–Boltzmann distribution assigns finite probabilities to find particles with speeds that exceed the speed of light). The real mathematical reason why f has to vanish faster than x^{-2} is that in the equivalent three-dimensional formulation $f(r)$ has to vanish faster than r^{-2} as $r \to \infty$ in order to be normalizable.

Problem 7.5. Show using $x = n\Delta x$ that Eqs (7.13) and (7.15) lead directly to Eqs. (7.18) and (7.23), respectively.

Together Eqs (7.18) and (7.23) express the essential features of *biased* random walk: A drift with speed v associated with the bias $k_r \neq k_l$, and a spread with a diffusion coefficient D. The linear dependence of the spread $\langle \delta x^2 \rangle$ on time is a characteristic feature of normal diffusion. Note that for a random walk in an isotropic three-dimensional space the corresponding relationship is

$$\langle \delta r^2 \rangle = \langle \delta x^2 \rangle + \langle \delta y^2 \rangle + \langle \delta z^2 \rangle = 6Dt \tag{7.24}$$

7.3.3 The probability distribution

Next consider the probability distribution itself. The solutions to the approximate Eqs (7.8) and (7.5) are the probability densities in Eqs (7.9) and (7.10), respectively, which are Gaussian functions. To gain insight on the nature of the approximation involved we consider, for simplicity, a model slightly different from that considered above, where jumps to the left or the right occur in every time-step, so that $p_r + p_l = 1$. Let the total number of steps taken by the particle be N. The probability for a particular walk with exactly n_r steps to the right (i.e. $n_l = N - n_r$ steps to the left, so that the final position relative to the origin is $n\Delta x$; $n = n_r - n_l = 2n_r - N$) is

$$W_N(n_r) = \frac{N!}{n_r!(N-n_r)!} p_r^{n_r} p_l^{N-n_r} \tag{7.25}$$

The coefficient $N!/[n_r!(N-n_r)!]$ is the number of distinct walks characterized by the given n_r. Note that the form (7.25) is normalized

$$\sum_{n_r=0}^{N} W_N(n_r) = \sum_{n_r=0}^{N} \frac{N!}{n_r!(N-n_r)!} p_r^{n_r} p_l^{N-n_r} = (p_r + p_l)^N = 1 \tag{7.26}$$

The distribution (7.25) is called *binomial.* Its most frequent textbook example is the outcome of flipping a coin with probabilities to win and lose given by p_r and p_l, respectively. The probability to have n_r successes out of N coin flips is then given by the binomial distribution (7.25).

The first moment of the distribution (7.25) can be computed according to

$$\begin{aligned} \langle n_r \rangle &= \sum_{n_r=1}^{N} \frac{N!}{n_r!(N-n_r)!} p_r^{n_r} p_l^{N-n_r} n_r = p_r \frac{\partial}{\partial p_r} \sum_{n_r=1}^{N} \frac{N!}{n_r!(N-n_r)!} p_r^{n_r} p_l^{N-n_r} \\ &= p_r \frac{\partial}{\partial p_r} (p_r + p_l)^N = p_r N (p_r + p_l)^{N-1} = p_r N \end{aligned} \tag{7.27}$$

Note that the identity $p_r + P_l = 1$ was used *after* the derivative with respect to p_r was taken.

Problem 7.6. Show that the second moment of the binomial distribution (7.25) is

$$\langle n_r^2 \rangle = \left(p_r \frac{\partial}{\partial p_r} \right)^2 (p_r + p_l)^N = (Np_r)^2 + Np_r p_l$$

so that the variance is

$$\left\langle \delta n_r^2 \right\rangle = Np_r p_l \qquad (7.28)$$

The net distance (in number of steps) that the particle walks from its original position is $n = n_r - n_l = 2n_r - N$. Obviously,

$$\langle n \rangle = (2p_r - 1)N \qquad (7.29)$$

while $\langle n^2 \rangle = 4\langle n_r^2 \rangle - 4\langle n_r \rangle N + N^2 = 4p_r^2 N^2 + 4Np_r p_l - 4p_r N^2 + N^2$. This leads to

$$\langle \delta n^2 \rangle = \langle n^2 \rangle - \langle n \rangle^2 = 4Np_r p_l = 4Np_r(1 - p_r) \qquad (7.30)$$

Note the difference between this result and Eq. (7.15). This difference reflects the fact that the model that leads to (7.15) is different from the present one that leads to (7.25) and consequently to (7.30): In the former $p_r + p_l = (k_r + k_l)\Delta t$ can be considerably smaller than 1, while in the latter $p_l + p_r = 1$. Equation (7.30) implies that if $p_r = 0$ or 1 there is no uncertainty in the walk so $\langle \delta n^2 \rangle = 0$. In contrast, Eq. (7.15) implies that uncertainty remains also when p_r or p_l (but not both) vanish because at each step the particle can move or not. For pure (unbiased) diffusion where $p_r = p_l = 1/2$ Eqs (7.15) and (7.30) yield identical results.

Consider now the distribution (7.25) in the limit $N \gg 1$. In this limit the length of the walk, $n\Delta x$ is much larger than the step size Δx, which was the basis for the expansion made in order to transform Eq. (7.4) to the form (7.5). In this limit the factorial factors in $W_N(n)$, Eq. (7.25) can be approximated by the Stirling formula, $\ln(N!) \approx N \ln N - N$, leading to

$$\begin{aligned} \ln[W_N(n_r)] &= N \ln N - n_r \ln n_r - (N - n_r) \ln(N - n_r) \\ &\quad + n_r \ln p_r + (N - n_r) \ln p_l \end{aligned} \qquad (7.31)$$

Further simplification is obtained if we expand $W_N(n_r)$ about its maximum at n_r^*. n_r^* is the solution of $\partial \ln[W_N(n_r)]/\partial n_r = 0$, which yields $n_r^* = Np_r = \langle n_r \rangle$. The nature of this extremum is identified as a maximum using

$$\frac{\partial^2 \ln W}{\partial n_r^2} = -\frac{1}{n_r} - \frac{1}{N-n_r} = -\frac{N}{n_r(N-n_r)} < 0 \tag{7.32}$$

When evaluated at n_r^* this gives $\partial^2 \ln W/\partial n_r^2 \mid_{n_r^*} = -1/(Np_rp_l)$. It is important to note that higher derivatives of ln W are negligibly small if evaluated at or near n_r^*. For example,

$$\left.\frac{\partial^3 \ln W}{\partial n_r^3}\right|_{n_r^*} = \frac{1}{n_r^{*2}} - \frac{1}{(N-n_r^*)^2} = \frac{1}{N^2}\left(\frac{1}{p_r^2} - \frac{1}{p_l^2}\right) \tag{7.33}$$

and, generally, derivatives of order k will scale as $(1/N)^{k-1}$. Therefore, for large N, W_N can be approximated by truncating the expansion after the first nonvanishing, second-order, term:

$$\ln W_N(n_r) \cong \ln W(n_r^*) - \frac{(n_r - n_r^*)^2}{2Np_rp_l} = \ln W(n_r^*) - \frac{(n_r - n_r^*)^2}{2\langle \delta n_r^2 \rangle} \tag{7.34}$$

where in the last step we have used Eq. (7.28). This leads to the Gaussian form

$$W_N(n_r) \cong \frac{1}{\sqrt{2\pi \langle \delta n_r^2 \rangle}} \exp\left[-\frac{(n_r - \langle n_r \rangle)^2}{2\langle \delta n_r^2 \rangle}\right]; \qquad \langle n_r \rangle = Np_r, \; \langle \delta n_r^2 \rangle = Np_rp_l \tag{7.35}$$

The pre-exponential term was taken to make the resulting Gaussian distribution normalized in the range $(-\infty, \infty)$ by replacing the sum over all n_r by an integral. In fact, n_r is bounded between 0 and N, however unless p_r is very close to 1 or 0, the distribution is vanishingly small near these boundaries and extending the limits of integration to $\pm\infty$ is an excellent approximation. For the variable $n = 2n_r - N$ we get

$$\begin{aligned} P_N(n) &\equiv W_N\left(\frac{N+n}{2}\right) \\ &\approx \exp\left\{-\frac{[n - N(p_r - p_l)]^2}{8Np_rp_l}\right\} \to \frac{1}{\sqrt{2\pi \langle \delta n^2 \rangle}} \exp\left\{-\frac{(n - \langle n \rangle)^2}{2\langle \delta n^2 \rangle}\right\} \end{aligned} \tag{7.36}$$

Again, in the last step we have calculated the normalization factor by replacing the sum over n by an integral in $(-\infty,\infty)$. The fact that starting from (7.25) we have obtained Gaussian distributions in the large N limit is an example of the central limit theorem of probability theory (see Section 1.1.1).

Finally, recalling that position and time are related to n and N by $x = n\Delta x$ and $t = N\Delta t$ we get from (7.36)

$$P(x,t) = \frac{1}{\sqrt{2\pi\langle\delta x^2\rangle}} \exp\left\{-\frac{(x-\langle x\rangle)^2}{2\langle\delta x^2\rangle}\right\} \tag{7.37}$$

with

$$\langle x\rangle = \frac{(p_r - p_l)\Delta x}{\Delta t}t = vt; \qquad \langle\delta x^2\rangle = 4p_r p_l \frac{(\Delta x)^2}{\Delta t}t \equiv 2Dt \tag{7.38}$$

Again, the result for the diffusion coefficient $D = 2p_r p_l(\Delta x)^2/\Delta t$ is the same as in Eq. (7.7) only when $p_r = p_l = 1/2$. The important observation is, again, that a Gaussian distribution was obtained as an approximation to the actual binomial one in the large N limit.

7.4 Some concepts from the general theory of stochastic processes

7.4.1 Distributions and correlation functions

In Section 7.1 we have defined a stochastic process as a time series, $z(t)$, of random variables. If observations are made at discrete times $0 < t_1 < t_2, \dots, < t_n$, then the sequence $\{z(t_i)\}$ is a discrete sample of the continuous function $z(t)$. In examples discussed in Sections 7.1 and 7.3 $z(t)$ was respectively the number of cars at time t on a given stretch of highway and the position at time t of a particle executing a one-dimensional random walk.

We can measure and discuss $z(t)$ directly, keeping in mind that we will obtain different *realizations* (stochastic trajectories) of this function from different experiments performed under identical conditions. Alternatively, we can characterize the process using the probability distributions associated with it. $P(z,t)dz$ is the probability that the realization of the random variable z at time t is in the interval between z and $z+dz$. $P_2(z_2t_2; z_1t_1)dz_1dz_2$ is the probability that z will have a value between z_1 and $z_1 + dz_1$ at t_1 and between z_2 and $z_2 + dz_2$ at t_2, etc. The time evolution of the process, if recorded in times $t_0, t_1, t_2, \dots, t_n$ is most generally represented by the joint probability distribution $P(z_nt_n; \dots; z_0t_0)$. Note that any such joint distribution function can be expressed as a reduced higher-order function, for example,

$$P(z_3t_3; z_1t_1) = \int dz_2 P(z_3t_3; z_2t_2; z_1t_1) \tag{7.39}$$

As discussed in Section 1.5.2, it is useful to introduce the corresponding conditional probabilities. For example,

$$P_1(z_1t_1 \mid z_0t_0)dz_1 = \frac{P_2(z_1t_1; z_0t_0)dz_1}{P_1(z_0t_0)} \tag{7.40}$$

is the probability that the variable z will have a value in the interval $z_1, \ldots, z_1 + dz_1$ at time t_1 given that it assumed the value z_0 at time t_0. Similarly,

$$P_2(z_4t_4; z_3t_3 \mid z_2t_2; z_1t_1)dz_3dz_4 = \frac{P_4(z_4t_4; z_3t_3; z_2t_2; z_1t_1)}{P_2(z_2t_2; z_1t_1)}dz_3dz_4 \tag{7.41}$$

is the conditional probability that z is in $z_4, \ldots, z_4+dz_4$ at t_4 and is in $z_3, \ldots, z_3+dz_3$ at t_3, given that its values were z_2 at t_2 and z_1 at t_1.

In the absence of time correlations, the values taken by $z(t)$ at different times are independent. In this case $P(z_nt_n; z_{n-1}t_{n-1}; \ldots; z_0t_0) = \prod_{k=0}^{n} P(z_k, t_k)$ and time correlation functions, for example, $C(t_2, t_1) = \langle z(t_2)z(t_1)\rangle$, are given by products of simple averages $C(t_2, t_1) = \langle z(t_2)\rangle\langle z(t_1)\rangle$, where $\langle z(t_1)\rangle = \int dz\, z\, P_1(z, t_1)$.This is often the case when the sampling times t_k are placed far from each other—farther than the longest *correlation time* of the process. More generally, the time correlation functions can be obtained from the joint distributions using the obvious expressions

$$C(t_2, t_1) = \int dz_1 \int dz_2 z_2 z_1 P_2(z_2t_2; z_1t_1) \tag{7.42a}$$

$$C(t_3, t_2, t_1) = \int dz_1 \int dz_2 \int dz_3 z_3 z_2 z_1 P_3(z_3t_3; z_2t_2; z_1t_1) \tag{7.42b}$$

In practice, numerical values of time correlations functions are obtained by averaging over an ensemble of realizations. Let $z^{(k)}(t)$ be the kth realization of the random function $z(t)$. Such realizations are obtained by observing z as a function of time in many experiments done under identical conditions. The correlation function $C(t_2, t_1)$ is then given by

$$C(t_2, t_1) = \lim_{N\to\infty} \frac{1}{N} \sum_{k=1}^{N} z^{(k)}(t_2)z^{(k)}(t_1) \tag{7.43}$$

If the stochastic process is stationary, the time origin is of no importance. In this case $P_1(z_1, t_1) = P_1(z_1)$ does not depend on time, while $P_2(z_2t_2; z_1t_1) = P_2(z_2, t_2 - t_1; z_1, 0)$ depends only on the time difference $\Delta t_{21} = t_2 - t_1$. In this case the correlation function $C(t_2, t_1) = C(\Delta t_{21})$ can be obtained by taking a time average over different origins along a single stochastic trajectory according to

$$C(t) = \lim_{N\to\infty} \frac{1}{N} \sum_{k=1}^{N} z(t_k + t)z(t_k) \tag{7.44}$$

Here the average is over a sample of reference times that span a region of time that is much larger than the longest correlation time of the process.

Further progress can be made by specifying particular kinds of processes of physical interest. In the following two sections we discuss two such kinds: Markovian and Gaussian.

7.4.2 Markovian stochastic processes

The process $z(t)$ is called Markovian if the knowledge of the value of z (say z_1) at a given time (say t_1) fully determines the probability of observing z at any later time

$$P(z_2t_2 \,|\, z_1t_1; z_0t_0) = P(z_2t_2 \,|\, z_1t_1); \qquad t_2 > t_1 > t_0 \tag{7.45}$$

Markov processes have no memory of earlier information. Newton equations describe deterministic Markovian processes by this definition, since knowledge of system state (all positions and momenta) at a given time is sufficient in order to determine it at any later time. The random walk problem discussed in Section 7.3 is an example of a stochastic Markov process.

The Markovian property can be expressed by

$$P(z_2t_2; z_1t_1; z_0t_0) = P(z_2t_2 \,|\, z_1t_1)P(z_1t_1; z_0t_0); \qquad \text{for } t_0 < t_1 < t_2 \tag{7.46}$$

or

$$P(z_2t_2; z_1t_1 \,|\, z_0t_0) = P(z_2t_2 \,|\, z_1t_1)P(z_1t_1 \,|\, z_0t_0); \qquad \text{for } t_0 < t_1 < t_2 \tag{7.47}$$

because, by definition, the probability to go from (z_1,t_1) to (z_2,t_2) is independent of the probability to go from (z_0,t_0) to (z_1,t_1). The above relation holds for any intermediate point between (z_0,t_0) and (z_2,t_2). As with any joint probability, integrating the left-hand side of Eq. (7.47) over z_1 yields $P(z_2t_2 \,|\, z_0t_0)$. Thus for a Markovian process

$$P(z_2t_2 \,|\, z_0t_0) = \int dz_1 P(z_2t_2 \,|\, z_1t_1)P(z_1t_1 \,|\, z_0t_0) \tag{7.48}$$

This is the *Chapman–Kolmogorov equation.*

Problem 7.7. Show that for a Markovian process

$$P_N(z_N t_N; z_{N-1} t_{N-1}; \ldots ; z_1 t_1; z_0 t_0) = P_1(z_0, t_0) \prod_{n=1}^{N} P_2(z_n t_n \,|\, z_{n-1} t_{n-1}) \tag{7.49}$$

The time evolution in a Markovian stochastic process is therefore fully described by the *transition probability* $P_2(zt \,|\, z't')$.

What is the significance of the Markovian property of a physical process? Note that the Newton equations of motion as well as the time-dependent Schrödinger equation are Markovian in the sense that the future evolution of a system described by these equations is fully determined by the present ("initial") state of the system. Non-Markovian dynamics results from reduction procedures used in order to focus on a "relevant" subsystem as discussed in Section 7.2, the same procedures that led us to consider stochastic time evolution. To see this consider a "universe" described by two variables, z_1 and z_2, which satisfy the Markovian equations of motion

$$\frac{dz_1}{dt} = F_1\left(z_1(t), z_2(t), t\right) \tag{7.50a}$$

$$\frac{dz_2}{dt} = F_2\left(z_1(t), z_2(t), t\right) \overset{\text{assumed}}{\longrightarrow} F_2\left(z_1(t), t\right) \tag{7.50b}$$

For simplicity we have taken F_2 to depend only on $z_1(t)$. If z_1 is the "relevant" subsystem, a description of the dynamics in the subspace of this variable can be achieved if we integrate Eq. (7.50b) to get

$$z_2(t) = z_2(t=0) + \int_0^t dt' F_2(z_1(t'), t') \tag{7.51}$$

Inserting this into (7.50a) gives

$$\frac{dz_1}{dt} = z_1(t=0) + F_1\left(z_1(t), z_2(t=0) + \int_0^t dt' F_2\left(z_1(t'), t'\right), t\right) \tag{7.52}$$

This equation describes the dynamics in the z_1 subspace, and its non-Markovian nature is evident. Starting at time t, the future evolution of z_1

is seen to depend not only on its value at time t, but also on its past history, since the right-hand side depends on all values of $z_1(t')$ starting from $t' = 0$.[2]

Why has the Markovian time evolution (7.50) of a system with two degrees of freedom become a non-Markovian description in the subspace of one of them? Equation (7.51) shows that this results from the fact that $z_2(t)$ responds to the historical time evolution of z_1, and therefore depends on past values of z_1, not only on its value at time t. More generally, consider a system A + B made of a part (subsystem) A that is relevant to us as observers, and another part, B, that affects the relevant subsystem through mutual interaction but is otherwise uninteresting. The non-Markovian behavior of the reduced description of the physical subsystem A reflects the fact that at any time t subsystem A interacts with the rest of the total system, that is, with B, whose state is affected by its past interaction with A. In effect, the present state of B carries the memory of past states of the relevant subsystem A.

This observation is very important because it points to a way to consider this memory as a qualitative attribute of system B (the environment or the bath) that determines the physical behavior of system A. In the example of Eq. (7.50), where system B comprises one degree of freedom z_2, its dynamics is solely determined by its interaction with system A represented by the coordinate z_1, and the memory can be as long as our observation. In practical applications, however, system A represents only a few degrees of freedom, while B is the macroscopic surrounding environment. B is so large relative to A that its dynamics may be dominated by interactions between B particles. Our physical experience tells us that if we disturb B then leave it to itself, it relaxes back to thermal equilibrium with a characteristic relaxation time τ_B. In other words, B "forgets" the disturbance it underwent on this timescale. If this remains true also in the case where A and B interact continuously (which says that also in this case τ_B is dominated by the internal dynamics of B), then the state of B at time t does not depend on disturbances in B that were caused by A at times earlier than $t' = t - \tau_B$. Consequently, dynamics in the A subspace at time t will depend on the history of A at earlier times going back only as far as this t'. The relaxation time τ_B can be therefore identified with the *memory time* of the environment B.

---

[2] Equation (7.52) shows also the origin of the stochastic nature of reduced descriptions. Focusing on z_1, we have no knowledge of the initial state $z_2(t = 0)$ of the "rest of the universe." At most we may know the distribution (e.g. Boltzmann) of different initial states. Different values of $z_2(t = 0)$ correspond to different realizations of the "relevant" trajectory $z_1(t)$. When the number of "irrelevant" degrees of freedom increases, this trajectory assumes an increasingly stochastic character in the sense that we can infer less and less about its evolution from the knowledge of its behavior along any given time segment.

We can now state the condition for the reduced dynamics of subsystem A to be Markovian: This will be the case if the characteristic timescale of the evolution of A is slow relative to the characteristic relaxation time associated with the environment B. When this condition holds, measurable changes in the A subsystem occur slowly enough so that on this relevant timescale B appears to be always at thermal equilibrium, and independent of its historical interaction with A. To reiterate, denoting the characteristic time for the evolution of subsystem A by τ_A, the condition for the time evolution within the A subspace to be Markovian is

$$\tau_B \ll \tau_A \tag{7.53}$$

While Markovian stochastic processes play important role in modeling molecular dynamics in condensed phases, their applicability is limited to processes that involve relatively slow degrees of freedom. Most intramolecular degrees of freedom are characterized by timescales that are comparable or faster than characteristic environmental times, so that the inequality (7.53) often does not hold. Another class of stochastic processes that are amenable to analytic descriptions also in non-Markovian situations is discussed next.

7.4.3 Gaussian stochastic processes

The special status of the Gaussian ("normal") distribution in reduced descriptions of physical processes was discussed in Section 1.4.4. It stems from the central limit theorem of probability theory and the fact that random variables that appear in coarse-grained descriptions of physical processes are themselves combinations of many more or less independent random variables. The same argument can be made for the transition probability associated with the time evolution step in a stochastic description of coarse-grained systems, assuming that the corresponding probability is affected by many random events. It leads to the conclusion that taking a Gaussian form for this probability is in many cases a reasonable model. A succession of such evolution steps, whether Markovian or not, constitutes a Gaussian stochastic process. As a general definition, a stochastic process $z(t)$ is Gaussian if the probability distribution of its observed values $z_1, z_2, \ldots, z_n$ at any n time points $t_1, t_2, \ldots, t_n$ (for any value of the integer n) is an n-dimensional Gaussian distribution.

$$P_n(z_1t_1; z_2t_2; \ldots; z_nt_n) = ce^{-(1/2)\sum_{j=1}^{n}\sum_{k=1}^{n} a_{jk}(z_j-m_j)(z_k-m_k)}; \qquad -\infty < z_j < \infty \tag{7.54}$$

where $m_j = (j = 1, \ldots n)$ are constants and the matrix $(a_{jk}) = \mathbf{A}$ is symmetric and positive definite (i.e. $\mathbf{u}^\dagger\mathbf{A}\mathbf{u} > 0$ for any vector $\mathbf{u}$) and where c is a normalization factor.

A Gaussian process can be Markovian. As an example consider the Markovian process characterized by the transition probability

$$P(z_k t_k \,|\, z_l t_l) = \frac{1}{\sqrt{2\pi}\,\Delta_{kl}} \exp\left[-\frac{(z_k - z_l)^2}{2\Delta_{kl}^2}\right] \tag{7.55}$$

This process is Gaussian by definition, since (7.54) is satisfied for any pair of times. The distribution (7.55) satisfies

$$\int dz_1 P(z_2 t_2 \,|\, z_1 t_1) P(z_1 t_1 \,|\, z_0 t_0) = \frac{1}{\sqrt{2\pi(\Delta_{21}^2 + \Delta_{10}^2)}} \exp\left[-\frac{(z_2 - z_0)^2}{2(\Delta_{21}^2 + \Delta_{10}^2)}\right] \tag{7.56}$$

Therefore the Markovian property (7.48) is satisfied provided that

$$\Delta_{20}^2 = \Delta_{21}^2 + \Delta_{10}^2 \tag{7.57}$$

If we further assume that Δ_{kl} is a function only of the time difference $t_k - t_l$, that is, $\Delta_{kl} = \Delta(t_k - t_l)$, it follows that its form must be

$$\Delta(t) = \sqrt{2Dt} \tag{7.58}$$

where D is some constant. Noting that Δ_{10}^2 is the variance of the probability distribution in Eq. (7.55), we have found that in a Markovian process described by (7.55) this variance is proportional to the elapsed time. Comparing this result with (7.9) we see that we have just identified regular diffusion as a Gaussian Markovian stochastic process.

Taken independently of the time ordering information, the distribution (7.54) is a multivariable, n-dimensional, Gaussian distribution

$$P_n(z_1, z_2, \ldots, z_n) = ce^{-(1/2)\sum_{j=1}^{n}\sum_{k=1}^{n} a_{jk}(z_j - m_j)(z_k - m_k)}; \qquad -\infty < z_j < \infty \tag{7.59}$$

In Appendix 7A we show that this distribution satisfies

$$\langle z_j \rangle = m_j; \qquad \langle \delta z_j \delta z_k \rangle = \left[(\mathbf{A})^{-1}\right]_{j,k} \qquad (\text{where } \delta z = z - \langle z \rangle) \tag{7.60}$$

This shows that a Gaussian distribution is completely characterized by the first two moments of its variables. Furthermore, we show in Appendix 7A that the so-called *characteristic function* of the n-variable Gaussian distribution, $\Phi_n(x_1,x_2,\ldots,x_n)$, defined as the Fourier transform of this distribution

$$\begin{aligned}\Phi_n(x_1,\ldots,x_n) &= \int_{-\infty}^{\infty} dz_1,\ldots,\int_{-\infty}^{\infty} dz_n P_n(z_1,z_2,\ldots,z_n)e^{i\sum_{j=1}^{n} x_j z_j} \\ &= \langle e^{i\sum_{j=1}^{n} x_j z_j}\rangle \end{aligned} \tag{7.61}$$

is given by

$$\Phi_n(\mathrm{x}) = \langle e^{i\mathbf{x}\cdot\mathbf{z}}\rangle = e^{i\mathbf{m}\cdot\mathbf{x}-(1/2)\mathbf{x}\cdot\mathbf{A}^{-1}\cdot\mathbf{x}} \tag{7.62}$$

where the vectors $\mathbf{z}$, $\mathbf{x}$, and $\mathbf{m}$ are defined as $\mathbf{z} = (z_1, z_2,\ldots,z_n)$ $\mathbf{m} = (m_1, m_2,\ldots,m_n)$, and $\mathbf{x} = (x_1, x_2,\ldots,x_n)$. More explicitly, this implies the following identity for the multivariable $(z_1, z_2,\ldots,z_n)$ Gaussian distribution

$$\left\langle e^{i\sum_j x_j z_j}\right\rangle = e^{i\sum_j x_j\langle z_j\rangle-(1/2)\sum_j\sum_k x_j\langle\delta z_j\delta z_k\rangle x_k} \tag{7.63}$$

where $\{x_j, j = 1,\ldots,n\}$ are any constants.

Problem 7.8. For a two-variable distribution of the type (7.54)

$$P_2(z_1,z_2) = ce^{-(1/2)\sum_{j=1}^{2}\sum_{k=1}^{2} a_{jk}(z_j-m_j)(z_k-m_k)}; \quad -\infty < z_j < \infty$$

show that

$$\langle e^{z_1}\rangle = \exp\left[\langle z_1\rangle + \frac{1}{2}\langle(\delta z_1)^2\rangle\right]$$

and

$$\langle e^{z_1+z_2}\rangle = \exp\left\{\langle z_1\rangle + \langle z_2\rangle + \frac{1}{2}[\langle(\delta z_1)^2\rangle + \langle(\delta z_2)^2\rangle + 2\langle(\delta z_1)(\delta z_2)\rangle]\right\}$$

Equations (7.60)–(7.63) describe general properties of many-variable Gaussian distributions. For a Gaussian random process the set $\{z_j\}$ corresponds to a sample $\{z_j,t_j\}$ from this process. This observation can be used to convert Eq. (7.63) to a general identity for a Gaussian stochastic process $z(t)$ and a general function of time $x(t)$ (see Appendix 7B)

$$\left\langle \exp\left(i\int_{t_0}^{t} dt' x(t')z(t')\right)\right\rangle = \exp\left(i\int_{t_0}^{t} dt' x(t')m(t') - \frac{1}{2}\int_{t_0}^{t} dt_1 \times \int_{t_0}^{t} dt_2 C_z(t_1,t_2)x(t_1)x(t_2)\right) \quad (7.64)$$

where

$$\begin{aligned} m(t) &= \langle z(t)\rangle \\ C_z(t_1,t_2) &= \langle \delta z(t_1)\delta z(t_2)\rangle \xrightarrow{\text{stationary process}} C_z(t_1-t_2) \\ \delta z(t) &= z(t) - m(t) \end{aligned} \quad (7.65)$$

Equation (7.64) is a general identity for a Gaussian stochastic process characterized by its average $m(t)$ and the time correlation functions $C_z(t_1, t_2)$. In many applications the stochastic process under study is stationary. In such cases $\langle z\rangle = m$ does not depend on time while $C_z(t_1,t_2) = C_z(t_1 - t_2)$ depends only on the time difference.

7.4.4 A digression on cumulant expansions

The identities (7.63) and (7.64) are very useful because exponential functions of random variables of the forms that appear on the left sides of these identities are frequently encountered in practical applications. For example, we have seen (cf. Eq. (1.5)) that the average $\langle e^{\alpha z}\rangle$, regarded as a function of α, is a generating function for the moments of the random variable z (see also Section 7.5.4 for a physical example). In this respect it is useful to consider extensions of (7.63) and (7.64) to non-Gaussian random variables and stochastic processes. Indeed, the identity (compare Problem 7.8)

$$\langle e^{\alpha z}\rangle = \exp[\alpha\langle z\rangle + (1/2)\alpha^2\langle(\delta z)^2\rangle] \quad (7.66)$$

that holds for a Gaussian distribution is a special case of the so-called cumulant expansion (valid for any distribution)

$$\langle e^{\alpha z}\rangle = \exp[\alpha\langle z\rangle_c + (1/2)\alpha^2\langle z^2\rangle_c + \cdots + (1/n!)\alpha^n\langle z^n\rangle_c + \cdots] \quad (7.67)$$

where the *cumulants* $\langle z^n\rangle_c$can be expressed in terms of the moments $\langle z^n\rangle$ such that the cumulant of order n is given by a linear combinations of moments of order n

and lower. For example, the first three cumulants are given by

$$\begin{aligned} \langle z \rangle_c &= \langle z \rangle \\ \langle z^2 \rangle_c &= \langle z^2 \rangle - \langle z \rangle^2 = \langle \delta z^2 \rangle \\ \langle z^3 \rangle_c &= \langle z^3 \rangle - 3\langle z \rangle \langle z^2 \rangle + 2\langle z \rangle^3 \end{aligned} \tag{7.68}$$

and for a Gaussian distribution all cumulants of order higher than 2 can be shown to vanish (which leads to Eq. (7.66)). For further discussion of cumulant expansions see Appendix 7C.

Problem 7.9. Use the procedure described in Appendix 7C to express the fourth cumulant $\langle z^4 \rangle_c$ in terms of the moments $\langle z^n \rangle; n = 1, 2, 3, 4$. Show that the third and fourth cumulants of the Gaussian distribution function $P(z) = \sqrt{(\alpha/\pi)} \exp[-\alpha z^2]; z = -\infty, \ldots, \infty$ vanish.

7.5 Harmonic analysis

Just as a random variable is characterized by the moments of its distribution, a stochastic process is characterized by its time correlation functions of various orders. In general, there are an infinite number of such functions, however we have seen that for the important class of Gaussian processes the first moments and the two-time correlation functions, simply referred to as time correlation functions, fully characterize the process. Another way to characterize a stationary stochastic process is by its spectral properties. This is the subject of this section.

7.5.1 The power spectrum

Consider a general stationary stochastic process $x(t)$, a sample of which is observed in the interval $0 \le t \le T$. Expand it in Fourier series

$$x(t) = \sum_{n=-\infty}^{\infty} x_n e^{i\omega_n t} \qquad \omega_n = \frac{2\pi n}{T}, \quad n = 0, \pm 1 \ldots \tag{7.69}$$

where x_n are determined from

$$x_n = \frac{1}{T} \int_0^T dt x(t) e^{-i\omega_n t} \tag{7.70}$$

If $x(t)$ is real then $x_n = x_{-n}^*$. Equation (7.69) resolves $x(t)$ into its spectral components, and associates with it a set of coefficients $\{x_n\}$ such that $|x_n|^2$ is the strength or intensity of the spectral component of frequency ω_n. However, since each realization of $x(t)$ in the interval $0, \ldots, T$ yields a different set $\{x_n\}$, the variables x_n are themselves random, and characterized by some (joint) probability function $P(\{x_n\})$. This distribution in turn is characterized by its moments, and these can be related to properties of the stochastic process $x(t)$. For example, the averages $\langle x_n \rangle$ satisfy

$$\langle x_n \rangle = \frac{1}{T} \int_0^T dt \, \langle x(t) \rangle \, e^{-i\omega_n t} \tag{7.71}$$

and since $\langle x(t) \rangle = \langle x \rangle$ does not depend on t ($x(t)$ being a stationary process), this implies

$$\langle x_n \rangle = \frac{\langle x \rangle}{T} \int_0^T dt e^{-i(2\pi n/T)t} = \langle x \rangle \, \delta_{n,0} \tag{7.72}$$

Note that from Eq. (7.70) $x_0 = (1/T) \int_0^T x(t) \equiv \bar{x}^T$ is the time average of $x(t)$ for any particular sampling on the interval T. For an ergodic process $\lim_{T \to \infty} \bar{x}^T = \langle x \rangle$ we thus find that $x_0 \overset{T \to \infty}{\longrightarrow} \langle x_0 \rangle = \langle x \rangle$.

For our purpose the important moments are $\langle |x_n|^2 \rangle$, sometimes referred to as the *average strengths* of the Fourier components ω_n. The power spectrum $I(\omega)$ of the stochastic process is defined as the $T \to \infty$ limit of the average intensity at frequency ω:

$$I(\omega) = \lim_{T \to \infty} \left(\frac{\sum_{n \in W_{\Delta\omega}} \langle |x_n|^2 \rangle}{\Delta\omega} \right);$$

$$W_{\Delta\omega} = \{ n \mid \omega - \Delta\omega/2 < (2\pi n/T) \le \omega + \Delta\omega/2 \} \tag{7.73}$$

where $n \in W_{\Delta\omega}$ encompasses all n with corresponding frequencies $\omega_n = (2\pi/T)n$ in the interval $\omega, \ldots, \omega \pm \Delta\omega/2$. If T is large enough we can use frequency intervals $\Delta\omega$ that are large enough so that they contain many Fourier components $\Delta\omega/(2\pi/T) = (T/2\pi)\Delta\omega \gg 1$, but small enough so that the strengths $|x_n|^2$ do not appreciably change within the interval. In this case the sum on the right-hand side of (7.73) may be represented by $\langle |x_n|^2 \rangle (T/2\pi)\Delta\omega$. This implies

$$I(\omega) = \lim_{T \to \infty} \frac{T}{2\pi} \langle |x_n|^2 \rangle; \qquad n = \frac{\omega T}{2\pi} \tag{7.74}$$

Note that, as defined, $I(\omega)$ is a real function that satisfies $I(-\omega) = I(\omega)$.

Problem 7.10. Show that another expression for the power spectrum is

$$I(\omega) = \lim_{T\to\infty} \sum_n \langle |x_n|^2 \rangle \delta(\omega - \omega_n); \qquad \omega_n = \frac{2\pi}{T} n \tag{7.75}$$

7.5.2 The Wiener–Khintchine theorem

An important relationship between $I(\omega)$ and the time correlation function $C(t) = \langle x(\tau)x(t+\tau)\rangle = \langle x(0)x(t)\rangle$ of $x(t)$ is the *Wiener–Khintchine theorem*, which states that

$$I(\omega) = \frac{1}{2\pi} \int_{-\infty}^{\infty} dt e^{-i\omega t} C(t) \qquad \text{or} \qquad C(t) = \int_{-\infty}^{\infty} d\omega e^{i\omega t} I(\omega) \tag{7.76}$$

If x is a complex function, this theorem holds for $C(t) = \langle x^*(0)x(t)\rangle$. The proof of this relationship is given in Appendix 7D. The power spectrum of a given stochastic process is thus identified as the Fourier transform of the corresponding time correlation function.

The power spectrum was defined here as a property of a given stochastic process. In the physics literature it is customary to consider a closely related function that focuses on the properties of the thermal environment that couples to the system of interest and affects the stochastic nature of its evolution. This is the *spectral density* that was discussed in Section 6.5.2. (see also Section 8.2.6). To see the connection between these functions we recall that in applications of the theory of stochastic processes to physical phenomena, the stochastic process $x(t)$ represents a physical observable A, say a coordinate or a momentum of some observed particle. Suppose that this observable can be expanded in harmonic normal modes $\{u_j\}$ as in Eq. (6.79)

$$A(t) = \sum_j c_j^{(A)} u_j(t) \tag{7.77}$$

where $c_j^{(A)}$ are the corresponding weights. The correlation function $C_{AA}(t) = \langle A(t)A(0)\rangle$ is therefore given by (cf. Eq. (6.88))

$$C_{AA}(t) = k_B T \sum_j \frac{(c_j^{(A)})^2}{\omega_j^2} \cos(\omega_j t) = \frac{2k_B T}{\pi} \int_0^{\infty} d\omega \frac{J_A(\omega)}{\omega} \cos(\omega t) \tag{7.78}$$

where (cf. Eq. (6.92))

$$J_A(\omega) = \pi g(\omega)(c^{(A)}(\omega))^2/(2\omega) \tag{7.79}$$

was identified as the bath spectral density, and we have used the subscript A to emphasize the fact that $J_A(\omega)$ characterizes both the bath and the variable A. Because (cf. Eq. (6.93)) $J_A(-\omega) = -J_A(\omega)$, Eq. (7.78) may be rewritten as

$$C_{AA}(t) = \frac{k_B T}{\pi} \int_{-\infty}^{\infty} d\omega \frac{J_A(\omega)}{\omega} e^{i\omega t} \tag{7.80}$$

so that

$$I_A(\omega) = \frac{k_B T}{\pi} \frac{J_A(\omega)}{\omega} \tag{7.81}$$

The two functions $I_A(\omega)$ and $J_A(\omega)$ are seen to convey the same physical information and their coexistence in the literature just reflects traditions of different scientific communities. The important thing is to understand their physical contents: we have found that the power spectrum is a function that associates the dynamics of an observable A with the dynamics of a reference harmonic system with density of modes given by (7.81) and (7.79).

7.5.3 Application to absorption

More insight into the significance of the power spectrum/spectral function concept can be gained by considering the rate at which a system absorbs energy from an external driving field. Assume that our system is driven by an external periodic force, $F \cos \omega t$ that is coupled to some system coordinate A, that is, it is derived from a potential $-AF \cos \omega t$. Taking A again to be the superposition (7.77) of system normal modes, the equation of motion of each normal mode is

$$\ddot{u}_j = -\omega_j^2 u_j + c_j F \cos \omega t \tag{7.82}$$

To simplify notation we have removed the superscript (A) from the coefficients c_j. Consider the rate at which such mode absorbs energy from the external field. It is convenient to assume the presence of a small damping term $\eta \dot{u}_j$, taking η to zero at the end of the calculation. This makes it possible to treat the energy absorption as a steady-state process. The equation of motion is then

$$\ddot{u}_j = -\omega_j^2 u_j - \eta \dot{u}_j + c_j F \cos \omega t \tag{7.83}$$

Multiplying Eq. (7.83) by $\dot{u}_j$ leads to

$$\frac{dE_j}{dt} = -\eta \dot{u}_j^2 + c_j F \dot{u}_j \cos \omega t \tag{7.84}$$

where

$$E_j = \frac{1}{2}(\dot{u}_j^2 + \omega_j^2 u_j^2) \tag{7.85}$$

is the oscillator energy (see Eq. (6.76)). At steady state, the average oscillator energy does not change in time, so Eq. (7.85) yields

$$\frac{dE_j}{dt} = 0 = -\eta\overline{\dot{u}_j^2} + c_jF\overline{\dot{u}_j\cos(\omega t)} \tag{7.86}$$

where the overbars represent time averages. The two terms on the right, the first representing the time-averaged dissipation and the second corresponding to the time-averaged pumping, must balance each other. The pumping rate expressed as a function of the pumping frequency ω is the contribution, $L_j(\omega)$, of the mode j to the absorption lineshape. Thus,

$$L_j(\omega) \sim \eta\overline{\dot{u}_j^2} \tag{7.87}$$

At steady state u_j oscillates with the same frequency of the driving field. Its motion is obtained from (7.83) by looking for a solution of the form $u_j(t) = Re(U_je^{i\omega t})$ and solving for U_j. We get[3]

$$u_j(t) = \mathrm{Re}\left[\frac{c_jF}{\omega_j^2 - \omega^2 + i\omega\eta}e^{i\omega t}\right] \tag{7.88a}$$

$$\dot{u}_j(t) = \mathrm{Re}\left[\frac{i\omega c_jF}{\omega_j^2 - \omega^2 + i\omega\eta}e^{i\omega t}\right] \tag{7.88b}$$

Using $\overline{\cos^2(\omega t)} = \overline{\sin^2(\omega t)}$ =(1/2) and $\overline{\sin(\omega t)\cos(\omega t)} = 0$, Eqs (7.87) and (7.88b) yield

$$L_j(\omega) \sim \eta\overline{\dot{u}_j^2} = \frac{c_j^2F^2}{2}\frac{\eta\omega^2}{(\omega_j^2 - \omega^2)^2 + (\omega\eta)^2} \tag{7.89}$$

When $\eta \ll \omega_j$ the absorption is dominated by frequencies ω close to ω_j and we may approximate $(\omega_j^2 - \omega^2) = (\omega_j - \omega)(\omega_j + \omega) \cong 2\omega(\omega_j - \omega)$. This leads to

$$L_j(\omega) \sim \frac{c_j^2F^2}{4}\frac{\eta/2}{(\omega_j - \omega)^2 + (\eta/2)^2} \xrightarrow{\eta\to 0} \frac{\pi c_j^2F^2}{4}\delta(\omega_j - \omega) \tag{7.90}$$

Summing over all modes yields

$$L(\omega) = \sum_j L_j(\omega) = \frac{\pi F^2c^2(\omega)}{4}g(\omega) \sim \omega^2 I_A(\omega) \tag{7.91}$$

We have found that up to constant factors, the absorption lineshape is determined by the power spectrum that characterizes the coordinate that couples to the external

---

[3] The result (7.88) is most easily obtained by solving variants of (7.83) with driving terms $(1/2)C_jFe^{\pm i\omega t}$ then combining the corresponding solutions

field. Note that this power spectrum is associated with the motion of this coordinate in the absence of the driving field.

7.5.4 The power spectrum of a randomly modulated harmonic oscillator

Having found a relationship between the absorption lineshape associated with a periodically modulated system variable and the power spectrum of this variable, we now consider a specific example. Consider a harmonic oscillator which is randomly perturbed so that its frequency changes in time as:

$$\omega(t) = \omega_0 + \delta\omega(t) \tag{7.92}$$

where $\delta\omega(t)$ is a stochastic process. This is a model for a system interacting with its thermal environment, where we assume that this interaction is expressed by Eq. (7.92). It is convenient to use the complex amplitude $a(t)$ defined in Eq. (6.38)

$$a(t) = x(t) + \frac{i}{m\omega}p(t) \tag{7.93}$$

so that the variables x and p can be replaced by a and a^*. In what follows we will calculate the power spectrum of these variables.[4]

Problem 7.11. Show that $|a(t)|^2 = 2E/(m\omega^2)$, where E is the oscillator energy.

The equation of motion for $a(t)$ is

$$\dot{a}(t) = -i\omega(t)a(t) \tag{7.94}$$

whose solution is (putting $a_0 = a(t = 0)$)

$$a(t) = a_0 \exp\left[-i\int_0^t dt'\omega(t')\right] \tag{7.95}$$

Since $\delta\omega\ (t)$ is a stochastic process, so are $a(t)$ and $x(t) = (1/2)[a(t) + a^*(t)]$. Consider the time correlation function

$$\langle a^*(0)a(t)\rangle = \left\langle |a_0|^2 \exp\left[-i\int_0^t dt'\omega(t')\right]\right\rangle \tag{7.96}$$

[4] Since $x = (1/2)(a + a^*)$ we have $\langle x(0)x(t)\rangle \sim 2Re\left(\langle a(0)a(t)\rangle + \langle a^*(0)a(t)\rangle\right)$. The term $\langle a(0)a(t)\rangle = \langle (a(0))^2\rangle\langle \exp\left[-i\int_0^t dt'\omega(t')\right]\rangle$ can be disregarded because (using Eq. (7.93) and assuming thermal equilibrium at $t = 0$)$\langle (a(0))^2\rangle = 0$. Therefore $\langle x(0)x(t)\rangle \sim 2Re\left(\langle a^*(0)a(t)\rangle\right)$ and the power spectra of $x(t)$ and of $a(t)$ are essentially the same.

The initial value $a(0)$ is assumed independent of the process $\omega(t)$, so

$$\frac{\langle a^*(0)a(t)\rangle}{\langle |a(0)|^2\rangle} = \left\langle \exp\left[-i\int_0^t dt'\omega(t')\right]\right\rangle = e^{-i\omega_0 t}\phi(t) \tag{7.97}$$

where

$$\phi(t) = \left\langle \exp\left[-i\int_0^t dt'\,\delta\omega(t')\right]\right\rangle \tag{7.98}$$

Once we evaluate $\phi(t)$ we can obtain the power spectrum of the randomly modulated harmonic oscillator using the Wiener–Khintchine theorem (7.76)

$$I_a(\omega) = \frac{1}{2\pi}\int_{-\infty}^{\infty} \langle a^*(0)a(t)\rangle e^{-i\omega t}dt = \frac{|a_0|^2}{2\pi}\int_{-\infty}^{\infty} \phi(t)e^{-i\omega' t}dt,$$
$$\omega' = \omega + \omega_0 \tag{7.99}$$

We now assume that the stochastic frequency modulation $\delta\omega(t)$ is a stationary Gaussian process with $\langle\delta\omega(t)\rangle = 0$ and $\langle\delta\omega(t_0)\delta\omega(t_0+t)\rangle = \langle\delta\omega^2\rangle s(t)$, where $s(t) = s(-t)$ is defined by this relationship. The parameter $\langle\delta\omega^2\rangle$ and the function $s(t)$ characterize the physics of the random frequency modulations and are assumed known. From Eq. (7.64) we get

$$\phi(t) = \left\langle \exp\left(-i\int_0^t dt'\,\delta\omega(t')\right)\right\rangle = e^{-(1/2)\langle\delta\omega^2\rangle\int_0^t dt_1\int_0^t dt_2 s(t_1-t_2)} \tag{7.100}$$

The integral in the exponent may be transformed as follows:

$$\frac{1}{2}\int_0^t dt_1\int_0^t dt_2\, s(t_1-t_2) = \int_0^t dt_1\int_0^{t_1} dt_2\, s(t_1-t_2) = \int_0^t dt_1\int_0^{t_1} d\tau\, s(\tau)$$
$$= \int_0^t d\tau(t-\tau)\,s(\tau) \tag{7.101}$$

The last equality is obtained by changing the order of integration as in the transition from (7.127) to (7.128) in Appendix 7D. Equation (7.100) then becomes

$$\phi(t) = \exp\left(-\langle\delta\omega^2\rangle\int_0^t d\tau(t-\tau)\,s(\tau)\right) \tag{7.102}$$

The resulting physical behavior is determined by the interplay between two physical parameters. First, $\Omega = \langle \delta\omega^2 \rangle^{1/2}$ measures the amplitude of the random frequency modulations. Second, $\tau_c = \int_0^\infty d\tau\, s(\tau) = \langle \delta\omega^2 \rangle^{-1} \int_0^\infty d\tau \langle \delta\omega(0)\delta\omega(\tau) \rangle$ measures the correlation time of these modulations. Depending on their relative magnitude we can get qualitatively different spectra.

To see this in detail consider the simple model

$$s(t) = e^{-t/\tau_c} \tag{7.103}$$

Using (7.103) in (7.102) results in

$$\phi(t) = \exp\left[-\alpha^2 \left(\frac{t}{\tau_c} - 1 + e^{-t/\tau_c}\right)\right] \tag{7.104}$$

where

$$\alpha = \tau_c \Omega \tag{7.105}$$

In the limit $\alpha \to \infty$ $\phi(t)$ vanishes unless t is very small. We can therefore expand the exponent in (7.104) in power of t up to order t^2. This leads to

$$\phi(t) = e^{-(1/2)\Omega^2 t^2} \tag{7.106}$$

and $I_a(\omega)$ is a Gaussian[5]

$$I_a(\omega) = \frac{1}{2\pi} \int_{-\infty}^{\infty} dt \phi(t) e^{-i\omega t} = \frac{1}{\Omega\sqrt{2\pi}} \exp\left(-\frac{\omega^2}{2\Omega^2}\right) \tag{7.107}$$

In the opposite limit, $\alpha \to 0$, the main contribution to the integral (7.99) will come from large t. In this case $\phi(t)$ may be approximated by

$$\phi(t) = \exp\left(-\alpha^2 \frac{t}{\tau_c}\right) = \exp\left(-\tau_c \Omega^2 t\right) \tag{7.108}$$

and the spectral function (7.99) is a Lorentzian, $(\gamma/\pi)/(\omega^2 + \gamma^2)$, with $\gamma = \tau_c \Omega^2$.

We have seen in Section 7.5.3 that the power spectrum of a given system is closely associated with the absorption lineshape in that system. The analysis presented above indicates that the spectral lineshape of a stochastically modulated oscillator assumes qualitatively different forms depending on the amplitude and timescale of the modulation. We will return to these issues in Chapter 18.

---

[5] Equation (7.107) holds for ω large relative to Ω, but not in the asymptotic limit $\omega \to \infty$. It can be shown that the Fourier transform of (7.104) approaches asymptotically a ω^{-6} behavior.

Appendix 7A: Moments of the Gaussian distribution

Consider the n-dimensional Gaussian distribution (7.54). Here it is shown that this distribution satisfies

$$\langle z_j \rangle = m_j; \qquad \langle \delta z_j \delta z_k \rangle = [(A)^{-1}]_{j,k} \quad (\text{where } \delta z = z - \langle z \rangle) \tag{7.109}$$

The time ordering information in (7.54) is not relevant here. Equation (7.109) obviously holds for $n = 1$, where $W(z) = ce^{-(1/2)a(z-m)^2}$ gives $\langle z \rangle = m$ and $\langle (z-m)^2 \rangle = -2(d/da)(\ln \int_{-\infty}^{\infty} dz e^{-(1/2)az^2}) = a^{-1}$. In the general n-variable case we introduce the *characteristic function* of n variables $\Phi_n(x_1, \ldots, x_n)$, essentially the Fourier transform of the probability function

$$\begin{aligned} \Phi_n(x_1 \ldots x_n) &= \int_{-\infty}^{\infty} dz_1 \ldots \int_{-\infty}^{\infty} dz_n P_n(z_1, z_2, \ldots, z_n)\, e^{i\sum_{j=1}^n x_j z_j} \\ &= \left\langle e^{i\sum_{j=1}^n x_j z_j} \right\rangle \end{aligned} \tag{7.110}$$

The characteristic function can be used to generate the moments of the distribution W_n according to (compare Eqs (1.5)–(1.7))

$$\langle z_j \rangle = -i\left(\frac{\partial \Phi_n}{\partial x_j}\right)_{\mathbf{x}=0}; \qquad \langle z_j z_{j'} \rangle = -\left(\frac{\partial^2 \Phi_n}{\partial x_j \partial x_{j'}}\right)_{\mathbf{x}=0} \tag{7.111}$$

It is convenient to use a vector notation

$$\mathbf{z} = (z_1, \ldots, z_n); \qquad \mathbf{x} = (x_1, \ldots, x_n); \qquad \mathbf{m} = (m_1, \ldots, m_n) \tag{7.112}$$

define $\mathbf{y} = \mathbf{z} - \mathbf{m}$. Using also (7.54), the characteristic function takes the form

$$\Phi(\mathbf{x}) = c \int_{-\infty}^{\infty} d\mathbf{z} e^{-(1/2)\mathbf{y}\cdot\mathbf{A}\cdot\mathbf{y} + i\mathbf{x}\cdot\mathbf{z}} = c \int_{-\infty}^{\infty} d\mathbf{y} e^{-(1/2)\mathbf{y}\cdot\mathbf{A}\cdot\mathbf{y} + i\mathbf{x}\cdot\mathbf{y} + i\mathbf{x}\cdot\mathbf{m}} \tag{7.113}$$

Next we change variable $\mathbf{y} \to \mathbf{u} + i\mathbf{b}$, where $\mathbf{b}$ is a constant vector to be determined below. The expression in the exponent transforms to

$$\begin{aligned} &-(1/2)\mathbf{y}\cdot\mathbf{A}\cdot\mathbf{y} + i\mathbf{x}\cdot\mathbf{y} + i\mathbf{x}\cdot\mathbf{m} \to \\ &\quad -(1/2)\mathbf{u}\cdot\mathbf{A}\cdot\mathbf{u} \underbrace{-i\mathbf{u}\cdot\mathbf{A}\cdot\mathbf{b}}_{1} \underbrace{+(1/2)\mathbf{b}\cdot\mathbf{A}\cdot\mathbf{b}}_{2} \underbrace{+i\mathbf{x}\cdot\mathbf{u}}_{1} \underbrace{-\mathbf{x}\cdot\mathbf{b}}_{2} + i\mathbf{x}\cdot\mathbf{m} \end{aligned} \tag{7.114}$$

Now choose $\mathbf{b}$ so that $\mathbf{A}\cdot\mathbf{b}=\mathbf{x}$ or $\mathbf{b}=\mathbf{A}^{-1}\cdot\mathbf{x}$. Then the terms marked 1 in (7.114) cancel while the terms marked 2 give $-(1/2)\mathbf{x}\cdot\mathbf{A}^{-1}\mathbf{x}$. This finally results in

$$\Phi(x)=e^{i\mathbf{m}\cdot\mathbf{x}-(1/2)\mathbf{x}\cdot\mathbf{A}^{-1}\cdot\mathbf{x}}c\int_{-\infty}^{\infty}d\mathbf{u}e^{-(1/2)\mathbf{u}\cdot\mathbf{A}\cdot\mathbf{u}}=e^{i\mathbf{m}\cdot\mathbf{x}-(1/2)\mathbf{x}\cdot\mathbf{A}^{-1}\cdot\mathbf{x}} \qquad (7.115)$$

The second equality results from the fact that $W_n=ce^{-(1/2)\mathbf{y}\cdot\mathbf{A}\cdot\mathbf{y}}$ is normalized to unity. We found

$$\Phi(x)=\left\langle e^{i\mathbf{x}\cdot\mathbf{z}}\right\rangle=e^{i\mathbf{m}\cdot\mathbf{x}-(1/2)\mathbf{x}\cdot\mathbf{A}^{-1}\cdot\mathbf{x}} \qquad (7.116)$$

A Taylor expansion about $\mathbf{x}$=0 yields

$$1+i\mathbf{x}\cdot\langle\mathbf{z}\rangle-\frac{1}{2}\mathbf{x}\cdot\langle\mathbf{zz}\rangle\cdot\mathbf{x}+\cdots=1+i\mathbf{x}\cdot\mathbf{m}-\frac{1}{2}\mathbf{x}\cdot(\mathbf{mm})\cdot\mathbf{x}-\frac{1}{2}\mathbf{x}\cdot\mathbf{A}^{-1}\cdot\mathbf{x}+\cdots \quad (7.117)$$

Here $\mathbf{zz}$ is a short-hand notation for the matrix $\mathbf{Z}_{ij}=\mathbf{z}_i\mathbf{z}_j$, and similarly for $\mathbf{mm}$.

By equating coefficients of equal powers of $\mathbf{x}$ in (7.117) we find (see also (7.109))

$$\langle\mathbf{z}\rangle=\mathbf{m}, \qquad \text{that is, } \mathbf{z}_j=m_j$$

and

$$\langle\mathbf{zz}\rangle=\mathbf{mm}+\mathbf{A}^{-1}, \qquad \text{that is, } \langle(z_j-m_j)(z_{j'}-m_{j'})\rangle=(\mathbf{A}^{-1})_{j,j'} \qquad (7.118)$$

which is equivalent to (7.109).

Appendix 7B: Proof of Eqs (7.64) and (7.65)

Here we prove, for a Gaussian stochastic processes $z(t)$ and a general function of time $x(t)$ the results (7.64) and (7.65). Our starting point is (cf. Eq. (7.63))

$$\left\langle e^{i\sum_j x_j z_j}\right\rangle=e^{i\sum_j x_j\langle z_j\rangle-(1/2)\sum_j\sum_k x_j\langle\delta z_j\delta z_k\rangle x_k} \qquad (7.119)$$

where the sums are over the n random variables. Noting that this relationship holds for any set of constants $\{x_j\}$, we redefine these constants by setting

$$x_j\to x(t_j)\Delta t_j \qquad (7.120)$$

and take the limit $\Delta t_j\to 0$ and $n\to\infty$ in the interval $t_0<t_1<t_2<\cdots<t_n$. Then

$$\sum_{j=1}^{n}x_j\langle z_j\rangle\to\int_{t_0}^{t}dt'x(t')\langle z(t')\rangle$$

and

$$\sum_{j=1}^{n}\sum_{k=1}^{n} x_j\langle z_j z_k\rangle x_k \to \int_{t_0}^{t}\int_{t_0}^{t} dt_1 dt_2 \langle z(t_1)z(t_2)\rangle x(t_1)x(t_2)$$

This immediately yields (7.64) and (7.65).

Appendix 7C: Cumulant expansions

Let z be a random variable and consider the function $e^{\alpha z}$. The average $\langle e^{\alpha z}\rangle$ is a generating function (see Eqs (1.5)–(1.7)) for the moments $\langle z^n\rangle$:

$$\langle e^{\alpha z}\rangle = 1 + \alpha\langle z\rangle + (1/2)\alpha^2\langle z^2\rangle + \cdots + (1/n!)\alpha^n\langle z^n\rangle + \cdots \tag{7.121}$$

The cumulants $\langle z^n\rangle_c$ are defined by

$$\langle e^{\alpha z}\rangle = \exp[\alpha\langle z\rangle_c + (1/2)\alpha^2\langle z^2\rangle_c + \cdots + (1/n!)\alpha^n\langle z^n\rangle_c + \cdots] \tag{7.122}$$

We can express the cumulant of order n as a linear combinations of moments of order $m \le n$ by expanding the right hand side of Eq. (7.122) in a Taylor series, and equating equal powers of α in the resulting expansion and in (7.121). This leads to

$$\begin{aligned} \langle z\rangle_c &= \langle z\rangle \\ \langle z^2\rangle_c &= \langle z^2\rangle - \langle z\rangle^2 = \langle \delta z^2\rangle \\ \langle z^3\rangle_c &= \langle z^3\rangle - 3\langle z\rangle\langle z^2\rangle + 2\langle z\rangle^3 \end{aligned} \tag{7.123}$$

We can generalize this to many random variables and even to a continuous array of such variable s. Starting from the left-hand side of (7.64) and using a Taylor expansion we have

$$\left\langle \exp\left(i\int_{t_0}^{t} dt' x(t')z(t')\right)\right\rangle = 1 + i\int_{t_0}^{t} dt' x(t')\langle z(t')\rangle - \frac{1}{2}\int_{t_0}^{t} dt' \times \int_{t_0}^{t} dt'' x(t')x(t'')\langle z(t')z(t'')\rangle + \cdots \tag{7.124}$$

The cumulant expansion is defined in analogy to (7.122)

$$\left\langle \exp\left(i \int_{t_0}^{t} dt' x(t') z(t') \right) \right\rangle = \exp\left(i \int_{t_0}^{t} dt' x(t') \langle z(t') \rangle_c - \frac{1}{2} \int_{t_0}^{t} dt' \right.$$
$$\left. \times \int_{t_0}^{t} dt'' x(t') x(t'') \langle z(t') z(t'') \rangle_c + \cdots \right) \tag{7.125}$$

Again, expanding the exponent in (7.125) in a power series and comparing similar orders of $x(t)$ in the resulting series with Eq. (7.124) we find

$$\begin{aligned} &\langle z(t) \rangle_c = \langle z(t) \rangle \\ &\langle z(t') z(t'') \rangle_c = \langle \delta z(t') \delta z(t'') \rangle; \qquad \delta z(t) = z(t) - \langle z(t) \rangle \end{aligned} \tag{7.126}$$

A common approximation is to truncate the cumulant expansion at some order, usually the second. Comparing to Eqs (7.63) and (7.64) we see that for a Gaussian process this approximation is exact. In fact, it may be shown that for a Gaussian process not only does the sum of all higher cumulants vanish, but every cumulant higher than the second is zero.

Appendix 7D: Proof of the Wiener–Khintchine theorem

Starting from Eq. (7.70) we have

$$\begin{aligned} \langle |x_n|^2 \rangle &= \frac{1}{T^2} \int_0^T dt_1 \int_0^T dt_2 \langle x(t_2) x(t_1) \rangle e^{-i\omega_n (t_1 - t_2)} \\ &= \frac{1}{T^2} \int_0^T dt_1 \left(\int_0^{t_1} dt_2 + \int_{t_1}^{T} dt_2 \right) C(t_1 - t_2) e^{-i\omega_n (t_1 - t_2)} \\ &= \frac{1}{T^2} \int_0^T dt_1 \int_0^{t_1} dt C(t) e^{-i\omega_n t} + \frac{1}{T^2} \int_0^T dt_1 \int_0^{T - t_1} dt C(-t) e^{i\omega_n t} \end{aligned} \tag{7.127}$$

Note that if $x(t)$ is a complex function the same relationships hold for $C(t) = \langle x^*(0) x(t) \rangle$. The integration regions corresponding to the two integrals in (7.127) are shown in Fig. 7.3, where the arrows show the direction taken by the inner

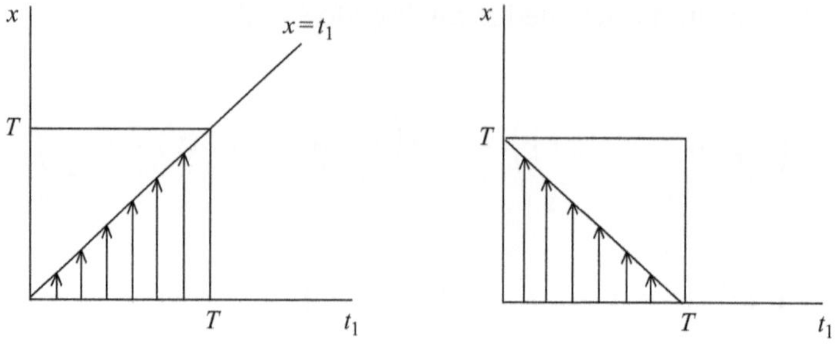

FIG. 7.3 The integration regions and procedures taken for the two integrals in Eq. (7.127)

integrals. By changing the integration direction we can change the orders of these integrations to get

$$\langle |x_n|^2 \rangle = \frac{1}{T^2}\left[\int_0^T dt(T-t)C(t)e^{-i\omega_n t} + \int_0^T dt(T-t)C(-t)e^{i\omega_n t}\right] \tag{7.128}$$

Using the definition (7.74) of $I(\omega)$, we get

$$I(\omega_n) = \lim_{T\to\infty}\frac{1}{2\pi T}\left[\int_0^T dt(T-t)C(t)e^{-i\omega_n t} + \int_0^T dt(T-t)C(-t)e^{i\omega_n t}\right] \tag{7.129}$$

assuming that the integrals $\int_0^\infty dtC(t)e^{i\omega_n t}$ and $\int_0^\infty dt\, tC(t)e^{i\omega_n t}$ are finite, this yields

$$I(\omega) = \frac{1}{2\pi}\left(\int_0^\infty dtC(t)e^{-i\omega t} + \int_0^\infty dtC(-t)e^{i\omega t}\right) = \frac{1}{2\pi}\int_{-\infty}^{\infty} dtC(t)e^{-i\omega t} \tag{7.130}$$

This concludes the proof.

Further reading

See end of Chapter 8.

8

STOCHASTIC EQUATIONS OF MOTION

Never suppose the atoms had a plan,
Not with wise intelligence imposed
An order on themselves, nor in some pact
Agreed what movements each should generate.
No, it was all fortuitous...

Lucretius (c.99–c.55 BCE*) "The way things are" translated by*
Rolfe Humphries, Indiana University Press, 1968

We have already observed that the full phase space description of a system of N particles (taking all $6N$ coordinates and velocities into account) requires the solution of the deterministic Newton (or Schrödinger) equations of motion, while the time evolution of a small subsystem is stochastic in nature. Focusing on the latter, we would like to derive or construct appropriate equations of motion that will describe this stochastic motion. This chapter discusses some methodologies used for this purpose, focusing on classical mechanics as the underlying dynamical theory. In Chapter 10 we will address similar issues in quantum mechanics.

8.1 Introduction

The time evolution of stochastic processes can be described in two ways:

1. *Time evolution in probability space.* In this approach we seek an equation (or equations) for the time evolution of relevant probability distributions. In the most general case we deal with an infinite hierarchy of functions, $P(z_n t_n; z_{n-1} t_{n-1}; \ldots; z_1 t_1)$ as discussed in Section 7.4.1, but simpler cases exist, for example, for Markov processes the evolution of a single function, $P(z, t; z_0 t_0)$, fully characterizes the stochastic dynamics. Note that the stochastic variable z stands in general for all the variables that determine the state of our system.
2. *Time evolution in variable space.* In this approach we seek an equation of motion that describes the evolution of the stochastic variable $z(t)$ itself (or

equations of motion for several such variables). Such equations of motions will yield stochastic trajectories $z(t)$ that are realizations of the stochastic process under study. The stochastic nature of these equations is expressed by the fact that for any initial condition z_0 at $t = t_0$ they yield infinitely many such realizations in the same way that measurements of $z(t)$ in the laboratory will yield different such realizations.

Two routes can be taken to obtain such stochastic equations of motions, of either kind:

1. Derive such equations from first principles. In this approach, we start with the deterministic equations of motion for the entire system, and derive equations of motion for the subsystem of interest. The stochastic nature of the latter stems from the fact that the state of the complementary system, "the rest of the world," is not known precisely, and is given only in probabilistic terms.
2. Construct phenomenological stochastic equations using physical arguments, experimental observations, and intuition.

In this chapter we will usually take the second route (see Section 8.2.5 for an example of first principle derivation).

In Chapter 7 we saw examples of evolution equations in probability space that were constructed via the phenomenological route. Equation (7.3) for the nearest neighbor random walk problem,

$$\frac{\partial P(n,t)}{\partial t} = k_r(P(n-1,t) - P(n,t)) + k_l(P(n+1,t) - P(n,t)) \tag{8.1}$$

which describes the time evolution of the probability distribution in terms of the *transition rates* between different "states" of the system is one example. Another is the diffusion equation (the three-dimensional analog of Eq. (7.8))

$$\frac{\partial P(\mathbf{r},t)}{\partial t} = D\nabla^2 P(\mathbf{r},t) \tag{8.2}$$

The fact that under these equations the probability distribution at time t is fully determined by the distribution at any earlier time implies that these processes are Markovian.

Equations (8.1) and (8.2) should be solved under given initial conditions, $P(n, 0)$ and $P(\mathbf{r}, 0)$, respectively. If these are given by $P(n, t = 0) = \delta_{n,n_0}$ and $P(\mathbf{r}, t) = \delta(\mathbf{r} - \mathbf{r}_0)$ the resulting solutions are the conditional probabilities $P(n, t \mid n_0, t_0)$ and $P(\mathbf{r}, t \mid \mathbf{r}_0, t_0)$ to be at n or $\mathbf{r}$ given that the system started at n_0 or $\mathbf{r}_0$, respectively. In the present context these can be identified as the transition probabilities of the corresponding stochastic processes—from n_0 to n or from $\mathbf{r}_0$ to $\mathbf{r}$. *A Markovian process is completely determined by these transition probabilities.* We can thus rewrite Eq. (8.2), for example, in a more specific way as

$$\frac{\partial P(\mathbf{r},t\,|\,\mathbf{r}_0,t_0)}{\partial t} = D\nabla^2 P(\mathbf{r},t\,|\,\mathbf{r}_0,t_0); \qquad P(\mathbf{r},t_0\,|\,\mathbf{r}_0,t_0) = \delta(\mathbf{r}-\mathbf{r}_0) \tag{8.3}$$

Problem 8.1. Show that for the initial condition $P(\mathbf{r},t=t_0)=\delta(\mathbf{r}-\mathbf{r_0})$ the solution of Eq. (8.2) is

$$P(\mathbf{r},t\,|\,\mathbf{r}_0,t_0) = \frac{1}{(4\pi D(t-t_0))^{3/2}}\exp\left(-\frac{(\mathbf{r}-\mathbf{r}_0)^2}{4D(t-t_0)}\right); \qquad t>t' \tag{8.4}$$

A stochastic process whose transition probability $P(\mathbf{r},t\,|\,\mathbf{r}_0,t_0)$ satisfies Eq. (8.3) is called a *Wiener process*. Another well-known Markovian stochastic process is the *Orenstein–Uhlenbeck process*, for which the transition probability satisfies the equation (in one-dimension)

$$\frac{\partial P(x,t\,|\,x_0,t_0)}{\partial t} = \gamma\frac{\partial}{\partial x}(xP(x,t\,|\,x_0,t_0)) + D\frac{\partial^2}{\partial x^2}P(x,t\,|\,x_0,t_0) \tag{8.5}$$

with $\gamma>0$. The solution of this equation (again with $P(\mathbf{r},t_0\,|\,\mathbf{r}_0,t_0)=\delta(\mathbf{r}-\mathbf{r}_0)$) is the Gaussian distribution

$$P(xt\,|\,x_0t_0) = \sqrt{\frac{\gamma}{2\pi D(1-a^2)}}\exp\left[-\frac{\gamma(x-ax_0)^2}{2D(1-a^2)}\right]; \qquad a=e^{-\gamma(t-t_0)} \tag{8.6}$$

In the limit $\gamma\to 0$ this becomes a Wiener process. Both Eqs (8.3) and (8.5) are special cases of the *Fokker–Planck equation*(see Section 8.4).

Problem 8.2. Calculate the equilibrium correlation function $\langle x(t)x(0)\rangle$ for a system undergoing the Orenstein–Uhlenbeck process.

Solution: The equilibrium distribution implied by Eq. (8.5) is the $t-t_0\to\infty$ limit of Eq. (8.6). In this limit Eq. (8.6) becomes

$$P_{\text{eq}}(x) = P(x,t\to\infty\,|\,x_0,t_0) = \sqrt{\frac{\gamma}{2\pi D}}\exp[-\gamma x^2/(2D)] \tag{8.7}$$

Therefore the equilibrium joint probability distribution for the stochastic variable to take the value x at time t and x' at time t' in equilibrium is (from Eqs (7.40) and (8.7))

$$P_2(x,t;x',t') = P(x,t\,|\,x',t')P_{eq}(x') = \frac{\gamma}{2\pi D\sqrt{1-a^2}}\exp\left[-\gamma\frac{x^2+x'^2-2xx'a}{2D(1-a^2)}\right] \tag{8.8}$$

Here we may treat t and t' on equal footing by taking $a = e^{-\gamma|t-t'|}$. The correlation function can be obtained from (cf. Eq. (7.42a))

$$\langle x(t)x(0)\rangle = \int dx \int dx' xx' P_2(x,t;x',t') \tag{8.9}$$

which yields after some algebra

$$\langle x(t)x(0)\rangle = \frac{Da}{\gamma} = \langle x^2\rangle e^{-\gamma|t-t'|} \tag{8.10}$$

Equations (8.1) and (8.2) are two examples of equations that describe a Markovian stochastic process in terms of the time evolution of its transition probability, $P(n,t|n_0,t_0)$ or $P(\mathbf{r},t \mid \mathbf{r}_0,t_0)$ given the initial conditions $P(n,t_0 \mid n_0,t_0) = \delta_{n,n_0}$ and $P(\mathbf{r},t_0|\mathbf{r}_0,t_0) = \delta(\mathbf{r}-\mathbf{r}_0)$. Apart from the actual form (that depends on the physical nature of the process) they differ from each other in that the system described by (8.1) has a discrete set of states $\{n\}$ while in (8.2) the state space is continuous. In correspondence $P(n,t \mid n_0,t_0)$ is a probability, while $P(\mathbf{r},t \mid \mathbf{r}_0,t_0)$ is a probability density. As seen in Section 7.3.1 these equations may describe the same physical process, with (8.2) obtained from a coarse-graining procedure applied to (8.1). Many times however, the use of discrete distributions appears in descriptions of physical processes in the space of the energy states of systems with discrete spectra, while continuous distributions appear when describing processes in position-momentum space. More important is to note that because continuous formulations usually involve coarse-graining, that is, collapsing many observables within the resolution window of our observation into a single coarse-grained variable, it follows from the central limit theorem of probability theory (Section 1.1.1) that the distributions involved are Gaussian, which is why modeling of physical processes in terms of Wiener or Orenstein–Uhlenbeck processes is often useful. Even when the process is not Gaussian, a continuous representation often leads to a time evolution equation, called a Fokker–Planck equation (see Section 8.4), whose form (a generalization of Eq. (8.5)) stems from the common situation where transitions involving a given state are dominated by states in its immediate neighborhood. A more general description of time evolution in probability space is often referred to as a *master equation* (Section 8.3). Before addressing these types of stochastic equations of motion we consider in the next section an alternative description in the space of the stochastic variable itself, the so called *Langevin equation.*

8.2 The Langevin equation

8.2.1 General considerations

Sometimes we find it advantageous to focus our stochastic description not on the probability but on the random variable itself. This makes it possible to address more directly the source of randomness in the system and its effect on the time evolution of the interesting subsystem. In this case the basic stochastic input is not a set of transition probabilities or rates, but the actual effect of the "environment" on the "interesting subsystem." Obviously this effect is random in nature, reflecting the fact that we do not have a complete microscopic description of the environment.

As discussed in Section 8.1, we could attempt to derive these stochastic equations of motion from first principles, that is, from the full Hamiltonian of the system+environment. Alternatively we can attempt to construct the equation of motion using intuitive arguments and as much of the available physical information as possible. Again, this section takes the second route. As an example consider the equation of motion of a particle moving in a one-dimensional potential,

$$\ddot{x} = -\frac{1}{m}\frac{\partial V(x)}{\partial x} \tag{8.11}$$

and consider the effect on this particle's dynamics of putting it in contact with a "thermal environment." Obviously the effect depends on the strength of interaction between the particle and this environment. A useful measure of the latter within a simple intuitive model is the friction force, proportional to the particle velocity, which acts to slow down the particle:

$$\ddot{x} = -\frac{1}{m}\frac{\partial V(x)}{\partial x} - \gamma\dot{x} \tag{8.12}$$

The effect of friction is to damp the particle energy. This can most easily be seen by multiplying Eq. (8.12) by $m\dot{x}$, using $m\dot{x}\ddot{x} + \dot{x}(\partial V(x)/\partial x) = (d/dt)[E_K + E_P] = \dot{E}$

to get $\dot{E} = -2\gamma E_k$. Here E, E_K, and E_P are respectively the total particle energy and its kinetic and potential components. Equation (8.12) thus describes a process of energy dissipation, and leads to zero energy, measured from a local minimum on the potential surface, at infinite time. It therefore cannot in itself describe the time evolution of a particle in thermal equilibrium. What is missing is the random "kicks" that the particle occasionally receives from the surrounding thermal particles. These kicks can be modeled by an additional random force in Eq. (8.12)

$$\ddot{x} = -\frac{1}{m}\frac{\partial V(x)}{\partial x} - \gamma\dot{x} + \frac{1}{m}R(t) \tag{8.13}$$

The function $R(t)$ describes the effects of random collisions between our subsystem (henceforth referred to as "system"), that may sometimes be a single particle or a single degree of freedom, and the molecules of the thermal environment ("bath"). This force is obviously a stochastic process, and a full stochastic description of our system is obtained once we define its statistical nature.

What can be said about the statistical character of the stochastic process $R(t)$? First, from symmetry arguments valid for stationary systems, $\langle R(t)\rangle = 0$, where the average can be either time or ensemble average. Second, since Eq. (8.12) seems to describe the relaxation of the system at temperature $T = 0$, R should be related to the finite temperature of the thermal environment. Next, at $T = 0$, the time evolution of x according to Eq. (8.12) is Markovian (knowledge of x and $\dot{x}$ fully determines the future of x), so the system-bath coupling introduced in (8.12) is of Markovian nature. This implies that the action of the bath on the system at time t does not depend on history of the system or the bath; in particular, the bath has no memory of what the system did in the past (see Section 7.4.2). The additional finite temperature term $R(t)$ has to be consistent with the Markovian form of the damping term. Finally, in the absence of further knowledge and because R is envisioned as a combined effect of many environmental motions, it makes sense to assume that, for each time t, $R(t)$ is a Gaussian random variable, and that the stochastic process $R(t)$ is a Gaussian process (Section 7.4.3).

We have already argued (Section 7.4.2) that the Markovian nature of the system evolution implies that the relaxation dynamics of the bath is much faster than that of the system. *The bath loses its memory on the timescale of interest for the system dynamics.* Still the timescale for the bath motion is not unimportant. If, for example, the sign of $R(t)$ changes infinitely fast, it makes no effect on the system. Indeed, in order for a finite force R to move the particle it has to have a finite duration. It is convenient to introduce a timescale τ_B, which characterizes the bath motion, and to consider an approximate picture in which $R(t)$ is constant in the interval $[t, t+\tau_B]$, while $R(t_1)$ and $R(t_2)$ are independent Gaussian random variables if

$|t_1 - t_2| \geq (1/2)\tau_B$. Accordingly,

$$\langle R(t_1)R(t_1 + t)\rangle = CS(t) \tag{8.14}$$

where $S(t)$ is 1 if $|t| < (1/2)\tau_B$, and is 0 otherwise. Since $R(t)$ was assumed to be a Gaussian process, the first two moments specify completely its statistical nature. The assumption that the bath is fast relative to the timescales that characterize the system implies that τ_B is much shorter than all timescales (inverse frequencies) derived from the potential $V(x)$ and much smaller than the relaxation time γ^{-1} for dissipation of the system energy.

In Eqs (8.13) and (8.14), both γ and C originate from the system–bath coupling, and should therefore be somehow related to each other. In order to obtain this relation it is sufficient to consider Eq. (8.13) for the case where V does not depend on position, whereupon

$$\dot{v} = -\gamma v + \frac{1}{m}R(t) \tag{8.15}$$

($v = \dot{x}$ is the particle velocity). This equation can be solved as a first-order inhomogeneous differential equation, to yield

$$v(t) = v(t = 0)e^{-\gamma t} + \frac{1}{m}\int_0^t dt' e^{-\gamma(t-t')}R(t') \tag{8.16}$$

For long times, as the system reaches equilibrium, only the second-term on the right of (8.16) contributes. For the average $\langle u\rangle$ at thermal equilibrium this gives zero, while for $\langle v^2\rangle$ we get

$$\langle v^2\rangle = \frac{1}{m^2}\int_0^t dt' \int_0^t dt'' e^{-\gamma(t-t')-\gamma(t-t'')} CS(t' - t'') \tag{8.17}$$

Since the integrand is negligible unless $|t' - t''| \leq \tau_B \ll 1/\gamma$, $\langle v^2\rangle$ in Eq. (8.17) can be approximated by

$$\langle v^2\rangle = \frac{1}{m^2}\int_0^t dt' e^{-2\gamma(t-t')} \int_0^t dt'' CS(t' - t'') = \frac{1}{2m^2\gamma}C\tau_B \tag{8.18}$$

To get the final result we took the limit $t \to \infty$. Since in this limit the system should be in thermal equilibrium we have $\langle v^2\rangle = k_B T/m$, whence

$$C = \frac{2m\gamma k_B T}{\tau_B} \tag{8.19}$$

Using this result in Eq. (8.14) we find that the correlation function of the Gaussian random force R has the form

$$\langle R(t_1)R(t_1+t)\rangle = 2m\gamma k_B T \frac{S(t)}{\tau_\mathrm{B}} \xrightarrow{\tau_\mathrm{B}\to 0} 2m\gamma k_B T\delta(t) \tag{8.20}$$

For the system's motion to be Markovian τ_B has to be much shorter than the relevant system's timescales. Equation (8.20) indicates that its actual magnitude is not important and the random force may be thought of as δ-correlated. The limiting process described above indicates that mathematical consistency requires that as $\tau_\mathrm{B} \to 0$ the second moment of the random force diverge, and the proper limiting form of the correlation function is a Dirac δ function in the time difference. Usually in analytical treatments of the Langevin equation this limiting form is convenient. In numerical solutions however, the random force is generated at time intervals Δt, determined by the integration routine. The random force is then generated as a Gaussian random variable with zero average and variance equal to $2m\gamma k_B T/\Delta t$.

We have thus seen that the requirement that the friction γ and the random force $R(t)$ together act to bring the system to thermal equilibrium at long time, naturally leads to a relation between them, expressed by Eq. (8.20). This is a relation between fluctuations and dissipation in the system, which constitutes an example of the *fluctuation–dissipation theorem* (see also Chapter 11). In effect, the requirement that Eq. (8.20) holds is equivalent to the condition of detailed balance, imposed on transition rates in models described by master equations, in order to satisfy the requirement that thermal equilibrium is reached at long time (see Section 8.3).

8.2.2 The high friction limit

The friction coefficient γ defines the timescale, γ^{-1} of thermal relaxation in the system described by (8.13). A simpler stochastic description can be obtained for a system in which this time is shorter than any other characteristic timescale of our system.[1] This high friction situation is often referred to as the *overdamped limit.* In this limit of large γ, the velocity relaxation is fast and it may be assumed to quickly reaches a steady state for any value of the applied force, that is, $\dot{v} = \ddot{x} = 0$. This statement is not obvious, and a supporting (though not rigorous) argument is provided below. If true then Eqs (8.13) and (8.20) yield

$$\frac{dx}{dt} = \frac{1}{\gamma m}\left(-\frac{dV}{dx} + R(t)\right); \qquad \langle R\rangle = 0; \qquad \langle R(0)R(t)\rangle = 2m\gamma k_B T\delta(t) \tag{8.21}$$

[1] But, as discussed in Section 8–2.1, not relative to the environmental relaxation time.

This is a Langevin type equation that describes strong coupling between the system and its environment. Obviously, the limit $\gamma \to 0$ of deterministic motion cannot be identified here.

Why can we, in this limit, neglect the acceleration term in (8.13)? Consider a particular realization of the random force in this equation and denote $-dV/dx + R(t) = F(t)$. Consider then Eq. (8.13) in the form

$$\dot{v} = -\gamma v + \frac{1}{m}F(t) \tag{8.22}$$

If F is constant then after some transient period (short for large γ) the solution of (8.22) reaches the constant velocity state

$$v = \frac{F}{m\gamma} \tag{8.23}$$

The neglect of the $\dot{v}$ term in (8.22) is equivalent to the assumption that Eq. (8.23) provides a good approximation for the solution of (8.22) also when F depends on time. To find the conditions under which this assumption holds consider the solution of (8.22) for a particular Fourier component of the time-dependent force

$$F(t) = F_\omega e^{i\omega t} \tag{8.24}$$

Disregarding any initial transient amounts to looking for a solution of (8.22) of the form

$$v(t) = v_\omega e^{i\omega t} \tag{8.25}$$

Inserting (8.24) and (8.25) into (8.22) we find

$$v_\omega = \frac{F_\omega/m}{i\omega + \gamma} = \frac{F_\omega}{m\gamma}\frac{1}{1 + i\omega/\gamma} \tag{8.26}$$

which implies

$$v(t) = \frac{F(t)}{m\gamma}(1 + O(\omega/\gamma)) \tag{8.27}$$

We found that Eq. (8.23) holds, with corrections of order ω/γ. It should be emphasized that this argument is not rigorous because the random part of $F(t)$ is in principle fast, that is, contain Fourier components with large ω. More rigorously, the transition from Eq. (8.13) to (8.21) should be regarded as coarse-graining in time to get a description in which the fast components of the random force are averaged to zero and velocity distribution is assumed to follow the remaining instantaneous applied force.

8.2.3 Harmonic analysis of the Langevin equation

If $R(t)$ satisfies the Markovian property (8.20), it follows from the Wiener–Khintchine theorem (7.76) that its power spectrum is constant

$$I_R(\omega) = \text{constant} \equiv I_R \tag{8.28}$$

$$C_R(t) = \int_{-\infty}^{\infty} d\omega e^{i\omega t} I_R(\omega) = 2\pi I_R \delta(t) \tag{8.29}$$

$$I_R = \frac{m\gamma k_B T}{\pi} \tag{8.30}$$

Equation (8.28) implies that all frequencies are equally presented in this random force spectrum. A stochastic process of this type is called *a white noise.*

From the spectral density of $R(t)$ we can find the spectral density of stochastic observables that are related to R via linear Langevin equations. For example, consider the Langevin equation (8.13) with $V(x) = (1/2)m\omega_0^2 x^2$ (the so called Brownian harmonic oscillator)

$$\frac{d^2x}{dt^2} + \gamma \frac{dx}{dt} + \omega_0^2 x = \frac{1}{m} R(t) \tag{8.31}$$

and apply the Fourier expansion (7.69) to R, x, and $v = dx/dt$

$$R(t) = \sum_{n=-\infty}^{\infty} R_n e^{i\omega_n t}; \qquad x(t) = \sum_{n=-\infty}^{\infty} x_n e^{i\omega_n t};$$

$$v(t) = \sum_{n=-\infty}^{\infty} v_n e^{i\omega_n t}; \qquad v_n = i\omega_n x_n \tag{8.32}$$

Using these expansions in (8.31) yields

$$x_n = \frac{1}{\left(\omega_0^2 - \omega_n^2 + i\omega_n \gamma\right) m} R_n; \qquad v_n = \frac{i\omega_n}{\left(\omega_0^2 - \omega_n^2 + i\omega_n \gamma\right) m} R_n \tag{8.33}$$

The power spectrum of any of these stationary processes is given by Eq. (7.74). Therefore, Eq. (8.33) implies a relation between these spectra

$$I_x(\omega) = \frac{1}{\left|\omega_0^2 - \omega^2 + i\gamma\omega\right|^2} \frac{I_R(\omega)}{m^2} = \frac{1}{\left(\omega_0^2 - \omega^2\right)^2 + \gamma^2\omega^2} \frac{I_R(\omega)}{m^2} \tag{8.34a}$$

$$I_v(\omega) = \frac{\omega^2}{\left(\omega_0^2 - \omega^2\right)^2 + \gamma^2\omega^2} \frac{I_R(\omega)}{m^2} \tag{8.34b}$$

In the absence of an external potential (free Brownian motion; $\omega_0 = 0$) Eq. (8.34b) becomes

$$I_v(\omega) = \frac{1}{m^2\left(\omega^2+\gamma^2\right)} I_R(\omega) \tag{8.35}$$

In the Markovian case $I_R(\omega) = I_R$ is independent of ω.[2] Together with Eq. (8.30) these are explicit expressions for the corresponding power spectra.

As an application consider the velocity time correlation function for the simple Brownian motion. Using Eqs (7.76), (8.30), and (8.35) we get

$$\begin{aligned} C_v(t_2-t_1) \equiv \langle v(t_1)v(t_2)\rangle &= \frac{I_R}{m^2}\int_{-\infty}^{\infty} d\omega\, e^{i\omega(t_2-t_1)}\frac{1}{\omega^2+\gamma^2} \\ &= \frac{\pi I_R}{m^2\gamma}e^{-\gamma|t_1-t_2|} = \frac{k_BT}{m}e^{-\gamma|t_1-t_2|} \end{aligned} \tag{8.36}$$

—an exponential decay with a pre-exponential coefficient given by the equilibrium value of $\langle v^2\rangle$, as expected. Similarly, for the harmonic Brownian motion, we get using Eqs (7.76) and (8.34a)

$$C_x(t) \equiv \langle x(0)x(t)\rangle = \frac{I_R}{m^2}\int_{-\infty}^{\infty} d\omega\, e^{i\omega t}\frac{1}{(\omega_0^2-\omega^2)^2+\gamma^2\omega^2} \tag{8.37}$$

This integral is most easily done by complex integration, where the poles of the integrand are $\omega = \pm(i/2)\gamma \pm \omega_1$ with $\omega_1 = \sqrt{\omega_0^2 - \gamma^2/4}$. It leads to

$$C_x(t) = \frac{\pi I_R}{m^2\gamma\omega_0^2}\left(\cos\omega_1 t + \frac{\gamma}{2\omega_1}\sin\omega_1 t\right)e^{-\gamma t/2} \quad \text{for } t>0 \tag{8.38}$$

For $t = 0$ we have $\langle x^2\rangle = \pi I_R(m^2\gamma\omega_0^2)^{-1}$ and using (8.30) we get $m\omega_0^2\langle x^2\rangle = k_BT$, again as expected.

8.2.4 The absorption lineshape of a harmonic oscillator

The Langevin equation (8.31), with $R(t)$ taken to be a Gaussian random force that satisfies $\langle R\rangle = 0$ and $\langle R(0)R(t)\rangle = 2m\gamma k_BT\delta(t)$, is a model for the effect of a thermal environment on the motion of a classical harmonic oscillator, for example, the nuclear motion of the internal coordinate of a diatomic molecule in solution.

[2] It is important to note that Eq. (8.31), with a constant γ, is valid only in the Markovian case. Its generalization to non-Markovian situations is discussed in Section 8.2.6.

A standard experimental probe of this motion is infrared spectroscopy. We may use the results of Sections 7.5 and 8.2.3 to examine the effect of interaction with the thermal environment on the absorption lineshape. The simplest model for the coupling of a molecular system to the radiation field is expressed by a term $-\boldsymbol{\mu} \cdot \boldsymbol{\mathcal{E}}$ in the Hamiltonian, where $\boldsymbol{\mu}$ is the molecular dipole, and $\boldsymbol{\mathcal{E}}(t)$ is the oscillating electric field (see Section 3.1). For a one-dimensional oscillator, assuming that μ is proportional to the oscillator displacement from its equilibrium position and taking $\mathcal{E}(t) \sim \cos(\omega t)$, we find that the coupling of the oscillator to the thermal environment and the radiation field can be modeled by Eq. (8.31) supplemented by a term $(F/m)\cos(\omega t)$ where F denotes the radiation induced driving force. We can use the resulting equation to compute the radiation energy absorbed by the oscillator following the procedure of Section 7.5.3. Alternatively, Eq. (8.31) implies that our oscillator can be described as a superposition of normal modes of the overall system including the bath (see Sect. 8.2.5). In this sense the coordinate x that couples to the radiation field is equivalent to the coordinate A (Eq. (7.77)) used in Section 7.5.3. This implies, using Eq. (7.91), the absorption lineshape

$$L(\omega) \sim \omega^2 I_x(\omega) = \frac{\omega^2}{(\omega_0^2 - \omega^2)^2 + (\gamma\omega)^2} \frac{\gamma k_B T}{\pi m} \tag{8.39}$$

In the underdamped limit $\gamma \ll \omega_0$, which is relevant for molecules in condensed phases, $L(\omega)$ is strongly peaked about $\omega = \omega_0$. Near the peak we can approximate the denominator in (8.39) by $(\omega_0^2 - \omega^2)^2 + (\omega\gamma)^2 \cong 4\omega_0^2(\omega - \omega_0)^2 + \omega_0^2\gamma^2$, so that

$$\tilde{L}(\omega) = L(\omega) \Big/ \int_{-\infty}^{\infty} d\omega L(\omega) = \frac{1}{\pi} \frac{\gamma/2}{(\omega_0 - \omega)^2 + (\gamma/2)^2} \tag{8.40}$$

This is a Lorentzian lineshape whose width is determined by the friction. The latter, in turn, corresponds to the rate of energy dissipation. It is significant that the normalized lineshape (characterized by its center and width) does not depend on the temperature. This result is associated with the fact that the harmonic oscillator is characterized by an energy level structure with constant spacing, or classically–with and energy independent frequency.

In Section 6.2.3 we have seen that a simple quantum mechanical theory based on the golden rule yields an expression for the absorption lineshape that is given essentially by the Fourier transform of the relevant dipole correlation function $\langle \mu(0)\mu(t) \rangle$. Assuming again that μ is proportional to the displacement x of the oscillator from its equilibrium position we have

$$L(\omega) = \alpha \int_{-\infty}^{\infty} dt e^{-i\omega t} \langle x(0)x(t) \rangle \tag{8.41}$$

that, using (7.76) and (8.37) leads again to the result (8.40).

Another point of interest is the close similarity between the lineshapes associated with the quantum damped two-level system, Eq. (9.40), and the classical damped harmonic oscillator. We will return to this issue in Section 9.3.

Problem 8.3. Show that

$$\tilde{L}(\omega) = L(\omega)/\int_{-\infty}^{\infty} d\omega L(\omega) = \left(2\pi\left\langle x^2\right\rangle\right)^{-1}\int_{-\infty}^{\infty} dt e^{-i\omega t}\langle x(0)x(t)\rangle \tag{8.42}$$

Problem 8.4.

(1) If $z(t)$ is a real stationary stochastic process so that $\langle z(t_1)z(t_2)\rangle = C_z(t_1-t_2)$ show that $z(\omega) = \int_{-\infty}^{\infty} dt e^{-i\omega t} z(t)$ satisfies

$$\langle z(\omega_1)z(\omega_2)\rangle = 2\pi\delta(\omega_1+\omega_2)C_z(\omega_2) \tag{8.43}$$

$$C_z(\omega) = \int_{-\infty}^{\infty} dt e^{-i\omega t} C_z(t) = C_z(-\omega) = C_z^*(\omega) \tag{8.44}$$

In particular, verify that $\langle R(\omega_1)R(\omega_2)\rangle = 4\pi m k_B T\gamma\delta(\omega_1+\omega_2)$.

(2) For the position correlation function of a harmonic oscillator use these results together with (cf. Eq. (8.33))

$$x(\omega) = \frac{m^{-1}R(\omega)}{\omega_0^2 - \omega^2 + i\omega\gamma} \tag{8.45}$$

to show that

$$C_x(\omega) = \frac{2k_B T\gamma/m}{(\omega_0^2-\omega^2)^2 + (\omega\gamma)^2} \tag{8.46}$$

This is another route to the corresponding absorption lineshape.

8.2.5 Derivation of the Langevin equation from a microscopic model

The stochastic equation of motion (8.13) was introduced as a phenomenological model based on the combination of experience and intuition. We shall now attempt to derive such an equation from "first principles," namely starting from the Newton equations for a particular microscopic model.[3] In this model the "system" is a one-dimensional particle of mass m moving in a potential $V(x)$, and the bath is a collection of independent harmonic oscillators. The Hamiltonian is taken to be

$$H = \frac{p^2}{2m} + V(x) + H_{\text{bath}} + H_{\text{int}} \tag{8.47}$$

with the bath and system–bath interaction Hamiltonians given by

$$H_{\text{bath}} + H_{\text{int}} = \frac{1}{2}\sum_j m_j \left[\dot{q}_j^2 + \omega_j^2 \left(q_j + \frac{c_j}{m_j\omega_j^2}x \right)^2 \right] \tag{8.48}$$

The "interaction" that appears in (8.48) contains, in addition to a linear coupling term $x\sum_j c_j q_j$, also a "compensating term" $\sum_j (c_j x)^2/(2m_j\omega_j^2)$ that has the effect that the minimum potential experienced by the particle at any point x along the x-axis is $V(x)$. This minimum is achieved when all bath coordinates q_j adjust to the position x of the particle, that is, take the values $-[c_j/(m_j\omega_j^2)]x$.

The equations of motion for the "system" and the bath particles are

$$\ddot{x} = -\frac{1}{m}\frac{\partial V}{\partial x} - \frac{1}{m}\sum_j c_j \left(\frac{c_j}{m_j\omega_j^2}x + q_j \right) \tag{8.49}$$

and

$$\ddot{q}_j = -\omega_j^2 q_j - \frac{c_j}{m_j}x \tag{8.50}$$

Equation (8.50) is an inhomogeneous differential equation for $q_j(t)$, whose solution can be written as

$$q_j(t) = Q_j(t) + \tilde{q}_j(t) \tag{8.51}$$

where

$$Q_j(t) = q_{j0}\cos(\omega_j t) + \frac{\dot{q}_{j0}}{\omega_j}\sin(\omega_j t) \tag{8.52}$$

[3] For the equivalent quantum mechanical derivation of the "quantum Langevin equation" see G. W. Ford and M. Kac, J. Stat. Phys. **46**, 803 (1987); G. W. Ford, J. T. Lewis, and R. F. O'Connell, Phys. Rev. A **37**, 4419 (1988).

is the solution of the corresponding homogeneous equation in which q_{j0} and $\dot{q}_{j0}$ should be sampled from the equilibrium distribution of the free bath, and where $\tilde{q}_j(t)$ is a solution of the inhomogeneous equation. A rather nontrivial such solution is[4]

$$\tilde{q}_j(t) = -\frac{c_j}{m_j\omega_j^2}x(t) + \frac{c_j}{m_j\omega_j^2}\int_0^t d\tau \cos\left(\omega_j(t-\tau)\right)\dot{x}(\tau) \tag{8.53}$$

Using (8.51)–(8.53) in (8.49) now leads to

$$\ddot{x} = -\frac{1}{m}\frac{\partial V(x)}{\partial x} - \int_0^t d\tau Z(t-\tau)\dot{x}(\tau) + \frac{1}{m}R(t) \tag{8.54}$$

$$Z(t) = \frac{1}{m}\sum_j \frac{c_j^2}{m_j\omega_j^2}\cos(\omega_j t) \tag{8.55}$$

$$R(t) = -\sum_j c_j\left(q_{j0}\cos(\omega_j t) + \frac{\dot{q}_{j0}}{\omega_j}\sin(\omega_j t)\right) \tag{8.56}$$

The following points are noteworthy:

1. The function $R(t)$, which is mathematically identical to the variable A of Section 6.5.1,[5] represents a stochastic force that acts on the system coordinate x. Its stochastic nature stems from the lack of information about q_{j0} and $\dot{q}_{j0}$. All we know about these quantities is that, since the thermal bath is assumed to remain in equilibrium throughout the process, they should be sampled from

---

[4] Eqs. (8.52) and (8.53) imply that the initial state of the bath modes is sampled from a thermal distribution in presence of the system. To check that (8.53) satisfies (8.50) write it in the form

$$\tilde{q}_j(t) = -\frac{c_j}{m_j\omega_j^2}x(t) + \frac{c_j}{m_j\omega_j^2}\mathrm{Re}F;\ F = e^{i\omega_j t}\int_0^t d\tau e^{-i\omega_j\tau}\dot{x}(\tau)$$

so that $\ddot{\tilde{q}}_j = -(c_j/(m_j\omega_j^2))(\ddot{x} - \mathrm{Re}\ddot{F})$, and prove the identity $\ddot{F} = i\omega\dot{F} + \ddot{x}$. This, together with the fact that x and its time derivatives are real lead to $\ddot{\tilde{q}}_j = (c_j/(m_j\omega_j^2))\mathrm{Re}(i\omega_j\dot{F})$. Using also $\dot{F} = i\omega_j F + \dot{x}$ leads to $\ddot{\tilde{q}}_j = -(c_j/m_j)\mathrm{Re}F$, which (using the equation that relates q_j to $\mathrm{Re}F$ above) is identical to (8.50)

[5] See Eqs (6.79), (6.81a), where the mass weighted normal coordinates u_j was used instead of q_j.

an equilibrium Boltzmann distribution,[6] that is, they are Gaussian random variables that satisfy

$$\begin{aligned}\langle q_{j0}\rangle &= \langle \dot{q}_{j0}\rangle = 0\\ (1/2)m_j\langle \dot{q}_{j0}\dot{q}_{j'0}\rangle &= (1/2)m_j\omega_j^2\langle q_{j0}q_{j'0}\rangle = (1/2)k_BT\delta_{j,j'}\\ \langle q_{j0}\dot{q}_{j'0}\rangle &= 0\end{aligned} \tag{8.57}$$

2. The system–bath interaction term in (8.48) is xf, where $f = \sum_j c_j q_j$ is the force exerted by the thermal environment on the system. The random force $R(t)$, Eq. (8.56) is seen to have a similar form,

$$R(t) = \sum_j c_j q_j^{(0)}(t) \tag{8.58}$$

where $q_j^{(0)}(t) = q_{j0}\cos(\omega_j t) + \omega_j^{-1}\dot{q}_{j0}\sin(\omega_j t)$ represents the motion of a free bath mode, undisturbed by the system.
3. Using Eqs (8.56) and (8.57) we can easily verify that

$$\langle R(0)R(t)\rangle = mk_BTZ(t) \tag{8.59}$$

Comparing Eqs (7.77)–(7.79) we see that $Z(t)$ is essentially the Fourier transform of the spectral density associated with the system–bath interaction. The differences are only semantic, originating from the fact that in Eqs (7.77)–(7.79) we used mass renormalized coordinates while here we have associated a mass m_j with each harmonic bath mode j.

Equation (8.54) is a stochastic equation of motion similar to Eq. (8.13). However, we see an important difference: Eq. (8.54) is an integro-differential equation in which the term $\gamma\dot{x}$ of Eq. (8.13) is replaced by the integral $\int_0^t d\tau Z(t-\tau)\dot{x}(\tau)$. At the same time the relationship between the random force $R(t)$ and the damping, Eq. (8.20), is now replaced by (8.59). Equation (8.54) is in fact the *non-Markovian generalization* of Eq. (8.13), where the effect of the thermal environment on the system is not instantaneous but characterized by a memory—at time t it depends on the past interactions between them. These past interactions are important during a *memory time*, given by the lifetime of the *memory kernel* $Z(t)$. The Markovian limit is obtained when this kernel is instantaneous

$$\text{Markovian limit:}\quad Z(t) = 2\gamma\delta(t) \tag{8.60}$$

[6] This is in fact a subtle point, because by choosing the solution (8.53) we affect the choice of $q_j(0)$ and $\dot{q}_j(0)$. For further discussion of this point see P. Hanggi, in Stochastic Dynamics, edited by L. Schimansky-Geier and T. Poschel (Springer Verlag, Berlin, 1997), Lecture notes in Physics Vol. 484, p. 15.

in which case Eqs (8.13) and (8.20) are recovered from Eqs (8.54) and (8.59).

8.2.6 The generalized Langevin equation

As discussed in Section 8.2.1, the Langevin equation (8.13) describes a Markovian stochastic process: The evolution of the stochastic system variable $x(t)$ is determined by the state of the system *and the bath* at the same time t. The instantaneous response of the bath is expressed by the appearance of a constant damping coefficient γ and by the white-noise character of the random force $R(t)$.

The microscopic model described in the previous section leads to Eqs (8.54)–(8.56) as precursors of this Markovian picture. The latter is obtained in the limit where the timescale for relaxation of the thermal environment is short relative to all characteristic system times, as expressed mathematically by Eq. (8.60). This limit, however, is far from obvious. The characteristic times in molecular systems are associated with electronic processes (typical timescale 10^{-15}–10^{-16} s), vibrational motions (10^{-14}–10^{-15} s), librations, rotations, and center of mass motions ($>10^{-12}$ s). This should be compared with typical thermal relaxation times in condensed phases that can be estimated in several ways. The simplest estimate, obtained from dividing a typical intermolecular distance (10^{-8} cm) by a typical thermal velocity (10^4 cm s^{-1}) give a result, 10^{-12} s that agrees with other estimates. Obviously this timescale is longer than characteristic vibrational and electronic motions in molecular systems. A similar picture is obtained by comparing the characteristic frequencies (spacing between energy levels) associated with molecular electronic motions (1–4 eV) and intramolecular vibrational motions ($\sim$0.1 eV) with characteristic cutoff (Debye) frequencies that are of order 0.01–0.1 eV for molecular environments. One could dismiss electronic processes as unimportant for room temperature systems in the absence of light, still intramolecular motions important in describing the dynamics of chemical reaction processes are also often considerably faster than typical environmental relaxation times.

The Markovian picture cannot be used to describe such motions. The generalized Langevin equation

$$\ddot{x} = -\frac{1}{m}\frac{\partial V(x)}{\partial x} - \int_0^t d\tau Z(t-\tau)\dot{x}(\tau) + \frac{1}{m}R(t) \tag{8.61}$$

with $R(t)$ being a Gaussian random force that satisfies

$$\langle R \rangle = 0; \qquad \langle R(0)R(t)\rangle = mk_BTZ(t) \tag{8.62}$$

is a useful model for such situations. While its derivation in the previous section has invoked a harmonic model for the thermal bath, this model is general enough for most purposes (see Section 6.5). The simple damping term $-\gamma\dot{x}$ in Eq. (8.13) is now replaced by the non-Markovian friction term $-\int_0^t d\tau Z(t-\tau)\dot{x}(\tau)$ with the memory kernel $Z(t)$ that satisfies Eq. (8.62). The time dependence of Z characterizes the memory of the bath—the way its response is affected by past influences. A characteristic "memory time" can be defined by

$$\tau_{\text{mem}} = \frac{1}{Z(0)} \int_0^\infty dt Z(t) \tag{8.63}$$

provided this integral converges.

It is important to point out that this does not imply that Markovian stochastic equations cannot be used in descriptions of condensed phase molecular processes. On the contrary, such equations are often applied successfully. The recipe for a successful application is to be aware of what can and what cannot be described with such approach. Recall that stochastic dynamics emerge when seeking coarse-grained or reduced descriptions of physical processes. The message from the timescales comparison made above is that Markovian descriptions are valid for molecular processes that are slow relative to environmental relaxation rates. Thus, with Markovian equations of motion we cannot describe molecular nuclear motions in detail, because vibrational periods (10^{-14} s) are short relative to environmental relaxation rates, but we should be able to describe vibrational relaxation processes that are often much slower, as is shown in Section 8.3.3.

Coming back to the non-Markovian equations (8.61) and (8.62), and their Markovian limiting form obtained when $Z(t)$ satisfies Eq. (8.60), we next seek to quantify the properties of the thermal environment that will determine its Markovian or non-Markovian nature.

Problem 8.5. Show that Eq. (8.55) can be written in the form

$$Z(t) = \frac{2}{\pi m} \int_0^\infty d\omega \frac{J(\omega)}{\omega} \cos(\omega t) \tag{8.64}$$

where

$$J(\omega) = \frac{\pi}{2} \sum_j \frac{c_j^2}{m_j \omega_j} \delta(\omega - \omega_j) \tag{8.65}$$

is the spectral density associated with the system–bath interaction.[7]

---

[7] Note the difference between Eqs (8.65) and (6.90) or (7.79). The mass m_j appears explicitly in (8.65) because here we did not use mass weighted normal mode coordinates as we did in Chapters 6 and 7. In practice this is just a redefinition of the coupling coefficient c_j.

The relaxation induced by the bath is seen to be entirely determined by the properties of this spectral function. In particular, a *Ohmic* bath is defined to have the property

$$J(\omega) = \eta\omega \tag{8.66}$$

where η is a constant. For such a bath Eq. (8.64) gives Eq. (8.60) with $\gamma = \eta/m$.

In reality, the Ohmic property, $J(\omega) \sim \omega$ can be satisfied only approximately because from (7.80) it follows that $\int_0^\infty d\omega(J(\omega)/\omega)$ has to be finite. A practical definition of Ohmic spectral density is

$$J(\omega) = \eta\omega e^{-\omega/\omega_c} \tag{8.67}$$

from which, using (8.64), it follows that

$$Z(t) = \frac{2\eta}{m}\frac{\omega_c/\pi}{1+(\omega_c t)^2} \tag{8.68}$$

ω_c represents a cutoff frequency beyond which the bath density of modes falls sharply. It is equivalent to the Debye frequency of Section 4.2.4, whose existence was implied by the discrete nature of atomic environments or equivalently by the finite density per unit volume of bath modes. Here it represents the fastest timescale associated with the thermal environment and the bath characteristic memory time (indeed Eqs (8.63) and (8.68) yield $\tau_{\text{mem}} = \pi/2\omega_c$). The Markovian requirement that the bath is fast relative to the system can be also expressed by requiring that ω_c is larger than all relevant system frequencies or energy spacings.

Problem 8.6. Show that the power spectrum (Section 7.5.1) of the stochastic process $R(t)$ is $I_R(\omega) = k_B T J(\omega)/(\pi\omega)$.

8.3 Master equations

As discussed in Section 8.1, a phenomenological stochastic evolution equation can be constructed by using a model to describe the relevant states of the system and the transition rates between them. For example, in the one-dimensional random walk problem discussed in Section 7.3 we have described the position of the walker by equally spaced points $n\Delta x$; $(n = -\infty, \ldots, \infty)$ on the real axis. Denoting by $P(n, t)$ the probability that the particle is at position n at time t and by k_r and k_l the probabilities per unit time (i.e. the rates) that the particle moves from a given site to the neighboring site on its right and left, respectively, we obtained a kinetic Eq. (7.3) for the time evolution of $P(n, t)$:

$$\frac{\partial P(n,t)}{\partial t} = k_r(P(n-1,t) - P(n,t)) + k_l(P(n+1,t) - P(n,t)) \tag{8.69}$$

This is an example of a master equation.[8] More generally, the transition rates can be defined between any two states, and the master equation takes the form

$$\frac{\partial P(m,t)}{\partial t} = \sum_{\substack{n \\ n\neq m}} k_{mn}P(n,t) - \sum_{\substack{n \\ n\neq m}} k_{nm}P(m,t) \tag{8.70}$$

where $k_{mn} \equiv k_{m\leftarrow n}$ is the rate to go from state n to state m. Equation (8.70) can be rewritten in the compact form

$$\frac{\partial P(m,t)}{\partial t} = \sum_n K_{mn}P(n,t); \quad \text{that is, } \frac{\partial \mathbf{P}}{\partial t} = \mathbf{KP} \tag{8.71}$$

provided we define

$$K_{mn} = k_{mn} \text{ for m} \neq \text{n}; \qquad K_{mm} = -\sum_{\substack{n \\ n\neq m}} k_{nm} \tag{8.72}$$

Note that (8.72) implies that $\sum_m K_{mn} = 0$ for all n. This is compatible with the fact that $\sum_m P(m,t) = 1$ is independent of time. The nearest neighbor random walk process is described by a special case of this master equation with

$$K_{mn} = k_l\delta_{n,m+1} + k_r\delta_{n,m-1} \tag{8.73}$$

In what follows we consider several examples.

8.3.1 The random walk problem revisited

The one-dimensional random walk problem described by Eq. (8.69) was discussed in Section 7.3. It was pointed out that summing either side of this equation over

[8] Many science texts refer to a 1928 paper by W. Pauli [W. Pauli, Festschrift zum 60. Geburtstage A. Sommerfelds (Hirzel, Leipzig, 1928) p. 30] as the first derivation of this type of Kinetic equation. Pauli has used this approach to construct a model for the time evolution of a many-sate quantum system, using transition rates obtained from quantum perturbation theory.

all n from $-\infty$ to ∞ yields zero, while multiplying this equation by n or n^2 then performing the summation yields (cf. Eqs (7.11), (7.12))

$$\frac{\partial \langle n \rangle}{\partial t} = k_r - k_l \tag{8.74}$$

and

$$\frac{\partial \langle n^2 \rangle}{\partial t} = 2\langle n \rangle (k_r - k_l) + k_r + k_l \tag{8.75}$$

For the initial conditions $\langle n \rangle (t = 0) = \langle n^2 \rangle (t = 0) = 0$, that is, for a particle that starts its walk from the origin, $n = 0$, these equations lead to (cf. (7.13), (7.15))

$$\langle n \rangle_t = (k_r - k_l)t = (p_r - p_l)N \tag{8.76}$$

$$\langle \delta n^2 \rangle_t = \langle n^2 \rangle_t - \langle n \rangle_t^2 = (k_r + k_l)t = (p_r + p_l)N \tag{8.77}$$

for a walker that has executed a total of N steps of duration Δt during time $t = N\Delta t$, with probabilities $p_r = k_r \Delta t$ and $p_l = k_l \Delta t$ to jump to the right and to the left, respectively, at each step.

More can be achieved by introducing the *generating function*, defined by[9]

$$F(s,t) = \sum_{n=-\infty}^{\infty} P(n,t)s^n; \qquad 0 < |s| < 1 \tag{8.78}$$

which can be used to generate all moments of the probability distribution according to:

$$\left[\left(s\frac{\partial}{\partial s} \right)^k F(s,t) \right]_{s=1} = \langle n^k \rangle \tag{8.79}$$

We can get an equation for the time evolution of F by multiplying the master equation (8.69) by s^n and summing over all n. Using $\sum_{n=-\infty}^{\infty} s^n P(n-1,t) = sF(s)$ and $\sum_{n=-\infty}^{\infty} s^n P(n+1,t) = F(s)/s$ leads to

$$\frac{\partial F(s,t)}{\partial t} = k_r s F(s,t) + k_l \frac{1}{s} F(s,t) - (k_r + k_l)F(s,t) \tag{8.80}$$

whose solution is

$$F(s,t) = Ae^{[k_r s + (k_l/s) - (k_r + k_l)]t} \tag{8.81}$$

[9] Note that (8.78) is a discrete analog of Eq. (1.5) with $s = e^{\alpha}$.

If the particle starts from $n = 0$, that is, $P(n, t = 0) = \delta_{n,0}$, Eq. (8.78) implies that $F(s, t = 0) = 1$. In this case the integration constant in Eq. (8.81) is $A = 1$. It is easily verified that using (8.81) in Eq (8.79) with $k = 1, 2$ leads to Eqs (8.76) and (8.77). Using it with larger k's leads to higher moments of the time-dependent distribution.

Problem 8.7. Equation (8.81) implies that $F(s = 1, t) = 1$ for all t. Using the definition of the generating function show that this result holds generally, not only for the generating function of Eq. (8.69).

8.3.2 Chemical kinetics

Consider the simple first-order chemical reaction, $A \xrightarrow{k} B$. The corresponding kinetic equation,

$$\frac{d\langle A\rangle}{dt} = -k\langle A\rangle \Rightarrow \langle A\rangle(t) = \langle A\rangle(t = 0)e^{-kt} \tag{8.82}$$

describes the time evolution of the *average* number of molecules A in the system.[10] Without averaging the time evolution of this number is a random process, because the moment at which a specific A molecule transforms into B is undetermined. The stochastic nature of radioactive decay, which is described by similar first-order kinetics, can be realized by listening to a Geiger counter. Fluctuations from the average can also be observed if we monitor the reaction in a small enough volume, for example, in a biological well.

Let $P(n, t)$ be the probability that the number of A molecules in the system at time t is n. We can derive a master equation for this probability by following a procedure similar to that used in Section 7.3.1 to derive Eq. (7.3) or (8.69):

$$P(n, t + \Delta t) = P(n, t) + k(n + 1)P(n + 1, t)\Delta t - knP(n, t)\Delta t$$

$$\Rightarrow \frac{\partial P(n, t)}{\partial t} = k(n + 1)P(n + 1, t) - knP(n, t) \tag{8.83}$$

Unlike in the random walk problem, the transition rate out of a given state n depends on n: The probability per unit time to go from $n+1$ to n is $k(n+1)$, and the probability per unit time to go from n to $n - 1$ is kn. The process described by Eq. (8.83) is an example of a *birth-and-death* process. In this particular example there is no source feeding A molecules into the system, so only death steps take place.

[10] For detailed discussion and more examples see D. A. McQuarrie, *A Stochastic Approach to Chemical Kinetics*, J. Appl. Probability **4**, 413 (1967).

Problem 8.8. How should Eq. (8.83) be modified if molecules A are inserted into the system with the characteristic constant insertion rate k_a (i.e. the probability that a molecule A is inserted during a small time interval Δt is $k_a \Delta t$)?

The solution of Eq. (8.83) is easily achieved using the generating function method. The random variable n can take only non-negative integer values, and the generating function is therefore

$$F(s,t) = \sum_{n=0}^{\infty} s^n P(n,t) \tag{8.84}$$

Multiplying (8.83) by s^n and doing the summation leads to

$$\frac{\partial F(s,t)}{\partial t} = k\frac{\partial F}{\partial s} - ks\frac{\partial F}{\partial s} = k(1-s)\frac{\partial F}{\partial s} \tag{8.85}$$

where we have used identities such as

$$\sum_{n=0}^{\infty} s^n n P(n,t) = s\frac{\partial}{\partial s}F(s,t) \tag{8.86}$$

and

$$\sum_{n=0}^{\infty} s^n (n+1) P(n+1,t) = \sum_{n=1}^{\infty} s^{n-1} n P(n,t) = \frac{\partial F}{\partial s} \tag{8.87}$$

If $P(n,t=0) = \delta_{n,n_0}$ then $F(s,t=0) = s^{n_0}$. It is easily verified by direct substitution that for this initial condition the solution of Eq. (8.85) is

$$F(s,t) = \left[1 + (s-1)e^{-kt}\right]^{n_0} \tag{8.88}$$

This will again give all the moments using Eq. (8.79).

Problem 8.9. Show that for this process

$$\langle n \rangle_t = n_0 e^{-kt} \tag{8.89}$$

$$\langle \delta n^2 \rangle_t = n_0 e^{-kt}(1 - e^{-kt}) \tag{8.90}$$

The first moment, (8.89), gives the familiar evolution of the average A population. The second moment describes fluctuations about this average. It shows that the variance of these fluctuations is zero at $t = 0$ and $t = \infty$, and goes through a maximum at some intermediate time.

8.3.3 The relaxation of a system of harmonic oscillators

In this example the master equation formalism is applied to the process of vibrational relaxation of a diatomic molecule represented by a quantum harmonic oscillator.[11] In a reduced approach we focus on the dynamics of just this oscillator, and in fact only on its energy. The relaxation described on this level is therefore a particular kind of random walk in the space of the energy levels of this oscillator. It should again be emphasized that this description is constructed in a phenomenological way, and should be regarded as a model. In the construction of such models one tries to build in all available information. In the present case the model relies on quantum mechanics in the weak interaction limit that yields the relevant transition matrix elements between harmonic oscillator levels, and on input from statistical mechanics that imposes a certain condition (detailed balance) on the transition rates.

We consider an ensemble of such oscillators contained in a large excess of chemically inert gas which acts as a constant temperature heat bath throughout the relaxation process. We assume that these oscillators are far from each other and do not interact among themselves, so that the energy exchange which controls the relaxation takes place primarily between the oscillators and the "solvent" gas.

The most important physical inputs into the stochastic model are the transition probabilities per unit time between any two vibrational levels. Naturally these transition rates will be proportional to the number Z of collisions undergone by the molecule per unit time. For each collision we assume that the transition probability between oscillator states n and m is proportional to Q_{mn}, the absolute square of the matrix element of the oscillator coordinate q between these states,[12] given by (cf. Eq. (2.141)):

$$Q_{nm} = Q_{mn} = |q_{nm}|^2 = |q_{01}|^2[n\delta_{n,m+1} + m\delta_{n,m-1}] \tag{8.91}$$

Finally, the transition probability between levels n and m must contain a factor that depends on the temperature and on the energy difference between these states. This factor, denoted below by $f(E_n - E_m)$, conveys information about the energy available for the transition, for example, telling us that a transition from a lower

[11] This section is based on E. W. Montroll and K. E. Shuler, J. Chem. Phys. **26**, 454 (1957).

[12] This assumption relies on the fact that the amplitude q of molecular vibrations about the equilibrium nuclear configuration x_{eq} is small. The interaction $V(x_{\text{eq}} + q, \text{B})$ between the oscillator and the surrounding bath B can then be expanded in powers of q, keeping terms up to first order. This yields $V = C - Fq$ where $C = V(x_{\text{eq}}, \text{B})$ is a constant and $F = -(\partial V/\partial q)_{q=0}$. When the effective interaction $-Fq$ is used in the golden rule formula (9.25) for quantum transition rates, we find that the rate between states i and j is proportional to $|\, q_{ij}\, |^2$. This is true also for radiative transition probabilities, therefore the same formalism can be applied to model the interaction of the oscillator with the radiation field.

energy to a higher energy state is impossible at zero temperature. The transition probability per unit time between two levels n and m can now be written in the form:

$$k_{nm} = ZQ_{nm}f(E_{nm}); \qquad E_{nm} = E_n - E_m \tag{8.92}$$

These rates can be used in the master equation (8.70) for the probability $P(n, t)$ (denoted below $P_n(t)$) to find the oscillator in its nth level at time t:

$$\frac{\partial P_m(t)}{\partial t} = \sum_n (k_{mn}P_n(t) - k_{nm}P_m(t)) \tag{8.93}$$

Equation (8.91) implies that the transitions occur only between levels adjacent to each other. More information about the rates (8.92) is obtained from the condition of detailed balance: *At thermal equilibrium any two levels must be in thermal equilibrium with respect to each other.* Therefore

$$k_{nm}P_m^{\text{eq}} - k_{mn}P_n^{\text{eq}} = 0 \tag{8.94}$$

so that (since $q_{nm} = q_{mn}$)

$$\frac{f(E_{nm})}{f(E_{mn})} = \frac{P_n}{P_m} = \exp[-\beta(E_n - E_m)]; \qquad \beta = (k_B T)^{-1} \tag{8.95}$$

If we assume that the probability of going down in energy does not depend on the temperature (since no activation is needed) we can denote $f(E_{n,n+1}) = \kappa$ so that $f(E_{n+1,n}) = \kappa e^{-\beta\varepsilon}$, where $\varepsilon = \hbar\omega$ is the energy spacing between adjacent levels. Using also Eqs (8.91) and (8.92) we can write

$$k_{n,n+1} = ZQ_{01}\kappa(n+1) \tag{8.96a}$$

$$k_{n+1,n} = ZQ_{01}\kappa(n+1)e^{-\beta\varepsilon} \tag{8.96b}$$

$$k_{n,m} = 0 \quad \text{unless} \quad m = n \pm 1 \tag{8.96c}$$

Using these rates in the master equation (8.93) we have

$$\frac{\partial P_n}{\partial t} = k_{n,n+1}P_{n+1} + k_{n,n-1}P_{n-1} - k_{n+1,n}P_n - k_{n-1,n}P_n \tag{8.97}$$

and redefining $ZQ_{01}\kappa t \equiv \tau$, we get

$$\frac{\partial P_n}{\partial \tau} = (n+1)P_{n+1} + ne^{-\beta\varepsilon}P_{n-1} - [(n+1)e^{-\beta\varepsilon} + n]P_n \tag{8.98}$$

Equation (8.98) describes the thermal relaxation of the internal nuclear motion of a diatomic molecule modeled as a harmonic oscillator. It is interesting to note that in addition to the physical parameter $\beta\varepsilon$ that appears explicitly in (8.98), the time evolution associated with this relaxation is given explicitly in terms of only one additional parameter, the product $ZQ_{01}\kappa$ that relates the time variable τ to the real physical time t.

The full solution of Eq. (8.98) is described in the paper by E. W. Montroll and K. E. Shuler (see footnote 11). Here we focus on the time evolution of the first moment $\langle n\rangle(t)$. Multiplying Eq. (8.98) by n and summing over all n between 0 and ∞ leads to

$$\frac{\partial\langle n\rangle}{\partial\tau} = A + e^{-\beta\varepsilon}B \tag{8.99}$$

with

$$A = \sum_{n=0}^{\infty}(n(n+1)P_{n+1} - n^2P_n) = -\sum_{n=0}^{\infty}(n+1)P_{n+1} = -\langle n\rangle \tag{8.100}$$

and

$$\begin{aligned} B &= \sum_{n=0}^{\infty}(n^2P_{n-1} - n(n+1)P_n) = \sum_{n=0}^{\infty}((n-1)nP_{n-1} + nP_{n-1} - n(n+1)P_n) \\ &= \sum_{n=0}^{\infty} nP_{n-1} = \langle n\rangle + 1 \end{aligned} \tag{8.101}$$

Using (8.100) and (8.101) leads to

$$\frac{\partial\langle n\rangle}{\partial\tau} = -\bar{k}_v\langle n\rangle + c \tag{8.102a}$$

$$\bar{k}_v = 1 - e^{-\beta\varepsilon}; \qquad c = e^{-\beta\varepsilon} \tag{8.102b}$$

The solution of (8.102) for the initial condition $\langle n\rangle = \langle n\rangle_0$ at $t = 0$ is easily found to be

$$\begin{aligned} \langle n\rangle_t &= \langle n\rangle_0 e^{-\bar{k}_v\tau} + \frac{c}{\bar{k}_v}(1 - e^{-\bar{k}_v\tau}) \\ &= \langle n\rangle_0 e^{-k_v t} + \frac{c}{\bar{k}_v}(1 - e^{-k_v t}) \end{aligned} \tag{8.103}$$

where

$$k_v = \left(1 - e^{-\beta\varepsilon}\right) ZQ_{01}\kappa \tag{8.104}$$

Noting that

$$\frac{c}{\bar{k}_v} = \frac{1}{e^{\beta\varepsilon} - 1} = \langle n \rangle_{\text{eq}} \tag{8.105}$$

is the equilibrium thermal population of the oscillator, we can write (8.103) in the physically appealing form

$$\langle n \rangle_t = \langle n \rangle_0 e^{-k_v t} + \langle n \rangle_{\text{eq}}(1 - e^{-k_v t}) \tag{8.106}$$

The relaxation to thermal equilibrium is seen to be exponential, with a rate given by (8.104). It is interesting to note that in the infinite temperature limit, where $k_v = 0$, Eq. (8.102) describes a constant heating rate of the oscillator. It is also interesting to compare the result (8.106) to the result (9.65) of the very different quantum formalism presented in Section 9.4; see the discussion at the end of Section 9.4 of this point.

For completeness we also cite from the same paper (see footnote 11) the expression for the variance $\sigma(t) = \langle n^2 \rangle_t - \langle n \rangle_t^2$

$$\sigma(t) = \sigma_{\text{eq}} + [\sigma_0 - \sigma_{\text{eq}}]e^{-2k_v t} + [\langle n \rangle_0 - \langle n \rangle_{\text{eq}}][1 + 2\langle n \rangle_{\text{eq}}]e^{-k_v t}(1 - e^{-k_v t}) \tag{8.107}$$

where

$$\sigma_{\text{eq}} = \langle n \rangle_{\text{eq}}(1 + \langle n \rangle_{\text{eq}}) \tag{8.108}$$

The result (8.107) shows that in the course of the relaxation process the width of the energy level distribution increases (due to the last term in (8.107)) before decreasing again. This effect is more pronounced for larger $[\langle n \rangle_0 - \langle n \rangle_{\text{eq}}]$, that is, when the initial excitation energy is much larger than $k_B T$.

8.4 The Fokker–Planck equation

In many practical situations the random process under observation is continuous in the sense that (1) the space of possible states is continuous (or it can be transformed to a continuous-like representation by a coarse-graining procedure), and (2) the change in the system state during a small time interval is small, that is, if the system is found in state x at time t then the probability to find it in state $y \neq x$ at time $t + \delta t$ vanishes when $\delta t \to 0$.[13] When these, and some other conditions detailed below, are satisfied, we can derive a partial differential equation for the probability distribution, the Fokker–Planck equation, which is discussed in this Section.

[13] In fact we will require that this probability vanishes faster than δt when $\delta t \to 0$.

8.4.1 A simple example

As an example without rigorous mathematical justification consider the master equation for the random walk problem

$$\begin{aligned}\frac{\partial P(n,t)}{\partial t} &= k_r P(n-1,t) + k_l P(n+1,t) - (k_r + k_l)P(n,t) \\ &= -k_r(P(n,t) - P(n-1,t)) - k_l(P(n,t) - P(n+1,t)) \\ &= -k_r(1 - e^{-(\partial/\partial n)})P(n,t) - k_l(1 - e^{(\partial/\partial n)})P(n,t) \end{aligned} \tag{8.109}$$

In the last step we have regarded n as a continuous variable and have used the Taylor expansion

$$e^{a(\partial/\partial n)}P(n) = 1 + a\frac{\partial P}{\partial n} + \frac{1}{2}a^2\frac{\partial^2 P}{\partial n^2} + \cdots = P(n+a) \tag{8.110}$$

In practical situations n is a very large number—it is the number of microscopic steps taken on the timescale of a macroscopic observation. This implies that $\partial^k P/\partial n^k \gg \partial^{k+1}P/\partial n^{k+1}$.[14] We therefore expand the exponential operators according to

$$1 - e^{\pm(\partial/\partial n)} = \mp\frac{\partial}{\partial n} - \frac{1}{2}\frac{\partial^2}{\partial n^2} \tag{8.111}$$

and neglect higher-order terms, to get

$$\frac{\partial P(n,t)}{\partial t} = -A\frac{\partial P(n,t)}{\partial n} + B\frac{\partial^2 P(n,t)}{\partial n^2} \tag{8.112}$$

where $A = k_r - k_l$ and $B = (k_r + k_l)/2$. We can give this result a more physical form by transforming from the number-of-steps variable n to the position variable $x = n\Delta x$ where Δx is the step size. At this point we need to distinguish between $P_n(n)$, the probability in the space of position indices, which is used without the subscript n in (8.112), and the probability density on the x-axis, $P_x(x) = P_n(n)/\Delta x$, that is used without the subscript x below. We omit these subscripts above and below because the nature of the distribution is clear from the text. This transformation leads to

$$\frac{\partial P(x,t)}{\partial t} = -v\frac{\partial P(x,t)}{\partial x} + D\frac{\partial^2 P(x,t)}{\partial x^2} \tag{8.113}$$

[14] For example if $f(n) = n^a$ then $\partial f/\partial n = an^{a-1}$ which is of order f/n. The situation is less obvious in cases such as the Gausssian distribution $f(n) \sim \exp((n - \langle n\rangle)^2/2\langle \delta n^2\rangle)$. Here the derivatives with respect to n adds a factor $\sim(n - \langle n\rangle)/\langle \delta n^2\rangle$ that is much smaller than 1 as long as $n - \langle n\rangle \ll \langle n\rangle$ because $\langle \delta n^2\rangle$ is of order $\langle n\rangle$.

where $v = \Delta x A$ and $D = \Delta x^2 B$. Note that we have just repeated, using a somewhat different procedure, the derivation of Eq. (7.5). The result (8.113) (or (7.5)) is a Fokker–Planck type equation.

As already discussed below Eq. (7.5), Eq. (8.113) describes a drift diffusion process: For a symmetric walk, $k_r = k_l$, $v = 0$ and (8.113) becomes the diffusion equation with the diffusion coefficient $D = \Delta x^2(k_r + k_l)/2 = \Delta x^2/2\tau$. Here τ is the hopping time defined from $\tau = (k_r + k_l)^{-1}$. When $k_r \neq k_l$ the parameter v is nonzero and represents the drift velocity that is induced in the system when an external force creates a flow asymmetry in the system. More insight into this process can be obtained from the first and second moment of the probability distribution $P(x, t)$ as was done in Eqs (7.16)–(7.23).

8.4.2 The probability flux

Additional insight can be obtained by rewriting Eq. (8.113) in the form:

$$\frac{\partial P(x,t)}{\partial t} = -\frac{\partial J(x,t)}{\partial x} \tag{8.114a}$$

$$J(x,t) = vP(x,t) - D\frac{\partial P(x,t)}{\partial x} \tag{8.114b}$$

Equations (8.114a) and (8.114b) represent a simple example of the continuity equation for conserved quantities discussed in Section 1.1.4. In particular Eq. (8.114a) expresses the fact that the probability distribution P is a conserved quantity and therefore its time dependence can stem only from boundary fluxes. Indeed, from (8.114a) it follows that $P_{ab}(t) = \int_a^b dxP(x,t); a < b$ satisfies $dP_{ab}(t)/dt = J(a,t) - J(b,t)$, which identifies $J(x,t)$ as the *probability flux* at point x: $J(a,t)$ is the flux entering (for positive J) at a, $J(b,t)$—the flux leaving (if positive) at b. In one-dimension J is of dimensionality t^{-1}, and when multiplied by the total number of walkers gives the number of such walkers that pass the point x per unit time in the direction determined by the sign of J. Equation (8.114b) shows that J is a combination of the drift flux, vP, associated with the net local velocity v, and the diffusion flux, $D\partial P/\partial x$ associated with the spatial inhomogeneity of the distribution. In a three-dimensional system the analog of Eq. (8.114) is

$$\frac{\partial P(\mathbf{r},t)}{\partial t} = -\nabla \cdot \mathbf{J}(\mathbf{r},t)$$

$$\mathbf{J}(\mathbf{r},t) = \mathbf{v}P(\mathbf{r},t) - D\nabla P(\mathbf{r},t) \tag{8.115}$$

Now $P(\mathbf{r}, t)$ is of dimensionality l^{-3}. The flux vector $\mathbf{J}$ has the dimensionality $l^{-2}t^{-1}$ and expresses the passage of walkers per unit time and area in the $\mathbf{J}$ direction.

It is important to emphasize that, again, the first of Eqs (8.115) is just a conservation law. Integrating it over some volume Ω enclosed by a surface S_Ω and denoting

$$P_\Omega(t) = \int_\Omega d\Omega P(\mathbf{r}, t) \tag{8.116}$$

we find, using the divergence theorem of vector calculus, Eq. (1.36),

$$dP_\Omega(t)/dt = -\int_{S_\Omega} d\mathbf{S} \cdot \mathbf{J}(\mathbf{r}, t) \tag{8.117}$$

where $d\mathbf{S}$ is a vector whose magnitude is a surface element and its direction is a vector normal to this element in the direction outward of the volume Ω.[15] Equation (8.117) states that the change in P inside the region Ω is given by the balance of fluxes that enter and leave this region.

8.4.3 Derivation of the Fokker–Planck equation from the Chapman–Kolmogorov equation

The derivation of the Fokker–Planck (FP) equation described above is far from rigorous since the conditions for neglecting higher-order terms in the expansion of $\exp(\pm\partial/\partial x)$ were not established. Appendix 8A outlines a rigorous derivation of the FP equation for a Markov process that starts from the Chapman–Kolmogorov equation

$$P(\mathbf{x}_3 t_3 \mid \mathbf{x}_1 t_1) = \int d\mathbf{x}_2 P(\mathbf{x}_3 t_3 \mid \mathbf{x}_2 t_2) P(\mathbf{x}_2 t_2 \mid \mathbf{x}_1 t_1) \qquad t_3 \geq t_2 \geq t_1 \tag{8.118}$$

In the most general case $\mathbf{x} = \{x_j; j = a, b, \ldots\}$ is a multivariable stochastic process. This derivation requires that the following conditions should be satisfied:

(a) The Markov process is continuous, that is, for any $\varepsilon > 0$

$$\lim_{\Delta t \to 0} \frac{1}{\Delta t} \int_{|x-y|>\varepsilon} d\mathrm{x} P(\mathbf{x}, t + \Delta t \mid \mathbf{y}, t) = 0 \tag{8.119}$$

Namely, the probability for the final state $\mathbf{x}$ to be different from the initial state $\mathbf{y}$ vanishes faster then Δt as $\Delta t \to 0$.

[15] The minus sign in (8.117) enters because, by convention, a vector (e.g. the flux) normal to a surface that defines a closed sub-space is taken positive when it points in the outward direction.

(b) The following functions

$$A_i(\mathbf{x},t) = \lim_{\Delta t \to 0} \frac{1}{\Delta t} \int d\mathbf{z}(z_i - x_i) P(\mathbf{z}, t + \Delta t \mid \mathbf{x}, t) \quad (8.120a)$$

$$B_{i,j}(\mathbf{x},t) = \lim_{\Delta t \to 0} \frac{1}{\Delta t} \int d\mathbf{z}(z_i - x_i)(z_j - x_j) P(\mathbf{z}, t + \Delta t \mid \mathbf{x}, t) \quad (8.120b)$$

exist for all $\mathbf{x}$. Note that the integral in (8.120a) is the average vector-distance that the random variable makes during time Δt, which is indeed expected to be linear in Δt for any systematic motion (it vanishes for pure diffusion). The integral in (8.120b) on the other hand is expected to be of order $(\Delta t)^2$ for systematic motion, in which case $B_{i,j} = 0$, but can be linear in Δt (implying nonzero $B_{i,j}$) for stochastic motion such as diffusion.

In Appendix 8A we show that when these conditions are satisfied, the Chapman–Kolmogorov integral equation (8.118) leads to two partial differential equations. The Fokker–Planck equation describes the future evolution of the probability distribution

$$\frac{\partial}{\partial t} P(\mathbf{x}, t \mid \mathbf{y}, t_0) = -\sum_i \frac{\partial}{\partial x_i} [A_i(\mathbf{x}, t) P(\mathbf{x}, t \mid \mathbf{y}, t_0)] + \frac{1}{2} \sum_{ij} \frac{\partial^2}{\partial x_i \partial x_j} [B_{ij}(\mathbf{x}, t) P(\mathbf{x}, t \mid \mathbf{y}, t_0)] \quad (8.121)$$

and the "backward" Fokker–Planck equation describes its evolution towards the past

$$\frac{\partial P(\mathbf{x}, t \mid \mathbf{y}, t_0)}{\partial t_0} = -\sum_i A_i(\mathbf{y}, t_0) \frac{\partial P(\mathbf{x}, t \mid \mathbf{y}, t_0)}{\partial y_i} - \frac{1}{2} \sum_{ij} B_{ij}(\mathbf{y}, t_0) \frac{\partial^2 P(\mathbf{x}, t \mid \mathbf{y}, t_0)}{\partial y_i \partial y_j} \quad (8.122)$$

Each of Eqs (8.121) and (8.122) is fully equivalent, under the conditions specified, to the Chapman–Kolmogorov equation. Furthermore, if the functions A_i and B_{ij} are time independent, the conditional probability $P(\mathbf{x}, t \mid \mathbf{y}, t_0)$ depends only on the time interval $t - t_0$ and therefore $\partial P(\mathbf{x}, t \mid \mathbf{y}, t_0)/\partial t_0 = -\partial P(\mathbf{x}, t \mid \mathbf{y}, t_0)/\partial t$. In this case Eqs (8.121) and (8.122) relate to each other in the following way. Writing the former in the form

$$\frac{\partial}{\partial t} P(\mathbf{x}, t \mid \mathbf{y}, t_0) = \hat{L}(\mathbf{x}) P(\mathbf{x}, t \mid \mathbf{y}, t_0) \quad (8.123a)$$

where the operator $\hat{L}(\mathbf{x})$ is defined by the right-hand side of (8.121), then Eq. (8.122) is

$$\frac{\partial}{\partial t}P(\mathbf{x},t \mid \mathbf{y},t_0) = \hat{L}^{\dagger}(\mathbf{y})\,P(\mathbf{x},t \mid \mathbf{y},t_0) \tag{8.123b}$$

where the operator $\hat{L}^{\dagger}$ is the adjoint of $\hat{L}$.

To gain some insight into the physical significance of these equations consider the case where $B_{i,j} = 0$ for all i and j. Equation (8.121) then becomes

$$\frac{\partial}{\partial t}P(\mathbf{x},t \mid \mathbf{y},t_0) = -\sum_i \frac{\partial}{\partial x_i}[A_i(\mathbf{x},t)P(\mathbf{x},t \mid \mathbf{y},t_0)] \tag{8.124}$$

It is easily realized that this equation describes the completely deterministic motion

$$\frac{dx_i}{dt} = A_i(\mathbf{x}(t),t); \qquad x_i(t=t_0) = y_i \text{ (all } i) \tag{8.125}$$

To see this note that if (8.125) holds then

$$J_i = (dx_i/dt)P(\mathbf{x},t \mid \mathbf{y},t_0) = A_i(\mathbf{x},t)P(\mathbf{x},t \mid \mathbf{y},t_0)$$

is the probability flux in the direction x_i.[16] Equation (8.124) can therefore be written as

$$\frac{\partial P(\mathbf{x},t \mid \mathbf{y},t_0)}{\partial t} = -\sum_i \frac{\partial}{\partial x_i}J_i(\mathbf{x},t) = -\nabla\cdot\mathbf{J} \tag{8.126}$$

which, as discussed above Eq. (8.115), is a statement on the conservation of probability. The one-dimensional analog to this is Eq. (8.114) for the case $D = 0$. In that case $\dot{x}P(x,t) = vP(x,t)$ became the probability flux, and the rate of change of P in time is given by $(\partial/\partial x)[\dot{x}P(x)] = v(\partial/\partial x)P(x,t)$.

We may conclude that Eq. (8.124) is a probabilistic reformulation of the information contained in the deterministic time evolution. This implies that not only $P(\mathbf{x},t_0 \mid \mathbf{y},t_0) = \delta(\mathbf{x}-\mathbf{y})$, but for a later time

$$P(\mathbf{x},t \mid \mathbf{y},t_0) = \delta(\mathbf{x}-\mathbf{x}(t \mid \mathbf{y},t_0)) \tag{8.127}$$

where $\mathbf{x}(t \mid \mathbf{y},t_0)$ is the (deterministic) solution to Eq. (8.125). Under Eq. (8.124) the conditional probability distribution $P(\mathbf{x},t \mid \mathbf{y},t_0)$ remains a δ function at all time.

[16] For example, if $\mathbf{x} = (x_1,x_2,x_3)$ denotes a position in space and $P(\mathbf{x},t \mid \mathbf{y},t_0)$ is the probability to find a particle at this position given that it was at position $\mathbf{y}$ at time t_0, then for a total particle number N, $N(dx_i/dt)P(\mathbf{x},t \mid \mathbf{y},t_0)$ is the particle flux in the direction x_i (number of particles moving per second through a unit cross-sectional area normal to x_i) which, when divided by N, yields the probability flux in that direction.

The stochastic spread of the distribution about the deterministic path results from the B terms in Eqs (8.121) and (8.122) in analogy with the D term in Eq. (8.113).

Problem 8.10. Show that $P(\mathbf{x}, t \mid \mathbf{y}, t_0)$ given by Eq. (8.127) satisfies Eq. (8.124)

Problem 8.11. Let the vector $\mathbf{x}$ in (8.124) be $(\mathbf{p}^N, \mathbf{r}^N)$, that is, a point in the phase space of a Hamiltonian system with N particles, and let Eqs (8.125) be the Hamilton equations of motion (this is a statement about the functions $A(\mathbf{x})$). Show that in this case Eq. (8.124) becomes the Liouville equation for the phase space density $f(\mathbf{p}^N, \mathbf{r}^N; t)$, that is, (c.f. Eq. (1.104))

$$\frac{\partial f(\mathbf{r}^N, \mathbf{p}^N; t)}{\partial t} = -\left(\frac{\partial f}{\partial \mathbf{r}^N}\frac{\partial H}{\partial \mathbf{p}^N} - \frac{\partial f}{\partial \mathbf{p}^N}\frac{\partial H}{\partial \mathbf{r}^N}\right) \tag{8.128}$$

8.4.4 Derivation of the Smoluchowski equation from the Langevin equation: The overdamped limit

Another route to the Fokker–Planck equation starts from the Langevin equation. Since the latter describes a continuous stochastic process, a Fokker–Planck equation is indeed expected in the Markovian case. We note in passing that using generalized Langevin equations such as Eq. (8.54) as starting points makes it possible to consider also non-Markovian situations, however, we shall limit ourselves to the one-dimensional Markovian case. The general case, which starts from Eqs (8.13) and (8.20), is taken up in the next section. Here we consider the simpler high friction limit, where the Langevin equation takes the form (cf. Eq. (8.21))

$$\frac{dx}{dt} = \frac{1}{\gamma m}\left(-\frac{dV}{dx} + R(t)\right) \tag{8.129}$$

with

$$\langle R \rangle = 0; \qquad \langle R(0)R(t) \rangle = 2m\gamma k_B T\delta(t) \tag{8.130}$$

Our aim is to find the corresponding equation for $P(x, t)$, the probability density to find the particle position at x; the velocity distribution is assumed equilibrated on the timescale considered. Note that in Section 8.1 we have distinguished between stochastic equations of motion that describe the time evolution of a system in state space (here x), and those that describe this evolution in probability space. We now deal with the transformation between such two descriptions.

The starting point for this task is an expression of the fact that the integrated probability is conserved. As already discussed, this implies that the time derivative of P should be given by the gradient of the flux $\dot{x}P$, that is,

$$\frac{\partial P(x,t)}{\partial t} = -\frac{\partial}{\partial x}(\dot{x}P) = \hat{\Omega}P \tag{8.131a}$$

where, using (8.129), the operator $\hat{\Omega}$ is given by

$$\hat{\Omega} = -\frac{1}{\gamma m}\frac{\partial}{\partial x}\left[\left(-\frac{dV}{dx} + R(t)\right)\right] \tag{8.131b}$$

The essence of the calculation that leads to the desired Fokker–Planck equation, known in this limit as the Smoluchowski equation, is a coarse-grained average of the time evolution (8.131) over the fast variation of $R(t)$. This procedure, described in Appendix 8B, leads to

$$\frac{\partial P(x,t)}{\partial t} = D\frac{\partial}{\partial x}\left(\beta\frac{\partial V}{\partial x} + \frac{\partial}{\partial x}\right)P(x,t); \qquad \beta = \frac{1}{k_B T} \tag{8.132}$$

$$D = \frac{k_B T}{m\gamma} \tag{8.133}$$

which is the desired Smoluchowski equation. When the potential V is constant it becomes the well-known diffusion equation. Equation (8.133) is a relation between the diffusion constant D and the friction coefficient γ, which in turn is related to the fluctuations in the system via the fluctuation–dissipation relation (8.130). We discuss this relation further in Section 11.2.4.

Next we consider some properties of Eq. (8.132). First note that it can be rewritten in the form

$$\frac{\partial P(x,t)}{\partial t} = -\frac{\partial}{\partial x}J(x,t) \tag{8.134}$$

where the probability flux J is given by

$$J = -D\left(\frac{\partial}{\partial x} + \beta\frac{\partial V}{\partial x}\right)P(x,t) \tag{8.135}$$

As discussed above (see Eq. (8.114) and the discussion below it), Eq. (8.134) has the form of a conservation rule, related to the fact that the overall probability is conserved.[17] The three-dimensional generalization of (8.132)

$$\frac{\partial P(x,t)}{\partial t} = D\nabla \cdot (\beta \nabla V + \nabla)P(\mathbf{r},t) \tag{8.136}$$

can similarly be written as a divergence of a flux

$$\frac{\partial P(\mathbf{r},t)}{\partial t} = -\nabla \cdot \mathbf{J} \tag{8.137a}$$

$$\mathbf{J} = -D(\beta \nabla V + \nabla)P(\mathbf{r},t) \tag{8.137b}$$

Again, as discussed in Section 8.4.2, Eq. (8.137a) is just a conservation law, equivalent to the integral form (8.117)

$$\frac{dP_\Omega}{dt} = -\int_{S_\Omega} \mathbf{J}(\mathbf{r}) \cdot d\mathbf{S} \tag{8.138}$$

where P_Ω, the probability that the particle is in the volume Ω, is given by (8.116).

Second, the flux is seen to be a sum of two terms, $J = J_D + J_F$, where $J_D = -D\partial P/\partial x$ (or, in three dimensions, $\mathbf{J}_D = -D\nabla P$) is the diffusion flux, while $J_F = D\beta(-\partial V/\partial x)P$ (or , in three dimensions, $\mathbf{J}_F = \beta D(-\nabla V)P$) is the flux caused by the force $F = -\partial V/\partial x$ (or $\mathbf{F} = -\nabla V$). The latter corresponds to the term $\mathbf{v}P$ in (8.114b), where the drift velocity $\mathbf{v}$ is proportional to the force, that is, $J_F = uFP$. This identifies the mobility u as

$$u = \beta D = (m\gamma)^{-1} \tag{8.139}$$

Again, this relation is discussed further in Section 11.2.4.

Finally note that at equilibrium the flux should be zero. Equation (8.135) then leads to a Boltzmann distribution.

$$\frac{\partial P}{\partial x} = -\beta \frac{\partial V}{\partial x} P \Rightarrow P(x) = \text{const} \cdot e^{-\beta V(x)} \tag{8.140}$$

---

[17] If N is the total number of particles then $NP(x)$ is the particles number density. The conservation of the integrated probability, that is, $\int dxP(x,t) = 1$ is a statement that the total number of particles is conserved: In the process under discussion particles are neither destroyed nor created, only move in position space.

8.4.5 Derivation of the Fokker–Planck equation from the Langevin equation: General case

Next consider the general one-dimensional Langevin Eq. (8.13)

$$\dot{x} = v$$
$$\dot{v} = -\frac{1}{m}\frac{\partial V(x)}{\partial x} - \gamma v + \frac{1}{m}R(t) \tag{8.141}$$

with a Gaussian random force $R(t)$ that again satisfies (8.130). Here x and $v = \dot{x}$ are respectively the position and velocity of a Brownian particle. We now seek an equation for $P(x, v, t)$, the joint probability density that the particle position and velocity at time t are x and v, respectively. The starting point is the two-dimensional analog of Eq. (8.131)

$$\frac{\partial P(x, v, t)}{\partial t} = -\frac{\partial}{\partial x}(\dot{x}P) - \frac{\partial}{\partial v}(\dot{v}P) \tag{8.142}$$

Again, this is just a statement about the conservation of probability. To show this multiply both sides by the phase space volume element $dxdv$. On the left the term $\partial/\partial t[P(x, v)dxdv]$ is the rate of change of the probability that the particle occupies this infinitesimal phase space volume. The two terms on the right represent the two contributions to this rate from fluxes in the x and v directions: For example, $-\dot{x}\partial P/\partial x \times dxdv = -[\dot{x}P(x + dx, v) - \dot{x}P(x, v)]dv$ is a contribution to the change in $Pdxdv$ per unit time due to particles that enter (when $v > 0$) the element $dxdv$ at position x and exit the same volume element at position $x + dx$. Similarly, $-\dot{v}\partial P/\partial v \times dxdv = [\dot{v}(x, v)P(x, v) - \dot{v}(x, v + dv)P(x, v + dv)]dx$ is the change per unit time arising from particles changing their velocity (see Fig. 8.1).

Using Eqs (8.141) and (8.142) we now have

$$\frac{\partial P(x, v, t)}{\partial t} = \hat{\Omega}(t)P$$
$$\hat{\Omega}(t) = -v\frac{\partial}{\partial x} + \frac{1}{m}\frac{\partial V}{\partial x}\frac{\partial}{\partial v} + \frac{\partial}{\partial v}\left(\gamma v - \frac{1}{m}R(t)\right) \tag{8.143}$$

which has a form similar to (8.131), only with a different operator $\hat{\Omega}(t)$ and can be treated in an analogous way. Repeating the procedure that lead to Eq. (8.132) (see further details in Appendix 8C) now leads to the Fokker–Planck equation

$$\frac{\partial P(x, v, t)}{\partial t} = \left[-v\frac{\partial}{\partial x} + \frac{1}{m}\frac{\partial V}{\partial x}\frac{\partial}{\partial v} + \gamma\frac{\partial}{\partial v}\left(v + \frac{k_B T}{m}\frac{\partial}{\partial v}\right)\right]P(x, v, t) \tag{8.144}$$

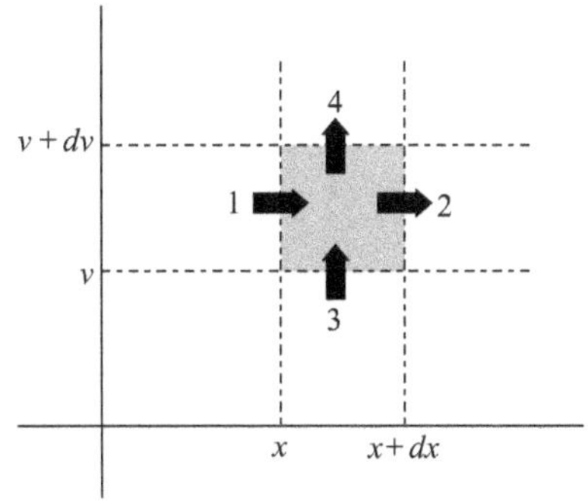

FIG. 8.1 The probability that the particle is at the position range $x,\ldots,x+dx$ and at the velocity range $v,\ldots,v+dv$ (the shaded area of phase space shown in the figure) is $P(x,v)dxdv$. It is changing by the four fluxes shown in the figure. The rate at which probability flows into the shaded area through the left boundary is $J_1dv = vP(x,v)dv$ where J_1 is the flux entering (or leaving if $v < 0$) at the left boundary. This change reflects particles changing their position near position x. Similarly $J_2dv = vP(x+dx,v)dv$ is the rate at which probability flows by particles changing position near the $x+dx$ boundary. $J_3dx = \dot{v}(x,v)P(x,v)dx$ is the rate at which probability flows by particles changing their velocity near v, and $J_4dx = \dot{v}(x,v+dv)P(x,v+dv)dx$ is the corresponding rate at $v+dv$.

Problem 8.12. Show that the Boltzmann distribution $P \sim e^{-\beta((1/2)mv^2+V(x))}$ satisfies Eq. (8.144) with $\partial P/\partial t = 0$.

In order to understand the physical content of this equation consider first the case where γ vanishes. In this case the Langevin equation (8.141) becomes the deterministic Newton equation $\dot{x} = v; \dot{v} = -(1/m)\partial V/\partial x$, and Eq. (8.144) with $\gamma = 0$ is just Eq. (8.142), that is, an expression for the conservation of probability, written for the deterministic Newtonian case. The reader may note that in this case Eq. (8.142) is in fact the Liouville equation (1.104) for this one-dimensional single particle system.

In the general case, the conservation of probability is still expressed by an equation that identifies the time derivative of P as a divergence of a probability flux

$$\frac{\partial P(x,v,t)}{\partial t} = -\nabla \cdot \mathbf{J} = -\frac{\partial}{\partial x}J_x - \frac{\partial}{\partial v}J_v \tag{8.145}$$

where

$$J_x = vP(x,v,t) \tag{8.146}$$

and

$$J_v = J_v^{\text{nwt}} + J_v^{\text{dis}} \tag{8.147a}$$

$$J_v^{\text{nwt}} = -\frac{1}{m}\frac{\partial V}{\partial x}P(x, v, t) \tag{8.147b}$$

$$J_v^{\text{dis}} = -\gamma\left(v + \frac{k_B T}{m}\frac{\partial}{\partial v}\right)P(x, v, t) \tag{8.147c}$$

The flux in the x direction is associated with the particles velocity, as in the deterministic case. The flux in the v direction consists of two parts. The deterministic Newtonian part, J_v^{nwt}, results from the acceleration associated with the potential V, and the dissipative part, J_v^{dis}, results from the coupling to the thermal environment. Note that this dissipative flux does not depend on the potential V.

Problem 8.13. Show that for the Boltzmann distribution $P \sim e^{-\beta((1/2)mv^2+V(x))}$ the dissipative flux $J_v^{\text{dis}}(x, v)$ vanishes at every position x.

8.4.6 The multidimensional Fokker–Planck equation

The above analysis was done for a single particle moving in one dimension, but can be extended to higher dimensions using the same procedure. The starting point is the multidimensional analog of Eq. (8.141), given by

$$\ddot{x}_j = -\frac{1}{m}\frac{\partial V(x^N)}{\partial x_j} - \sum_l \gamma_{jl}\dot{x}_l + \frac{1}{m}R_j(t)$$
$$\langle R_j \rangle = 0; \qquad \langle R_j(0)R_l(t)\rangle = 2m\gamma_{jl}k_B T\delta(t) \tag{8.148}$$

Problem 8.14. Show that the second of Eqs (8.148) is indeed the correct relationship between fluctuation and dissipation for such a system.
(To do this consider the case in which $V = 0$, and use the transformation that diagonalizes γ_{jl}.)

Note that the form of Eq. (8.148) expresses the possibility that the different degrees of freedom x_j are coupled to each other not only via their interaction potential V, but in principle also through their mutual coupling to the environment.

Constructing the corresponding Fokker–Planck equation now proceeds as before. A particularly simple model is the one in which the coupling between the variables $\{x_j\}$ through their interaction with the environment is neglected. In this case $\gamma_{jl} = \gamma_j \delta_{jl}$ and a straightforward application of the procedure described above leads to

$$\frac{\partial P(x^N, v^N, t)}{\partial t} = \sum_j \left[-v_j \frac{\partial}{\partial x_j} + \frac{1}{m} \frac{\partial V}{\partial x_j} \frac{\partial}{\partial v_j} + \gamma_j \frac{\partial}{\partial v_j} \left(v_j + \frac{k_B T}{m} \frac{\partial}{\partial v_j} \right) \right] P(x^N, v^N, t) \tag{8.149}$$

8.5 Passage time distributions and the mean first passage time

We have already noted the difference between the Langevin description of stochastic processes in terms of the stochastic variables, and the master or Fokker–Planck equations that focus on their probabilities. Still, these descriptions are equivalent to each other when applied to the same process and variables. It should be possible to extract information on the dynamics of stochastic variables from the time evolution of their probability distribution, for example, the Fokker–Planck equation. Here we show that this is indeed so by addressing the *passage time distribution* associated with a given stochastic process. In particular we will see (problem 14.3) that the first moment of this distribution, the *mean first passage time*, is very useful for calculating rates.

We consider a system described in terms of a stochastic variable x whose probability distribution evolves according to

$$\frac{\partial P(x, t|x_0)}{\partial t} = \hat{L}(x) P(x, t|x_0) \tag{8.150}$$

$\hat{L}(x)$ can be the Fokker–Planck operator, the difference operator in a master equation, etc., and x, that may stand for a group of variables, represents a point (state) in the state space of the system. $P(x, t|x_0)$ is the probability density to find the system in state x at time t given that it started in state x_0 at time $t = 0$. We seek an answer to the following question: *Given this initial condition (particle starts at state* x_0 at $t = 0$), *what is the probability* $\Pi(x_1, t|x_0)dt$ *that it will reach the state* x_1 *for the first time between times t and t+dt*? When the problem is multidimensional, that is, when x represents several stochastic variables, the language should be modified somewhat: We will usually ask about reaching a surface, not a point in the space of these variables. In what follows we focus for simplicity on the single variable case and continue this discussion using the language of a particle moving along the x-axis.

For specificity we take $x_1 > x_0$. Because the question involves the *first* time of the particle arrival to x_1, we may impose absorbing boundary conditions at this point, that is, $P(x_1, t) = 0$ for all t. Given this boundary condition, the integrated probability to make this first arrival at x_1 between times 0 and t is equal to the probability not to remain in the interval $(-\infty, x_1)$ at time t, that is,

$$\int_0^t dt' \Pi(x_1, t'|x_0) = 1 - \int_{-\infty}^{x_1} dx P(x, t|x_0) \tag{8.151}$$

This in turn implies that

$$\Pi(x_1, t|x_0) = -\int_{-\infty}^{x_1} dx \frac{\partial}{\partial t} P(x, t|x_0) \tag{8.152}$$

Note that for such absorbing boundary problem $P(x, t \to \infty|x_0) \to 0$ for x in $(-\infty, x_1)$.

Problem 8.15. Show that $\Pi(x_1, t|x_0)$ is normalized, that is, $\int_0^\infty dt \Pi(x_1, t|x_0) = 1$.

Equation (8.152) is an expression for the passage time distribution $\Pi(x_1, t|x_0)$. The mean first passage time $\tau(x_1, x_0)$ is its first moment

$$\tau(x_1, x_0) = \int_0^\infty dt\, t \Pi(x_1, t|x_0) \tag{8.153}$$

Inserting Eq. (8.152) and integrating by parts then leads to

$$\tau(x_1, x_0) = \int_{-\infty}^{x_1} dx \int_0^\infty dt P(x, t|x_0) \tag{8.154}$$

Obviously, if $P(x, t|x_0)$ is known we can compute the mean first passage time from Eq. (8.154). We can also find an equation for this function, by operating with backward evolution operator $\hat{L}^\dagger(x_0)$ on both sides of Eq. (8.154). Recall that when the operator $\hat{L}$ is time independent the backward equation (8.122) takes the form (cf. Eq. (8.123b)) $\partial P(x, t|x_0)/\partial t = \hat{L}^\dagger(x_0) P(x, t|x_0)$ where $\hat{L}^\dagger$ is the adjoint operator. Applying it to (8.154) leads to

$$L^{\dagger}(x_0)\tau(x_1,x_0) = \int_{-\infty}^{x_1} dx \int_0^{\infty} dt \frac{\partial P}{\partial t} = \int_{-\infty}^{x_1} dx(-P(x,t=0)) = -1 \qquad (8.155)$$

where we have used the fact that $P(x,t=0)$ is normalized in $(-\infty,x_1)$. Thus we have a differential equation for the mean first passage time

$$L^{\dagger}(x_0)\tau(x_1,x_0) = -1 \qquad (8.156)$$

that should be solved with the boundary condition $\tau(x_1,x_0=x_1)=0$.

As an example consider the following form for the operator $\hat{L}$

$$\hat{L}(x) = -\frac{d}{dx}\left[a(x) - b(x)\frac{d}{dx}\right] \qquad (8.157)$$

which, for $a(x) = -\beta D\partial V/\partial x$ and $b(x) = D$ is the Smoluchowski operator (8.132). The equation for $\tau(x_1,x_0)$ is

$$\hat{L}^{\dagger}(x_0)\tau(x_1,x_0) = a(x_0)\frac{d\tau(x_1,x_0)}{dx_0} + \frac{d}{dx_0}\left(b(x_0)\frac{d\tau(x_1,x_0)}{dx_0}\right) = -1 \qquad (8.158)$$

This differential equation can be easily solved,[18] or it can be checked by direct substitution that its solution is

$$\tau(x_1,x_0) = -\int_{c_2}^{x_0} dx'[b(x')f(x')]^{-1}\int_{c_1}^{x'} dx''f(x'') \qquad (8.159a)$$

$$f(x) = \exp\left(\int^{x} dx'\frac{a(x')}{b(x')}\right) \qquad (8.159b)$$

where c_1 and c_2 are integration constants that should be determined from the boundary conditions. In particular, the choice $c_2 = x_1$ has to be made in order to satisfy the requirement that τ should vanish if $x_0 = x_1$. c_1 is the point where $d\tau(x_1,x_0)/dx_1$ vanishes. Note that $f(x)$ is the equilibrium solution of Eq. (8.150) with the operator $\hat{L}$ given by (8.157).

Passage time distributions and the mean first passage time provide a useful way for analyzing the time evolution of stochastic processes. An application to chemical reactions dominated by barrier crossing is given in Section 14.4.2 and Problem 14.3.

[18] To solve this equation define $y(x) = b(x)[d\tau(x)/dx]$ and solve $(a(x)/b(x))y(x)+dy(x)/dx = -1$ by making the substitution $y(x) = u(x)\exp[-\int^{x} dx'a(x')/b(x')]$.

Appendix 8A: Obtaining the Fokker–Planck equation from the Chapman–Kolmogorov equation

We start from the Chapman–Kolmogorov equation

$$P(\mathbf{x}_3t_3 \mid \mathbf{x}_1t_1) = \int d\mathbf{x}_2 P(\mathbf{x}_3t_3 \mid \mathbf{x}_2t_2)P(\mathbf{x}_2t_2 \mid \mathbf{x}_1t_1) \qquad t_3 \geq t_2 \geq t_1 \tag{8.160}$$

where in general $\mathbf{x}$ is a multivariable stochastic process. Recall that this is a general property (in fact can be viewed as the definition) of Markovian stochastic processes. We further assume that the following conditions are satisfied:

1. The Markov process is continuous, that is, for any $\varepsilon > 0$

$$\lim_{\Delta t \to 0} \frac{1}{\Delta t} \int\limits_{|x-y|>\varepsilon} d\mathbf{x}\, P(\mathbf{x}, t + \Delta t \mid \mathbf{y}, t) = 0 \tag{8.161}$$

   Namely, the probability for the final state $\mathbf{x}$ to be different from the initial state $\mathbf{y}$ vanishes faster than Δt as $\Delta t \to 0$.

2. The following functions

$$\begin{aligned} A_i(\mathbf{x}, t) &= \lim_{\Delta t \to 0} \frac{1}{\Delta t} \int d\mathbf{z}(z_i - x_i)P(\mathbf{z}, t + \Delta t \mid \mathbf{x}, t) \\ B_{i,j}(\mathbf{x}, t) &= \lim_{\Delta t \to 0} \frac{1}{\Delta t} \int d\mathbf{z}(z_i - x_i)(z_j - x_j)P(\mathbf{z}, t + \Delta t \mid \mathbf{x}, t) \end{aligned} \tag{8.162}$$

   exist for all $\mathbf{x}$. Note that since the process is continuous, the contributions to these integrals come from regions of $\mathbf{z}$ infinitesimally close to $\mathbf{x}$. Also note that higher moments of the form

$$C_{ijk} \equiv \lim_{\Delta t \to 0} \frac{1}{\Delta t} \int d\mathbf{z}(z_i - x_i)(z_j - x_j)(z_k - x_k)P(\mathbf{z}, t + \Delta t \mid \mathbf{x}, t) \tag{8.163}$$

   (and higher) *must* be zero. To show this define

$$C(\mathbf{a}) = \sum_{ijk} a_i a_j a_k C_{ijk} \tag{8.164}$$

   Knowing $C(\mathbf{a})$ we can get C_{ijk} from

$$C_{ijk} = \frac{1}{3!} \frac{\partial^3}{\partial a_i \partial a_j \partial a_k} C(\mathbf{a}) \tag{8.165}$$

We will show that $C(\mathbf{a}) = 0$ for all $\mathbf{a}$. This is because

$$\begin{aligned}
C(\mathbf{a}) &= \lim_{\Delta t \to 0} \frac{1}{\Delta t} \int d\mathbf{z}[(\mathbf{z} - \mathbf{x}) \cdot \mathbf{a}]^3 P(\mathbf{z}, t + \Delta t \mid \mathbf{x}, t) \\
&\leq \lim_{\Delta t \to 0} \frac{1}{\Delta t} \int d\mathbf{z} \mid (\mathbf{z} - \mathbf{x}) \cdot \mathbf{a} \mid ((\mathbf{z} - \mathbf{x}) \cdot \mathbf{a})^2 P(\mathbf{z}, t + \Delta t \mid \mathbf{x}, t) \\
&\leq \varepsilon \mid \mathbf{a} \mid \lim_{\Delta t \to 0} \frac{1}{\Delta t} \int d\mathbf{z}((\mathbf{z} - \mathbf{x}) \cdot \mathbf{a})^2 P(\mathbf{z}, t + \Delta t \mid \mathbf{x}, t) \qquad (8.166)
\end{aligned}$$

where $\varepsilon \to 0$ when $\Delta t \to 0$. In the last expression $\varepsilon \mid \mathbf{a} \mid$ is multiplied by $\mathbf{a} \cdot \mathbf{B} \cdot \mathbf{a}$ which is finite, so $C(\mathbf{a})$ is 0 for all $\mathbf{a}$ and therefore $C_{ijk} = 0$. The same argument holds for any moment of order 3 or higher.[19]

The forward equation: The Fokker–Planck equation is now derived as the differential form of the Chapman–Kolmogorov equation: For any function $f(\mathbf{x})$

$$\begin{aligned}
\frac{\partial}{\partial t} \int d\mathbf{z} f(\mathbf{z}) P(\mathbf{z}, t \mid \mathbf{y}, t_0) &= \lim_{\Delta t \to 0} \frac{1}{\Delta t} \int d\mathbf{z} f(\mathbf{z}) \left[P(\mathbf{z}, t + \Delta t \mid \mathbf{y}, t_0) - P(\mathbf{z}, t \mid \mathbf{y}, t_0)\right] \\
&= \lim_{\Delta t \to 0} \frac{1}{\Delta t} \left\{ \int d\mathbf{z} \int d\mathbf{x} f(\mathbf{z}) P(\mathbf{z}, t + \Delta t \mid \mathbf{x}, t) P(\mathbf{x}, t \mid \mathbf{y}, t_0) \right. \\
&\quad \left. - \int d\mathbf{x} f(\mathbf{x}) P(\mathbf{x}, t \mid \mathbf{y}, t_0) \right\} \qquad (8.167)
\end{aligned}$$

In the first integral replace $f(\mathbf{z})$ by its Taylor expansion about $\mathbf{x}$,

$$f(\mathbf{z}) = f(\mathbf{x}) + \sum_i \frac{\partial f(\mathbf{x})}{\partial x_i}(z_i - x_i) + \sum_{i,j} \frac{1}{2} \frac{\partial^2 f}{\partial x_i \partial x_j}(z_i - x_i)(z_j - x_j) \qquad (8.168)$$

We have seen above that higher-order terms do not contribute. The term arising from $f(\mathbf{x})$ and the last integral in (8.167) cancels because $\int d\mathbf{z} P(\mathbf{z}, t + \Delta t \mid \mathbf{x}, t) = 1$,

[19] In Eq. (8.161) we could write $A_i(x,t) \leq \lim_{\Delta t \to 0} \frac{1}{\Delta t} \int d\mathbf{z} \mid z_i - x_i \mid P$, but we cannot go further because both $|z_i - x_i|$ and Δt are infinitesimal.

so (8.167) becomes

$$\frac{\partial}{\partial t}\int d\mathbf{z}f(\mathbf{z})P(\mathbf{z},t\mid \mathrm{y},t_0)=\lim_{\Delta t\to 0}\frac{1}{\Delta t}\left\{\int d\mathbf{z}\int d\mathbf{x}\right.$$

$$\times\left[\sum_i (z_i-x_i)\frac{\partial f}{\partial x_i}+\frac{1}{2}\sum_{i,j}(z_i-x_i)(z_j-x_j)\frac{\partial^2 f}{\partial x_i\partial x_j}\right]$$

$$\left.\times\, P(\mathbf{z},t+\Delta t\mid \mathbf{x},t)\,P(\mathbf{x},t\mid \mathbf{y},t_0)\right\}$$

$$=\int d\mathbf{x}\left[\sum_i A_i(\mathbf{x})\frac{\partial f}{\partial x_i}+\frac{1}{2}\sum_{i,j}B_{ij}(\mathbf{x})\frac{\partial^2 f}{\partial x_i\partial x_j}\right]P(\mathbf{x},t\mid \mathbf{y},t_0) \qquad (8.169)$$

Next we integrate the right-hand side of Eq. (8.169) by parts. Since f was an arbitrary function we can assume that f and its first and second derivatives vanish on the surface of our system. Hence

$$\frac{\partial}{\partial t}\int d\mathbf{x}f(\mathbf{x})P(\mathbf{x},t\mid \mathbf{y},t_0)=\int d\mathbf{x}f(\mathbf{x})\Bigg[-\sum_i\frac{\partial}{\partial x_i}[A_i(\mathbf{x},t)P(\mathbf{x},t\mid \mathbf{y},t_0)]$$

$$+\sum_i\sum_j\frac{1}{2}\frac{\partial^2}{\partial x_i\partial x_j}[B_{ij}(\mathbf{x},t)P(\mathbf{x},t\mid \mathbf{y},t_0)]\Bigg] \qquad (8.170)$$

So *for a continuous Markov process,*

$$\frac{\partial}{\partial t}P(\mathbf{x},t\mid \mathbf{y},t_0)=-\sum_i\frac{\partial}{\partial x_i}[A_i(\mathbf{x},t)P(\mathbf{x},t\mid \mathbf{y},t_0)]$$

$$+\frac{1}{2}\sum_{i,j}\frac{\partial^2}{\partial x_i\partial x_j}\left[B_{ij}(\mathbf{x},t)P(\mathbf{x},t\mid \mathbf{y},t_0)\right] \qquad (8.171)$$

This is the Fokker–Planck equation that corresponds, under the conditions specified, to the Chapman–Kolmogorov equation (8.118).

The backward equation. Now consider

$$\frac{\partial}{\partial t_0}P(\mathbf{x},t\mid \mathbf{y},t_0)=\lim_{\Delta t_0\to 0}\frac{1}{\Delta t_0}[P(\mathbf{x},t\mid \mathbf{y},t_0+\Delta t_0)-P(\mathbf{x},t\mid \mathbf{y},t_0)] \qquad (8.172)$$

Multiplying the first term on the right by $1=\int d\mathbf{z}P(\mathbf{z},t_0+\Delta t_0\mid \mathbf{y},t_0)$ and writing the second term in the form $\int d\mathbf{z}P(\mathbf{x},t\mid \mathbf{z},t_0+\Delta t_0)P(\mathbf{z},t_0+\Delta t_0\mid \mathbf{y},t_0)$ yields the right-hand side of (8.172) in the form

$$\lim_{\Delta t_0 \to 0} \frac{1}{\Delta t_0} \left\{ \int d\mathbf{z} P(\mathbf{z}, t_0 + \Delta t_0 \mid \mathbf{y}, t_0) \left[P(\mathbf{x}, t \mid \mathbf{y}, t_0 + \Delta t_0) - P(\mathbf{x}, t \mid \mathbf{z}, t_0 + \Delta t_0) \right] \right\} \tag{8.173}$$

Inside the square brackets we may put $\Delta t_0 = 0$ and use a Taylor expansion to get

$$P(\mathbf{x}, t \mid \mathbf{y}, t_0) - P(\mathbf{x}, t \mid \mathbf{z}, t_0) = -\sum_i \frac{\partial P(\mathbf{x}, t \mid \mathbf{y}, t_0)}{\partial y_i}(z_i - y_i) - \frac{1}{2}\sum_i \sum_j \frac{\partial^2 P(\mathbf{x}, t \mid \mathbf{y}, t_0)}{\partial y_i \partial y_j}(z_i - y_i)(z_j - y_j) - \cdots \tag{8.174}$$

This again leads to limits of the form

$$\lim_{\Delta t_0 \to 0} (1/\Delta t_0) \int d\mathbf{z} P(\mathbf{z}, t_0 + \Delta t_0 \mid \mathbf{y}, t_0) \prod_j (z_j - y_j)$$

that are evaluated as before. Using the definitions (8.162) and the fact that moments of this kind of order higher than 2 vanish, we finally get the *backward Fokker–Planck equation*

$$\frac{\partial P(\mathbf{x}, t \mid \mathbf{y}, t_0)}{\partial t_0} = -\sum_i A_i(\mathbf{y}, t_0) \frac{\partial P(\mathbf{x}, t \mid \mathbf{y}, t_0)}{\partial y_i} - \frac{1}{2}\sum_{ij} B_{ij}(\mathbf{y}, t_0) \frac{\partial^2 P(\mathbf{x}, t \mid \mathbf{y}, t_0)}{\partial y_i \partial y_j} \tag{8.175}$$

Appendix 8B: Obtaining the Smoluchowski equation from the overdamped Langevin equation

Our starting point is Eqs (8.129) and (8.130). It is convenient to redefine the timescale

$$\tau = t/(\gamma m) \tag{8.176}$$

Denoting the random force on this timescale by $\rho(\tau) = R(t)$, we have $\langle \rho(\tau_1)\rho(\tau_2) \rangle = 2m\gamma k_B T \delta(t_1 - t_2) = 2k_B T \delta(\tau_1 - \tau_2)$. The new Langevin equation becomes

$$\frac{dx}{d\tau} = -\frac{dV(x)}{dx} + \rho(\tau) \tag{8.177a}$$

$$\langle \rho \rangle = 0 \qquad \langle \rho(0)\rho(\tau) \rangle = 2k_B T \delta(\tau) \tag{8.177b}$$

The friction γ does not appear in these scaled equations, but any rate evaluated from this scheme will be inversely proportional to γ when described on the real (i.e. unscaled) time axis.

In these scaled time variable Eqs (8.131) take the forms

$$\frac{\partial P(x,\tau)}{\partial \tau} = \hat{\Omega}(\tau)P$$
$$\hat{\Omega}(\tau) = \frac{\partial}{\partial x}\left(\frac{\partial V}{\partial x} - \rho(\tau)\right) \tag{8.178}$$

Integrate between τ and $\tau + \Delta\tau$ to get

$$P(x,\tau+\Delta\tau) = P(x,\tau) + \int_{\tau}^{\tau+\Delta\tau} d\tau_1 \hat{\Omega}(\tau_1)P(x,\tau_1) \tag{8.179}$$

The operator $\hat{\Omega}$ contains the random function $\rho(\tau)$. Repeated iterations in the integral and averaging over all realizations of ρ lead to

$$P(x,\tau+\Delta\tau) - P(x,\tau) = \left[\int_{\tau}^{\tau+\Delta\tau} d\tau_1 \langle\hat{\Omega}(\tau_1)\rangle \right.$$
$$\left. + \int_{\tau}^{\tau+\Delta\tau} d\tau_1 \int_{\tau}^{\tau_1} d\tau_2 \langle\hat{\Omega}(\tau_1)\hat{\Omega}(\tau_2)\rangle + \cdots \right] P(x,\tau) \tag{8.180}$$

our aim now is to take these averages using the statistical properties of ρ and to carry out the required integrations keeping only terms of order $\Delta\tau$. To this end we note that $\hat{\Omega}$ is of the form $\hat{\Omega}(\tau) = \hat{A} + \hat{B}\rho(\tau)$ where $\hat{A}$ and $\hat{B}$ are the deterministic operators $\partial/\partial x(\partial V(x)/\partial x)$ and $\partial/\partial x$, respectively. Since $\langle\rho\rangle = 0$ the first term in the square bracket is simply

$$\hat{A}\Delta\tau = \frac{\partial}{\partial x}\frac{\partial V(x)}{\partial x}\Delta\tau \tag{8.181}$$

where the operator $\partial/\partial x$ is understood to operate on everything on its right. The integrand in the second term inside the square brackets contains terms of the forms AA, $AB\langle\rho\rangle = 0$, and $B^2\langle\rho(\tau_1)\rho(\tau_2)\rangle$. The double time integrals with the deterministic AA integrand are of order $\Delta\tau^2$ and may be neglected. The only contributions of order $\Delta\tau$ come from the *BB* terms, which, using Eq. (8.177b), lead to

$$\int_{\tau}^{\tau+\Delta\tau} d\tau_1 \int_{\tau}^{\tau_1} d\tau_2 \langle\rho(\tau_1)\rho(\tau_2)\rangle \frac{\partial^2}{\partial x^2} = \int_{\tau}^{\tau+\Delta\tau} d\tau_1 k_B T \frac{\partial^2}{\partial x^2} = k_B T \frac{\partial^2}{\partial x^2}\Delta\tau \quad (8.182)$$

With a little effort we can convince ourselves that higher-order terms in the expansion (8.180) contribute only terms of order Δt^2 or higher. Consider for example the third-order term $\int_{\tau}^{\tau+\Delta\tau} d\tau_1 \int_{\tau}^{\tau_1} d\tau_2 \int_{\tau}^{\tau_2} d\tau_3 \langle\hat{\Omega}(\tau_1)\hat{\Omega}(\tau_2)\hat{\Omega}(\tau_3)\rangle$ that yields integrals involving *AAA*, *AAB*, *ABB*, and *BBB* terms. The integral with the deterministic *AAA* term is of order $\Delta\tau^3$ and can be disregarded. The *AAB* and *BBB* terms lead to results that contain $\langle\rho\rangle$ and $\langle\rho\rho\rho\rangle$ which are zero. The only terms that may potentially contribute are of the type *ABB*. However, they do not: The integrands that involve such terms appear with functions such as $\langle\rho(\tau_1)\rho(\tau_2)\rangle$, which yields a δ-function that eliminates one of the three time integrals. The remaining two time integrals yield a $\Delta\tau^2$ term and do not contribute to order $\Delta\tau$.

Similar considerations show that all higher-order terms in Eq. (8.180) may be disregarded. Equations (8.181) and (8.182) finally lead to

$$\frac{\partial P(x,\tau)}{\partial \tau} = \left(\frac{\partial}{\partial x}\frac{dV}{dx} + k_B T \frac{\partial^2}{\partial x^2}\right) P(x,\tau) \quad (8.183)$$

Transforming back to the original time variable $t = \gamma m \tau$ yields the *Smoluchowski equation* (8.132) and (8.133).

Appendix 8C: Derivation of the Fokker–Planck equation from the Langevin equation

Our starting point is Eq. (8.143)

$$\frac{\partial P(x,v,t)}{\partial t} = \hat{\Omega}P = -v\frac{\partial P}{\partial x} + \frac{1}{m}\frac{\partial V}{\partial x}\frac{\partial}{\partial v}P - \frac{\partial}{\partial v}\left[\left(-\gamma v + \frac{1}{m}R(t)\right)P\right] \quad (8.184)$$

As in (8.178), the operator $\hat{\Omega}$ is of the form $\hat{\Omega}(\tau) = \hat{A} + \hat{B}R(t)$ in which $\hat{A}$ and $\hat{B}$ are deterministic operators and $R(t)$ is a random function of known statistical properties. We can therefore proceed in exactly the same way as in Appendix 8B. In what follows we will simplify this task by noting that the right-hand side of (8.184) contains additive contributions of Newtonian and dissipative terms. The former is just the Liouville equation

$$\left(\frac{\partial P(x,v,t)}{\partial t}\right)_{\text{nwt}} = -v\frac{\partial P}{\partial x} + \frac{1}{m}\frac{\partial V}{\partial x}\frac{\partial}{\partial v}P \tag{8.185}$$

The dissipative part (terms that vanish when $\gamma = 0$, including the $R(t)$ term) does not depend on the potential V, and we should be able to derive its contribution to the Fokker–Planck equation for a system in which V is constant, that is, $\partial V/\partial x = 0$. In this case we can focus on the Langevin equation for the velocity

$$\dot{v} = -\gamma v + \frac{1}{m}R(t) \tag{8.186}$$

and look for an equation for the probability $P(v,t)$ associated with this stochastic differential equation. In analogy to (8.131) we now have

$$\left(\frac{\partial P(v,t)}{\partial t}\right)_{\text{dis}} = -\frac{\partial}{\partial v}[\dot{v}P] = -\frac{\partial}{\partial v}\left[\left(-\gamma v + \frac{1}{m}R(t)\right)P(v,t)\right] \tag{8.187}$$

that we rewrite in the form

$$\begin{aligned}\frac{\partial P(v,t)}{\partial t} &= \hat{\Omega}(t)P \\ \hat{\Omega}(t) &= \frac{\partial}{\partial v}\left(\gamma v - \frac{1}{m}R(t)\right)\end{aligned} \tag{8.188}$$

Integrating between t and $t + \Delta t$ and iterating leads to

$$\begin{aligned}P(v,t+\Delta t) = P(v,t) + \int_t^{t+\Delta t} dt_1 \hat{\Omega}(t_1)P(v,t_1) = \Bigg[1 + \int_t^{t+\Delta t} dt_1 \hat{\Omega}(t_1) \\ + \int_t^{\Delta t} dt_1 \hat{\Omega}(t_1) \int_t^{t_1} dt_2 \hat{\Omega}(t_2) + \cdots \Bigg] P(v,t)\end{aligned} \tag{8.189}$$

The rest of the calculation is done in complete analogy to the transition from (8.178) and (8.180) to (8.183). In the present case we get

$$\begin{aligned}P(v,t+\Delta t) &= \left(1 + \frac{\partial}{\partial v}(\gamma v)\Delta t + \frac{\gamma k_B T}{m}\frac{\partial^2}{\partial v^2}\Delta t\right)P(v,t) \\ \Rightarrow \left(\frac{\partial P(v,t)}{\partial t}\right)_{\text{dis}} &= \frac{\partial}{\partial v}\gamma\left[v + \frac{k_B T}{m}\frac{\partial}{\partial v}\right]P(v,t)\end{aligned} \tag{8.190}$$

which is the dissipative part of the time evolution. The full Fokker–Planck equation is obtained by adding the Liouville terms (8.185), leading to Eq. (8.144).

Further Reading (Chapters 7 and 8)

C. W. Gardiner, *Handbook of Stochastic Methods*, 3rd edn (Springer, Berlin, 2004).

R. Kubo, M. Toda, and N. Hashitsume, *Statistical Physics II* , 2nd edn (Springer, Berlin, 2003).

H. Risken, *The Fokker–Planck Equation*, 2nd edn (Springer, Berlin, 1989).

Z. Schuss, *Theory and Applications of Stochastic Differential Equations* (Wiley, New York, 1980).

N. G. van Kampen, *Stochastic Processes in Physics and Chemistry* (North-Holland, Amsterdam, 1992).

R. Zwanzig, *Non Equilibrium Statistical Mechanics* (Oxford University Press, Oxford, 2001).

N. Wiener, J. Math. Phys. *2*, 131–174 (1923) (on the foundations of the theory of stochastic processes – for the mathematically oriented reader)

9

INTRODUCTION TO QUANTUM RELAXATION PROCESSES

Since earth and water,
Air and fire, those elements which form
The sums of things are, all of them, composed
Of matter that is born and dies, we must
Conclude that likewise all the universe
Must be of mortal nature. Any time
We see that parts are transient substances
We know that their total is as fugitive,
And when the main components of the world
Exhaust themselves or come to birth again
Before our very eyes, we may be sure
That heaven and earth will end, as certainly
As ever they once began…

Lucretius (c.99–c.55 BCE) "The way things are"
translated by Rolfe Humphries, Indiana University Press, 1968

The first question to ask about the phenomenon of relaxation is why it occurs at all. Both the Newton and the Schrödinger equations are symmetrical under time reversal: The Newton equation, $dx/dt = v\,; dv/dt = -\partial V/\partial x$, implies that particles obeying this law of motion will retrace their trajectory back in time after changing the sign of both the time t and the particle velocities v. The Schrödinger equation, $\partial\psi/\partial t = -(i/\hbar)\hat{H}\psi$, implies that if $(\psi(t)$ is a solution then $\psi^*(-t)$ is also one, so that observables which depend on $|\psi|^2$ are symmetric in time. On the other hand, nature clearly evolves asymmetrically as asserted by the second law of thermodynamics. *How does this asymmetry arise in a system that obeys temporal symmetry in its time evolution?* Readers with background in thermodynamics and statistical mechanics have encountered the intuitive answer: Irreversibility in a system with many degrees of freedom is essentially a manifestation of the system "getting lost in phase space": A system starts from a given state and evolves in time. If the number of accessible states is huge, the probability that the system will find its way back to the initial state in finite time is vanishingly small, so that an observer who monitors properties associated with the initial state will see an irreversible evolution. The question is how is this irreversible behavior manifested through the

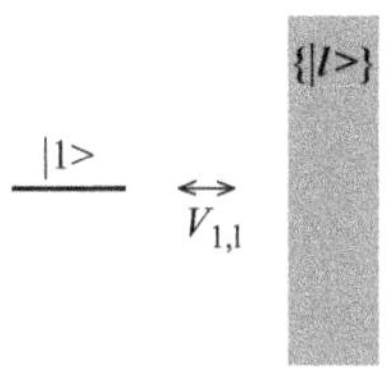

FIG. 9.1 A model for quantum mechanical relaxation: A single zero-order level $|1\rangle$ is initially populated. This level is coupled to, and energetically overlaps with, a continuous manifold of other zero-order levels represented by the shaded area. This manifold (here $\{|l\rangle\}$) is sometimes denoted in the text by the corresponding capital letter L.

reversible equations of motion, and how does it show in the quantitative description of the time evolution. This chapter provides an introduction to this subject using the time-dependent Schrödinger equation as a starting point. Chapter 10 discusses more advanced aspects of this problem within the framework of the quantum Liouville equation and the density operator formalism.

9.1 A simple quantum-mechanical model for relaxation

In what follows we consider a simple quantum-mechanical model for irreversibility. In addition to providing a simple demonstration of how irreversibility arises in quantum mechanics, we will see that this model can be used as a prototype of many physical situations, showing not only the property of irreversible relaxation but also many of its observable consequences.

We consider a Hamiltonian written as a sum

$$\hat{H} = \hat{H}_0 + \hat{V} \tag{9.1}$$

and use the set of eigenstates of $\hat{H}_0$ as a basis. We assume that this set is given by a single state $|1\rangle$ of zero-order energy E_1 and a manifold of states $\{|l\rangle\}$ $(l = 2, 3, \ldots)$ with zero-order energies E_l, see Fig. 9.1. The set $|1\rangle, \{|l\rangle\}$ is taken to be orthonormal, that is, $\langle 1|1\rangle = 1$, $\langle 1|l\rangle = 0$ for all l and $\langle l|l'\rangle = \delta_{l,l'}$ for all l and l'. These states are coupled by the "perturbation" V. We consider a model in which $V_{1,1} = V_{ll} = 0$ for all l, however $V_{1,l} \neq 0$ so the state 1 is coupled to all states in the manifold $\{l\}$. This information is contained in the following expressions for $\hat{H}_0$ and $\hat{V}$:

$$\begin{aligned}
\hat{H}_0 &= E_1|1\rangle\langle 1| + \sum_l E_l|l\rangle\langle l| \\
\hat{V} &= \sum_l (V_{1,l}|1\rangle\langle l| + V_{l,1}|l\rangle\langle 1|)
\end{aligned} \tag{9.2}$$

Another way to express this information is by writing the matrix representation of $\hat{H}$ in the given basis.

$$\mathbf{H} = \begin{pmatrix} E_1 & V_{1,2} & V_{1,3} & V_{1,4} & \cdots \\ V_{2,1} & E_2 & 0 & 0 & \cdots \\ V_{3,1} & 0 & E_3 & 0 & \cdots \\ V_{4,1} & 0 & 0 & E_4 & 0 \\ \vdots & \vdots & \vdots & 0 & \ddots \end{pmatrix} \tag{9.3}$$

We want to solve the time-dependent Schrödinger equation

$$\frac{d\psi(t)}{dt} = -\frac{i}{\hbar}\hat{H}\psi(t) \tag{9.4}$$

under the assumption that the system is initially in state $|1\rangle$. In particular, we want to evaluate the probability $P_1(t)$ to find the system in state 1 at time t.

Before setting to solve this mathematical problem, we should note that while the model is mathematically sound and the question asked is meaningful, it cannot represent a complete physical system. If the Hamiltonian was a real representation of a physical system we could never prepare the system in state $|1\rangle$. Still, we shall see that this model represents a situation which is ubiquitous in molecular systems, not necessarily in condensed phase. Below we outline a few physical problems in which our model constitutes a key element:

1. Consider the generic two-level model, Eq. (2.13), with the levels now denoted g and s with energies $E_s > E_g$. An extended system that includes also the environment may be represented by states that will be denoted $|s,\{e\}\rangle$, $|g,\{e\}\rangle$ where $\{e\}$ defines states of the environment. A common phrase is to say that these molecular states are "dressed" by the environment. Now consider this generic molecule in state s and assume that the environment is at zero temperature. In this case $\{e\} = \{e\}_g$ is the ground state of the environment. Obviously the initial state $|s,\{e\}_g\rangle$ is energetically embedded in a continuum of states $|g,\{e\}_x\rangle$ where $\{e\}_x$ are excited states of the environment. This is exactly the situation represented in Fig. 9.1, where level $|1\rangle$ represents the state $|s,\{e\}_g\rangle$ while levels $|l\rangle$ are the states $|g,\{e\}_x\rangle$ with different excited state of the environment. An important aspect common to all models of this type is that the continuous manifold of states $\{|l\rangle\}$ is bound from below: State $|g,\{e\}_g\rangle$ is obviously its lowest energy state.

2. The generality of this picture is emphasized by the observation that even for a single atom or molecule in vacuum the ever present radiation field constitutes such an environment (see Section 9.2.3 below). Any excited molecular state is coupled to lower molecular states dressed by photons.

3. In an isolated large molecule, each excited electronic state is coupled to a dense manifold of vibrational levels associated with lower electronic states. This

can lead to the appearance of *radiationless relaxation* (no photon emitted) in single isolated molecules (see Section 9.2.2).

All these physical examples can be described by the model of Fig. 9.1 and Eq. (9.2). Let us now return to this simple model and address the probability $P_1(t)$ to find the system in state 1 at time t given that $P_1(t=0)=1$.

We start by writing the general solution of the time-dependent Schrödinger equation in the form

$$\psi(t) = C_1(t)|1\rangle + \sum_l C_l(t)|l\rangle \tag{9.5}$$

Insert (9.5) into (9.4), then multiply the resulting equation by $\langle 1|$ or $\langle l|$ to get

$$\hbar\frac{d}{dt}C_1 = -iE_1C_1 - i\sum_l V_{1,l}C_l \tag{9.6}$$

$$\hbar\frac{d}{dt}C_l = -iE_lC_l - iV_{l,1}C_1; \qquad \text{for each } l \tag{9.7}$$

This set of equations should be solved under the initial condition $C_1(t=0)=1$; $C_l(t=0)=0$ for all l. We want to find the probability $P_1(t) = |C_1(t)|^2$ that the system is still in state 1 at time t.

Equations (9.6) and (9.7) constitute a linear initial value problem that can conveniently be solved using Laplace transforms as described in Section 2.6. The formal answer to our problem has already been obtained, Eqs (2.60) and (2.61), which imply

$$C_1(t) = -\frac{1}{2\pi i}\int_{-\infty}^{\infty} dE e^{-iEt/\hbar} G_{1,1}(E+i\varepsilon); \qquad \varepsilon \to 0+ \tag{9.8a}$$

$$G_{1,1}(z) = \langle 1|\frac{1}{z-H}|1\rangle \tag{9.8b}$$

This is a Fourier transform of the diagonal 1, 1 matrix element of the Green's operator $\hat{G}(E+i\varepsilon)$ where

$$\hat{G}(z) = \frac{1}{z-\hat{H}} = \frac{1}{z-\hat{H}_0-\hat{V}} \tag{9.9}$$

A convenient way to evaluate this matrix element starts by defining also

$$G_0(z) = \frac{1}{z-H_0} \tag{9.10}$$

so that $(G_0)_{1,1} = (z - E_1)^{-1}$, $(G_0)_{l,l'} = (z - E_l)^{-1}\delta_{l,l'}$, and $(G_0)_{1,l} = 0$. $\hat{G}(z)$ satisfies the so-called *Dyson identities*.[1]

$$\hat{G}(z) = \hat{G}_0(z) + \hat{G}_0(z)\hat{V}\hat{G}(z) = \hat{G}_0(z) + \hat{G}(z)\hat{V}\hat{G}_0(z) \tag{9.11}$$

Starting from the first identity in (9.11) we take its 1,1 and l, 1 matrix elements, that is, $G_{1,1} = (G_0)_{1,1} + (\hat{G}_0\hat{V}\hat{G})_{1,1}$ and $G_{l,1} = (\hat{G}_0\hat{V}\hat{G})_{l,1}$. Using the resolution of the identity operator

$$|1\rangle\langle 1| + \sum_l |l\rangle\langle l| = 1 \tag{9.12}$$

leads to

$$G_{1,1} = (G_0)_{1,1} + (G_0)_{1,1}\sum_l V_{1,l}G_{l,1} \tag{9.13}$$

$$G_{l,1} = (G_0)_{l,l}V_{l,1}G_{1,1} \tag{9.14}$$

Inserting (9.14) into (9.13) and using the identities below (9.10) it follows that

$$G_{1,1}(z) = \frac{1}{z - E_1} + \frac{1}{z - E_1}\left(\sum_l \frac{|V_{1,l}|^2}{z - E_l}\right)G_{1,1}(z) \tag{9.15}$$

that is, (putting $z = E + i\epsilon$ and taking the limit $\epsilon \to 0$),

$$G_{1,1}(E) = \lim_{\varepsilon\to 0}\frac{1}{E + i\varepsilon - E_1 - \sum_l |V_{1,l}|^2/(E - E_l + i\varepsilon)} \tag{9.16}$$

This is the function to be Fourier transformed according to Eq. (9.8).

Before continuing with this task we make the following observation: Our problem deals with Hamiltonian whose spectrum spans infinitely many energy levels, however the physics of interest focuses on a small local (in energy) part of this infinite Hilbert space—the energetic neighborhood of the initially prepared level $|1\rangle$, that affects its future evolution. The Green function element $G_{1,1}$ contains the information on level $|1\rangle$ in an explicit way, while the effect of all other (infinitely many!) levels appears only in a sum

$$B_1(E) = \sum_l \frac{|V_{1,l}|^2}{E - E_l + i\varepsilon} \tag{9.17}$$

[1] For example, starting from $\hat{G}_0^{-1} = \hat{G}^{-1} + \hat{V}$ and multiplying it by $\hat{G}$ from the left and by $\hat{G}_0$ from the right yields the second identity of Eq. (9.11).

that is often characterized (as will be seen below) by just a few parameters.

Focusing on the interesting subspace of an overall system and attempting to characterize its behavior while discarding uninteresting information on the rest of the system has been a repeating motif in our discussions. Mathematically, this is often done by "projecting" the dynamics encoded in our equations of motion, here the Schrödinger equation, onto the interesting subspace.[2] Techniques based on *projection operators* are very useful in this respect. In Appendix 9A we repeat the derivation of Eq. (9.16) using this technique.

As a prelude to evaluating the Fourier transform (9.8) let us consider first the function $B_1(E)$ and assume that the manifold $\{l\}$ constitutes a continuum of states. In this case the summation over l corresponds to the integral

$$\sum_l \rightarrow \int_{-\infty}^{\infty} dE_l \rho_L(E_l) \tag{9.18}$$

where $\rho_L(E)$ denotes the density of states in the $\{l\}$ manifold. Note that the fact that we took the integration limits to be $(-\infty \ldots \infty)$ does not necessarily mean that the eigenvalues $\{E_l\}$ extend between these limits. The actual information concerning this eigenvalue spectrum is in the density of states $\rho_L(E_l)$ that can be zero below some threshold. Equation (9.17) now takes the form

$$B_1(E) = \int_{-\infty}^{\infty} dE_l \frac{\overline{(|V_{1,l}|^2)_{E_l}} \rho_L(E_l)}{E - E_l + i\varepsilon} = \frac{1}{2\pi} \int_{-\infty}^{\infty} dE_l \frac{\Gamma_1(E_l)}{E - E_l + i\varepsilon} \tag{9.19}$$

where $\overline{(|V_{1,l}|^2)_E}$ is the average of the squared coupling over all continuum levels l that have energy E,[3] and where

$$\Gamma_1(E) \equiv 2\pi \overline{(|V_{1,l}|^2)_E} \rho_L(E) \tag{9.20}$$

Consider first the particularly simple case where the manifold $\{|l\rangle\}$ extends in energy from $-\infty$ to ∞ and where $\Gamma_1(E)$ does not depend on E. In this case

$$B_1(E) = (\Gamma_1/2\pi) \int_{-\infty}^{\infty} dx \frac{1}{E - x + i\varepsilon} = (\Gamma_1/2\pi) \int_{-\infty}^{\infty} dx \frac{E - x - i\varepsilon}{(E - x)^2 + \varepsilon^2} \tag{9.21}$$

---

[2] Much like projecting forces acting on a given body onto the direction of interest.

[3] A formal definition is $\overline{(|V_{1,l}|^2)_E} = \sum_l |V_{1l}|^2 \delta(E - E_l) / \sum_l \delta(E - E_l) = (\rho_L(E))^{-1} \sum_l |V_{1l}|^2 \delta(E - E_l)$.

The real part vanishes by symmetry and, using $\int_{-\infty}^{\infty} dx\varepsilon/(x^2+\varepsilon^2) = \pi$, we get

$$B_1(E) = -\frac{1}{2}i\Gamma_1 \tag{9.22}$$

Using this with (9.16) in (9.8) yields[4] (see Sections 1.1.6 and 2.6)

$$C_1(t) = -\frac{1}{2\pi i}\int_{-\infty}^{\infty} dE \frac{e^{-i(E+i\varepsilon)t/\hbar}}{E - E_1 + (1/2)i\Gamma_1} = e^{-iE_1t/\hbar-(1/2)\Gamma_1 t/\hbar} \tag{9.23}$$

So finally

$$C_1(t) = C_1(0)\exp(-iE_1t/\hbar - (1/2)\Gamma_1 t/\hbar), \tag{9.24a}$$

and

$$|C_1(t)|^2 = e^{-\Gamma_1 t/\hbar} = e^{-k_1 t} \tag{9.24b}$$

Our model assumptions lead to exponential decay of the probability that the system remains in the initial state, where the decay rate k_1 is given by the so-called Fermi "golden rule" formula,

$$k_1 \equiv \frac{\Gamma_1}{\hbar} = \frac{2\pi}{\hbar}\overline{|V_{1,l}|^2}\rho_L \tag{9.25}$$

Note that k_1 has the dimension [time]$^{-1}$ while Γ_s is of dimensionality [energy].

It is important to emphasize that the assumptions that $|V|^2\rho$ is a constant and that the $\{l\}$ manifold extends from $-\infty$ to ∞ are not essential for irreversibility but only for the simple single exponential decay (9.24). In fact, as discussed above, the spectrum $\{E_l\}$ never extends to $-\infty$ because it is bounded from below by the ground state. A more general evaluation starts from and uses the identity (cf. Eq. (1.71))

$$\frac{1}{E - E_l + i\varepsilon} \xrightarrow{\varepsilon\to 0+} \text{PP}\frac{1}{E - E_l} - i\pi\delta(E - E_l) \tag{9.26}$$

where PP is the principal part of the integral (Section 1.1.6). This identity is meaningful only inside an integral. Using it in (9.17) or (9.19) leads to

$$B_1(E) = \Lambda_1(E) - (1/2)i\Gamma_1(E) \tag{9.27}$$

[4] Note that the infinitesimal term $i\varepsilon$ in Eq. (9.16) can be disregarded relative to $(1/2)i\Gamma_1$.

where $\Lambda_1(E)$ and $\Gamma_1(E)$ are the real functions

$$\Gamma_1(E) = 2\pi \sum_l |V_{1,l}|^2 \delta(E - E_l) = 2\pi (\overline{|V_{1,l}|^2 \rho_L(E_l)})_{E_l=E} \tag{9.28}$$

and

$$\Lambda_1(E) = \text{PP} \sum_l \frac{|V_{1,l}|^2}{E - E_l} = \text{PP} \int_{-\infty}^{\infty} dE_l \frac{\overline{|V_{1,l}|^2 \rho_L(E_l)}}{E - E_l} \tag{9.29}$$

The structure of the integrand in Eq. (9.8), as implied by Eqs. (9.16), (9.17) and (9.27), suggests that Λ_1 corresponds to a shift in the unperturbed energy E_1, while the presence of an imaginary term in the denominator of this integrand is the origin of the resulting relaxation behavior. A strong dependence of these functions on E will lead to a relaxation process characterized by a nonexponential decay. In practice, exponential decay is observed in many situations, suggesting that assumptions similar to those made above are good approximations to reality.[5]

More insight into the nature of the result obtained above may be gained by making the following observation: The Green function element

$$G_{1,1}(E) = \lim_{\varepsilon \to 0} \frac{1}{E + i\varepsilon - E_1 - B_1(E)} \tag{9.30}$$

was seen to be an instrument for studying the time evolution in the subspace of the Hilbert space spanned by the state $|1\rangle$– starting from $|1\rangle$, the probability amplitude to remain in this state is given by (9.8). When $|1\rangle$ is an eigenstate of the Hamiltonian (i.e. when $\hat{V} = 0$), $B_1(E) = 0$ and $G_{1,1}(E)$ is really a property of the state $|1\rangle$ alone. When this is not so the function $B_1(E)$ is seen to represent the effect of the rest of the Hilbert space on the time evolution within the 1 subspace. This function is referred to as the *self energy* associated with the level 1. In particular, we have seen that when it is approximately independent of E, the real part of B_1 contributes a shift in E_1, while its imaginary part represents the decay rate of the probability that the system remains in this state. In a sense the complex number $E_1 + \text{Re}(B_1) + i\text{Im}(B_1) = \tilde{E}_1 - (1/2)i\Gamma_1$ may be thought of as a renormalized (complex) energy eigenvalue associated with the state $|1\rangle$. Indeed, from the point of view of the "interesting state" $|1\rangle$, the effect of adding the coupling $\hat{V}$ to the

[5] One of these assumption was that the continuum $\{l\}$ extends from $-\infty$ to ∞. This is often a good approximation to the situation where the edge(s) of the continuum is(are) far from the energetic region of interest, in this case the energy E_1. In the solid state physics literature this is sometimes referred to as the *wide band approximation*.

Hamiltonian $\hat{H}_0$ was to affect the following change on the corresponding Green function element

$$\frac{1}{E - E_1 + i\varepsilon} \Rightarrow \frac{1}{E - \tilde{E}_1 + (1/2)i\Gamma_1}; \qquad (\varepsilon \to 0) \tag{9.31}$$

We will see in Section 9.3 that in addition to representing a decay rate of state $|1\rangle$, Γ_1 also defines an energy width of this state.

9.2 The origin of irreversibility

Did the analysis in Section 9.1 demonstrate irreversibility? It should be emphasized that while the dynamics of the system is completely reversible, as implied by the underlying equations of motion, the appearance of irreversibility has resulted from the particular question asked. This section focuses on understanding this and other aspects of quantum irreversibility.

9.2.1 Irreversibility reflects restricted observation

By their nature, the dynamics governed by either the Newton or the Schrödinger equations are fully reversible. The fact that the probability (9.24b) to find the system in the initial state $|1\rangle$ decays with time reflects the restricted character of the observation. In many situations such restricted observations are associated naturally with the physics of the system: We are interested in the state of a small part of a large system, and the evolution of this small part appears irreversible. We often use the term "system" to denote those degrees of freedom that we are specifically interested in, and the term "bath" for the rest of the (much larger) system. In a classical analogy, a small subsystem at temperature T_1 in contact with a large "thermal bath" with temperature T_2 will relax irreversibly until T_1 becomes equal to T_2, while a state of the overall system given in terms of the position and momentum of every atom will evolve in a systematic (and reversible) way.

9.2.2 Relaxation in isolated molecules

We are quite used to these observations in macroscopic phenomena. What may appear as a surprise is that such situations are also encountered in microscopic systems, including single molecules. For example, an optical transition of a large molecule into an excited electronic state is often followed by relaxation of the electronic energy due to coupling to nuclear (vibrational) levels associated with lower electronic states, in a way which appears to be "radiationless" (no photon emitted) and "collisionless" (take place on a timescale shorter than collision times at the given pressure). Figure 12.3 shows a schematic example. In Fig. 9.1 level 1 can represent one of the vibronic levels in the excited electronic state 2 of Fig. 12.3, while the manifold $\{l\}$ corresponds to the manifold $\{|1, \mathbf{v}\rangle\}$ of vibronic states associated with the ground electronic state 1 of that figure. In a large molecule the density of such levels can be enormous (see Section 12.4.1 and Problem 12.2), making this manifold an effective continuum.

Another relaxation process encountered in isolated molecules is the phenomenon of intramolecular vibrational relaxation. Following excitation of a high-lying vibrational level associated with a particular molecular mode, the excitation energy can rapidly spread to other nuclear modes. This is again a case of an initially prepared single state decaying into an effective continuum.

In both cases, because of restrictions imposed on the excitation process (e.g. optical selection rules), the initially excited state is not an exact eigenstate of the molecular Hamiltonian (see below). At the same time, if the molecule is large enough, this initially prepared zero-order excited state is embedded in a "bath" of a very large number of other states. Interaction between these zero-order states results from residual molecular interactions such as corrections to the Born Oppenheimer approximation in the first example and anharmonic corrections to nuclear potential surfaces in the second. These exist even in the absence of interactions with other molecules, giving rise to relaxation even in isolated (large) molecules. The quasi-continuous manifolds of states are sometimes referred to as "molecular heat baths." The fact that these states are initially not populated implies that these "baths" are at zero temperature.

Problem 9.1. In the analysis that led to the result (9.24) for the decay of the initially prepared state $|1\rangle$ we have used the representation defined by the eigenstates of $\hat{H}_0$. In the alternative representation defined by the full set of eigenstates $\{|j\rangle\}$ of $\hat{H}$ the initial state is given by

$$\Psi(t=0) = |1\rangle = \sum_j C_j|j>; \qquad C_j = \langle 1|j\rangle \tag{9.32}$$

Show that in terms of the coefficients C_j the probability $P_1(t)$ that the system remains in the initial state is given by

$$P_1(t) = \left| \int_{-\infty}^{\infty} dE\,\overline{(\rho_J(E_j)|C_j|^2)}_{E_j=E}\, e^{-iEt/\hbar} \right|^2 \tag{9.33}$$

where $\rho_J(E_j)$ is the density of states in the manifold of eigenstates of $\hat{H}$ and where $\overline{(\rho_J(E_j)|C_j|^2)}_{E_j=E} \equiv L(E)$ is a coarse-grained average (see Section (1.4.4)) of $\rho_J(E_j)|C_j|^2$ in the neighborhood of the energy E. What can you infer from this result on the functional form of the function $L(E)$. Can you offer a physical interpretation of this function?

9.2.3 Spontaneous emission

Even the excited states of a single atom are embedded in a continuum of other states. As discussed in Section 3.2.3 this continuum corresponds to the states of the radiation field sitting on lower atomic states. Casting that discussion in our present notation we have (cf. Eqs (3.21)–(3.24)) $\hat{H}_0 = \hat{H}_M + \hat{H}_R$, $\hat{H} = \hat{H}_0 + \hat{H}_{MR}$, where $\hat{H}_M$ and $\hat{H}_R$ are the Hamiltonians of the molecule and of the free radiation field, respectively, and $\hat{H}_{MR}$ is their mutual interaction. The Hamiltonian $\hat{H}_R$ was shown to represent a collection of modes—degrees of freedom that are characterized by a frequency ω, a polarization vector $\boldsymbol{\sigma}$, and a wavevector $\mathbf{k}$, which satisfy the relations $\boldsymbol{\sigma} \cdot \mathbf{k} = 0$ and $\omega = ck$ with c being the speed of light.

To simplify our notation, we will suppress in what follows the polarization vector $\boldsymbol{\sigma}$, that is, the vector $\mathbf{k}$ will be taken to denote both wavevector and polarization. The time evolution of a mode $\mathbf{k}$ of frequency $\omega_{\mathbf{k}}$ is determined by a harmonic oscillator Hamiltonian, $\hat{h}_{\mathbf{k}} = \hbar\omega_{\mathbf{k}}\hat{a}_{\mathbf{k}}^{\dagger}\hat{a}_{\mathbf{k}}$, and its quantum state—by the corresponding occupation number $n_{\mathbf{k}}$, the numbers of photons in this mode. The state of the radiation field is determined by the set $\{n_{\mathbf{k}}\}$ of occupation numbers of the different modes, and the ground ("vacuum") state of the field is given by $\{n\} = \{0\} = (0, \ldots, 0)$. The eigenstates of $\hat{H}_0$ may be denoted $|j; \{n\}\rangle$ where the index j denotes the molecular state. We refer to such states as "dressed," for example the state $|j; (0, \ldots, 0, 1_{\mathbf{k}}, 0, \ldots, 0)\rangle$ is the molecular state j dressed by one photon in mode $\mathbf{k}$. Again, to simplify notation we will often represent such one-photon states by $|j; 1_{\mathbf{k}}\rangle$ or $|j; \mathbf{k}\rangle$, and sometimes, if our concern is only the photon frequency, by $|j; \omega\rangle$. The corresponding zero-photon state $|j; \{0\}\rangle$ will usually be written simply as $|j\rangle$.

The model of Fig. 9.1 may thus represent the decay of an excited molecular state with no photons, $|1\rangle = |x, \{0\}\rangle$, to the continuum of states $\{|l\rangle\} = \{|g, 1_{\mathbf{k}}\rangle\}$ that combine the ground molecular state with a continuum of single photon states of the radiation field. The relaxation $|1\rangle \to \{|l\rangle\}$ is then the process of spontaneous emission, and the rate will then yield the radiative relaxation rate of the corresponding excited molecular state, as discussed in detail in Section 3.2.3.

9.2.4 Preparation of the initial state

An important ingredient in our analysis was the characterization of the initial state of the system $|1\rangle$ as a nonstationary state. Otherwise, if $|1\rangle$ was an eigenstate of $\hat{H}$, its time evolution would satisfy $\psi(t) = e^{-i\hat{H}t/\hbar}\psi(0) = e^{-iE_1t/\hbar}|1\rangle$, and $|C_1(t)|^2 = |\langle 1|e^{-i\hat{H}t/\hbar}|1\rangle|^2 = 1$. How can the system be put into such a nonstationary state?

The answer is not unique, but a general statement can be made: *A short time external perturbation exerted on a system in a stationary state (i.e. an eigenstate of the system's Hamiltonian) will generally move the system into a nonstationary state provided that the duration of this perturbation is short relative to $\hbar/\Delta E$, where ΔE is a typical spacing between the system's energy levels in the spectral range of interest.* In what follows we describe a particular example.

Consider a molecule in its ground state ψ_g, an exact eigenstate of the molecular Hamiltonian, subjected to the very short external perturbation $\hat{M}\delta(t)$ (such as caused by a very short radiation pulse, in which case $\hat{M}$ is proportional to the dipole moment operator). From Eq. (2.74) truncated at the level of first-order perturbation theory

$$\Psi_I(t) = \Psi_I(0) - \frac{i}{\hbar}\int_0^t dt_1 \hat{V}_I(t_1)\Psi_I(0) \tag{9.34}$$

we find, using also $\Psi_I(t) = \exp(i\hat{H}_M t/\hbar)\Psi(t)$ (see Eq. 2.70) that[6]

$$\begin{aligned}\Psi(t) &= e^{-(i/\hbar)\hat{H}_M t}\left(\psi_g - \frac{i}{\hbar}\hat{M}\psi_g\right) \\ &= e^{-(i/\hbar)E_g t}\psi_g - \frac{i}{\hbar}e^{-(i/\hbar)\hat{H}_M t}\hat{M}\psi_g; \qquad (t > 0)\end{aligned} \tag{9.35}$$

Therefore, the excited component in the resulting state arises from $\hat{M}\psi_g$. Now, if, because of selection rules, $\langle 1|\hat{M}|g\rangle \neq 0$ but $\langle l|\hat{M}|g\rangle = 0$ for all l, the excited state of the system following this sudden excitation will be the non stationary state $|1\rangle$.

[6] A reader keen on technical details may wonder about an apparently missing factor of $\frac{1}{2}$, since $\int_0^\infty dt\delta(t) = 1/2$. However the proper integral to take starts infinitesimally below zero, since we want the state obtained after the system that started in ψ_g before the onset of the pulse, has experienced the full pulse.

9.3 The effect of relaxation on absorption lineshapes

A very important result of the theory of quantum dynamics is the connection between the time evolution in a given spectral region and the absorption lineshape into the same region. That such a connection exists is to be expected, because the time evolution is determined by the distribution of initial amplitudes among exact eigenstates according to Eq. (2.6), while the absorption process, in principle, prepares these initial amplitudes in the spectral region of interest.

To see this connection in more detail we extend the model of Figs 9.1 and Eq. (9.2) to include two discrete states, the ground state $|g\rangle$ and an excited state $|s\rangle$, and a continuum of states $\{|l\rangle\}$ that may represent the ground state dressed by environmental or radiation field states. We assume that $|s\rangle$ is the only excited state in the relevant spectral region that is radiatively coupled to the ground state $|g\rangle$ so it can be initially prepared as explained in Section 9.2.4. In the subspace that encompasses the state $|s\rangle$ and the continuum $\{|l\rangle\}$, the former plays the same role as state $|1\rangle$ in Fig. 9.1. We now focus on the excitation from g to s; specifically we pose the question: What is the corresponding absorption lineshape?

The molecular model, shown in Fig. 9.2, is now given by

$$\hat{H}_{\text{M}} = \hat{H}_{0\text{M}} + \hat{V} \tag{9.36a}$$

$$H_{0M} = E_g|g\rangle\langle g| + E_s|s\rangle\langle s| + \sum_l E_l|l\rangle\langle l| \tag{9.36b}$$

$$V = \sum_l (V_{s,l}|s\rangle\langle l| + V_{l,s}|l\rangle\langle s|) \tag{9.36c}$$

It should be stated at the outset that the models of Figs 9.1 and 9.2 are too simple for most cases of interest for the simple reason that, following the excitation of any system, at least two relaxation channels are available. We have already argued that every excited molecular state can interact with the continuum of photon-dressed states associated with lower molecular states, leading to spontaneous emission. This is a radiative relaxation channel. In addition there are usually several nonradiative channels where the molecule relaxes to lower states by transferring energy to nonradiative modes such as intramolecular and intermolecular nuclear motions.[7] We will see (see Problem 9.2 below) that extending the model of Fig. 9.1 to more relaxation channels is a simple matter as long as different relaxation processes are independent of each other. We consider first the simple, single channel model, but

[7] However, for excited atoms in collisionless conditions only the radiative relaxation channel is open. Here "collisionless" means that the time between collisions is much longer than the radiative relaxation time.

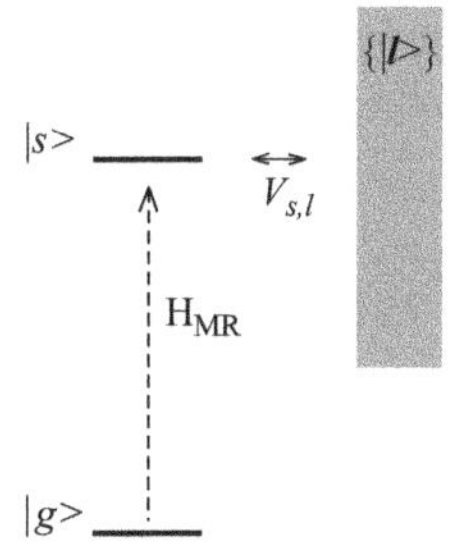

FIG. 9.2 The model of Fig. 9.1, extended to show how a state $|s\rangle$ (equivalent to $|1\rangle$ in Fig. 9.1) may be prepared from the ground state $|g\rangle$ by optical excitation. $\hat{H}_{MR}$, the molecule–radiation field interaction is assumed to couple states $|g\rangle$ and $|s\rangle$ but not states $|g\rangle$ and $|l\rangle$. $\hat{V}$ couples between $|s\rangle$ and states in the manifold $\{|l\rangle\}$ so that if $|s\rangle$ was initially prepared the ensuing time evolution is obtained from the formalism of Section 9.1.

keep in mind that the $|s\rangle \to \{|l\rangle\}$ relaxation can describe different type of relaxation depending on the physical nature of the manifold $\{|l\rangle\}$.

Coming back to the model of Fig. 9.2, we already know that the coupling $\hat{V}$ between the state $|s\rangle$ and the manifold $\{|l\rangle\}$ leads to the decay of $|s\rangle$ following an initial preparation of the system in this state. We are also given that the ground state $|g\rangle$ is coupled radiatively only to $|s\rangle$ but not to $\{|l\rangle\}$, that is, $\langle s|\hat{\mu}|g\rangle \neq 0$ and $\langle l|\hat{\mu}|g\rangle = 0$, where $\hat{\mu}$ is the molecular dipole moment operator. When such situations arise, the state $|s\rangle$ is sometimes referred to as a *doorway state*.

The complete system under consideration now comprises both the molecule and the radiation field, and the corresponding Hamiltonian is

$$\hat{H} = \hat{H}_{0M} + \hat{V} + \hat{H}_R + \hat{H}_{MR} = \hat{H}_0 + \hat{V} + \hat{H}_{MR}; \quad \hat{H}_0 = \hat{H}_{0M} + \hat{H}_R \tag{9.37}$$

As was indicated above, a state of the radiation field is defined by specifying population of each mode, and in particular single photon states (one photon in mode $\mathbf{k}$ of frequency ω, no photons in other modes) will be denoted by $|1_{\mathbf{k}}\rangle$, $|\mathbf{k}\rangle$, or $|\omega\rangle$ as will be convenient. We will sometimes use $|\text{vac}\rangle = |0, \ldots, 0, 0, 0 \ldots\rangle$ to denote the "vacuum" or ground state of the radiation field, that is, the state with no photons.

The *absorption lineshape* corresponds to the photon-energy dependence of the rate at which the photon is absorbed by the molecule. We consider absorption under conditions where it is a linear process, that is, where the rate at which the molecular system absorbs energy from the radiation field at frequency ω is proportional to the radiation intensity (number of photons) at this frequency.[8] Under such conditions it is enough to consider the rate of absorption from a single photon state and to use the

[8] This is the condition of validity of the Beer–Lambert law of absorption.

basis of zero and one-photon eigenstates of the Hamiltonian $\hat{H}_0 = \hat{H}_{0M} + \hat{H}_R$. In particular we are interested in the rate at which the initial 1-photon state $|0\rangle \equiv |g, \mathbf{k}\rangle$ (the molecule in the ground state and the radiation field in the state $|\mathbf{k}\rangle$) of energy

$$E_0 = E_g + \hbar\omega_{\mathbf{k}} \tag{9.38}$$

disappears due to coupling via $\hat{H}_{MR}$ to the state $|s, vac\rangle$, the excited molecular state s with no photons.[9] For simplicity of notation we will use $|s\rangle$ both for the excited molecular state (eigenstate of $\hat{H}_{0M}$) and as a shorthand notation for $|s, vac\rangle$ (eigenstate of $\hat{H}_0 = \hat{H}_{0M} + \hat{H}_R$); the distinction between these entities should be clear from the text.

The full Hamiltonian for the process of interest, written in the dressed state basis is

$$\hat{H} = \hat{H}_0 + \hat{V} + \hat{H}_{MR} \tag{9.39a}$$

$$\hat{H}_0 = E_0|0\rangle\langle 0| + E_s|s\rangle\langle s| + \sum_l E_l|l\rangle\langle l| + \hat{H}_R \tag{9.39b}$$

$$\hat{V} = \sum_l (V_{s,l}|s\rangle\langle l| + V_{l,s}|l\rangle\langle s|) \tag{9.39c}$$

$$\hat{H}_{MR} = \alpha(\mu_{s,g}|s\rangle\langle 0| + \mu_{g,s}|0\rangle\langle s|) \tag{9.39d}$$

and the corresponding level scheme is shown in Fig. 9.3. In Eq. (9.39d) we have used the fact that matrix element of $\hat{H}_{MR}$ between the dressed states $|0\rangle = |g, \mathbf{k}\rangle$ and $|s, vac\rangle$ are proportional to matrix elements of the molecular dipole moment operator $\hat{\mu}$ between the corresponding molecular states $|g\rangle$ and $|s\rangle$, and have written α for the proportionality coefficient. Also for simplicity we disregard the vector nature of $\hat{\mu}$.

Note that in Fig. 9.2 $|s\rangle$ represents a molecular state, while in Fig. 9.3 it stands for the dressed state $|s, \text{vac}\rangle$. Note also that the physical nature of the continuum $\{|l\rangle\}$ and the coupling $V_{s,l}$ depends on the physical process under consideration. In the dressed state picture of Fig. 9.3 this continuum may represent the radiative channel $\{|g, \mathbf{k}\rangle\}$ or a nonradiative channel, for example, $\{|g, \mathbf{v}; \text{vac}\rangle\}$ of vibrational levels $\mathbf{v}$ associated with the electronic ground state g. In the former case the coupling

[9] If, alternatively, we take $|0\rangle = |g, n_k\rangle$, a state with n_k photons in the mode $\mathbf{k}$, then $|s\rangle$ is a state with one less photon than in $|0\rangle$. n_k is a measure of the intensity of the incident beam. One can then show, using Eqs (3.1), (3.70), and (2.157), that α in Eq. (9.39d) is proportional to $\sqrt{n_k}$, so that the rate of absorbing photons, Eq. (9.40), is proportional to n_k. Keeping this in mind it is sufficient to consider the transition from one-photon ground state to zero-photon excited state.

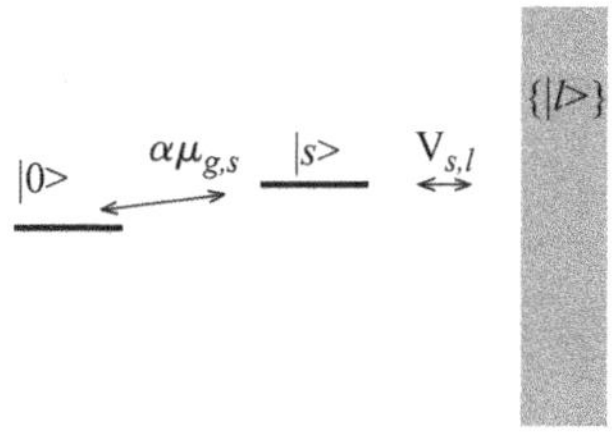

FIG. 9.3 Same as Fig. 9.2, now cast in the dressed states (eigenstates of $\hat{H}_M + \hat{H}_R$) form. $|0\rangle = |g; \mathbf{k}\rangle$ corresponds to the molecule in the ground state with a single photon of mode **k**. $|s\rangle$ describes the molecule in an excited state and the radiation field in its vacuum state. The coupling between $|0\rangle$ and $|s\rangle$ is proportional to the dipole matrix element $\mu_{g,s}$ between the corresponding molecular states.

$\hat{V}$ is identical to $\hat{H}_{MR}$.[10] The exact nature of $\{|l\rangle\}$ is however unimportant for the continuation of our discussion.

We are interested in the rate at which the dressed state $|0\rangle = |g, \mathbf{k}\rangle$, or rather the probability that the system remains in this state, decays because of its coupling to the state $|s, \text{vac}\rangle$ and through it to the continuum $\{|l\rangle\}$. The absorption lineshape is this rate, displayed as a function of $\omega = kc$. This rate is evaluated in Appendix 9B and leads to the following expression for the absorption lineshape

$$L(\omega) \propto \frac{\alpha^2 |\mu_{g,s}|^2 (\Gamma_s/2)}{(E_g + \hbar\omega - \tilde{E}_s)^2 + (\Gamma_s/2)^2} \tag{9.40}$$

This is a Lorentzian centered at a shifted energy of the state s, $\tilde{E}_s = E_s + \Lambda_s$, whose width at half height is Γ_s, where $\Lambda_s = \Lambda_s(E_s)$ and $\Gamma_s = \Gamma_s(E_s)$ are given by Eqs (9.29) and (9.28), respectively (with the subscript 1 replaced by s everywhere).

Problem 9.2. Consider the model where the doorway state $|s\rangle$ is coupled to two different continua, R and L (see Fig. 9.4).

Show that under the same model assumptions used above the absorption lineshape is Lorentzian and the decay rate of state $|s\rangle$ after it is initially prepared is exponential. Also show that the decay rate is $\Gamma_s/\hbar$ and the width of the Lorentzian is Γ_s with

$$\Gamma_s = \Gamma_{s,R} + \Gamma_{s,L} = 2\pi[|V_{s,R}|^2 \rho_R + |V_{s,L}|^2 \rho_L]_{E_s} \tag{9.41}$$

[10] Note the subtle difference between this radiative coupling which is a sum over all modes of the radiation field, and the coupling (9.39d) which involves only the particular mode that enters in state $|0\rangle$.

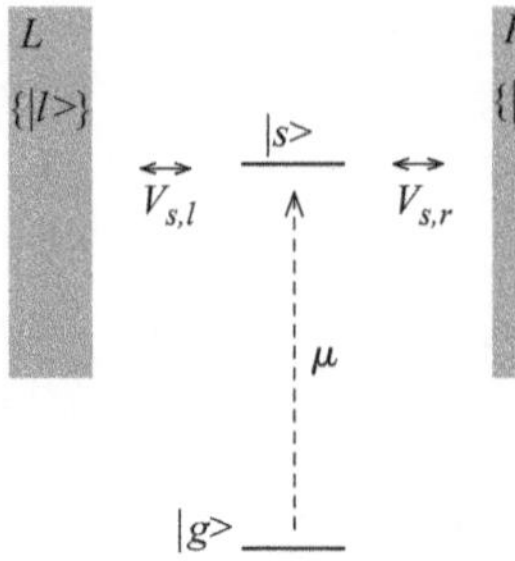

FIG. 9.4 Same as Fig. 9.2, except that the doorway state $|s\rangle$ is now coupled to two relaxation channels represented by the continua L and R.

where R and L stand for the corresponding manifolds and where $[]_{Es}$ has the same meaning with respect to each manifold as in Eq. (9.28).

Also show that, under these conditions, the yield of the "L product" is

$$Y_L = \frac{\Gamma_{s,L}}{\Gamma_{s,L} + \Gamma_{s,R}} \tag{9.42}$$

We end this discussion by noting the following points:

1. The simple exponential decay and Lorentzian lineshape obtained above result from the simple model assumptions used, in particular the infinite energy extent of the continuum $\{|l\rangle\}$ and the weak dependence on the energy E of $\Gamma_s(E)$ and $\Lambda_s(E)$.
2. In the procedure (Appendix 9B) to evaluate the lineshape (9.40) we use the representation defined by the states $\{|j\rangle\}$ that diagonalize the Hamiltonian in the ($|s\rangle$, $\{|l\rangle\}$) subspace. Of course any basis can be used for a mathematical analysis. It was important and useful to state the *physical* problem in terms of the zero-order states $|s\rangle$ and $\{|l\rangle\}$ because an important attribute of the model was that in the latter representation the ground state $|g\rangle$ is coupled by the radiation field only to the state $|s\rangle$, which therefore has the status of a *doorway* state. This state is also referred to as a *resonance* state, a name used for the spectral feature associated with an underlying picture of a discrete zero-order state embedded in and coupled to a continuous manifold of such states.
3. For the initial value problem with $\psi(t = 0) = |s\rangle$ we got an exponential decay with the characteristic relaxation rate $k_s = \Gamma_s/\hbar$. For the absorption lineshape into state $|s\rangle$ we got a Lorentzian with linewidth given by the same Γ_s. There appears to be a fundamental relationship between the lifetime

and linewidth associated with such resonance state. More specifically, these results, Eqs (9.25) and (9.40), have the characteristic of the Heisenberg uncertainty principle: The lifetime $(k_s)^{-1}$ of a level (in a sense the minimum uncertainty in the time at which it may be observed) and the linewidth Γ_s associated with this level (the minimum uncertainty in its energy) are related by $(k_s)^{-1}\Gamma_s = \hbar$.

4. Is the last observation an inherently quantum-mechanical result? A negative answer is obtained from the calculation (Section 8.2.4) of the lineshape associated with a classical underdamped harmonic oscillator. The normalized lineshape is obtained (see Eq. (8.40) in the form

$$\tilde{L}(\omega) = \frac{\gamma/\pi}{(\omega_0 - \omega)^2 + \gamma^2} \tag{9.43}$$

where γ, the linewidth in this classical result is the friction coefficient, that is, the rate at which the oscillator loses energy by friction with it environment. In both quantum and classical systems the linewidth is seen to be just the inverse of the relaxation time. In fact, the only quantum element seen in Eq. (9.40) is the association of the energy E with the frequency ω through the Planck relationship, $E = \hbar\omega$. Otherwise these systems have common characteristics, and intuition obtained from classical harmonic oscillator problems is often useful for the corresponding quantum two-level problems. This useful analogy breaks when the dynamics of the two-level system involves saturation, a common situation in pumped two-level systems that does not have an equivalent in the harmonic oscillator case.

5. We have seen (Section 6.2.3) that a Lorentzian lineshape corresponds to an exponentially decaying dipole autocorrelation function. For the Hamiltonian of Eqs (9.36) and (9.39) this correlation function is $C_\mu(t) = \langle g|e^{i\hat{H}_\mathrm{M}t/\hbar}\hat{\mu}e^{-i\hat{H}_\mathrm{M}t/\hbar}\hat{\mu}|g\rangle = e^{iE_g t/\hbar}\langle g|\hat{\mu}e^{-i\hat{H}_\mathrm{M}t/\hbar}\hat{\mu}|g\rangle = \sum_j e^{i(E_g-E_j)t/\hbar}|\langle g|\hat{\mu}|j\rangle|^2 = |\langle g|\hat{\mu}|s\rangle|^2 \sum_j |\langle s|j\rangle|^2 e^{i(E_g-E_j)t/\hbar}$, where the states $|j\rangle$ are exact eigenstates of $\hat{H}_\mathrm{M}$. The reader may attempt to show that the same conditions that lead to exponential relaxation of state $|s\rangle$ after it is initially prepared also imply that $|C_\mu(t)|$ is an exponentially decaying function, both with the same decay rate $\Gamma_s/\hbar$.

The quantum relaxation problems discussed above correspond to zero temperature situations. This is seen from the fact that the initial population of level $|s\rangle$ was not obtained thermally, otherwise the levels $\{l\}$ in the same energy neighborhood would be equally populated. The fact that these levels carry zero probability at $t = 0$ is a manifestation of zero temperature. In the next section we consider another quantum relaxation problem, the relaxation of a quantum oscillator coupled to a thermal bath of similar oscillators, where finite temperature effects are taken into account.

9.4 Relaxation of a quantum harmonic oscillator

We next consider another example of quantum-mechanical relaxation. In this example an isolated harmonic mode, which is regarded as our system, is weakly coupled to an infinite bath of other harmonic modes. This example is most easily analyzed using the boson operator formalism (Section 2.9.2), with the Hamiltonian

$$\hat{H} = \hbar\omega_0 \hat{a}^\dagger \hat{a} + \hbar \sum_j \omega_j \hat{b}_j^\dagger \hat{b}_j + \hbar \sum_j \{u_j \hat{a}^\dagger \hat{b}_j + u_j^* \hat{a} \hat{b}_j^\dagger\} \tag{9.44}$$

The first two terms on the right describe the system and the "bath", respectively, and the last term is the system–bath interaction. This interaction consists of terms that annihilate a phonon in one subsystem and simultaneously create a phonon in the other.[11] The creation and annihilation operators in Eq. (9.44) satisfy the commutation relations:

$$[\hat{a}, \hat{a}^\dagger] = 1; \quad [\hat{a}, \hat{a}] = 0; \quad [\hat{a}^\dagger, \hat{a}^\dagger] = 0; \tag{9.45a}$$

$$\hat{a}, \hat{a}^\dagger \text{ commutes with all } \hat{b}, \hat{b}^\dagger$$

$$[\hat{b}_j, \hat{b}_j^\dagger] = 1; \quad [\hat{b}_j, \hat{b}_j] = 0; \quad [\hat{b}_j^\dagger, \hat{b}_j^\dagger] = 0; \tag{9.45b}$$

$$\hat{b}_j, \hat{b}_j^\dagger \text{ commutes with all } \hat{b}_{j'}, \hat{b}_{j'}^\dagger \text{ for } j \neq j'$$

The Heisenberg equations of motion, $\dot{\hat{A}} = (i/\hbar)[\hat{H}, \hat{A}]$ for the Heisenberg-representation operators $a(t)$ and $b(t)$ are derived using these commutations

[11] Transforming to coordinate and momentum operators using Eqs (2.152), the interaction term in (9.44) is seen to depend on the momenta. A more standard interaction expressed in terms of the coordinates only, say $x_1 x_2$, when transformed into the creation and annihilation operator representation will contain the four products $a_1^\dagger a_2, a_1 a_2^\dagger, a_1^\dagger a_2^\dagger$, and $a_1 a_2$. The neglect of the last two terms in Eq. (9.44) is known as the *rotating wave approximation* (RWA). (See also Section 3.2.2 and the derivation of Eq. (3.28).) It is justified for weak coupling by the observation that such terms cannot conserve energy in low order. The use of this approximation in the present context should be exercised with caution: It can be shown that for small ω_0 the lowest eigenvalues of this Hamiltonian imply imaginary frequencies. Still, the treatment presented here should serve as an introduction to the somewhat more involved treatment needed if the RWA is avoided (see K. Lindenberg, and B. J. West, Phys. Rev. A, **30**, 568–582 (1984) and G. W. Ford, , J. T. Lewis, et al., Phys. Rev. A , **37**, 4419–4428 (1988). These references treat the same problem without resorting to the RWA).

relations. We get:

$$\dot{\hat{a}}(t) = -i\omega_0\hat{a}(t) - i\sum_j u_j\hat{b}_j(t) \tag{9.46a}$$

$$\dot{\hat{b}}_j(t) = -i\omega_j\hat{b}_j(t) - iu_j^*\hat{a}(t) \tag{9.46b}$$

The initial conditions at $t = 0$ are the corresponding Schrödinger operators. This model is seen to be particularly simple: All operators in Eq. (9.46) commute with each other, therefore this set of equations can be solved as if these operators are scalars.

Note that Eqs (9.46) are completely identical to the set of equations (9.6) and (9.7). The problem of a single oscillator coupled linearly to a set of other oscillators that are otherwise independent is found to be isomorphic, in the rotating wave approximation, to the problem of a quantum level coupled to a manifold of other levels. There is one important difference between these problems though. Equations (9.6) and (9.7) were solved for the initial conditions $C_0(t = 0) = 1$, $C_l(t = 0) = 0$, while here $\hat{a}(t = 0)$ and $\hat{b}_j(t = 0)$ are the Schrödinger representation counterparts of $\hat{a}(t)$ and $\hat{b}_j(t)$. Still, Eqs (9.46) can be solved by Laplace transform following the route used to solve (9.6) and (9.7).

In what follows we take a different route (that can be also applied to (9.6) and (9.7)) that sheds more light on the nature of the model assumptions involved. We start by writing the solution of Eq. (9.46b) in the form

$$\hat{b}_j(t) = \hat{b}_j(0)e^{-i\omega_j t} - iu_j^* \int_0^t d\tau e^{-i\omega_j(t-\tau)}\hat{a}(\tau) \tag{9.47}$$

Inserting this into the equation for a yields

$$\dot{\hat{a}}(t) = -i\omega_0\hat{a} - i\sum_j u_j\hat{b}_j(0)e^{-i\omega_j t} - \sum_j |u_j|^2 \int_0^t d\tau e^{-i\omega_j(t-\tau)}\hat{a}(\tau) \tag{9.48}$$

which, by transforming according to $\hat{a} = \hat{\bar{a}}e^{-i\omega_0 t}$, becomes

$$\dot{\hat{\bar{a}}}(t) = -i\sum_j u_j\hat{b}_j(0)e^{-i(\omega_j-\omega_0)t} - \int_0^t d\tau\hat{\bar{a}}(\tau)S(\omega_0, t-\tau) \tag{9.49}$$

with

$$S(\omega_0, t) = \sum_j |u_j|^2 e^{-i(\omega_j-\omega_0)t} \tag{9.50}$$

Note that interchanging the order of the finite integration and the, in principle, infinite series S is permitted provided the latter converges, and we assume that it does. In fact, we will assume that $S(t)$ vanishes everywhere except near $t = 0$ and that this range near $t = 0$ is small enough so that (1) in (9.49) $a(\tau)$ may be taken as a constant, $a(t)$, out of the integral and (2) the lower limit of integration can be extended to $-\infty$. This leads to

$$\dot{\hat{a}}(t) = -i\sum_j u_j\hat{b}_j(0)e^{-i(\omega_j-\omega_0)t} - \hat{a}(t)\int_0^\infty d\tau S(\omega_0,\tau) \tag{9.51}$$

where we have further used $\int_{-\infty}^{t} d\tau S(t-\tau) = \int_0^\infty d\tau S(\tau)$.

What is the justification for the assumption that $S(t)$ vanishes unless t is very close to zero? For an answer let us rewrite Eq. (9.50) in the form

$$\begin{aligned} S(\omega_0,t) &= \int_{-\infty}^{\infty} d\omega e^{-i(\omega-\omega_0)t}\sum_j |u_j|^2\delta(\omega-\omega_j) \equiv \int_{-\infty}^{\infty} d\omega e^{-i(\omega-\omega_0)t}C(\omega) \\ &= \int_{-\infty}^{\infty} d\omega e^{i\omega t}C(\omega_0-\omega) \end{aligned} \tag{9.52}$$

The function $S(\omega_0,t)$ is seen to be the Fourier transform of the *coupling density*[12]

$$C(\omega) \equiv \sum_j |u_j|^2\delta(\omega-\omega_j) = |u(\omega)|^2 g(\omega) \tag{9.53}$$

where $|u(\omega)|^2 = \overline{(|u_j|^2)_{\omega_j=\omega}}$, with the bar denoting an average over intervals of ω that are large relative to the spacing between subsequent ω_j's, and where $g(\omega)$ is the density of modes at frequency ω, defined by Eq. (4.32). The second equality in (9.53) becomes exact in the infinite bath limit where the spectrum of normal mode frequencies is continuous.

Consider now Eq. (9.52). The behavior of S as a function of time depends on the behavior of $C(\omega)$ about ω_0. If $C(\omega)$ was constant in all range $-\infty < \omega < \infty$ we could take it out of the integral in (9.52) to get $S(\omega_0,t) = 2\pi C\delta(t)$. This constitutes the *wide band approximation*. In reality $C(\omega)$ may be different from zero, and approximated by a constant, only in some finite frequency interval about

[12] In Section 6.5.2 we introduce the closely related *spectral density* function, $J(\omega) = \pi g(\omega)u^2(\omega)/(2\omega)$.

ω_0 characterized by a width ω_c. This leads to a function $S(\omega_0, t)$ that vanishes only beyond some finite value of t, of order ω_c^{-1}. For example, if $C(\omega) \sim e^{-((\omega-\omega_0)/\omega_c)^2}$ then $S(\omega_0, t) \sim e^{-(\omega_c t/2)^2}$. Therefore, to be able to approximate $\tilde{a}(\tau)$ in (9.49) by $\tilde{a}(t)$ we need to assume that $\tilde{a}(\tau)$ does not change appreciably during the time interval of order $(\omega_c)^{-1}$. What helps at this point is the fact that we have already eliminated the fast oscillations $e^{-i\omega_0 t}$ by the transformation $a \rightarrow \tilde{a}$. The remaining time dependence of $\tilde{a}(t)$ stems from the relaxation induced by the bath of b modes, so what we assume in effect is that this relaxation is slow relative to $(\omega_c)^{-1}$—a weak coupling approximation. When this assumption holds, Eq. (9.49) may indeed be approximated by (9.51).

Next consider the function

$$F(\omega_0) = \int_0^\infty d\tau S(\omega_0, \tau) = \int_0^\infty d\tau \sum_j |u_j|^2 e^{-i(\omega_j-\omega_0)\tau} \tag{9.54}$$

Since the integrand is strongly peaked about $\tau = 0$, we may multiply it by a factor $e^{-\eta\tau}$ with a very small positive η without affecting the result. This, however, makes it possible to perform the τ integral before the summation, leading to (in analogy to the treatment that leads from (9.17) to (9.27)–(9.29))

$$F(\omega_0) = \lim_{\eta\to 0}\left(i\sum_j \frac{|u_j|^2}{\omega_0 - \omega_j + i\eta}\right) = i\delta\omega_0 + \frac{1}{2}\gamma \tag{9.55}$$

where

$$\delta\omega_0 \equiv \mathrm{PP}\int d\omega \frac{|u(\omega)|^2 g(\omega)}{\omega_0 - \omega} \tag{9.56}$$

is the principal part integral defined in Section 1.1.6 and where

$$\gamma \equiv 2\pi C(\omega) = 2\pi(|u(\omega)|^2 g(\omega))_{\omega=\omega_0} \tag{9.57}$$

Equation (9.52) now becomes

$$\dot{\tilde{a}}(t) = (-i\delta\omega_0 - (1/2)\gamma)\tilde{a}(t) - i\sum_j u_j \hat{b}_j(0) e^{-i(\omega_j-\omega_0)t} \tag{9.58}$$

which is equivalent to (putting $\tilde{\omega}_0 = \omega_0 + \delta\omega_0$)

$$\dot{\hat{a}}(t) = -i(\tilde{\omega}_0 - (1/2)i\gamma)\hat{a}(t) - i\sum_j u_j \hat{b}_j(0) e^{-i\omega_j t} \tag{9.59}$$

We have found that the linear coupling of the mode ω_0 to the continuum of modes $\{\omega_j\}$ leads to a shift $\delta\omega_0$ in ω_0 as well as to an imaginary contribution, $(1/2)i\gamma$ to this frequency. The latter amounts to a damping effect. In addition the coupling leads to an inhomogeneous time-dependent term in the equation of motion for $\hat{a}(t)$ that brings in the effect of the free time evolution of all the bath modes. These three effects: A frequency shift, damping, and a time-dependent function of the free bath motion (to be later interpreted as "noise" exerted by the bath) constitute the essence of the effect of coupling to the bath on the motion of our system. We have seen (Section 8.2) that similar effects characterize the behavior of the equivalent classical system in a formalism that leads to the *Langevin equation* for the evolution of a system interacting with its thermal environment. Indeed, Eq. (9.59) is an example of a *quantum Langevin equation.*

Using the solution of $\dot{y}(t) = -ky + f(t)$ in the form $y(t) = y(0)e^{-kt} + \int_0^t dt' e^{-k(t-t')} f(t')$ we get the solution of Eq. (9.59)

$$\hat{a}(t) = e^{-i\tilde{\omega}_0 t-(1/2)\gamma t}\hat{a}(0) + \sum_j u_j \frac{e^{-i\tilde{\omega}_0 t-(1/2)\gamma t} - e^{-i\omega_j t}}{\tilde{\omega}_0 - \omega_j - (1/2)i\gamma}\hat{b}_j(0) \tag{9.60a}$$

and, taking the complex conjugate

$$\hat{a}^\dagger(t) = e^{i\tilde{\omega}_0 t-(1/2)\gamma t}\hat{a}^\dagger(0) + \sum_j u_j^* \frac{e^{i\tilde{\omega}_0 t-(1/2)\gamma t} - e^{i\omega_j t}}{\tilde{\omega}_0 - \omega_j + (1/2)i\gamma}\hat{b}_j^\dagger(0) \tag{9.60b}$$

This is our final result. In comparison with the result obtained for the decay of a prepared state, Eq. (9.24a), we see that the essential difference lies in the fact that in that problem the initial condition $C_l(t = 0) = 0$ was naturally used and therefore did not appear in the final result, while here, equally naturally, $b_j(t = 0) \neq 0$.

Problem 9.3. Show that $\hat{a}(t)$ and $a^\dagger(t)$ give by Equations (9.60) satisfy the commutation relations (9.45a)

To see the significance of this result, consider the time evolution of the average population of the system oscillator, $\langle n(t)\rangle = \langle a^\dagger(t)a(t)\rangle$. In the spirit of the Heisenberg representation of time-dependent quantum mechanics, this average is over the initial state of the system. From Eqs (9.60) we see that four averages are encountered. First $\langle a^\dagger(t = 0)b_j(t = 0)\rangle = \langle b_j^\dagger(t = 0)a(t = 0)\rangle = 0$ express an assumption that initially the system and bath are uncorrelated, so that, for example, $\langle a^\dagger(t = 0)b(t = 0)\rangle = \langle a^\dagger(t = 0)\rangle\langle b(t = 0)\rangle = 0$. (The equalities $\langle b(t = 0)\rangle = \langle b^\dagger(t = 0)\rangle = 0$ reflect the fact that the bath is initially at thermal equilibrium (see Eq. (2.197)). Second, $\langle a^\dagger(t=0)a(t=0)\rangle = n_0$ where n_0 is the initial state of the system oscillator. Finally, $\langle b_j^\dagger(t=0)b_j(t=0)\rangle = \langle n_j\rangle_T$, where

$$\langle n_j\rangle_T = \langle n(\omega_j)\rangle_T = \frac{1}{e^{(\hbar\omega_j/k_BT)}-1} \tag{9.61}$$

also expresses the model assumption that the bath is initially at thermal equilibrium.

Using Eqs (9.60) we now get

$$\begin{aligned}\langle n(t)\rangle =& n_0 e^{-\gamma t} + \sum_j \frac{|u_j|^2\langle n_j\rangle_T}{(\tilde{\omega}_0-\omega_j)^2+((1/2)\gamma)^2} \\ &\times (1+e^{-\gamma t}-2e^{-(1/2)\gamma t}\cos[(\tilde{\omega}_0-\omega_j)t])\end{aligned} \tag{9.62}$$

Consider first the $t\to\infty$ limit. In this case

$$\begin{aligned}\langle n(t\to\infty)\rangle &= \sum_j \frac{|u_j|^2\langle n_j\rangle}{(\tilde{\omega}_0-\omega_j)^2+((1/2)\gamma)^2} \\ &= \int d\omega_j \langle n(\omega_j)\rangle \frac{\gamma/2\pi}{(\tilde{\omega}_0-\omega_j)^2+((1/2)\gamma)^2} \\ &\cong \langle n(\tilde{\omega}_0)\rangle_T \int d\omega \frac{\gamma/2\pi}{(\tilde{\omega}_0-\omega)^2+((1/2)\gamma)^2} = \langle n(\tilde{\omega}_0)\rangle_T \\ &= \frac{1}{e^{(\hbar\tilde{\omega}_0/k_BT)}-1}\end{aligned} \tag{9.63}$$

In the last steps we have again used the assumption that γ is small so that the Lorentzian in the integrand is strongly peaked about ω_0.

The same approximation can be applied to the second term in Eq. (9.62). Using

$$\int_{-\infty}^{\infty} d\omega \frac{\gamma/2\pi}{\omega^2+((1/2)\gamma)^2}\cos(\omega t) = e^{-(1/2)\gamma|t|} \tag{9.64}$$

then leads to

$$\langle n(t)\rangle = n_0 e^{-\gamma t} + \langle n\rangle_T(1-e^{-\gamma t}) \tag{9.65}$$

We have found that due to its coupling with the thermal bath our system relaxes to a final thermal equilibrium at temperature T irrespective of its initial state. The relaxation process is exponential and the rate is given by γ, Eq. (9.57). Note that in the model studied this rate does not depend on the temperature.[13]

It is remarkable that with relatively modest theoretical tools we have been able to account for the problem of thermal relaxation from a microscopic approach. Still, one should keep in mind the approximations made above when trying to relate these results to the real world. Our main assumption (made in the paragraph above Eq. (9.54)) was that $C(\omega)$ is finite and fairly constant in a sizable neighborhood about ω_0. In particular, we have assumed that γ is much smaller than the extent of this neighborhood. $C(\omega)$ in turn is dominated by the mode density $g(\omega)$ which in crystals can be quite structured as a function of ω. Our theory will fail if ω_0 is very close to 0 or to a sharp feature in $g(\omega)$.

A most important observation for molecular vibrational relaxation is the fact that molecular frequencies are usually larger than the upper cutoff ω_D beyond which $g(\omega)$ and subsequently $C(\omega)$ vanish (see Section 4.2.4). The existence of such a cutoff, which is a direct consequence of the discrete nature of matter, implies that by the theory presented above the relaxation rates of most molecular vibrations in monoatomic environments vanish.[14] Indeed, it is found experimentally that relaxation processes in which the "system" frequency is smaller than the host cutoff frequency are much faster than those in which the opposite is true. However, it is also found that the rate of the latter processes is not zero. This implies the existence of relaxation mechanisms not described by the model presented by Eq. (9.44). We will come back to this issue in Chapter 13.

Finally, it is also interesting to compare the result (9.65) to the result (8.106) of the very different semiclassical formalism presented in Section (8.3.3). If we identify γ of the present treatment with the factor $ZQ_{01}\kappa$ Eq. (8.96)[15] the two results are identical for $\varepsilon = \hbar\omega \gg k_BT$. The rotating wave approximation used in the model (9.44) cannot reproduce the correct result in the opposite, classical, limit. Most studies of vibrational relaxation in molecular systems are done at temperatures considerably lower than ε/k_B, where both approaches predict temperature-independent relaxation. We will see in Chapter 13 that temperature-dependent rates that are often observed experimentally are associated with anharmonic interactions that often dominate molecular vibrational relaxation.

---

[13] This holds as long as $\hbar\omega > k_BT$. In the opposite, classical, limit the rotating wave approximation invoked here cannot be used. This can be seen by comparing Eq. (9.65) to Eqs (8.104) and (8.106).

[14] Polyatomic solids have of course high frequencies associated with their intramolecular motions.

[15] Indeed, both represent, in their corresponding models, the zero temperature transition rate from level $n = 1$ to level $n = 0$ of the harmonic oscillator.

9.5 Quantum mechanics of steady states

Both experimentally and theoretically, the study of dynamical processes can proceed along two main routes. We can either monitor the time evolution of a system after it starts at $t = 0$ in some nonequilibrium state and follow its relaxation to equilibrium, or we can observe the system under the influence of some force (or forces)[16] and monitor fluxes that develop in response to these forces. Equation (9.24) is an answer to a problem of the first kind, mathematically a solution to a given initial value problem. Even though not formulated in this way, Eq. (9.40) is an answer to a problem of the second kind, giving the flux going from the ground to an excited molecular state that results from driving by an external electromagnetic field. The purpose of this section is to formalize the treatment of quantum dynamical problems of the second kind.

9.5.1 Quantum description of steady-state processes

The time-dependent Schrödinger equation can be evaluated to yield stationary solutions of the form

$$\Psi(\mathbf{r}, t) = \psi_k(\mathbf{r}) \exp(-(i/\hbar)E_k t) \tag{9.66}$$

leading to the time-independent Schrödinger equation for the eigenfunctions ψ_k and the eigenvalues E_k. Alternatively, it can be solved as an initial value problem that yields $\psi(r, t)$ given $\psi(r, t = 0)$. In both cases the solutions are obtained under given boundary conditions. Note that the word "stationary" applied to Eq. (9.66) does not imply that this solution is time-independent, only that observables associated with it are constant in time. For closed systems in the absence of external forces, another important attribute of the states is that they carry no flux. Both attributes also characterize classical equilibrium states.

In classical physics we are familiar with another kind of stationary states, so-called steady states, for which observables are still constant in time however fluxes do exist. A system can asymptotically reach such a state when the boundary conditions are not compatible with equilibrium, for example, when it is put in contact with two heat reservoirs at different temperatures or matter reservoirs with different chemical potentials. Classical kinetic theory and nonequilibrium statistical mechanics deal with the relationships between given boundary conditions and the resulting steady-state fluxes. The time-independent formulation of scattering theory is in fact a quantum theory of a similar nature (see Section 2.10).

[16] A "force" should be understood here in a generalized way as any influence that drives the system away from equilibrium.

In addition to studying actual steady-state phenomena, it is sometime useful to use them as routes for evaluating rates. Consider, for example, the first-order reaction $A \to P$ and suppose we have a theory that relates $A(t)$ to $A(t=0)$. A rate coefficient can then be defined by $k(t) = -t^{-1}\ln\{[A(t=0)]^{-1}A(t)\}$, though its usefulness is usually limited to situations where k is time-independent, that is, when A obeys first-order kinetics, $A(t) \sim \exp(-kt)$, at least for long times. In the latter case we may consider the steady state that is established when A is restricted to be constant while P is restricted to be zero (these restrictions may be regarded as boundary conditions), implying that a constant current, $J = kA$, exists in the system. A theory that relates the constants A and J in such a steady state is therefore a route for finding k. The approximate evaluation of the rate associated with the Lindemann mechanism of chemical reactions (see Section 14.2) is a simple example of such a procedure. Less trivial applications of the same idea are found in Section 14.4.

What is the quantum mechanical analog of this approach? Consider the simple example that describes the decay of a single level coupled to a continuum, Fig. 9.1 and Eq. (9.2). The time-dependent wavefunction for this model is $\psi(t) = C_1(t)|1\rangle + \sum_l C_l(t)|l\rangle$, where the time-dependent coefficients satisfy (cf. Eqs (9.6) and (9.7))

$$\begin{aligned} \hbar\frac{d}{dt}C_1 &= -iE_1C_1 - i\sum_l V_{1,l}C_l \\ \hbar\frac{d}{dt}C_l &= -iE_lC_l - iV_{l,1}C_1; \qquad \text{all } l \end{aligned} \tag{9.67}$$

The result (9.24) is obtained by solving this as an initial value problem, given that $C_1(t=0) = 1$. Alternatively, suppose that the population in state $|1\rangle$ remains always constant so that $C_1(t) = c_1\exp(-(i/\hbar)E_1t)$. In this case the first equation in (9.67) is replaced by Eq. (9.68a) below, where we have also supplemented the second equation by an infinitesimal absorbing term, so that

$$\hbar\frac{d}{dt}C_1 = -iE_1C_1 \;\Rightarrow\; C_1(t) = c_1\exp(-(i/\hbar)E_1t) \tag{9.68a}$$

$$\begin{aligned} \hbar\frac{d}{dt}C_l &= -iE_lC_l - iV_{l,1}C_1(t) - (1/2)\eta C_l \\ &= -iE_lC_l - iV_{l,1}c_1\exp(-(i/\hbar)E_1t) - (1/2)\eta C_l \end{aligned} \tag{9.68b}$$

η will be put to zero at the end of the calculation. Equation (9.68b) admits a steady-state solution of the form

$$C_l(t) = c_le^{-(i/\hbar)E_1t} \tag{9.69}$$

where

$$c_l = \frac{V_{l,1}c_1}{E_1 - E_l + i\eta/2} \tag{9.70}$$

This steady-state solution results from the balance in Eq. (9.68b) between the "driving term" $V_{l,1}c_1 \exp(-(i/\hbar)E_1 t)$ that pumps the amplitude of state l and the term $(1/2)\eta C_l$ that damps it. Note that at steady state it is the observable $|C_l|^2$, not the amplitude $C_l(t)$, which remains constant in time. The total flux through the system in this steady state can be calculated by observing that it must be equal to the rate at which population disappears in the continuum

$$\begin{aligned} J =&(\eta/\hbar)\sum_l |C_l|^2 = |C_1|^2 \sum_l |V_{l,1}|^2 \frac{\eta/\hbar}{(E_1 - E_l)^2 + (\eta/2)^2} \\ &\xrightarrow{\eta \to 0} |C_1|^2 \frac{2\pi}{\hbar} \sum_l |V_{l,1}|^2 \delta(E_1 - E_l) \end{aligned} \tag{9.71}$$

This flux corresponds to the steady-state rate

$$k = \frac{J}{|C_1|^2} = \frac{2\pi}{\hbar} \sum_l |V_{l,1}|^2 \delta(E_1 - E_l) = \frac{2\pi}{\hbar} (\overline{|V_{l,1}|^2 \rho_L})_{E_l = E_1} = \Gamma_1/\hbar \tag{9.72}$$

This simple steady-state argument thus leads to the same golden rule rate expression, Eq. (9.25), obtained before for this model.

Let us now repeat the same derivation for the slightly more complicated example described by the Hamiltonian

$$\hat{H} = \hat{H}_0 + \hat{V} \tag{9.73}$$

$$\hat{H}_0 = E_0|0\rangle\langle 0| + E_1|1\rangle\langle 1| + \sum_l E_l|l\rangle\langle l| + \sum_r E_r|r\rangle\langle r| \tag{9.74}$$

$$\begin{aligned} \hat{V} =&V_{0,1}|0\rangle\langle 1| + V_{1,0}|1\rangle\langle 0| + \sum_l (V_{l,1}|l\rangle\langle 1| + V_{1,l}|1\rangle\langle l|) \\ &+ \sum_r (V_{r,1}|r\rangle\langle 1| + V_{1,r}|1\rangle\langle r|) \end{aligned} \tag{9.75}$$

Here it is the level $|0\rangle$ that is taken as the "driving state," and the flux is carried through another level $|1\rangle$ coupled to two continua, $L = \{l\}$ and $R = \{r\}$. Looking for a solution to the time-dependent Schrödinger equation of the form

$$\psi(t) = C_0(t)|0\rangle + C_1(t)|1\rangle + \sum_l C_l(t)|l\rangle + \sum_r C_r(t)|r\rangle \tag{9.76}$$

the equations equivalent to (9.67) are

$$\begin{aligned}\hbar\dot{C}_0 &= -iE_0C_0 - iV_{0,1}C_1\\ \hbar\dot{C}_1 &= -iE_1C_1 - iV_{1,0}C_0 - i\textstyle\sum_l V_{1,l}C_l - i\sum_r V_{1,r}C_r\\ \hbar\dot{C}_l &= -iE_lC_l - iV_{l,1}C_1\\ \hbar\dot{C}_r &= -iE_rC_r - iV_{r,1}C_1\end{aligned} \tag{9.77}$$

while those corresponding to (9.67) are

$$\hbar\dot{C}_0 = -iE_0C_0 \;\Rightarrow\; C_0(t) = c_0e^{-(i/\hbar)E_0t} \tag{9.78a}$$

$$\hbar\dot{C}_1 = -iE_1C_1 - iV_{1,0}c_0e^{-(i/\hbar)E_0t} - i\sum_l V_{1,l}C_l - i\sum_r V_{1,r}C_r \tag{9.78b}$$

$$\hbar\dot{C}_l = -iE_lC_l - iV_{l,1}C_1 - (\eta/2)C_l \tag{9.78c}$$

$$\hbar\dot{C}_r = -iE_rC_r - iV_{r,1}C_1 - (\eta/2)C_r \tag{9.78d}$$

At $t \to \infty$ we again reach a steady state where the amplitudes are $C_j(t) = c_j\exp(-(i/\hbar)E_0t)$ ($j = 0, 1, \{l\}, \{r\}$), and the coefficients c_j satisfy

$$0 = i(E_0 - E_1)c_1 - iV_{1,0}c_0 - i\sum_l V_{1,l}c_l - i\sum_r V_{1,r}c_r \tag{9.79a}$$

$$0 = i(E_0 - E_l)c_l - iV_{l,1}c_1 - (\eta/2)c_l \tag{9.79b}$$

$$0 = i(E_0 - E_r)c_r - iV_{r,1}c_1 - (\eta/2)c_r \tag{9.79c}$$

The solution of (9.79c)

$$c_r = \frac{V_{r,1}c_1}{E_0 - E_r + i\eta/2} \tag{9.80}$$

is now substituted in the last term of (9.79a). Repeating procedures from Section 9.1 (compare (9.17), (9.19), (9.27)–(9.29)), we have

$$-i\sum_r V_{1,r}c_r \equiv -iB_{1R}(E_0)c_1 \tag{9.81}$$

$$\begin{aligned}B_{1R}(E) &\equiv \lim_{\eta\to 0}\sum_r \frac{|V_{1r}|^2}{E - E_r + i\eta/2} = \Lambda_{1R}(E) - (1/2)i\Gamma_{1R}(E)\\ \Gamma_{1R}(E) &= 2\pi\left(\overline{|V_{1r}|^2\rho_R(E_r)}\right)_{E_r=E}\\ \Lambda_{1R}(E) &= \mathrm{PP}\int_{-\infty}^{\infty} dE_r\frac{\overline{|V_{1r}|^2}\rho_R(E_r)}{E - E_r}\end{aligned} \tag{9.82}$$

$B_{1R}(E)$ is the self energy of level $|1\rangle$ due to its interaction with the continuum R. Similar results, with L, l replacing R, r, are obtained by inserting the solution for c_l from Eq. (9.79b) into (9.79a), leading to an additional contribution $B_{1L}(E)$ to the self energy of this level due to its interaction with the continuum L. Using these results in (9.79a) leads to

$$c_1 = \frac{V_{1,0}c_0}{E_0 - \tilde{E}_1 + (i/2)\Gamma_1(E_0)} \tag{9.83}$$

where

$$\Gamma_1(E) = \Gamma_{1L}(E) + \Gamma_{1R}(E) \tag{9.84}$$

and

$$\tilde{E}_1 = \tilde{E}_1(E) = E_1 + \Lambda_{1R}(E) + \Lambda_{1L}(E) \tag{9.85}$$

Using (9.83) in (9.79b) and (9.79c) now yields

$$|c_r|^2 = |C_r|^2 = \frac{|V_{r,1}|^2}{(E_0 - E_r)^2 + (\eta/2)^2} \frac{|V_{1,0}|^2|c_0|^2}{(E_0 - \tilde{E}_1)^2 + (\Gamma_1(E_0)/2)^2} \tag{9.86}$$

Equation (9.86), and the equivalent expression (with r replaced by l everywhere) for $|c_l|^2 = |C_l|^2$ give the steady-state population of individual levels, r and l, in the continua. This steady state was obtained by assigning to each such level a decay rate η. Therefore, the total steady-state flux out of the system through the continuum (channel) R is, in analogy to (9.71)

$$J_{0\to R} = (\eta/\hbar)\sum_r |c_r|^2 \xrightarrow{\eta\to 0} \frac{|V_{1,0}|^2}{(E_0 - \tilde{E}_1)^2 + (\Gamma_1(E_0)/2)^2} \frac{\Gamma_{1R}(E_0)}{\hbar}|c_0|^2 \tag{9.87}$$

and the corresponding steady-state rate is $J_{0\to R}/|c_0|^2$. Again similar results are obtained for the channel L, so finally

$$k_{0\to K} = \frac{J_{0\to K}}{|c_0|^2} = \frac{|V_{1,0}|^2}{(E_0 - \tilde{E}_1)^2 + (\Gamma_1(E_0)/2)^2} \frac{\Gamma_{1K}(E_0)}{\hbar}; \qquad K = L, R \tag{9.88}$$

Problem 9.4. Using Eq. (9.78d) to derive an equation for $(d/dt)|C_r|^2$. Show that the flux $J_{0\to R}$ is also given by

$$J_{0\to R} = \frac{2}{\hbar}\text{Im}\left(\sum_r V_{r1}C_r^*C_1\right)$$

The result (9.88) gives the decay rates of a state $|0\rangle$ that is coupled to two relaxation channels L and R via an intermediate state $|1\rangle$. Viewed as functions of the initial energy E_0, these rates peak when this energy is equal to the (shifted) energy $\tilde{E}_1$ of the intermediate state. In the present context the intermediate level $|1\rangle$ is sometimes referred to as a resonance level. Note that the decay rates Γ in these expressions are defined at the energy E_0 of the driving state, not E_1 of the resonance state.

9.5.2 Steady-state absorption

The model (9.73)–(9.75) was presented as an initial value problem: We were interested in the rate at which a system in state $|0\rangle$ decays into the continua L and R and have used the steady-state analysis as a trick. The same approach can be more directly applied to genuine steady state processes such as energy resolved (also referred to as "continuous wave") absorption and scattering. Consider, for example, the absorption lineshape problem defined by Fig. 9.4. We may identify state $|0\rangle$ as the photon-dressed ground state, state $|1\rangle$ as a zero-photon excited state and the continua R and L with the radiative and nonradiative decay channels, respectively. The interactions $V_{1,0}$ and $V_{1,r}$ correspond to radiative (e.g. dipole) coupling elements between the zero photon excited state $|1\rangle$ and the ground state (or other lower molecular states) dressed by one photon. The radiative quantum yield is given by the flux ratio $Y_R = J_{0\to R}/(J_{0\to R} + J_{0\to L}) = \Gamma_{1R}/(\Gamma_{1R} + \Gamma_{1L})$.

Note that in such spectroscopic or scattering processes the "pumping state" $|0\rangle$ represents a particular state of energy E_0 out of a continuous manifold. In most cases this state belongs to one of the manifolds R and L. For example, in the absorption lineshape problem this photon-dressed ground state is one particular state of the radiative (R) continuum of such states.

9.5.3 Resonance tunneling

Consider a one-dimensional tunneling problem where a particle coming from the left encounters a double barrier potential as seen in Fig. 9.5. This is a potential scattering problem, usually analyzed in terms of scattering functions that are naturally traveling waves. However, when the tunneling process is dominated by resonance state(s) in the barrier region, or in other words the scattering is dominated by quasi-bound states in the scattering region, it is sometimes advantageous to formulate the problem in terms of basis states that are confined to different regions of space.[17]

In this "local basis" approach the zero-order problem is defined in terms of states localized on the left side of the barrier (the L continuum), the right side (the

[17] Such an approach to quantum tunneling was first formulated by Bardeen, Phys. Rev. Letters, **6**, 59 (1961).

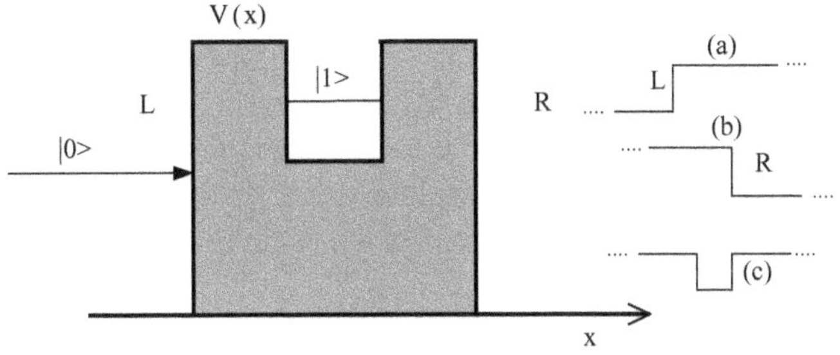

FIG. 9.5 A double barrier model of resonance tunneling. Starting from state $|0\rangle$ on the left, our problem is to compute the fluxes into the continua L and R, defined in terms of basis states that are restricted to the left and right sides of the barrier, respectively. Below the barrier energy such states can be taken as eigenstates of a particle moving in the potentials (a) and (b) respectively, which are shown on the right. The basis is supplemented by state $|1\rangle$, taken as the bound eigenstate of a particle in the potential (c)

R continuum), and in the intermediate well. For example, for energies lower than the top of the barrier these can be taken as eigenstates of Hamiltonians defined with the potentials (a), (b), and (c) shown on the right of Fig. 9.5. The results obtained above then correspond to the case where direct coupling between the L and R states can be disregarded and where it is assumed that the intermediate well between the two barriers can support only one state $|1\rangle$. $J_{0\to R}$, Eq. (9.87), and $J_{0\to L}$ (Eq. (9.87) with L replacing R everywhere) are then the transmitted and reflected fluxes, respectively, associated with one state $|0\rangle$ in the L continuum.[18,19]

Before turning to analyze the solution , it is important to keep in mind that this is not a general solution to the scattering problem represented in Fig. 9.5. Rather, we are interested in the tunneling flux in energy regions where it is dominated

[18] The eigenstates of the Hamiltonians defined by the potentials (a), (b), and (c) in Fig. 9.5 constitute in fact a non-orthogonal basis, because they are derived from different Hamiltonians. The time-dependent Schrödinger equation can be easily represented with such a basis and it may easily be verified that Eqs (9.79) remain of the same form, except that each V_{ij} is replaced by $\tilde{V}_{ij} = V_{ij} - ES_{ij}$, where $V_{ij} = \langle i|\hat{H}|j\rangle$ ($\hat{H}$ being the full Hamiltonian) and $S_{ij} = \langle i|j\rangle$. With this substitution all results obtained in this section can be extended to this more general case.

[19] Readers familiar with scattering theory may be confused by the lack of distinction between the so-called incoming and outgoing states. Indeed, the present formulation of the transmission problem is expressed in terms of states that are localized on the two sides of the barrier, essentially standing waves that cannot be labeled as incoming or outgoing but are combinations of both. This local state representation is convenient for resonance transmission problems because it allows for a natural description of the resonance state as a state that is localized on the barrier in the zero-order level of description. Note, however, that this formulation is useful only for energies well below the barrier where the localized basis states provide a good physical starting point.

by a particular resonance state. We can then limit ourselves to energies close to the zero-order energy of this state (the eigenvalue of the corresponding bound state of the potential (c) in Fig. 9.5). When this state is well below the barrier top, the corresponding continuum states (eigenfunctions of Hamiltonian with the potentials (a) and (b) in Fig. 9.5) are well localized on their respective sides of the barrier, and the local basis approach provides a useful description of the tunneling problem.

Consider then the transmitted flux $J_{0\to R}$, given by Eq. (9.87) or (9.88) with $K = R$. The flux per unit initial energy is obtained by multiplying it by $\rho_L(E_0)$—the density of states in the L continuum. Using

$$\Gamma_{1L}(E_0) = 2\pi |V_{1,0}|^2 \rho_L(E_0) \tag{9.89}$$

we get the transmission flux per unit energy in the form

$$\left(\frac{dJ_{L\to R}(E)}{dE}\right)_{E=E_0} = \frac{1}{2\pi\hbar}\mathcal{T}(E_0)|c_0|^2 \tag{9.90}$$

with

$$\mathcal{T}(E) = \frac{\Gamma_{1L}(E)\Gamma_{1R}(E)}{(E-\tilde{E}_1(E))^2 + (\Gamma_1(E)/2)^2} \tag{9.91}$$

To see the physical significance of the function $\mathcal{T}(E)$ return to Eq. (9.87) which expresses the transmitted flux associated with a single state of energy E_0 and momentum $p_0 = \sqrt{2mE_0}$ where m is the mass of the transmitted particle. The incident flux per particle is $p_0/2m\bar{L}$ where $\bar{L}$ is the normalization length of the single particle wavefunction (so that $\bar{L}^{-1}$ is the corresponding single particle density). The factor two in the denominator reflects the fact that in the standing wave representation that is used here only half the particles of energy E_0 move in the direction of the barrier. Now, Eq. (9.87) can be written in the form

$$J_{0\to R} = (\text{incident flux}) \times \mathcal{T}(E_0) = |c_0|^2 \frac{p_0}{2m\bar{L}}\mathcal{T}(E_0) \tag{9.92}$$

To see this note that $p_0/(2mL)$ can be cast in terms of the one-dimensional density of states in the form (from Eq. (2.96) using $E = p^2/2m$)

$$\frac{p_0}{2m\bar{L}} = (2\pi\hbar\rho_L(E_0))^{-1} \tag{9.93}$$

Using (9.89), (9.91), and (9.93) in (9.92) indeed leads to (9.87). Equation (9.92) implies that $\mathcal{T}(E_0)$ is the transmission coefficient (ratio between transmitted and incident fluxes) at energy E_0. This coefficient is seen to be symmetric in both directions, as required by microscopic reversibility.

Some more observations can be made at this point:

1. The transmission coefficient at energy E_0 has been expressed in terms of this incident energy, the (shifted) energy of the intermediate barrier state $|1\rangle$, and the widths associated with the decay rates of that barrier state into the left and right sides of the barrier. It is interesting to note that these widths should be calculated not at energy E_1 (that will normally be used in calculating the decay rate of this resonance state if initially prepared) but at the incident energy E_0.
2. For a symmetric barrier $\Gamma_{1L} = \Gamma_{1R}$ Eq. (9.91) shows that the maximum transmission, obtained on resonance ($E_0 = \tilde{E}_1$), is 1, irrespective of the coupling strengths that determine Γ_{1L} and Γ_{1R}. These couplings, in turn, determine the linewidth of the peak observed when the transmission is monitored as a function of E_0.
3. The population $|c_0|^2$ of the state that pumps the system does not appear in the rate expression (9.88), however it determines the observed flux through Eqns (9.90) or (9.92). In the particular application (Section 17.2.2) to the problem of electronic conduction, when the left and right continua represent the free electron states of metal electrodes $|c_0|^2$ will be identified as the Fermi–Dirac equilibrium occupation probability at energy E_0, $f(E_0) = [\exp((E_0 - \mu)/k_B T) + 1]^{-1}$, where μ is the chemical potential of the corresponding electrode.
4. The transmission problem analyzed above is one-dimensional: Incident and transmitted states were characterized only by their energy. In three dimensions these states may be characterized in terms of the total energy E_0 and by the wavevectors $\mathbf{k}_y$, $\mathbf{k}_z$ in the directions perpendicular to the incident (x) direction, so that the incident energy and the magnitude of the momentum in the incident direction are

$$E_x = E_0 - (\hbar^2/2m)(k_y^2 + k_z^2); \qquad p_x = \sqrt{2mE_x} \tag{9.94}$$

A particular transmission event at energy E_0 can involve the incident states $E_0, \mathbf{k}_y, \mathbf{k}_z$ and the transmitted state $E_0, \mathbf{k}'_y, \mathbf{k}'_z$, (referred to as *channels* in the present context. Note that any such channel corresponds to the continuum of kinetic energy states associated with motion in the x direction). In Appendix 9C it is shown that the total transmitted flux per unit energy between all possible channels is again given by Eq. (9.90) where $\Gamma_{1L}(E)$ and $\Gamma_{1R}(E)$ are, as before, the decay rates of the resonance state $|1\rangle$ into the left and right (three-dimensional) continua. Also derived in Appendix 9C is a generalization of the present treatment to the situation where transmission is promoted through N barrier states. The result is again of the form for the transmitted flux density, however with a generalized expression for the transmission coefficient

$$\mathcal{T}(E) = \text{Tr}[\hat{\Gamma}^{(L)}(E)\hat{G}^{(B)\dagger}(E)\hat{\Gamma}^{(R)}(E)\hat{G}^{(B)}(E)] \tag{9.95}$$

where the barrier Green's operator is

$$\hat{G}^{(B)}(E) = (E\hat{I}^{(B)} - \hat{H}^{(B)})^{-1} \tag{9.96}$$

In (9.96) $\hat{I}^{(B)}$ is an $N \times N$ unit matrix and $\hat{H}^{(B)}$ is the barrier Hamiltonian, an $N \times N$ matrix defined by $H^{(B)}_{n,n'} = H_{n,n'} + B_{n,n'}$, where $\hat{B} = \hat{B}^{(L)} + \hat{B}^{(R)}$ is the self energy matrix, a generalization of the function $B_{1R}(E) + B_{1L}(E)$ (e.g. Eq. (9.82)) now defined by Eq. (9.133), and where $\hat{\Gamma}^{(K)} = -2\text{Im}\hat{B}^{(K)}$, $K = L, R$.

Problem 9.5. Show that the barrier Hamiltonian can be written in the form $\hat{H}^{(B)} = \hat{P}\hat{H}\hat{P} + \hat{B}$ where $\hat{B}$ is defined as above and $\hat{P}$ is a projection operator on the subspace of the barrier, that is, $\hat{P} = \sum_{n=1}^{N} |n\rangle\langle n|$ where $|n\rangle$, $n = 1, \ldots, N$ is a basis that spans the barrier's subspace.

Appendix 9A: Using projection operators

The mathematical analysis of the dynamics of systems interacting with encompassing reservoirs, whose detailed dynamics is of no direct interest, is facilitated by the use of projection operators or projectors. A simple example is provided by the use of such projectors to rearrange a system of linear equations. Let

$$\mathbf{A} = \begin{pmatrix} a_{11} & \ldots & a_{1N} \\ & & \\ a_{N1} & \ldots & a_{NN} \end{pmatrix}; \quad \mathbf{x} = \begin{pmatrix} x_1 \\ \vdots \\ x_N \end{pmatrix}; \quad \mathbf{u} = \begin{pmatrix} u_1 \\ \vdots \\ u_N \end{pmatrix} \tag{9.97}$$

be and $N \times N$ matrix and N-vectors. Consider the system of linear equations

$$\mathbf{Ax} = \mathbf{u} \tag{9.98}$$

and define the projector matrices

$$\mathbf{P} = \begin{pmatrix} \mathbf{1}_n & 0 \\ 0 & 0_{N-n} \end{pmatrix}; \quad \mathbf{Q} = \begin{pmatrix} 0_n & 0 \\ 0 & \mathbf{1}_{N-n} \end{pmatrix} \tag{9.99}$$

where $\mathbf{1}_n$ is an $n \times n$ identity matrix (and similarly $\mathbf{1}_{N-n}$ is an $(N-n) \times (N-n)$ identity matrix). **P** projects onto the upper left $n \times n$ part of **A** while **Q** projects onto the lower right $(N-n) \times (N-n)$ part. Obviously the relationship $\mathbf{P}+\mathbf{Q}=\mathbf{1}$ (the $N \times N$ identity matrix), $\mathbf{P}^2 = \mathbf{P}, \mathbf{Q}^2 = \mathbf{Q}$, and $\mathbf{PQ} = 0$ are all satisfied. We can use identities such as

$$\mathbf{x} = (\mathbf{P}+\mathbf{Q})\mathbf{x} \tag{9.100}$$

for a vector **x**, and

$$\mathbf{A} = (\mathbf{P}+\mathbf{Q})\mathbf{A}(\mathbf{P}+\mathbf{Q}) \tag{9.101}$$

for a matrix **A** to separate the set of coupled linear equations (9.98) into two distinct sets

$$\mathbf{PAP}\cdot\mathbf{Px} + \mathbf{PAQ}\cdot\mathbf{Qx} = \mathbf{Pu} \tag{9.102}$$

$$\mathbf{QAP}\cdot\mathbf{Px} + \mathbf{QAQ}\cdot\mathbf{Qx} = \mathbf{Qu} \tag{9.103}$$

We may now formally solve Eq. (9.103) for **Qx** and insert in (9.102). This leads to a set of equations in the "P subspace." A common situation is where **u** is in the P subspace, that is, $\mathbf{Qu} = 0$.[20] In this case we find $\mathbf{Qx} = -(\mathbf{QAQ})^{-1}\mathbf{QAP}\cdot\mathbf{Px}$ which leads to

$$\mathbf{Px} = (\mathbf{PAP} - \mathbf{PAQ}(\mathbf{QAQ})^{-1}\mathbf{QAP})^{-1}\mathbf{Pu} \tag{9.104}$$

Since we have explicit forms for all terms on the right, Eq. (9.104) provides an explicit solution for the "interesting" part, P-part, of our system.

Problem 9.6. Show that a matrix **A** can be formally written in the form

$$\mathbf{A} = \begin{pmatrix} \mathbf{PAP} & \mathbf{PAQ} \\ \mathbf{QAP} & \mathbf{QAQ} \end{pmatrix} \tag{9.105}$$

---

[20] The mathematical problem was to find the vector **x** given the vector **u**. The physical problem may have identified the subspace **P** as the "interesting subspace" because the initial information **u** is given in that space (i.e. $\mathbf{Qu} = 0$) and information about **x** is needed also in that space (i.e. we require only s**Px**).

Problem 9.7. Show that

$$(\mathbf{A}^{-1})_{\mathrm{PP}} \equiv \mathbf{P}\mathbf{A}^{-1}\mathbf{P} = (\mathbf{A}_{\mathrm{PP}} - \mathbf{A}_{\mathrm{PQ}}(\mathbf{A}_{\mathrm{QQ}})^{-1}\mathbf{A}_{\mathrm{QP}})^{-1} \tag{9.106}$$

To show this use the identity $\mathbf{A}\mathbf{A}^{-1} = \mathbf{1}$ in the form

$$\begin{pmatrix} \mathbf{A}_{\mathrm{PP}} & \mathbf{A}_{\mathrm{PQ}} \\ \mathbf{A}_{\mathrm{QP}} & \mathbf{A}_{\mathrm{QQ}} \end{pmatrix} \begin{pmatrix} (\mathbf{A}^{-1})_{\mathrm{PP}} & (\mathbf{A}^{-1})_{\mathrm{PQ}} \\ (\mathbf{A}^{-1})_{\mathrm{QP}} & (\mathbf{A}^{-1})_{\mathrm{QQ}} \end{pmatrix} = \mathbf{I} \tag{9.107}$$

Equation (9.107) constitutes a set of the four equations. Two of them are

$$\begin{aligned} \mathbf{A}_{\mathrm{PP}}(\mathbf{A}^{-1})_{\mathrm{PP}} + \mathbf{A}_{\mathrm{PQ}}(\mathbf{A}^{-1})_{\mathrm{QP}} &= \mathbf{I}_P \\ \mathbf{A}_{\mathrm{QP}}(\mathbf{A}^{-1})_{\mathrm{PP}} + \mathbf{A}_{\mathrm{QQ}}(\mathbf{A}^{-1})_{\mathrm{QP}} &= 0 \end{aligned} \tag{9.108}$$

Solving these for $\mathbf{P}\mathbf{A}^{-1}\mathbf{P}$ yields (9.106).

As an example of an application of this formalism consider again the problem of the decay of an initially prepared level $|1\rangle$ coupled to a continuum $\{|l\rangle\}$ as in Eqs (9.2)–(9.7) and Fig. 9.1. Let

$$\hat{P} = |1\rangle\langle 1|; \qquad \hat{Q} = 1 - P = \sum_l |l\rangle\langle l| \tag{9.109}$$

so that $(H_0)_{\mathrm{PQ}} = (H_0)_{\mathrm{QP}} = 0$ and $V_{\mathrm{PP}} = V_{\mathrm{QQ}} = 0$. Let $\hat{A} = EI - \hat{H}_0 - \hat{V}$, so that $\hat{A}^{-1} = \hat{G}$ is the Green operator. Use Eq. (9.106) to obtain

$$G_{\mathrm{PP}} = \left(EP - H_0P - V_{\mathrm{PQ}} \frac{1}{EQ - H_0Q} V_{\mathrm{QP}} \right)^{-1} \tag{9.110}$$

Using (9.109) this may be written explicitly

$$G_{1,1} = \frac{1}{E - E_1 - \sum_l V_{1,l}(E - E_l)^{-1} V_{l,1}} \tag{9.111}$$

which is what was found before (see Eq. (9.16)). Note that we now have a more general and powerful way to obtain the Green function for any given subsystem, because $\hat{P}$ and $\hat{Q}$ can be chosen in more general ways.

Appendix 9B: Evaluation of the absorption lineshape for the model of Figs 9.2 and 9.3

The model under consideration is given in the dressed state representation in Fig. 9.3. It comprises a level $|0\rangle$ coupled to a level $|s\rangle$ which in turn is coupled to a continuum $\{|l\rangle\}$. As far as the relaxation of $|0\rangle$ is concerned the representation used for the subsystem of states ($|s\rangle$, $\{|l\rangle\}$) is immaterial; we could equally well use a basis of states $\{j\}$, each a linear combination of $|s\rangle$ and $\{|l\rangle\}$ states, that diagonalize the Hamiltonian in this subspace. Explicitly

$$\begin{aligned} |j\rangle &= C_{j,s}|s\rangle + \sum_l C_{j,l}|l\rangle \\ (\hat{H}_0 + \hat{V})|j\rangle &= E_j|j\rangle \end{aligned} \tag{9.112}$$

The calculation of the decay rate of $|0\rangle$ in this representation amounts to repeating the problem represented by the model of Fig. 9.1 and Eq. (9.2), where states 1 and $\{l\}$ are now replaced by 0 and $\{j\}$, respectively. One needed element is the coupling between states $|0\rangle$ and $|j\rangle$. Using (9.112) and the fact that $\langle 0|\hat{H}|l\rangle = 0$ we find that

$$\langle 0|\hat{H}|j\rangle = \alpha\mu_{g,s}C_{j,s} \tag{9.113}$$

where the constant α was introduced in Eq. (9.39d).

As discussed in Section 9.1, the required decay rate is obtained under certain conditions as $(-)$ the imaginary part of the self energy $B_0(E_0)$ of the state $|0\rangle$. The latter is defined from

$$G_{0,0}(E) \equiv \langle 0|\frac{1}{E - \hat{H} + i\varepsilon}|0\rangle = \frac{1}{E - E_0 - B_0(E)} \tag{9.114}$$

An approximate expression for $B_0(E)$ is found using the Dyson equation (9.11) and the coupling scheme of Fig. 9.3. Noting that the Dyson equation is valid for any separation of the Hamiltonian to two additive contributions, we now use the separation

$$\hat{H} = [\hat{H} - \alpha(\mu_{g,s}|0\rangle\langle s| + \mu_{s,g}|s\rangle\langle 0|)] + \alpha(\mu_{g,s}|0\rangle\langle s| + \mu_{s,g}|s\rangle\langle 0|) \tag{9.115}$$

and use the term in square bracket as "$\hat{H}_0$" and the last term on the right as the "coupling," to write

$$G_{0,0} = \bar{G}_{0,0} + \bar{G}_{0,0}\alpha\mu_{g,s}G_{s,0} \tag{9.116a}$$

$$G_{s,0} = \bar{G}_{s,s}\alpha\mu_{s,g}G_{0,0} \tag{9.116b}$$

Here $\bar{G} \equiv [E - (\hat{H} - \alpha(\mu_{g,s}|0\rangle\langle s| + \mu_{s,g}|s\rangle\langle 0|)) + i\varepsilon]^{-1}$ is the Green's function associated with the Hamiltonian without the term (9.39d) that couples the molecule to the incident mode of the radiation field. Eliminating $G_{s,0}$ from (9.116) we get

$$G_{0,0}(E) = \frac{1}{E - E_0 - \alpha^2|\mu_{g,s}|^2\bar{G}_{s,s}(E)} \tag{9.117}$$

It is easily realized that $\bar{G}_{s,s}$ here is the same as $G_{s,s}$ defined by Eqs (9.30) and (9.27) (where state 1 plays the same role as state s here). This implies that, if the assumptions leading to Eq. (9.31) hold then,

$$B_0(E_0) = \frac{\alpha^2|\mu_{g,s}|^2}{E_0 - \tilde{E}_s + (1/2)i\Gamma_s} \tag{9.118}$$

Recalling that $E_0 = E_g + \hbar\omega$ we find that the absorption lineshape is given by

$$L(\omega) \propto -\mathrm{Im}B_0(E_0) = \frac{\alpha^2|\mu_{g,s}|^2(\Gamma_s/2)}{(E_g + \hbar\omega - \tilde{E}_s)^2 + (\Gamma_s/2)^2} \tag{9.119}$$

which is Eq. (9.40).

Appendix 9C: Resonance tunneling in three dimensions

Here we generalize the transmission problem of Section 9.5.3 to three dimensions and to many barrier states. Consider first the three-dimensional problem with a single barrier state. The barrier is taken to be rectangular and of a finite width in the transmission (x) direction, so it divides our infinite system into two semi-infinite parts, right (R) and left (L). The transmission is again assumed to result from interactions between free particle states in the L and R subspaces and a single state $|1\rangle$ localized in the barrier, as seen in Fig. 9.5. These free particle states are plane waves in the y and z directions, and can be characterized by the total energy E_0 and by the wavevectors $\mathbf{k}_y$, $\mathbf{k}_z$, so that the incident energy and the magnitude of the momentum in the direction x normal to the barrier are

$$E_x = E_0 - (\hbar^2/2m)(k_y^2 + k_z^2); \qquad p_x = \sqrt{2mE_x} \tag{9.120}$$

With these notations, Eq. (9.87) can be rewritten in the form

$$J_{(\mathbf{k}_y,\mathbf{k}_z,E_0)\to R} = \frac{|V_{1,(\mathbf{k}_y,\mathbf{k}_z,E_0)}|^2}{(E_0 - \tilde{E}_1)^2 + (\Gamma_1(E_0)/2)^2}\frac{\Gamma_{1R}(E_0)}{\hbar}|c_0|^2 \tag{9.121}$$

This is an expression for the steady-state rate of population transfer from a driving state in a particular one-dimensional *channel* $(\mathbf{k}_y, \mathbf{k}_z, E_0)$ on the left of the barrier into the right (now three-dimensional) continuum. The corresponding density of such one-dimensional states is $\rho_x(E_x) = mL/(\pi\hbar p_x)$. The flux *per unit energy* associated with this channel is thus[21]

$$J_{(\mathbf{k}_y,\mathbf{k}_z,E_0)\to R}\rho_x(E_x) = \frac{1}{(2\pi\hbar)}\frac{\Gamma_{1L}^{(\mathbf{k}_y,\mathbf{k}_z)}(E_0)\Gamma_{1R}(E_0)}{(E_0-\tilde{E}_1)^2+(\Gamma_1(E_0)/2)^2}|c_0|^2 \tag{9.122}$$

where

$$\Gamma_{1L}^{(\mathbf{k}_y,\mathbf{k}_z)}(E_0) = 2\pi|V_{1,(\mathbf{k}_y,\mathbf{k}_z,E_0)}|^2\rho_x(E_x) \tag{9.123}$$

Finally, the total flux from all channels of total energy E_0 is obtained by summing over all $\mathbf{k}_y$, $\mathbf{k}_z$ for which E_x of Eq. (9.120) is nonnegative. This yields the total flux per unit energy at energy E_0 in a form similar to (9.90)

$$\left(\frac{dJ_{L\to R}(E)}{dE}\right)_{E=E_0} = \frac{1}{2\pi\hbar}\mathcal{T}(E_0)|c_0|^2 = \frac{1}{(2\pi\hbar)}\frac{\Gamma_{1L}(E_0)\Gamma_{1R}(E_0)}{(E_0-\tilde{E}_1)^2+(\Gamma_1(E_0)/2)^2}|c_0|^2 \tag{9.124}$$

where

$$\Gamma_{1L}(E_0) = \sum_{\mathbf{k}_y,\mathbf{k}_z}\Gamma_{1L}^{(\mathbf{k}_y,\mathbf{k}_z)}(E_0) \tag{9.125}$$

is the leftward decay rate of state $|1\rangle$ into the left continuum, here expressed as sum of rates of decay into the individual one-dimensional channels. Note that Γ_{1R} can be expressed in a similar way, though we did not need to use it in the derivation above.

Further insight into the structure of this solution may be obtained by denote these channels by the collective indices α so that Eq. (9.125) takes the form

$$\Gamma_{1L}(E) = \sum_{\alpha}\Gamma_{1L}^{\alpha}(E) \qquad \text{(and same for } \Gamma_{1R}) \tag{9.126}$$

This implies that

$$\mathcal{T}_1(E_0) = \sum_{\alpha,\alpha'}\mathcal{T}_{\alpha,\alpha'}(E_0) \tag{9.127}$$

[21] We assume that all states of energy E_0 are associated with the same probability $|c_0|^2$, as expected if the continua represented bulk system at thermal equilibrium.

where

$$\mathcal{T}_{\alpha,\alpha'}(E) = \frac{\Gamma^{\alpha}_{1L}(E)\Gamma^{\alpha'}_{1R}(E)}{(E - \tilde{E}_1(E))^2 + (\Gamma_1(E)/2)^2} \tag{9.128}$$

Expressions such as (9.124) are sometimes called "all to all" microcanonical transition rates (or, in this case, differential transition fluxes), since they express that total flux due to all states of energy E.

Next we generalize these results to the case where transmission is induced by many barrier-localized states, $|1\rangle, |2\rangle, \dots, |N\rangle$. The transition under consideration is then between a left manifold of states $\{l\}$ and a right manifold $\{r\}$ due to their mutual coupling with barrier states $\{n\}, n = 1, 2, \dots, N$. These states are not necessarily eigenstates of the Hamiltonian, so in the derivation below we will encounter matrix elements such as $H_{n,n'} = \langle n|\hat{H}|n'\rangle$. We will however assume for simplicity that all states are orthogonal to each other, $\langle n|n'\rangle = \delta_{n,n'}$ (see footnote 18 about the more general case). Again we consider a driving state $|0\rangle$ that couples to the barrier states, and the fluxes following from this into the left and right continua. The generalization of Eqs (9.78) into the present case reads

$$\begin{aligned}\hbar\dot{C}_n &= -iH_{n,n}C_n - iH_{n,0}C_0(t) - i\sum_{n'\neq n} H_{n,n'}C_{n'} - i\sum_l H_{n,l}C_l - i\sum_r H_{n,r}C_r; \\ &\quad n, n' = 1, \dots, N\end{aligned} \tag{9.129a}$$

$$\hbar\dot{C}_k = -iH_{k,k}C_k - \sum_{n=1}^{N} iH_{k,n}C_n - (\eta/2)C_k; \qquad k = l, r \tag{9.129b}$$

with $C_0(t) = c_0 \exp(-(i/\hbar)E_0 t)$. Again we look for steady-state solutions of the forms $C_j(t) = c_j \exp(-(i/\hbar)E_0 t)$ $(j = \{n\}, \{l\}, \{r\})$, and get the equations equivalent to (9.79)

$$\begin{aligned}0 = i(E_0 - H_{n,n})c_n - iH_{n,0}c_0 - i\sum_{n'\neq n} H_{n,n'}c_{n'} \\ -i\sum_l H_{n,l}c_l - i\sum_r H_{n,r}c_r \qquad n, n' = 1, \dots, N\end{aligned} \tag{9.130a}$$

$$0 = i(E_0 - H_{k,k})c_k - \sum_{n=1}^{N} iH_{k,n}c_n - (\eta/2)c_k; \qquad k = l, r \tag{9.130b}$$

Equation (9.130b) yields

$$c_k = \frac{\sum_{n=1}^{N} H_{k,n}c_n}{E_0 - H_{k,k} + i(\eta/2)}; \qquad k = l, r \tag{9.131}$$

and inserting this into (9.130a) leads to

$$0 = i(E_0 - H_{n,n})c_n - iH_{n,0}c_0 - i\sum_{n'\neq n} H_{n,n'}c_{n'} - i\sum_{n'} B_{n,n'}(E_0)c_{n'} \tag{9.132}$$

with

$$\begin{aligned}
&B_{n,n'}(E) = B^{(L)}_{n,n'}(E) + B^{(R)}_{n,n'}(E) \\
&B^{(K)}_{n,n'}(E) \equiv \sum_{k\in K} \frac{H_{n,k}H_{k,n'}}{E - H_{k,k} + i\eta/2} = \Lambda^{(K)}_{n,n'}(E) - \frac{1}{2}i\Gamma^{(K)}_{n,n'}(E); \qquad K = L, R \\
&\Gamma^{(K)}_{n,n'}(E) = 2\pi(H_{n,k}H_{k,n'}\rho_K(E_k))_{E_k=E}; \qquad k \in K, \quad K = L, R \\
&\Lambda^{(K)}_{n,n'}(E) = \mathrm{PP}\int_{-\infty}^{\infty} dE_k \frac{H_{n,k}H_{k,n'}\rho_K(E_k)}{E - E_k}; \qquad k \in K, \quad K = L, R
\end{aligned} \tag{9.133}$$

Note that the self energy defined above, Eq. (9.82), has now become a non-diagonal matrix.

We now define an effective Hamiltonian matrix $\hat{H}^{(B)}$ in the subspace of the barrier states (a $N \times N$ matrix)

$$H^{(B)}_{n,n'} = H_{n,n'} + B_{n,n'} \tag{9.134}$$

and a corresponding barrier Green's operator

$$G^{(B)}(E) = (E\hat{I}^{(B)} - \hat{H}^{(B)})^{-1} \tag{9.135}$$

and obtain a solution for the coefficients c_n ($n = 1, \ldots, N$) in the form ($\hat{I}^{(B)}$ is an $N \times N$ unit matrix)

$$c_n = c_0 \sum_{n'} G^{(B)}_{n,n'} H_{n',0} \tag{9.136}$$

Using this in (9.131) yields, for example, for the R continuum

$$c_r = \frac{\sum_{n=1}^{N}\sum_{n'=1}^{N} H_{r,n}G^{(B)}_{n,n'}H_{n',0}}{E_0 - H_{r,r} + i(\eta/2)} c_0 \tag{9.137}$$

The flux through the R continuum is now obtained in the form (compare (9.87))

$$J_{0\to R} = (\eta/\hbar)\sum_r |c_r|^2 \xrightarrow{\eta\to 0} \hbar^{-1}|c_0|^2$$

$$\times \sum_r 2\pi\delta(E_0 - H_{rr}) \sum_{n_1=1}^{N}\sum_{n_1'=1}^{N}\sum_{n=1}^{N}\sum_{n'=1}^{N} H_{0,n_1'} G^{(B)\dagger}_{n_1',n_1} H_{n_1,r} H_{r,n} G^{(B)}_{n,n'} H_{n',0}$$

$$= \hbar^{-1}|c_0|^2 \sum_{n_1=1}^{N}\sum_{n_1'=1}^{N}\sum_{n=1}^{N}\sum_{n'=1}^{N} H_{0,n_1'} G^{(B)\dagger}_{n_1',n_1} \Gamma^{(R)}_{n_1,n} G^{(B)}_{n,n'} H_{n',0} \qquad (9.138)$$

Finally, repeating the steps that lead from (9.121) to (9.124) we now find

$$\left(\frac{dJ_{L\to R}(E)}{dE}\right)_{E=E_0} = \frac{|c_0|^2}{2\pi\hbar} \sum_{n_1=1}^{N}\sum_{n_1'=1}^{N}\sum_{n=1}^{N}\sum_{n'=1}^{N} \Gamma^{(L)}_{n',n_1'} G^{(B)\dagger}_{n_1',n_1} \Gamma^{(R)}_{n_1,n} G^{(B)}_{n,n'}$$

$$= \frac{|c_0|^2}{2\pi\hbar} \mathrm{Tr}[\hat{\Gamma}^{(L)}(E_0)\hat{G}^{(B)\dagger}(E_0)\hat{\Gamma}^{(R)}(E_0)\hat{G}^{(B)}(E_0)] \qquad (9.139)$$

where the trace operation denotes a sum over barrier states $\{n\}$ of the diagonal elements of the matrix $\hat{\Gamma}^{(L)}\hat{G}^{\dagger}\hat{\Gamma}^{(R)}\hat{G}$. This is a product of four $N \times N$ matrices, all defined in the subspace of the barrier states. This concludes our derivation of Eqs (9.90) and (9.95) for the present model.

10

THE QUANTUM MECHANICAL DENSITY OPERATOR AND ITS TIME EVOLUTION: QUANTUM DYNAMICS USING THE QUANTUM LIOUVILLE EQUATION

Surely the atoms never began by forming
A conscious pact, a treaty with each other,
Where they should stay apart, where they come together.
More likely, being so many, in many ways
Harassed and driven through the universe
From an infinity of time, by trying
All kind of motion, every combination,
They came at last into such disposition
As now establishes the sum of things...

Lucretius (c.99–c.55 BCE*) "The way things are"*
translated by Rolfe Humphries, Indiana University Press, 1968

The starting point of the classical description of motion is the Newton equations that yield a phase space trajectory $(\mathbf{r}^N(t), \mathbf{p}^N(t))$ for a given initial condition $(\mathbf{r}^N(0), \mathbf{p}^N(0))$. Alternatively one may describe classical motion in the framework of the Liouville equation (Section (1.2.2)) that describes the time evolution of the phase space probability density $f(\mathbf{r}^N, \mathbf{p}^N; t)$. For a closed system fully described in terms of a well specified initial condition, the two descriptions are completely equivalent. Probabilistic treatment becomes essential in reduced descriptions that focus on parts of an overall system, as was demonstrated in Sections 5.1–5.3 for equilibrium systems, and in Chapters 7 and 8 that focus on the time evolution of classical systems that interact with their thermal environments.

This chapter deals with the analogous quantum mechanical problem. Within the limitations imposed by its nature as expressed, for example, by Heisenberg-type uncertainty principles, the Schrödinger equation is deterministic. Obviously it describes a deterministic evolution of the quantum mechanical wavefunction. The analog of the phase space probability density $f(\mathbf{r}^N, \mathbf{p}^N; t)$ is now the quantum mechanical density operator (often referred to as the "density matrix"), whose time evolution is determined by the quantum Liouville equation. Again, when the system is fully described in terms of a well specified initial wavefunction, the two descriptions are equivalent. The density operator formalism can, however, be carried over to situations where the initial state of the system is not well characterized and/or a reduced description of part of the overall system is desired. Such situations are considered later in this chapter.

10.1 The density operator and the quantum Liouville equation

10.1.1 The density matrix for a pure system

Consider a system characterized by a given Hamiltonian operator $\hat{H}$, an orthonormal basis $\{\phi_n\}$ (also denoted $\{|n\rangle\}$) that spans the corresponding Hilbert space and a time dependent wavefunction $\Psi(t)$—a normalized solution of the Schrödinger equation. The latter may be represented in terms of the basis functions as

$$\Psi(t) = \sum_n C_n(t)\phi_n \tag{10.1}$$

The normalization of $\Psi(t)$ implies that $\sum_n |C_n|^2 = 1$. When the state of the system is given in terms of such wavefunction we say that the system is in a *pure state.*

Consider also a dynamical variable A that is represented by an operator $\hat{A}$. Its expectation value at time t is given by

$$\langle A\rangle_t = \langle\Psi(t)|\hat{A}|\Psi(t)\rangle = \sum_{n,n'} C_{n'}(t)C_n^*(t)A_{n,n'} \equiv \sum_{n,n'} \rho_{n',n}(t)A_{n,n'} \tag{10.2}$$

The coefficients $\rho_{n,n'}$ in (10.2) define the matrix elements of the *density operator* $\hat{\rho}$ in the given basis. For a pure state $\hat{\rho}$ can be written explicitly as

$$\hat{\rho}(t) \equiv |\Psi(t)\rangle\langle\Psi(t)| = \sum_{n,n'} C_n(t)C_{n'}^*(t)|n\rangle\langle n'| \tag{10.3}$$

so that indeed

$$\rho_{n,n'}(t) = \langle\phi_n|\hat{\rho}(t)|\phi_{n'}\rangle = C_n(t)C_{n'}^*(t) \tag{10.4}$$

Using the completeness of the basis, that is, $\sum_n |n\rangle\langle n| = 1$, Eq. (10.2) is seen to be equivalent to

$$\langle A\rangle_t = \mathrm{Tr}[\hat{\rho}\hat{A}] \tag{10.5}$$

which is the quantum analog of Eq. (1.100) if $\hat{\rho}$ is perceived as a quantum analog of the distribution function f. Another element in this analogy is provided by the equivalence

$$\int d\mathbf{r}^N \int d\mathbf{p}^N f(\mathbf{r}^N, \mathbf{p}^N; t) = 1 \quad \Leftrightarrow \quad \mathrm{Tr}[\hat{\rho}] = \sum_n \rho_{nn} = 1 \tag{10.6}$$

which follows from (10.3).

The time evolution of the density operator can be found from the time evolution of $\Psi(t)$ and the definition (10.3):

$$\begin{aligned}\frac{d}{dt}\hat{\rho}(t) &= \frac{d}{dt}(|\Psi(t)\rangle\langle\Psi(t)|) = \left(\frac{d}{dt}|\Psi(t)\rangle\right)\langle\Psi(t)| + |\Psi(t)\rangle\left(\frac{d}{dt}\langle\Psi(t)|\right) \\ &= -\frac{i}{\hbar}\hat{H}|\Psi(t)\rangle\langle\Psi(t)| + \frac{i}{\hbar}|\Psi(t)\rangle\langle\Psi(t)|\hat{H} \end{aligned} \tag{10.7}$$

or

$$\frac{d}{dt}\hat{\rho}(t) = -\frac{i}{\hbar}[\hat{H}, \hat{\rho}(t)] \equiv -i\mathcal{L}\hat{\rho}(t) \tag{10.8}$$

where $\mathcal{L} \equiv \hbar^{-1}[\hat{H},]$ is the quantum Liouville operator.

Equation (10.8) may be compared with the Heisenberg equation of motion (2.66) for the Heisenberg representation $\hat{A}_{\mathrm{H}}(t) = \exp((i/\hbar)\hat{H}t)\hat{A}\exp(-(i/\hbar)\hat{H}t)$ of the operator $\hat{A}$

$$\frac{d}{dt}\hat{A}_{\mathrm{H}}(t) = \frac{i}{\hbar}[\hat{H}, \hat{A}_{\mathrm{H}}(t)] = i\mathcal{L}\hat{A}_{\mathrm{H}}(t) \tag{10.9}$$

We see that the density operator $\hat{\rho}(t)$, the quantum analog of the classical phase space distribution $f(\mathbf{r}^N, \mathbf{p}^N; t)$, is different from other operators that represent dynamical variables. The same difference in time evolution properties was already encountered in classical mechanics between dynamical variables and the distribution function, as can be seen by comparing Eq. (10.8) with (1.104) and Eq. (10.9) with (1.99). This comparison also emphasizes the correspondence between the classical and quantum Liouville operators.

Two other properties of the density operators follow from its definition (10.3). First, it is Hermitian, that is, $\hat{\rho}^\dagger(t) = \hat{\rho}(t)$. Second it is idempotent, that is, satisfies the property

$$\hat{\rho}^2 = \hat{\rho} \tag{10.10}$$

10.1.2 Statistical mixtures

As defined above, the density operator provides an alternative but equivalent description of the information contained in a pure quantum mechanical state. Its real advantage emerges when we encounter systems whose state is not known completely. For example, we may know the probabilities $P_n = |C_n|^2$ to be in the different states n (defined in terms of some basis set $\{\phi_n\}$), without knowing the actual state $\psi = \sum_n C_n\phi_n$ that requires knowledge of the phases of the complex numbers C_n. In the extreme case of such ignorance all phases are equally possible and should be averaged upon in any calculation. In this case Eq. (10.4) becomes

$$\rho_{nn'} = |C_n|^2\delta_{n,n'} = P_n\delta_{n,n'} \tag{10.11}$$

We refer to such a state as a *statistical mixture*. An example is a system in thermal equilibrium whose eigenfunctions constitute the basis $\{\phi_n\}$. The probabilities P_n are then given by the Boltzmann distribution and all microscopic states compatible with these probabilities (i.e. all phases of the complex coefficients C_n) are assumed to be equally probable.

It should be obvious that a pure state and the statistical mixture with the same $|C_n|$ are not equivalent. For example, the average value of an observable A in the pure state is $\langle\psi|\hat{A}|\psi\rangle = \sum_{n,n'} C_n^* C_{n'} A_{n,n'}$, where $A_{n,n'} = \langle n|A|n'\rangle$, while in the corresponding statistical mixture $\langle A\rangle = \sum_n P_n A_{n,n} = \sum_n |C_n|^2 A_{n,n}$.

Above we have contrasted the pure state with a statistical mixture represented by a diagonal density matrix. We now make these statements more general:

1. In the representation defined with a basis $\{\phi_n\}$, $\hat{\rho}$ is a matrix with elements $\rho_{n,n'}$. The diagonal elements, $\rho_{nn} = P_n$, are the probabilities that the system is in states n. In the pure state $\psi = \sum_n C_n \phi_n$ we found that

$$P_n = |C_n|^2 \quad \text{and} \quad \rho_{n,n'} = C_n C_{n'}^* \tag{10.12}$$

2. In the statistical mixture the last equality is not satisfied. This does not necessarily mean that in a statistical mixture $\hat{\rho}$ has to be diagonal. If it is diagonal in the basis $\{\phi_n\}$, that is, $\hat{\rho} = \sum_n P_n |\phi_n\rangle\langle\phi_n|$, we can go to another representation $\{\psi_k\}$ such that $\phi_n = \sum_k a_{nk}\psi_k$ in which $\hat{\rho}$ takes the non-diagonal form

$$\hat{\rho} = \sum_n P_n \sum_{k,k'} a_{nk} a_{nk'}^* |\psi_k\rangle\langle\psi_{k'}| = \sum_{k,k'} \rho_{kk'} |\psi_k\rangle\langle\psi_{k'}| \tag{10.13}$$

where $\rho_{kk'} = \sum_n P_n a_{nk} a_{nk'}^*$.

3. So how is a pure state different from a mixed state? In the former the elements of the corresponding density matrix are related to each other in a specific way, Eq. (10.12), resulting from their association with the amplitudes of the expansion of the pure state in terms of the given basis. In a mixed state such a relationship does not exist.
4. An operational statement of the difference between a pure state and a statistical mixture can be made with respect to their diagonal representations. In such representation the pure state density matrix will have only one nonzero element on its diagonal, that will obviously take the value 1. A diagonal density matrix representing a statistical mixture must have at least two elements (whose sum is 1) on its diagonal.

An important observation is that several, but not all, properties of the pure state density operator remain also in mixed states:

1 $\hat{\rho}$ is Hermitian and its diagonal elements ρ_{nn} are real and positive. These elements represent the average probabilities to find the system in the corresponding state (see also 6 below).
2. $\mathrm{Tr}\hat{\rho} = 1$.
3. For any observable represented by an operator $\hat{A}$

$$\langle A \rangle = \mathrm{Tr}(\hat{\rho}\hat{A}) \tag{10.14}$$

This follows from the fact that this relation has to hold for the diagonal representation and from the fact that the Trace operation does not depend on the representation used.
4. The time evolution of the density operator is described by the Liouville equation

$$\frac{d}{dt}\hat{\rho}(t) = -\frac{i}{\hbar}[\hat{H}, \hat{\rho}(t)] = -i\mathcal{L}\hat{\rho}(t) \tag{10.15}$$

To show that this remains true for mixed states consider the diagonal representation of the initial state, $\hat{\rho}(t = 0) = \sum_n P_n |\psi_n\rangle\langle\psi_n|$. This describes a mixed state in which the probability to be in the pure state ψ_n is P_n. In the following time evolution each pure state evolves according to the time-dependent Schrödinger equation so that $\Psi_n(t) = \exp(-(i/\hbar)\hat{H}t)\psi_n(0)$ and therefore P_n also represents the probability to be in state $\Psi_n(t)$ at time t. It follows that $\hat{\rho}(t) = \sum_n P_n |\Psi_n(t)\rangle\langle\Psi_n(t)|$ so that

$$\hat{\rho}(t) = e^{-(i/\hbar)\hat{H}t}\hat{\rho}(0)e^{(i/\hbar)\hat{H}t} \tag{10.16}$$

from which (10.15) follows. Eq. (10.16) is sometimes written as a formal solution of Eq. (10.15), that is,

$$\hat{\rho}(t) = e^{-i\mathcal{L}t}\rho(0) \equiv e^{-(i/\hbar)\hat{H}t}\hat{\rho}(0)e^{(i/\hbar)\hat{H}t} \tag{10.17}$$

5. In general $\hat{\rho}$ does not satisfy $\hat{\rho}^2 = \hat{\rho}$. This identity holds only for a pure state.
6. Given that $\hat{\rho}$ is diagonal in the representation $\{\psi_j\}$, then for any state ψ we have $\langle\psi|\hat{\rho}|\psi\rangle = \sum_j P_j |\langle\psi|\psi_j\rangle|^2 \geq 0$. $\hat{\rho}$ is therefore *a positive operator*. $\langle\psi|\hat{\rho}|\psi\rangle$ is seen to be the *average probability* to find the system in state ψ.

10.1.3 Representations

The time evolution of the density operator, Eqs (10.15) and (10.16), stems from the time dependence of the wavefunctions, and describes the time evolution of $\hat{\rho}$

in the Schrödinger representation. (As usual, we omit the subscript "S" from the Schrödinger $\hat{\rho}$ as long as its identity may be inferred from the text. Below, however, we sometimes write this subscript explicitly). In the corresponding Heisenberg representation

$$\hat{\rho}_{\rm H} = \hat{\rho}_{\rm S}(t=0) = e^{(i/\hbar)\hat{H}t}\hat{\rho}_{\rm S}(t)e^{-(i/\hbar)\hat{H}t} = e^{i\mathcal{L}t}\hat{\rho}_{\rm S}(t) \tag{10.18}$$

$\hat{\rho}_{\rm H}$ does not depend on t just as the wavefunction $\psi_{\rm H}$ does not. It will prove useful to consider also the time evolution of the density operator in the interaction representation associated with some suitable decomposition of $\hat{H}$, $\hat{H} = \hat{H}_0 + \hat{V}$. This representation of $\hat{\rho}$ is defined by

$$\hat{\rho}_{\rm I}(t) \equiv e^{i\hat{H}_0t/\hbar}\hat{\rho}_{\rm S}(t)e^{-i\hat{H}_0t/\hbar} \tag{10.19a}$$

$$= e^{i\hat{H}_0t/\hbar}e^{-i\hat{H}t/\hbar}\hat{\rho}(t=0)e^{i\hat{H}t/\hbar}e^{-i\hat{H}_0t/\hbar} \tag{10.19b}$$

To obtain the corresponding equation of motion take time derivative of (10.19a):

$$\begin{aligned}\frac{d\hat{\rho}_{\rm I}}{dt} &= e^{(i/\hbar)\hat{H}_0t}\left[\frac{i}{\hbar}\hat{H}_0\hat{\rho}_{\rm S}(t) - \frac{i}{\hbar}\hat{\rho}_{\rm S}(t)\hat{H}_0 + \frac{d\hat{\rho}_{\rm S}}{dt}\right]e^{-(i/\hbar)\hat{H}_0t} \\ &= e^{(i/\hbar)\hat{H}_0t}\left[\frac{i}{\hbar}\hat{H}_0\hat{\rho}_{\rm S}(t) - \frac{i}{\hbar}\hat{\rho}_{\rm S}(t)\hat{H}_0 - \left(\frac{i}{\hbar}\hat{H}\hat{\rho}_{\rm S}(t) - \frac{i}{\hbar}\hat{\rho}_{\rm S}(t)\hat{H}\right)\right]e^{-(i/\hbar)\hat{H}_0t} \\ &= e^{(i/\hbar)\hat{H}_0t}\left[-\frac{i}{\hbar}\hat{V}\hat{\rho}_{\rm S}(t) + \frac{i}{\hbar}\hat{\rho}_{\rm S}(t)\hat{V}\right]e^{-(i/\hbar)\hat{H}_0t} = -\frac{i}{\hbar}[\hat{V}_{\rm I}(t), \hat{\rho}_{\rm I}(t)]\end{aligned} \tag{10.20}$$

Thus

$$\frac{d\hat{\rho}_{\rm I}}{dt} = -\frac{i}{\hbar}[\hat{V}_{\rm I}(t), \hat{\rho}_{\rm I}(t)] \tag{10.21}$$

Note that an equation similar to (10.19a) that relates the interaction representation of any other operator to the Schrödinger representation

$$\hat{A}_{\rm I}(t) = e^{(i/\hbar)\hat{H}_0t}A_{\rm S}e^{-(i/\hbar)\hat{H}_0t} \tag{10.22}$$

leads to

$$\frac{d\hat{A}_{\rm I}}{dt} = \frac{i}{\hbar}[\hat{H}_0, \hat{A}_{\rm I}] \tag{10.23}$$

which is obviously different from (10.21). The origin of this difference is the fact that in the Schrödinger representation $A_{\rm S}$ is time independent (unless A has intrinsic time dependence) while $\rho_{\rm S}(t)$ depends on time.

Table 10.1 summarizes our findings, as well as those from Sections 2.7.1 and 2.7.2, by comparing the different transformations and the corresponding equations

TABLE 10.1 The Schrödinger, Heisenberg, and interaction representations of the quantum time evolution.

Schrödinger representation	Heisenberg representation	Interaction representation
$\Psi_S(t) = e^{-i\hat{H}t/\hbar}\Psi(t=0)$	$\Psi_H(t) = e^{i\hat{H}t/\hbar}\Psi_S(t) = \Psi_S(t=0)$	$\Psi_I(t) = e^{i\hat{H}_0 t/\hbar}\Psi_S(t)$
$\frac{\partial \Psi_S}{\partial t} = -(i/\hbar)\hat{H}\Psi_S$	(time independent)	$\frac{\partial \Psi_I}{\partial t} = -(i/\hbar)\hat{V}_I(t)\Psi_I$
$\hat{A}_S = \hat{A}(t=0)$	$\hat{A}_H(t) = e^{(i/\hbar)\hat{H}t}\hat{A}_S e^{-(i/\hbar)\hat{H}t}$	$\hat{A}_I(t) = e^{(i/\hbar)\hat{H}_0 t}A_S e^{-(i/\hbar)\hat{H}_0 t}$
(time independent)	$\frac{d\hat{A}_H(t)}{dt} = \frac{i}{\hbar}[\hat{H}, \hat{A}_H(t)]$	$\frac{d\hat{A}_I(t)}{dt} = \frac{i}{\hbar}[\hat{H}_0, \hat{A}_I(t)]$
$\hat{\rho}_S(t) = e^{-(i/\hbar)\hat{H}t}\hat{\rho}(0)e^{(i/\hbar)\hat{H}t}$	$\hat{\rho}_H = e^{(i/\hbar)\hat{H}t}\hat{\rho}_S(t)e^{-(i/\hbar)\hat{H}t}$	$\hat{\rho}_I(t) = e^{i\hat{H}_0 t/\hbar}\hat{\rho}_S(t)e^{-i\hat{H}_0 t/\hbar}$
$\frac{d}{dt}\hat{\rho}_S(t) = -\frac{i}{\hbar}[\hat{H}, \hat{\rho}_S(t)]$	$= \hat{\rho}(t=0)$ (time independent)	$\frac{d\hat{\rho}_I}{dt} = -\frac{i}{\hbar}[\hat{V}_I(t), \hat{\rho}_I(t)]$

of motion for the quantum mechanical wavefunction, the density operator, and a "regular" operator (i.e. an operator that represent a dynamical variable) in the absence of explicit time dependence, that is, in the absence of time-dependent external forces.

Note that the time-dependent average of an operator $\hat{A}$ is the same in all representations

$$\langle A\rangle_t = \text{Tr}[\hat{\rho}_H\hat{A}_H(t)] = \text{Tr}[\hat{\rho}_S(t)\hat{A}_S] = \text{Tr}[\hat{\rho}_I(t)\hat{A}_I(t)] \tag{10.24}$$

as of course should be.

Problem 10.1.

(1) Show that the trace of $\hat{\rho}^2$ is 1 for a pure state and smaller than 1 for a mixed state.
(2) Show that for a pure state $\text{Tr}(\hat{\rho}^2)$ cannot be larger than 1; that in fact if $\text{Tr}(\hat{\rho}) = 1$ and $\text{Tr}(\hat{\rho}^2) > 1$ then $\hat{\rho}$ has negative eigenvalues, that is, is unphysical.

Problem 10.2. Let $\hat{H} = \hat{H}_0 + \hat{H}_1(t)$. Show by direct differentiation with respect to t that the solution to the Liouville equation $(d/dt)\hat{\rho}(t) = -(i/\hbar)[\hat{H}_0 + \hat{H}_1(t), \hat{\rho}(t)] = -i(\mathcal{L}_0 + \mathcal{L}_1(t))\hat{\rho}(t)$ may be written in the form

$$\hat{\rho}(t) = e^{-i(t-t_0)\mathcal{L}_0}\hat{\rho}(t_0) - i\int_{t_0}^{t} dt' e^{-i(t-t')\mathcal{L}_0}\mathcal{L}_1(t')\hat{\rho}(t') \tag{10.25}$$

(Hint: Multiply both sides by $\exp(it\mathcal{L}_0)$ before taking derivative).

10.1.4 Coherences

In a general representation $\hat{\rho}$ is non-diagonal. In terms of the basis $\{\psi_j\}$ that diagonalizes ρ we may write

$$\rho_{nn'} = \sum_j P_j \langle n|\psi_j\rangle\langle\psi_j|n'\rangle \tag{10.26}$$

If $|n\rangle = \sum_j C_{nj}\psi_j$ and $|n'\rangle = \sum_j C_{n'j}\psi_j$ we find

$$\rho_{nn'} = \sum_j P_j C_{nj}^* C_{n'j} = \langle C_{nj}^* C_{n'j}\rangle \tag{10.27}$$

We see that the non-diagonal element $\rho_{nn'}$ of the density matrix is the averaged product of cross terms between states n and n'. These elements, which appear in calculations of interference effects between these states, are referred to as "*coherences*." If $\rho_{nn'}$ in Eq. (10.27) is found to vanish, the corresponding interference is averaged out.

Problem 10.3. Show that in the basis of eigenstates of H

$$\frac{d}{dt}\rho_{nn} = 0; \qquad \frac{d}{dt}\rho_{nm} = -\frac{i}{\hbar}(E_n - E_m)\rho_{nm} \tag{10.28}$$

Problem 10.4. Consider a system for which the Hamiltonian is $\hat{H} = \hat{H}_0 + \hat{V}$, or, in the representation defined by the eigenstates of $\hat{H}_0$, $\hat{H} = \sum_m E_m|m\rangle\langle m| + \sum\sum_{m\neq n} V_{m,n}|m\rangle\langle n|$. In the same representation the density operator is $\hat{\rho} = \sum_m\sum_n \rho_{m,n}|m\rangle\langle n|$. Show that in this representation the Liouville equation is

$$\frac{d\rho_{m,n}}{dt} = -\frac{i}{\hbar}E_{m,n}\rho_{m,n} - \frac{i}{\hbar}\sum_l (V_{m,l}\rho_{l,n} - V_{l,n}\rho_{m,l}) \tag{10.29}$$

where $E_{m,n} = E_m - E_n$.

Problem 10.5. Show that in any basis

$$\rho_{nn}\rho_{mm} \geq |\rho_{nm}|^2 \qquad \text{(equality holds for a pure state)} \tag{10.30}$$

Solution: To prove the inequality (10.30) we first note that for a pure state $\hat{\rho} = |\psi\rangle\langle\psi|$, and Eq. (10.30) is satisfied as an identity, with each side equal to

$|\langle n|\psi\rangle|^2\,|\langle m|\psi\rangle|^2$. In the more general case $\hat{\rho}$ may still be expressed in the diagonal representation,

$$\rho = \sum_j a_j|\psi_j\rangle\langle\psi_j|; \qquad a_j \geq 0; \qquad \sum_j a_j = 1 \tag{10.31}$$

The inequality (10.30) then takes the form

$$\left(\sum_j a_j|\langle n|\psi_j\rangle|^2\right)\left(\sum_j a_j|\langle m|\psi_j\rangle|^2\right) \geq \left|\sum_j a_j\langle n|\psi_j\rangle\langle\psi_j|m\rangle\right|^2 \tag{10.32}$$

This, however, is just the Schwarz inequality (Section 1.1.8). Indeed, Eq. (10.32) is identical to the inequality satisfied by two complex vectors, (cf. Eq. (1.81)), $|\mathbf{e}|^2\,|\mathbf{f}|^2 \geq |\mathbf{e}^* \cdot \mathbf{f}|^2$, if we identify $e_j = \sqrt{a_j}\langle n|\psi_j\rangle$ and $f_j = \sqrt{a_j}\langle m|\psi_j\rangle$. Another proof of (10.30) is obtained by defining the wavefunctions $|\psi\rangle = \hat{\rho}^{1/2}|n\rangle$; $|\phi\rangle = \hat{\rho}^{1/2}|m\rangle$ and realizing the Eq. (10.30) can then be rewritten in the form $\langle\psi|\psi\rangle\langle\phi|\phi\rangle \geq |\langle\psi|\phi\rangle|^2$ which is another expression, Eq. (1.85), of the Schwarz inequality.

10.1.5 Thermodynamic equilibrium

The expression for the classical distribution function in thermodynamic equilibrium reflects the Boltzmann equilibrium property

$$f(\mathbf{r}^N, \mathbf{p}^N) = \frac{e^{-\beta H(\mathbf{r}^N,\mathbf{p}^N)}}{\int d\mathbf{r}^N \int d\mathbf{p}^N e^{-\beta H(\mathbf{r}^N,\mathbf{p}^N)}} \tag{10.33}$$

Similarly, for a quantum system in thermal equilibrium, the populations of stationary states are given by the Boltzmann factors $P_k \sim e^{-\beta E_k}$, and coherences between such states are zero. This implies that *in the basis of eigenstates of* the system Hamiltonian $\hat{H}$

$$\hat{\rho}_{\text{eq}} = \frac{\sum_j e^{-\beta E_j}|\psi_j\rangle\langle\psi_j|}{\sum_j e^{-\beta E_j}} \tag{10.34}$$

and more generally,

$$\hat{\rho}_{\text{eq}} = Z^{-1}e^{-\beta\hat{H}}; \qquad Z = \text{Tr}[\hat{\rho}_{\text{eq}}] \tag{10.35}$$

The thermal average of an observable represented by an operator $\hat{A}$ is, according to Eq. (10.14)

$$\langle A \rangle_T = \mathrm{Tr}[\hat{\rho}_{eq}\hat{A}] = \frac{\mathrm{Tr}[e^{-\beta\hat{H}}\hat{A}]}{\mathrm{Tr}[e^{-\beta\hat{H}}]} \tag{10.36}$$

Problem 10.6. Show that

$$\langle A \rangle_T = \langle A^\dagger \rangle_T^* \tag{10.37}$$

For future reference we cite here without proof a useful identity that involves the harmonic oscillator Hamiltonian $\hat{H} = \hat{p}^2/2m + (1/2)m\omega^2\hat{q}^2$ and an operator of the general form $\hat{A} = \exp[\alpha_1\hat{p} + \alpha_2\hat{q}]$ with constant parameters α_1 and α_2, that is, the exponential of a linear combination of the momentum and coordinate operators. The identity, known as the *Bloch theorem*, states that the thermal average $\langle \hat{A} \rangle_T$ (under the harmonic oscillator Hamiltonian) is related to the thermal average $\langle (\alpha_1\hat{p} + \alpha_2\hat{q})^2 \rangle_T$ according to

$$\langle e^{\alpha_1\hat{p}+\alpha_2\hat{q}} \rangle_T = e^{(1/2)\langle(\alpha_1\hat{p}+\alpha_2\hat{q})^2\rangle_T} \tag{10.38}$$

Problem 10.7. Prove the classical analog of (10.38), that is,

$$\langle \exp(\alpha_1 p + \alpha_2 q) \rangle_T = \exp[(1/2)\langle(\alpha_1 p + \alpha_2 q)^2\rangle_T]$$

where $\langle A(p,q) \rangle_T = \int dp \int dq A(p,q) \exp(-\beta H(p,q)) / \int dp \int dq \exp(-\beta H(p,q))$ and $H(p,q)$ is the classical harmonic oscillator Hamiltonian. (Note: the needed two-dimensional integrations can be done directly, or you can use the general relationship (7.63)).

10.2 An example: The time evolution of a two-level system in the density matrix formalism

In Section 2.2 we have used the two coupled states model as a simple playground for investigating time evolution in quantum mechanics. Here we reformulate this problem in the density matrix language as an example for using the quantum Liouville equation

$$\frac{d\hat{\rho}}{dt} = -\frac{i}{\hbar}[\hat{H}, \hat{\rho}(t)] \tag{10.39}$$

The Hamiltonian is taken in the form

$$\hat{H} = H_{11}|1\rangle\langle 1| + H_{22}|2\rangle\langle 2| + H_{12}|1\rangle\langle 2| + H_{21}|2\rangle\langle 1| \tag{10.40}$$

and the density operator is

$$\hat{\rho} = \rho_{11}(t)|1\rangle\langle 1| + \rho_{22}(t)|2\rangle\langle 2| + \rho_{12}(t)|1\rangle\langle 2| + \rho_{21}(t)|2\rangle\langle 1| \tag{10.41}$$

Using

$$\begin{aligned}[\hat{H}, \hat{\rho}] = &H_{11}(\rho_{12}|1\rangle\langle 2| - \rho_{21}|2\rangle\langle 1|) + H_{22}(\rho_{21}|2\rangle\langle 1| - \rho_{12}|1\rangle\langle 2|) \\ &+ H_{12}[-\rho_{11}|1\rangle\langle 2| + \rho_{22}|1\rangle\langle 2| + \rho_{21}|1\rangle\langle 1| - \rho_{21}|2\rangle\langle 2|] \\ &+ H_{21}[\rho_{11}|2\rangle\langle 1| - \rho_{22}|2\rangle\langle 1| + \rho_{12}|2\rangle\langle 2| - \rho_{12}|1\rangle\langle 1|]\end{aligned} \tag{10.42}$$

we get

$$\hbar\frac{d\rho_{11}}{dt} = -iH_{12}\rho_{21} + iH_{21}\rho_{12} \tag{10.43a}$$

$$\hbar\frac{d\rho_{22}}{dt} = iH_{12}\rho_{21} - iH_{21}\rho_{12} \tag{10.43b}$$

$$\begin{aligned}\hbar\frac{d\rho_{12}}{dt} &= -iH_{11}\rho_{12} + iH_{22}\rho_{12} + iH_{12}\rho_{11} - iH_{12}\rho_{22} \\ &= -iE_{12}\rho_{12} + iH_{12}\rho_{11} - iH_{12}\rho_{22}\end{aligned} \tag{10.43c}$$

$$\begin{aligned}\hbar\frac{d\rho_{21}}{dt} &= iH_{11}\rho_{21} - iH_{22}\rho_{21} - iH_{21}\rho_{11} + iH_{21}\rho_{22} \\ &= -iE_{21}\rho_{21} - iH_{21}\rho_{11} + iH_{21}\rho_{22}\end{aligned} \tag{10.43d}$$

where we have denoted $E_{21} = -E_{12} = H_{22} - H_{11}$. Note that the sum of Eqs (10.43a) and (10.43b) vanishes, because $\rho_{11} + \rho_{22} = 1$. There are therefore only three independent variables. Defining

$$\sigma_z(t) \equiv \rho_{11}(t) - \rho_{22}(t) \tag{10.44}$$

$$\sigma_+(t) \equiv \rho_{21}(t) \tag{10.45}$$

$$\sigma_-(t) \equiv \rho_{12}(t) \tag{10.46}$$

we obtain from Eqs (10.43) the following equations of motion for these new variables

$$\frac{d\sigma_z}{dt} = 2\frac{i}{\hbar}H_{21}\sigma_- - 2\frac{i}{\hbar}H_{12}\sigma_+$$
$$= \frac{i}{\hbar}[(H_{12}+H_{21})-(H_{12}-H_{21})]\sigma_- - \frac{i}{\hbar}[(H_{12}+H_{21})+(H_{12}-H_{21})]\sigma_+ \tag{10.47a}$$

$$\frac{d\sigma_+}{dt} = i\omega\sigma_+ - \frac{i}{\hbar}H_{21}\sigma_z \tag{10.47b}$$

$$\frac{d\sigma_-}{dt} = -i\omega\sigma_- + \frac{i}{\hbar}H_{12}\sigma_z \tag{10.47c}$$

where $\omega = E_{12}/\hbar$.

Problem 10.8. Show that $\sigma_z(t)$, $\sigma_+(t)$, and $\sigma_-(t)$ are the expectation values of the operators

$$\hat{\sigma}_z = |1\rangle\langle 1| - |2\rangle\langle 2| \tag{10.48a}$$

$$\hat{\sigma}_+ = |1\rangle\langle 2| \tag{10.48b}$$

$$\hat{\sigma}_- = |2\rangle\langle 1| \tag{10.48c}$$

For example, $\sigma_z(t) = \mathrm{Tr}(\hat{\rho}(t)\hat{\sigma}_z)$, etc.

In terms of

$$\sigma_x = \sigma_+ + \sigma_-; \qquad \sigma_y = -i(\sigma_+ - \sigma_-) \tag{10.49}$$

these evolution equations take the forms

$$\frac{d\sigma_x}{dt} = \omega\sigma_y + \frac{i}{\hbar}(H_{12}-H_{21})\sigma_z \tag{10.50a}$$

$$\frac{d\sigma_y}{dt} = -\omega\sigma_x - \frac{1}{\hbar}(H_{12}+H_{21})\sigma_z \tag{10.50b}$$

$$\frac{d\sigma_z}{dt} = -\frac{i}{\hbar}(H_{12}-H_{21})\sigma_x + \frac{1}{\hbar}(H_{12}+H_{21})\sigma_y \tag{10.50c}$$

Equations (10.47) or (10.50) do not have a mathematical or numerical advantage over Eqs (10.43), however, they show an interesting analogy with another physical system, a spin $\frac{1}{2}$ particle in a magnetic field. This is shown in Appendix 10A. A more important observation is that as they stand, Eqs (10.43) and their equivalents (10.47) and (10.50) do not contain information that was not available in the regular time-dependent Schrödinger equation whose solution for this problem was discussed in Section 2.2. The real advantage of the Liouville equation appears in the description of processes in which the time evolution of the density matrix cannot be associated with that of wavefunctions. Such cases are discussed below.

10.3 Reduced descriptions

Nothing is less real than realism ... Details are confusing. It is only by selection, by elimination, by emphasis, that we get at the real meaning of things. (Georgia O'Keeffe)

In Chapter 7 (see in particular Section 7.2) we have motivated the use of reduced descriptions of dynamical processes, where we focus on the dynamics of the subsystem of interest under the influence of its environment. This leads to reduced descriptions of dynamical processes whose stochastic nature stems from the incomplete knowledge of the state of the bath. The essence of a reduction process is exemplified by the relationship

$$P(x_1) = \int dx_2 P(x_1, x_2) \tag{10.51}$$

between the joint probability distribution for two variables x_1 and x_2 and the probability distribution of the variable x_1 alone, irrespective of the value of x_2. Extensive use of such reduction procedures was done in Section 5.3 in conjunction with the theory of classical liquids. Obviously, the same concept and the same need exist also in quantum mechanics, and the density operator, the quantum analog of the classical phase space distribution function is the natural starting point for such considerations. In what follows we discuss such reduction procedures in the quantum mechanical framework.

10.3.1 General considerations

Let S be the quantum system of interest and let B be the surrounding bath, also a quantum system. The Hamiltonian is

$$\hat{H} = \hat{H}_{\text{S}} + \hat{H}_{\text{B}} + \hat{H}_{\text{SB}} = \hat{H}_0 + \hat{V} \tag{10.52}$$

where $\hat{V} = \hat{H}_{\text{SB}}$ will denote here the system–bath interaction. Let $\{|s\rangle\}$ and $\{|b\rangle\}$ be the (assumed orthonormal) sets of eigenstates of $\hat{H}_{\text{S}}$ and $\hat{H}_{\text{B}}$, respectively. Then the density operator $\hat{\rho}$ of the overall system–bath super-system may be written in the representation defined by the product states $|sb\rangle = |s\rangle|b\rangle$ as

$$\begin{aligned} \hat{\rho} &= \sum_{s,b}\sum_{s',b'} \rho_{sb,s'b'} |sb\rangle\langle s'b'| \\ \rho_{sb,s'b'} &= \langle sb|\hat{\rho}|s'b'\rangle \end{aligned} \tag{10.53}$$

A reduced description of the subsystem S alone will provide a density operator

$$\hat{\sigma} = \sum_{s,s'} \sigma_{s,s'} |s\rangle\langle s'| \qquad (10.54)$$
$$\sigma_{s,s'} = \langle s|\hat{\sigma}|s'\rangle$$

in the system sub-space. Such a density operator has to have the property that the average of any system operator $\hat{A} = \sum_{s,s'} A_{s,s'} |s\rangle\langle s'|$ is given by

$$\langle A\rangle = \text{Tr}_\text{S}[\hat{\sigma}\hat{A}] \qquad (10.55)$$

The same average can be taken in the overall system

$$\langle A\rangle = \text{Tr}_\text{S+B}[\hat{\rho}\hat{A}] = \sum_{sb} \langle sb|\hat{\rho}\hat{A}|sb\rangle = \sum_{sb}\sum_{s'b'} \langle sb|\hat{\rho}|s'b'\rangle\langle s'b'|\hat{A}|sb\rangle \qquad (10.56)$$

However, $\hat{A}$, being an operator on the system only satisfies $\langle s'b'|\hat{A}|sb\rangle = A_{s,s'}\delta_{b,b'}$. Equation (10.56) therefore implies

$$\langle A\rangle = \sum_{s,s'}\sum_{b} \langle sb|\hat{\rho}|s'b\rangle\langle s'|\hat{A}|s\rangle = \text{Tr}_\text{S}[(\text{Tr}_\text{B}\hat{\rho})\hat{A}] \qquad (10.57)$$

Comparing (10.57) with (10.55) we conclude

$$\hat{\sigma} = \text{Tr}_\text{B}\hat{\rho} \qquad (10.58)$$

Equation (10.58) is the quantum mechanical analog of Eq. (10.51).

Problem 10.9. A system that comprises a two-level sub-system and a bath is found in a pure state, characterized by finite probabilities to exist in states 1 and 2 of the two-level subsystem where each of them is associated with a different bath state, b_1 and b_2, respectively. Show that the corresponding reduced density matrix of the 2-level subsystem does not describe a pure state but a statistical mixture.

Solution: In the basis of direct products $|j, b\rangle = |j\rangle|b\rangle$ of eigenstates of the isolated 2-level subsystem and the bath, the given pure state is $\psi = C_1|1, b_1\rangle + C_2|2, b_2\rangle$, so that the probabilities that the subsystem is in states $|1\rangle$ or $|2\rangle$ are $|C_1|^2$ and $|C_2|^2$ respectively. The corresponding density operator is

$$\hat{\rho} = |C_1|^2|1, b_1\rangle\langle 1, b_1| + |C_2|^2|2, b_2\rangle\langle 2, b_2| + C_1C_2^*|1, b_1\rangle\langle 2, b_2| \\ + C_1^*C_2|2, b_2\rangle\langle 1, b_1|$$

The reduced density matrix of the subsystem alone is (using $\langle b_1|b_2\rangle = 0$)

$$\hat{\sigma} = Tr_B\hat{\rho} = \langle b_1|\hat{\rho}|b_1\rangle + \langle b_2|\hat{\rho}|b_2\rangle = |C_1|^2|1\rangle\langle 1| + |C_2|^2|2\rangle\langle 2|$$

which obviously describes a mixed state.

Problem 10.10. Show that Eq. (10.58) is satisfied for $\hat{\rho}$ and $\hat{\sigma}$ defined by Eqs (10.53) and (10.54) provided that

$$\sigma_{s,s'} = \sum_b \rho_{sb,s'b} \tag{10.59}$$

Note that the same results apply to time-dependent density operators, that is, Eq. (10.59) holds for the corresponding $\sigma_{s,s'}(t)$ and $\rho_{sb,s'b}(t)$, whether their time dependence is intrinsic as in Eq. (10.16) or stems from external perturbations.

Using this reduction operation we may obtain interesting relationships by taking traces over bath states of the equations of motion (10.15) and (10.21). Consider for example the Liouville equation in the Schrödinger representation, Eq. (10.15). (Note: below, an operator $\hat{A}$ in the interaction representation is denoted $\hat{A}_\mathrm{I}$ while in the Schrödinger representation it carries no label. Labels S and B denote system and bath.)

$$\frac{d\hat{\rho}}{dt} = -\frac{i}{\hbar}[\hat{H}, \hat{\rho}] = -\frac{i}{\hbar}[\hat{H}_\mathrm{S}, \hat{\rho}] - \frac{i}{\hbar}[\hat{H}_\mathrm{B}, \hat{\rho}] - \frac{i}{\hbar}[\hat{V}, \hat{\rho}] \tag{10.60}$$

Taking Tr_B of both sides we note that $\mathrm{Tr}_\mathrm{B}[\hat{H}_\mathrm{S}, \hat{\rho}] = [\hat{H}_\mathrm{S}, \mathrm{Tr}_\mathrm{B}\hat{\rho}] = [\hat{H}_\mathrm{S}, \hat{\sigma}]$ while $\mathrm{Tr}_\mathrm{B}([\hat{H}_\mathrm{B}, \hat{\rho}]) = \sum_b [E_b, \hat{\rho}] = 0$. This leads to

$$\frac{d\hat{\sigma}}{dt} = -\frac{i}{\hbar}[\hat{H}_\mathrm{S}, \hat{\sigma}] - \frac{i}{\hbar}\mathrm{Tr}_\mathrm{B}([\hat{V}, \hat{\rho}]) \tag{10.61}$$

Next consider the same time evolution in the interaction representation. For the overall system we have (cf. Eq. (10.21))

$$\frac{d\hat{\rho}_{\mathrm{I}}}{dt} = -\frac{i}{\hbar}[\hat{V}_{\mathrm{I}}(t), \hat{\rho}_{\mathrm{I}}(t)] \tag{10.62}$$

where

$$\hat{\rho}_{\mathrm{I}}(t) = e^{(i/\hbar)\hat{H}_0 t}\hat{\rho}(t)e^{-(i/\hbar)\hat{H}_0 t} \tag{10.63a}$$

$$\hat{V}_{\mathrm{I}}(t) = e^{(i/\hbar)\hat{H}_0 t}\hat{V}e^{-(i/\hbar)\hat{H}_0 t} \tag{10.63b}$$

Defining, in analogy to (10.58)

$$\hat{\sigma}_{\mathrm{I}}(t) \equiv \mathrm{Tr}_{\mathrm{B}}\hat{\rho}_{\mathrm{I}}(t) \tag{10.64}$$

it follows from (10.63a) that

$$\hat{\sigma}_{\mathrm{I}}(t) = e^{(i/\hbar)\hat{H}_{\mathrm{S}} t}\hat{\sigma}(t)e^{-(i/\hbar)\hat{H}_{\mathrm{S}} t} = e^{(i/\hbar)\hat{H}_{\mathrm{S}} t}(\mathrm{Tr}_{\mathrm{B}}\hat{\rho}(t))e^{-(i/\hbar)\hat{H}_{\mathrm{S}} t} \tag{10.65}$$

Problem 10.11. (1) Show that (10.65) follows from (10.63a). (2) Use Eqs (10.65) and (10.61) together with the definitions (10.63) to prove the following identity

$$\frac{d\hat{\sigma}_{\mathrm{I}}(t)}{dt} = -\frac{i}{\hbar}\mathrm{Tr}_{\mathrm{B}}([\hat{V}_{\mathrm{I}}(t), \hat{\rho}_{\mathrm{I}}(t)]) \tag{10.66}$$

Proof of (10.66): Take the time derivative of (10.65) to get

$$\frac{d\hat{\sigma}_{\mathrm{I}}(t)}{dt} = \frac{i}{\hbar}[\hat{H}_{\mathrm{S}}, \hat{\sigma}_{\mathrm{I}}(t)] + e^{(i/\hbar)\hat{H}_{\mathrm{S}} t}\frac{d\hat{\sigma}(t)}{dt}e^{-(i/\hbar)\hat{H}_{\mathrm{S}} t} \tag{10.67}$$

then use Eq. (10.61) to find that the second term on the right in (10.67) is $-(i/\hbar)[\hat{H}_{\mathrm{S}}, \hat{\sigma}_{\mathrm{I}}(t)] - (i/\hbar)e^{(i/\hbar)\hat{H}_{\mathrm{S}} t}\mathrm{Tr}_{\mathrm{B}}([\hat{V}, \hat{\rho}])e^{-(i/\hbar)\hat{H}_{\mathrm{S}} t}$. Equation (10.67) can therefore be written as

$$\begin{aligned}\frac{d\hat{\sigma}_{\mathrm{I}}(t)}{dt} &= -\frac{i}{\hbar}e^{(i/\hbar)\hat{H}_{\mathrm{S}} t}\mathrm{Tr}_{\mathrm{B}}([\hat{V}, \hat{\rho}])e^{-(i/\hbar)\hat{H}_{\mathrm{S}} t} \\ &= -\frac{i}{\hbar}\mathrm{Tr}_{\mathrm{B}}(e^{(i/\hbar)\hat{H}_{\mathrm{S}} t}[\hat{V}, \hat{\rho}]e^{-(i/\hbar)\hat{H}_{\mathrm{S}} t}) = -\frac{i}{\hbar}\mathrm{Tr}_{\mathrm{B}}(e^{(i/\hbar)\hat{H}_0 t}[\hat{V}, \hat{\rho}]e^{-(i/\hbar)\hat{H}_0 t})\end{aligned} \tag{10.68}$$

which, using (10.63) gives (10.66).

Our goal is to describe the dynamics of our subsystem by constructing an equation of motion for $\hat{\sigma}$. This equation should show the influence of coupling to the thermal reservoir, however, we desire that our dynamical description will be self contained in the sense that elements of the overall density operator $\hat{\rho}$ will not explicitly appear in it. Equation (10.66) is obviously not yet of such a form, but will be used as a starting point for our later discussion (see Section 10.4.3).

10.3.2 A simple example—the quantum mechanical basis for macroscopic rate equations

Consider two coupled multilevel systems L and R, characterized by their spectrum of eigenvectors and eigenvalues. The Hamiltonian without the intersystem coupling is

$$\hat{H}_0 = \hat{H}_L + \hat{H}_R \tag{10.69}$$

$$\hat{H}_L = \sum_l E_l |l\rangle\langle l|; \qquad \hat{H}_R = \sum_r E_r |r\rangle\langle r| \tag{10.70}$$

We assume that $\hat{V}$, the operator that couples systems L and R to each other, mixes only l and r states, that is, $V_{l,l'} = V_{r,r'} = 0$. We are interested in the transition between these two subsystems, induced by $\hat{V}$. We assume that: (1) the coupling $\hat{V}$ is weak coupling in a sense explained below, and (2) the relaxation process that brings each subsystem by itself (in the absence of the other) into thermal equilibrium is much faster that the transition induced by $\hat{V}$ between them. Note that assumption (2), which implies a separation of timescales between the $L \rightleftarrows R$ transition and the thermal relaxation within the L and R subsystems, is consistent with assumption (1).

In the absence of $\hat{V}$ the subsystems reach their own thermal equilibrium so that their density matrices are diagonal, with elements given by

$$\rho_{k,k} \equiv P_k = f_K(E_k); \quad f_K(E) = \frac{e^{-\beta E}}{\mathrm{Tr}(e^{-\beta H_K})}; \qquad k = l, r; \quad K = L, R \tag{10.71}$$

When $\hat{V} \neq 0$ transitions between L and R can take place, and their populations evolve in time. Defining the total L and R populations by $P_K(t) = \sum_k P_k(t)$, our goal is to characterize the kinetics of the $L \rightleftharpoons R$ process. This is a reduced description because we are not interested in the dynamics of individual level $|l\rangle$ and $|r\rangle$, only in the overall dynamics associated with transitions between the L and R "species." Note that "reduction" can be done on different levels, and the present focus is on P_L and P_R and the transitions between them. This reduction is not done by limiting attention to a small physical subsystem, but by focusing on a subset of density-matrix elements or, rather, their combinations.

We start from the Liouville equation (10.29) written in the basis of the $|l\rangle$ and $|r\rangle$ states,

$$\frac{d\rho_{k,k'}}{dt} = -\frac{i}{\hbar}E_{k,k'}\rho_{k,k'} - \frac{i}{\hbar}\sum_{k''}(V_{k,k''}\rho_{k'',k'} - V_{k'',k'}\rho_{k,k''}); \qquad k,k',k'' = l,r \tag{10.72}$$

where $E_{k,k'} = E_k - E_{k'}$, and write it separately for the diagonal and non-diagonal elements of $\hat{\rho}$. Recalling that $\hat{V}$ couples only between states of different subsystems we get

$$\frac{d\rho_{l,l}}{dt} = \frac{dP_l}{dt} = -\frac{i}{\hbar}\sum_r(V_{l,r}\rho_{r,l} - V_{r,l}\rho_{l,r}) = -\frac{2}{\hbar}\mathrm{Im}\sum_r V_{r,l}\rho_{l,r} \tag{10.73}$$

(and a similar equation with $l \leftrightarrow r$)

$$\frac{d\rho_{l,r}}{dt} = -\frac{i}{\hbar}E_{l,r}\rho_{l,r} - \frac{i}{\hbar}(V_{l,r}\rho_{r,r} - V_{l,r}\rho_{l,l}) + \begin{bmatrix} V \times \text{terms containing} \\ \text{non-diagonal } \rho \text{ elements} \end{bmatrix} \tag{10.74}$$

In what follows we will disregard the terms containing non-diagonal elements of $\hat{\rho}$ multiplying elements of $\hat{V}$ on the right-hand side of (10.74). The rationale for this approximation is that provided assumptions (1) and (2) above are valid, $\hat{\rho}$ remains close to the diagonal form obtained when $\hat{V} = 0$; with non-diagonal terms of order $\hat{V}$.

Below we will use the timescale separation between the (fast) thermal relaxation within the L and R subsystems and the (slow) transition between them in one additional way: We will assume that relative equilibrium within each subsystem is maintained, that is,

$$\begin{aligned} P_l(t) &= P_L(t)f_L(E_l) \\ P_r(t) &= P_R(t)f_R(E_r) \end{aligned} \tag{10.75}$$

Assume now that at the distant past, $t \to -\infty$, the two systems L and R were uncoupled from each other and at their internal thermal equilibrium states (10.71). This also implies that $\rho_{r,l}(t = -\infty) = 0$. At that point in the distant past the intersystem coupling $\hat{V}$ was switched on. Propagation according to (10.74) yields

$$\rho_{l,r}(t) = -\frac{i}{\hbar}V_{l,r}\int_{-\infty}^{t} d\tau e^{-i(E_{l,r}/\hbar)(t-\tau)}(P_r(\tau) - P_l(\tau)) \tag{10.76}$$

where $E_{l,r} = E_l - E_r$. Upon inserting into (10.73) this leads to

$$\frac{dP_l}{dt} = \sum_r \int_{-\infty}^{t} d\tau S_{l,r}(t-\tau)(P_r(\tau) - P_l(\tau)) \tag{10.77}$$

where

$$S_{l,r}(t-\tau) \equiv \frac{2}{\hbar^2}\mathrm{Re}|V_{l,r}|^2 e^{-i(E_{l,r}/\hbar)(t-\tau)} = \frac{2}{\hbar^2}|V_{l,r}|^2 \cos\left(\frac{E_{l,r}}{\hbar}(t-\tau)\right) \tag{10.78}$$

Finally, summing Eq. (10.77) over all l and using (10.75) we get

$$\frac{dP_L}{dt} = \int_{-\infty}^{t} d\tau K_{L\leftarrow R}(t-\tau)P_R(\tau) - \int_{-\infty}^{t} d\tau K_{R\leftarrow L}(t-\tau)P_L(\tau) \tag{10.79}$$

with

$$K_{L\leftarrow R}(t) = \frac{2}{\hbar^2}\sum_r f_R(E_r)\sum_l |V_{l,r}|^2 \cos\left(\frac{1}{\hbar}E_{l,r}t\right) \tag{10.80}$$

and

$$K_{R\leftarrow L}(t) = \frac{2}{\hbar^2}\sum_l f_L(E_l)\sum_r |V_{l,r}|^2 \cos\left(\frac{1}{\hbar}E_{l,r}t\right) \tag{10.81}$$

Problem 10.12. Show that

$$K_{R\leftarrow L}(t=0) = (2/\hbar^2)\sum_l f_L(E_l)\langle l|\hat{V}^2|l\rangle \tag{10.82}$$

(and a similar relation in which $r \leftrightarrow l$ for $K_{L\leftarrow R}(t=0)$).

The kinetic Eq. (10.79) is non-Markovian: the rate at which $P_L(t)$ changes depends on its earlier values, going back over a time period characterized by the "memory time"—times above which $K_{R\leftarrow L}(t)$, $K_{L\leftarrow R}(t) \sim 0$. To understand the microscopic origin of this memory consider, for example, the function $K_{R\leftarrow L}(t)$.

We assume that the manifold of states $\{|r\rangle\}$ is a continuum, that is, that the corresponding eigenvalues $\{E_r\}$ span a continuous energy range. Let E_{r0} be the center of this range, then

$$\sum_r |V_{l,r}|^2 \cos\left(\frac{1}{\hbar}E_{l,r}t\right) = \mathrm{Re}\int dE_r \rho(E_r)|V_{l,r}|^2 \exp\left(\frac{i}{\hbar}E_{l,r}t\right)$$
$$= \mathrm{Re}\left\{e^{(i/\hbar)(E_l-E_{r0})t}\int_{-\infty}^{\infty} dE_r\rho(E_r)|V_{l,r}|^2 \times \exp\left(-\frac{i}{\hbar}(E_r - E_{r0})t\right)\right\} \tag{10.83}$$

Provided that $\rho(E_r)|V_{l,r}|^2$ is a relatively smooth function of E_r, the Fourier transform in (10.83) decays to zero on the timescale $\hbar W_R^{-1}$, where W_R is the spectral width of the manifold $\{E_r\}$. A similar argument holds for $K_{L\leftarrow R}(t)$, which decays to zero on a timescale $\hbar W_L^{-1}$. If these timescales are much smaller than the characteristic $L \rightleftarrows R$ transition time, that is, if the spectral widths of the L and R manifolds are large relative to the inverse reaction rate (multiplied by $\hbar$) we can replace $P_R(\tau)$ and $P_L(\tau)$ in (10.79) by $P_R(t)$ and $P_L(t)$, respectively, to get

$$\frac{dP_L}{dt} = k_{L\leftarrow R}P_R - k_{R\leftarrow L}P_L \tag{10.84}$$

where

$$k_{R\leftarrow L} = \int_0^{\infty} d\tau K_{R\leftarrow L}(\tau); \qquad k_{L\leftarrow R} = \int_0^{\infty} d\tau K_{L\leftarrow R}(\tau) \tag{10.85}$$

The approximation that leads to from Eq. (10.79) to (10.84), which relies on the large spectral width and the smooth spectral functions of the state manifolds involved, is sometimes referred to as the *wide band approximation*. Similar arguments were used in Section 9.1 in treating the decay of a single state coupled to a continuous manifold of states, in order to obtain a constant decay rate given by the golden rule formula, Eq. (9.25). Also in the present case, under the approximations invoked above, the rates (10.85) can be written as thermally averaged golden-rule expressions (see Problem 10.13 below).

Problem 10.13. Show that

$$k_{R\leftarrow L} = \int_0^\infty d\tau K_{R\leftarrow L}(\tau) = \sum_l f_L(E_l) k_{\{r\}\leftarrow l} \tag{10.86}$$

where $k_{\{r\}\leftarrow l} = (2\pi/\hbar)\sum_r |V_{l,r}|^2 \delta(E_l - E_r)$ has the form of a simple golden-rule type rate to go from state l into a continuous manifold $\{r\}$. Therefore $k_{R\leftarrow L}$ is just the thermal average of this rate over the thermal distribution of l states. $k_{L\leftarrow R}$ may be interpreted in an equivalent way.

Equations (10.79)–(10.85) provide the basis for many macroscopic rate theories, for example, the kinetic theory of chemical reaction rates. Obviously it was formulated above for a very simple situation that in the language of chemical reaction rates corresponds to unimolecular inter-conversion. Still, the concepts that were introduced are general, and can be used in more complex situations. We have relied on two key ideas: First, the separation of timescales between the (small) transition rate under discussion and the (fast) thermal relaxation rate has made it possible to focus on the transition between two subsystems defined by manifolds of energy states, and avoid the need to address individual transitions between all microscopic levels. The use of the density matrix formalism was critical at this stage, as it has made it possible to consider separately the diagonal and non-diagonal elements of the density matrix and to invoke the consequence of the timescale separation discussed above with regard to their relative sizes. This leads to Eq. (10.79). Second, we have used arguments similar to those encountered in our discussion of the decay of a level coupled to a broad continuum in order to go over to the Markovian limit, Eq. (10.84). These arguments again rely on timescale separation, now between the (relatively short) time, W^{-1}, associated with the spectral structure of the continuous level manifolds that affect irreversible decay, and the (relatively long) time that characterizes the process of interest.

The above derivation has relied in an essential way on the smallness of the non-diagonal elements of the density matrix in the energy representation that was chosen in accord with our physical picture of the system. Without explicitly stating the fact, we have assumed that *dephasing*, that is, the damping of coherences reflected in the non-diagonal density matrix elements, is fast. In what follows we explore more general applications of the density matrix formalism, where the existence of the thermal environment and the coupling of system of interest to this environment are considered explicitly. This will make it possible address directly population and phase relaxation and the dependence of their rates on the physical characteristics of the system.

10.4 Time evolution equations for reduced density operators: The quantum master equation

Let us state our goal again. The system of interest is in contact with its thermal environment. This environment is interesting to us only as much as it affects the dynamics of our system. We want to derive closed equations of motion for the system, where "closed" implies that only relevant variables, those belonging to the system's subspace, appear explicitly. The density matrix formalism provides a convenient quantum mechanical framework for this task, where we seek an equation of motion for $\hat{\sigma} = \mathrm{Tr}_B \hat{\rho}$, the so called *quantum master equation*. An analogous methodology that starts from the classical distribution function $f(\mathbf{r}^N, \mathbf{p}^N; t)$ is equally useful in classical mechanics, however, with the exception of deriving a classical Langevin equation for a system interacting with a harmonic bath (Section 8.2.5), the reduced equations of motion advanced in Chapter 8 were constructed phenomenologically. The derivation of equation (10.84) can be seen as a microscopic basis for the phenomenological master equations used in Section 8.3. Now we aim for a more general microscopic derivation which, as we will see, not only provides the foundation for such reduced descriptions, but can also identify new dynamical issues not easily come by in a phenomenological approach. Projection operators, operators that project onto the subspace of interest, are very useful for carrying out this task.

10.4.1 Using projection operators

We have already encountered the projection operator formalism in Appendix 9A, where an application to the simplest system–bath problem—a single level interacting with a continuum, was demonstrated. This formalism is general can be applied in different ways and flavors. In general, a projection operator (or projector) $\hat{P}$ is defined with respect to a certain sub-space whose choice is dictated by the physical problem. By definition it should satisfy the relationship $\hat{P}^2 = \hat{P}$ (operators that satisfy this relationship are called *idempotent*), but other than that can be chosen to suit our physical intuition or mathematical approach. For problems involving a system interacting with its equilibrium thermal environment a particularly convenient choice is the *thermal projector*: An operator that projects the total system–bath density operator on a product of the system's reduced density operator and the equilibrium density operator of the bath, $\hat{\rho}_{\mathrm{eq}}^{(B)}$

$$\hat{P}\hat{\rho} = \hat{\rho}_{\mathrm{eq}}^{(B)} \mathrm{Tr}_B \hat{\rho} = \hat{\rho}_{\mathrm{eq}}^{(B)} \hat{\sigma} \tag{10.87}$$

Since $\mathrm{Tr}_B \hat{\rho}_{\mathrm{eq}}^{(B)} = 1$ $\hat{P}$ is indeed idempotent, $\hat{P}^2 = \hat{P}$. The complementary projector $\hat{Q}$ is defined simply by $\hat{Q} = 1 - \hat{P}$.

Problem 10.14. Using the definitions (10.63a), (10.64) and (10.58) show that

$$\hat{P}\hat{\rho}_{\mathrm{I}}(t) = \hat{\rho}_{\mathrm{eq}}^{(B)}\hat{\sigma}_{\mathrm{I}}(t) \tag{10.88}$$

The projection operator $\hat{P}$ is chosen according to our stated need: We want an equation of motion that will describe the time evolution of a system in contact with a thermally equilibrated bath. $\hat{P}\hat{\rho}$ of Eq. (10.87) is the density operator of just this system, and its dynamics is determined by the time evolution of the system's density operator $\hat{\sigma}$. Finding an equation of motion for this evolution is our next task.

10.4.2 The Nakajima–Zwanzig equation

It is actually simple to find a formal time evolution equation "in P space." This formal simplicity stems from the fact that the fundamental equations of quantum dynamics, the time-dependent Schrödinger equation or the Liouville equation, are linear. Starting from the quantum Liouville equation (10.8) for the overall system—system and bath,

$$\frac{d}{dt}\hat{\rho}(t) = -\frac{i}{\hbar}[\hat{H}, \hat{\rho}(t)] \equiv -i\mathcal{L}\hat{\rho}(t); \qquad \mathcal{L} \equiv \hbar^{-1}[\hat{H},] \tag{10.89}$$

we want to find an equation of motion for the density matrix of a chosen subsystem. Let $\hat{P}$ be a projector on this relevant part of the overall system and let $\hat{Q} = 1 - \hat{P}$. Then (10.89) trivially leads to

$$\frac{d}{dt}\hat{P}\hat{\rho} = -i\hat{P}\mathcal{L}\hat{\rho} = -i\hat{P}\mathcal{L}\hat{P}\hat{\rho} - i\hat{P}\mathcal{L}\hat{Q}\hat{\rho} \tag{10.90}$$

$$\frac{d}{dt}\hat{Q}\hat{\rho} = -i\hat{Q}\mathcal{L}\hat{\rho} = -i\hat{Q}\mathcal{L}\hat{P}\hat{\rho} - i\hat{Q}\mathcal{L}\hat{Q}\hat{\rho} \tag{10.91}$$

These equations look complicated, however, *in form* (as opposed to *in physical contents*) they are very simple. We need to remember that in any representation that uses a discrete basis set $\hat{\rho}$ is a vector and $\mathcal{L}$ is a matrix. The projectors $\hat{P}$ and $\hat{Q}$ are also matrices that project on parts of the vector space (see Appendix 9A). For example, in the simplest situation

$$\hat{P}\hat{\rho} = \begin{pmatrix} \rho_P \\ 0 \end{pmatrix}; \qquad \hat{Q}\hat{\rho} = \begin{pmatrix} 0 \\ \rho_Q \end{pmatrix}; \qquad \hat{\rho} = \hat{P}\hat{\rho} + \hat{Q}\hat{\rho} = \begin{pmatrix} \rho_P \\ \rho_Q \end{pmatrix} \tag{10.92}$$

where ρ_P is just that part of $\hat{\rho}$ that belongs to the P space, etc. Similarly

$$\hat{P}\mathcal{L}\hat{P} = \begin{pmatrix} \mathcal{L}_{PP} & 0 \\ 0 & 0 \end{pmatrix}; \qquad \hat{P}\mathcal{L}\hat{Q} = \begin{pmatrix} 0 & \mathcal{L}_{PQ} \\ 0 & 0 \end{pmatrix}, \text{etc.} \tag{10.93}$$

where we should keep in mind that $\mathcal{L}_{PQ}$ is not necessarily a square matrix because the two complementary subspaces defined by $\hat{P}$ and $\hat{Q}$ are not usually of equal dimensions. Writing Eq. (10.89) in the form

$$\frac{d}{dt}\begin{pmatrix}\rho_P\\ \rho_Q\end{pmatrix} = -i\begin{pmatrix}\mathcal{L}_{PP} & \mathcal{L}_{PQ}\\ \mathcal{L}_{QP} & \mathcal{L}_{QQ}\end{pmatrix}\begin{pmatrix}\rho_P\\ \rho_Q\end{pmatrix} \tag{10.94a}$$

Equations (10.90) and (10.91) are seen to be just the equivalent set of equations for ρ_P and ρ_Q. One word of caution is needed in the face of possible confusion: To avoid too many notations it has become customary to use $\hat{P}\hat{\rho}$ also to denote ρ_P, $\hat{P}\mathcal{L}\hat{P}$ also to denote $\mathcal{L}_{PP}$, etc., and to let the reader decide from the context what these structures mean. With this convention Eq. (10.94a) is written in the form

$$\frac{d}{dt}\begin{pmatrix}\hat{P}\hat{\rho}\\ \hat{Q}\hat{\rho}\end{pmatrix} = -i\begin{pmatrix}\hat{P}\mathcal{L}\hat{P} & \hat{P}\mathcal{L}\hat{Q}\\ \hat{Q}\mathcal{L}\hat{P} & \hat{Q}\mathcal{L}\hat{Q}\end{pmatrix}\begin{pmatrix}\hat{P}\hat{\rho}\\ \hat{Q}\hat{\rho}\end{pmatrix} \tag{10.94b}$$

Indeed, Eqs (10.90) and (10.91) are written in this form.

We proceed by integrating Eq. (10.91) and inserting the result into (10.90). Again the procedure is *simple in form*. If $\hat{P}\hat{\rho} = x$ and $\hat{Q}\hat{\rho} = y$ were two scalar variables and all other terms were scalar coefficients, this would be a set of two coupled first order differential equations

$$\frac{d}{dt}x = Ax + By \tag{10.95}$$

$$\frac{d}{dt}y = Cx + Dy \tag{10.96}$$

and we could proceed by solving (10.96) to get (as can be verified by taking the time derivative)

$$y(t) = e^{D(t-t_0)}y(t_0) + \int_{t_0}^{t} d\tau e^{D(t-\tau)}Cx(\tau) \tag{10.97}$$

and inserting into (10.95) to get

$$\frac{d}{dt}x = Ax + B\int_{t_0}^{t} d\tau e^{D(t-\tau)}Cx(\tau) + Be^{D(t-t_0)}y(t_0) \tag{10.98}$$

We will be doing exactly the same thing with Eqs. (10.90) and (10.91). Integration of (10.91) yields

$$\hat{Q}\hat{\rho}(t) = e^{-i\hat{Q}\mathcal{L}(t-t_0)}\hat{Q}\hat{\rho}(t_0) - i\int_{t_0}^{t} d\tau e^{-i\hat{Q}\mathcal{L}(t-\tau)}\hat{Q}\mathcal{L}\hat{P}\hat{\rho}(\tau) \tag{10.99}$$

(this can again be verified by taking time derivative) and inserting this into (10.90) leads to

$$\frac{d}{dt}\hat{P}\hat{\rho}(t) = -i\hat{P}\mathcal{L}\hat{P}\hat{\rho}(t) - \int_{t_0}^{t} d\tau \hat{P}\mathcal{L}e^{-i\hat{Q}\mathcal{L}(t-\tau)}\hat{Q}\mathcal{L}\hat{P}\hat{\rho}(\tau) - i\hat{P}\mathcal{L}e^{-i\hat{Q}\mathcal{L}(t-t_0)}\hat{Q}\hat{\rho}(t_0) \tag{10.100}$$

The identity (10.100) is the Nakajima–Zwanzig equation. It describes the time evolution of the "relevant part" $\hat{P}\hat{\rho}(t)$ of the density operator. This time evolution is determined by the three terms on the right. Let us try to understand their physical contents. In what follows we refer to the relevant and irrelevant parts of the overall system as "system" and "bath" respectively.

The first term, $-i\hat{P}\mathcal{L}\hat{P}\hat{\rho}(t)$ describes the time evolution that would be observed if the system was uncoupled from the bath throughout the process (i.e. if $\hat{Q}\mathcal{L}\hat{P} = \hat{P}\mathcal{L}\hat{Q} = 0$). The second term is the additional contribution to the time evolution of the system that results from its coupling to the bath. This contribution appears as a memory term that depends on the past history, $\hat{P}\hat{\rho}(\tau)$, of the system. Consider the integrand in this term, written in the form[1]

$$\underbrace{\hat{P}\mathcal{L}\hat{Q}}_{4} \times \underbrace{e^{-i\hat{Q}\mathcal{L}\hat{Q}(t-\tau)}}_{3} \times \underbrace{\hat{Q}\mathcal{L}\hat{P}}_{2} \times \underbrace{\hat{P}\hat{\rho}(\tau)}_{1}$$

It shows the relevant (system) part of the density operator at time τ (1) coupled to the bath (2), propagated in the bath subspace from time τ to time t (3) and affecting again the system via the system-bath coupling (4). This is a mathematical expression of what we often refer to as a *reaction field* effect: The system at some time τ appears to act on itself at some later time t, and the origin of this action is the reaction of the system at time t to the effect made by the same system on the bath at some earlier time τ.

The last term on the right-hand side of Eq. (10.100) also has a clear physical interpretation: This is a contribution to force exerted on the system at time t, associated with the initial ($t = t_0$) correlations between the system and the bath embedded in the term $\hat{Q}\hat{\rho}(t_0)$. There are many situations where this contribution to the relevant time evolution can be disregarded, at least at long time, and it is identically zero if $\hat{Q}\hat{\rho}(t_0) = 0$. The last situation appears when $\hat{P}$ is the thermal projector (10.87) if we *assume* that until time t_0 the system and bath were uncoupled with the bath kept at thermal equilibrium.

[1] Note that $\exp(\hat{Q}\mathcal{L}\hat{Q}t)\hat{Q}\hat{\rho} = \exp(\hat{Q}\mathcal{L}t)\hat{Q}\hat{\rho}$.

We end this discussion with two comments. First, we note that the Nakajima–Zwanzig equation (10.100) is exact; no approximations whatever were made in its derivation. Second, this identity can be used in many ways, depending on the choice of the projection operator $\hat{P}$. The thermal projector (10.87) is a physically motivated choice. In what follows we present a detailed derivation of the quantum master equation using this projector and following steps similar to those taken above, however, we will sacrifice generality in order to get practical usable results.

10.4.3 Derivation of the quantum master equation using the thermal projector

Practical solutions of dynamical problems are almost always perturbative. We are interested in the effect of the thermal environment on the dynamical behavior of a given system, so a natural viewpoint is to assume that the dynamics of the system alone is known and to take the system–bath coupling as the perturbation. We have seen (Section 2.7.3) that time dependent perturbation theory in Hilbert space is most easily discussed in the framework of the interaction representation. Following this route[2] we start from the Liouville equation in this representation (cf. Eq. (10.21))

$$\frac{d\hat{\rho}_{\mathrm{I}}}{dt} = -\frac{i}{\hbar}[\hat{V}_{\mathrm{I}}(t), \hat{\rho}_{\mathrm{I}}(t)] \tag{10.101}$$

and, using $\hat{P} + \hat{Q} = 1$, write the two projected equations

$$\frac{d}{dt}\hat{P}\hat{\rho}_{\mathrm{I}} = -\frac{i}{\hbar}\hat{P}[\hat{V}_{\mathrm{I}}, (\hat{P} + \hat{Q})\hat{\rho}_{\mathrm{I}}] \tag{10.102}$$

$$\frac{d}{dt}\hat{Q}\hat{\rho}_{\mathrm{I}} = -\frac{i}{\hbar}\hat{Q}[\hat{V}_{\mathrm{I}}, (\hat{P} + \hat{Q})\hat{\rho}_{\mathrm{I}}] \tag{10.103}$$

Now use Eqs (10.88) and (10.64) in (10.102) to get

$$\begin{aligned}\frac{d}{dt}\hat{\sigma}_{\mathrm{I}} &= -\frac{i}{\hbar}\mathrm{Tr}_{\mathrm{B}}[\hat{V}_{\mathrm{I}}, \hat{\rho}_{\mathrm{eq}}^{(B)}\hat{\sigma}_{\mathrm{I}}] - \frac{i}{\hbar}\mathrm{Tr}_{\mathrm{B}}[\hat{V}_{\mathrm{I}}, \hat{Q}\hat{\rho}_{\mathrm{I}}] \\ &= -\frac{i}{\hbar}[\hat{\bar{V}}_{\mathrm{I}}, \hat{\sigma}_{\mathrm{I}}] - \frac{i}{\hbar}\mathrm{Tr}_{\mathrm{B}}[\hat{V}_{\mathrm{I}}, \hat{Q}\hat{\rho}_{\mathrm{I}}]\end{aligned} \tag{10.104}$$

where

$$\hat{\bar{V}} = \mathrm{Tr}_{\mathrm{B}}(\hat{V}\hat{\rho}_{\mathrm{eq}}^{(B)}) \tag{10.105}$$

and

$$\hat{\bar{V}}_{\mathrm{I}} = e^{(i/\hbar)\hat{H}_{\mathrm{S}}t}\hat{\bar{V}}e^{-(i/\hbar)\hat{H}_{\mathrm{S}}t} \tag{10.106}$$

We shall see below that the last term on the right in (10.104) is second order and higher in the system–bath interaction $\hat{V}$. The time evolution obtained by

[2] This derivation follows that of V. May and O. Kühn, *Charge and Energy Transfer Dynamics in Molecular Systems* (Wiley-VCH, Berlin, 2000).

disregarding it

$$\frac{d\hat{\sigma}_\mathrm{I}}{dt} = -\frac{i}{\hbar}[\hat{\bar{V}}_\mathrm{I}, \hat{\sigma}_\mathrm{I}] \tag{10.107}$$

corresponds to a modified system Hamiltonian, where $\hat{H}_\mathrm{S}$ is replaced by $\hat{H}_\mathrm{S} + \hat{\bar{V}}$. Indeed, the equivalent equation in the Schrödinger representation is

$$\frac{d\hat{\sigma}}{dt} = -\frac{i}{\hbar}[\hat{H}_\mathrm{S} + \hat{\bar{V}}, \hat{\sigma}] \tag{10.108}$$

The operator $\hat{\bar{V}}$ has a very simple interpretation: it is a mean potential that corrects the system Hamiltonian for the average effect of the bath. Such corrections are very important, for example, in determining solvent shifts of spectral lines. Such shifts result from the fact that the average solvent interaction often influences differently the energies of the ground and excited states of a solvent molecule. At the same time it is clear that such average interactions can only affect the system eigenstates and energy levels, but cannot cause relaxation. We see that relaxation phenomena must be associated with the last term of (10.104) that was neglected in (10.107). Moreover, when addressing relaxation, we will often disregard the $\hat{\bar{V}}$ term: This amounts to including it in $\hat{H}_\mathrm{S}$ thus working with a renormalized system Hamiltonian that includes the energy shifts associated with the average effect of the solvent.

Problem 10.15. Show that $\mathrm{Tr}_\mathrm{B}[\hat{V}_\mathrm{I}, \hat{\rho}_\mathrm{eq}^{(B)}\hat{\sigma}_\mathrm{I}] = [\hat{\bar{V}}_\mathrm{I}, \hat{\sigma}_\mathrm{I}]$ where $\hat{\bar{V}}_\mathrm{I}$ is defined by (10.105) and (10.106).

Consider now this last term, $(i/\hbar)\mathrm{Tr}_\mathrm{B}[\hat{V}_\mathrm{I}, \hat{\mathcal{Q}}\hat{\rho}_\mathrm{I}]$, in (10.104). It contains $\hat{\mathcal{Q}}\hat{\rho}_\mathrm{I}$ whose time evolution is given by Eq. (10.103). We rewrite this equation in the form

$$\frac{d}{dt}\hat{\mathcal{Q}}\hat{\rho}_\mathrm{I} = -\frac{i}{\hbar}\hat{\mathcal{Q}}[\hat{V}_\mathrm{I}, \hat{\rho}_\mathrm{eq}^{(B)}\hat{\sigma}_\mathrm{I}] - \frac{i}{\hbar}\hat{\mathcal{Q}}[\hat{V}_\mathrm{I}, \hat{\mathcal{Q}}\hat{\rho}_\mathrm{I}] \tag{10.109}$$

and formally integrate it to get

$$\hat{\mathcal{Q}}\hat{\rho}_\mathrm{I}(t) = \hat{\mathcal{Q}}\hat{\rho}_\mathrm{I}(0) - \frac{i}{\hbar}\int_0^t dt'\hat{\mathcal{Q}}[\hat{V}_\mathrm{I}(t'), \hat{\rho}_\mathrm{eq}^{(B)}\hat{\sigma}_\mathrm{I}(t')] - \frac{i}{\hbar}\int_0^t dt'\hat{\mathcal{Q}}[\hat{V}_\mathrm{I}(t'), \hat{\mathcal{Q}}\hat{\rho}_\mathrm{I}(t')] \tag{10.110}$$

This equation can be iterated by inserting this expression for $\hat{\mathcal{Q}}\hat{\rho}_\mathrm{I}(t)$ into the integrand in the second term on the right, and a perturbative expansion in increasing powers of $\hat{V}$ can be obtained by continuing this procedure repeatedly. This is the analog of the perturbative expansion of the time-dependent wavefunction, Eq. (2.76). The resulting infinite series for $\hat{\mathcal{Q}}\hat{\rho}_\mathrm{I}(t)$ can be inserted into Eq. (10.104) to yield a formally exact equation in P-space. Of course this equation contains the effect of the system–thermal bath coupling and is generally very difficult to simplify and to solve.

Fortunately a substantial amount of relevant physics can be extracted by considering the low-order terms in this expansion. The lowest order is the mean potential approximation (10.107). The next order is obtained by neglecting the last term on the right-hand side of (10.110) and inserting the remaining expression for $\hat{Q}\hat{\rho}_\text{I}$ into Eq. (10.104). The resulting approximate time evolution equation for the system density operator $\hat{\sigma}$ is what is usually referred to as the quantum master equation.

Problem 10.16. Starting from Eq. (10.66), show that replacing $\hat{\rho}_\text{I}(t)$ on the right-hand side by $\hat{\rho}_\text{I}(t) = \hat{\sigma}_\text{I}(t)\hat{\rho}_\text{I}^{(B)}(t)$, where $\hat{\rho}_\text{I}^{(B)}(t) = e^{i\hat{H}_\text{B}t/\hbar}\hat{\rho}^{(B)}(t)e^{-i\hat{H}_\text{B}t/\hbar}$ is the density operator of the thermal reservoir and $\hat{\sigma}_\text{I}(t) = e^{i\hat{H}_\text{S}t/\hbar}\hat{\sigma}(t)e^{-i\hat{H}_\text{S}t/\hbar}$, then taking Tr_B of both sides, leads to

$$\frac{d\hat{\sigma}_\text{I}(t)}{dt} = -\frac{i}{\hbar}[\text{Tr}_\text{B}(\hat{V}_\text{I}(t)\hat{\rho}_\text{I}^{(B)}(t)), \hat{\sigma}_\text{I}(t)] = -\frac{i}{\hbar}[\hat{\bar{V}}_\text{I}(t), \hat{\sigma}_\text{I}(t)] \tag{10.111}$$

where $\hat{\bar{V}}(t) = \text{Tr}_\text{B}(\hat{V}\hat{\rho}^{(B)}(t))$ and $\hat{\bar{V}}_\text{I}(t) = e^{(i/\hbar)\hat{H}_\text{S}t}\hat{\bar{V}}(t)e^{-(i/\hbar)\hat{H}_\text{S}t}$. Equation (10.111) has the same form as Eq. (10.107), however, the definition (10.105) is replaced by a more general definition involving the time-dependent density operator of the bath.

10.4.4 The quantum master equation in the interaction representation

As just stated, we henceforth use the term "quantum master equation" (QME) to denote the approximate time evolution equation for the system's density matrix $\hat{\sigma}$ obtained in second order in the system–bath coupling $\hat{V}$. To obtain this equation we start from Eq. (10.104) and use a simplified version of Eq. (10.110)

$$\hat{Q}\hat{\rho}_\text{I}(t) = -\frac{i}{\hbar}\int_0^t dt'\hat{Q}[\hat{V}_\text{I}(t'), \hat{\rho}_\text{eq}^{(B)}\hat{\sigma}_\text{I}(t')] \tag{10.112}$$

in which we have truncated the right-hand side after the term that is lowest order in $\hat{V}$ and also disregarded the initial correlation term $\hat{Q}\hat{\rho}_\text{I}(0)$. The latter approximation amounts to assuming that $\hat{\rho}_\text{I}(0)$ is in P space, that is, that initially the system and the bath are uncorrelated and that the bath is in thermal equilibrium, or at least to assuming that the effect of initial correlations decays fast relative to the timescale at which the system is observed. Inserting (10.112) into (10.104) leads to

$$\frac{d}{dt}\hat{\sigma}_\text{I} = -\frac{i}{\hbar}[\hat{\bar{V}}_\text{I}, \hat{\sigma}_\text{I}] - \frac{1}{\hbar^2}\int_0^t d\tau\,\text{Tr}_\text{B}[\hat{V}_\text{I}(t), \hat{Q}[\hat{V}_\text{I}(\tau), \hat{\rho}_\text{eq}^{(B)}\hat{\sigma}_\text{I}(\tau)]] \tag{10.113}$$

We note in passing that had we included the bath–average interaction (10.105) as part of the system's Hamiltonian, then the first term on the right of (10.113) would not appear. This is indeed the recommended practice for system-thermal bath interactions, however, we keep this term explicitly in (10.113) and below because, as will be seen, an equivalent time-dependent term plays an important role in describing the interaction of such system with an external electromagnetic field.

Next consider the integrand in (10.113)

$$\begin{aligned}
&\mathrm{Tr_B}[\hat{V}_{\mathrm{I}}(t), \hat{Q}[\hat{V}_{\mathrm{I}}(\tau), \hat{\rho}_{\mathrm{eq}}^{(B)}\hat{\sigma}_{\mathrm{I}}(\tau)]] \\
&\quad = \mathrm{Tr_B}\{\hat{V}_{\mathrm{I}}(t)\hat{Q}(\hat{V}_{\mathrm{I}}(\tau)\hat{\rho}_{\mathrm{eq}}^{(B)}\hat{\sigma}_{\mathrm{I}}(\tau)) - \hat{V}_{\mathrm{I}}(t)\hat{Q}(\hat{\rho}_{\mathrm{eq}}^{(B)}\hat{\sigma}_{\mathrm{I}}(\tau)\hat{V}_{\mathrm{I}}(\tau)) \\
&\quad -\hat{Q}(\hat{V}_{\mathrm{I}}(\tau)\hat{\rho}_{\mathrm{eq}}^{(B)}\hat{\sigma}_{\mathrm{I}}(\tau))\hat{V}_{\mathrm{I}}(t) + \hat{Q}(\hat{\rho}_{\mathrm{eq}}^{(B)}\hat{\sigma}_{\mathrm{I}}(\tau)\hat{V}_{\mathrm{I}}(\tau))\hat{V}_{\mathrm{I}}(t)\}
\end{aligned} \tag{10.114}$$

Further simplification is achieved if we assume that the interaction $\hat{V}$ is a product of system and bath operators, that is

$$\hat{V} = \hat{V}^{\mathrm{S}}\hat{V}^{\mathrm{B}} \tag{10.115}$$

so that $\hat{\bar{V}} = \hat{V}^{\mathrm{S}}\bar{V}^{\mathrm{B}}$ and $\hat{V}_{\mathrm{I}}(t) = \hat{V}_{\mathrm{I}}^{\mathrm{S}}\hat{V}_{\mathrm{I}}^{\mathrm{B}}$; $\hat{V}_{\mathrm{I}}^{\mathrm{S}} = e^{(i/\hbar)\hat{H}_{\mathrm{S}}t}\hat{V}^{\mathrm{S}}e^{-(i/\hbar)\hat{H}_{\mathrm{S}}t}$; $\hat{V}_{\mathrm{I}}^{\mathrm{B}} = e^{(i/\hbar)\hat{H}_{\mathrm{B}}t}\hat{V}^{\mathrm{B}}e^{-(i/\hbar)\hat{H}_{\mathrm{B}}t}$. To see how the simplification works consider for example the second term on the right of Eq. (10.114)

$$\begin{aligned}
&\mathrm{Tr_B}\{-\hat{V}_{\mathrm{I}}(t)\hat{Q}(\hat{\rho}_{\mathrm{eq}}^{(B)}\hat{\sigma}_{\mathrm{I}}(\tau)\hat{V}_{\mathrm{I}}(\tau))\} \\
&\quad = \mathrm{Tr_B}\{-\hat{V}_{\mathrm{I}}(t)\hat{\rho}_{\mathrm{eq}}^{(B)}\hat{\sigma}_{\mathrm{I}}(\tau)\hat{V}_{\mathrm{I}}(\tau)\} - \mathrm{Tr_B}\{-\hat{V}_{\mathrm{I}}(t)\hat{P}(\hat{\rho}_{\mathrm{eq}}^{(B)}\hat{\sigma}_{\mathrm{I}}(\tau)\hat{V}_{\mathrm{I}}(\tau))\}
\end{aligned} \tag{10.116}$$

Using (10.115) and the cyclic property of the trace, the first term on the right of (10.116) takes the form

$$\begin{aligned}
\mathrm{Tr_B}\{-\hat{V}_{\mathrm{I}}(t)\hat{\rho}_{\mathrm{eq}}^{(B)}\hat{\sigma}_{\mathrm{I}}(\tau)\hat{V}_{\mathrm{I}}(\tau)\} &= -\mathrm{Tr_B}(\hat{V}_{\mathrm{I}}^{\mathrm{B}}(\tau)\hat{V}_{\mathrm{I}}^{\mathrm{B}}(t)\hat{\rho}_{\mathrm{eq}}^{(B)}) \cdot \hat{V}_{\mathrm{I}}^{\mathrm{S}}(t)\hat{\sigma}_{\mathrm{I}}(\tau)\hat{V}_{\mathrm{I}}^{\mathrm{S}}(\tau) \\
&= -\langle\hat{V}_{\mathrm{I}}^{\mathrm{B}}(\tau)\hat{V}_{\mathrm{I}}^{\mathrm{B}}(t)\rangle \cdot \hat{V}_{\mathrm{I}}^{\mathrm{S}}(t)\hat{\sigma}_{\mathrm{I}}(\tau)\hat{V}_{\mathrm{I}}^{\mathrm{S}}(\tau)
\end{aligned} \tag{10.117}$$

and, using (10.87), the second is

$$\begin{aligned}
&\mathrm{Tr_B}\{-\hat{V}_{\mathrm{I}}(t)\hat{P}(\hat{\rho}_{\mathrm{eq}}^{(B)}\hat{\sigma}_{\mathrm{I}}(\tau)\hat{V}_{\mathrm{I}}(\tau))\} \\
&\quad = -\mathrm{Tr_B}[\hat{V}_{\mathrm{I}}^{\mathrm{B}}(t)\hat{V}_{\mathrm{I}}^{\mathrm{S}}(t)\hat{\rho}_{\mathrm{eq}}^{(B)}\mathrm{Tr_B}(\hat{\rho}_{\mathrm{eq}}^{(B)}\hat{V}_{\mathrm{I}}^{\mathrm{B}}(\tau)\hat{\sigma}_{\mathrm{I}}(\tau)\hat{V}_{\mathrm{I}}^{\mathrm{S}}(\tau))] \\
&\quad = -\mathrm{Tr_B}(\hat{V}_{\mathrm{I}}^{\mathrm{B}}(t)\hat{\rho}_{\mathrm{eq}}^{(B)})\mathrm{Tr_B}(\hat{\rho}_{\mathrm{eq}}^{(B)}\hat{V}_{\mathrm{I}}^{\mathrm{B}}(\tau))\hat{V}_{\mathrm{I}}^{\mathrm{S}}(t)\hat{\sigma}_{\mathrm{I}}(\tau)\hat{V}_{\mathrm{I}}^{\mathrm{S}}(\tau) \\
&\quad = -(\bar{V}^{\mathrm{B}})^2\hat{V}_{\mathrm{I}}^{\mathrm{S}}(t)\hat{\sigma}_{\mathrm{I}}(\tau)\hat{V}_{\mathrm{I}}^{\mathrm{S}}(\tau)
\end{aligned} \tag{10.118}$$

where

$$\bar{V}^{\mathrm{B}} \equiv \langle \hat{V}^{\mathrm{B}} \rangle = \mathrm{Tr}_{\mathrm{B}}(\hat{V}_{\mathrm{I}}^{\mathrm{B}}(t)\hat{\rho}_{\mathrm{eq}}^{(B)}) \tag{10.119}$$

is time independent. Together, Eqs (10.116)–(10.118) yield

$$\mathrm{Tr}_{\mathrm{B}}\{-\hat{V}_{\mathrm{I}}(t)\hat{Q}(\hat{\rho}_{\mathrm{eq}}^{(B)}\hat{\sigma}_{\mathrm{I}}(\tau)\hat{V}_{\mathrm{I}}(\tau))\} = -C(\tau - t)\hat{V}_{\mathrm{I}}^{\mathrm{S}}(t)\hat{\sigma}_{\mathrm{I}}(\tau)\hat{V}_{\mathrm{I}}^{\mathrm{S}}(\tau) \tag{10.120}$$

where

$$\begin{aligned} C(t-\tau) &= \langle \hat{V}_{\mathrm{I}}^{\mathrm{B}}(t)\hat{V}_{\mathrm{I}}^{\mathrm{B}}(\tau)\rangle - (\bar{V}^{\mathrm{B}})^2 = \langle \delta\hat{V}_{\mathrm{I}}^{\mathrm{B}}(t)\delta\hat{V}_{\mathrm{I}}^{\mathrm{B}}(\tau)\rangle \\ \delta\hat{V}^{\mathrm{B}} &\equiv \hat{V}^{\mathrm{B}} - \bar{V}^{\mathrm{B}} \end{aligned} \tag{10.121}$$

is a bath correlation function. Time correlation functions (Chapter 6) involving bath operators are seen to emerge naturally in our development. This is the way by which information about the bath appears in the reduced description of our system.

Problem 10.17. Repeat the procedure used above to get Eq. (10.120), now using instead of Eq. (10.115) a sum of products of systems and bath operators

$$\hat{V} = \sum_n \hat{V}_n^{\mathrm{S}}\hat{V}_n^{\mathrm{B}} \tag{10.122}$$

Show that the result equivalent to Eq. (10.120) is in this case

$$\mathrm{Tr}_{\mathrm{B}}\{-\hat{V}_{\mathrm{I}}(t)\hat{Q}(\hat{\rho}_{\mathrm{eq}}^{(B)}\hat{\sigma}_{\mathrm{I}}(\tau)\hat{V}_{\mathrm{I}}(\tau))\} = -\sum_{n,m} C_{nm}(\tau - t)\hat{V}_{\mathrm{I}m}^{\mathrm{S}}(t)\hat{\sigma}_{\mathrm{I}}(\tau)\hat{V}_{\mathrm{I}n}^{\mathrm{S}}(\tau) \tag{10.123}$$

where

$$C_{nm}(t) = \langle \delta\hat{V}_{\mathrm{I}n}^{\mathrm{B}}(t)\delta\hat{V}_{\mathrm{I}m}^{\mathrm{B}}(0)\rangle \tag{10.124}$$

and

$$\hat{V}_{\mathrm{I}n}^{\mathrm{S}} = e^{(i/\hbar)\hat{H}_{\mathrm{S}}t}\hat{V}_n^{\mathrm{S}}e^{-(i/\hbar)\hat{H}_{\mathrm{S}}t}; \qquad \hat{V}_{\mathrm{I}n}^{\mathrm{B}} = e^{(i/\hbar)\hat{H}_{\mathrm{B}}t}\hat{V}_n^{\mathrm{B}}e^{-(i/\hbar)\hat{H}_{\mathrm{B}}t} \tag{10.125}$$

Repeating the procedure that led to (10.120) for all terms in Eq. (10.114) and collecting the resulting expression leads to the final result

$$\frac{d\hat{\sigma}_{\mathrm{I}}}{dt} = -\frac{i}{\hbar}[\hat{\bar{V}}_{\mathrm{I}}, \hat{\sigma}_{\mathrm{I}}] - \frac{1}{\hbar^2}\int_0^t d\tau \{C(t-\tau)[\hat{V}_{\mathrm{I}}^{\mathrm{S}}(t), \hat{V}_{\mathrm{I}}^{\mathrm{S}}(\tau)\hat{\sigma}_{\mathrm{I}}(\tau)] - C^*(t-\tau)[\hat{V}_{\mathrm{I}}^{\mathrm{S}}(t), \hat{\sigma}_{\mathrm{I}}(\tau)\hat{V}_{\mathrm{I}}^{\mathrm{S}}(\tau)]\} \tag{10.126}$$

where the first term in the integrand results from the first and third terms in Eq. (10.114), while the second term is obtained from the second and forth terms in that equation. The equivalent result for the more general case (10.122) is

$$\begin{aligned}\frac{d\hat{\sigma}_{\mathrm{I}}}{dt} = &-\frac{i}{\hbar}[\hat{\bar{V}}_{\mathrm{I}}, \hat{\sigma}_{\mathrm{I}}] \\ &- \frac{1}{\hbar^2}\sum_{n,m}\int_0^t d\tau (C_{mn}(t-\tau)[\hat{V}_{\mathrm{I}m}^{\mathrm{S}}(t), \hat{V}_{\mathrm{I}n}^{\mathrm{S}}(\tau)\hat{\sigma}_{\mathrm{I}}(\tau)] \\ &- C_{mn}^*(t-\tau)[\hat{V}_{\mathrm{I}m}^{\mathrm{S}}(t), \hat{\sigma}_{\mathrm{I}}(\tau)\hat{V}_{\mathrm{I}n}^{\mathrm{S}}(\tau)])\end{aligned} \tag{10.127}$$

where

$$C_{mn}(t) = \langle \delta\hat{V}_{\mathrm{I}m}^{\mathrm{B}}(t)\delta\hat{V}_{\mathrm{I}n}^{\mathrm{B}}(0)\rangle = C_{nm}^*(-t) \tag{10.128}$$

The second equality results from the general identity (6.64)

$$C_{AB}(t) = C_{BA}^*(-t); \qquad \text{where } C_{AB}(t) = \langle \hat{A}(t)\hat{B}(0)\rangle \tag{10.129}$$

for any two Hermitian bath operators $\hat{A}$ and $\hat{B}$ with $\hat{X}(t) = e^{i\hat{H}_{\mathrm{B}}t/\hbar}\hat{X}e^{-i\hat{H}_{\mathrm{B}}t/\hbar}$ $(\hat{X} = \hat{A}, \hat{B})$.

10.4.5 The quantum master equation in the Schrödinger representation

Equations (10.126) and (10.127) represent the quantum master equation in the interaction representation. We now transform it to the Schrödinger picture using

$$\hat{\sigma}(t) = e^{-(i/\hbar)\hat{H}_{\mathrm{S}}t}\hat{\sigma}_{\mathrm{I}}(t)e^{(i/\hbar)\hat{H}_{\mathrm{S}}t} \tag{10.130}$$

which yields

$$\frac{d\hat{\sigma}}{dt} = -\frac{i}{\hbar}[\hat{H}_{\mathrm{S}}, \hat{\sigma}] + e^{-(i/\hbar)\hat{H}_{\mathrm{S}}t}\frac{d\hat{\sigma}_{\mathrm{I}}(t)}{dt}e^{(i/\hbar)\hat{H}_{\mathrm{S}}t} \tag{10.131}$$

Using (10.126) we find

$$e^{-(i/\hbar)\hat{H}_S t}\frac{d\hat{\sigma}_I(t)}{dt}e^{(i/\hbar)\hat{H}_S t} = -\frac{i}{\hbar}[\hat{\bar{V}},\hat{\sigma}(t)] - \frac{1}{\hbar^2}\int_0^t d\tau\{C(t-\tau)$$
$$\times [\hat{V}^S, e^{-(i/\hbar)\hat{H}_S(t-\tau)}\hat{V}^S\hat{\sigma}(\tau)e^{(i/\hbar)\hat{H}_S(t-\tau)}]$$
$$-C^*(t-\tau)[\hat{V}^S, e^{-(i/\hbar)\hat{H}_S(t-\tau)}\hat{\sigma}(\tau)\hat{V}^S e^{(i/\hbar)\hat{H}_S(t-\tau)}]\} \tag{10.132}$$

where all the operators on the right-hand side are in the Schrödinger representation. Using this in (10.131) and making the transformation $t-\tau \to \tau$ finally leads to

$$\frac{d\hat{\sigma}(t)}{dt} = \frac{-i}{\hbar}[\hat{H}_S + \hat{\bar{V}}, \hat{\sigma}] - \frac{1}{\hbar^2}\int_0^t d\tau\{C(\tau)[\hat{V}^S, e^{-(i/\hbar)\hat{H}_S\tau}\hat{V}^S\hat{\sigma}(t-\tau)e^{(i/\hbar)\hat{H}_S\tau}]$$
$$-C^*(\tau)[\hat{V}^S, e^{-(i/\hbar)\hat{H}_S\tau}\hat{\sigma}(t-\tau)\hat{V}^S e^{(i/\hbar)\hat{H}_S\tau}]\} \tag{10.133}$$

Problem 10.18. Show that the equivalent expression in the more general case (10.122) is

$$\frac{d\hat{\sigma}(t)}{dt} = \frac{-i}{\hbar}[\hat{H}_S + \hat{\bar{V}}, \hat{\sigma}] - \frac{1}{\hbar^2}\sum_{m,n}\int_0^t d\tau$$
$$\times \{C_{mn}(\tau)[\hat{V}_m^S, e^{-(i/\hbar)\hat{H}_S\tau}\hat{V}_n^S\hat{\sigma}(t-\tau)e^{(i/\hbar)\hat{H}_S\tau}]$$
$$- C_{mn}^*(\tau)[\hat{V}_m^S, e^{-(i/\hbar)\hat{H}_S\tau}\hat{\sigma}(t-\tau)\hat{V}_n^S e^{(i/\hbar)\hat{H}_S\tau}]\} \tag{10.134}$$

10.4.6 A pause for reflection

What did we achieve so far? We have an equation, (10.133) or (10.134), for the time evolution of the system's density operator. All terms in this equation are strictly defined in the system sub-space; the effect of the bath enters through correlation functions of bath operators that appear in the system–bath interaction. These correlation functions are properties of the unperturbed equilibrium bath. Another manifestation of the reduced nature of this equation is the appearance of memory: The time evolution of $\hat{\sigma}$ at time t is not determined just by $\hat{\sigma}(t)$ but by the past history of $\hat{\sigma}$. As explained in Sections 7.4.2 and 8.2.6, and in analogy to Eqs (10.95)–(10.98), this non-local temporal character (or non-Markovian behavior) stems from the fact that the system evolves at time t in response to the state of the bath at that time and the latter is determined by the history of the system–bath interaction. Equation (10.133) (or (10.134)) results from a low order expansion of the system–bath interaction, so its validity is expected to be limited to weak system–bath coupling. The neglect of initial system–bath correlations, expressed in dropping the term $\hat{Q}\hat{\rho}_{\mathrm{I}}(0)$ in Eq. (10.110) constitutes another approximation, or rather a restriction on the choice of the initial nonequilibrium state. There is a large class of problems, for example the study of nonequilibrium steady states, for which this approximation is of no consequence.

10.4.7 System-states representation

Next we express Eq. (10.133) in the $\hat{H}_{\mathrm{S}}$ representation, that is, the representation defined by the eigenstates of the system Hamiltonian $\hat{H}_{\mathrm{S}}$. Using relationships such as

$$[\hat{H}_{\mathrm{S}}, \hat{\sigma}]_{ab} = (E_a - E_b)\sigma_{ab} \tag{10.135}$$

$$[\hat{\bar{V}}, \hat{\sigma}]_{ab} = \sum_c (\bar{V}_{ac}\sigma_{cb} - \sigma_{ac}\bar{V}_{cb}) \tag{10.136}$$

$$\begin{aligned}&[\hat{V}^{\mathrm{S}}, e^{-(i/\hbar)\hat{H}_{\mathrm{S}}\tau}\hat{V}^{\mathrm{S}}\hat{\sigma}(t-\tau)e^{(i/\hbar)\hat{H}_{\mathrm{S}}\tau}]_{ab}\\ &\quad = \sum_{cd} V_{ac}^{\mathrm{S}}V_{cd}^{\mathrm{S}}\sigma_{db}(t-\tau)e^{(i/\hbar)(E_b-E_c)\tau} - \sum_{cd} V_{ac}^{\mathrm{S}}\sigma_{cd}(t-\tau)V_{db}^{\mathrm{S}}e^{(i/\hbar)(E_d-E_a)\tau}\end{aligned} \tag{10.137}$$

with $V_{ab}^{\mathrm{S}} = \langle a|\hat{V}_{\mathrm{S}}|b\rangle$ and a, b stand for eigenstates of $\hat{H}_{\mathrm{S}}$. We also define the coupling correlation function

$$M_{ab,cd}(t) \equiv \frac{1}{\hbar^2}C(t)V_{ab}^{\mathrm{S}}V_{cd}^{\mathrm{S}} \tag{10.138}$$

which, using (10.129), is easily shown to satisfy

$$M_{ab,cd}^*(t) = M_{ba,dc}(-t) = M_{dc,ba}(-t) \tag{10.139}$$

The second equality, which is trivial for this case, was added to show the correspondence to the more general case, Eq. (10.144) below. Using Eqs (10.135)–(10.138), Eq. (10.133) is expressed in the $\hat{H}_{\mathrm{S}}$ representation as

$$\frac{d\sigma_{ab}}{dt} = -i\omega_{ab}\sigma_{ab} - \frac{i}{\hbar}\sum_c(\bar{V}_{ac}\sigma_{cb} - \sigma_{ac}\bar{V}_{cb}) + \left(\frac{d\sigma_{ab}}{dt}\right)_{\mathrm{B}} \tag{10.140}$$

where $\omega_{ab} = (E_a - E_b)/\hbar$. The first two terms on the right-hand side constitute the system-states representation of the mean potential approximation, Eq. (10.108), and the last term is

$$\left(\frac{d\sigma_{ab}}{dt}\right)_B = -\sum_{c,d}\int_0^t d\tau\{M_{ac,cd}(\tau)e^{i\omega_{bc}\tau}\sigma_{db}(t-\tau)+M_{cd,db}(-\tau)e^{i\omega_{da}\tau}\sigma_{ac}(t-\tau)\}$$
$$+\sum_{c,d}\int_0^t d\tau(M_{db,ac}(\tau)e^{i\omega_{da}\tau}+M_{db,ac}(-\tau)e^{i\omega_{bc}\tau})\sigma_{cd}(t-\tau) \tag{10.141}$$

This last term will be seen to contain the physics of thermal relaxation.

Problem 10.19. For the function $R_{ab,cd}(\omega) \equiv \int_0^\infty d\tau M_{ab,cd}(\tau)e^{i\omega\tau}$ prove the identity

$$R_{ab,cd}(\omega) + R^*_{dc,ba}(\omega) = \int_{-\infty}^{\infty} dt e^{i\omega t} M_{ab,cd}(t) \tag{10.142}$$

Problem 10.20. Show that in the more general case (10.134) we get (10.141) with

$$M_{ab,cd}(t) = \frac{1}{\hbar^2}\sum_{m,n} C_{mn}(t)V^{Sm}_{ab}V^{Sn}_{cd} \tag{10.143}$$

that satisfies (using (10.128))

$$M^*_{ab,cd}(t) = \frac{1}{\hbar^2}\sum_{m,n} C_{nm}(-t)V^{Sm}_{ba}V^{Sn}_{dc} = M_{dc,ba}(-t) \tag{10.144}$$

Note that the result (10.141) satisfies the basic requirement of conservation of probability, that is

$$\frac{d}{dt}\mathrm{Tr_S}\hat{\sigma} = 0 \qquad \text{that is, } \frac{d}{dt}\sum_a \sigma_{aa} = 0 \tag{10.145}$$

at all time. Indeed, it is clearly satisfied by the first two terms (the mean potential approximation) in (10.140). To show that it is satisfied also by the last (relaxation) term put $a = b$ in (10.141) and sum over a to get

$$\sum_a \left(\frac{d\sigma_{aa}}{dt}\right)_B = \sum_{a,c,d} \int_0^t d\tau \{-M_{ac,cd}(\tau)e^{i\omega_{ac}\tau}\sigma_{da}(t-\tau) - M_{cd,da}(-\tau)e^{i\omega_{da}\tau}\sigma_{ac}(t-\tau) + M_{da,ac}(\tau)e^{i\omega_{da}\tau}\sigma_{cd}(t-\tau) + M_{da,ac}(-\tau)e^{i\omega_{ac}\tau}\sigma_{cd}(t-\tau)\} \tag{10.146}$$

It is easy to see by interchanging subscript notations that the first and third terms on the right of (10.146) cancel each other, as do the second and fourth terms. Other consistency issues are discussed in Section 10.4.10.

10.4.8 The Markovian limit—the Redfield equation

Simpler and more manageable expressions are obtained in the limit where the dynamics of the bath is much faster than that of the system. In this limit the functions $M(t)$ (i.e. the bath correlations functions $C(t)$ of Eq. (10.121) or $C_{mn}(t)$ of Eq. (10.124)) decay to zero as $t \to \infty$ much faster than any characteristic system timescale. One may be tempted to apply this limit by substituting τ in the elements $\sigma_{mn}(t-\tau)$ in Eq. (10.141) by zero and take these terms out of the integral over τ, however this would be wrong because, in addition to their relatively slow physical time evolution, non-diagonal elements of $\hat{\sigma}$ contain a fast phase factor. Consider for example the integral in first term on the right-hand side of (10.141),

$$I_1 = \int_0^t d\tau M_{ac,cd}(\tau)e^{i\omega_{bc}\tau}\sigma_{db}(t-\tau) \tag{10.147}$$

In the free system (i.e. without coupling to the reservoir) $\sigma_{db}(t) = e^{-i\omega_{db}t}\sigma_{db}(0)$. This fast phase oscillation, or its remaining signature in the presence of system–bath coupling, should not be taken out of the integral. We therefore use the interaction representation of $\hat{\sigma}$ (see Eq. (10.65))

$$\sigma_{db}(t) = e^{-i\omega_{db}t}\sigma^I_{db}(t) \tag{10.148}$$

and assume that the relaxation of $M_{ac,cd}(\tau)$ to zero as $\tau \to \infty$ is fast relative to the timescale on which σ^I_{db} changes. We then approximate $\sigma^I_{db}(t-\tau) \approx \sigma^I_{db}(t)$ and take it out of the integral. This yields

$$I_1 = e^{-i\omega_{db}t}\sigma^I_{db}(t) \int_0^t d\tau M_{ac,cd}(\tau)e^{i\omega_{bc}\tau + i\omega_{db}\tau} = \sigma_{db}(t) \int_0^t d\tau M_{ac,cd}(\tau)e^{i\omega_{dc}\tau} \tag{10.149}$$

In the second equality we have regained the Schrödinger representation of $\hat{\sigma}(t)$. A final approximation, valid for times longer than the relaxation time of $M(t)$, is to take the upper limit of the time integral in (10.149) to infinity, leading to

$$I_1 = R_{ac,cd}(\omega_{dc})\sigma_{db}(t) \tag{10.150}$$

where we denote here and henceforth

$$R_{ab,cd}(\omega) \equiv \int_0^\infty d\tau M_{ab,cd}(\tau)e^{i\omega\tau} \tag{10.151}$$

Proceeding along similar lines, the integral in the second term on the right of (10.141) is shown to be

$$\begin{aligned} I_2 &= \int_0^t d\tau M_{cd,db}(-\tau)e^{i\omega_{da}\tau}\sigma_{ac}(t-\tau) \rightarrow \sigma_{ac}(t)\int_0^\infty d\tau M_{cd,db}(-\tau)e^{-i\omega_{cd}\tau} \\ &= \sigma_{ac}(t)\int_0^\infty d\tau (M_{bd,dc}(\tau)e^{i\omega_{cd}\tau})^* = \sigma_{ac}(t)R^*_{bd,dc}(\omega_{cd}) \end{aligned} \tag{10.152}$$

where we have used the symmetry property, Eq. (10.139) or (10.144), of M. Similarly, the integrals in the third and forth terms are transformed to

$$\begin{aligned} I_3 &= \int_0^t d\tau M_{db,ac}(\tau)e^{i\omega_{da}\tau}\sigma_{cd}(t-\tau) \rightarrow \sigma_{cd}(t)\int_0^\infty d\tau M_{db,ac}(\tau)e^{i\omega_{ca}\tau} \\ &= \sigma_{cd}(t)R_{db,ac}(\omega_{ca}) \end{aligned} \tag{10.153}$$

and

$$\begin{aligned} I_4 &= \int_0^t d\tau M_{db,ac}(-\tau)e^{i\omega_{bc}\tau}\sigma_{cd}(t-\tau) \rightarrow \sigma_{cd}(t)\int_0^\infty d\tau M_{db,ac}(-\tau)e^{i\omega_{bd}\tau} \\ &= \sigma_{cd}(t)R^*_{ca,bd}(\omega_{db}) \end{aligned} \tag{10.154}$$

Combining all terms we get the so called *Redfield equation*

$$\begin{aligned}\frac{d\sigma_{ab}(t)}{dt} = & -i\omega_{ab}\sigma_{ab} - \frac{i}{\hbar}\sum_c (\bar{V}_{ac}\sigma_{cb} - \sigma_{ac}\bar{V}_{cb}) \\ & - \sum_{c,d}(R_{ac,cd}(\omega_{dc})\sigma_{db}(t) + R^*_{bd,dc}(\omega_{cd})\sigma_{ac}(t) \\ & - [R_{db,ac}(\omega_{ca}) + R^*_{ca,bd}(\omega_{db})]\sigma_{cd}(t))\end{aligned} \tag{10.155}$$

that was first introduced by Redfield in the nuclear magnetic resonance literature.[3] To summarize:

1. The Redfield equation describes the time evolution of the reduced density matrix of a system coupled to an equilibrium bath. The effect of the bath enters via the average coupling $\hat{\bar{V}} = \langle \hat{H}_{\text{SB}} \rangle$ and the "relaxation operator," the last sum on the right of Eq. (10.155). The physical implications of this term will be discussed below.
2 Equation (10.155) is written in the basis of eigenstates of the system Hamiltonian, $\hat{H}_{\text{S}}$. A better description is obtained by working in the basis defined by the eigenstates of $\hat{H}_{\text{S}} + \hat{\bar{V}}$. In the latter case the energy differences ω_{ab} will include shifts that result from the average system–bath coupling and second term on the right of (10.155) (or (10.156a) below) will not appear.
3. Equation (10.155) was obtained under three approximations. The first two are the neglect of initial correlations and the assumption of weak coupling that was used to approximate Eq. (10.110) by Eq. (10.112). The third is the assumption of timescale separation between the (fast) bath and the (slow) system used to get the final Markovian form.
4. The "kinetic coefficients" $R(\omega)$ that appear in the relaxation operator are given by Fourier–Laplace transforms $R_{ab,cd}(\omega) = \int_0^\infty d\tau M_{ab,cd}(\tau)e^{i\omega\tau}$ of the coupling correlation functions $M(t)$. These functions are defined by Eq. (10.138) and satisfy the symmetry property (10.139). In the more general case where the system–bath coupling is given by (10.122), these functions are given by Eq. (10.143) with the symmetry property (10.144).
5. The dynamics of the bath enters through the bath correlation functions, $C(t) = \langle \delta\hat{V}_{\text{I}}^{\text{B}}(t)\delta\hat{V}_{\text{I}}^{\text{B}}(0)\rangle = C^*(-t)$ or more generally $C_{mn}(t) = \langle \delta\hat{V}_{\text{I}m}^{\text{B}}(t)\delta\hat{V}_{\text{I}n}^{\text{B}}(0)\rangle = C^*_{nm}(-t)$. These functions are properties of the equilibrium bath only, independent of any system it might be coupled to. An important observation is that even though we have assumed that the bath is fast on the timescale of

[3] A. G. Redfield, IBM J. Res. Develop. **1**, 19 (1957); Adv. Magn. Reson. **1**, 1 (1965).

the system dynamics, the details of its dynamics do matter, that is, we could not have simply assumed that $C(t) \sim \delta(t)$. The reason for this is that the bath dynamics is usually slow relative to phase oscillations (related to the inverse spacing between energy levels) in the system. Indeed, we will see below that it is through details of the bath dynamics, expressed though relationships between Fourier transforms of bath correlation functions by equations like (6.73), that detailed balance enters in the reduced description of the system dynamics.

Problem 10.21. Verify that Eq. (10.155) can be rewritten in the form

$$\frac{d\sigma_{ab}(t)}{dt} = -i\omega_{ab}\sigma_{ab} - \frac{i}{\hbar}\sum_c (\bar{V}_{ac}\sigma_{cb} - \sigma_{ac}\bar{V}_{cb}) - \sum_{c,d} K_{ab,cd}\sigma_{cd}(t) \tag{10.156a}$$

with

$$K_{ab,cd} = \delta_{bd}\sum_e R_{ae,ec}(\omega_{ce}) + \delta_{ac}\sum_e R^*_{be,ed}(\omega_{de}) - [R_{db,ac}(\omega_{ca}) + R^*_{ca,bd}(\omega_{db})] \tag{10.156b}$$

10.4.9 Implications of the Redfield equation

Next we consider the physical implications of Eq. (10.155). We assume, as discussed above, that the terms involving $\bar{V}$ are included with the system Hamiltonian to produce renormalized energies so that the frequency differences ω_{ab} correspond to the spacings between these renormalized levels. This implies that the second term on the right of Eq. (10.155), the term involving $\bar{V}$, does not exist. We also introduce the following notation for the real and imaginary parts of the super-matrix R:

$$R_{ab,cd}(\omega) = \Gamma_{ab,cd}(\omega) + iD_{ab,cd}(\omega) \tag{10.157}$$

We will see below that it is the real part of R that dominates the physics of the relaxation process.

To get a feeling for the physical content of Eq. (10.155) let us consider first the time evolution of the diagonal elements of $\hat{\sigma}$ in the case where the non-diagonal elements vanish. In this case Eq. (10.155) (without the term involving $\bar{V}$)

becomes

$$\begin{aligned}\frac{d\sigma_{aa}(t)}{dt} &= -\sum_c [R_{ac,ca}(\omega_{ac}) + R^*_{ac,ca}(\omega_{ac})]\sigma_{aa}(t) \\ &\quad + \sum_c [R_{ca,ac}(\omega_{ca}) + R^*_{ca,ac}(\omega_{ca})]\sigma_{cc}(t) \\ &= -2\sum_c \Gamma_{ac,ca}(\omega_{ac})\sigma_{aa}(t) + 2\sum_c \Gamma_{ca,ac}(\omega_{ca})\sigma_{cc}(t) \qquad (10.158)\end{aligned}$$

This is a set of kinetic equations that describe transfer of populations within a group of levels. The transition rates between any two levels a and c are given by

$$k_{a\leftarrow c} = 2\Gamma_{ca,ac}(\omega_{ca}); \qquad k_{c\leftarrow a} = 2\Gamma_{ac,ca}(\omega_{ac}) \qquad (10.159)$$

Using Eqs (10.121), (10.138) and (10.151) we find that these rates are given by

$$k_{j\leftarrow l} = \frac{1}{\hbar^2}|V^S_{jl}|^2 \int_{-\infty}^{\infty} d\tau e^{i\omega_{lj}\tau}\langle \delta\hat{V}^B_I(\tau)\delta\hat{V}^B_I(0)\rangle \qquad (10.160)$$

Consequently

$$\frac{k_{a\leftarrow c}}{k_{c\leftarrow a}} = \frac{\int_{-\infty}^{\infty} d\tau e^{i\omega_{ca}\tau}\langle \delta\hat{V}^B_I(\tau)\delta\hat{V}^B_I(0)\rangle}{\int_{-\infty}^{\infty} d\tau e^{i\omega_{ac}\tau}\langle \delta\hat{V}^B_I(\tau)\delta\hat{V}^B_I(0)\rangle} = e^{\beta\hbar\omega_{ca}} \qquad (10.161)$$

where we have used Eq. (6.75). Here $\hbar\omega_{ca} = E_c - E_a$ where E_c and E_a are eigenvalues of $\hat{H}_S + \hat{\hat{V}}$. We see that the time evolution obtained within the specified approximations satisfies detailed balance with respect to these energies. This insures that the system will reach a Boltzmann distribution at equilibrium.

Problem 10.22. Show that $\mathrm{Re}(R_{ca,ac}(\omega)) = (1/2)\int_{-\infty}^{\infty} dt M_{ca,ac}(t)e^{i\omega t}$. Use this with Eqs (6.71) and (6.72) to show that the relationship (10.161) remains valid also in the more general case where M is given by (10.143), that is, when, for example, $k_{a\leftarrow c} = (\hbar^2)^{-1}\sum_{m,n} V^{Sm}_{ca}V^{Sn}_{ac}\int_{-\infty}^{\infty} dt e^{i\omega_{ca}t}C_{mn}(t)$.

Next, still within our drive to gain physical feeling for the solution of (10.155), assume that all but one element of σ is nonzero, and let this element be non-diagonal. How does it evolve? Taking $\sigma_{i,j} = \sigma_{a,b}\delta_{i,a}\delta_{j,b}$ everywhere on the right side of

Eq. (10.155) leads to

$$\frac{d\sigma_{ab}}{dt} = -i\omega_{ab}\sigma_{ab} - \left(\sum_c [R_{ac,ca}(\omega_{ac}) + R^*_{bc,cb}(\omega_{bc})]\right)\sigma_{ab} + (R_{bb,aa}(0) + R^*_{aa,bb}(0))\sigma_{ab} \tag{10.162}$$

or, alternatively,

$$\frac{d\sigma_{ab}}{dt} = -i\omega_{ab}\sigma_{ab} - \left(\sum_{c\neq a} R_{ac,ca}(\omega_{ac}) + \sum_{c\neq b} R^*_{bc,cb}(\omega_{bc})\right)\sigma_{ab} + (R_{bb,aa}(0) + R^*_{aa,bb}(0) - R_{aa,aa}(0) - R^*_{bb,bb}(0))\sigma_{ab} \tag{10.163}$$

Using (10.157) we see that the terms that involve the imaginary part of R just affect a (small in the weak coupling limit that was already assumed) normalization of the frequency ω_{ab}. The R terms with zero-frequency argument on the right of (10.163) are all real. Defining

$$\tilde{\omega}_{ab} = \omega_{ab} + \left(\sum_{c\neq a} D_{ac,ca}(\omega_{ac}) + \sum_{c\neq b} D_{bc,cb}(\omega_{bc})\right) \tag{10.164}$$

we can rewrite Eq. (10.163) in the form

$$\frac{d\sigma_{ab}}{dt} = -i\tilde{\omega}_{ab}\sigma_{ab} - \left(\sum_{c\neq a} \Gamma_{ac,ca}(\omega_{ac}) + \sum_{c\neq b} \Gamma_{bc,cb}(\omega_{bc})\right)\sigma_{ab} + (R_{bb,aa}(0) + R_{aa,bb}(0) - R_{aa,aa}(0) - R_{bb,bb}(0))\sigma_{ab} \tag{10.165}$$

In addition to the deterministic term $-i\tilde{\omega}_{ab}\sigma_{ab}$, we find on the right-hand side of Eq. (10.165) two relaxation terms. The first can be rewritten in terms of the transition rates k of Eq. (10.159), $\Gamma_{ac,ca}(\omega_{ac}) = (1/2)k_{c\leftarrow a}$. Using Eqs. (10.138) and (10.151) for the second, we find $R_{bb,aa}(0) + R_{aa,bb}(0) - R_{aa,aa}(0) - R_{bb,bb}(0) = -\hbar^{-2}\tilde{C}(0)(V^{\mathrm{S}}_{aa} - V^{\mathrm{S}}_{bb})^2$, where $\tilde{C}(\omega) = \int_0^\infty dt e^{i\omega t} C(t)$ and $\tilde{C}(0)$ is real and positive. We finally get

$$\frac{d\sigma_{ab}}{dt} = -i\tilde{\omega}_{ab}\sigma_{ab} - (1/2)\left(\sum_{c\neq a} k_{c\leftarrow a} + \sum_{c\neq b} k_{c\leftarrow b}\right)\sigma_{ab} - \hbar^{-2}\tilde{C}(0)(V^{\mathrm{S}}_{aa} - V^{\mathrm{S}}_{bb})^2\sigma_{ab} \tag{10.166}$$

Equation (10.166) shows two mechanisms for the relaxation of non-diagonal elements of the system density matrix (i.e. coherences). First, population relaxation out of states a and b with rates $k_a = \sum_{c \neq a} k_{c \leftarrow a}$ and $k_b = \sum_{c \neq b} k_{c \leftarrow b}$ manifests itself also here, giving rise to the relaxation rate $(1/2)(k_a + k_b)$. This may be intuitively expected as the contribution of population relaxation in states a and b to the relaxation of their mutual coherence. For example, in terms of the pure state amplitudes C_a and C_b we have $\sigma_{ab} = C_a C_b^*$, cf. Eq. (10.4), and if $|C_a|^2 \sim e^{-k_a t}$ then $C_a \sim e^{-(1/2)k_a t}$. This component of the coherence relaxation is sometimes referred to as "t_1 process." The corresponding rate, $k_1^{(ab)} = (t_1^{(ab)})^{-1}$ is given by

$$k_1^{(ab)} = (1/2)(k_a + k_b) \tag{10.167}$$

The other coherence relaxation process, with the rate

$$k_2^{(ab)} = (t_2^{(ab)})^{-1} = \hbar^{-2}\tilde{C}(0)(V_{aa}^{\mathrm{S}} - V_{bb}^{\mathrm{S}})^2 \tag{10.168}$$

is more interesting. To understand the origin of this relaxation we note that the difference $V_{aa}^{\mathrm{S}} - V_{bb}^{\mathrm{S}}$ is related to the fluctuations in the energy spacing between states a and b that result from the system coupling to its thermal environment. Indeed, the system–bath coupling that appears in Eqs (10.52) and (10.115) may be written in the form

$$\hat{V} = \hat{V}^{\mathrm{S}}\hat{V}^{\mathrm{B}} = \hat{V}^{\mathrm{S}}\bar{V}^{\mathrm{B}} + \hat{V}^{\mathrm{S}}\delta\hat{V}^{\mathrm{B}} \tag{10.169}$$

where $\bar{V}^{\mathrm{B}}$ and $\delta\hat{V}^{\mathrm{B}}$ were defined by (10.119) and (10.121), respectively, and we have included the $\hat{V}^{\mathrm{S}}\bar{V}^{\mathrm{B}}$ term in a renormalized system Hamiltonian. The remaining term, $\hat{V}^{\mathrm{S}}\delta\hat{V}^{\mathrm{B}}$, is responsible for the t_1 relaxation discussed above but also induces fluctuations in the system energy spacings that can be represented by

$$\hbar\delta\omega_{ab} = (V_{aa}^{\mathrm{S}} - V_{bb}^{\mathrm{S}})\delta\hat{V}^{\mathrm{B}} \tag{10.170}$$

This is because for any realization of the operator $\delta\hat{V}^{\mathrm{B}}$ (that satisfies $\langle\delta\hat{V}^{\mathrm{B}}\rangle = 0$) Equation (10.170) expresses a corresponding shift in the a–b energy spacing and because such realizations correspond to the changing instantaneous state of the thermal environment adjacent to our system. If $\delta\hat{V}^{\mathrm{B}}$ is replaced by a stochastic scalar function $r(t)$ ($\langle r(t)\rangle = 0$) then $\hbar\delta\omega_{ab}(t) = (V_{aa}^{\mathrm{S}} - V_{bb}^{\mathrm{S}})r(t)$ represents random modulations of this energy spacing. Indeed, the equation

$$d\sigma_{ab}/dt = -i(\omega_{ab} + \delta\omega_{ab}(t))\sigma_{ab} \tag{10.171}$$

was our starting point in analyzing the lineshape of a randomly modulated oscillator in Section 7.5.4. Equation (10.168) represents a relaxation process of the same type: $k_2^{(ab)}$ is a contribution to the relaxation rate of non-diagonal elements of the density matrix that results from random modulations of energy spacings that constitute the phases of these elements. This $k_2^{(ab)} = (t_2^{(ab)})^{-1}$ process is sometimes referred to as "pure dephasing" or "pure decoherence," or simply "t_2 relaxation." The word "pure" used in the present context implies a contribution to the phase relaxation that is not associated with population relaxation.

The relaxation terms in the time evolution of non-diagonal elements of the density matrix, Eq. (10.166), are thus comprised of population relaxation contributions embedded in rates such as $k_1^{(ab)}$, and pure phase relaxation expressed by rates of the $k_2^{(ab)}$ type. An important distinction between pure dephasing and population relaxation appears in their temperature dependence. At $T \to 0$ thermal relaxation can move population from higher to lower energy levels of the system but obviously not in the opposite direction, as implied by the detailed balance relation (10.161). The $T \to 0$ limit of the population relaxation rate depends on the energy gap, the bath density of states, and on details of the system–bath coupling. A particular example will be discussed in Chapter 13. Dephasing rates were seen to reflect the modulation of levels spacings in the system due to its interaction with the bath. Such modulations arise by repeated events of energy exchange between the system and the bath. A bath whose temperature remains zero at all time cannot affect such an exchange—it can only take energy out (t_1 relaxation) but not put it back into the system. This implies that pure dephasing vanishes at $T \to 0$, a conclusion that can be validated by proving that zero frequency transforms $\tilde{C}(0)$ of bath time correlation functions vanish at $T = 0$.[4]

10.4.10 Some general issues

The Redfield equation, Eq. (10.155) has resulted from combining a weak system–bath coupling approximation, a timescale separation assumption, and the energy state representation. Equivalent time evolution equations valid under similar weak coupling and timescale separation conditions can be obtained in other representations. In particular, the position space representation $\sigma(\mathbf{r}, \mathbf{r}')$ and the phase space representation obtained from it by the *Wigner transform*

$$\sigma(\mathbf{r}, \mathbf{p}) = \frac{1}{(\pi\hbar)^3} \int d\mathbf{r}' \sigma(\mathbf{r} - \mathbf{r}', \mathbf{r} + \mathbf{r}') e^{2i\mathbf{p}\cdot\mathbf{r}'/\hbar} \tag{10.172}$$

are often encountered in the condensed-phase literature; the expression (10.172) then serving as a convenient starting point for semiclassical approximations.

[4] The subject of zero temperature dephasing has some other subtle aspects that are not addressed here, and to some extent depends on the observable used to determine loss of phase. For more discussion of this issue see Y. Imry, arXiv:cond-mat/0202044.

Whatever representation is used, the reduced density operator $\hat{\sigma}$ should satisfy some basic requirements that can provide important consistency checks. The fact that $\mathrm{Tr}\hat{\rho}(t) = 1$ at all time implies that we should have $\mathrm{Tr}_S\hat{\sigma}(t) = 1$, as was indeed verified in Section 10.4.7. In addition, the reduced density operator $\hat{\sigma}$ should satisfy some other basic requirements:

1. Like any valid density operator it has to be semi-positive (i.e. no negative eigenvalues) at all time. This is implied by the fact that these eigenvalues correspond to state probabilities in the diagonal representation. Indeed, if the overall density operator $\hat{\rho}$ satisfies this requirement it can be shown that so does $\hat{\sigma} = \mathrm{Tr}_B\hat{\rho}$.
2. If the bath is kept at thermal equilibrium, the system should approach the same thermal equilibrium at long time. In practical situations we often address this distribution in the representation defined by the system eigenstates,[5] in which case the statement holds rigorously in the limit of zero coupling.[6] Detailed-balance relationships such as Eq. (10.161) indeed imply that a Boltzmann thermal distribution is a stationary ($d\sigma/dt = 0$) solution of the Redfield equation.

A third requirement is less absolute but still provide a useful consistency check for models that reduce to simple Brownian motion in the absence of external potentials: The dissipation should be invariant to translation (e.g. the resulting friction coefficient should not depend on position). Although it can be validated only in representations that depend explicitly on the position coordinate, it can be shown that Redfield-type time evolution described in such (position or phase space) representations indeed satisfies this requirement under the required conditions.

The main shortcoming of the Redfield time evolution is that it does not necessarily conserve the positivity property. In fact, it has been shown by Lindblad[7] that a linear Markovian time evolution that satisfies this condition has to be of the form

$$\dot{\hat{\sigma}} = -\frac{i}{\hbar}[\hat{H}_0, \hat{\sigma}] + \frac{1}{2}\sum_j ([\hat{V}_j\hat{\sigma}, \hat{V}_j^\dagger] + [\hat{V}_j, \hat{\sigma}\hat{V}_j^\dagger]) \tag{10.173}$$

where $\{\hat{V}_j\}$ is a set of system operators associated with the system–bath interaction. When constructing phenomenological relaxation models one often uses this form as a way to insure positivity. It can be shown that the general Redfield equation

[5] As discussed in Sections 10.4.8 and 10.4.9, these eigenstates may be defined in terms of a "system" Hamiltonian that contains the mean system–bath interaction.

[6] Otherwise the thermal distribution is approached with respect to the exact energies that may be shifted under this interaction.

[7] G. Lindblad, Commun. Math. Phys. **48**, 119 (1976).

is not of this form, and indeed it is possible to find, even for simple systems such as the damped harmonic oscillator, initial conditions for which positivity of the Redfield evolution is not satisfied. At the same time, it has been shown,[8] again for the damped harmonic oscillator, that the Lindblad equation (10.173) cannot satisfy together the conditions of translational invariance and detailed balance. It has to be concluded that no theory can yet yield a fully consistent master equation that describes thermal relaxation in the weak system–bath coupling. Nevertheless, on the practical level, Redfield equations and their analogs in different representations were found very useful for many applications, some of which are discussed below.

10.5 The two-level system revisited

10.5.1 The two-level system in a thermal environment

Further insight on the implications of the Redfield equation can be obtained by considering the special case of a two-level system. In the discussion of Section 10.4.10 we have included terms involving $\bar{V}$, which arise from the average effect of the bath on the system, in the system Hamiltonian $\hat{H}_S$. We will keep similar terms explicitly in the following discussion. They will be used for modeling the coupling between the system and time-dependent external forces as encountered for example, in the semiclassical treatment of a system interacting with a radiation field. The picture is then as follows: The average system/thermal–bath interaction is included in $\hat{H}_S$ so that the eigenstates and eigenvalues of this system Hamiltonian correspond to the renormalized system that includes the average effect of the thermal bath. At the same time a deterministic term appears in the Liouville equation in which a (generally time dependent) system operator $\hat{F}$ replaces the bath-average thermal interaction $\hat{\bar{V}}$. $\hat{F}$ will later represent the effect of an external electromagnetic field in the semiclassical level of description (Chapter 3), and for simplicity will be assumed to have no diagonal terms in the representation defined by the eigenstates of the thermally renormalized system Hamiltonian. Equation (10.155) then leads to

$$\begin{aligned}\frac{d\sigma_{11}}{dt} = & -\frac{i}{\hbar}(F_{12}\sigma_{21} - F_{21}\sigma_{12}) \\ & - 2\text{Re}R_{12,21}(\omega_{12})\,\sigma_{11} + 2\text{Re}R_{21,12}(\omega_{21})\,\sigma_{22} \\ & + (R_{21,11}(0) - R^*_{12,22}(0))\sigma_{12} - (R_{12,22}(0) - R^*_{21,11}(0))\sigma_{21} \qquad (10.174\text{a})\end{aligned}$$

[8] G. Lindblad, Rep. Math. Phys. **10**, 393 (1976).

$$
\begin{aligned}
\frac{d\sigma_{12}}{dt} = &-i\omega_{12}\sigma_{12} - \frac{i}{\hbar}F_{12}(\sigma_{22} - \sigma_{11}) \\
&+ (R_{22,11}(0) + R^*_{11,22}(0) - R_{11,11}(0) - R^*_{22,22}(0) \\
&\quad - R_{12,21}(\omega_{12}) - R^*_{21,12}(\omega_{21}))\sigma_{12} \\
&+ (R_{12,12}(\omega_{21}) + R^*_{21,21}(\omega_{12}))\sigma_{21} \\
&+ (R^*_{11,21}(\omega_{12}) - R^*_{22,21}(\omega_{12}) + R_{12,11}(0) - R^*_{21,11}(0))\,\sigma_{11} \\
&+ (R_{22,12}(\omega_{21}) - R_{11,12}(\omega_{21}) + R^*_{21,22}(0) - R_{12,22}(0))\,\sigma_{22} \qquad (10.174b)
\end{aligned}
$$

and equations for $d\sigma_{22}/dt$ and $d\sigma_{21}/dt$ obtained from (10.174(a,b)) by interchanging the indices 1 and 2 everywhere. Note that the structure of these equations is the same as before, except that F_{ij} has replaced $\bar{V}_{ij}$.

Problem 10.23. Using (10.151) and (10.138) show that the coefficients of σ_{11} and σ_{22} in Eq. (10.174a) satisfy

$$
2\mathrm{Re}R_{12,21}(\omega_{12}) = \frac{|V^S_{12}|^2}{\hbar^2}R_C(\omega_{12}); \qquad 2\mathrm{Re}R_{21,12}(\omega_{21}) = \frac{|V^S_{12}|^2}{\hbar^2}R_C(-\omega_{12})
$$

where $R_C(\omega) = \int_{-\infty}^{\infty} dt e^{i\omega t} C(t)$.

A useful simplified set of equations can be obtained by invoking an approximation based on the observation that in Eqs (10.174) there are two kinds of terms that transform between coherences and populations. The first involves the coupling F_{12}. The second are terms (last two terms in each of (10.174a) and (10.174b)) that involve the system–thermal bath interaction. In consistency with our weak thermal coupling model we will disregard the latter terms, that is, we assume that transitions between coherences and populations are dominated by the interaction with the external field. For a similar reason we will also drop the term involving σ_{21} on the right-hand side of Eq. (10.174b). Using the notation introduced in (10.159), (10.167), and (10.168) we then get

$$
\frac{d\sigma_{11}}{dt} = -\frac{i}{\hbar}(F_{12}\sigma_{21} - F_{21}\sigma_{12}) - k_{2\leftarrow 1}\sigma_{11} + k_{1\leftarrow 2}\sigma_{22} \qquad (10.175a)
$$

$$
\frac{d\sigma_{12}}{dt} = -i\omega_{12}\sigma_{12} - \frac{i}{\hbar}F_{12}(\sigma_{22} - \sigma_{11}) - k_d\sigma_{12} \qquad (10.175b)
$$

where k_d is the dephasing rate

$$k_d = k_1^{(12)} + k_2^{(12)} = \frac{1}{2}(k_{2\leftarrow 1} + k_{1\leftarrow 2}) + \hbar^{-2}\tilde{C}(0)(V_{11}^S - V_{22}^S)^2$$
$$\tilde{C}(0) = \int_0^\infty dtC(t) \tag{10.176}$$

and where, by (10.161),

$$\frac{k_{2\leftarrow 1}}{k_{1\leftarrow 2}} = e^{\beta\hbar\omega_{12}} \tag{10.177}$$

We have invoked the assumption that the interaction with the thermal environment is weak to disregard the difference between $\tilde{\omega}_{12}$ and ω_{12}. The corresponding equations for σ_{22} and σ_{21} are again obtained by interchanging $1 \leftrightarrow 2$.

10.5.2 The optically driven two-level system in a thermal environment—the Bloch equations

Equations (10.174) and (10.175) have the same mathematical structure as Eqs (10.155) except the specification to a two-level system and the replacement of $\bar{V}$ by $\hat{F}$. As discussed above, it makes sense to keep the F terms explicitly in these equations only when they depend on time. In what follows we consider one important problem of this kind, where $\hat{F}(t) \rightarrow \mathcal{E}(t)\hat{\mu}$ and $\mathcal{E}(t) = \mathcal{E}_0 \cos\omega t$, as a model for a two-level system interacting with an incident radiation field. We consider the special case where the light frequency is near resonance with the two-level system, that is,

$$\omega \sim \omega_{21} \quad \text{or} \quad \eta \equiv \omega - \omega_{21} \ll \omega \tag{10.178}$$

where we have denoted $\omega_{21} = (E_2 - E_1)/\hbar > 0$ and where η is the *detuning frequency*.

We are going to make one additional approximation. First make a transformation to new variables

$$\tilde{\sigma}_{12}(t) = e^{-i\omega t}\sigma_{12}(t); \qquad \tilde{\sigma}_{21}(t) = e^{i\omega t}\sigma_{21}(t); \qquad \tilde{\sigma}_{jj}(t) = \sigma_{jj}(t); \quad (j = 1, 2) \tag{10.179}$$

For the free two-level system $\sigma_{jk}(t) = \exp(-i\omega_{jk}t)\sigma_{jk}(0)$, so by (10.178) the transformed variables are slow functions of the time, where "slow" is measured against the timescale $\omega_{21}^{-1} \sim \omega^{-1}$. Equations (10.175) and the corresponding equations for

σ_{22} and σ_{21} become

$$\frac{d\sigma_{11}}{dt} = -\frac{i}{\hbar}\mathcal{E}_0 \cos(\omega t)(\mu_{12}e^{-i\omega t}\tilde{\sigma}_{21} - \mu_{21}e^{i\omega t}\tilde{\sigma}_{12}) - k_{2\leftarrow 1}\sigma_{11} + k_{1\leftarrow 2}\sigma_{22} \tag{10.180a}$$

$$\frac{d\sigma_{22}}{dt} = -\frac{i}{\hbar}\mathcal{E}_0 \cos(\omega t)(\mu_{21}e^{i\omega t}\tilde{\sigma}_{12} - \mu_{12}e^{-i\omega t}\tilde{\sigma}_{21}) - k_{1\leftarrow 2}\sigma_{22} + k_{2\leftarrow 1}\sigma_{11} \tag{10.180b}$$

$$\frac{d\tilde{\sigma}_{12}}{dt} = -i\eta\tilde{\sigma}_{12} - \frac{i}{\hbar}\mathcal{E}_0 \cos(\omega t)e^{-i\omega t}\mu_{12}(\sigma_{22} - \sigma_{11}) - k_d\tilde{\sigma}_{12} \tag{10.180c}$$

$$\frac{d\tilde{\sigma}_{21}}{dt} = i\eta\tilde{\sigma}_{21} - \frac{i}{\hbar}\mathcal{E}_0 \cos(\omega t)e^{i\omega t}\mu_{21}(\sigma_{11} - \sigma_{22}) - k_d\tilde{\sigma}_{21} \tag{10.180d}$$

In Eqs (10.180) the terms that depend explicitly on time originate from $\cos(\omega t)e^{\pm i\omega t} = (1/2)(1 + \exp(\pm 2i\omega t))$ and oscillate with frequency 2ω. The other rates in the problem are the detuning frequency and the thermal rates (population relaxation and dephasing). For optical transitions these rates are usually much smaller than ω, for example typical room temperature vibrational relaxation rates are of order $10^{12}\ \text{s}^{-1}$ while vibrational frequencies are in the range $10^{14}\ \text{s}^{-1}$. The effect of the fast terms, $\exp(\pm 2i\omega t)$, in Eqs (10.180) is therefore expected to be small provided that the field is not too strong, and they will be henceforth disregarded.[9] This is known as the *rotating wave approximation* (RWA).[10] Under this approximation Eqs (10.180) become

$$\frac{d\sigma_{11}}{dt} = -\frac{d\sigma_{22}}{dt} = -\frac{i}{2\hbar}\mathcal{E}_0\mu(\tilde{\sigma}_{21} - \tilde{\sigma}_{12}) - k_{2\leftarrow 1}\sigma_{11} + k_{1\leftarrow 2}\sigma_{22} \tag{10.181a}$$

$$\frac{d\tilde{\sigma}_{12}}{dt} = -i\eta\tilde{\sigma}_{12} - \frac{i}{2\hbar}\mathcal{E}_0\mu(\sigma_{22} - \sigma_{11}) - k_d\tilde{\sigma}_{12} \tag{10.181b}$$

[9] A formal way to do this is a coarse-graining procedure by which we take the average of Eqs (10.180) over the time interval $2\pi/2\omega$. If we assume that all terms except $\exp(\pm 2i\omega t)$ are constant on this timescale the result is equivalent to dropping out all terms containing these fast oscillating factors.

[10] The origin of this name can be understood by considering the product of two time-oscillating functions, $f(t) = f_1(t)f_2(t)$ with $f_j(t) = \cos(\omega_j t)$ and $\omega_1 \simeq \omega_2 > 0$. If we sit on a time-frame that rotates with $f_2(t)$ we find that in the product $f(t)$ there is a component that moves very slowly, at a frequency $\delta\omega = \omega_1 - \omega_2$, relative to this rotating frame, and another that moves very fast, with frequency $\omega_1 + \omega_2 \simeq 2\omega_2$, relative to it. Indeed, $\cos(\omega_1 t)\cos(\omega_2 t) = (1/2)[\cos((\omega_1 - \omega_2)t) + \cos((\omega_1 + \omega_2)t)]$. In the RWA we disregard the latter component.

$$\frac{d\tilde{\sigma}_{21}}{dt} = i\eta\tilde{\sigma}_{21} + \frac{i}{2\hbar}\mathcal{E}_0\mu(\sigma_{22} - \sigma_{11}) - k_{\mathrm{d}}\tilde{\sigma}_{21} \tag{10.181c}$$

We have also denoted $\mu_{12} = \mu_{21} \equiv \mu$, using the fact that the dipole moment operator is real. Equations (10.181), known as the *optical Bloch equations*, correspond to an effective two state system with energy spacing $\omega_{12} = \eta$ and a *time independent* interstate coupling $\mathcal{E}_0\hat{\mu}/2$, subjected to thermal relaxation that itself characterizes the original two state system and satisfies the detailed balance condition (10.177). We could derive the same set of equations using the dressed-state approach introduced in Chapter 9. In this approach applied to the present example we replace the original two-level system coupled to a radiation field with another two state system—a ground state dressed by a photon (more generally N photons) and an excited state without photons (or $N-1$ photons), and disregard the coupling of these dressed states to the infinite number of differently dressed states. This intuitive approach is now validated for near resonance conditions involving relatively weak fields.

A simpler form of Eqs (10.181) may be obtained by redefining variables according to Eqs (10.44)–(10.46) and (10.49)

$$\sigma_z(t) \equiv \sigma_{11}(t) - \sigma_{22}(t) \tag{10.182}$$

$$\tilde{\sigma}_x(t) = \tilde{\sigma}_{12}(t) + \tilde{\sigma}_{21}(t); \qquad \tilde{\sigma}_y(t) = i(\tilde{\sigma}_{12}(t) - \tilde{\sigma}_{21}(t)) \tag{10.183}$$

which leads to the optical Bloch equations in the forms[11]

$$\frac{d\sigma_z}{dt} = \frac{\mathcal{E}_0\mu}{\hbar}\tilde{\sigma}_y - k_r(\sigma_z - \sigma_{z,\mathrm{eq}}) \tag{10.184a}$$

$$\frac{d\tilde{\sigma}_x}{dt} = -\eta\tilde{\sigma}_y - k_{\mathrm{d}}\tilde{\sigma}_x \tag{10.184b}$$

$$\frac{d\tilde{\sigma}_y}{dt} = \eta\tilde{\sigma}_x - \frac{\mathcal{E}_0\mu}{\hbar}\sigma_z - k_{\mathrm{d}}\tilde{\sigma}_y \tag{10.184c}$$

where

$$\sigma_{z,\mathrm{eq}} \equiv \frac{k_{1\leftarrow 2} - k_{2\leftarrow 1}}{k_{1\leftarrow 2} + k_{2\leftarrow 1}} \tag{10.185}$$

is the equilibrium value of σ_z in the absence of radiation ($\mathcal{E}_0 = 0$),[12] and

$$k_r \equiv k_{1\leftarrow 2} + k_{2\leftarrow 1}. \tag{10.186}$$

---

[11] Note that, as defined, η is equivalent to $-\omega$ of Eqs (10.50).

[12] The corresponding values of σ_x and σ_y are $\sigma_{x,\mathrm{eq}} = \sigma_{y,\mathrm{eq}} = 0$.

The Bloch equations and related approximate models derived using similar principles are very useful as simple frameworks for analyzing optical response of material systems. Some examples for their use are provided in Chapter 18.

Appendix 10A: Analogy of a coupled 2-level system to a spin $\frac{1}{2}$ system in a magnetic field

A particle with a spin $\mathbf{S}$ has a magnetic moment $\mathbf{M} = G\mathbf{S}$ where G is the *gyromagnetic* constant. The energy of such a particle in a static magnetic field $\mathbf{B}$ is $E_S = -\mathbf{M} \cdot \mathbf{B}$. A particle of spin $\frac{1}{2}$ is a degenerate two-state system; the two states can be defined according to the spin direction with respect to an arbitrary axis. The degeneracy of this system is lifted in a magnetic field. Taking the field to be $\mathbf{B} = (0, 0, B_z)$, that is, in the z direction, it is convenient to work in the basis of the two eigenstates of the z component of the spin operator, denoted $|+\rangle$ and $|-\rangle$, with eigenvalues $+(1/2)\hbar$ and $-1/2\hbar$ respectively. These states are also eigenstates of the Hamiltonian, the corresponding energies are therefore

$$E_{\pm} = \mp \frac{1}{2} \hbar G B_z \tag{10.187}$$

Using a matrix-vector notation with this basis set the two eigenstates are

$$|+\rangle = \begin{pmatrix} 1 \\ 0 \end{pmatrix}; \qquad |-\rangle = \begin{pmatrix} 0 \\ 1 \end{pmatrix} \tag{10.188}$$

On this basis the Hamiltonian is represented by

$$\hat{H} = -G\mathbf{B} \cdot \hat{\mathbf{S}} = -\frac{1}{2} \hbar G B_z \hat{\sigma}_z \tag{10.189}$$

where the operator $\hat{\sigma}_z \equiv (2/\hbar)\hat{S}_z$ is

$$\hat{\sigma}_z = \begin{pmatrix} 1 & 0 \\ 0 & -1 \end{pmatrix} \tag{10.190}$$

$\hat{\sigma}_z$ is one of the three Pauli matrices whose mutual commutation relations correspond to angular momentum algebra. The other two are

$$\hat{\sigma}_x = \begin{pmatrix} 0 & 1 \\ 1 & 0 \end{pmatrix}; \qquad \hat{\sigma}_y = \begin{pmatrix} 0 & -i \\ i & 0 \end{pmatrix} \tag{10.191}$$

It is easily verified that

$$[\hat{\sigma}_x, \hat{\sigma}_y] = 2i\hat{\sigma}_z \qquad [\hat{\sigma}_y, \hat{\sigma}_z] = 2i\hat{\sigma}_x \qquad [\hat{\sigma}_z, \hat{\sigma}_x] = 2i\hat{\sigma}_y \tag{10.192}$$

Note that the Pauli matrices are the matrix representations, in the basis of Eq. (10.188), of the operators defined in Eq. (10.48), provided that we denote the states $|+\rangle$and $|-\rangle$ by $|1\rangle$ and $|2\rangle$, respectively.

$$\begin{aligned}\hat{\sigma}_z &= |+\rangle\langle+| - |-\rangle\langle-| = |1\rangle\langle 1| - |2\rangle\langle 2| \\ \hat{\sigma}_x &= |+\rangle\langle-| + |-\rangle\langle+| = |1\rangle\langle 2| + |2\rangle\langle 1| \\ \hat{\sigma}_y &= i(|-\rangle\langle+| - |+\rangle\langle-|) = i(|2\rangle\langle 1| - |1\rangle\langle 2|)\end{aligned} \tag{10.193}$$

In addition we define the operators

$$\hat{\sigma}_+ = |1\rangle\langle 2| = \frac{1}{2}(\hat{\sigma}_x + i\hat{\sigma}_y) \tag{10.194}$$

$$\hat{\sigma}_- = |2\rangle\langle 1| = \frac{1}{2}(\hat{\sigma}_x - i\hat{\sigma}_y) \tag{10.195}$$

whose matrix representations are

$$\hat{\sigma}_- = \begin{pmatrix} 0 & 0 \\ 1 & 0 \end{pmatrix} \qquad \hat{\sigma}_+ = \begin{pmatrix} 0 & 1 \\ 0 & 0 \end{pmatrix} \tag{10.196}$$

The operation of $\hat{\sigma}_+$ changes the spin from $-(1/2)\hbar$ to $+(1/2)\hbar$, that is, it moves the higher energy state to the lower energy state. $\hat{\sigma}_-$ acts in the opposite direction.

Consider now the Hamiltonian (10.40). Using the identities $|1\rangle\langle 1| = (1/2)(\hat{I} + \hat{\sigma}_z)$; $|2\rangle\langle 2| = (1/2)(\hat{I} - \hat{\sigma}_z)$, where $\hat{I}$ is the unity operator, we can rewrite it in the form

$$\begin{aligned}\hat{H} &= \frac{1}{2}(H_{11} - H_{22})\hat{\sigma}_z + \frac{1}{2}(H_{11} + H_{22}) + H_{12}\hat{\sigma}_+ + H_{21}\hat{\sigma}_- \\ &= (H_{11} - H_{22})\frac{1}{2}\hat{\sigma}_z + (H_{12} + H_{21})\frac{1}{2}\hat{\sigma}_x + i(H_{12} - H_{21})\frac{1}{2}\hat{\sigma}_y\end{aligned} \tag{10.197}$$

where, in the second line we have disregarded the constant $(1/2)(H_{11} + H_{22})$. We see that, up to the neglected constant energy, this Hamiltonian can be written as a spin $\frac{1}{2}$ system in a magnetic field,

$$\hat{H} = -G\mathbf{B}\cdot\hat{\mathbf{S}} = -(1/2)\hbar G(B_x\hat{\sigma}_x + B_y\hat{\sigma}_y + B_z\hat{\sigma}_z), \tag{10.198}$$

with the fictitious magnetic field whose components are

$$B_z = +\frac{1}{G\hbar}(H_{22} - H_{11}) \tag{10.199}$$

$$B_x = -\frac{1}{G\hbar}(H_{12} + H_{21}) = -\frac{2}{G\hbar}\text{Re}H_{12} \tag{10.200}$$

$$B_y = -\frac{i}{G\hbar}(H_{12} - H_{21}) = \frac{2}{G\hbar}\text{Im}H_{12} \tag{10.201}$$

In this representation the difference between the diagonal elements of $\hat{H}$ corresponds to the z component of the magnetic field, while the non-diagonal elements arise from the other components of $\mathbf{B}$, that is B_x and B_y. In this regard note that

$$B_\perp \equiv \sqrt{B_x^2 + B_y^2} = \frac{2}{\hbar}\left|\frac{H_{12}}{G}\right| \tag{10.202}$$

Using the Hamiltonian (10.197), the Heisenberg equations $\dot{\hat{\sigma}} = (i/\hbar)[\hat{H}, \hat{\sigma}]$ and the commutation relations (10.192), we can easily verify that Eqs (10.50) do not only stand for the averages $\langle\sigma\rangle$ but also for the Heisenberg operators $\hat{\sigma}_{\rm H}(t) = e^{(i/\hbar)\hat{H}t}\hat{\sigma}e^{-(i/\hbar)\hat{H}t}$. Another form of these equations is obtained using Eq. (10.198)

$$\frac{d\hat{\sigma}_x}{dt} = GB_z\hat{\sigma}_y - GB_y\hat{\sigma}_z \tag{10.203}$$

$$\frac{d\hat{\sigma}_y}{dt} = GB_x\hat{\sigma}_z - GB_z\hat{\sigma}_x \tag{10.204}$$

$$\frac{d\hat{\sigma}_z}{dt} = GB_y\hat{\sigma}_x - GB_x\hat{\sigma}_y \tag{10.205}$$

or

$$\frac{d\hat{\sigma}}{dt} = -G\mathbf{B} \times \hat{\sigma} \tag{10.206}$$

Equation (10.206) has the form of a classical time evolution equation of the magnetic moment associated with an orbiting charge in a magnetic field. Such a charge, circulating with an angular momentum $\mathbf{J}$, possesses a magnetic moment $\mathbf{m} = \gamma\mathbf{J}$. In a magnetic field $\mathbf{B}$ a torque $\mathbf{m} \times \mathbf{B}$ is exerted on the charge and the corresponding classical equation of motion is

$$\frac{d\mathbf{J}}{dt} = \mathbf{m} \times \mathbf{B} = \gamma\mathbf{J} \times \mathbf{B} \tag{10.207}$$

or

$$\frac{d\mathbf{m}}{dt} = \gamma\mathbf{m} \times \mathbf{B} \tag{10.208}$$

Since the scalar product of $\mathbf{m} \times \mathbf{B}$ with either $\mathbf{B}$ or $\mathbf{m}$ is zero, it follows that

$$\frac{d}{dt}(\mathbf{m})^2 = \frac{d}{dt}(\mathbf{m} \cdot \mathbf{B}) = 0 \tag{10.209}$$

This implies that $\mathbf{m}$ evolves such that its modulus is constant, maintaining a constant angle with the direction of $\mathbf{B}$. This motion is called precession. The angular velocity of this precession is $\omega = \gamma B$.

Further reading

K. Blum, Density Matrix Theory and Applications, 2nd edition (plenum, New York, 1996).

C. Cohen-Tannoudji, B. Diu and F. Laloe, Quantum Mechanics (Wiley, New York, 1977).

D. Kohen, C. C. Marston, and D. J. Tannor, J. Chem. Phys. **107**, 5326 (1997).

V. May and O. Kühn, *Charge and Energy Transfer Dynamics in Molecular Systems* (Wiley-VCH, Berlin, 2000).

R. Zwanzig, *Non-Equilibrium Statistical Mechanics* (Oxford University Press, Oxford, 2001).

11

LINEAR RESPONSE THEORY

If cause forever follows after cause
In infinite, undeviating sequence,
And a new motion always has to come
Out of an old one by fixed law; if atoms
Do not, by swerving, cause new moves which break
The Laws of fate; if cause forever follows,
In infinite sequence, cause—where would we get
This free will that we have, wrested from fate,
By which we go ahead...

Lucretius (c.99–c.55 BCE*) "The way things are" translated by Rolfe Humphries, Indiana University Press, 1968*

Equilibrium statistical mechanics is a first principle theory whose fundamental statements are general and independent of the details associated with individual systems. No such general theory exists for nonequilibrium systems and for this reason we often have to resort to ad hoc descriptions, often of phenomenological nature, as demonstrated by several examples in Chapters 7 and 8. Equilibrium statistical mechanics can however be extended to describe small deviations from equilibrium in a way that preserves its general nature. The result is Linear Response Theory, a statistical mechanical perturbative expansion about equilibrium. In a standard application we start with a system in thermal equilibrium and attempt to quantify its response to an applied (static- or time-dependent) perturbation. The latter is assumed small, allowing us to keep only linear terms in a perturbative expansion. This leads to a linear relationship between this perturbation and the resulting response.

Let us make these statements more quantitative. Consider a system characterized by the Hamiltonian $\hat{H}_0$. An external force acting on this system changes the Hamiltonian according to

$$\hat{H}_0 \longrightarrow \hat{H} = \hat{H}_0 + \hat{H}_1 \tag{11.1}$$

We take $\hat{H}_1$ to be of the form

$$\hat{H}_1(t) = -\hat{A}F(t) \tag{11.2}$$

where $F(t)$ is the external force that can depend on time and $\hat{A}$ is an operator that represents the dynamical variable $\mathcal{A}(\mathbf{r}^N, \mathbf{p}^N)$ (a function of the coordinates $\mathbf{r}^N$ and moments $\mathbf{p}^N$ of all particles in the system) that couples to this force. For example, for an external electric field $\mathcal{E}$ imposed on a one-dimensional classical system with one charged particle, $H_1 = -qx\mathcal{E}$, that is, $\mathcal{A} = x$ and $F = q\mathcal{E}$ where x and q are the particle position and charge, respectively. More generally $\hat{H}_1$ can be a sum of such products,

$$\hat{H}_1 = -\sum_j \hat{A}_j F_j \tag{11.3}$$

but for simplicity we consider below the simplest case (11.2). Next, we focus on another dynamical variable $\mathcal{B}(\mathbf{r}^N, \mathbf{p}^N)$, represented in quantum mechanics by an operator $\hat{B}$, and on the change $\langle \Delta B \rangle$ in the expectation value of this variable in response to the imposed perturbation. Linear response theory aims to characterize the linear relationship between the imposed small force F and the ensuing response $\langle \Delta B \rangle$.

It should be noted that in addition to mechanical forces such as electric or magnetic fields that couple to charges and polarization in our system, other kinds of forces exist whose effect cannot be expressed by Eq. (11.2). For example, temperature or chemical potential gradients can be imposed on the system and thermal or material fluxes can form in response. In what follows we limit ourselves first to linear response to mechanical forces whose effect on the Hamiltonian is described by Eqs (11.2) or (11.3).

11.1 Classical linear response theory

11.1.1 Static response

In previous chapters it was sometimes useful to use different notations for an observable A and the corresponding dynamical variable $\mathcal{A}(\mathbf{r}^N, \mathbf{p}^N)$. In this chapter we will not make this distinction because it makes the presentation somewhat cumbersome. The difference between these entities should be clear from the text.

Consider first the response to a static perturbation, that is, we take $F =$ constant in Eq. (11.2). In this case we are not dealing with a nonequilibrium situation, only comparing two equilibrium cases. In this case we need to evaluate $\Delta\langle B \rangle = \langle B \rangle - \langle B \rangle_0$ where

$$\langle B \rangle_0 = \frac{\int d\mathbf{r}^N \int d\mathbf{p}^N B e^{-\beta H_0}}{\int d\mathbf{r}^N \int d\mathbf{p}^N e^{-\beta H_0}} \tag{11.4a}$$

$$\langle B \rangle = \frac{\int d\mathbf{r}^N \int d\mathbf{p}^N B e^{-\beta (H_0 + H_1)}}{\int d\mathbf{r}^N \int d\mathbf{p}^N e^{-\beta (H_0 + H_1)}} \tag{11.4b}$$

where $\beta = (k_B T)^{-1}$. In what follows we use $\iint$ to denote $\int d\mathbf{r}^N \int d\mathbf{p}^N$. For a small perturbation, $\beta H_1 \ll 1$, we expand (11.4b) to linear order in this quantity using

$$e^{-\beta(H_0+H_1)} = e^{-\beta H_0}(1 - \beta H_1) + \mathrm{O}(H_1^2) \tag{11.5}$$

to get

$$\begin{aligned}\langle B\rangle &= \frac{\left[\iint e^{-\beta H_0}\right]\{\langle B\rangle_0 - \beta\langle H_1 B\rangle_0\}}{\left[\iint e^{-\beta H_0}\right]\{1 - \beta\langle H_1\rangle_0\}} = (\langle B\rangle_0 - \beta\langle H_1 B\rangle_0)(1 + \beta\langle H_1\rangle_0) \\ &= \langle B\rangle_0 + \beta\langle B\rangle_0\langle H_1\rangle_0 - \beta\langle H_1 B\rangle_0 + O[(\beta H_1)^2]\end{aligned} \tag{11.6}$$

So, to linear order

$$\begin{aligned}\Delta\langle B\rangle &= -\beta(\langle H_1 B\rangle_0 - \langle H_1\rangle_0\langle B\rangle_0) \\ &= \beta F(\langle AB\rangle_0 - \langle A\rangle_0\langle B\rangle_0) = \beta F\langle\delta A\delta B\rangle_0\end{aligned} \tag{11.7}$$

So we found the remarkable result

$$\begin{aligned}\Delta\langle B\rangle &= \chi_{BA} F \\ \chi_{BA} &= \beta\langle\delta A\delta B\rangle_0\end{aligned} \tag{11.8}$$

that is, the response function χ_{BA} (sometimes called admittance or susceptibility), the coefficient of the linear relationship between the applied force and the ensuing system response, is given in terms of a correlation function between the equilibrium fluctuations in A and B in the unperturbed system. Note that there are different susceptibilities, each associated with the way by which forcing one system variable invokes a response in another. Note also that χ_{BA} as defined is the isothermal susceptibility. We could also study the response of the system under other conditions, for example, the adiabatic susceptibility measures the linear response under the condition of constant system energy rather than constant temperature.

11.1.2 Relaxation

In (11.8) χ_{BA} is the response coefficient relating a *static response* in $\langle B\rangle$ to a static perturbation associated with a field F which couples to the system through an additive term $H_1 = -FA$ in the Hamiltonian. Consider next the dynamical experiment in which the system reached equilibrium with H_0+H_1 and then the field suddenly switched off. How does $\Delta\langle B\rangle$, the induced deviation of B from its original equilibrium value $\langle B\rangle_0$, relax to zero? The essential point in the following derivation is that the time evolution is carried out under the Hamiltonian H_0 (after the field has been switched off), while the thermal averaging over initial states is done for the Hamiltonian $H = H_0 + H_1$.

$$\langle B(t) \rangle = \frac{\iint e^{-\beta(H_0+H_1)} B_0(t \mid p^N, r^N)}{\iint e^{-\beta(H_0+H_1)}} \tag{11.9}$$

Here $B_0(t \mid p^N, r^N) \equiv B_0(p^N(t), r^N(t) \mid p^N, r^N)$ is the value of B at time t, after evolving under the Hamiltonian H_0 from the initial system's configuration $(\mathbf{r}^N, \mathbf{p}^N)$. Note that

$$\iint e^{-\beta H_0} B_0(t \mid p^N, r^N) = \langle B \rangle_0 \iint e^{-\beta H_0} \tag{11.10}$$

because we start from equilibrium associated with H_0 and propagate the system's trajectory with the same Hamiltonian. However,

$$\iint e^{-\beta(H_0+H_1)} B_0(t \mid p^N, r^N) \neq \langle B \rangle \iint e^{-\beta(H_0+H_1)} \tag{11.11}$$

because we start from equilibrium associated with $H_0 + H_1$, but the time evolution is done under H_0.

Starting from Eq. (11.9) we again expand the exponential operators. Once $\exp(-\beta(H_0 + H_1))$ is replaced by $\exp(-\beta H_0)(1 - \beta H_1)$ we get a form in which the time evolution and the averaging are done with the same Hamiltonian H_0. We encounter terms such as

$$\frac{\iint e^{-\beta H_0} B_0(p^N(t), r^N(t) \mid p^N, r^N)}{\iint e^{-\beta H_0}} = \langle B(t) \rangle_0 = \langle B \rangle_0 \tag{11.12}$$

and

$$\frac{\iint e^{-\beta H_0} H_1(p^N, r^N) B_0(p^N(t), r^N(t) \mid p^N, r^N)}{\iint e^{-\beta H_0}} = \langle H_1(0) B(t) \rangle_0 \tag{11.13}$$

With this kind of manipulations Eq. (11.9) becomes

$$\langle B(t) \rangle = \frac{\langle B \rangle_0 - \beta \langle H_1(0) B(t) \rangle_0}{1 - \beta \langle H_1 \rangle_0} \tag{11.14}$$

and to linear order in βH_1

$$\begin{aligned} \Delta \langle B(t) \rangle &= \langle B(t) \rangle - \langle B \rangle_0 = -\beta(\langle H_1(0) B(t) \rangle_0 - \langle H_1 \rangle_0 \langle B \rangle_0) \\ &= \beta F \langle \delta A(0) \delta B(t) \rangle_0 \end{aligned} \tag{11.15}$$

with

$$\delta A(0) = A(0) - \langle A \rangle_0; \qquad \delta B(t) = B(t) - \langle B \rangle_0 \tag{11.16}$$

On the left side of (11.15) we have the time evolution of a prepared deviation from equilibrium of the dynamical variable B. On the right side we have a time correlation function of spontaneous equilibrium fluctuations involving the dynamical variables A, which defined the perturbation, and B. The fact that the two time evolutions are the same has been known as the *Onsager regression hypothesis*. (The hypothesis was made before the formal proof above was known.)

11.1.3 Dynamic response

Consider next a more general situation where the weak external perturbation is time-dependent, $F = F(t)$. We assume again that the force is weak so that, again, the system does not get far from the equilibrium state assumed in its absence. In this case, depending on the nature of this external force, two scenarios are usually encountered at long time.

1. When a constant force is imposed on an open system, the system will eventually reach a nonequilibrium steady state where the response to the force appears as a time-independent flux. (A closed system in the same situation will reach a new equilibrium state, as discussed above.)
2. When the external force oscillates with a given frequency, the system will eventually reach a dynamic steady state in which system observables oscillate with the same frequency and often with a characteristic phase shift. The amplitude of this oscillation characterizes the response; the phase shift is associated with the imaginary part of this amplitude.

Linear response theory accounts for both scenarios by addressing the assumed linear relationship between the response of a dynamical variable B (i.e. the change in its average observed value) and the small driving field F. It is convenient to represent the long time behavior of the system under this driving by assuming that the external force has been switched on in the infinite past where the system was at its unperturbed equilibrium state. In terms of the density operator $\hat{\rho}(t)$ or the analogous classical distribution function $f\left(\mathbf{r}^N, \mathbf{p}^N; t\right)$ this implies

$$f\left(\mathbf{r}^N, \mathbf{p}^N; -\infty\right) = \frac{e^{-\beta \hat{H}_0}}{\int d\mathbf{r}^N \int d\mathbf{p}^N e^{-\beta \hat{H}_0}}; \ \hat{\rho}(-\infty) = \frac{e^{-\beta \hat{H}_0}}{Tr\left[e^{-\beta \hat{H}_0}\right]} \tag{11.17}$$

The most general linear relationship between the force $F(t)$ and the response $\langle \Delta B(t) \rangle$ is then

$$\langle \Delta B(t) \rangle = \int_{-\infty}^{t} dt' \chi_{BA}(t - t') F(t') \tag{11.18}$$

Note that causality, that is, the recognition that the response at time t can depend only on past perturbations and not on future ones, is built into (11.18). It is convenient to write

$$\langle \Delta B(t) \rangle = \int_{-\infty}^{\infty} dt' \chi_{BA}(t - t') F(t') \tag{11.19}$$

by defining the time-dependent susceptibility χ so that $\chi(t) = 0$ for $t < 0$. For the following special choice of $F(t)$:

$$\left.\begin{aligned} F(t) &= F \\ F(t) &= 0 \end{aligned}\right\} \qquad \begin{aligned} t &< 0 \\ t &\geq 0 \end{aligned} \tag{11.20}$$

Equation (11.19) should yield the result (11.15), that is, $\langle \Delta B(t) \rangle = \beta F \langle \delta A(0) \delta B(t) \rangle_0$. This implies that for $t > 0$

$$\beta \langle \delta A(0) \delta B(t) \rangle_0 = \int_{-\infty}^{0} dt' \chi_{BA}(t - t') = \int_{t}^{\infty} d\tau \chi_{BA}(\tau) \tag{11.21}$$

and by taking derivative with respect to time

$$\chi_{BA}(t) = -\theta(t) \beta \langle \delta A(0) \delta \dot{B}(t) \rangle_0 = \theta(t) \beta \langle \delta \dot{A}(0) \delta B(t) \rangle_0 \tag{11.22}$$

where the θ function is

$$\theta(t) = \begin{cases} 1 & \text{for } t > 0 \\ 0 & \text{otherwise} \end{cases} \tag{11.23}$$

The second equality in (11.22) follows from the symmetry property (6.32). We have found that the dynamic susceptibility is again given in terms of equilibrium correlation functions, in this case time correlation functions involving one of the variables A or B and the time derivative of the other. Note (c.f. Eq (1.99)) that if X is a dynamical variable, that is, a function of $(\mathbf{r}^N, \mathbf{p}^N)$ so is its time derivative.

11.2 Quantum linear response theory

The derivation of the quantum analog of the theory presented above follows essentially the same line, except that care must be taken with the operator algebra involved.

11.2.1 Static quantum response

The analogs of Eqs (11.4) are

$$\langle B\rangle_0 = \mathrm{Tr}[\hat{B}\hat{\rho}_{0,\mathrm{eq}}]; \quad \hat{\rho}_{0,\mathrm{eq}} = \frac{e^{-\beta\hat{H}_0}}{\mathrm{Tr}(e^{-\beta\hat{H}_0})} \tag{11.24a}$$

$$\langle B\rangle = \mathrm{Tr}[\hat{B}\hat{\rho}_{\mathrm{eq}}]; \quad \hat{\rho}_{\mathrm{eq}} = \frac{e^{-\beta(\hat{H}_0+\hat{H}_1)}}{\mathrm{Tr}(e^{-\beta(\hat{H}_0+\hat{H}_1)})} \tag{11.24b}$$

and the first-order perturbation expansion analogous to (11.5) is obtained from the operator identity (cf. Eq. (2.78))

$$\exp\left[-\beta\left(\hat{H}_0+\hat{H}_1\right)\right] = \exp\left(-\beta\hat{H}_0\right)\left(1-\int_0^\beta d\lambda e^{\lambda\hat{H}_0}\hat{H}_1 e^{-\lambda\left(\hat{H}_0+\hat{H}_1\right)}\right) \tag{11.25}$$

by replacing, as a lowest order approximation, $\exp\left[-\lambda\left(\hat{H}_0+\hat{H}_1\right)\right]$ by $\exp\left(-\lambda\hat{H}_0\right)$ inside the integral on the right to get

$$\exp\left[-\beta\left(\hat{H}_0+\hat{H}_1\right)\right] = \exp\left(-\beta\hat{H}_0\right)\left(1-\beta\hat{H}_1^{(\beta)}\right)+O\left(\hat{H}_1^2\right) \tag{11.26}$$

Here we have used the notation $\hat{X}^{(\beta)}$ for the *Kubo transform* of an operator $\hat{X}$ defined by

$$\hat{X}^{(\beta)} = \beta^{-1}\int_0^\beta d\lambda e^{\lambda\hat{H}_0}\hat{X}e^{-\lambda\hat{H}_0} \tag{11.27}$$

Problem 11.1. Show that in the basis of eigenstates of $\hat{H}_0$, $\hat{H}_0|j\rangle = \varepsilon_j|j\rangle$,

$$(\hat{X}^{(\beta)})_{ij} = \beta^{-1}\frac{e^{\beta(\varepsilon_i-\varepsilon_j)}-1}{\varepsilon_i-\varepsilon_j}X_{ij} \tag{11.28}$$

and that the high-temperature limit is $\lim_{\beta\to 0}\hat{X}^{(\beta)} = \hat{X}$.

When the external force F is small and constant we again seek a linear dependence of the form

$$\delta\langle B\rangle = \langle B\rangle - \langle B\rangle_0 = \chi_{BA}F \tag{11.29}$$

where χ_{BA} is the static quantum isothermal susceptibility. Using (11.26) in (11.24b) we get

$$\delta\langle\hat{B}\rangle = \beta F\left(\frac{\mathrm{Tr}[\hat{B}e^{-\beta\hat{H}_0}\hat{A}^{(\beta)}]}{\mathrm{Tr}[e^{-\beta\hat{H}_0}]} - \frac{\mathrm{Tr}[\hat{B}e^{-\beta\hat{H}_0}]}{\mathrm{Tr}[e^{-\beta\hat{H}_0}]}\frac{\mathrm{Tr}[e^{-\beta\hat{H}_0}\hat{A}^{(\beta)}]}{\mathrm{Tr}[e^{-\beta\hat{H}_0}]}\right)$$
$$= \beta\left(\langle\hat{A}^{(\beta)}\hat{B}\rangle_0 - \langle\hat{A}^{(\beta)}\rangle_0\langle\hat{B}\rangle_0\right)F = \beta\langle\delta\hat{A}^{(\beta)}\delta\hat{B}\rangle_0 F \tag{11.30}$$

So that

$$\chi_{BA} = \beta\left\langle\delta\hat{A}^{(\beta)}\delta\hat{B}\right\rangle_0 \tag{11.31}$$

where for any operator $\hat{X}$ we define $\delta\hat{X} = \hat{X} - \langle\hat{X}\rangle_0$. We see that the classical limit, Eq. (11.8), is obtained from Eq. (11.31) by replacing $\hat{X}^{(\beta)}$ by the corresponding dynamical variable X, irrespective of β.

Problem 11.2. Show that for any two operators $\hat{A}$ and $\hat{B}$

$$\langle\hat{A}^{(\beta)}\hat{B}\rangle = \langle\hat{B}^{(\beta)}\hat{A}\rangle \tag{11.32}$$

Proof: Using Eq. (11.27) we have

$$\langle\hat{A}^{(\beta)}\hat{B}\rangle = \beta^{-1}\frac{\int_0^\beta d\lambda\mathrm{Tr}[e^{-\beta\hat{H}_0}e^{\lambda\hat{H}_0}\hat{A}e^{-\lambda\hat{H}_0}\hat{B}]}{\mathrm{Tr}[e^{-\beta\hat{H}_0}]} \tag{11.33}$$

Consider the numerator expressed in the basis of eigenstates of $\hat{H}_0$

$$\int_0^\beta d\lambda\mathrm{Tr}[e^{-\beta\hat{H}_0}e^{\lambda\hat{H}_0}\hat{A}e^{-\lambda\hat{H}_0}\hat{B}] = \int_0^\beta d\lambda\sum_j\sum_k e^{-\beta\varepsilon_j}e^{\lambda(\varepsilon_j-\varepsilon_k)}\langle j|\hat{A}|k\rangle\langle k|\hat{B}|j\rangle$$
$$= \sum_j\sum_k\frac{e^{-\beta\varepsilon_k}-e^{-\beta\varepsilon_j}}{\varepsilon_j-\varepsilon_k}\langle j|\hat{A}|k\rangle\langle k|\hat{B}|j\rangle \tag{11.34}$$

Interchanging j and k gives

$$\int_0^\beta d\lambda\mathrm{Tr}\left[e^{-\beta\hat{H}_0}e^{\lambda\hat{H}_0}\hat{A}e^{-\lambda\hat{H}_0}\hat{B}\right] = \sum_j\sum_k\frac{e^{-\beta\varepsilon_k}-e^{-\beta\varepsilon_j}}{\varepsilon_j-\varepsilon_k}\langle j|\hat{B}|k\rangle\langle k|\hat{A}|j\rangle$$
$$= \int_0^\beta d\lambda\mathrm{Tr}\left[e^{-\beta\hat{H}_0}e^{\lambda\hat{H}_0}\hat{B}e^{-\lambda\hat{H}_0}\hat{A}\right] \tag{11.35}$$

This, together with (11.33) imply that (11.32) holds.

11.2.2 Dynamic quantum response

Now let the external force $F(t)$ be time-dependent. We will repeat the procedure followed in the classical case, assuming that $F(t)$ is given by the step function (11.20), that is, following the onset of a perturbation $\hat{H}_1 = -F\hat{A}$ at $t = -\infty$, F is switched off at $t = 0$. We want to describe the subsequent relaxation process of the system as expressed by the evolution of the expectation of an observable $\hat{B}$ from its value $\langle \hat{B} \rangle$ (equilibrium average under $\hat{H} = \hat{H}_0 + \hat{H}_1$) at $t = 0$ to the final value $\langle \hat{B} \rangle_0$ at $t \to \infty$. In what follows we will follow two routes for describing this relaxation.

Our first starting point is the equation equivalent to (11.9)

$$\langle B \rangle(t) = \frac{\mathrm{Tr}[e^{-\beta(\hat{H}_0+\hat{H}_1)} e^{(i/\hbar)\hat{H}_0 t} \hat{B} e^{-(i/\hbar)\hat{H}_0 t}]}{\mathrm{Tr}[e^{-\beta(\hat{H}_0+\hat{H}_1)}]} \tag{11.36}$$

Using the identity (11.25) we expand the thermal operators in (11.36) to first order in $\hat{H}_1$: $\exp[-\beta(\hat{H}_0 + \hat{H}_1)] = \exp(-\beta\hat{H}_0)(1 - \int_0^\beta d\lambda e^{\lambda\hat{H}_0}\hat{H}_1 e^{-\lambda\hat{H}_0})$. Using the definition of the interaction representation

$$\hat{B}_I(t) = \exp(i\hat{H}_0 t/\hbar)\hat{B}\exp(-i\hat{H}_0 t/\hbar) \tag{11.37}$$

and the identity $\mathrm{Tr}[\hat{\rho}_{0,\mathrm{eq}}\hat{B}_I(t)] = \langle B \rangle_0$ that holds for all t, we find

$$\begin{aligned}\langle B \rangle(t) &= \frac{\mathrm{Tr}[e^{-\beta\hat{H}_0}\hat{B}_I(t)] - \mathrm{Tr}[e^{-\beta\hat{H}_0}\int_0^\beta d\lambda e^{\lambda\hat{H}_0}\hat{H}_1 e^{-\lambda\hat{H}_0}\hat{B}_I(t)]}{\mathrm{Tr}[e^{-\beta\hat{H}_0}](1 - \mathrm{Tr}[e^{-\beta\hat{H}_0}\int_0^\beta d\lambda e^{\lambda\hat{H}_0}\hat{H}_1 e^{-\lambda\hat{H}_0}]/\mathrm{Tr}[e^{-\beta\hat{H}_0}])} \\ &= \langle B \rangle_0 - \left\langle \int_0^\beta d\lambda e^{\lambda\hat{H}_0}\hat{H}_1 e^{-\lambda\hat{H}_0}\hat{B}_I(t) \right\rangle_0 + \langle B \rangle_0 \left\langle \int_0^\beta d\lambda e^{\lambda\hat{H}_0}\hat{H}_1 e^{-\lambda\hat{H}_0} \right\rangle_0\end{aligned} \tag{11.38}$$

or

$$\langle B \rangle(t) - \langle B \rangle_0 = \beta F\left(\left\langle \hat{A}^{(\beta)}\hat{B}_I(t)\right\rangle_0 - \langle B \rangle_0 \left\langle \hat{A}^{(\beta)} \right\rangle\right) = \beta \left\langle \delta\hat{A}^{(\beta)}\delta\hat{B}_I(t) \right\rangle_0 F \tag{11.39}$$

where the Kubo transform $\hat{A}^{(\beta)}$ is defined by Eq. (11.27). In the classical limit $\delta\hat{A}^{(\beta)} = \delta A$ and Eq. (11.39) becomes identical to (11.15). The rest of the development follows the same steps as those leading to Eq. (11.22) and results in the linear response equation (11.19) with the quantum expression for the isothermal susceptibility

$$\chi_{BA}(t) = -\theta(t)\beta\langle \delta\hat{A}^{(\beta)}\delta\dot{\hat{B}}_I(t)\rangle_0 = \theta(t)\beta\langle \delta\dot{\hat{A}}^{(\beta)}\delta\hat{B}_I(t)\rangle_0 \tag{11.40}$$

As in the classical case, the important physical aspect of this result is that the time dependence in $\delta\hat{B}_I(t)$ as well as the equilibrium thermal average are evaluated with respect to the Hamiltonian $\hat{H}_0$ of the unperturbed system.

Let us now follow a different route, starting from the quantum Liouville equation for the time evolution of the density operator $\hat{\rho}$

$$\frac{\partial\hat{\rho}(t)}{\partial t} = -i(\mathcal{L}_0 + \mathcal{L}_1(t))\hat{\rho}(t)$$
$$\mathcal{L}_0\hat{\rho} = \hbar^{-1}[\hat{H}_0, \hat{\rho}]; \qquad \mathcal{L}_1(t)\hat{\rho} = \hbar^{-1}[\hat{H}_1, \hat{\rho}] = -\hbar^{-1}F(t)[\hat{A}, \hat{\rho}] \tag{11.41}$$

Assume as in Eq. (11.17) that the system is at its unperturbed equilibrium state in the infinite past

$$\hat{\rho}(-\infty) = \hat{\rho}_{0,\text{eq}} = \frac{e^{-\beta\hat{H}_0}}{\text{Tr}[e^{-\beta\hat{H}_0}]} \tag{11.42}$$

Using an integral equation representation of (11.41) given by Eq. (10.25) and keeping only the lowest-order correction to $\hat{\rho}_{0,\text{eq}}$ leads to

$$\hat{\rho}(t) = \hat{\rho}_{0,\text{eq}} - i\int_{-\infty}^{t} dt' e^{-i(t-t')\mathcal{L}_0}\mathcal{L}_1(t')\hat{\rho}_{0,\text{eq}}$$
$$= \hat{\rho}_{0,\text{eq}} - \frac{i}{\hbar}\int_{-\infty}^{t} dt' e^{-(i/\hbar)(t-t')\hat{H}_0}[\hat{H}_1(t'), \hat{\rho}_{0,\text{eq}}]e^{(i/\hbar)(t-t')\hat{H}_0} \tag{11.43}$$

where we have also used the identity $\exp(-i\mathcal{L}_0 t)\hat{\rho}_{0,\text{eq}} = \hat{\rho}_{0,\text{eq}}$. The deviation of an observable B from its equilibrium value under $\hat{\rho}_{0,\text{eq}}$ is given by

$$\delta\langle B\rangle(t) = \langle B(t)\rangle - \langle B\rangle_0 = \text{Tr}[\hat{\rho}(t)\hat{B}] - \text{Tr}[\hat{\rho}_{0,\text{eq}}\hat{B}]$$
$$= -\frac{i}{\hbar}\int_{-\infty}^{t} dt'\text{Tr}[\hat{B}e^{-(i/\hbar)(t-t')\hat{H}_0}[\hat{H}_1(t'), \hat{\rho}_{0,\text{eq}}]e^{(i/\hbar)(t-t')\hat{H}_0}]$$
$$= -\frac{i}{\hbar}\int_{-\infty}^{t} dt'\text{Tr}[\hat{B}_I(t-t')[\hat{H}_1(t'), \hat{\rho}_{0,\text{eq}}]] \tag{11.44}$$

where the last equality is obtained by cyclically changing the order of operators inside the trace and by using the interaction representation (11.37) of the operator

$\hat{B}$. Note that since $\mathrm{Tr}[\hat{H}_1(t'), \hat{\rho}_{0,\mathrm{eq}}] = 0$ we can replace $\hat{B}_I$ by $\delta\hat{B}_I = \hat{B}_I - \langle B\rangle_0$ in (11.44) to get

$$\delta\langle B\rangle(t) = -\frac{i}{\hbar}\int_{-\infty}^{t} dt' \mathrm{Tr}[\delta\hat{B}_I(t-t')[\hat{H}_1(t'), \hat{\rho}_{0,\mathrm{eq}}]] \tag{11.45}$$

Next we use $\hat{H}_1(t) = -F(t)\hat{A}$ and the cyclic property of the trace to rewrite Eq. (11.45) in the form

$$\begin{aligned}\delta\langle B\rangle(t) &= \frac{i}{\hbar}\int_{-\infty}^{t} dt' F(t')\mathrm{Tr}[\delta\hat{B}_I(t-t')[\hat{A}, \hat{\rho}_{0,\mathrm{eq}}]] \\ &= \frac{i}{\hbar}\int_{-\infty}^{t} dt' F(t')[\langle\delta\hat{B}_I(t-t')\hat{A}\rangle_0 - \langle\hat{A}\delta\hat{B}_I(t-t')\rangle_0] \\ &= \frac{i}{\hbar}\int_{-\infty}^{t} dt' F(t')\langle[\delta\hat{B}_I(t-t'), \hat{A}]\rangle_0\end{aligned} \tag{11.46}$$

Finally we note that under the commutator we can replace $\hat{A}$ by $\delta\hat{A} = \hat{A} - \langle\hat{A}\rangle_0$. We have found that in the linear response approximation

$$\delta\langle B\rangle(t) = \int_{-\infty}^{t} dt' \chi_{BA}(t-t')F(t') \tag{11.47}$$

where

$$\chi_{BA}(t) = \frac{i}{\hbar}\langle[\delta\hat{B}_I(t), \delta\hat{A}]\rangle_0 \tag{11.48}$$

or

$$\delta\langle B\rangle(t) = \int_{-\infty}^{\infty} dt' \chi_{BA}(t-t')F(t') \tag{11.49}$$

with $\chi_{BA}(t)$ given by

$$\chi_{BA}(t) = \begin{cases} \frac{i}{\hbar}\langle[\delta\hat{B}_I(t), \delta\hat{A}]\rangle_0 & t > 0 \\ 0 & t < 0 \end{cases} \tag{11.50}$$

Problem 11.3. Use the Kubo's identity (Appendix 11A, Eq. (11.81)) with $\hat{H}$ replaced by $\hat{H}_0$ to show that Eqs (11.50) and (11.40) are equivalent

Solution: Indeed

$$\chi_{BA}(t) = \frac{i}{\hbar}\langle[\delta\hat{B}_I(t), \delta\hat{A}]\rangle_0 = \frac{i}{\hbar}\mathrm{Tr}\{\hat{\rho}_{0,\mathrm{eq}}[\delta\hat{B}_I(t), \delta\hat{A}]\} = \frac{i}{\hbar}\mathrm{Tr}\{[\delta\hat{A}, \hat{\rho}_{0,\mathrm{eq}}]\delta\hat{B}_I(t)\}$$

$$= \mathrm{Tr}\{\hat{\rho}_{0,\mathrm{eq}}\int_0^\beta d\lambda\,\delta\dot{\hat{A}}_I(-i\hbar\lambda)\delta\hat{B}_I(t)\} = \int_0^\beta d\lambda\langle\delta\dot{\hat{A}}_I(-i\hbar\lambda)\delta\hat{B}_I(t)\rangle_0$$

$$= \beta\langle\delta\dot{\hat{A}}^{(\beta)}\delta\hat{B}_I(t)\rangle_0 = -\beta\langle\delta\hat{A}^{(\beta)}\delta\dot{\hat{B}}_I(t)\rangle_0 \tag{11.51}$$

11.2.3 Causality and the Kramers–Kronig relations

The fact that the response functions $\chi(t)$ in Eqs (11.19) and (11.49) vanish for $t < 0$ is physically significant: It implies that an "effect" cannot precede its "cause," that is, the "effect" at time t cannot depend on the value of the "cause" at a later time. This obvious physical property is also reflected in an interesting and useful mathematical property of such response functions. To see this we express the response function $\chi_{BA}(t) = \frac{i}{\hbar}\theta(t)\langle[\delta\hat{B}_I(t), \delta\hat{A}]\rangle_0$ in terms of the eigenstates (and corresponding energies) of $\hat{H}_0$

$$\chi_{BA}(t) = \frac{i}{\hbar}\theta(\mathrm{t})\mathrm{Tr}\{\hat{\rho}_{0,\mathrm{eq}}(e^{i\hat{H}_0 t/\hbar}\delta\hat{B}e^{-i\hat{H}_0 t/\hbar}\delta\hat{A} - \delta\hat{A}e^{i\hat{H}_0 t/\hbar}\delta\hat{B}e^{-i\hat{H}_0 t/\hbar})\}$$

$$= \frac{i}{\hbar}\theta(t)\sum_k\sum_l P_k(e^{i\omega_{kl}t}\delta B_{kl}\delta A_{lk} - e^{-i\omega_{kl}t}\delta B_{lk}\delta A_{kl})$$

$$= \frac{i}{\hbar}\theta(t)\sum_k\sum_l (P_k - P_l)e^{i\omega_{kl}t}\delta B_{kl}\delta A_{lk} \tag{11.52}$$

where $P_k = \exp(-\beta\varepsilon_k)/\sum_k \exp(-\beta\varepsilon_k)$, $\omega_{kl} = (\varepsilon_k - \varepsilon_l)/\hbar$, and $X_{kl} = \langle k|\hat{X}|l\rangle$ for any operator $\hat{X}$.

Now consider the Fourier transform of this function, $\tilde{\chi}(\omega) = \int_{-\infty}^{\infty} dte^{i\omega t}\chi(t)$. We use the identity[1]

$$\int_{-\infty}^{\infty} dte^{ixt}\theta(t)e^{ix't} = \lim_{\eta\to 0+} \frac{i}{x + x' + i\eta} \tag{11.53}$$

to write

$$\tilde{\chi}_{BA}(\omega) = -\frac{1}{\hbar}\lim_{\eta\to 0+}\sum_k\sum_l (P_k - P_l)\frac{1}{\omega + \omega_{kl} + i\eta}\delta B_{kl}\delta A_{lk} \tag{11.54}$$

The appearance of $i\eta$ in the denominators here defines the analytical properties of this function: The fact that $\tilde{\chi}(\omega)$ is analytic on the upper half of the complex ω plane and has simple poles (associated with the spectrum of $\hat{H}_0$) on the lower half is equivalent to the casual nature of its Fourier transform—the fact that it vanishes for $t < 0$. An interesting mathematical property follows. For any function $\chi(\omega)$ that is (1) analytic in the half plane $\text{Re}\omega > 0$ and (2) vanishes fast enough for $|\omega| \to \infty$ we can write (see Section 1.1.6)

$$\lim_{\eta\to 0}\int_{-\infty}^{\infty} d\omega' \frac{\chi(\omega')}{\omega' - \omega - i\eta} = 2\pi i\chi(\omega) \tag{11.55}$$

Using (cf. Eq. (1.71)) $\lim_{\eta\to 0}(\omega' - \omega - i\eta)^{-1} = \text{PP}(\omega' - \omega)^{-1} + i\pi\delta(\omega' - \omega)$ we find for the real and imaginary parts of $\chi = \chi_1 + i\chi_2$

$$\chi_1(\omega) = \frac{1}{\pi}\text{PP}\int_{-\infty}^{\infty} d\omega' \frac{\chi_2(\omega')}{\omega' - \omega}$$

$$\chi_2(\omega) = -\frac{1}{\pi}\text{PP}\int_{-\infty}^{\infty} d\omega' \frac{\chi_1(\omega')}{\omega' - \omega} \tag{11.56}$$

---

[1] Note that the existence of the θ function is important in this identity. The inverse Fourier transform is

$$\lim_{\eta\to 0+}\frac{1}{2\pi}\int_{-\infty}^{\infty} dxe^{-ixt}\frac{i}{x + x' + i\eta} = \theta(t)e^{ix't}$$

as is easily shown by contour integration.

The transformation defined by (11.56) is called *Hilbert transform*, and we have found that the real and imaginary parts of a function that is analytic in half of the complex plane and vanishes at infinity on that plane are Hilbert transforms of each other. Thus, causality, by which response functions have such analytical properties, also implies this relation. On the practical level this tells us that if we know the real (or imaginary) part of a response function we can find its imaginary (or real) part by using this transform.

Problem 11.4. Note that if $\chi_1(\omega)$ is symmetric under sign inversion of ω, that is, $\chi_1(\omega) = \chi_1(-\omega)$, then $\chi_2(\omega)$ is antisymmetric, $\chi_2(\omega) = -\chi_2(-\omega)$. Show that in this case Eqs (11.56) can be rewritten in the form

$$\chi_1(\omega) = \frac{2}{\pi} PP \int_0^\infty d\omega' \frac{\omega' \chi_2(\omega')}{\omega'^2 - \omega^2}$$

$$\chi_2(\omega) = -\frac{2\omega}{\pi} PP \int_0^\infty d\omega' \frac{\chi_1(\omega')}{\omega'^2 - \omega^2} \tag{11.57}$$

In this form the equations are known as the Kramers–Kronig relations.

11.2.4 Examples: mobility, conductivity, and diffusion

Consider a homogeneous and isotropic system of classical noninteracting charged particles under an external, position-independent electric field $E_x(t)$ in the x direction. In this case

$$H_1 = -q\left(\sum_j x_j\right) E_x(t) \tag{11.58}$$

where x_j is the displacement of particle j in the x direction and q is the particle charge. In the notation of Eq. (11.2) we now have

$$F = E_x \quad \text{and} \quad A = q\sum\nolimits_j x_j \tag{11.59}$$

We want to calculate the response to this force as expressed by the average speed of a given particle l, and since in equilibrium $\langle \mathbf{v}_l \rangle_0 = 0$ we can write

$$\Delta B = \dot{x}_l = v_{lx} \tag{11.60}$$

In focusing on the x component of the response we anticipate that the response in the orthogonal directions vanishes, as can be easily verified using the procedure below. Equation (11.18) takes the form

$$\langle v_{lx}(t)\rangle = \int_{-\infty}^{t} dt' \chi(t-t')E_x(t') \tag{11.61}$$

where, using Eq. (11.22) in the form $\chi_{BA} = \beta\langle\delta\dot{A}(0)\delta B(t)\rangle_0$

$$\chi(t) = \beta q\left\langle\left(\sum_j v_{jx}(0)\right)v_{lx}(t)\right\rangle_0 = \beta q\langle v_{lx}(0)v_{lx}(t)\rangle_0 \tag{11.62}$$

In the last equality we have used the fact that the equilibrium velocities of different particles are uncorrelated. For $E_x(t) = E_x =$ constant Eq. (11.61) gives

$$\langle v_x\rangle = \left(\beta q\int_0^{\infty} dt\langle v_x(0)v_x(t)\rangle\right)E_x \tag{11.63}$$

Here and below we have dropped the subscript 0 from the correlation functions. Indeed, to this lowest order we could take the thermal averages using either H or H_0. We have also dropped the subscript l because this result is obviously the same for all identical particles. The equivalent quantum result is

$$\langle v_x\rangle = \left(\beta q\int_0^{\infty} dt\langle \hat{v}_x^{(\beta)}(0)\hat{v}_x(t)\rangle\right)E_x \tag{11.64}$$

For simplicity we continue to consider the classical case. We can now discuss several equivalent transport functions (or "transport coefficients"):

11.2.4.1 Mobility

The coefficient that multiples the force, qE_x, in (11.63) is the *mobility u*,

$$u = \beta\int_0^{\infty} dt\langle v_x(0)v_x(t)\rangle = \frac{\beta}{3}\int_0^{\infty} dt\langle \mathbf{v}(0)\cdot\mathbf{v}(t)\rangle \tag{11.65}$$

In the last equality we have used the fact that an isotropic system u does not depend on direction.

11.2.4.2 *Diffusion*

Equation (11.65) is closely related to the expression (6.14) for the diffusion coefficient D, so that

$$u = \frac{D}{k_B T} \tag{11.66}$$

This result is known as the Stokes–Einstein relation. It can also be derived from elementary considerations: Let a system of noninteracting charged particles be in thermal equilibrium with a uniform electric field $E_x = -\partial\Phi/\partial x$ in the x direction, so that the density of charged particles satisfies $\rho(x) \sim \exp(-\beta q \Phi(x))$. In equilibrium, the diffusion flux, $-D\partial\rho/\partial x = -\beta D q \rho E_x$ and the drift flux, $uqE_x\rho$ should add to zero. This yields (11.66).

11.2.4.3 *Friction*

The diffusion coefficient and therefore the mobility are closely related to the friction coefficient γ that determines the energy loss by the moving particles. Writing the acceleration $\dot{v}_x$ of the moving charged particle as a sum of electrostatic and friction forces

$$\langle \dot{v}_x \rangle = \frac{1}{m} q E_x - \gamma \langle v_x \rangle \tag{11.67}$$

and putting at steady state $\langle \dot{v}_x \rangle = 0$, leads to $u = (m\gamma)^{-1}$ or

$$D = \frac{k_B T}{m\gamma} \tag{11.68}$$

11.2.4.4 *Conductivity*

The conductivity σ connects the external electric field E_x and the observed current $\langle J_x \rangle$ via $\langle J_x \rangle = \sigma E_x$. The average current is $\langle J_x \rangle = \rho q \langle v_x \rangle$ where ρ is the carrier density. The average velocity v_x is obtained from (11.66) in the form $\langle v_x \rangle = uqE_x = DqE_x/k_B T$. This yields the *Nernst–Einstein* equation

$$\sigma = \frac{q^2 \rho}{k_B T} D \tag{11.69}$$

Consider now the time-dependent case. From Eq. (11.63) we get

$$J_x(t) = \rho q \langle v_x(t) \rangle = \int_{-\infty}^{\infty} dt \sigma(t - t') E_x(t) \tag{11.70}$$

$$\sigma(t) = \begin{cases} \rho q^2 \beta \langle \delta v_x(0) \delta v_x(t) \rangle = \dfrac{\rho q^2}{3 k_B T} \langle \mathbf{v}(0) \cdot \mathbf{v}(t) \rangle & t > 0 \\ 0 & t < 0 \end{cases} \tag{11.71}$$

In Fourier space, $J_x(\omega) = \int_{-\infty}^{\infty} dte^{i\omega t} J_x(t)$, etc. we get

$$\mathbf{J}(\omega) = \sigma(\omega)\mathbf{E}(\omega) \tag{11.72}$$

with the frequency-dependent conductivity, $\sigma(\omega)$, which is given by the Fourier–Laplace transform of the velocity–time correlation function,[2]

$$\sigma(\omega) = \int_{-\infty}^{\infty} dt\sigma(t)e^{i\omega t} = \frac{\rho q^2 \beta}{3} \int_{0}^{\infty} dt \langle \mathbf{v}(0) \cdot \mathbf{v}(t) \rangle e^{i\omega t} \tag{11.73}$$

The generalization of these results to anisotropic system is easily done within the same formalism, see for example, the book by Kubo et al. referred to at the end of this chapter. The result (11.73) (or the more general expression below) is often referred to as Kubo's formula of conductivity.

11.2.4.5 Conductivity and diffusion in a system of interacting particles

If carrier–carrier interactions are not disregarded we cannot obtain the above transport coefficients from the single particle response (11.63). Linear response theory should now be used for collective variables. Starting from Eqs (11.58) and (11.59) we seek the response in the current density

$$J_x(\mathbf{r}, t) = q \sum_j \dot{x}_j(t)\delta(\mathbf{r} - \mathbf{r}_j) \tag{11.74}$$

Classical linear response theory now yields $\sigma(t) = \beta\langle \delta\dot{A}\delta B(t)\rangle_0$ (for $t > 0$) with $\delta B = J_x(\mathbf{r}, t)$ and $\delta\dot{A} = q\sum_j \dot{x}_j = \int d\mathbf{r}' J_x(\mathbf{r}', 0)$. This leads to Eq. (11.70) with

$$\begin{aligned}\sigma(t) = \beta \int d\mathbf{r}' \langle J_x(\mathbf{r}', 0) J_x(\mathbf{r}, t)\rangle_0 &= \beta \int d\mathbf{r} \langle J_x(0,0) J_x(\mathbf{r}, t)\rangle_0 \\ &= \frac{\beta}{3} \int d\mathbf{r} \langle \mathbf{J}(0,0) \cdot \mathbf{J}(\mathbf{r}, t)\rangle_0\end{aligned} \tag{11.75}$$

while the DC conductivity is given by

$$\sigma = \frac{\beta}{3} \int d\mathbf{r} \int_0^{\infty} dt \langle \mathbf{J}(0,0) \cdot \mathbf{J}(\mathbf{r}, t)\rangle_0 \tag{11.76}$$

[2] Note that σ of Eq. (11.69) (or $\sigma(\omega)$ of Eq. (11.73)) have dimensionalities of $(\text{time})^{-1}$ while $\sigma(t)$ of Eq. (11.71) has dimensionality $(\text{time})^{-2}$.

In Eqs (11.75) and (11.76) the current density vector is

$$\mathbf{J}(\mathbf{r},t) = q\sum_j \dot{\mathbf{r}}_j(t)\delta(\mathbf{r}-\mathbf{r}_j) \tag{11.77}$$

These equations relate the static and dynamic conductivities to the time and space correlation functions of equilibrium fluctuations in the local current density.

Turning now to the diffusion coefficient, the single particle expression (Section 6.2.1)

$$D_{\text{tr}} = \int_0^\infty dt\langle v_x(0)v_x(t)\rangle = \frac{1}{3}\int_0^\infty dt\langle \mathbf{v}(0)\cdot\mathbf{v}(t)\rangle \tag{11.78}$$

still has a meaning in a system of interacting particles, because it is possible to follow the motion of single particles in such a system. This is done by marking a small fraction of such particles by, for example, isotope substitution and following the marked particles. The diffusion coefficient that characterizes the motion observed in this way is called the *tracer diffusion coefficient.*

The tracer diffusion coefficient however is not the transport coefficient to be used in the linear relationship (Fick's law) $\mathbf{J}_c = -D\nabla c$ between the diffusion current $\mathbf{J}_c$ and the particle concentration c or in the diffusion equation $\partial c/\partial t = D\nabla^2 c$. The coefficient D in these equations does depend on correlations between the motions of different molecules. We have used the notation D_{tr} above to distinguish the tracer diffusion coefficient from the so-called *chemical diffusion coefficient D* that appears in the diffusion equation.

A fully microscopic theory of chemical diffusion can be constructed, however, it requires a careful distinction between the motions of the observed species and the underlying host, and is made complicated by the fact that, as defined, the diffusion coefficient relates flux to the concentration gradient while the actual force that drives diffusion is gradient of the chemical potential. An alternative useful observable is the so-called *conductivity diffusion coefficient*, which is *defined* for the motion of charged particles by the Nernst–Einstein equation (11.69)

$$D_\sigma = \frac{k_B T}{cq^2}\sigma \tag{11.79}$$

More generally, any force could be used to move the particles, so a more general definition of this type of transport coefficient will be the "mobility diffusion coefficient," $D_u = k_B T u$ (cf. Eq. (11.66)). Note that while this relationship between the conductivity and the diffusion coefficient was *derived* for noninteracting carriers, we now use this equation as a definition also in the presence of interparticle interactions, when σ is given by Eq. (11.76).

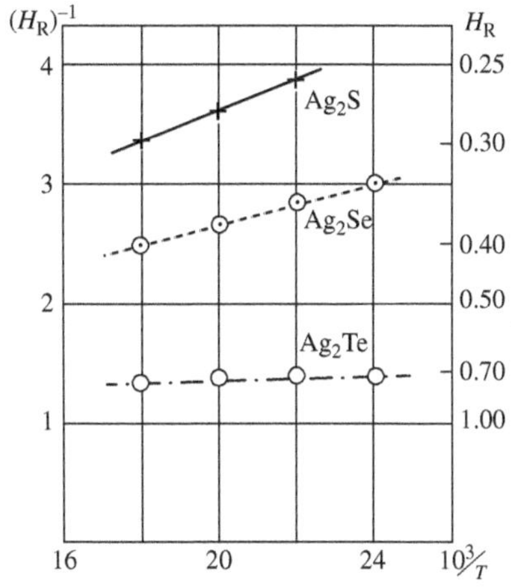

Fig. 11.1 The Haven ratios, plotted against inverse temperature, for the diffusion of silver ions in three solid ionic conductors: Ag_2S, Ag_2Se, and Ag_2Te. (From H. Okazaki, J. Phys. Soc. Jpn, **43**, 213 (1977).)

It should be realized that D_σ and D_u just express the properties of the conductivity and mobility from which they are derived, and using them in the Fick's law $\mathbf{J}_c = -D\nabla c$ is at best a crude approximation. On the other hand they contain information about interparticle correlations that result from carrier–carrier interactions. A useful quantity that gauge the importance of such correlations is the *Haven ratio*

$$H_R \equiv \frac{D_{tr}}{D_\sigma} \tag{11.80}$$

which is unity in a system of noninteracting particles and deviates from 1 when carrier–carrier interactions affect the observable D_σ. An example is shown in Fig. 11.1.

Appendix 11A: The Kubo identity

Here we prove the *Kubo's identity* for any operator $\hat{A}$ and Hamiltonian $\hat{H}$. It states

$$[e^{-\beta\hat{H}}, \hat{A}] = e^{-\beta\hat{H}} \int_0^\beta d\lambda e^{\lambda\hat{H}} [\hat{A}, \hat{H}] e^{-\lambda\hat{H}} = i\hbar e^{-\beta\hat{H}} \int_0^\beta d\lambda e^{\lambda\hat{H}} \dot{\hat{A}} e^{-\lambda\hat{H}}$$

$$= i\hbar e^{-\beta\hat{H}} \int_0^\beta d\lambda \dot{\hat{A}}_H(-i\hbar\lambda) \tag{11.81}$$

where $\hat{A}_{\mathrm{H}}(x) = e^{(i/\hbar)\hat{H}x}\hat{A}e^{-(i/\hbar)\hat{H}x}$ is the Heisenberg representation of $\hat{A}$. The first equality in (11.81) is verified by taking the ij matrix element of the two sides in the basis of eigenstates of $\hat{H}$. On the left we have $\langle i|[e^{-\beta\hat{H}},\hat{A}]|j\rangle = A_{ij}(e^{-\beta\varepsilon_i} - e^{-\beta\varepsilon_j})$. The same matrix element on the right is

$$\langle i|e^{-\beta\hat{H}}\int_0^\beta d\lambda e^{\lambda\hat{H}}[\hat{A},\hat{H}]e^{-\lambda\hat{H}}|j\rangle = e^{-\beta\varepsilon_i}\int_0^\beta d\lambda e^{\lambda\varepsilon_i}(A_{ij}\varepsilon_j - \varepsilon_i A_{ij})e^{-\lambda\varepsilon_j}$$

$$= A_{ij}(\varepsilon_j - \varepsilon_i)e^{-\beta\varepsilon_i}\int_0^\beta d\lambda e^{\lambda(\varepsilon_i-\varepsilon_j)} \tag{11.82}$$

which is easily shown to be the same. The second equality in (11.81) is based on the identities

$$\frac{d\hat{A}_{\mathrm{H}}}{dt} = \frac{i}{\hbar}[\hat{H},\hat{A}_{\mathrm{H}}(t)] = \frac{i}{\hbar}e^{(i/\hbar)\hat{H}t}[\hat{H},\hat{A}]e^{-(i/\hbar)\hat{H}t} = e^{(i/\hbar)\hat{H}t}\dot{\hat{A}}e^{-(i/\hbar)\hat{H}t} \tag{11.83}$$

Here $\hat{A}_{\mathrm{H}}$ and $\hat{A}$ denote respectively the Heisenberg and Schrödinger representations of the operator. Equation (11.83) implies that $\dot{\hat{A}}_{\mathrm{H}} = d\hat{A}_{\mathrm{H}}/dt$ and $\dot{\hat{A}}$ are respectively time derivatives of $\hat{A}$ in the Heisenberg and the Schrödinger representations. Eq. (11.81) is a relationship between these representations in which t is replaced by $-i\hbar\lambda$.

Further reading

B. J. Berne and R. Pecora, *Dynamic Light Scattering* (Wiley, New York, 1976).
D. Chandler, *Introduction to Modern Statistical Mechanics* (Oxford University Press, Oxford, 1986).
R. Kubo, M. Toda, and N. Hashitsume, *Statistical Physics* II, *Springer Series in Solid State Sciences*, 2nd ed. (Springer, Berlin, 1995).
R. Zwanzig, *Non Equilibrium Statistical Mechanics* (Oxford University Press, Oxford, 2001).

12

THE SPIN–BOSON MODEL

Sometimes, you know, we can not see dazzling objects
Through an excess of light; whoever heard
Of doorways, portals, outlooks, in such trouble?
Besides, if eyes are doorways, might it not
Be better to remove them, sash, jamb, lintel,
And let the spirit have a wider field?

Lucretius (c.99–c.55 BCE*) "The way things are"*
translated by Rolfe Humphries, Indiana University Press, 1968

In a generic quantum mechanical description of a molecule interacting with its thermal environment, the molecule is represented as a few level system (in the simplest description just two, for example, ground and excited states) and the environment is often modeled as a bath of harmonic oscillators (see Section 6.5). The resulting theoretical framework is known as *the spin–boson model*,[1] a term that seems to have emerged in the Kondo problem literature (which deals with the behavior of magnetic impurities in metals) during the 1960s, but is now used in a much broader context. Indeed, it has become one of the central models of theoretical physics, with applications in physics, chemistry, and biology that range far beyond the subject of this book. Transitions between molecular electronic states coupled

[1] The term "spin–boson model" seems to have emerged in the Kondo problem literature (which deals with the interactions between an impurity spin in a metal and the surrounding electron bath) during the 1960s, however fundamental works that use different aspects of this model were published earlier. In 1953, Wangsness and Bloch (R. K. Wangsness and F. Bloch, Phys. Rev. **89**, 728 (1953)) presented a framework for the theoretical discussion of spin relaxation due to environmental interactions, that evolved into theory of the Bloch equations in a later paper by Bloch (F. Bloch, Phys. Rev. **105**, 1206 (1957)) and the more rigorous description by Redfield (A. G. Redfield, IBM J. Res. Develop. **1**, 19 (1957)); see Section 10.4.8). Marcus (R. A. Marcus, J. Chem. Phys. **24**, 966; 979 (1956); see Chapter 16) has laid the foundation of the theory of electron transfer in polar solvents and Holstein (T. Holstein, Ann. Phys. (NY), **8**, 325; 343 (1959)) published his treatise of polaron formation and dynamics in polar crystals. Much of the later condensed phase literature has been reviewed by Leggett et al. (A. J. Leggett, S. Chakravarty, A. T. Dorsey, M. P. A. Fisher, A. Garg, W. Zwerger, Rev. Mod. Phys. **59**, 1 (1987)), see also H. Grabert and A. Nitzan, editors, Chem. Phys. **296**(2–3) (2004). In many ways the problem of a few level system interacting with the radiation field (Chapter 18) also belong to this class of problems.

to nuclear vibrations, environmental phonons, and photon modes of the radiation field fall within this class of problems. The present chapter discusses this model and some of its mathematical implications. The reader may note that some of the subjects discussed in Chapter 9 are reiterated here in this more general framework.

12.1 Introduction

In Sections 2.2 and 2.9 we have discussed the dynamics of the two-level system and of the harmonic oscillator, respectively. These exactly soluble models are often used as prototypes of important classes of physical system. The harmonic oscillator is an exact model for a mode of the radiation field (Chapter 3) and provides good starting points for describing nuclear motions in molecules and in solid environments (Chapter 4). It can also describe the short-time dynamics of liquid environments via the instantaneous normal mode approach (see Section 6.5.4). In fact, many linear response treatments in both classical and quantum dynamics lead to harmonic oscillator models: Linear response implies that forces responsible for the return of a system to equilibrium depend linearly on the deviation from equilibrium—a harmonic oscillator property! We will see a specific example of this phenomenology in our discussion of dielectric response in Section 16.9.

The two-level model is the simplest prototype of a quantum mechanical system that has no classical analog. A spin $\frac{1}{2}$ particle is of course an example, but the same model is often used also to describe processes in multilevel systems when the dynamics is dominated by two of the levels. The dynamics of an anharmonic oscillator at low enough temperatures may be dominated by just the two lowest energy levels. The electronic response of a molecular system is often dominated by just the ground and the first excited electronic states. Low temperature tunneling dynamics in a double well potential can be described in terms of an interacting two-level system, each level being the ground state on one of the wells when it is isolated from the other. Finally, as a mathematical model, the two-level dynamics is often a good starting point for understanding the dynamics of a few level systems.

The prominence of these quantum dynamical models is also exemplified by the abundance of theoretical pictures based on the spin–boson model—a two (more generally a few) level system coupled to one or many harmonic oscillators. Simple examples are an atom (well characterized at room temperature by its ground and first excited states, that is, a two-level system) interacting with the radiation field (a collection of harmonic modes) or an electron spin interacting with the phonon modes of a surrounding lattice, however this model has found many other applications in a variety of physical and chemical phenomena (and their extensions into the biological world) such as atoms and molecules interacting with the radiation field, polaron formation and dynamics in condensed environments, electron transfer processes, quantum solvation phenomena, spin–lattice relaxation, molecular vibrational relaxation, exciton dynamics in solids, impurity relaxation in solids, interaction of magnetic moments with their magnetic environment, quantum computing (the need to understand and possibly control relaxation effects in quantum bits, or qubits), and more. In addition, the spin–boson model has been extensively used as a playground for developing, exploring, and testing new theoretical methods, approximations, and numerical schemes for quantum relaxation processes, including perturbation methods, exactly solvable models, quantum-numerical methodologies and semiclassical approximations. A few of these results and applications are presented below.

12.2 The model

We consider a two-level system coupled to a bath of harmonic oscillators that will be referred to as a *boson field*. Two variations of this model, which differ from each other by the basis used to describe the two-level system, are frequently encountered. In one, the basis is made of the eigenstates of the two-state Hamiltonian that describes the isolated system. The full Hamiltonian is then written

$$\hat{H} = \hat{H}_0 + \hat{V}_{\mathrm{SB}} \tag{12.1}$$

where the zero-order Hamiltonian describes the separated subsystems (see Fig. 12.1)

$$\hat{H}_0 = \hat{H}_M + \hat{H}_B = E_1|1\rangle\langle 1| + E_2|2\rangle\langle 2| + \sum_\alpha \hbar\omega_\alpha \hat{a}_\alpha^\dagger \hat{a}_\alpha \tag{12.2a}$$

and the coupling is taken in the form

$$\hat{V}_{\mathrm{SB}} = \sum_{j,j'=1}^{2} \sum_\alpha V_{j,j'}^\alpha |j\rangle\langle j'|(\hat{a}_\alpha^\dagger + \hat{a}_\alpha) \tag{12.2b}$$

The rationale behind this choice of system–bath interaction is that it represents the first term in the expansion of a general interaction between the

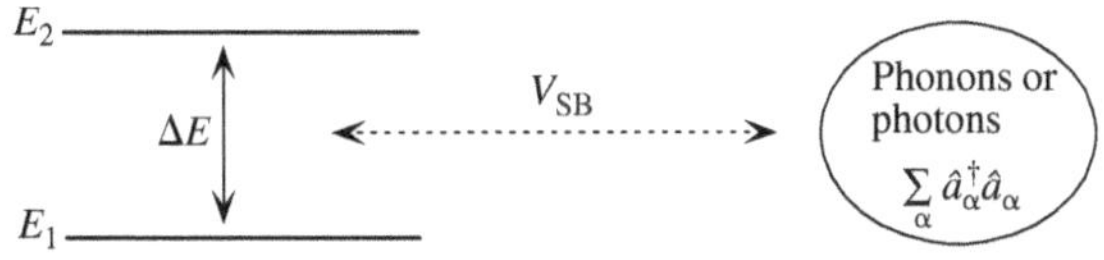

FIG. 12.1 The spin–boson model for a two-level molecule coupled to a system of harmonic oscillators.

two-level system and the harmonic bath in the bath normal mode coordinates, $\hat{x}_\alpha = \sqrt{\hbar/(2m_\alpha\omega_\alpha)}(\hat{a}_\alpha^\dagger + \hat{a}_\alpha)$, that express deviations from the minimum energy configuration.[2] In other situations the coupling takes place through the momentum operator which is linear in $(\hat{a}_\alpha^\dagger - \hat{a}_\alpha)$. An example is the important case of system–radiation field coupling. If the system does not have a permanent dipole moment the coupling $\hat{V}_{SB}$ is non-diagonal in the system states and takes the form (cf. Eq. (3.27))

$$\hat{V}_{SB} = -i\sum_{\mathbf{k}}\sum_{\sigma_{\mathbf{k}}}\sqrt{\frac{2\pi\hbar\omega_k}{\varepsilon\Omega}}[(\hat{\mu}_{12}\cdot\sigma_{\mathbf{k}})|1\rangle\langle 2| + (\hat{\mu}_{21}\cdot\sigma_{\mathbf{k}})|2\rangle\langle 1|](\hat{a}_{\mathbf{k},\sigma_{\mathbf{k}}} - \hat{a}_{\mathbf{k},\sigma_{\mathbf{k}}}^\dagger) \tag{12.3}$$

where $\hat{\mu}$ is the system dipole operator and where the harmonic modes are characterized in terms of the wavevector $\mathbf{k}$ and the polarization vector $\sigma_{\mathbf{k}}$.

In the second model, the basis chosen to describe the two-level (or few level) system is not made of the system eigenstates. In what follows we denote these states $|L\rangle$ and $|R\rangle$

$$\hat{H} = \hat{H}_0 + \hat{V}_S + \hat{V}_{SB} \tag{12.4}$$

$\hat{H}_0$ and $\hat{V}_{SB}$ have the forms (12.2)

$$\hat{H}_0 = \hat{H}_{0M} + \hat{H}_{0B} = E_L|L\rangle\langle L| + E_R|R\rangle\langle R| + \sum_\alpha \hbar\omega_\alpha\hat{a}_\alpha^\dagger\hat{a}_\alpha \tag{12.5a}$$

$$\hat{V}_{SB} = \sum_{j,j'=L}^{R}\sum_\alpha V_{j,j'}^\alpha|j\rangle\langle j'|(\hat{a}_\alpha^\dagger + \hat{a}_\alpha) \tag{12.5b}$$

and the additional term is the non-diagonal part of the system Hamiltonian

$$\hat{V}_S = V_{LR}^S|L\rangle\langle R| + V_{RL}^S|R\rangle\langle L| \tag{12.6}$$

Sometimes this is done as a computational strategy, for example, atomic orbitals are used as a basis set in most molecular computations. In other cases this choice reflects our physical insight. Consider, for example, tunneling in a double well potential U, Fig. 12.2(a), where the barrier between the two wells is high relative to both the thermal energy k_BT and the zero-point energy in each well. We have already indicated that a two-level model can be useful for describing the low-temperature dynamics of this system. Denoting by ψ_L and ψ_R the wavefunctions that represent the ground states of the separated potentials U_L, Fig. 12.2(b), and U_R, Fig. 12.2(c),

[2] The zero-order term of this expansion, which is independent of x_α, just redefines the zero-order Hamiltonian. Disregarding higher order reflects the expectation that in condensed phases the deviations x_α from the minimum energy configuration are small.

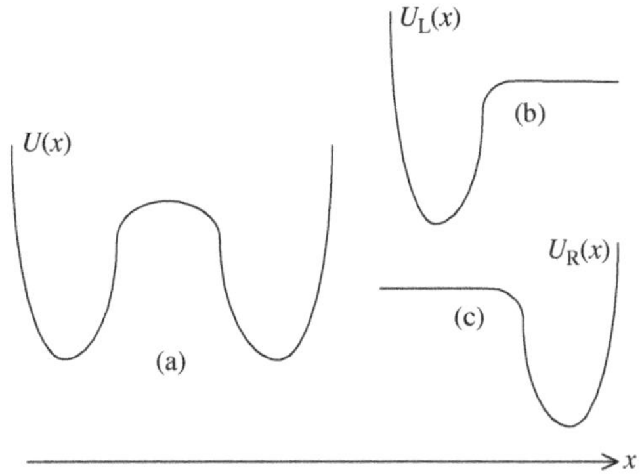

FIG. 12.2 A double well potential $U(x)$ (a) and the potentials $U_L(x)$ and $U_R(x)$, (b) and (c), which define the states $|L\rangle$ and $|R\rangle$ used in the local state representation of a 2-level system.

respectively, the two lowest states of the potential U are approximated well by the even and odd linear combinations, $\psi_{1,2} = \psi_L \pm \psi_R$. While $\psi_{1,2}$ represent the exact ground states of the potential $U(x)$, tunneling is more readily described in terms of transitions between the local states ψ_R and ψ_L.

It should be emphasized that, while the two models, Eqs (12.1), (12.2) and (12.4)–(12.6) are mathematically just different representations of what may be seen as the same Hamiltonian, they are used in different physical contexts. The former model is used to describe transitions between system eigenstates that are induced by the interaction of a two-level system with a boson field, as exemplified by the interaction between a system and a radiation field, Eq. (12.3). In contrast, the latter model is used to examine the effect of a boson bath on the transition between states of the system that are (1) coupled to each other also in the absence of this field and (2) associated with distinctly different polarizations of the boson environment in the different system states. This is exemplified by the electron transfer problem discussed in Chapter 16, where states L and R correspond to charge localization at different spatial positions in a polar solvent. Obviously, a two-level system may be described in terms of its eigenstates or any other basis, and the dynamics caused by its coupling to an external field or a thermal bath can be studied in any representation. Physical reality often guides us to choose a particular representation. In the tunneling example discussed above and in the electron transfer problem of Chapter 16 the local state representation is convenient because the system can be initially prepared such a local state. We have encountered a similar example in Chapter 9, where the study of the decay of a prepared "doorway" state coupled to a continuous manifold of background states was studied in the representation defined by these states and not by the eigenstates of the system Hamiltonian, because such a doorway state could be experimentally prepared and monitored.

Choosing a physically motivated representation is useful in developing physically guided approximation schemes. A commonly used approximation for the model (12.4)–(12.6) is to disregard terms with $j \neq j'$ in the system–bath interaction (12.5b). The overall Hamiltonian then takes the form

$$\hat{H} = \left[E_L + \sum_\alpha V_L^\alpha (\hat{a}_\alpha^\dagger + \hat{a}_\alpha)\right] |L\rangle\langle L| + \left[E_R + \sum_\alpha V_R^\alpha (\hat{a}_\alpha^\dagger + \hat{a}_\alpha)\right] |R\rangle\langle R|$$
$$+ V_{LR}|L\rangle\langle R| + V_{RL}|R\rangle\langle L| + \sum_\alpha \hbar\omega_\alpha \hat{a}_\alpha^\dagger \hat{a}_\alpha \tag{12.7}$$

The spin–boson coupling in this Hamiltonian is diagonal in the local state basis. The rationale for this model is that in this local state representation bath induced coupling between different local states is small relative to the interstates coupling V_{RL} because the corresponding local wavefunctions almost do not overlap. However the bath affects the system in states L and R in a substantial way. Its effect in the Hamiltonian (12.7) appears as fluctuations in the local state energies associated with the instantaneous configurations of the harmonic bath (again expanded to first order in the bath coordinates). Interestingly, the Hamiltonian (12.7) can be transformed to a form similar to (12.3) but with a nonlinear coupling to the boson field. This is shown in the next section.

12.3 The polaron transformation

Consider the n-level equivalent of the Hamiltonian (12.7)

$$\hat{H} = \sum_n \left(E_n + \sum_\alpha g_{n\alpha}\hat{x}_\alpha\right) |n\rangle\langle n| + \sum_{n\neq n'} V_{n,n'}|n\rangle\langle n'| + \hat{H}_B(\{\hat{p}_\alpha, \hat{x}_\alpha\}) \tag{12.8}$$

where (using (2.153))

$$g_{n\alpha} = V_n^\alpha \sqrt{\frac{2m_\alpha\omega_\alpha}{\hbar}} \tag{12.9}$$

and where

$$H_B(\{\hat{p}_\alpha, \hat{x}_\alpha\}) = \sum_\alpha \left(\frac{\hat{p}_\alpha^2}{2m_\alpha} + \frac{1}{2}m_\alpha\omega_\alpha^2\hat{x}_\alpha^2\right) \tag{12.10}$$

is the harmonic bath Hamiltonian, with $\hat{x}_\alpha$, $\hat{p}_\alpha$, ω_a, and m_α denoting the position and momentum operators, the frequency and mass, respectively, of the harmonic bath modes. We now carry the unitary transformation, known as the *polaron transformation*

$$\hat{\tilde{H}} = \hat{U}\hat{H}\hat{U}^{-1}; \qquad \hat{U} \equiv \prod_n \hat{U}_n \tag{12.11a}$$

$$\hat{U}_n = \exp(-i|n\rangle\langle n|\hat{\Omega}_n); \qquad \hat{\Omega}_n = \sum_\alpha \hat{\Omega}_{n,\alpha} \tag{12.11b}$$

$$\hat{\Omega}_{n,\alpha} = \frac{g_{n\alpha}\hat{p}_\alpha}{\hbar m_\alpha \omega_\alpha^2} \Rightarrow i\hat{\Omega}_{n,\alpha} = \frac{g_{n\alpha}}{m_\alpha \omega_\alpha^2}\frac{\partial}{\partial x_\alpha} \tag{12.11c}$$

Now use Eq. (2.178) to find

$$\begin{aligned}\hat{U}\hat{H}_B\hat{U}^{-1} &= \sum_\alpha \left(\frac{\hat{p}_\alpha^2}{2m_\alpha} + \frac{1}{2}m_\alpha\omega_\alpha^2\left(\hat{x}_\alpha - \sum_n \frac{g_{n\alpha}}{m_\alpha\omega_\alpha^2}|n\rangle\langle n|\right)^2\right)\\ &= \hat{H}_B + \frac{1}{2}\sum_n\left(\sum_\alpha \frac{g_{n\alpha}^2}{m_\alpha\omega_\alpha^2}\right)|n\rangle\langle n| - \sum_n\left(\sum_\alpha g_{n\alpha}\hat{x}_\alpha\right)|n\rangle\langle n|\end{aligned} \tag{12.12}$$

$$\begin{aligned}\hat{U}\sum_n\left(E_n + \sum_\alpha g_{n\alpha}\hat{x}_\alpha\right)|n\rangle\langle n|\hat{U}^{-1} &= \sum_n\left(E_n + \sum_\alpha g_{n\alpha}\hat{x}_\alpha\right)|n\rangle\langle n|\\ &\quad - \sum_n\sum_\alpha \frac{g_{n\alpha}^2}{m_\alpha\omega_\alpha^2}|n\rangle\langle n|\end{aligned} \tag{12.13}$$

in deriving (12.12) and (12.13) we have used $(|n\rangle\langle n|)^2 = |n\rangle\langle n|$, and the fact that when evaluating transformations such as $\hat{U}\hat{x}_\alpha\hat{U}^{-1}$ the operator $|n\rangle\langle n|$ in $\hat{U}_n$ can be regarded as a scalar. In addition it is easily verified that

$$\begin{aligned}e^{-i\sum_n|n\rangle\langle n|\hat{\Omega}_n}|n_1\rangle\langle n_2|e^{i\sum_n|n\rangle\langle n|\hat{\Omega}_n} &= e^{-i|n_1\rangle\langle n_1|\hat{\Omega}_{n_1}}|n_1\rangle\langle n_2|e^{i|n_2\rangle\langle n_2|\hat{\Omega}_{n_2}}\\ &= e^{-i\hat{\Omega}_{n_1}}|n_1\rangle\langle n_2|e^{i\hat{\Omega}_{n_2}} = |n_1\rangle\langle n_2|e^{i(\hat{\Omega}_{n_2}-\hat{\Omega}_{n_1})}\end{aligned} \tag{12.14}$$

Equations (12.8) and (12.11)–(12.14) yield

$$\begin{aligned}\hat{\tilde{H}} &= \sum_n\left(E_n - \left(\sum_\alpha \frac{g_{n\alpha}^2}{2m_\alpha\omega_\alpha^2}\right)\right)|n\rangle\langle n| + \sum_{n\neq n'} V_{n,n'}e^{-i(\hat{\Omega}_n-\hat{\Omega}_{n'})}|n\rangle\langle n'| + \hat{H}_B(\{p_\alpha, x_\alpha\})\\ &\equiv \hat{\tilde{H}}_0 + \sum_{n\neq n'} V_{n,n'}e^{-i(\hat{\Omega}_n-\hat{\Omega}_{n'})}|n\rangle\langle n'|\end{aligned} \tag{12.15}$$

In this transformed Hamiltonian $\hat{\hat{H}}_0$ again describes uncoupled system and bath; the new element being a shift in the state energies resulting from the system–bath interactions. In addition, the interstate coupling operator is transformed to

$$\hat{V} = V_{n,n'}|n\rangle\langle n'| \rightarrow \hat{\hat{V}} = V_{n,n'}e^{-i(\hat{\Omega}_n - \hat{\Omega}_{n'})}|n\rangle\langle n'| = V_{n,n'}e^{-\sum_\alpha \lambda_\alpha^{n,n'}(\partial/\partial x_\alpha)}|n\rangle\langle n'| \tag{12.16a}$$

with

$$\lambda_\alpha^{n,n'} = \lambda_\alpha^{(n)} - \lambda_\alpha^{(n')}; \qquad \lambda_\alpha^{(n)} = \frac{g_{n\alpha}}{m_\alpha \omega_\alpha^2} \tag{12.16b}$$

To see the significance of this result consider a typical matrix element of this coupling between eigenstates of $\hat{\hat{H}}_0$. These eigenstates may be written as $|n, \mathbf{v}\rangle = |n\rangle\chi_\mathbf{v}(\{x_\alpha\})$, where the elements v_α of the vector $\mathbf{v}$ denote the states of different modes α, that is, $\chi_{n,\mathbf{v}}(\{x_\alpha\}) = \prod_\alpha \chi_{v_\alpha}(x_\alpha)$ are eigenstates of $\hat{H}_B$ and $\chi_{v_\alpha}(x_\alpha)$ is the eigenfunction that corresponds to the v_αth state of the single harmonic mode α. A typical coupling matrix element is then

$$\langle n, \mathbf{v}|\hat{\hat{V}}|n', \mathbf{v}'\rangle = V_{n,n'}\langle\chi_\mathbf{v}(\{x_\alpha\})|e^{-i(\hat{\Omega}_n - \hat{\Omega}_{n'})}|\chi_{\mathbf{v}'}(\{x_\alpha\})\rangle \tag{12.17}$$

that is, the coupling between two vibronic states $|n, \mathbf{v}\rangle$ and $|n', \mathbf{v}'\rangle$ is given by the bare interstate coupling $V_{n,n'}$ "renormalized" by the term

$$\begin{aligned}\langle\chi_\mathbf{v}(\{x_\alpha\})|e^{-i(\hat{\Omega}_n - \hat{\Omega}_{n'})}|\chi_{\mathbf{v}'}(\{x_\alpha\})\rangle &= \prod_\alpha \langle\chi_{v_\alpha}(x_\alpha)|e^{-\lambda_\alpha^{n,n'}(\partial/\partial x_\alpha)}|\chi_{v'_\alpha}(x_\alpha)\rangle \\ &= \prod_\alpha \langle\chi_{v_\alpha}(x_\alpha)|\chi_{v'_\alpha}(x_\alpha - \lambda_\alpha^{n,n'})\rangle\end{aligned} \tag{12.18}$$

The absolute square of these term, which depend on $\mathbf{v}$, $\mathbf{v}'$, and the set of shifts $\{\lambda_\alpha^{n,n'}\}$, are known as *Franck–Condon factors*.

12.3.1 The Born Oppenheimer picture

The polaron transformation, executed on the Hamiltonian (12.8)–(12.10) was seen to yield a new Hamiltonian, Eq. (12.15), in which the interstate coupling is "renormalized" or "dressed" by an operator that shifts the position coordinates associated with the boson field. This transformation is well known in the solid-state physics literature, however in much of the chemical literature a similar end is achieved via a different route based on the Born–Oppenheimer (BO) theory of molecular vibronic structure (Section 2.5). In the BO approximation, molecular vibronic states are of the form $\phi_n(\mathbf{r}, \mathbf{R})\chi_{n,\mathbf{v}}(\mathbf{R})$ where $\mathbf{r}$ and $\mathbf{R}$ denote electronic and nuclear coordinates, respectively, $\phi_n(\mathbf{r}, \mathbf{R})$ are eigenfunctions of the electronic Hamiltonian (with corresponding eigenvalues $E_{\mathrm{el}}^{(n)}(\mathbf{R})$) obtained at fixed nuclear coordinates $\mathbf{R}$ and

$\chi_{n,\mathbf{v}}(\mathbf{R})$ are nuclear wavefunctions associated, for each electronic state n, with a nuclear potential surface given by $E_{\text{el}}^{(n)}(\mathbf{R})$. These nuclear potential surfaces are therefore different for different electronic states, and correspond within the harmonic approximation to different sets of normal modes. Mathematically, for any given potential surface, we first find the corresponding equilibrium position, that is, the minimum energy configuration $E_{\text{el,eq}}^{(n)}$, and make the harmonic approximation by disregarding higher than quadratic terms in the Taylor expansion of the potentials about these points. The eigenvectors and eigenvalues of the Hessian matrices of the nth surface, $\mathcal{H}_{\alpha,\alpha'}^{(n)} = (\partial^2 E_{\text{el}}^{(n)}(\mathbf{R})/\partial \mathbf{R}_\alpha \partial \mathbf{R}_{\alpha'})_{\text{eq}}$, yield the normal-mode coordinates, $\mathbf{x}^{(n)} \equiv \{x_\alpha^{(n)}\}$ and the corresponding frequencies $\{\omega_\alpha^{(n)}\}$ of the nuclear motion. In this harmonic approximation the potential surfaces are then

$$E_{\text{el}}^{(n)}(\mathbf{R}) = E_n + \frac{1}{2}\sum_\alpha m_\alpha \omega_\alpha^{(n)2} x_\alpha^{(n)2} \tag{12.19}$$

where $E_n \equiv E_{\text{el,eq}}^{(n)}$.

The sets of normal modes obtained in this way are in principle different for different potential surfaces and can be related to each other by a unitary rotation in the nuclear coordinate space (see further discussion below). An important simplification is often made at this point: We assume that the normal modes associated with the two electronic states are the same, $\{x_\alpha^{(n)}\} = \{x_\alpha\}$, except for a shift in their equilibrium positions. Equation (12.19) is then replaced by

$$\begin{aligned} E_{\text{el}}^{(n)}(\mathbf{R}) &= E_n + \frac{1}{2}\sum_\alpha m_\alpha \omega_\alpha^2 (x_\alpha - \lambda_\alpha^{(n)})^2 \\ &= E_n + \sum_\alpha \hbar\omega_\alpha (\bar{x}_\alpha - \bar{\lambda}_\alpha^{(n)})^2 \end{aligned} \tag{12.20}$$

where the dimensionless coordinates and shifts are defined by

$$\bar{x}_\alpha \equiv x_\alpha \sqrt{\frac{m_\alpha \omega_\alpha}{2\hbar}}; \qquad \bar{\lambda}_\alpha^{(n)} \equiv \lambda_\alpha^{(n)} \sqrt{\frac{m_\alpha \omega_\alpha}{2\hbar}} \tag{12.21}$$

A schematic view of the two potential surfaces projected onto a single normal mode is seen in Fig. 12.3. The normal mode shifts $\lambda_\alpha^{(n)}$ express the deviation of the equilibrium configuration of electronic state n from some specified reference configuration (e.g. the ground state equilibrium), projected onto the normal mode directions. Other useful parameters are the single mode reorganization energy E_{r}^{α}, defined by the inset to Fig. 12.3,

$$E_{\text{r}}^{\alpha} = \hbar\omega_\alpha \bar{\lambda}_\alpha^2 \tag{12.22a}$$

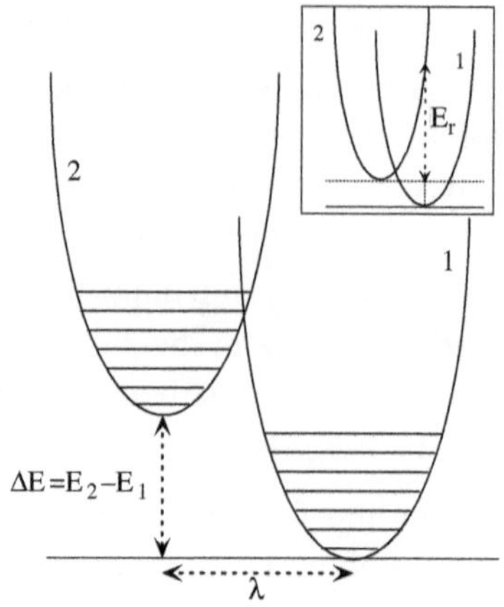

FIG. 12.3 A schematic view of the shifted potential surfaces model, shown for simplicity as a one-dimensional (single mode) representation. The inset is a similar figure on a different scale that shows the reorganization energy E_{r}.

and the corresponding total reorganization energy

$$E_{\mathrm{r}} = \sum_{\alpha} E_{\mathrm{r}}^{\alpha} \tag{12.22b}$$

What is the justification for this approximation? Our accumulated experience in molecular spectroscopy involving low-lying electronic states teaches us that many optical spectra can be interpreted *approximately* using model nuclear potential surfaces in which the *identities* of the normal-mode coordinates do not change in the electronic transition. A geometrical picture of this observation is that the harmonic surfaces shift in parallel with each other. Mixing the normal modes will amount in this picture to a relative rotation of the potential surfaces between the different electronic states, and the assumption is that this rotation is small and may be disregarded to first approximation. Note that this does not mean that the molecular shape remains constant in the transition. Any change in the equilibrium position of a normal mode that is not totally symmetric in the molecular symmetry group will lead to a change in molecular shape.

To summarize, the Born–Oppenheimer states are of the form $\phi_n(\mathbf{r},\mathbf{R})\chi_{n,\mathbf{v}}(\mathbf{R})$ where the vibrational wavefunction $\chi_{n,\mathbf{v}}(\mathbf{R})$ is an eigenstate of the nuclear Hamiltonian $\hat{H}_B^{(n)}$ associated with the electronic state n. In the harmonic approximation these Hamiltonians are separable, $\hat{H}_B^{(n)} = \sum_\alpha \hat{h}_{n\alpha}$, so that $\chi_{n,\mathbf{v}}(\mathbf{R}) = \prod_\alpha \chi_{n,v_\alpha}(x_\alpha)$ where χ_{n,v_α} are eigenfunctions of the mode Hamiltonians $\hat{h}_{n\alpha}$. In the shifted harmonic surfaces model these normal modes keep their identity in different electronic states, except that their equilibrium positions depend on the electronic state. Formally this can be expressed as

$$\hat{h}_{2\alpha} = \hat{U}_\alpha \hat{h}_{1\alpha} \hat{U}_\alpha^{-1} \tag{12.23}$$

where $\hat{U}_\alpha$ is the unitary position shift operator (Eqs (2.173) and (2.175))

$$\hat{U}_\alpha = e^{-\lambda_\alpha(\partial/\partial x_\alpha)} = e^{\bar{\lambda}_\alpha(\hat{a}_\alpha^\dagger - \hat{a}_\alpha)} \tag{12.24}$$

and λ_α is the shift associated with mode α between the two electronic states (same as $\lambda_\alpha^{1,2}$ in the notation of Eq. (12.16)).

Consider now transition between vibronic levels associated with different electronic states that are described in the Born–Oppenheimer approximation. Any residual coupling $\hat{V}(\mathbf{r}, \mathbf{R})$ not taken into account under the BO approximation, as well as coupling induced by external fields, can cause such transitions. For allowed optical transitions this is the electronic dipole operator. Electronic radiationless relaxation following optical excitation in molecular processes is best described in the full BO picture, whereupon perturbations that lead to interstate coupling between states of the same spin multiplicity stem from corrections to this picture (Eq. (2.53)). Charge transfer processes (Chapter 16) are usually described within a diabatic local state picture, where the dominant interaction is associated with electrons on one center feeling the other. In either case, a general coupling matrix element between two vibronic states $\phi_n(\mathbf{r}, \mathbf{R})\chi_{n,\mathbf{v}}(\mathbf{R})$ and $\phi_{n'}(\mathbf{r}, \mathbf{R})\chi_{n',\mathbf{v}'}(\mathbf{R})$ is of the form

$$V_{n,\mathbf{v};n',\mathbf{v}'} = \langle \chi_{n,\mathbf{v}} | \langle \phi_n | \hat{V}(\mathbf{r}, \mathbf{R}) | \phi_{n'} \rangle_\mathbf{r} | \chi_{n',\mathbf{v}'} \rangle_\mathbf{R} \tag{12.25}$$

where $\langle\rangle_\mathbf{r}$ and $\langle\rangle_\mathbf{R}$ indicate integrations in the electronic and nuclear subspaces, respectively. In the so-called *Condon approximation* the dependence of the electronic matrix element on the nuclear configuration is disregarded, that is, $\langle \phi_n | \hat{V}(\mathbf{r}, \mathbf{R}) | \phi_{n'} \rangle_\mathbf{r} \rightarrow V_{n,n'}$ is taken to be independent of $\mathbf{R}$, whereupon

$$V_{n,\mathbf{v};n',\mathbf{v}'} = V_{n,n'} \langle \chi_{n,\mathbf{v}} | \chi_{n',\mathbf{v}'} \rangle_\mathbf{R} \tag{12.26}$$

In the shifted harmonic surfaces model $\chi_{n,\mathbf{v}}(\mathbf{R}) = \prod_\alpha \chi_{n,v_\alpha}(x_\alpha)$ and $\chi_{n',\mathbf{v}'}(\mathbf{R}) = \prod_\alpha \chi_{n',v'_\alpha}(x_\alpha - \lambda_\alpha^{n,n'})$, so Eq. (12.26) is identical to (12.17) and (12.18).

We have thus found that the interstate coupling (12.17) associated with the Hamiltonian (12.15) is the same as that inferred from the Born–Oppenheimer picture in the Condon approximation, under the assumption that different potential surfaces are mutually related by only rigid vertical and horizontal shifts.[3] In spite

[3] Note that the term $\sum_\alpha g_{n\alpha}\hat{x}_\alpha$ in Eq. (12.8) contributes both horizontal and vertical shift: Limiting ourselves to the contribution of a single mode α we have: $E_n + g_\alpha x_\alpha = (1/2)m\omega_\alpha^2 x_\alpha^2 + g_\alpha x_\alpha = (1/2)m\omega_\alpha^2(x_\alpha - \lambda_\alpha)^2 - (1/2)m\omega_\alpha^2\lambda_\alpha^2$ where $\lambda_\alpha = g_\alpha/(m\omega_\alpha^2)$. The associated vertical shift is $(1/2)m\omega_\alpha^2\lambda_\alpha^2 = g_\alpha^2/(2m\omega_\alpha^2)$ which is indeed the vertical shift contribution that enters in Eq. (12.15).

of the approximate nature of the shifted harmonic surfaces picture, this model is very useful both because of its inherent simplicity and because it can be sometimes justified on theoretical grounds as in the electron transfer problem (Chapter 16). The parallel shift parameters λ can be obtained from spectroscopic data or, as again exemplified by the theory of electron transfer, by theoretical considerations.

12.4 Golden-rule transition rates

12.4.1 The decay of an initially prepared level

Let us now return to the two model Hamiltonians introduced in Section 12.2, and drop from now on the subscripts S and SB from the coupling operators. Using the polaron transformation we can describe both models (12.1), (12.2) and (12.4)–(12.6) in a similar language, where the difference enters in the form of the coupling to the boson bath

$$\hat{H} = \hat{H}_0 + \hat{V} \tag{12.27a}$$

$$\begin{aligned} \hat{H}_0 &= E_1|1\rangle\langle 1| + E_2|2\rangle\langle 2| + \hat{H}_B \\ \hat{H}_B &= \sum_\alpha \hbar\omega_\alpha \hat{a}_\alpha^\dagger \hat{a}_\alpha \end{aligned} \tag{12.27b}$$

In one coupling model we use (12.2b) where, for simplicity, terms with $j = j'$ are disregarded

$$\hat{V} = |1\rangle\langle 2| \sum_\alpha V_{1,2}^\alpha (\hat{a}_\alpha^\dagger + \hat{a}_\alpha) + |2\rangle\langle 1| \sum_\alpha V_{2,1}^\alpha (\hat{a}_\alpha^\dagger + \hat{a}_\alpha) \tag{12.28a}$$

Sometimes an additional approximation is invoked by disregarding in (12.28a) terms that cannot conserve energy in the lowest order treatment. Under this so-called *rotating wave approximation*[4] the coupling (12.28a) is replaced by (for $E_2 > E_1$)

$$\hat{V}_{\text{RWA}} = |1\rangle\langle 2| \sum_\alpha V_{1,2}^\alpha \hat{a}_\alpha^\dagger + |2\rangle\langle 1| \sum_\alpha V_{2,1}^\alpha \hat{a}_\alpha \tag{12.28b}$$

[4] The rationale for this approximation can be seen in the interaction picture in which $\hat{V}$ becomes

$$\begin{aligned} \hat{V}_I(t) &= \exp((i/\hbar)\hat{H}_0 t)\hat{V} \exp(-(i/\hbar)\hat{H}_0 t) \\ &= |1\rangle\langle 2| \exp((i/\hbar)(E_1 - E_2)t) \sum_\alpha V_{1,2}^\alpha (\hat{a}_\alpha^\dagger \exp(i\omega_\alpha t) + \hat{a}_\alpha \exp(-i\omega_\alpha t)) + h.c. \end{aligned}$$

The RWA keeps only terms for which $E_1 - E_2 \pm \hbar\omega_\alpha$ can be small, the argument being that strongly oscillating terms make only a small contribution to the transition rate.

In the other model, Eq. (12.15) written for a two-state system, $\hat{H}_0$ is given again by (12.27b), however now

$$\hat{V} = V_{1,2}e^{i(\hat{\Omega}_2-\hat{\Omega}_1)}|1\rangle\langle 2| + V_{2,1}e^{-i(\hat{\Omega}_2-\hat{\Omega}_1)}|2\rangle\langle 1| \tag{12.29a}$$

$$\hat{\Omega}_2 - \hat{\Omega}_1 = \sum_\alpha \frac{(g_{2\alpha} - g_{1\alpha})\hat{p}_\alpha}{\hbar m_\alpha \omega_\alpha^2} = \sum_\alpha i\bar{\lambda}_\alpha(\hat{a}_\alpha^\dagger - \hat{a}_\alpha) \tag{12.29b}$$

where we have dropped the tilde notation from the Hamiltonian, denoted $\lambda_{2,1}^\alpha$ simply by λ_α and have redefined the energies E_n to include the shifts $\sum_\alpha g_{n\alpha}^2/(2m_\alpha\omega_\alpha^2)$ that were written explicitly in Eq. (12.15). We have also defined (compare Eq. (2.176))

$$\bar{\lambda}_\alpha \equiv \lambda_\alpha\sqrt{\frac{m_\alpha\omega_\alpha}{2\hbar}} = \frac{g_{2\alpha} - g_{1\alpha}}{m_\alpha\omega_\alpha^2}\sqrt{\frac{m_\alpha\omega_\alpha}{2\hbar}} \tag{12.30}$$

Equations (12.28) and (12.29) describe different spin-boson models that are commonly used to describe the dynamics of a two-level system interacting with a boson bath. Two comments are in order:

(a) The word "bath" implies here two important attributes of the boson subsystem: First, the boson modes are assumed to constitute a continuum, characterized by a density of modes function $g(\omega)$, so that the number of modes in a frequency range between ω and $\omega + d\omega$ is given by $g(\omega)d\omega$. Second, the boson field is large and relaxes fast relative to the dynamics of the two-level system. It can therefore be assumed to maintain its equilibrium state throughout the process.

(b) The couplings terms (12.28a) and (12.29a) that characterize the two models differ from each other in an essential way: When the spin-boson coupling vanishes ($V_{1,2}^\alpha = 0$ for all α in (12.28); $g_{1\alpha} = g_{2\alpha}$ for all α in (12.29)) the exact system Hamiltonian becomes $\hat{H}_0$ in the first case and $\hat{H}_0 + V$ in the second. The basis states $|1\rangle$ and $|2\rangle$ are therefore eigenstates of the free system in the first case, but can be taken as local states (still coupled by V therefore equivalent to $|L\rangle$ and $|R\rangle$ in Eq. (12.5)) in the second.

As an example consider, within the model (12.29), the time evolution of the system when it starts in a specific state of $\hat{H}_0$, for example, $\Psi(t = 0) = |2, \mathbf{v}\rangle = |2\rangle\prod_\alpha |v_\alpha\rangle$ where $|v_\alpha\rangle$ is an eigenfunctions of the harmonic oscillator Hamiltonian that represents mode α of the boson bath, with the energy $(v_\alpha + (1/2))\hbar\omega_\alpha$. In the absence of coupling to the boson field, namely when $\bar{\lambda}_\alpha = 0$, that is, $\hat{\Omega}_2 - \hat{\Omega}_1 = 0$ in (12.29), the remaining interstate coupling V_{12} cannot change the state $\mathbf{v}$ of the bath and the problem is reduced to the dynamics of two coupled levels ($|2, \mathbf{v}\rangle$ and $|1, \mathbf{v}\rangle$)

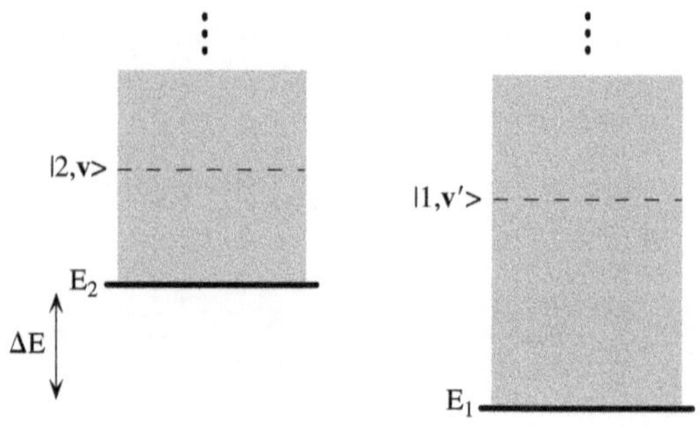

FIG. 12.4 A dressed-states representation of the model of Fig. 12.1.

discussed in Section 2.2, yielding the solution (for $\Psi(t=0) = |2\rangle$; cf. Eq. (2.32))

$$P_2(t) = 1 - P_1(t)$$
$$P_1(t) = \frac{4|V_{12}|^2}{(E_1 - E_2)^2 + 4|V_{12}|^2} \sin^2\left[\frac{\Omega_R}{2}t\right] \tag{12.31}$$

where Ω_R is the Rabi frequency, $\Omega_R = (1/\hbar)\sqrt{(E_2 - E_1)^2 + 4|V_{12}|^2}$. Two facts are evident: (1) The specification of the bath state $|\mathbf{v}\rangle$ is immaterial here, and (2) in this case we cannot speak of a *rate* that characterizes the $1 \leftrightarrow 2$ transition.

The coupling to the boson bath can change this in a dramatic way because initial levels of the combined spin–boson system are coupled to a *continuum* of other levels. Indeed Fig. 12.1 can be redrawn in order to display this feature, as seen in Fig. 12.4. Two continuous manifolds of states are seen, "seating" on level 1 and 2, that encompass the states $|1,\mathbf{v}'\rangle = |1\rangle \prod_\alpha |v'_\alpha\rangle$ and $|2,\mathbf{v}\rangle = |2\rangle \prod_\alpha |v_\alpha\rangle$ with zero-order energies $E_{1,\mathbf{v}'}$ and $E_{2,\mathbf{v}}$, respectively, where

$$E_{n,\mathbf{v}} = E_n + E_{\text{boson}}(\mathbf{v}) = E_n + \sum_\alpha \hbar\omega_\alpha(v_\alpha + (1/2)) \tag{12.32}$$

The continuous nature of these manifolds stems from the continuous distribution of boson modes. The picture is altogether similar to the dressed state picture discussed in Sections 9.2 and 9.3, where states 1 and 2 were ground and excited molecular electronic states, while the boson subsystem was the radiation field, and where we have considered a specific case where level 2 with no photons interacts with the continuum of 1-photon states seating on the ground state 1.

The initial state, $\Psi(t = 0) = |2,\mathbf{v}\rangle = |2\rangle \prod_\alpha |v_\alpha\rangle$, is a state of the overall system—a particular level in the state manifold 2. The general considerations of Section 9.1 (see also Section 10.3.2) have taught us that under certain fairly general conditions the probability to remain in this level decays exponentially by transfer to states in the manifold 1, with a rate given by the golden-rule formula

$$k_{1\leftarrow 2\mathbf{v}} = \frac{2\pi}{\hbar}\sum_{\mathbf{v}'} |\langle 2,\mathbf{v}|\hat{V}|1,\mathbf{v}'\rangle|^2 \delta(E_{2,\mathbf{v}} - E_{1,\mathbf{v}'}) \tag{12.33}$$

The assumption that thermal relaxation in the bath is fast relative to the timescale determined by this rate (see statement (2) of Problem 12.1) makes it possible to define also the thermally averaged rate to go from state 2 to state 1

$$k_{1\leftarrow 2} = \frac{2\pi}{\hbar}\sum_{\mathbf{v}} P_{\mathbf{v}} \sum_{\mathbf{v}'} |\langle 2,\mathbf{v}|\hat{V}|1,\mathbf{v}'\rangle|^2 \delta(E_{2,\mathbf{v}} - E_{1,\mathbf{v}'}) \tag{12.34}$$

where (denoting $E_{\mathbf{v}} = E_{\text{boson}}(\mathbf{v})$)

$$P_{\mathbf{v}} = \frac{e^{-\beta E_{2,\mathbf{v}}}}{Q_2} = \frac{e^{-\beta E_{\mathbf{v}}}}{Q_{\text{boson}}}; \qquad \beta = (k_B T)^{-1} \tag{12.35}$$

$$Q_2 = \sum_{\mathbf{v}} e^{-\beta E_{2,\mathbf{v}}}; \qquad Q_{\text{boson}} = \sum_{\mathbf{v}} e^{-\beta E_{\mathbf{v}}} \tag{12.36}$$

is the canonical distribution that characterizes the boson bath.

Problem 12.1. Refer to the general discussions of Sections 9.1 and 10.3.2 in order to explain the following two statements: (1) Eq. (12.33) for the partial rates and hence Eq. (12.34) for the thermally averaged rate are valid rate expressions only if the partial rates (12.33) are larger than the inverse of $\hbar\rho_1(E_{2,\mathbf{v}}) = \hbar\sum_{\mathbf{v}'}\delta(E_{2,\mathbf{v}} - E_{1,\mathbf{v}'})$. (2) The thermally averaged rate, $k_{2\to 1}$ of Eq. (12.34), is meaningful only if it is much smaller than the rate of thermal relaxation between the levels **v** of the initial "2" manifold.

An important observation, made in statement (1) of Problem 12.1, is that we do not really need a continuous distribution of modes in the boson field in order for the manifold ($1\mathbf{v}'$) to be *practically* continuous in the sense that the rate expressions (12.33) and (12.34) are valid. A large finite number, $N \gg 1$, of modes can provide a sufficiently large density of states in manifold 1, $\rho_1(E_{2,\mathbf{v}})$, with energies $E_{1\mathbf{v}'} = \sum_{\alpha=1}^{N}(v'_\alpha + 1/2)\hbar\omega_\alpha$ in the neighborhood of the energy $E_{2,\mathbf{v}}$, provided the energy gap $E_2 - E_1$ is large enough (a reasonable criterion is $E_2 - E_1 \gg \hbar\langle\omega\rangle$ where ω is the average mode frequency). This stems from the huge number of possible combinations of occupation numbers v_α that will yield an energy $E_{1,\mathbf{v}'}$ in a finite neighborhood of any energy $E_{2,\mathbf{v}}$. This is demonstrated by Problem 12.2.

Problem 12.2. Obtain a rough estimate of the density of vibrational states $\rho(E)$ as a function of energy, for a molecule that contains 30 harmonic modes of average frequency $\omega = 500\ \text{cm}^{-1}$ using the following procedure: Assume first that all the modes have the same frequency, $500\ \text{cm}^{-1}$. Then the only possible energy levels (relative to the ground vibrational state) are integer products of this number, $E(L) = 500 \times L$. Calculate the degeneracy $D(L)$ of the energy level $E(L)$ and *estimate* the actual density of states from the results obtained. How fast does $k_{1\leftarrow 2}$ need to be for an expression like (12.33) to be valid if $E_{21} = 10,000\ \text{cm}^{-1}$?

Solution: For N modes and $E = \hbar\omega L$ we have L indistinguishable quanta that should be distributed among these N modes. The number of possibilities is a standard problem in combinatorics and the result, $(N + L - 1)!/[(N - 1)!L!]$, is the degeneracy of a level of energy E. The density of states can be roughly estimated to be $\rho(E) = [(N + E/\hbar\omega - 1)!]/[(N - 1)!(E/\hbar\omega)!]/\hbar\omega$. For $\omega = 500\ \text{cm}^{-1}$ and $E = 10\,000\ \text{cm}^{-1}$ this is $49!/(29!20!)/500 \simeq 5.7 \times 10^{10}$ cm, that is, $\rho \sim 5.7 \times 10^{10}$ states per wavenumber or $\sim 2.8 \times 10^{26}$ states per erg. This translates into the time $t = \hbar\rho \sim 0.28$ s. The rate therefore has to be faster than $3.6\ \text{s}^{-1}$ for expression (12.33) to hold.

The interstate energy E_{21}, the number of modes N, and the frequency ω used in the estimate made in Problem 12.2 are typical for moderately large molecules. This rationalizes the observation that electronically excited large molecules can relax via *radiationless* pathways in which population is transferred from the excited electronic state to higher vibrational levels of lower electronic states. We may conclude that large isolated molecules can, in a sense, provide their own boson bath and relax accordingly. In such cases, however, the validity of the assumption that thermal relaxation in the boson bath is faster than the $1 \leftrightarrow 2$ transition dynamics may not hold. Radiationless transition rates between electronic states of the same spin multiplicity can be as fast as 10^9–$10^{15}\ \text{s}^{-1}$,[5] while thermal relaxation rates vary. For large molecules in condensed phases thermal equilibrium of nuclear motion is usually achieved within 1–10 ps. For small molecules and for molecules in the gas phase this time can be much longer. In such situations the individual rates (12.33) may have to be considered specifically. We will not consider such cases here.

[5] Nonradiative rates that are considerably slower than that will not be observed if the $2 \rightarrow 1$ transition is optically allowed. In the latter case radiative relaxation (i.e. fluorescence) on timescales of 10^{-8}–10^{-9} s will be dominant.

12.4.2 The thermally averaged rate

We now proceed with the thermally averaged 2 → 1 rate, Eq. (12.34), rewritten in the form

$$k_{1\leftarrow 2} = \frac{2\pi}{\hbar}\sum_{\mathbf{v}} P_{\mathbf{v}} \sum_{\mathbf{v}'} |\langle \mathbf{v}|\hat{V}_{2,1}|\mathbf{v}'\rangle|^2 \delta(E_2 - E_1 + E_{\text{boson}}(\mathbf{v}) - E_{\text{boson}}(\mathbf{v}')) \tag{12.37}$$

We will evaluate this rate for the two models considered above. In the model of (12.28a)

$$\hat{V}_{1,2} = \sum_{\alpha} V_{1,2}^{\alpha}(\hat{a}_{\alpha}^{\dagger} + \hat{a}_{\alpha}); \qquad \hat{V}_{2,1} = \sum_{\alpha} V_{2,1}^{\alpha}(\hat{a}_{\alpha}^{\dagger} + \hat{a}_{\alpha}); \qquad V_{2,1}^{\alpha} = (V_{1,2}^{\alpha})^* \tag{12.38}$$

while from (12.29) in

$$\hat{V}_{1,2} = V_{1,2} e^{-\sum_{\alpha} \bar{\lambda}_{\alpha}(\hat{a}_{\alpha}^{\dagger} - \hat{a}_{\alpha})}; \qquad \hat{V}_{2,1} = V_{2,1} e^{\sum_{\alpha} \bar{\lambda}_{\alpha}(\hat{a}_{\alpha}^{\dagger} - \hat{a}_{\alpha})}; \qquad V_{2,1} = V_{1,2}^* \tag{12.39}$$

Now use the identity $\delta(x) = (2\pi\hbar)^{-1}\int_{-\infty}^{\infty} dt e^{ixt/\hbar}$ to rewrite Eq. (12.37) in the form

$$\begin{aligned} k_{1\leftarrow 2} &= \frac{1}{\hbar^2}\sum_{\mathbf{v}} P_{\mathbf{v}} \sum_{\mathbf{v}'} \langle \mathbf{v}|\hat{V}_{2,1}|\mathbf{v}'\rangle\langle \mathbf{v}'|\hat{V}_{1,2}|\mathbf{v}\rangle \int_{-\infty}^{\infty} dt e^{i(E_{2\mathbf{v}} - E_{1\mathbf{v}'})t/\hbar} \\ &= \frac{1}{\hbar^2}\int_{-\infty}^{\infty} dt e^{iE_{2,1}t/\hbar} \sum_{\mathbf{v}} P_{\mathbf{v}} \langle \mathbf{v}| e^{i\hat{H}_B t/\hbar} \hat{V}_{2,1} e^{-i\hat{H}_B t/\hbar} \sum_{\mathbf{v}'} (|\mathbf{v}'\rangle\langle \mathbf{v}'|)\hat{V}_{1,2}|\mathbf{v}\rangle \\ &= \frac{1}{\hbar^2}\int_{-\infty}^{\infty} dt e^{iE_{2,1}t/\hbar} \langle \hat{V}_{2,1}(t)\hat{V}_{1,2}\rangle \end{aligned} \tag{12.40}$$

where $E_{2,1} = E_2 - E_1$, $\hat{V}_{2,1}(t) = e^{i\hat{H}_B t/\hbar}\hat{V}_{2,1}e^{-i\hat{H}_B t/\hbar}$ is the interaction operator in the Heisenberg representation and where $\langle \cdots \rangle$ denotes a thermal average in the boson subspace. To get this result we have used the fact that $\sum_{\mathbf{v}'}(|\mathbf{v}'\rangle\langle \mathbf{v}'|)$ is a unit operator in the boson subspace.

Problem 12.3. Explain the difference in the forms of Eqs (12.40) and (6.20).

We have thus found that the $k_{1\leftarrow 2}$ is given by a Fourier transform of a quantum time correlation function computed at the energy spacing that characterizes the two-level system,

$$k_{1\leftarrow 2} = \frac{1}{\hbar^2}\tilde{C}_{21}(E_{2,1}/\hbar); \qquad \tilde{C}_{21}(\omega) = \int_{-\infty}^{\infty} dte^{i\omega t} C_{21}(t);$$
$$C_{21}(t) \equiv \langle \hat{V}_{2,1}(t)\hat{V}_{1,2}\rangle \tag{12.41}$$

Problem 12.4.

(1) Assume that an expression analogous to (12.33) holds also for the transition $|1, \mathbf{v}'\rangle \rightarrow 2$, that is,

$$k_{2\leftarrow 1\mathbf{v}'} = \frac{2\pi}{\hbar}\sum_{\mathbf{v}} |\langle 2, \mathbf{v}|\hat{V}|1, \mathbf{v}'\rangle|^2 \delta(E_{2,\mathbf{v}} - E_{1,\mathbf{v}'}) \tag{12.42}$$

(when would you expect this assumption to be valid?), so that the thermally averaged $1 \rightarrow 2$ rate is

$$k_{2\leftarrow 1} = \frac{2\pi}{\hbar}\sum_{\mathbf{v}'} P_{\mathbf{v}'} \sum_{\mathbf{v}} |\langle 2, \mathbf{v}|\hat{V}|1, \mathbf{v}'\rangle|^2 \delta(E_{2\mathbf{v}} - E_{1,\mathbf{v}'}) \tag{12.43}$$

Using the same procedure as above, show that this leads to

$$k_{2\leftarrow 1} = \frac{1}{\hbar^2}\int_{-\infty}^{\infty} dte^{-iE_{2,1}t/\hbar} C_{12}(t); \qquad C_{12}(t) \equiv \langle \hat{V}_{1,2}(t)\hat{V}_{2,1}\rangle \tag{12.44}$$

(2) Use Eqs (12.41), (12.44), and (6.73) to show that

$$k_{2\leftarrow 1} = k_{1\leftarrow 2}e^{-E_{2,1}/k_B T} \tag{12.45}$$

that is, the rates calculated from the golden-rule expression satisfy detailed balance.

12.4.3 Evaluation of rates

For the model (12.28a) $\hat{V}_{1,2}$ and $\hat{V}_{2,1}$ are given by Eq. (12.38) and the corresponding Heisenberg representation (with respect to $\hat{H}_B$) operator is $\hat{V}_{j,k}(t) = \sum_\alpha V_{j,k}^\alpha(\hat{a}_\alpha^\dagger e^{i\omega_\alpha t} + \hat{a}_\alpha e^{-i\omega_\alpha t})$ where $j = 1, k = 2$ or $j = 2, k = 1$. Using this in (12.41) yields

$$C_{21}(t) = C_{12}(t) = \sum_\alpha |V_{2,1}^\alpha|^2 \langle \hat{a}_\alpha^\dagger \hat{a}_\alpha e^{i\omega_\alpha t} + \hat{a}_\alpha \hat{a}_\alpha^\dagger e^{-i\omega_\alpha t}\rangle$$
$$= \int_0^\infty d\omega g(\omega)|V_{1,2}(\omega)|^2 (n(\omega)e^{i\omega t} + (n(\omega)+1)e^{-i\omega t}) \quad (12.46)$$

where $n(\omega) = (e^{\beta\hbar\omega} - 1)^{-1}$ is the thermal boson occupation number and $g(\omega) = \sum_\alpha \delta(\omega - \omega_\alpha)$ is the density of modes in the boson field.[6] We note in passing that the function $\sum_\alpha |V_{1,2}^\alpha|^2 \delta(\omega - \omega_\alpha) = g(\omega)|V_{1,2}(\omega)|^2$ is essentially the spectral density, the coupling weighted density of modes (see Sections 6.5.2, 7.5.2, and 8.2.6), associated with the system–bath coupling. We have discussed several models for such functions in Sections 6.5.2 and 8.2.6.

Using Eq. (12.46) in (12.41) we find that for $E_{2,1} > 0$ the term containing $\exp(i\omega t)$ in (12.46) does not contribute to $k_{1\leftarrow 2}$. We get

$$k_{1\leftarrow 2} = \frac{2\pi}{\hbar^2} g(\omega_{2,1})|V_{1,2}(\omega_{2,1})|^2 (n(\omega_{2,1}) + 1); \qquad \omega_{2,1} = E_{2,1}/\hbar \quad (12.47)$$

Similarly, Eq. (12.44) yields

$$k_{2\leftarrow 1} = \frac{2\pi}{\hbar^2} g(\omega_{2,1})|V_{1,2}(\omega_{2,1})|^2 n(\omega_{2,1}) \quad (12.48)$$

Note that for a model characterized by an upper cutoff in the boson density of states, for example, the Debye model, these rates vanish when the level spacing of the two-level system exceeds this cutoff. Note also that the rates (12.47) and (12.48) satisfy the detailed balance relationship (12.45).

Next consider the model defined by Eqs (12.27) and (12.29). The correlation functions $C_{21}(t)$ and $C_{12}(t)$ are now

$$C_{21}(t) = \langle \hat{V}_{2,1}(t)\hat{V}_{1,2}\rangle = |V_{2,1}|^2 \prod_\alpha C_{21}^\alpha(t) \quad (12.49a)$$

where

$$C_{21}^\alpha(t) = \langle e^{\bar{\lambda}_\alpha(\hat{a}_\alpha^\dagger e^{i\omega_\alpha t} - \hat{a}_\alpha e^{-i\omega_\alpha t})} e^{-\bar{\lambda}_\alpha(\hat{a}_\alpha^\dagger - \hat{a}_\alpha)}\rangle \quad (12.49b)$$

[6] The function $V_{1,2}(\omega)$ is defined by a coarse-graining procedure, $\sum_{\omega_\alpha \in \Delta\omega} |V_{1,2}^\alpha|^2 = \Delta\omega g(\omega)|V_{1,2}(\omega)|^2$ where $\omega_\alpha \in \Delta\omega$ denotes $\omega + \Delta\omega/2 \geq \omega_\alpha \geq \omega - \Delta\omega/2$ and $\Delta\omega$ is large relative to $(g(\omega))^{-1}$. A formal definition is $|V_{12}(\omega)^2| = g^{-1}(\omega)\sum_\alpha |V_{12}^\alpha|^2 \delta(\omega - \omega_\alpha)$.

and

$$C_{12}(t) \equiv \langle \hat{V}_{1,2}(t)\hat{V}_{2,1}\rangle = |V_{2,1}|^2 \prod_{\alpha} C_{12}^{\alpha}(t) \tag{12.50a}$$

$$C_{12}^{\alpha}(t) = \langle e^{-\bar{\lambda}_\alpha(\hat{a}_\alpha^\dagger e^{i\omega_\alpha t} - \hat{a}_\alpha e^{-i\omega_\alpha t})} e^{\bar{\lambda}_\alpha(\hat{a}_\alpha^\dagger - \hat{a}_\alpha)}\rangle \tag{12.50b}$$

These quantum thermal averages over an equilibrium boson field can be evaluated by applying the raising and lowering operator algebra that was introduced in Section 2.9.2.

Problem 12.5. Use the identities

$$e^{\hat{A}}e^{\hat{B}} = e^{\hat{A}+\hat{B}}e^{(1/2)[\hat{A},\hat{B}]} \tag{12.51}$$

(for operators $\hat{A}$, $\hat{B}$ which commute with their commutator $[\hat{A}, \hat{B}]$) and

$$\langle e^{\hat{A}}\rangle_T = e^{(1/2)\langle \hat{A}^2\rangle_T} \quad \text{(Bloch theorem)} \tag{12.52}$$

(for harmonic oscillator Hamiltonian and an operator $\hat{A}$ that is linear in $\hat{a}$ and $\hat{a}^\dagger$) to show that

$$K \equiv \langle e^{\alpha_1\hat{a}+\beta_1\hat{a}^\dagger} e^{\alpha_2\hat{a}+\beta_2\hat{a}^\dagger}\rangle_T = e^{(\alpha_1+\alpha_2)(\beta_1+\beta_2)(n+1/2)+(1/2)(\alpha_1\beta_2-\beta_1\alpha_2)} \tag{12.53}$$

where $n = \langle \hat{a}^\dagger\hat{a}\rangle = (e^{\beta\hbar\omega} - 1)^{-1}$.

Using (12.53) to evaluate (12.49b) and (12.50b) we get

$$C_{21}^{\alpha}(t) = C_{12}^{\alpha}(t) = e^{-\bar{\lambda}_\alpha^2(2n_\alpha+1)+\bar{\lambda}_\alpha^2((n_\alpha+1)e^{-i\omega_\alpha t}+n_\alpha e^{i\omega_\alpha t})} \tag{12.54}$$

So that

$$k_{1\leftarrow 2} = k(\omega_{21}); \qquad \omega_{21} = (E_2 - E_1)/\hbar \tag{12.55a}$$

$$k(\omega_{21}) = \frac{|V_{12}|^2}{\hbar^2} e^{-\sum_\alpha \bar{\lambda}_\alpha^2(2n_\alpha+1)} \int_{-\infty}^{\infty} dt e^{i\omega_{21}t+\sum_\alpha \bar{\lambda}_\alpha^2(n_\alpha e^{i\omega_\alpha t}+(n_\alpha+1)e^{-i\omega_\alpha t})} \tag{12.55b}$$

Equations (12.55), sometime referred to as multiphonon transition rates for reasons that become clear below, are explicit expressions for the golden-rule transitions rates between two levels coupled to a boson field in the shifted parallel harmonic potential surfaces model. The rates are seen to depend on the level spacing E_{21}, the normal mode spectrum $\{\omega_\alpha\}$, the normal mode shift parameters $\{\bar{\lambda}_\alpha\}$, the temperature (through the boson populations $\{n_\alpha\}$) and the nonadiabatic coupling parameter $|V_{12}|^2$. More insight on the dependence on these parameters is obtained by considering different limits of this expression.

Problem 12.6. Show that in the limit where both $\bar{\lambda}_\alpha^2$ and $\bar{\lambda}_\alpha^2 n_\alpha$ are much smaller than 1, Eqs (12.55a) and (12.55b) yield the rates (12.47) and (12.48), respectively, where $|V_{1,2}^\alpha|^2$ in Eq (12.46) is identified with $|V_{12}|^2\bar{\lambda}_\alpha^2$ so that $|V_{12}(\omega)|^2$ in (12.48) is identified with $|V_{12}|^2\bar{\lambda}^2(\omega)$.

The rate expressions (12.47) and (12.48) are thus seen to be limiting forms of (12.55), obtained in the low-temperature limit provided that $\bar{\lambda}_\alpha^2 \ll 1$ for all α. On the other hand, the rate expression (12.55) is valid if V_{12} is small enough, irrespective of the temperature and the magnitudes of the shifts $\bar{\lambda}_\alpha$.

12.5 Transition between molecular electronic states

Transitions between molecular electronic states are often described by focusing on the two electronic states involved, thus leading to a two-state model. When such transitions are coupled to molecular vibrations, environmental phonons or radiation-field photons the problem becomes a spin–boson-type. The examples discussed below reiterate the methodology described in this chapter in the context of physical applications pertaining to the dynamics of electronic transitions in molecular systems.

12.5.1 The optical absorption lineshape

A direct consequence of the observation that Eqs. (12.55) provide also golden-rule expressions for transition rates between molecular electronic states in the shifted parallel harmonic potential surfaces model, is that the same theory can be applied to the calculation of optical absorption spectra. The electronic absorption lineshape expresses the photon-frequency dependent transition rate from the molecular ground state dressed by a photon, $|\bar{g}\rangle \equiv |g, \hbar\omega\rangle$, to an electronically excited state without a photon, $|x\rangle$. This absorption is broadened by electronic–vibrational coupling, and the resulting spectrum is sometimes referred to as the Franck–Condon envelope of the absorption lineshape. To see how this spectrum is obtained from the present formalism we start from the Hamiltonian (12.7) in which states L and R are replaced by $|\bar{g}\rangle$ and $|x\rangle$ and V_{LR} becomes $V_{\bar{g}x}$—the coupling between molecule and radiation field. The modes $\{\alpha\}$ represent intramolecular as well as intermolecular vibrational motions that couple to the electronic transition in the way defined by this Hamiltonian,

$$\hat{H} = \left[E_g + \hbar\omega + \sum_\alpha V_g^\alpha (\hat{a}_\alpha^\dagger + \hat{a}_\alpha) \right] |\bar{g}\rangle\langle\bar{g}| + \left[E_x + \sum_\alpha V_x^\alpha (\hat{a}_\alpha^\dagger + \hat{a}_\alpha) \right] |x\rangle\langle x|$$
$$+ V_{\bar{g}x}|\bar{g}\rangle\langle x| + V_{x\bar{g}}|x\rangle\langle\bar{g}| + \sum_\alpha \hbar\omega_\alpha \hat{a}_\alpha^\dagger \hat{a}_\alpha \tag{12.56}$$

We have already seen that this form of electron–phonon coupling expresses shifts in the vibrational modes equilibrium positions upon electronic transition, a standard model in molecular spectroscopy. Applying the polaron transformation to get a Hamiltonian equivalent to (12.27) and (12.29), then using Eq. (12.34) with $E_2 = E_{\bar{g}} = E_g + \hbar\omega$ and $E_1 = E_x$ leads to the electronic absorption lineshape in the form

$$L_{\text{abs}}(\omega) \sim \sum_{\mathbf{v}} P_{\mathbf{v}} \sum_{\mathbf{v}'} |\langle g\mathbf{v}|\hat{\mu}e^{i(\hat{\Omega}_x - \hat{\Omega}_g)}|x\mathbf{v}'\rangle|^2 \delta(E_g + \hbar\omega - E_x + E_{\text{vib}}(\mathbf{v}) - E_{\text{vib}}(\mathbf{v}'))$$
$$= |\mu_{gx}|^2 \sum_{\mathbf{v}} P_{\mathbf{v}} \sum_{\mathbf{v}'} |\langle \mathbf{v}|e^{i(\hat{\Omega}_x - \hat{\Omega}_g)}|\mathbf{v}'\rangle|^2 \delta(E_g + \hbar\omega - E_x + E_{\text{vib}}(\mathbf{v}) - E_{\text{vib}}(\mathbf{v}')) \tag{12.57}$$

where $\hat{\mu}$ is the electronic dipole operator, the molecular–electronic part of the molecule–radiation field coupling, and where in the last expression we have invoked the Condon approximation. As already discussed, the operator $e^{i(\hat{\Omega}_x - \hat{\Omega}_g)}$ affects a rigid displacement of the nuclear wavefunctions. The matrix elements

$$(\text{FC})_{\mathbf{v},\mathbf{v}'} = |\langle \mathbf{v}|e^{i(\hat{\Omega}_x - \hat{\Omega}_g)}|\mathbf{v}'\rangle|^2 \tag{12.58}$$

called *Franck–Condon factors*, are absolute squares of overlap integrals between nuclear wavefunctions associated with parallel-shifted nuclear potential surfaces.

A word of caution is needed here. The golden-rule expression, Eq. (12.33) or (12.43), was obtained for the rate of decay of a level interacting with a continuous manifold (Section 9.1), not as a perturbation theory result[7] but under certain conditions (in particular a dense manifold of final states) that are not usually satisfied for optical absorption. A similar expression is obtained in the weak coupling limit using time-dependent perturbation theory, in which case other conditions are not

[7] This statement should be qualified: The treatment that leads to the golden-rule result for the exponential decay rate of a state interacting with a continuum is not a short-time theory and in this sense nonperturbative, however we do require that the continuum will be "broad." In relaxation involving two-level systems this implies $E_{21} \gg \Gamma = 2\pi V^2 \rho$, that is, a relatively weak coupling.

needed. It is in the latter capacity that we apply it here. The distinction between these applications can be seen already in Eq. (12.33) which, for the zero temperature case (putting $\mathbf{v} = 0$ for the ground vibrational level in the dressed electronic state $|\bar{g}\rangle$), yields

$$k_{x\leftarrow\bar{g}0} = \frac{2\pi}{\hbar}\sum_{\mathbf{v}'} |\langle\bar{g},0|\hat{V}|x,\mathbf{v}'\rangle|^2\delta(E_{\bar{g}0} - E_{x\mathbf{v}'}) \tag{12.59}$$

This expression can be interpreted as a decay rate of level $|\bar{g},0\rangle$ into the manifold $\{|x,\mathbf{v}'\rangle\}$ only if this manifold is (1) continuous or at least dense enough, and (2) satisfies other requirements specified in Section 9.1. Nevertheless, Eq. (12.59) can be used as a lineshape expression even when that manifold is sparse, leading to the zero temperature limit of (12.57)

$$L_{\text{abs}}(\omega) \sim |\mu_{gx}|^2 \sum_{\mathbf{v}'} |\langle 0|e^{i(\hat{\Omega}_x-\hat{\Omega}_g)}|\mathbf{v}'\rangle|^2\delta(E_g + \hbar\omega - E_x - E_{\text{vib}}(\mathbf{v}')) \tag{12.60}$$

It displays a superposition of lines that correspond to the excitation of different numbers of vibrational quanta during the electronic transition (hence the name multiphonon transition rate). The relative line intensities are determined by the corresponding Franck–Condon factors. The fact that the lines appear as δ functions results from using perturbation theory in the derivation of this expression. In reality each line will be broadened and simplest theory (see Section 9.3) yields a Lorentzian lineshape.

Consider now the $T \to 0$ limit of Eq. (12.55b) written for the absorption lineshape of a diatomic molecule with a single vibrational mode α,

$$\begin{aligned} L_{\text{abs}}(\omega) &\sim |\mu_{gx}|^2 e^{-\bar{\lambda}_\alpha^2} \int_{-\infty}^{\infty} dt e^{-i(\omega_{xg}-\omega)t+\bar{\lambda}_\alpha^2 e^{-i\omega_\alpha t}} \\ &= |\mu_{gx}|^2 e^{-\bar{\lambda}_\alpha^2} \int_{-\infty}^{\infty} dt e^{-i(\omega_{xg}-\omega)t} \sum_{v=0}^{\infty} \frac{1}{v!}\bar{\lambda}_\alpha^{2v} e^{-iv\omega_\alpha t} \\ &\sim |\mu_{gx}|^2 \sum_{v=0}^{\infty} \frac{e^{-\bar{\lambda}_\alpha^2}}{v!}\bar{\lambda}_\alpha^{2v}\delta(\omega_{xg} + v\omega_\alpha - \omega) \end{aligned} \tag{12.61}$$

We have seen (Eqs (2.185) and (2.186)) that the coefficients in front of the δ-functions are the corresponding Franck–Condon factors, so Eq. (12.61) is just another way to write Eq. (12.60) with the Franck–Condon factors explicitly evaluated.

Equations (12.60) and (12.61) are expressions for the low temperature (i.e. $k_BT < \hbar\omega_\alpha$) electronic absorption lineshape. The frequency dependence originates from the individual transition peaks, that in reality are broadened by intramolecular and intermolecular interactions and may overlap, and from the Franck–Condon envelope

$$(FC_{0,v}(\omega))_{v=(\omega-\omega_{xg})/\omega_\alpha} = \begin{cases} 0 & \omega < \omega_{xg} \\ e^{-\bar{\lambda}_\alpha^2}\bar{\lambda}_\alpha^{2[(\omega-\omega_{xg})/\omega_\alpha]}/[(\omega-\omega_{xg})/\omega_\alpha]! & \omega > \omega_{xg} \end{cases} \tag{12.62}$$

This Franck–Condon envelope characterizes the broadening of molecular electronic spectra due to electronic–vibrational coupling.

12.5.2 Electronic relaxation of excited molecules

When a molecule is prepared in an excited electronic state, the subsequent time evolution should eventually take the molecule back to the ground state. This results from the fact that electronic energy spacings ΔE_{el} between lower molecular electronic states are usually much larger than k_BT. The corresponding relaxation process may be radiative—caused by the interaction between the molecule and the radiation field and accompanied by photon emission, or nonradiative, resulting from energy transfer from electronic to nuclear degrees of freedom, that is, transition from an excited electronic state to higher vibrational levels associated with a lower electronic state. The excess vibrational energy subsequently dissipates by interaction with the environment (vibrational relaxation, see Chapter 13), leading to dissipation of the initial excess energy as heat.[8] The terms radiative and nonradiative (or radiationless) transitions are used to distinguish between these two relaxation routes. Both processes can be described within the spin–boson model: In the radiative case the radiation field can be represented as a set of harmonic oscillators—the photons, while in the nonradiative case the underlying nuclear motion associated with intramolecular and intermolecular vibrations is most simply modeled by a set of harmonic oscillators.

In what follows we focus on the nonradiative relaxation process (the treatment of radiative relaxation, namely fluorescence, is similar to that of absorption discussed in the previous section). An important observation is that *the mechanism and consequently the rate of the electronic transition depend critically on how the nuclei behave during its occurrence.* Figure 12.5 depicts a schematic view of this process,

[8] It is also possible that the molecule will dispose of excess vibrational energy radiatively, that is, by infrared emission, however this is not very likely in condensed phases because relaxation to solvent degrees of freedom is usually much faster. Even in low-pressure samples the relaxation due to collisions with the walls is usually more efficient than the infrared emission route.

showing two extreme possibilities for this nuclear motion. In the low-temperature limit route *a* has to be taken. This is a nuclear tunneling process that accompanies the electronic transition. In the opposite, high-temperature case pathway *b* dominates. This is an activated process, characterized by an activation energy E_A shown in the figure.

We should keep in mind that the two routes: Tunneling in case *a* and activation in case *b* refer to the nuclear motion that underlines the electronic transition. In fact, the mathematical difference between the rates of these routes stems from the corresponding Franck–Condon factors that determine the overlap between the nuclear wavefunctions involved in the transition. The nuclear wavefunctions associated with process *a* are localized in wells that are relatively far from each other and their mutual overlap in space is small—a typical tunneling situation. In contrast, in process *b* the electronic transition takes place at the crossing between the nuclear potential surfaces where the overlap between the corresponding nuclear wavefunctions is large. This route will therefore dominate if the temperature is high enough to make this crossing region energetically accessible.

We will see below that the relative importance of these routes depends not only on the temperature but also on the nuclear shift parameters λ, the electronic energy gap ΔE, and the vibrational frequencies. We should also note that these two routes represent extreme cases. Intermediate mechanisms such as thermally activated tunneling also exist. Mixed situations, in which some nuclear degrees of freedom have to be activated and others, characterized by low nuclear masses and small shifts λ, can tunnel, can also take place.

A particular kind of electronic relaxation process is electron transfer. In this case (see Chapter 16) the electronic transition is associated with a large rearrangement of the charge distribution and consequently a pronounced change of the nuclear configuration, which translate into a large λ. Nuclear tunneling in this case is a very low-probability event and room temperature electron transfer is usually treated as an activated process.

12.5.3 The weak coupling limit and the energy gap law

Consider now the application of Eq. (12.55) to the transition rate from an excited electronic state 2 to a lower state 1 in the absence of any external field. For simplicity we focus on the low-temperature limit, $k_B T < \hbar\omega_{\min}$ where $\omega_{\min}$ is the lowest phonon frequency. This implies $n_\alpha = 0$ for all α, so (12.55b) becomes

$$k_{1\leftarrow 2}(\omega_{21}) = \frac{|V_{12}|^2}{\hbar^2} e^{-\sum_\alpha \bar{\lambda}_\alpha^2} \int_{-\infty}^{\infty} dt e^{i\omega_{21}t + \sum_\alpha \bar{\lambda}_\alpha^2 e^{-i\omega_\alpha t}} \tag{12.63}$$

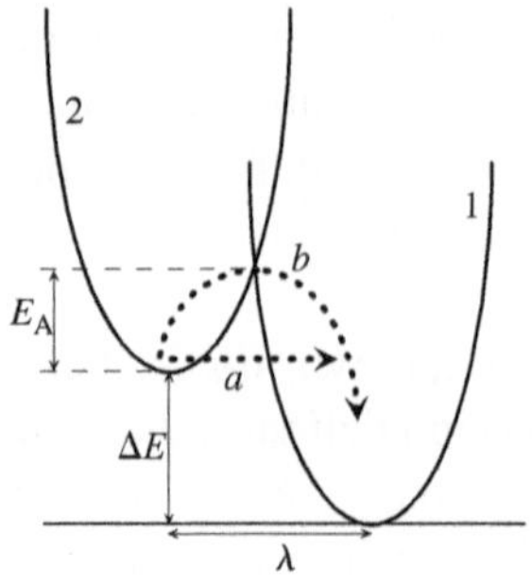

FIG. 12.5 The nuclear tunneling (a) and nuclear activation (b) pathways to nonradiative electronic relaxation.

where $E_{21} = \hbar\omega_{21}$ has been denoted ΔE in Fig. 12.5. For a single mode model, the analog of Eq. (12.61) is

$$k_{1\leftarrow 2}(\omega_{21}) = \frac{2\pi|V_{12}|^2}{\hbar^2}\sum_{\nu=0}^{\infty}\frac{e^{-\bar{\lambda}_\alpha^2}}{\nu!}\bar{\lambda}_\alpha^{2\nu}\delta(\omega_{21} - \nu\omega_\alpha)$$
$$\approx \frac{2\pi|V_{12}|^2}{\hbar^2}e^{-\bar{\lambda}_\alpha^2}\frac{\bar{\lambda}_\alpha^{2\bar{\nu}}}{\bar{\nu}!}; \qquad \bar{\nu} = \frac{\omega_{21}}{\omega_\alpha} \tag{12.64}$$

For $\bar{\nu} \gg 1$, that is, large energy gap, $E_2 - E_1 \gg \hbar\omega_\alpha$, the ω_{21} dependence is given by

$$k_{1\leftarrow 2}(\omega_{21}) \sim \exp(\bar{\nu}\ln\bar{\lambda}_\alpha^2 - \bar{\nu}\ln\bar{\nu}) \tag{12.65}$$

For $\bar{\lambda}_\alpha < 1$ this function decreases exponentially or faster with the energy gap.

The same observation can be made also for the general many-mode case, for which we get

$$k_{1\leftarrow 2}(\omega_{21}) = \frac{2\pi|V_{12}|^2}{\hbar^2}e^{-\sum_\alpha\bar{\lambda}_\alpha^2}\sum_{\{\nu_\alpha\}}\delta\left(\omega_{21} - \sum_\alpha \nu_\alpha\omega_\alpha\right)\prod_\alpha\frac{\bar{\lambda}_\alpha^{2\nu_\alpha}}{\nu_\alpha!} \tag{12.66}$$

where, as usual, $\{\nu_\alpha\}$ denotes a set of vibrational quantum numbers, ν_α for mode α, that specify a molecular vibrational state. Again we consider the large energy gap limit, $\omega_{21} \gg \omega_c$ where ω_c is the highest phonon frequency. We also focus on the weak electron–phonon coupling limit, $\bar{\lambda}_\alpha^2 \ll 1$ for all modes. In this case the sum in (12.66) is dominated by the terms with the smallest ν_α, that is, by modes with $\omega_\alpha \simeq \omega_c$. For an order of magnitude estimate we may replace $\bar{\lambda}_\alpha^2$ for these modes by an average value $\bar{\lambda}^2$, so

$$k_{1\leftarrow 2}(\omega_{21}) \sim \bar{\lambda}^{2(\omega_{21}/\omega_c)}S; \qquad S = \sum_{\substack{\{\nu_\alpha\} \\ \sum_\alpha \nu_\alpha = \omega_{21}/\omega_c}}\prod_\alpha\frac{1}{\nu_\alpha!} \tag{12.67}$$

where the terms that contribute to S are associated with the group of modes with frequencies close to ω_c. Equation (12.67) shows an exponential decrease (since $\bar{\lambda} < 1$) of the rate with increasing $\bar{v} \equiv \omega_{21}/\omega_c$, that is, with larger electronic energy gap E_{21}, with corrections that arise from the dependence of S on ω_{21}. $\bar{v}$ is the number of vibrational quanta that the n_c high-frequency modes must accept from the electronic motion. If, for example, $n_c \gg \bar{v}$, the most important contributions to (12.67) are the $n_c!/(\bar{v}!(n_c - \bar{v})!)$ terms with $v_\alpha = 1$ or 0, so that $k_{1\leftarrow 2}(\omega_{21}) \sim \bar{\lambda}^{2\bar{v}} n_c!/(\bar{v}!(n_c - \bar{v})!)$ is the analog of (12.64) and (12.65).

This "energy gap law": Inverse exponential decrease of the rate on the gap between the electronic origins of the two states involved, characterizes the nuclear tunneling route to electronic relaxation. As discussed above, it stems from the energy gap dependence of the Franck–Condon factors that determine the magnitudes of the dominant contributions to the rate at the given gap. As seen in Section 2.10, tunneling processes depend exponentially on parameters related to the potential barrier. The result obtained here has a similar character but is different in specifics because the electronic energy gap does not reflect a barrier height for the nuclear tunneling process.

12.5.4 The thermal activation/potential-crossing limit

Consider now the opposite case where the shift parameter $\bar{\lambda}^2$ and/or the temperature are large. Specifically we assume $\sum_\alpha \bar{\lambda}_\alpha^2 n_\alpha \gg 1$. In this case the integrand in Eq. (12.55b) is very short-lived and can be approximated by expanding the exponential up to second order in the $\omega_\alpha t$ factors.[9] This short-time approximation leads to

$$k_{1\leftarrow 2} = \frac{|V_{12}|^2}{\hbar^2} e^{-\sum_\alpha \bar{\lambda}_\alpha^2 (2n_\alpha+1)} \int_{-\infty}^{\infty} dt e^{i\omega_{21}t + \sum_\alpha \bar{\lambda}_\alpha^2((n_\alpha+1)e^{-i\omega_\alpha t} + n_\alpha e^{i\omega_\alpha t})}$$

$$k_{1\leftarrow 2} = \frac{|V_{12}|^2}{\hbar^2} \int_{-\infty}^{\infty} dt e^{i\omega_{21}t - it\sum_\alpha \bar{\lambda}_\alpha^2 \omega_\alpha - (1/2)t^2 \sum_\alpha \bar{\lambda}_\alpha^2 \omega_\alpha^2 (2n_\alpha+1)}$$

$$= \frac{|V_{12}|^2}{\hbar^2} \sqrt{\frac{\pi}{a}} e^{-(\omega_{21} - E_r/\hbar)^2/4a}; \qquad a = \frac{1}{2}\sum_\alpha (2n_\alpha + 1)\bar{\lambda}_\alpha^2 \omega_\alpha^2 \qquad (12.68)$$

[9] Applying the short time approximation should be done with care. The physics has to be such that the correlation function remains zero (or at least unimportant) after it became zero at short time. The result does not depend on what is the physical process (e.g., broadening by the surrounding liquid) that makes this happen.

where E_r is the reorganization energy defined by Eqs. (12.22). A simpler equation is obtained in the classical limit where $n_\alpha = k_B T/\hbar\omega_\alpha$ for all modes α, so $a = k_B T E_r/\hbar^2$:

$$k_{1\leftarrow 2} = \frac{|V_{12}|^2}{\hbar}\sqrt{\frac{\pi}{k_B T E_r}}e^{-(E_{21}-E_r)^2/4k_B T E_r} \tag{12.69}$$

This result has the form of a thermally activated rate, with activation energy given by

$$E_A = \frac{(E_{21}-E_r)^2}{4E_r} \tag{12.70}$$

Problem 12.7. Show that E_A, Eq. (12.70) is equal to the height of the *minimum-energy crossing point* of the two potential surfaces $E_{el}^{(1)}(\mathbf{R})$ and $E_{el}^{(2)}(\mathbf{R})$ (Eq. (12.20)) above the bottom of the $E_{el}^{(2)}(\mathbf{R})$ potential surface.

Solution: The two N-dimensional potential surfaces (N—number of phonon modes) cross on an $(N-1)$—dimensional surface defined by $E_{el}^{(1)}(\{x_\alpha\}) = E_{el}^{(2)}(\{x_\alpha\})$ where $E_{el}^{(1)}(\{x_\alpha\}) = \sum_\alpha \hbar\omega_\alpha(\bar{x}_\alpha - \bar{\lambda}_\alpha)^2 - E_{21}$ and $E_{el}^{(2)}(\{x_\alpha\}) = \sum_\alpha \hbar\omega_\alpha \bar{x}_\alpha^2$. Using Eq. (12.22) the equation for this surface is

$$-2\sum_\alpha \hbar\omega_\alpha \bar{x}_\alpha \bar{\lambda}_\alpha + E_r - E_{21} = 0 \tag{12.71}$$

The crossing point of minimum energy can be found as the minimum of $E_{el}^{(2)}(\{x_\alpha\})$ under the condition (12.71). Defining the Lagrangian $F = \sum_\alpha \hbar\omega_\alpha \bar{x}_\alpha^2 + B(-2\sum_\alpha \hbar\omega_\alpha \bar{x}_\alpha \bar{\lambda}_\alpha + E_r - E_{21})$ the condition for extremum is found as $\bar{x}_\alpha = B\bar{\lambda}_\alpha$. Using this in (12.71) yields the Lagrange multiplier $B = -(E_{21}-E_r)/2E_r$, whence

$$x_\alpha^{(\min)} = -\frac{(E_{21}-E_r)\lambda_\alpha}{2E_r} \tag{12.72}$$

The energy at the lowest energy crossing point is $E_A = E_{el}^{(2)}(\{x_\alpha^{\min}\})$. Using (12.72) and (12.22) leads indeed to (12.70).

We have thus found that in this high temperature, strong electron–phonon coupling limit the electronic transition is dominated by the lowest crossing point of the two potential surfaces, that is, the system chooses this pathway for the electronic transition. It is remarkable that this result, with a strong characteristic of classical rate theory, was obtained as a limiting form of the quantum golden-rule expression for the transition rate. Equation (12.69) was first derived by Marcus in the context of electron transfer theory (Chapter 16).

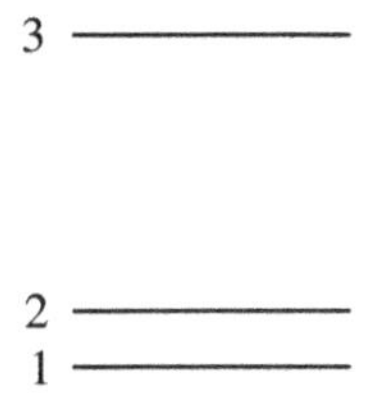

FIG. 12.6 A three-level system used in the text to demonstrate the promotion of effective coupling between spin levels 1 and 2 via their mutual coupling with level 3.

12.5.5 Spin–lattice relaxation

The evaluation of relaxation rates in the previous sections was based on the assumption that the energy spacing in the two-level system under consideration is large relative to the maximum (cutoff) phonon frequency. This is the common state of affairs for nonradiative electronic relaxation processes, resulting in the multiphonon character of these processes. When the two-level energy spacing is smaller than the boson cutoff frequency, the relaxation is dominated by one-boson processes and may be studied with the Hamiltonian (12.27) and (12.28). Radiative relaxation associated with the interaction (12.3) is an example. The rates associated with such one-boson relaxation processes are given by Eqs (12.47) and (12.48).[10] They are proportional to $g(\omega_{21})$, the boson mode density at the two-level frequency. In the high-temperature limit, $k_BT \gg E_{21}$, they are also proportional to T.

Consider now the case where the energy spacing E_{21} is very small. Such cases are encountered in the study of relaxation between spin levels of atomic ions embedded in crystal environments, so called *spin–lattice relaxation*. The spin level degeneracy is lifted by the local crystal field and relaxation between the split levels, caused by coupling to crystal acoustical phonons, can be monitored. The relaxation as obtained from (12.47) and (12.48) is very slow because the density of phonon modes at the small frequency ω_{21} is small (recall that $g(\omega) \sim \omega^2$).

In fact, when $\omega_{21} \to 0$, other relaxation mechanisms should be considered. It is possible, for example, that the transition from level 2 to 1 is better routed through a third, higher-energy level, 3, as depicted in Fig. 12.6, because the transition rates $k_{3\leftarrow 2}$ and $k_{1\leftarrow 3}$ are faster than $k_{1\leftarrow 2}$. In this case, sometimes referred to as the *Orbach mechanism*, the transition $2 \to 3$ is the rate-determining step and, if $k_BT \ll E_{32}$, the observed relaxation will depend exponentially on temperature, $k_{1\leftarrow 2,\text{apparent}} = \exp(-E_{32}/k_BT)$. Another relaxation mechanism is a two-phonon process analogous to Raman light scattering. We note that the one-phonon coupling

---

[10] The measured relaxation rate in a process $A \underset{k_2}{\overset{k_1}{\leftrightarrows}} B$ is $k_1 + k_2$.

(12.2b) is a first-order term in an expansion in bath normal mode coordinate. The second-order term in this expansion leads to interaction terms such as

$$\hat{V}_{\mathrm{SB}} = \sum_{\alpha,\beta} (V_{1,2}^{\alpha,\beta}|1\rangle\langle 2| + V_{2,1}^{\alpha,\beta}|2\rangle\langle 1|)(\hat{a}_\alpha^\dagger + \hat{a}_\alpha)(\hat{a}_\beta^\dagger + \hat{a}_\beta) \tag{12.73}$$

According to Eq. (12.40) the golden-rule rate is now given by

$$k_{1\leftarrow 2} = \frac{1}{\hbar^2}\sum_{\alpha,\beta}|V_{1,2}^{\alpha,\beta}|^2 \int_{-\infty}^{\infty} dt e^{i\omega_{21}t}\langle(\hat{a}_\alpha^\dagger e^{i\omega_\alpha t} + \hat{a}_\alpha e^{-i\omega_\alpha t})(\hat{a}_\beta^\dagger e^{i\omega_\beta t} + \hat{a}_\beta e^{-i\omega_\beta t})$$
$$\times (\hat{a}_\alpha^\dagger + \hat{a}_\alpha)(\hat{a}_\beta^\dagger + \hat{a}_\beta)\rangle \tag{12.74}$$

Only terms with $\alpha \neq \beta$ that satisfy $\omega_\alpha - \omega_\beta = \pm\omega_{21}$ survive the time integration. We get

$$\begin{aligned} k_{1\leftarrow 2} &= \frac{2\pi}{\hbar^2}\sum_{\alpha,\beta}|V_{1,2}^{\alpha,\beta}|^2(\langle\hat{a}_\alpha^\dagger\hat{a}_\beta\hat{a}_\alpha\hat{a}_\beta^\dagger\rangle\delta(\omega_{21}+\omega_\alpha-\omega_\beta) \\ &\quad + \langle\hat{a}_\alpha\hat{a}_\beta^\dagger\hat{a}_\alpha^\dagger\hat{a}_\beta\rangle\delta(\omega_{21}-\omega_\alpha+\omega_\beta)) \\ &= \frac{2\pi}{\hbar^2}\sum_{\alpha,\beta}|V_{1,2}^{\alpha,\beta}|^2[n(\omega_\alpha)(n(\omega_\beta)+1)\delta(\omega_{21}+\omega_\alpha-\omega_\beta) \\ &\quad + (n(\omega_\alpha)+1)n(\omega_\beta)\delta(\omega_{21}-\omega_\alpha+\omega_\beta)] \\ &\simeq \frac{4\pi}{\hbar^2}\int_0^{\omega_{\mathrm{D}}} d\omega g^2(\omega)|V_{1,2}(\omega)|^2 n(\omega)(n(\omega)+1) \end{aligned} \tag{12.75}$$

where $n(\omega) = (e^{\hbar\omega/k_BT} - 1)^{-1}$, ω_{D} is the Debye frequency and where in the last step we have converted the double sum to a double integral over the normal modes, have used the δ functions to do one integration and have approximated the resulting integrand by taking the limit $\omega_{21} \to 0$. We have also assumed for simplicity that $|V_{1,2}^{\alpha,\beta}|^2$ depends on α and β only through $\omega_\alpha \simeq \omega_\beta$.

Further analysis is possible only if more information on $V_{1,2}(\omega)$ is available. The theory of spin lattice relaxation leads to $V_{1,2}(\omega) \sim \omega$. At low temperature the integral in (12.75) is dominated by the low-frequency regime where we can use $g(\omega) \sim \omega^2$ (see Section 4.2.4). We then have

$$k_{1\leftarrow 2} \sim \int_0^{\omega_{\mathrm{D}}} d\omega\omega^6 n(\omega)(n(\omega)+1) \tag{12.76}$$

and provided that $T \ll \hbar\omega_D/k_B$ this leads to (compare to the analysis of the low-temperature behavior of Eq. (4.52))

$$k_{1\leftarrow 2} \sim T^7 \tag{12.77}$$

Indeed, both the exponential temperature dependence that characterize the Orbach process and the T^7 behavior associated with the Raman type process have been observed in spin lattice relaxation.[11]

12.6 Beyond the golden rule

The golden-rule rate expressions obtained and discussed above are very useful for many processes that involve transitions between individual levels coupled to boson fields, however there are important problems whose proper description requires going beyond this simple but powerful treatment. For example, an important attribute of this formalism is that it focuses on the rate of a given process rather than on its full time evolution. Consequently, a prerequisite for the success of this approach is that the process will indeed be dominated by a single rate. In the model of Figure 12.3, after the molecule is excited to a higher vibrational level of the electronic state 2 the relaxation back into electronic state 1 is characterized by the single rate (12.34) only provided that thermal relaxation within the vibrational subspace in electronic state 2 is faster than the 2 $\rightarrow$ 1 electronic transition. This is often the case in condensed phase environments but exceptions have been found increasingly often since picosecond and femtosecond timescales became experimentally accessible. Generalized golden-rule approaches may still be useful in such cases.[12]

In many cases, reliable theoretical descriptions of multi-rate processes can be obtained by using master equations in which individual rates are obtained from golden-rule type calculations (see Sections 8.3.3 and 10.4). A condition for the validity of such an approach is that individual rate processes will proceed independently. For example, after evaluating the rates $k_{1\leftarrow 2}$ and $k_{2\leftarrow 1}$, Eqs (12.55a) and (12.55b), a description of the overall dynamics of the coupled two-level system by the kinetic scheme $1 \underset{k_{1\leftarrow 2}}{\overset{k_{2\leftarrow 1}}{\rightleftarrows}} 2$ relies on the assumption that after each transition, say from 1 to 2, the system spends a long enough time in state 2, become equilibrated

[11] P. L. Scott and C. D. Jeffries, Phys. Rev. **127**, 32 (1962); G. H. Larson and C. D. Jeffries, Phys. Rev. **141**, 461 (1966).

[12] R. D. Coalson, D. G. Evans, and A. Nitzan, J. Chem. Phys. **101**, 436 (1994); M. Cho and R. J. Silbey, J. Chem. Phys. **103**, 595 (1995).

in this state (or, more poetically, forgets its past), so that the reverse transition occurs independently. When this is not the case such simple kinetic schemes fail. Generalized quantum master equations (e.g. Section 10.4.2) can be used in such cases, however they are often hard to implement. Note that situations in which successive processes are not decoupled from each other occur also in classical systems.

Other cases that require going beyond the golden-rule involve transitions which by their nature are of high order in the interaction. Processes studied in conjunction with nonlinear spectroscopy (see Section 18.7) are obvious examples.

Finally, the golden-rule fails when the basic conditions for its validity are not satisfied. For a general review of these issues see Leggett et al.[13] Conditions for the validity of the golden-rule involve relatively uniform coupling to a relatively wide continuum, and one consistency check is that the decaying level, broadened by the decay width Γ, is still wholly contained within the continuum. For example, referring to Fig. 12.4, this is not satisfied for a level near the origin of state 2 if $\Gamma = 2\pi V^2 \rho > \Delta E$. Such "overdamped" cases have to be handled by more advanced methodologies, for example, path integral methods[14] that are beyond the scope of this text.

---

[13] A. J. Leggett, S. Chakravarty, A. T. Dorsey, M. P. A. Fisher, A. Garg, and W. Zwerger, Rev. Mod. Phys. **59**, 1 (1987).

[14] N. Makri and D. E. Makarov, J. Chem. Phys. **102**, 4600 (1995); **102**, 4611 (1995); D. G. Evans, A. Nitzan, and M. A. Ratner, J. Chem. Phys. **108**, 6387 (1998).

PART III

APPLICATIONS

13

VIBRATIONAL ENERGY RELAXATION

You see that stones are worn away by time,
Rocks rot, and towers topple, even the shrines
And images of the gods grow very tired,
Develop crack or wrinkles, their holy wills
Unable to extend their fated term,
To litigate against the Laws of Nature...

Lucretius (c.99–c.55 BCE*) "The way things are" translated by Rolfe Humphries, Indiana University Press, 1968*

An impurity molecule located as a solute in a condensed solvent, a solid matrix or a liquid, when put in an excited vibrational state will loose its excess energy due to its interaction with the surrounding solvent molecules. Vibrational energy accumulation is a precursor to all thermal chemical reactions. Its release by vibrational relaxation following a reactive barrier crossing or optically induced reaction defines the formation of a product state. The direct observation of this process by, for example, infrared emission or more often laser induced fluorescence teaches us about its characteristic timescales and their energetic (i.e. couplings and frequencies) origin. These issues are discussed in this chapter.

13.1 General observations

Before turning to our main task, which is constructing and analyzing a model for vibrational relaxation in condensed phases, we make some general observations about this process. In particular we will contrast condensed phase relaxation with its gas phase counterpart and will comment on the different relaxation pathways taken by diatomic and polyatomic molecules.

First, vibrational relaxation takes place also in low density gases. Collisions involving the vibrationally excited molecule may result in transfer of the excess vibrational energy to rotational and translational degrees of freedom of the overall system. Analysis based on collision theory, with the intermolecular interaction potential as input, then leads to the *cross-section* for inelastic collisions in which vibrational and translational/rotational energies are exchanged. If C^* is the concentration of vibrationally excited molecules and ρ is the overall gas density, the relaxation rate coefficient k_{gas} is defined from the bimolecular rate law

$$\frac{dC^*}{dt} = -k_{\text{gas}} C^* \rho \tag{13.1}$$

so that the relaxation time is

$$(\tau^{(\text{gas})})^{-1} \equiv -\frac{1}{C^*}\frac{dC^*}{dt} = k_{\text{gas}}\rho \tag{13.2}$$

When comparing this relaxation to its condensed phase counterpart one should note a technical difference between the ways relaxation rates are defined in the two phases. In contrast to the bimolecular rate coefficient k_{gas}, in condensed environments the density is high and is not easily controlled, so the relaxation rate is conventionally defined in terms of a unimolecular rate coefficient k_{cond}, defined from $dC^*/dt = -k_{\text{cond}}C^* = -(\tau^{(\text{cond})})^{-1}C^*$. This difference between the two definitions should be taken into account in any meaningful comparison between rates in the two phases.

Next, consider the relaxation of a diatomic molecule, essentially a single oscillator of frequency ω_0 interacting with its thermal environment, and contrast its behavior with a polyatomic molecule placed under similar conditions. The results of the simple harmonic relaxation model of Section 9.4 may give an indication about the expected difference. The harmonic oscillator was shown to relax to the thermal equilibrium defined by its environment, Eq. (9.65), at a rate given by (cf. Eq. (9.57))

$$\gamma = 2\pi(|u(\omega)|^2 g(\omega))_{\omega=\omega_0} \tag{13.3}$$

where u is the coupling strength and $g(\omega)$ is the density of modes characterizing the environment that was taken as a bath of harmonic oscillators. For harmonic environments this mode density function is characterized by an upper cutoff, the Debye frequency ω_{D}, beyond which modes are not allowed (the cutoff wavelength is of the order of the nearest interatomic distance). This implies that in harmonic relaxation models the rate vanishes for $\omega_0 > \omega_{\text{D}}$. In realistic environments this translates, as we will see, into an exponential falloff of the rate when ω_0 increases beyond ω_{D}. Since typical Debye frequencies are smaller than vibrational frequencies of many diatomic species, we expect vibrational relaxation of such species to be slow.

In polyatomic molecules, however, other relaxation pathways can show up in such cases, using combination of intermode energy transfer with relaxation to circumvent the "Debye restriction." Consider for example a pair of molecular modes

1 and 2, of frequencies ω_1 and ω_2, respectively, such that ω_1 is larger than ω_D while both ω_2 and $\omega_1 - \omega_2$ are smaller than ω_D. An initial excitation of mode 1 can relax by successive steps of transferring quanta of its vibrational energy to mode 2, with subsequent relaxation of the latter. The energy transferred to the bath at each relaxation step is either $\hbar(\omega_1 - \omega_2)$ or $\hbar\omega_2$, both below the Debye threshold. Such pathways are characterized by bath-assisted redistribution of intramolecular vibrational energy into low-frequency modes followed by relaxation of the latter into the thermal environment. They are facilitated by the fact that polyatomic molecules always have some low-frequency modes associated with relatively heavy segments of the molecule vibrating against each other. Such pathways multiply and become more efficient for larger molecules.

It is indeed observed that vibrational relaxation of large molecules is invariably fast, at the picosecond time range. In contrast, the relaxation of high-frequency diatomic molecules can be very slow and is often overtaken by competing relaxation processes including energy transfer to other molecules that may be present as impurities or infrared emission which is a relatively slow (typically in the millisecond time regime for allowed transitions) process by itself. To get a feeling for what should be considered "slow" or "fast" in these type of processes note that a typical molecular vibrational period is of the order 10^{-13}–10^{-14} s. In solution or in a solid matrix such a molecule is surrounded by molecules of the host environment at contact distances, and beats against them 10^{13}–10^{14} times per second.[1] The observation that excess vibrational energy can sometimes survive for milliseconds or more makes us appreciate the associated relaxation process as very slow indeed.

An example that shows the sometimes intricate role of competing relaxation mechanisms is given in Fig. 13.1, which depicts the relaxation behavior of several vibrational levels of the CN radical embedded in Ne matrix at 4 K. Here the process that involves repeated solvent assisted transitions between two electronic manifolds provides a faster relaxation route than pure vibrational transitions. Such pure transitions do show up on a timescale slower by three orders of magnitude when the former pathway is not available.

In the rest of this chapter we focus on the "pure" vibrational relaxation problem, that is, on the relaxation of an oscillator (usually modeled as harmonic) coupled to its thermal environment. From the theoretical point of view this problem supplements the spin boson model considered in Chapter 12. Indeed, these are the $N = 2$ and the $N = \infty$ limits of an N level system. For a harmonic oscillator there is an added feature, that these states are equally spaced. In this case individual

[1] In addition to this vibrational beating there are additional encounters due to the center of mass motion. For intermolecular distances of order $\sim$1 Å this would imply a thermal collision rate of $\sim 10^{12}$ s^{-1} at room temperature.

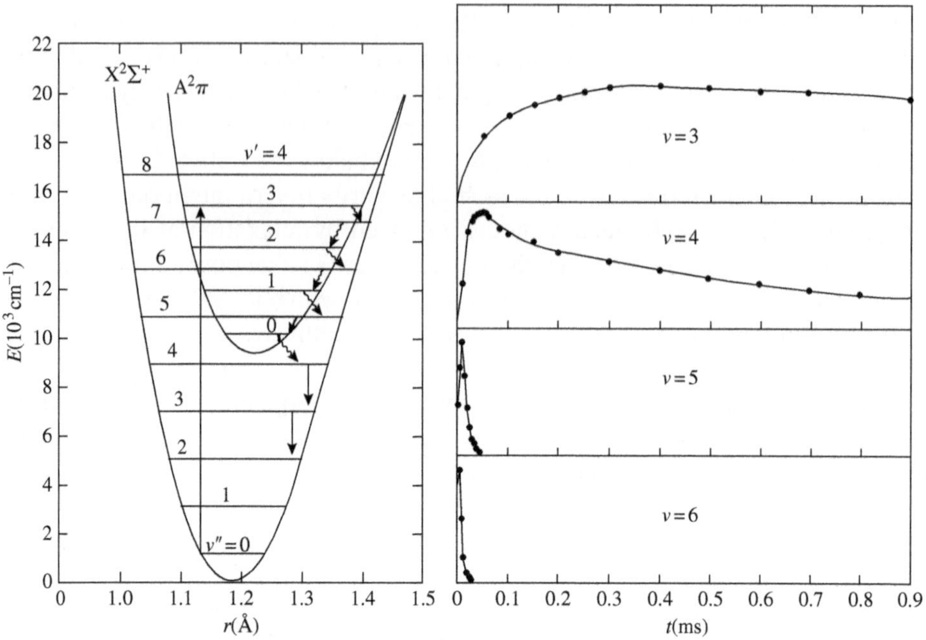

FIG. 13.1 Relaxation in the $X^2\Sigma^+$ (ground electronic state) and $A^2\Pi$ (excite electronic state) vibrational manifolds of the CN radical in Ne host matrix at $T = 4$ K, following excitation into the third vibrational level of the Π state. Populations in individual vibrational levels in both electronic states are monitored independently by fluorescence (for the Π state) and by laser induced fluorescence (for the Σ state). The preferred relaxation pathway for energies above the origin of the Π state is found to be medium assisted internal conversion[2] as indicated by arrows in the left panel. The right panel shows the dynamics of population and subsequent decays of the vibrational levels 6, 5, 4, and 3 of the ground Σ state. Levels 6 and 5 relax much faster (lifetimes in the order of 1–3 μs) than levels 4 and 3 (lifetimes in the ms range). For the latter the internal conversion-assisted pathway is closed as seen in the state diagram on the left, so these long lifetimes correspond to pure vibrational transitions. (From V. E. Bondybey and A. Nitzan, Phys. Rev. Lett. **38**, 889 (1977).)

transitions cannot be resolved (though their rate can be calculated) and the only observable relaxation is that of the total energy. In realistic situations the relaxing oscillator is anharmonic and individual transitions can be monitored. The choice between applying a two-level or an $N = \infty$ level (i.e., a harmonic oscillator) model to a particular observation depends on whether it involves observable individual transitions between pairs of vibrational levels (that can be resolved because

[2] *Internal conversion* is a term used to describe radiationless transition between electronic states of the same spin multiplicity. *Intersystem crossing* is the term used of the similar process in which the spin state changes.

of the anharmonicity) or whether we follow the relaxation of the overall vibrational energy. It should be noticed, however, that an important aspect of the two problems is common to both: they involve the exchange of a well-defined quantum of energy with the host medium and both are equally sensitive to the issue of whether this quantum is smaller or larger than that implied by the host Debye frequency.

13.2 Construction of a model Hamiltonian

Our aim is to consider the main physical factors affecting the vibrational relaxation of a diatomic molecule, embedded as an impurity or a solute in a condensed hosting environment, by considering a simple model that captures these factors. To this end we will first impose a drastic simplification (whose failure in some cases will be discussed later) by avoiding a detailed consideration of local motions, oscillations and rotations, that may couple to the vibration under study. This is achieved by (1) considering a mono-atomic solvent, (2) assuming that the molecule resides in a spherical cavity within this solvent, and (3) making a *breathing sphere* model of the diatomic molecule. In this model the molecule is taken to be a sphere with the molecular mass M, whose radius $a + q$ vibrates about the equilibrium radius a with an amplitude q. The center $\mathbf{R}$ of this sphere corresponds to the position of the molecular center of mass. The coordinate q is modeled as a harmonic oscillator of mass m and frequency ω that corresponds to the reduced mass and the intramolecular frequency of the diatomic molecule. The interaction between this impurity molecule and a bath atom j at position $\mathbf{r}_j$ is, in this model, a function of a single distance parameter, and may be written in the form $V(|\mathbf{r}_j - \mathbf{R}| - a - q)$. This is obviously a highly simplified picture because in reality the interaction depends also on the molecular orientation relative to the molecule–atom distance vector. It appears, however, to contain the important physical ingredients of the process: the oscillator mass and frequency, a representative local mode (the center of mass motion of the molecule in its solvent cage) and the host dynamics through the time-dependent positions $\mathbf{r}_j(t)$.

An important simplification can be made at this point by noting that the amplitude q of a molecular vibration is usually much smaller than intermolecular and intramolecular distances. Therefore the interaction potential may be expanded according to

$$V(|\mathbf{r}_j - \mathbf{R}| - a - q) \cong V(|\mathbf{r}_j - \mathbf{R}| - a) + f(|\mathbf{r}_j - \mathbf{R}| - a)q \tag{13.4}$$

where

$$f(|\mathbf{r}_j - \mathbf{R}| - a) = -\left(\frac{\partial V(x)}{\partial x}\right)_{x=|\mathbf{r}_j-\mathbf{R}|-a} \tag{13.5}$$

is the force exerted on the intramolecular bond, frozen in its equilibrium length a, by the solvent. Note that this force depends on the instantaneous solvent configuration $\{\mathbf{r}_j\}$.

The resulting system–bath Hamiltonian can therefore be written in the form

$$H = H_{\mathrm{S}} + H_{\mathrm{B}} + H_{\mathrm{SB}} \tag{13.6}$$

where the "system" Hamiltonian, H_{S}, describes the intramolecular motion which is assumed to be harmonic

$$H_{\mathrm{S}} = \frac{p^2}{2m} + \frac{1}{2}m\omega^2 q^2 \tag{13.7}$$

The bath consists of all solvent atoms and the spherical impurity with its internal motion frozen at $q = 0$,

$$H_{\mathrm{B}} = \frac{P^2}{2M} + \sum_j \frac{p_j^2}{2m_j} + \sum_j V(|\mathbf{r}_j - \mathbf{R}| - a) + U_{\mathrm{B}}(\{\mathbf{r}_j\}) \tag{13.8}$$

where $U_{\mathrm{B}}(\{\mathbf{r}_j\})$ is the intermolecular solvent potential. When the bath is a solid environment all atoms execute small amplitude motions about their equilibrium positions. We can therefore repeat the procedure of Section 4.2.1: Expand the potential (the two last terms in (13.8)) in the atomic deviations from these equilibrium positions, truncate the expansion at the quadratic level, and diagonalize the resulting coupled oscillator equations. This leads to the harmonic bath model and the representation of H_{B} as a sum of normal mode Hamiltonians (cf. Eq. (6.76))

$$H_{\mathrm{B}} = \frac{1}{2}\sum_k (p_k^2 + \omega_k^2 u_k^2) \tag{13.9}$$

These normal modes evolve independently of each other. Their classical equations of motion are $\ddot{u}_k = -\omega_k^2 u_k$, whose general solution is given by Eqs (6.81). This bath is assumed to remain in thermal equilibrium at all times, implying the phase space probability distribution (6.77), the thermal averages (6.78), and equilibrium time correlation functions such as (6.82). The quantum analogs of these relationships were discussed in Section 6.5.3.

Finally, the oscillator–bath coupling is

$$H_{\mathrm{SB}} = F(\{r_j\})q = q\sum_j f(|\mathbf{r}_j - \mathbf{R}| - a) \tag{13.10}$$

where F is the total force exerted on the intramolecular bond by the solvent. It is in principle a function of the position of all solvent atoms, but because intermolecular forces are relatively short range, the dominant contributions are from solvent atoms close to the solute molecule, usually atoms in the first solvation shell.

Equations (13.6), (13.7), (13.9), and (13.10) provide the general structure of our model. The relaxation process depends on details of this model mainly through the form and the magnitude of H_{SB}. We will consider in particular two forms for this interaction. One, which leads to an exactly soluble model, is the bilinear interaction model in which the force $F(\{r_j\})$ is expanded up to first order in the deviations δr_j of the solvent atoms from their equilibrium positions, $F(\{r_j\}) = \sum_j F'(\{r_j^{eq}\})\delta r_j$, then these deviations are expanded in the normal modes. This eventually leads to a normal mode expansion of the force F with expansion coefficients determined by the interactions (13.4) and the transformation that relate the atomic deviations δr_j to the normal modes.

$$F = \sum_k A_k u_k \tag{13.11}$$

The other, more realistic interaction model is derived from assuming an exponentially repulsive interaction, $V(r) = A\exp(-\alpha r)$, in (13.4). Putting $r = |\mathbf{r}_j - \mathbf{R}| - a - q$ and expanding to first order in q yields

$$H_{SB} = \alpha A q \sum_j e^{-\alpha(|\mathbf{r}_j - \mathbf{R}| - a)} \tag{13.12}$$

This choice is appropriate for the common case where the frequency ω of our impurity vibration is very high relative to typical bath frequencies (as estimated, for example, by the bath Debye frequency). In this case only close encounters between the impurity and the bath molecules can couple efficiently to this high-frequency motion,[3] and such close encounters are indeed dominated by an exponential repulsion.

In our application of the model (13.12) we will make another simplifying approximation, which is valid for solid environments but can be rationalized also for liquid solvents.[4] Consider one of the contributions to the interaction potential (13.10). For

[3] This results from the fact that a time-dependent potential can efficiently affect the motion of an oscillator of frequency ω only if its spectrum contains similar frequencies, that is, if it varies with time at rates comparable to these frequencies. Our oscillator can experience such a potential only at close encounters that explore the steep regimes of the intermolecular interaction.

[4] The essence of the simplification discussed here is the assumption that the oscillator–host interaction is dominated by the host atoms nearest to the atom, the so-called "first solvation shell" of host atoms. In a solid host these atoms are fixed, while in a liquid they may interchange with other host atoms on the timescale of the observed relaxation. However, the argument used here relies mainly on the number of nearest atoms, not on their identity, and while this number may fluctuate somewhat in a liquid its average provides a reasonable measure for the interaction experienced by the oscillator.

this term we can take the x-axis in the direction from $\mathbf{R}$ to $\mathbf{r}_j$, so that the exponential function can be written as $e^{-\alpha(x_j - X - a)}$. Now assume that (1) x_j can be expressed in terms of deviations from some static equilibrium value $x_j^{(\mathrm{eq})}$, that is $x_j = x_j^{(\mathrm{eq})} + \delta x_j$; (2) $x_j^{(\mathrm{eq})} - X = |\mathbf{r}_j^{(\mathrm{eq})} - \mathbf{R}|$ is the same for all solvent atoms which are nearest neighbors to the impurity molecule; the number of these atoms will be denoted N_{nn}; (3) the contributions of all other solvent atoms to H_{SB} is small and can be disregarded. Under these approximations Eq. (13.12) takes the form

$$H_{\mathrm{SB}} = Be^{-\alpha\delta x}q \tag{13.13}$$

where B is a constant given by

$$B = \alpha AN_{nn}e^{-\alpha\left(|\mathbf{r}^{(\mathrm{eq})} - \mathbf{R}| - a\right)} \tag{13.14}$$

and δx is the (radial) deviation of an atom in the first coordination shell about the solute molecule from its equilibrium position.[5]

The model defined above will be used below to investigate vibrational relaxation (VR) of an impurity molecule in a condensed host. This simplified model cannot be expected to account quantitatively for the observed VR rate, but we can use it in order to understand the mechanism of this relaxation and the dependence of the relaxation rate on the nature of the hosting environment, the temperature, the molecular spectrum (i.e. the energy separation between the levels involved) and the molecule–solvent interaction.

13.3 The vibrational relaxation rate

The model Hamiltonian (13.6)–(13.8) and (13.13) and (13.14) can be used as a starting point within classical or quantum mechanics. For most diatomic molecules of interest $\hbar\omega > k_B T$, which implies that our treatment must be quantum mechanical. In this case all dynamical variables in Eqs (13.6)–(13.8) and (13.13)–(13.14) become operators.

---

[5] The rest of the discussion in this chapter just uses B as a constant parameter. Still, it should be noted that the linear relationship $B \sim N_{nn}$ is highly questionable. Since the calculated rate is proportional to B^2 it would imply that the rate goes like N_{nn}^2. Another equally reasonable model assumption is that each nearest neighbor acts independently, therefore contributes additively to the rate, in which case the resulting rate goes like N_{nn}. One can even envision situations in which different nearest neighbors interfere destructively, in which case the dependence on N_{nn} will be sublinear. More than anything, this uncertainty reflects a shortcoming of the spherical breathing sphere model that disregards the fact that the interaction of the molecule with its surrounding neighbors depends on their mutual orientations.

Consider first the transition rate between two molecular levels i and f with energies E_i and E_f—eigenstates and eigenvalues of $\hat{H}_S$, Eq. (13.7). We use the golden rule expression (12.34) to evaluate this rate on the basis of eigenstates of $\hat{H}_S + \hat{H}_B$. Denoting by ε_α and $|\alpha\rangle$ the eigenvalues and eigenfunctions of the bath Hamiltonian, the transition under consideration is between the group of states $\{|i,\alpha\rangle\}$ and $\{|f,\alpha'\rangle\}$, averaged over the thermal distribution in the $\{\alpha\}$ manifold and sum over all final states α', and the rate is

$$k_{f\leftarrow i} = \frac{2\pi}{\hbar}\sum_\alpha \frac{e^{-\beta\varepsilon_\alpha}}{Q}\sum_{\alpha'}|(\hat{H}_{SB})_{i\alpha,f\alpha'}|^2\delta(E_i + \varepsilon_\alpha - E_f - \varepsilon_{\alpha'}) \tag{13.15}$$

Q is the bath partition function, $Q = \sum_\alpha e^{-\beta\varepsilon_\alpha}$. Following the procedure that leads from Eq. (12.34) or (12.37) to (12.44) we can recast this rate expression in the time correlation form

$$k_{f\leftarrow i} = \frac{1}{\hbar^2}\int_{-\infty}^{\infty} dt e^{iE_{i,f}t/\hbar}\langle(\hat{H}_{SB})_{i,f}(t)(\hat{H}_{SB})_{f,i}\rangle_T; \qquad E_{i,f} = E_i - E_f \tag{13.16}$$

Here the matrix elements in the molecular subspace, for example $(\hat{H}_{SB})_{i,f}$ are operators in the bath space, $(\hat{H}_{SB})_{i,f}(t) = \exp(i\hat{H}_B t/\hbar)(\hat{H}_{SB})_{i,f}\exp(-i\hat{H}_B t/\hbar)$ and $\langle\ldots\rangle_T = \mathrm{Tr}[e^{-\beta\hat{H}_B}\ldots]/\mathrm{Tr}[e^{-\beta\hat{H}_B}]$, where Tr denotes a trace over the eigenstates of $\hat{H}_B$. At this point expression (13.16) is general. In our case $\hat{H}_{SB} = \hat{F}\hat{q}$ where $\hat{F}$ is an operator in the bath sub-space and $\hat{q}$ is the coordinate operator of our oscillator. This implies that $(\hat{H}_{SB})_{i,f} = q_{i,f}\hat{F}$ and consequently

$$k_{f\leftarrow i} = \frac{1}{\hbar^2}|q_{i,f}|^2\int_{-\infty}^{\infty} dt e^{i\omega_{i,f}t}\langle\hat{F}(t)\hat{F}(0)\rangle_T \tag{13.17}$$

with $\omega_{i,f} = (E_i - E_f)/\hbar$. The relaxation rate is seen to be given by the $\omega_{i,f}$ Fourier component of the time correlation function of the force exerted by the environment on the vibrating coordinate when frozen at its equilibrium value.

Note that, using Eq. (12.45) (or applying Eq. (6.75)) we find that this result satisfies detailed balance, that is, $k_{f\leftarrow i} = e^{\beta\hbar\omega_{i,f}}k_{i\leftarrow f}$. Furthermore, this result confirms the assertion made above that for large $\omega_{i,f}$ only close encounters, for which the time-dependent force experienced by the vibrating coordinate has appreciable Fourier component in the corresponding frequency, contribute to the relaxation.

For a harmonic oscillator, a transition between levels $|n\rangle$ and $|n-1\rangle$ involves the matrix element $|q_{n,n-1}|^2 = (\hbar/2m\omega)n$, so that Eq. (13.17) becomes

$$k_{n-1\leftarrow n} = \frac{n}{2m\hbar\omega}\int_{-\infty}^{\infty} dte^{i\omega t}\langle\hat{F}(t)\hat{F}(0)\rangle_{\mathrm{T}} \equiv k_{\downarrow}(\omega)\, n \tag{13.18a}$$

$$k_{n\leftarrow n-1} = \frac{n}{2m\hbar\omega}\int_{-\infty}^{\infty} dte^{-i\omega t}\langle\hat{F}(t)\hat{F}(0)\rangle_{\mathrm{T}} \equiv k_{\uparrow}(\omega)n \tag{13.18b}$$

with $k_{\uparrow}(\omega) = k_{\downarrow}(\omega)\exp(-\beta\hbar\omega)$.

Problem 13.1. Use the detailed balance condition (6.75) to show that k in Eq. (13.18) is given by

$$k_{\downarrow}(\omega) = \frac{1}{2m\hbar\omega}(1+e^{-\beta\hbar\omega})^{-1}\int_{-\infty}^{\infty} dte^{i\omega t}\langle\{\hat{F}(t),\hat{F}(0)\}\rangle_{\mathrm{T}} \tag{13.19}$$

where the *anti-commutator* is defined by $\{\hat{F}(t),\hat{F}(0)\} \equiv \hat{F}(t)\hat{F}(0) + \hat{F}(0)\hat{F}(t)$

The fact that symmetrized time correlation functions of the form $\langle\{\hat{F}(t),\hat{F}(0)\}\rangle_{\mathrm{T}}$ are insensitive to the order of placing the operators $\hat{F}$ and $\hat{F}(t)$ suggests that they can be approximated by their classical counterparts,

$$\langle\{\hat{F}(t),\hat{F}(0)\}\rangle_{\mathrm{q}} \to 2\langle F(t)F(0)\rangle_{\mathrm{c}} \tag{13.20}$$

This leads to the approximation

$$\begin{aligned} k_{\downarrow\mathrm{sc}}(\omega) &= \frac{1}{m\hbar\omega}(1+e^{-\beta\hbar\omega})^{-1}\int_{-\infty}^{\infty} dte^{i\omega t}\langle F(t)F(0)\rangle_{\mathrm{c}} \\ &= \frac{2}{m\hbar\omega}(1+e^{-\beta\hbar\omega})^{-1}\int_{0}^{\infty} dt\cos(\omega t)\langle F(t)F(0)\rangle_{\mathrm{c}} \end{aligned} \tag{13.21a}$$

$$k_{\uparrow\mathrm{sc}}(\omega) = \frac{2}{m\hbar\omega}(1+e^{\beta\hbar\omega})^{-1}\int_{0}^{\infty} dt\cos(\omega t)\langle F(t)F(0)\rangle_{\mathrm{c}} \tag{13.21b}$$

In (13.20) and (13.21) the subscripts c and q correspond respectively to the classical and the quantum time correlation functions and k_{sc} denotes a "semi-classical" approximation. We refer to the form (13.21) as semiclassical because it carries aspects of the quantum thermal distribution even though the quantum time correlation function was replaced by its classical counterpart. On the face of it this approximation seems to make sense because one could anticipate that (1) the time correlation functions involved decay to zero on a short timescale (of order ~ 1 ps that characterizes solvent configuration variations), and (2) classical mechanics may provide a reasonable short-time approximation to quantum time correlation functions. Furthermore note that the rates in (13.21) satisfy detailed balance.

Note that in continuation of the same line of thought we may take the high temperature ($\beta \to 0$) limit of (13.21) and attempt to regard it as a classical approximation. We see, however, that the result (which no longer satisfies detailed balance)

$$k_{\downarrow\text{c}} = k_{\uparrow\text{c}} \equiv \lim_{k_B T/(\hbar\omega)\to\infty} k_{\text{sc}} = \frac{1}{m\hbar\omega}\int_0^\infty dt\cos(\omega t)\langle F(t)F(0)\rangle_{\text{c}} \tag{13.22}$$

is not classical in the sense that it depends on $\hbar$. This should not come as a surprise: The transition rate between two *levels* is a quantum mechanical concept! Note that the related rate of energy change, $k_{\text{c}} \times \hbar\omega$, does not depend on $\hbar$. Energy relaxation, however, should be discussed more carefully, as described below.

Consider now the overall relaxation process. As was done in Section 8.3.3, this process can be represented by a master equation for the probability P_n to be in quantum state n of the oscillator,

$$\frac{dP_n}{dt} = \frac{\partial P_n}{\partial t} = k_{n,n+1}P_{n+1} + k_{n,n-1}P_{n-1} - (k_{n+1,n} + k_{n-1,n})P_n \tag{13.23}$$

where $k_{i,j} = k_{i\leftarrow j}$. For the harmonic oscillator we found $k_{n-1\leftarrow n} = nk_\downarrow$ and $k_{n\leftarrow n-1} = nk_\downarrow e^{-\beta\hbar\omega}$, hence Eq. (13.23) has the same form as (8.98) and can be solved using the procedure described in Section 8.3.3. In particular, multiplying both sides of (13.23) by n and summing over all n yields

$$\frac{d\langle n\rangle}{dt} = -k_\downarrow(1 - e^{-\beta\hbar\omega})(\langle n\rangle - \langle n\rangle_{\text{eq}}) \tag{13.24}$$

where

$$\langle n\rangle_{\text{eq}} = \frac{1}{e^{\beta\hbar\omega} - 1} \tag{13.25}$$

Since the oscillator energy is $E = n\hbar\omega$, it follows from (13.24) that the rate coefficient for energy relaxation, k_E, is

$$k_E = k_{\downarrow}(1 - e^{-\beta\hbar\omega}) = \frac{1}{m\hbar\omega}\tanh(\beta\hbar\omega/2)\int_0^{\infty} dt\cos(\omega t)\langle\{\hat{F}(t), \hat{F}(0)\}\rangle_{\mathrm{T}} \tag{13.26}$$

Again, using the correspondence (13.20), we may consider the semiclassical and the classical approximations

$$k_{E,\mathrm{sc}} = \frac{2}{m\hbar\omega}\tanh(\beta\hbar\omega/2)\int_0^{\infty} dt\cos(\omega t)\langle F(t)F(0)\rangle_{\mathrm{c}} \tag{13.27}$$

$$k_{E,\mathrm{c}} = \lim_{k_BT/(\hbar\omega)\to\infty} k_{E,\mathrm{sc}} = \frac{\beta}{m}\int_0^{\infty} dt\cos(\omega t)\langle F(t)F(0)\rangle_{\mathrm{c}} \tag{13.28}$$

Next we turn to the evaluation of these expressions within specific interaction models.

13.4 Evaluation of vibrational relaxation rates

In this Section we apply the general formalism developed in Section 13.3 together with the interaction models discussed in Section 13.2 in order to derive explicit expressions for the vibrational energy relaxation rate. Our aim is to identify the molecular and solvent factors that determine the rate. We will start by analyzing the implications of a linear coupling model, than move on to study more realistic nonlinear interactions.

13.4.1 The bilinear interaction model

Consider first the model described by Eqs (13.6), (13.7), (13.10), and (13.11) where the harmonic oscillator under study is coupled bi-linearly to the harmonic bath. The relevant correlation functions $\langle\{\hat{F}(t), \hat{F}(0)\}\rangle$ and $\langle F(t)F(0)\rangle_{\mathrm{c}}$ should be calculated with $\hat{F} = \sum_j A_j\hat{u}_j$ and its classical counterpart, where u_j are coordinates of the bath normal modes. Such correlation functions were calculated in Section 6.5.

Problem 13.2. For a harmonic thermal bath show that

$$\int_{-\infty}^{\infty} dt e^{i\omega t}\langle \hat{F}(t)\hat{F}(0)\rangle_{\mathrm{q}} = \frac{\pi\hbar}{\omega(1-e^{-\beta\hbar\omega})}A^2(\omega)g(\omega) \tag{13.29a}$$

$$\int_{-\infty}^{\infty} dt e^{i\omega t}\langle\{\hat{F}(t),\hat{F}(0)\}\rangle_{\mathrm{q}} = 2\int_{0}^{\infty} dt \cos(\omega t)\langle\{\hat{F}(t),\hat{F}(0)\}\rangle_{\mathrm{q}}$$

$$= \frac{\pi\hbar}{\omega\tanh(\beta\hbar\omega/2)}A^2(\omega)g(\omega) \tag{13.29b}$$

while

$$\int_{-\infty}^{\infty} dt e^{i\omega t}\langle F(t)F(0)\rangle_{\mathrm{c}} = 2\int_{0}^{\infty} dt\cos(\omega t)\langle F(t)F(0)\rangle_{\mathrm{c}} = \frac{\pi k_B T}{\omega^2}A^2(\omega)g(\omega) \tag{13.30}$$

where the coupling density is defined by (cf. Eq. (6.91)) $A^2(\omega) \equiv \sum_k A_k^2\delta(\omega-\omega_k)/\sum_k\delta(\omega-\omega_k)$, and where $g(\omega)$ is the density of bath modes.

Solution: Using the fact that different normal modes are uncorrelated, the classical correlation function is computed according to

$$\langle F(t)F(0)\rangle = \sum_j A_j^2\langle u_j(t)u_j(0)\rangle \tag{13.31}$$

Using $u_j(t) = u_j(0)\cos(\omega_j t) + [\dot{u}_j(0)/\omega_j]\sin(\omega_j t)$, and $\langle u_j^2\rangle = k_B T/\omega^2$, $\langle u_j\dot{u}_j\rangle = 0$ we get (13.30). The quantum correlation function is also given by the form (13.31), with the classical u_j replaced by

$$\hat{u}_j(t) = \left(\frac{\hbar}{2\omega_j}\right)^{1/2}(\hat{a}_j^\dagger e^{i\omega_j t} + \hat{a}_j e^{-i\omega_j t}) \tag{13.32}$$

The position correlation function is

$$\langle\hat{u}_j(t)\hat{u}_j(0)\rangle = \frac{\hbar}{2\omega_j}(\langle n_j\rangle e^{i\omega_j t} + (\langle n_j\rangle + 1)e^{-i\omega_j t});\quad \langle n_j\rangle = [\exp(\beta\hbar\omega_j)-1]^{-1} \tag{13.33}$$

which leads to (13.29).

Note that the bath spectral density associated with the coupling coefficient A was defined as (cf. Eqs (6.90), (6.92))

$$J_A(\omega) = \frac{\pi g(\omega) A^2(\omega)}{2\omega} \tag{13.34}$$

Equations (13.26) and (13.29b) now provide an exact result, within the bilinear coupling model and the weak coupling theory that leads to the golden rule rate expression, for the vibrational energy relaxation rate. This result is expressed in terms of the oscillator mass m and frequency ω and in terms of properties of the bath and the molecule–bath coupling expressed by the coupling density $A^2(\omega)g(\omega)$ at the oscillator frequency

$$k_E = \frac{\pi A^2(\omega) g(\omega)}{2m\omega^2} \tag{13.35}$$

This rate has two remarkable properties: First, it does not depend on the temperature and second, it is proportional to the bath density of modes $g(\omega)$ and therefore vanishes when the oscillator frequency is larger than the bath cutoff frequency (Debye frequency). Both features were already encountered (Eq. (9.57)) in a somewhat simpler vibrational relaxation model based on bilinear coupling and the rotating wave approximation. Note that temperature independence is a property of the energy relaxation rate obtained in this model. The inter-level transition rate, Eq. (13.19), satisfies (cf. Eq. (13.26)) $k_\downarrow = k_E(1 - e^{-\beta\hbar\omega})^{-1}$ and does depend on temperature.

Equation (13.35) is the exact golden-rule rate expression for the bilinear coupling model. For more realistic interaction models such analytical results cannot be obtained and we often resort to numerical simulations (see Section 13.6). Because classical correlation functions are much easier to calculate than their quantum counterparts, it is of interest to compare the approximate rate $k_{E,\mathrm{sc}}$, Eq. (13.27), with the exact result k_E. To this end it is useful to define the *quantum correction factor*

$$\xi = \frac{\int_0^\infty dt \cos(\omega t)\langle\{\hat{F}(t), \hat{F}(0)\}\rangle_{\mathrm{q}}}{\int_0^\infty dt \cos(\omega t)\langle\{F(t), F(0)\}\rangle_{\mathrm{c}}} = \frac{(1 + e^{-\beta\hbar\omega})\int_{-\infty}^{\infty} dt e^{i\omega t}\langle\hat{F}(t)\hat{F}(0)\rangle_{\mathrm{q}}}{2\int_{-\infty}^{\infty} dt e^{i\omega t}\langle F(t)F(0)\rangle_{\mathrm{c}}} \tag{13.36}$$

If we had a theory for this factor, we could calculate quantum relaxation rates using computed classical correlation functions. For the bilinear model (13.11) we get, using (13.29) and (13.30)

$$\xi = \frac{\hbar\omega}{2k_BT}\frac{1}{\tanh(\beta\hbar\omega/2)} \tag{13.37}$$

One could be tempted to use this factor also for general interaction models, that is to assume that the following relationship

$$\int_{-\infty}^{\infty} dte^{i\omega t}\langle \hat{F}(t)\hat{F}(0)\rangle_q \approx \frac{\beta\hbar\omega}{(1-e^{-\beta\hbar\omega})}\int_{-\infty}^{\infty} dte^{i\omega t}\langle F(t)F(0)\rangle_c \tag{13.38}$$

holds approximately also in more general cases. We will see however, Eq. (13.68) that this ansatz fails badly at low temperatures in the interesting cases where ω is considerably larger than the medium's Debye frequency.

Problem 13.3. Show that for the bilinear coupling model, the classical limit rate, and the exact quantum result are identical, and are both given by[6]

$$k_E = k_{E,c} = \frac{1}{mk_BT}\int_0^{\infty} dt\cos(\omega t)\langle F(t)F(0)\rangle_c \tag{13.39}$$

Equation (13.39) implies that *in the bilinear coupling*, the vibrational energy relaxation rate for a "quantum harmonic oscillator in a quantum harmonic bath" is the same as that obtained from a fully classical calculation ("a classical harmonic oscillator in a classical harmonic bath"). In contrast, the semiclassical approximation (13.27) gives an error that diverges in the limit $T \to 0$. Again, this result is specific to the bilinear coupling model and fails in models where the rate is dominated by the nonlinear part of the impurity–host interaction.

13.4.2 Nonlinear interaction models

We have seen that vibrational relaxation rates can be evaluated analytically for the simple model of a harmonic oscillator coupled linearly to a harmonic bath. Such model may represent a reasonable approximation to physical reality if the frequency of the oscillator under study, that is the mode that can be excited and monitored, is well embedded within the spectrum of bath modes. However, many processes of interest involve molecular vibrations whose frequencies are higher than the solvent Debye frequency. In this case the linear coupling rate (13.35) vanishes, reflecting the fact that in a linear coupling model relaxation cannot take place in the absence of modes that can absorb the dissipated energy. The harmonic Hamiltonian

[6] J. S. Bader and B. J. Berne, J. Chem. Phys. **100**, 8359 (1994).

described by Eqs (13.6), (13.7) and (13.9)–(13.11) is, however, an approximation based on expanding a realistic potential up to quadratic order about the minimum energy configuration. Relaxation of a high-frequency impurity oscillator can only take place by exploring regions of the potential surface far from the equilibrium configuration. This observation stands behind the choice of the interaction form (13.12) that leads to (13.13).

With this understanding, we can continue in two ways. First we can use the interaction (13.13) in the golden-rule rate expression—approach we take in Section 13.4.4. Alternatively, we may use the arguments that (1) transitions between states of the high-frequency impurity oscillator can occur with appreciable probability only during close encounters with a bath atom (see footnote 3), and (2) during such encounters, the interactions of the oscillators with other bath atoms is relatively small and can be disregarded, in order to view such encounters as binary collision events. This approach is explored in the next section.

13.4.3 The independent binary collision (IBC) model

Consider first the assumption that those oscillator–bath encounters that lead to vibrational relaxation can be described as *uncorrelated* binary collisions.[7] It is sufficient to focus on the rate expression (13.17)

$$k_{f\leftarrow i} = \frac{|q_{i,f}|^2}{\hbar^2} \int_{-\infty}^{\infty} dt e^{i\omega_{i,f}t} \langle \hat{F}(t)\hat{F}(0)\rangle_T \tag{13.40}$$

which is the basis to all other rates derived in Section 13.3. We proceed by writing the force operator $\hat{F}$ as a sum of contributions from all bath atoms

$$\hat{F} = \sum_\alpha \hat{f}_\alpha \tag{13.41}$$

the uncorrelated nature of successive encounters implies that in the force autocorrelation function

$$\langle \hat{F}(t)\hat{F}(0)\rangle = \sum_\alpha \langle \hat{f}_\alpha(t)\hat{f}_\alpha(0)\rangle + \sum_\alpha \sum_{\alpha'\neq\alpha} \langle \hat{f}_\alpha(t)\hat{f}_{\alpha'}(0)\rangle, \tag{13.42}$$

we can disregard the second term. The resulting autocorrelation function is now expressed as a sum of contributions from individual bath atoms. Next we want to

[7] Further reading: J. Chesnoy and G. M. Gale, Adv. Chem. Phys. **70**, 298–355 (1988).

express each such contribution as resulting from a single collision event. This is based on the realization that repeated collisions with the same atoms, if uncorrelated, are no different than collisions with different atoms and can be treated as such as long as correct counting of the overall collision rate is made. Also, for simplicity we will use henceforth classical mechanics language but keep in mind that the correlation functions below should be treated quantum mechanically if needed. To calculate the correlation function $\langle f_\alpha(t) f_\alpha(0)\rangle$ we write the classical force $f_\alpha(t)$ in the form

$$f_\alpha(t) = f_\alpha(\mathbf{r}_\alpha(t)) = f_\alpha(t; \mathbf{r}_\alpha(0), \mathbf{p}_\alpha(0)) \tag{13.43}$$

where $\mathbf{r}_\alpha(t)$ is the "collision coordinate," that is, position at time t of the colliding bath atom relative to the relaxing oscillator, and $\mathbf{p}_\alpha(t)$ is the associated momentum. The initial values $\mathbf{r}_\alpha(0)$ and $\mathbf{p}_\alpha(0)$ are to be sampled from a Boltzmann distribution. The collision event is described here loosely: In actual application more than one coordinate may be needed to describe it. The important feature in the form (13.43) is that the time-dependent force resulting from interaction with bath particle α is expressed in terms of an isolated collision, that is, the magnitude of the force exerted by particle α at time t depends only on its position and momentum relative to the target. Therefore

$$\begin{aligned}\langle F(t)F(0)\rangle &= \sum_\alpha \langle f_\alpha(t; \mathbf{r}_\alpha, \mathbf{p}_\alpha) f_\alpha(\mathbf{r}_\alpha)\rangle \\ &= \rho \int d^3 r g(\mathbf{r}) \langle f(t; \mathbf{r}, \mathbf{p}) f(\mathbf{r})\rangle_{\mathrm{p}}\end{aligned} \tag{13.44}$$

and

$$k_{f\leftarrow i} = \rho \frac{|q_{i,f}|^2}{\hbar^2} \int d^3 r g(\mathbf{r}) \int_{-\infty}^{\infty} dt e^{i\omega_{i,f} t} \langle f(t; \mathbf{r}, \mathbf{p}) f(\mathbf{r})\rangle_{\mathrm{p}} \equiv \rho \int d^3 r g(\mathbf{r}) B_{i,f}(\mathbf{r}, T) \tag{13.45a}$$

$$B_{i,f}(\mathbf{r}, T) = \frac{|q_{i,f}|^2}{\hbar^2} \int_{-\infty}^{\infty} dt e^{i\omega_{i,f} t} \langle f(t; \mathbf{r}, \mathbf{p}) f(\mathbf{r})\rangle_{\mathrm{p}} \tag{13.45b}$$

In the second line of (13.44) and in (13.45) $f(\mathbf{r})$ denotes the force between the relaxing coordinate and a single bath atom at distance $\mathbf{r}$ away and $f(t; \mathbf{r}, p)$ is the same force, a time t later, given that the initial momentum of the bath atom was $\mathbf{p}$. All bath atoms are assumed identical and their bulk density is denoted ρ. $g(\mathbf{r})$ is the impurity–solvent pair distribution function, $\langle\,\rangle_p$ denotes thermal average over initial momenta and T is the temperature. We have used the fact (see Section 5.3) that the local density of bath atoms near the relaxing impurity is $\rho g(\mathbf{r})$.

The important feature in the final result (13.45) is the fact that all the needed dynamical information is associated with the time course of a single collision event. To calculate the correlation function that appears here it is sufficient to consider a single binary collision with thermal initial conditions. The result is given in terms of the function B, a two-body property that depends only on the relative position of the particles (the initial configuration for the collision event) and the temperature. The host structure as determined by the N-body force enters only through the configurational average that involves the pair distribution $g(\mathbf{r})$.

Within the IBC approximation the result (13.45) is general. It is particularly useful for interpretation of vibrational relaxation in pure liquids, because it can be related immediately to measurement of the same property in the corresponding low pressure gas phases. To show this define

$$y(\mathbf{r}) = \frac{g(\mathbf{r})}{g_{\text{gas}}(\mathbf{r})} \tag{13.46}$$

Then

$$k_{f\leftarrow i} = \rho \int d^3r g_{\text{gas}}(\mathbf{r}) B_{i,f}(\mathbf{r}, T) y(\mathbf{r}) \tag{13.47}$$

The advantage of this form is that $g_{\text{gas}}(\mathbf{r})$ already contains the short-range structural features of $g(\mathbf{r})$, therefore $y(\mathbf{r})$ depends relatively weakly on $\mathbf{r}$. We may define a mean collision distance $\mathbf{R}^*$ by

$$\int d^3r g_{\text{gas}}(\mathbf{r}) B_{i,f}(\mathbf{r}, T)(\mathbf{r} - \mathbf{R}^*) = 0 \tag{13.48}$$

and expand $y(\mathbf{r})$s about $\mathbf{R}^*$

$$y(\mathbf{r}) = y(\mathbf{R}^*) + (\mathbf{r} - \mathbf{R}^*) y'(\mathbf{R}^*) \tag{13.49}$$

to get[8]

$$k_{f\leftarrow i} = \rho y(\mathbf{R}^*) k_{f\leftarrow i,\text{gas}} \tag{13.50}$$

[8] P. K. Davis and I. Oppenheim, J. Chem. Phys. **57**, 507 (1972).

Here k_{gas} is the gas phase VR rate

$$k_{f\leftarrow i,\text{gas}} = \int d^3 r g_{\text{gas}}(\mathbf{r}) B_{i,f}(\mathbf{r}, T) \cong \int d^3 r e^{-\beta u(\mathbf{r})} B_{i,f}(\mathbf{r}, T) \tag{13.51}$$

where $u(\mathbf{r})$ is the two-body potential between the relaxing oscillator and the bath particle and where we have used Eq. (5.54). The important outcome of this analysis lies in the fact that k_{gas} is relatively easy to compute or to simulate, and also in the fact that much of the important physics of the relaxation process is already contained in k_{gas}: For example, the ratio $k/(k_{\text{gas}}\rho)$ is of the order of $0.5 \ldots 2$ for many simple fluids (see footnote 7), while k itself varies over several orders of magnitudes between these fluids.

13.5 Multi-phonon theory of vibrational relaxation

A different approach to the calculation of the rates $k_{f\to i}$, Eq. (13.40), and the related rates $k_\uparrow$ and k_E is to evaluate the force autocorrelation function associated with the interaction (13.13) and (13.14) and the corresponding force

$$\hat{F} = Be^{-\alpha\delta\hat{x}} = Be^{-\sum_k (\alpha_k a_k + \alpha_k^* a_k^\dagger)} \tag{13.52}$$

In the second equality we have expanded the coordinate deviation δx in normal modes coordinates, and expressed the latter using raising and lowering operators. The coefficients α_k are defined accordingly and are assumed known. They contain the parameter α, the coefficients of the normal mode expansion and the transformation to raising/lowering operator representation. Note that the inverse square root of the volume Ω of the overall system enters in the expansion of a local position coordinate in normal modes scales,[9] hence the coefficients α_k scale like $\Omega^{-1/2}$.

Recall that the interaction form (13.52) was chosen to express the close encounter nature of a molecule–bath interaction needed to affect a transition in which the molecular energy change is much larger than $\hbar\omega_{\text{D}}$ where ω_{D} is the Debye cutoff frequency of the thermal environment. This energy mismatch implies that many bath phonons are generated in such transition, as will be indeed seen below.

Using (13.52) in (13.40) the transition rate is obtained in the form

$$k_{f\leftarrow i} = \frac{|q_{i,f}|^2}{\hbar^2}|B|^2 \int_{-\infty}^{\infty} dt e^{i\omega_{i,f}t} \prod_k \langle \hat{b}_k(t)\hat{b}_k(0)\rangle_{\text{T}} \tag{13.53}$$

[9] As a one-dimensional example this was seen in Eq. (4.30) where the transformation coefficients contain the inverse square root of the total number N of lattice atoms which is proportional to the lattice volume.

where

$$\hat{b}_k(t) = e^{-[\alpha_k \hat{a}_k(t) + \alpha_k^* \hat{a}_k^\dagger(t)]} \tag{13.54}$$

with

$$\begin{aligned} \hat{a}_k(t) &= e^{i\hat{H}_{\mathrm{B}}t/\hbar} a_k e^{-i\hat{H}_{\mathrm{B}}t/\hbar} = a_k e^{-i\omega_k t} \\ \hat{a}_k^\dagger(t) &= \hat{a}_k^\dagger e^{i\omega_k t} \end{aligned} \tag{13.55}$$

The structure of Eqs. (13.53)–(13.55) is similar to that encountered in the evaluation of rates in the spin–boson problem, Sections 12-4.2 and 12-4.3 (see Eqs (12.44), (12.49)–(12.50)) and our evaluation proceeds in the same way. We need to calculate the thermal average

$$C_k \equiv \langle e^{-(\alpha_k \hat{a}_k e^{-i\omega_k t} + \alpha_k^* \hat{a}_k^\dagger e^{i\omega_k t})} e^{-(\alpha_k \hat{a}_k + \alpha_k^* \hat{a}_k^\dagger)} \rangle_{\mathrm{T}} \tag{13.56}$$

This will be accomplished in two stages described in Problem 12-5 and Eq. (12.53). First we bring all operators onto a single exponent. This leads to

$$C_k = \langle e^{-\alpha_k \hat{a}_k (1+e^{-i\omega_k t}) - \alpha_k^* \hat{a}_k^\dagger (1+e^{i\omega_k t})} e^{-i|\alpha_k|^2 \sin \omega_k t} \rangle \tag{13.57}$$

Second we use the *Bloch theorem* (cf. Eq. (10.38)) that states that for a harmonic system, if $\hat{A}$ is linear in the coordinate and momentum operators, then $\langle e^{\hat{A}} \rangle_{\mathrm{T}} = \exp[(1/2)\langle A^2 \rangle_{\mathrm{T}}]$. In our case

$$\hat{A} = -\alpha_k (1 + e^{-i\omega_k t}) \hat{a}_k - \alpha_k^* (1 + e^{i\omega_k t}) \hat{a}_k^\dagger \tag{13.58}$$

so

$$\begin{aligned} \langle \hat{A}^2 \rangle_{\mathrm{T}} &= |\alpha_k|^2 |1 + e^{i\omega_k t}|^2 (\langle \hat{a}_k \hat{a}_k^\dagger \rangle_{\mathrm{T}} + \langle \hat{a}_k^\dagger \hat{a}_k \rangle_{\mathrm{T}}) \\ &= |\alpha_k|^2 |1 + e^{i\omega_k t}|^2 (2n_k + 1) \end{aligned} \tag{13.59}$$

where $n_k = [\exp(\beta\hbar\omega_k) - 1]^{-1}$ is the phonon thermal occupation number. Using Eqs. (13.58) and (13.59) in (13.57) leads to

$$C_k = e^{|\alpha_k|^2 (2n_k+1)} e^{|\alpha_k|^2 ([n_k+1] e^{-i\omega_k t} + n_k e^{i\omega_k t})} \tag{13.60}$$

Inserting in Eq. (13.53) we find the relaxation rate

$$k_{f\leftarrow i} = \frac{|q_{i,f}|^2}{\hbar^2} |B|^2 e^{\sum_k |\alpha_k|^2 (2n_k+1)} \int_{-\infty}^{\infty} dt e^{i\omega_{i,f} t + \sum_k |\alpha_k|^2 [(n_k+1) e^{-i\omega_k t} + n_k e^{i\omega_k t}]} \tag{13.61}$$

The result (13.61) has the characteristic form of a multiphonon relaxation rate that was already encountered in Chapter 12 (compare Eq. (12.55)). Note the appearance in this result of the important parameters that determine this rate: The coupling matrix element $q_{i,f}$ between the two system states, the parameters α_k that determine the strength of the system–bath coupling, the energy gap $\omega_{i,f}$ between the two levels for which the transition is considered, the phonon frequencies and their thermal occupation numbers. Several more points are noteworthy:

1. The mathematical derivation of Eqs (13.61) and (12.55) is identical. The physical origins of the models that were used in these cases were different: The coupling parameters λ that appear in (12.55) express the strength of electron–phonon coupling that appear in Eqs (12.8), (12.16b) and were shown to reflect parallel horizontal shifts of Born–Oppenheimer potential surfaces between different electronic states. The coupling parameters α in (13.61) originate from the inverse length that characterizes the range of the exponential repulsion (13.52) associated with the molecule–bath interaction.

2. In Eq. (13.61) the coupling parameters α_k appear in sums of the form $\sum_k |\alpha_k|^2$ over the bath normal modes. This sum is equivalent to the integral form $\int d\omega\, g(\omega)\alpha^2(\omega)$. As was explained below Eq (13.52), $\alpha(\omega) \sim \Omega^{-1/2}$ where Ω is the overall volume (essentially the volume of the bath). Since the density of bath modes is linear in Ω (see, for example, Eq. (4.40) or (4.47)) the final result does not depend on this volume, as expected.

3. As discussed before with respect to Eq. (12.55), also the rate expression (13.61) incorporates the restriction of energy conservation: To make the $i \to f$ transition we need to generate or eliminate enough phonons with the correct combinations of numbers and frequencies to exactly account for this energy transfer. To see this consider one of the terms in the Fourier integral of Eq. (13.61) that is obtained by making a Taylor expansion of the exponent:

$$\frac{1}{N!}\int_{-\infty}^{\infty} dt e^{i\omega_{i,f}t}\left[\sum_k |\alpha_k|^2\{(n_k+1)e^{-i\omega_k t} + n_k e^{i\omega_k t}\}\right]^N \tag{13.62}$$

Following this by a multinomial expansion of $[]^N$ we see that this (and all other terms) are combinations of terms of the form $\delta(\omega_{i,f} - \sum_k l_k\omega_k + \sum_{k'} l_{k'}\omega_{k'})$, where l_k and $l_{k'}$ are integers—all possible energy conserving combinations of phonons created and destroyed. These energy conserving terms are weighted by factors containing powers of n_k and (n_k+1) because squared matrix elements for phonon creation or annihilation operators are proportional to these factors. The temperature dependence of the vibrational relaxation rate results from the presence of these terms.

4. A more transparent representation of the temperature dependence can be obtained in simple models. Consider for example an Einstein-type model where the phonon spectrum is represented by a single frequency ω_a. The rate is loosely written in this case in the form

$$k_{f\leftarrow i} = \frac{|q_{i,f}|^2}{\hbar^2}|B|^2 e^{|\alpha|^2(2n_a+1)} \int_{-\infty}^{\infty} dt e^{i\omega_{i,f}t+|\alpha|^2(n_a+1)e^{-i\omega_a t}+|\alpha|^2 n_a e^{i\omega_a t}} \tag{13.63}$$

with $n_a = [\exp(\beta\hbar\omega_a) - 1]^{-1}$. For convenience we specify henceforth to downwards transitions, whereupon $\omega_{i,f} > 0$. The Fourier integral can again be evaluated by making a Taylor expansion followed by a binomial expansion:

$$\begin{aligned} I &\equiv \int_{-\infty}^{\infty} dt e^{i\omega_{i,f}t+|\alpha|^2(n_a+1)e^{-i\omega_a t}+|\alpha|^2 n_a e^{i\omega_a t}} \\ &= \int_{-\infty}^{\infty} dt e^{i\omega_{i,f}t} \sum_{l=0}^{\infty} \frac{|\alpha|^{2l}}{l!}((n_a+1)e^{-i\omega_a t} + n_a e^{i\omega_a t})^l \\ &= 2\pi \sum_{l=0}^{\infty} \frac{|\alpha|^{2l}}{l!} \sum_{s=0}^{l} \frac{l!}{(l-s)!s!}(n_a+1)^s n_a^{l-s} \delta(\omega_{i,f} - (2s-l)\omega_a) \end{aligned} \tag{13.64}$$

Consider now the weak coupling limit where $|\alpha|^2$ is very small so that: (1) the exponent that multiplies the integral I in (13.63) can be disregarded, and (2) only the term with the smallest l that is compatible with energy conservation in (13.64) contributes. This is the term

$$s = l = \frac{\omega_{i,f}}{\omega_a} \tag{13.65}$$

that yields

$$k_{\downarrow} \sim \frac{|V_{i,f}|^2}{\hbar^2} \frac{|\alpha|^{2(\omega_{i,f}/\omega_a)}}{(\omega_{i,f}/\omega_a)!}(n_a+1)^{(\omega_{i,f}/\omega_a)} \tag{13.66}$$

where $V_{if} = Bq_{i,f}$. At low temperature this rate becomes temperature independent, while as $(k_B T/\hbar\omega_a) \to \infty$ it diverges like $(k_B T/\hbar\omega_a)^{(\omega_{i,f}/\omega_a)}$. The ratio $(\omega_{i,f}/\omega_a)$ is the number of phonons participating in the transition, and the onset of temperature dependence as T increases is at $k_B T \sim \hbar\omega_a$. Beyond this threshold, the rate increases very rapidly for large $(\omega_{i,f}/\omega_a)$.

5. As in the weak coupling limit of the rate (12.55), analyzed in Section 12.5.3, the weak coupling limit (13.66) of the vibrational relaxation rate also has the characteristic form of an energy-gap law: It decreases exponentially, like $|\alpha|^{2(\omega_{i,f}/\omega_a)}$, with increasing dimensionless energy gap $\omega_{i,f}/\omega_a$.

6. For a harmonic oscillator of frequency ω, the ratio $k_\downarrow(T)/k_\downarrow(T \to \infty) = [(n_a(T)+1)(\hbar\omega_a/k_BT)]^{\omega/\omega_a}$ is, according to (13.19) and (13.22), equal to $2(1+e^{-\beta\hbar\omega})^{-1}\zeta$ where ζ is the quantum correction factors defined in (13.36). Using (13.66) this implies that for the exponential repulsion model in the weak coupling approximation

$$\zeta = \frac{1}{2}(1+e^{-\beta\hbar\omega})\left[\frac{\hbar\omega_a}{k_BT}(n_a+1)\right]^{(\omega/\omega_a)} \tag{13.67}$$

In the low temperature limit this becomes

$$\zeta \overset{T\to 0}{\longrightarrow} \frac{1}{2}\left(\frac{\hbar\omega_a}{k_BT}\right)^{\omega/\omega_a} \tag{13.68}$$

indicating the failure of the ansatz (13.38).

Figure 13.2 shows experimental relaxation data for different vibrational levels of the ground electronic state of Oxygen molecules embedded in a solid Argon

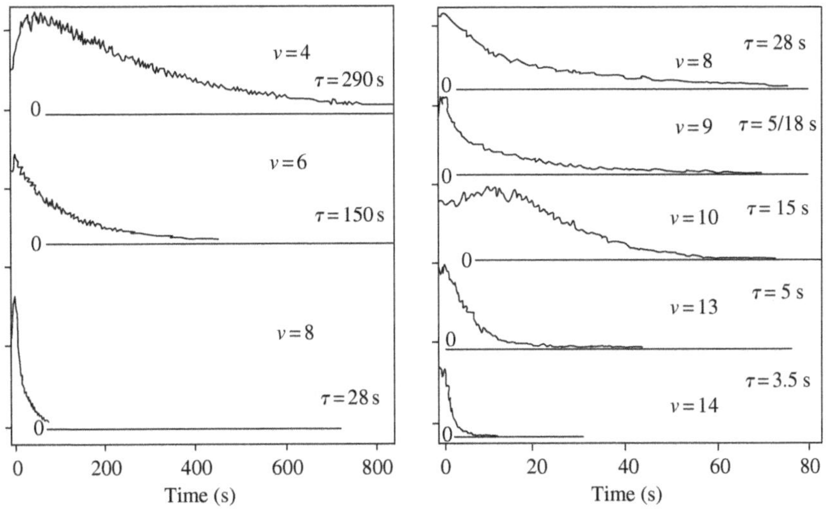

FIG. 13.2 The relaxation of different vibrational levels of the ground electronic state of $^{16}O_2$ in a solid Ar matrix. Analysis of these results indicates that the relaxation of the $v < 9$ levels is dominated by radiative decay and possible transfer to impurities. The relaxation of the upper levels probably takes place by the multiphonon mechanism discussed here. (From A. Salloum and H. Dubust, Chem. Phys. **189**, 179 (1994).)

matrix. Oxygen is vibrationally excited by energy transfer from IR pumped carbon monoxide, and its subsequent relaxation is followed by pump-probe spectroscopy, that is, exciting into a higher electronic state and probing the ensuing fluorescence. The vibrational frequency of molecular oxygen is $\sim$ 1500 cm^{-1} and the Debye frequency of solid Ar is of the order of 65 cm^{-1}. The high phonon order, $1500/65 = 23$, of the process explains the long times seen and the sensitivity to competing relaxation pathways.

13.6 Effect of supporting modes

As already discussed in Section 13.1, the multiphonon pathway for vibrational relaxation is a relatively slow relaxation process, and, particularly at low temperatures the system will use other relaxation routes where accessible. Polyatomic molecules take advantage of the existence of relatively small frequency differences, and relax by subsequent medium assisted vibrational energy transfer between molecular modes. Small molecules often find other pathways as demonstrated in Section 13.1 for the relaxation of the CN radical. When the concentration of impurity molecules is not too low, intermolecular energy transfer often competes successfully with local multiphonon relaxation. For example, when a population of CO molecules in low temperature rare gas matrices is excited to the $v = 1$ level, the system takes advantage of the molecular anharmonicity by undergoing an intermediate relaxation of the type

$$\text{CO}(v = 1) + \text{CO}(v) \rightarrow \text{CO}(v = 0) + \text{CO}(v + 1) + \Delta E \tag{13.69}$$

where the excess energy $\Delta E = E_1 - E_0 - (E_{v+1} - E_v) > 0$ is smaller than the matrix Debye frequency and is easily deposited into the matrix. When this route is no longer available, the relaxation from the lower vibrational levels of CO proceeds radiatively.[10]

As another interesting example consider vibrational relaxation of an HCl molecules embedded in an Argon matrix.[11] At $T = 9$ K it is found that transition rates between the lowest vibrational levels are $k_{0\leftarrow 1} = 8 \times 10^2$ s^{-1} ($\omega =$ 2871 cm^{-1}) and $k_{1\leftarrow 2} = 3.8 \times 10^3$ s^{-1} ($\omega = 2768$ cm^{-1}). These transitions are resolved because of the molecular anharmonicity, and the fact that $k_{0\leftarrow 1} < k_{1\leftarrow 2}$ seems to agree with expectations based on the "energy gap law." On the other

[10] H. Dubost in *Chemistry and Physics of Matrix Isolated Species*, edited by L. Andrews and M. Moscovits (Elsevier, Amsterdam, 1989).

[11] F. Legay, Vibrational Relaxation in Matrices, in *Chemical and Biochemical Applications of Lasers*, Vol II (1977), p. 43.

hand, the observation that for DCl under identical conditions $k_{1\leftarrow 2} = 4 \times 10^2\ \mathrm{s}^{-1}$ ($\omega = 2078\ \mathrm{cm}^{-1}$) is in contradiction with this picture. Similarly, vibrational relaxation rates for NH radicals in Ar matrix ($T = 4.2$ K) are[12] $k_{0\leftarrow 1} = 1.2 \times 10^6\ \mathrm{s}^{-1}$ ($\omega = 2977\ \mathrm{cm}^{-1}$) and $k_{1\leftarrow 2} = 6.2 \times 10^6\ \mathrm{s}^{-1}$ ($\omega = 2718\ \mathrm{cm}^{-1}$), however for the deuterated species ND it is found that $k_{0\leftarrow 1} \leq 2 \times 10^4\ \mathrm{s}^{-1}(\omega = 2217\ \mathrm{cm}^{-1})$.

How can we explain such apparent breakdown of the energy gap law? Again, we should bear in mind the possibility that the observed relaxation proceeds via alternative routes such as transfer to impurities, as discussed above. These possibilities should be checked and suppressed by carefully adjusting experimental conditions while carrying control experiments. Assuming that the results described above represent "real" vibrational relaxation we should turn for answers to the physical model considered. To this end note that the simple "Einstein" model considered above fails if the process is dominated by a relatively strong coupling of the relaxing mode to a local phonon mode. Local vibrations often emerge when an impurity molecule is inserted into an otherwise pure host matrix. A simple easy-to-visualize example is the center of mass motion of the impurity in its solvation cavity. Suppose that such a mode exists, and suppose further that its frequency is considerably higher than thermal energy, $\hbar\omega_1 \gg k_B T$, so its thermal occupation may be assumed to vanish. In this case the integral I, Eq. (13.64), will be replaced by

$$I \equiv \int_{-\infty}^{\infty} dt e^{i\omega_{i,f}t + |\alpha_1|^2 e^{-i\omega_1 t} + |\alpha_2|^2 (n+1) e^{-i\omega_2 t} + |\alpha_2|^2 n e^{i\omega_2 t}} \tag{13.70}$$

Here the subscript 1 denotes the local mode and the other modes are represented by the "Einstein frequency" ω_2, of the order of the solvent Debye frequency, and we have assumed that $\omega_1 > \omega_2$.

Now, if α_2 was zero, we can use the procedure that leads to Eq. (13.64) to get

$$I = \sum_l \frac{|\alpha_1|^{2l}}{l!} 2\pi\delta(\omega_{i,f} - l\omega_1) \tag{13.71}$$

However, having only one, or very few, local modes, the probability to match energy exactly, that is, to have $\omega_{i,f} - l\omega_1$ for some integer l is extremely small. Instead, the largest contribution to the rate comes from a process in which l local phonons are generated so that $\omega_{i,f} = l\omega_1 + \Delta\omega$, where l is the largest integer that satisfies $\omega_{i,f} - l\omega_1 > 0$. In other words, the local phonons "fill the energy gap" as much as possible. The residual energy $\hbar\Delta\omega$ is dissipated into lattice phonons. The

[12] V. E. Bondybey and L. E. Brus, J. Chem. Phys. **63**, 794 (1975).

integral I is therefore, essentially,

$$I \sim \frac{|\alpha_1|^{2l}}{l!} \int_{-\infty}^{\infty} dt e^{i(\omega_{i,f}-l\omega_1)t+|\alpha_2|^2[(n_2+1)e^{-i\omega_2 t}+n_2 e^{-i\omega_2 t}]}$$

$$\sim \frac{|\alpha_1|^{2l}}{l!} \frac{|\alpha_2|^{2(\omega_{i,f}-l\omega_1)/\omega_2}}{((\omega_{i,f}-l\omega_1)/\omega_2)!}(n_2+1)^{(\omega_{i,f}-l\omega_1)/\omega_2} \tag{13.72}$$

In this case the temperature dependence is expected to be much weaker than before, Eq. (13.66), because the temperature dependent factor (n_2+1) is raised to a much smaller power. Moreover, *the larger is the frequency of the local phonon the smaller is l, therefore, since $\alpha_1 \ll 1$, the larger is the rate.*

In the examples considered above it has been found that the principal local mode is the almost free, only slightly hindered, rotation of the H (or D) atom about its heavy partner, N for NH/ND or Cl for HCl/DCl. This local mode is not a harmonic oscillator, but the argument proceeds in the same way: Assuming free rotation, the rotational level l corresponds to the energy $\hbar Bl(l+1) \cong \hbar Bl^2$ where B is the rotational constant. The minimum number of rotational levels needed to fill the frequency gap is the maximum integer l that satisfies $\omega_{i,f} \geq Bl(l+1)$ that is, $l < \sqrt{\omega_{i,f}/B}$. When the H atom in the HX molecule is replaced by a deuterium, the molecular frequency decreases by a factor $\sqrt{2}$, however, the rotational constant B, which is inversely proportional to the moment of inertia, decreases by a factor of ~2. Therefore l is larger in the deuterated molecule, namely more rotational levels are needed to fill the frequency gap. The rate is therefore expected to be smaller in this case, as observed.[13]

13.7 Numerical simulations of vibrational relaxation

Numerical simulations have become a central tool in studies of condensed phase processes. The science, technique, and art of this tool are subjects of several excellent texts.[14] Here we assume that important problems such as choosing a

[13] For a detailed treatment of this model see R. B. Gerber and M. Berkowitz, Phys. Rev. Lett. **39**, 1000 (1977).

[14] See, for example, M. P. Allen and D. J. Tildesley, *Computer Simulation of Liquids* (Oxford University Press, Oxford, 1987); D. C. Rapaport, *The Art of Molecular Dynamics Simulation*, 2nd edn (Cambridge University Press, Cambridge, 2004); Daan Frenkel and B. Smit, *Understanding Molecular Simulation*, 2nd edn (Academic Press, San Diego, CA, 2002); J. M. Haile, *Molecular Dynamics Simulation: Elementary Methods* (Wiley, New York, 1997).

proper force field for the inter- and intramolecular interactions, using an adequately large system given the range of intermolecular forces and the phenomenon under study, avoiding artifacts associated with artificial boundary conditions and adequate sampling of initial conditions (needed, for example, to represent thermal equilibrium at given temperature and volume or pressure), have been handled. We thus assume that we can generate a table that gives the position and velocity of each atom in the system as a function of time. This table is a numerical representation of a phase space trajectory. Naturally, entries in such a numerical table will be given at discrete time points, determined by a pre-chosen time interval.

Two principal issues remain:

1. *Quantum versus classical mechanics*. Note that the previous paragraph uses the language of classical mechanics. The equivalent undertaking in quantum mechanics would be to generate a table that gives at subsequent time intervals the many-particle complex wavefunction or, more generally, the many-particle density matrix of the system. Such goals are far beyond our reach. Quantum dynamical simulations can be carried only for very small systems such as encountered in gas phase processes. Progress in quantum dynamical simulations of condensed phase processes is being made by developing tools for mixed quantum–classical simulations, in which one attempts to identify those degrees of freedom for which quantum mechanics is essential, and compute the quantum dynamics for these degrees of freedom when coupled to the rest of the system which is treated classically. Quantum mechanics is evidently essential when the process under study is strongly influenced by one or more of the following inherently quantum phenomena: (1) interference, (2) tunneling, (3) zero-point motion, and (4) energy level spacings larger than $k_B T$.

2. *Handling co-existing vastly different timescales*. The problem of vibrational relaxation is a good example for demonstrating this point. Disregarding electronic transitions, the shortest timescale in this problem is the period of the fast impurity oscillator which is the subject of our relaxation study. For the oxygen molecule in the experiment of Fig. 13.2 this is of the order of 20 fs. Decent numerical solution of the Newton equations should use a considerably smaller integration time interval, say 0.2 fs. On the other hand the measured relaxation time as seen in Fig. 13.2 is of the order of 1–100 s. This implies that a direct numerical observation of this vibrational relaxation, even within a classical mechanics based simulation, requires running trajectories of such lengths. Again, this is far beyond our present numerical abilities.

It should be noted that a similar problem exists also in systems with vastly different lengthscales. The smallest characteristic lengthscale in atomic-level condensed phase simulations is the interatomic distance which is of the same order as the atomic size. To simulate phenomena that involve much larger lengthscales we need to handle systems with huge number of atoms, again a numerical burden that can easily reach the limits of present abilities. This issue is not an important factor in numerical simulations of vibrational relaxation, where the dominant forces are short range. In fact, expecting that the process is dominated by close encounters of the relaxing molecule with its nearest neighbors, one could focus on a very small "inner system" that includes the molecule and its neighbors, and take the effect of the rest of the solvent by imposing random force and damping (i.e. using Langevin equations, Section 8.2) on the motion for these particles.[15] Indeed, the first simulations of vibrational relaxation where done using such Langevin or Brownian dynamics simulations.[16]

Coming back to the timescale issue, it is clear that direct observation of signals such as shown in Fig. 13.2 cannot be achieved with numerical simulations. Fortunately an alternative approach is suggested by Eq. (13.26), which provides a way to compute the vibrational relaxation rate directly. This calculation involves the autocorrelation function of the force exerted by the solvent atoms on the frozen oscillator coordinate. Because such correlation functions decay to zero relatively fast (on timescales in the range of pico to nano seconds depending on temperature), its numerical evaluation requires much shorter simulations. Several points should be noted:

1. Given a trajectory table, $(\mathbf{r}^N(t), \mathbf{v}^N(t))$, for all N particles in the system at successive time points t, the time correlation function of any dynamical variable is computed essentially as described by Eqs (7.43) and (7.44).
2. The procedure described here is an example for combining theory (that relates rates and currents to time correlation functions) with numerical simulations to provide a practical tool for rate evaluation. Note that this calculation *assumes* that the process under study is indeed a simple rate process characterized by a single rate. For example, this level of the theory cannot account for the nonexponential relaxation of the $v = 10$ vibrational level of O_2 in Argon matrix as observed in Fig. 13.2.
3. Difficulties associated with disparity of timescales may still be encountered even within this approach, in cases where the frequency of the impurity molecule is much larger than the cutoff (Debye) frequency, ω_D, of the host. Note that the rate (13.26) is given by the Fourier transform of the force auto-correlation function, taken at the frequency of the impurity oscillator. The time dependence of this correlation function reflects the time dependence of

[15] In fact, generalized Langevin equations (Section 8.2.6) need to be used in such applications to account for the fact that the "system" is usually faster than the "bath." Indeed, the spectral density, Eq. (8.65) should reflect the spectral characteristics of the bath, including its Debye frequency.

[16] M. Shugard, J. C. Tully, and A. Nitzan, J. Chem. Phys. **69**, 336, 2525 (1978).

the force, which in turn reflects the normal mode frequencies of the host. The Fourier transform ($t \to \omega$) of the force autocorrelation function is typically large for $\omega < \omega_{\text{D}}$ and decays exponentially for larger ω. For $\omega \gg \omega_{\text{D}}$, the evaluation of the required Fourier transform is a problem in numerical signal analysis—the need to extract a very small meaningful signal in a large noisy background.[17]

Finally, let us return to the issue of quantum mechanics. Our impurity vibration is usually such that $\hbar\omega > k_B T$ so this mode should obviously be handled quantum mechanically. Indeed, expression (13.26) was derived from the quantum mechanical golden rule for the transition rate between two levels of this oscillator. The host medium is characterized by a continuous spectrum, and since no interference or tunneling seem to be at play, nor does zero point motion appear to be an essential ingredient, one could perhaps anticipate that a classical description of the host could be at least a crude approximation expected to become better at higher temperatures. This is the rational for considering the semiclassical rate expressions (13.21) and (13.27). However, as indicated by Eq. (13.67) the quantum correction factor $\zeta(\omega)$ can be quite large when $\omega/\omega_{\text{D}} \gg 1$. Even though the solvent behaves classically in many respects, its ability to absorb a quantum of energy much larger than its natural frequencies shows a distinct quantum mechanical nature.

At low temperature the quantum correction factor can be huge, as seen in Eq. (13.68). Since quantum correlation functions are not accessible by numerical simulations, one may evaluate numerically the corresponding classical correlation function and estimate theoretically the quantum correction factor. Some attempts to use this procedure appear to give reasonable results,[18] however, it is not clear that the quantum correction factors applied in these works are of general applicability.

13.8 Concluding remarks

Our focus in this chapter was the rate at which a molecule interacting with its thermal environment releases its vibrational energy. Small diatomic molecules embedded in condensed monoatomic hosts provide natural choices for studying this phenomenon. We have found, however, that in many cases their direct vibrational relaxation is very slow and the system finds alternative routes for releasing its excess nuclear energy. On the other extreme end, relaxation is very fast, in the ps regime, for polyatomic molecules at room temperature. In such systems

[17] D. Rostkier-Edelstein, P. Graf, and A. Nitzan, J. Chem. Phys. **107**, 10470 (1997).

[18] See, for example, K. F. Everitt, J. L. Skinner, and B. M. Ladanyi, J. Chem. Phys. **116**, 179 (2002).

relaxation takes place through anharmonic interactions that couple two or more molecular modes with the environmental phonons, making it possible to interact with the main phonon spectrum of the environment rather than with its very small high-frequency tail.

An interesting comparison can be made between the relaxation of high-frequency molecular modes and the spin–lattice relaxation discussed in Section 12.5.5. In both cases direct relaxation is very slow because of incompatibility between energy spacings in the system and in the bath. Molecular frequencies are much higher, and spin energy spacings are much lower, than the spectral region where most of the host phonon spectrum is distributed. Both systems resolve their difficulties by invoking higher-order terms in their interaction with the environment.

As in the case of electronic relaxation (see Section 12.4.1 and problem 12-2), once a molecule becomes large enough it can provide enough density of vibrational modes to act as its own "heat bath." The ensuing relaxation process is referred to as *intramolecular vibrational relaxation* (or redistribution), in which a mode or a group of modes under observation exchange energy with the rest of the molecular nuclear space, even though the total molecular energy is unchanged.

Finally, while the focus of our discussion was energy relaxation, we should keep in mind that phase relaxation is also an important process of measurable spectroscopic consequences. We have discussed vibrational phase relaxation in Section 7.5.4 and the concept of dephasing in Section 10.4.9. We will come back to this issue in Section 18.5.

Further Reading

D. W. Oxtoby, Vibrational relaxation in liquids, Ann. Rev. Phys. Chem. *32*, 77–101 (1981).

J. C. Owrutsky, D. Raftery and R. M. Hochstrasser, Vibrational relaxation dynamics in solutions, Ann. Rev. Phys. Chem. *45*, 519–555 (1994).

J. L. Skinner, S. A. Egorov and K. F. Everitt, Vibrational relaxation in liquids and supercritical fluids, in: "Ultrafast infrared and Raman Spectroscopy", edited by M. Fayer (Marcel Dekker, New York City 2001).

14

CHEMICAL REACTIONS IN CONDENSED PHASES

Never suppose the atoms had a plan,
Nor with a wise intelligence imposed
An order on themselves, nor in some pact
Agreed what movements each should generate.
No, it was all fortuitous; for years,
For centuries, for eons all those motes
In infinite varieties of ways
Have always moved, since infinite time began,
Are driven by collisions, are borne on
By their own weight, in every kind of way
Meet and combine, try every possible,
Every conceivable pattern, till at length
Experiment culminates in that array
Which makes great things...

Lucretius (c.99–c.55 BCE*) "The way things are" translated by*
Rolfe Humphries, Indiana University Press, 1968

14.1 Introduction

Understanding chemical reactions in condensed phases is essentially the understanding of solvent effects on chemical processes. Such effects appear in many ways. Some stem from equilibrium properties, for example, solvation energies and free energy surfaces (see Section 14.3). Others result from dynamical phenomena: solvent effect on diffusion of reactants toward each other, dynamical cage effects, solvent-induced energy accumulation and relaxation, and suppression of dynamical change in molecular configuration by solvent induced friction.

In attempting to sort out these different effects it is useful to note that a chemical reaction proceeds by two principal dynamical processes that appear in three stages. In the first and last stages the reactants are brought together and products are separated from each other. In the middle stage the assembled chemical system undergoes the structural/chemical change. In a condensed phase the first and last stages involve diffusion, sometimes (e.g. when the species involved are charged) in a force field. The middle stage often involves the crossing of a potential barrier. When the barrier is high the latter process is rate-determining. In unimolecular reactions the species that undergoes the chemical change is already assembled and only the barrier crossing process is relevant.[1] On the other hand, in bi-molecular reactions with low barrier (of order $k_B T$ or less), the rate may be dominated by the diffusion process that brings the reactants together. It is therefore meaningful to discuss these two ingredients of chemical rate processes separately.

Most of the discussion in this chapter is based on a classical mechanics description of chemical reactions. Such classical pictures are relevant to many condensed phase reactions at and above room temperature and, as we shall see, can be generalized when needed to take into account the discrete nature of molecular states. In some situations quantum effects dominate and need to be treated explicitly. This is the case, for example, when tunneling is a rate determining process. Another important class is nonadiabatic reactions, where the rate determining process is hopping (curve crossing) between two electronic states. Such reactions are discussed in Chapter 16 (see also Section 14.3.5).

14.2 Unimolecular reactions

In addition to their inherent significance as an important class of chemical reactions, unimolecular rate processes in solution have attracted much attention because they provide convenient testing grounds to theories that describe solvent effect on barrier crossing irrespective of its additional role of modifying bi-molecular encounters. The study of such reactions therefore provides a convenient framework for analyzing solvent effects on barrier crossing phenomena.

A unimolecular process can take place once the reactant molecule has accumulated enough energy. The solvent environment controls energy transfer to and from this molecule, affects energy flow between different degrees of freedom within the molecule, and helps to dissipate excess energy in the product species. The way all these effects combine to yield the overall solvent effect on the unimolecular reaction rate is the subject of our study.

An important concept in chemical kinetics is the rate *coefficient*. For a unimolecular reaction involving species A, the rate coefficient k appears in the first-order kinetics law $d\mathbf{A}/dt = -k\mathbf{A}$, however we should bear in mind that even for unimolecular processes the existence of a rate coefficient as a time-independent constant requires additional assumptions. First, as the process proceeds, the system itself can change. For example, the solvent temperature may change due to the energy released or absorbed during the process. The solution composition may change due to disappearance of reactant and formation of product species. These potential difficulties can be avoided by using a large excess of solvent and taking proper care with the experimental design.

---

[1] In some photoinduced unimolecular reactions the barrier may be very small or altogether absent.

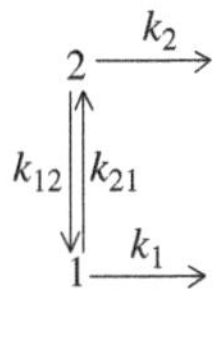

FIG. 14.1 A two-level model for thermal relaxation effect on chemical reactions.

More importantly, a molecular species A can exist in many quantum states; in fact the very nature of the required activation energy implies that several excited nuclear states participate. It is intuitively expected that individual vibrational states of the reactant will correspond to different reaction rates, so the appearance of a single macroscopic rate coefficient is not obvious. If such a constant rate is observed experimentally, it may mean that the process is dominated by just one nuclear state, or, more likely, that the observed macroscopic rate coefficient is an average over many microscopic rates. In the latter case $k = \sum_i P_i k_i$, where k_i are rates associated with individual states and P_i are the corresponding probabilities to be in these states. The rate coefficient k is therefore time-independent provided that the probabilities P_i remain constant during the process.[2] The situation in which the *relative* populations of individual molecular states remains constant even if the overall population declines is sometimes referred to as a *quasi steady state.* This can happen when the relaxation process that maintains thermal equilibrium between molecular states is fast relative to the chemical process studied. In this case $\{P_i\}$ remain thermal (Boltzmann) probabilities at all times. We have made such assumptions in earlier chapters; see Sections 10.3.2 and 12.4.2. We will see below that this is one of the conditions for the validity of the so-called *transition state theory* of chemical rates. We also show below that this can sometime happen also under conditions where the time-independent probabilities $\{P_i\}$ do not correspond to a Boltzmann distribution.

A simple example can clarify this issue. Suppose that the reactant A is a two-level molecule. Denote the levels by 1 and 2, the corresponding densities by A_1 and A_2 ($A_1 + A_2 = A$) and the microscopic rates out of these levels by k_1 and k_2. Let there be also an internal rate process with rates $k_{12} = k_{1\leftarrow 2}$ and $k_{21} = k_{2\leftarrow 1}$, that would maintain thermal equilibrium in a closed system. This implies $k_{12}/k_{21} = \exp(-\beta(E_1 - E_2))$. The overall kinetic scheme is shown in Fig. 14.1.

If we further take $k_1 = 0$ this becomes the *Lindemann mechanism* that is used to explain the observation that many gas-phase reactions of the type $A \to$ *product* that appear unimolecular at high pressure change their character to bimolecular at low pressure. Lindemann has postulated that such unimolecular reactions proceed

[2] Note that $\sum_i P_i = 1$, so the normalized state probabilities do not reflect the change in the overall number of reactant molecules.

via an activated state 2 of A which is obtained by collision with surrounding gas molecules M (so that the corresponding rates are proportional to the gas pressure):

$$\begin{aligned} A_1 + M &\underset{k_{12}}{\overset{k_{21}}{\rightleftharpoons}} A_2 + M \\ A_2 &\underset{k_2}{\longrightarrow} P \text{ (products)} \end{aligned} \tag{14.1}$$

The corresponding chemical rate equations are

$$\frac{dA_1}{dt} = -\bar{k}_{21}A_1 + \bar{k}_{12}A_2 \tag{14.2a}$$

$$\frac{dA_2}{dt} = \bar{k}_{21}A_1 - \bar{k}_{12}A_2 - k_2A_2 = -\frac{dA_1}{dt} - k_2A_2 \tag{14.2b}$$

$$\frac{dP}{dt} = k_2A_2 \tag{14.2c}$$

where $\bar{k}_{ij} = k_{ij}M$. This set of linear first-order differential equations can be solved exactly, but we will first take a simplified route in which we assume that a quasi steady state is established in this process. This implies that A_1/A_2 remains constant during the time evolution, which in turn implies that $(dA_2/dt)/(dA_1/dt) = A_2/A_1$. In the common case where $A_2 \ll A_1$ we can disregard dA_2/dt relative to dA_1/dt in (14.2b). This leads to

$$A_2 = \frac{\bar{k}_{21}A_1}{\bar{k}_{12} + k_2} \tag{14.3}$$

and the effective overall rate $k = (1/A)dP/dt$

$$k = \frac{\bar{k}_{21}k_2}{\bar{k}_{12} + k_2} \tag{14.4}$$

$\bar{k}_{21}$ and $\bar{k}_{12}$ express the strength of the interaction between the molecule and its thermal environment; in the present example they are proportional to the gas pressure. In the limit where these rates are large, specifically when $\bar{k}_{12} \gg k_2$, Eq. (14.4) yields $k = k_2(k_{21}/k_{12}) = k_2 \exp[-\beta(E_2 - E_1)]$, independent of the gas pressure. This reflects the fact that in this limit the thermal relaxation is much faster than the rate of product formation, therefore to a good approximation A_2 is given by its equilibrium value $A_2 = A_1 \exp[-\beta(E_2 - E_1)] \simeq A \exp[-\beta(E_2 - E_1)]$. In the opposite limit, when the thermal interaction is weak, $\bar{k}_{21} \ll \bar{k}_{12} \ll k_2$, the rate becomes $k = \bar{k}_{21} = k_{21}M$. In a pure gas A, (i.e. when M and A are the same species) this implies that the reaction appears to be bimolecular.

From the mathematical point of view, the dynamics of product formation in the reaction scheme (14.2) is characterized by two timescales (see below). Our attempt to identify a single reaction rate amounts to exploring the conditions under which one of these rates dominates the observed evolution. In the limit of fast thermal relaxation relative to the rate of product formation, it is the latter slow rate that dominates the observed process. In the opposite limit it is the excitation process $A_1 \to A_2$ that determines the rate. The latter is then dominated by the first of the reactions in (14.1)—accumulation of energy in A, and is proportional to the strength of the thermal interaction, for example, the bathing gas pressure.

The results obtained above are limiting cases of the full dynamics. The time evolution described by Eqs (14.2) is determined by the two roots, $\alpha_1 \le \alpha_2$ of the characteristic equation

$$\alpha^2 - \alpha(k_2 + \bar{k}_{21} + \bar{k}_{12}) + \bar{k}_{21}k_2 = 0 \tag{14.5}$$

and, for the initial condition $A_1(t=0) = A(t=0) = 1, A_2(t=0) = 0$ is given by

$$\begin{aligned} A_1(t) &= \frac{\bar{k}_{21}}{\alpha_2 - \alpha_1}\left[\left(\frac{k_2}{\alpha_1} - 1\right)e^{-\alpha_1 t} - \left(\frac{k_2}{\alpha_2} - 1\right)e^{-\alpha_2 t}\right] \\ A_2(t) &= \frac{\bar{k}_{21}}{\alpha_2 - \alpha_1}\left[e^{-\alpha_1 t} - e^{-\alpha_2 t}\right] \\ A(t) = A_1(t) + A_2(t) &= \frac{\alpha_2 e^{-\alpha_1 t} - \alpha_1 e^{-\alpha_2 t}}{\alpha_2 - \alpha_1} \end{aligned} \tag{14.6}$$

To obtain this result we have used (cf. Eq. (14.5)) $\alpha_1\alpha_2 = \bar{k}_{21}k_2$. The time evolution will appear unimolecular, with a single time-independent rate, only if $\alpha_1 \ll \alpha_2$. In this case, following a transient period of order $1/\alpha_2$, $A(t)$ proceeds to disappear exponentially

$$A(t) \simeq \left(\frac{\alpha_2}{\alpha_2 - \alpha_1}\right)e^{-\alpha_1 t} \simeq e^{-\alpha_1 t} \tag{14.7}$$

while the ratio A_2/A_1 remains time-independent

$$A_2(t)/A_1(t) \simeq \alpha_1/\left(k_2 - \alpha_1\right) \tag{14.8}$$

Note that the relationship $\alpha_1 \ll \alpha_2$ also implies that the amplitude of the fast component in the last of Eq. (14.6) is very small, implying that most of the process take place on the timescale determined by α_1 which therefore determines the reaction rate k. The nature of the constant distribution (14.8) depends on the actual rates.

Keeping in mind that $\bar{k}_{21} = \bar{k}_{12}e^{-\beta(E_2-E_1)} < \bar{k}_{12}$, we obtain for both $k_2 \ll \bar{k}_{12}$ (henceforth case 1) and $k_2 \gg \bar{k}_{12}$ (case 2), to a good approximation

$$\alpha_1 = k = \frac{\bar{k}_{21}k_2}{k_2 + \bar{k}_{12} + \bar{k}_{21}} \tag{14.9}$$

so that (from (14.8))

$$\frac{A_2}{A_1} = \frac{\bar{k}_{21}}{\bar{k}_{12} + k_2} \tag{14.10}$$

This implies that in case 1 a Boltzmann distribution is essentially maintained after the initial fast relaxation, while in case 2 the upper state is strongly depleted relative to the Boltzmann population. Furthermore, for $\beta(E_2 - E_1) \gg 1$ we have again $A_1 \simeq A$, and the effective unimolecular rate is $k = k_2A_2/A \simeq k_2\bar{k}_{21}/(\bar{k}_{12} + k_2)$, reproducing Eq. (14.4)

The result (14.4) tells us that as a function of the thermal relaxation rate (e.g. the gas pressure that determines M) the rate coefficient k grows until it saturates at the maximum value $k = k_2e^{-\beta(E_2-E_1)}$ which no longer depends on M. However, for very large densities (e.g. in liquids and solids) k_2 may depend on M and in fact it is expected to vanish as $M \to \infty$. This is because the process $A_2 \to P$ is a reaction in which atoms move to form new configurations a process that can be inhibited by friction or by lack of available volume needed for atomic motion. The Kramers model (Section 14.4) describes this effect by taking solvent-induced friction as an explicit parameter of the theory.

The observations made above emphasize two conditions for a linear multistep process to appear as a simple single-exponential relaxation: (1) One of the eigenvalues of the relaxation matrix has to be much smaller than the other. (2) This eigenvalue should dominate the relaxation, that is, the combined amplitudes of the relaxation modes associated with all other eigenvalues must be small relative to that associated with the smallest eigenvalue. Another useful observation is that the inequality $\alpha_1 \ll \alpha_2$ always holds if the reaction rate α_1 is much smaller from the thermal relaxation rate, here given by $\bar{k}_{12} + \bar{k}_{21}$, the rate at which a closed system ($k_2 = 0$) will approach equilibrium. Equation (14.9) shows that this will be satisfied if the activation energy is large enough, that is, $\beta(E_2 - E_1) \gg 1$.

The Lindemann model discussed above provides the simplest framework for analyzing the dynamical effect of thermal relaxation on chemical reactions. We will see that similar reasoning applies to the more elaborate models discussed below, and that the resulting phenomenology is, to a large extent, qualitatively the same. In particular, the Transition State Theory (TST) of chemical reactions, discussed in the next section, is in fact a generalization of the fast thermal relaxation limit of the Lindemann model.

14.3 Transition state theory

14.3.1 Foundations of TST

Consider a system of particles moving in a box at thermal equilibrium, under their mutual interactions. In the absence of any external forces the system will be homogenous, characterized by the equilibrium particle density. From the Maxwell velocity distribution for the particles, we can easily calculate the *equilibrium* flux in any direction inside the box, say in the positive x direction, $J_x = \rho\langle v_x\rangle$, where ρ is the density of particles and $\langle v_x\rangle = (\beta m/2\pi)^{1/2}\int_0^\infty dv_x v_x \exp(-\beta m v_x^2/2)$. Obviously, this quantity has no relation to the kinetic processes observed in the corresponding nonequilibrium system. For example, if we disturb the homogeneous distribution of particles, the rate of the resulting diffusion process is associated with the *net* particle flux (difference between fluxes in opposing directions) which is zero at equilibrium.

There are, however, situations where the equilibrium flux calculated as described above, *through a carefully chosen surface*, provides a good approximation for an observed nonequilibrium rate. The resulting *transition state theory* of rate processes is based on the calculation of just that equilibrium flux. In fact, for many chemical processes characterized by transitions through high-energy barriers, this approximation is so successful that *dynamical* effects to which most of this chapter is devoted lead to relatively small corrections. The essential ingredients of TST can be described by referring to the potential of Fig. 14.2, plotted against the *reaction coordinate* x. For specificity we regard the wells to the left and right of the barrier as "reactant" and "product" states, respectively. We consider an ensemble of systems prepared in the reactant well and examine the phase space trajectories that lead such a system to cross to the product side. For simplicity we use the word "particles" to

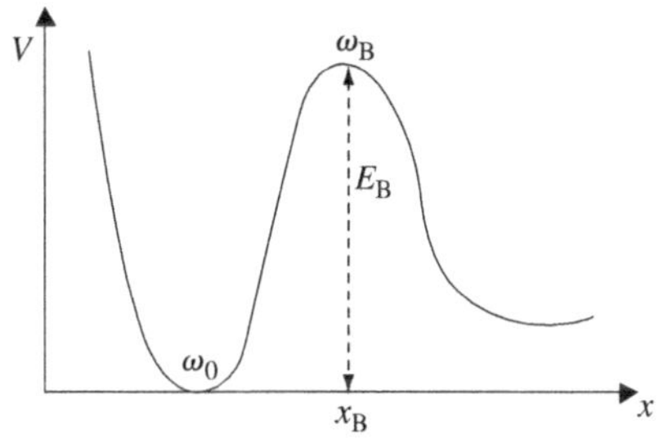

FIG. 14.2 A one-dimensional model for a barrier crossing process. The potential barrier is characterized by the well and barrier curvatures which determine the frequencies ω_0 and ω_B, and by the barrier height E_B.

describe these trajectories. TST is based on two critical assumptions:

1. *The reaction process does not disturb the thermal distribution within the reactant and product wells.* This assumption is based on the observation that the timescale of processes characterized by high barriers is much longer than the timescale of achieving *local* thermal equilibrium in the reactants and products regions. The only quantity which remains in a nonequilibrium state on this long timescale is the relative concentrations of reactants and products.
2. *The rate is determined by the equilibrium flux across the boundary separating reactants and products.* The fact that this boundary is characterized by a high-energy barrier is again essential here. Suppose that the barrier is infinite at first and we start with all particles in the "reactant state," say to the left of the barrier in Fig. 14.2. On a very short timescale thermal equilibrium is achieved in this state. Assumption (1) assures us that this thermal equilibrium is maintained also after we lower the barrier to its actual (large) height. Assumption (2) suggests that if we count the number of barrier-crossing events per unit time in the direction reactants $\rightarrow$ products using the (implied by assumption (1)) Maxwell distribution of velocities, we get a good representation of the rate. For this to be true, the event of barrier crossing has to be the deciding factor concerning the transformation of reactants to products.

This is far less simple than it sounds: after a particle has crossed the barrier, say from left to right, its fate is not yet determined. It is only after subsequent relaxation leads it toward the bottom of the right well that its identity as a product is established. If this happens before the particle is reflected back to the left, that crossing is *reactive*. Assumption (2) in fact states that all equilibrium trajectories crossing the barrier are reactive, that is, they go from reactants to products without being reflected. For this to be a good approximation to reality two conditions should be satisfied:

1. The barrier region should be small relative to the mean free path of the particles along the reaction coordinate, so that their transition from a well defined *left* to a well defined *right* is undisturbed and can be calculated from the thermal velocity.
2. Once a particle crosses the barrier, it relaxes quickly to the final equilibrium state before being reflected to its well of origin.

These conditions cannot be satisfied exactly. Indeed, they are incompatible with each other: the fast relaxation required by the latter implies that the mean free path is small, in contrast to the requirement of the former. In fact, assumption (2) must fail for processes without barrier. Such processes proceed by diffusion, which is defined over length scales large relative to the mean free path of the diffusing particle. A transition region that defines the "final" location of the particle to its left or to its right cannot be smaller than this mean free path.

For a transition region located at the top of a high barrier the situation is more favorable. Once the particle has crossed the barrier, it gains kinetic energy as it goes into the well region, and since the rate of energy loss due to friction is proportional to the kinetic energy, the particle may lose energy quickly and become identified as a product before it is reflected back to the reactant well. This fast energy loss from the reaction coordinate is strongly accelerated in real chemical systems where the reaction coordinate is usually strongly coupled, away from the barrier region, to other nonreactive molecular degrees of freedom. Thus, thermal relaxation may be disregarded in the small region at the barrier top and assumed fast just out of this region—exactly the conditions for validity of TST.

14.3.2 Transition state rate of escape from a one-dimensional well

Consider an activated rate process represented by the escape of a particle from a one-dimensional potential well. The Hamiltonian of the particle is

$$H = \frac{p^2}{2m} + V(x) \tag{14.11}$$

where $V(x)$ is characterized by a potential well with a minimum at $x = 0$ and a potential barrier peaked at $x = x_{\mathrm{B}} > 0$, separating reactants ($x < x_{\mathrm{B}}$) from products ($x > x_{\mathrm{B}}$) (see Fig. 14.2). Under the above assumptions the rate coefficient for the escape of a particle out of the reactant well is given by the forward flux at the *transition state* $x = x_{\mathrm{B}}$

$$k_{\mathrm{TST}} = \int_0^\infty dv v P(x_{\mathrm{B}}, v) = \langle v_{\mathrm{f}} \rangle P(x_{\mathrm{B}}) \tag{14.12}$$

where $P(x_{\mathrm{B}})dx$ is the equilibrium probability that the particle is within dx of x_{B},

$$P(x_{\mathrm{B}}) = \frac{\exp(-\beta E_{\mathrm{B}})}{\int_{-\infty}^{x_{\mathrm{B}}} dx \exp(-\beta V(x))}; \quad E_{\mathrm{B}} = V(x_{\mathrm{B}}) \tag{14.13}$$

and where $\langle v_{\mathrm{f}} \rangle$ is the average of the forward velocity

$$\langle v_{\mathrm{f}} \rangle = \frac{\int_0^\infty dv v e^{-(1/2)\beta m v^2}}{\int_{-\infty}^\infty dv e^{-(1/2)\beta m v^2}} = \frac{1}{\sqrt{2\pi \beta m}} \tag{14.14}$$

Note that the fact that only half the particles move in the forward direction is taken into account in the normalization of Eq. (14.14). For a high barrier, most of the contribution to the integral in the denominator of (14.13) comes from regions of the coordinate x for which $V(x)$ is well represented by an harmonic well, $V(x) = (1/2)m\omega_0^2x^2$. Under this approximation this denominator then becomes

$$\int_{-\infty}^{\infty} dx e^{-(1/2)\beta m\omega_0^2 x^2} = \sqrt{\frac{2\pi}{\beta m\omega_0^2}} \tag{14.15}$$

Inserting Eqs (14.13)–(14.15) into (14.12) leads to

$$k_{\mathrm{TST}} = \frac{\omega_0}{2\pi} e^{-\beta E_{\mathrm{B}}} \tag{14.16}$$

The transition state rate is of a typical Arrenius form: a product of a frequency factor that may be interpreted as the number of attempts, per unit time, that the particle makes to exit the well, and an activation term associated with the height of the barrier. It is important to note that it does not depend on the coupling between the molecule and its environment, *only on parameters that determine the equilibrium distribution.*

14.3.3 Transition rate for a multidimensional system

More insight into the nature of TST can be obtained from the generalization of the above treatment to a multidimensional system. Consider an $(N+1)$-dimensional system defined by the Hamiltonian

$$H = \sum_{i=0}^{N} \frac{\bar{p}_i^2}{2m_i} + \bar{V}(\bar{x}_0, \ldots, \bar{x}_N) \tag{14.17}$$

or, in terms of mass weighted coordinates and momenta

$$\begin{aligned} x_i &= \sqrt{m_i}\bar{x}_i; \qquad p_i = \bar{p}_i/\sqrt{m_i} = \dot{x}_i \\ V(x_0, \ldots, x_N) &= \bar{V}(\bar{x}_0, \ldots, \bar{x}_N) \end{aligned} \tag{14.18}$$

$$H = \sum_{i=0}^{N} \frac{p_i^2}{2} + V(x^{N+1}) \tag{14.19}$$

Here x^{N+1} denotes the set $(x_0, x_1, \ldots, x_N)$. It is assumed that the potential, $V(x^{N+1})$, has a well whose minimum is at some point x_A^{N+1} and which is surrounded by a domain of attraction, separated from the outside space by a potential barrier.

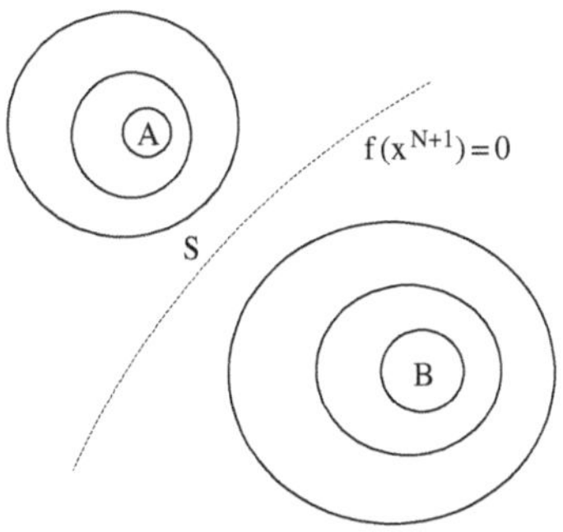

FIG. 14.3 A two dimensional analog of Fig. 14.2 that represents a multidimensional potential surface, showing two wells, A and B and the saddle point S that corresponds to the barrier. Also shown is the transition surface $f(x^{N+1}) = 0$.

Figure 14.3 is a contour plot that shows two such minima, representing the stable reactant and product configurations, and the transition surface (line in the figure) that separate their domains of attraction. The minimum of $V(x^{N+1})$ on this surface is the saddle point x_S^{N+1}. This dividing N-dimensional surface is defined in the $(N+1)$-dimensional space by the relation

$$f(x^{N+1}) = 0 \tag{14.20}$$

such that $f(x^{N+1}) < 0$ includes x_A^{N+1} and is defined as the reactant (say) part of the configuration space and $f(x^{N+1}) > 0$ is the product space defined in an analogous way. Note that, as defined, the dividing surface includes the saddle point x_S^{N+1}. This is a helpful but not formally required restriction, and in practical rate calculations we often use surfaces that do not satisfy this restriction.[3]

As before, TST assumes that (1) thermal equilibrium exists within the reactant space, and (2) trajectories that cross the dividing surface from the reactant to the product space do not recross on the timescale of thermal relaxation in the product space. A straightforward generalization of the calculation of Section 14.3.2 then leads to

$$k_{\mathrm{TST}} = Q_A^{-1} \int dp^{N+1} dx^{N+1} e^{-\beta H} \delta(f(x^{N+1}))(\nabla f \cdot p^{N+1})\Theta(\nabla f \cdot p^{N+1}) \tag{14.21}$$

where $p^{N+1} = (p_0, \ldots, p_N)$ is the velocity vector, Θ is the unit step function, and where the partition function of the reactant, Q_A, is defined by

$$Q_A = \int dp^{N+1} dx^{N+1} e^{-\beta H} \Theta(-f(\mathbf{x}^{N+1})) \tag{14.22}$$

[3] This happens for the practical reason that the position of the saddle point on a multidimensional potential surface is not always known.

While Eq. (14.21) seems much more complicated than its one-dimensional counterpart (14.12), a close scrutiny shows that they contain the same elements. The δ-function in (14.21) defines the dividing surface, the term $\nabla f \cdot p^{N+1}$ is the component of the momentum normal to this surface and the Θ function selects outwards going particles. If, for physical reasons, a particular direction, say x_0, is identified as the reaction coordinate, then a standard choice for f in the vicinity of the saddle point x_S^{N+1} is $f(x^{N+1}) = x_0 - x_{S0}$, where x_{S0} is the value of the reaction coordinate at that saddle point. This implies $\nabla f \cdot p^{N+1} = p_0$, that is, the component of the momentum along the reaction coordinate. With this choice Eq. (14.21) becomes

$$k_{TST} = Q_{0A}^{-1} \int dp_0 dx_0 e^{-\beta[p_0^2/2 + W(x_0)]} \delta(x_0 - x_{S0}) p_0 \Theta(p_0) \tag{14.23}$$

where

$$Q_{0A} = \int dp_0 dx_0 \Theta[-(x_0 - x_{S0})] e^{-\beta[p_0^2/2 + W(x_0)]} \tag{14.24}$$

and where

$$W(x_0) = -\beta^{-1} \ln \int dx^N e^{-\beta V(x^{N+1})}; \qquad \int dx^N = \int \ldots \int dx_1 dx_2 \ldots dx_N \tag{14.25}$$

is the potential of mean force along the reaction coordinate.

Problem 14.1. Show that Eqs (14.23) and (14.24) are equivalent to Eqs (14.12)–(14.14) if the one-dimensional potential in (14.12)–(14.14) is identified as $W(x)$.

This re-derivation of the one-dimensional TST result emphasizes the effective character of the potential used in one-dimensional treatments of barrier crossing problems. The one-dimensional model, Eq. (14.11), will yield the correct TST result provided that the potential $V(x)$ is taken as the effective potential of the reaction coordinate, that is, the potential of mean force along this coordinate where all other degrees of freedom are in thermal equilibrium at any given position of this coordinate.[4] It should be stressed, however, that this choice of the one-dimensional effective potential assumes that such a coordinate can be identified and that a point along this coordinate can be identified as the transition point that separates reactants from products.

[4] D. Chandler, J. Chem. Phys. **68**, 2959 (1978).

A more conventional form of the TST rate is obtained by inserting Eq. (14.24) into (14.23) and carrying out the integrations over p_0. This leads to

$$k_{\mathrm{TST}} = \frac{1}{\sqrt{2\pi\beta}} \frac{\int dx^{N+1} e^{-\beta V(x^{N+1})} \delta(x_0 - x_{\mathrm{S}0})}{\int dx^{N+1} e^{-\beta V(x^{N+1})} \Theta[-(x_0 - x_{\mathrm{S}0})]} \tag{14.26}$$

For a high barrier, $\beta E_{\mathrm{B}} \gg 1$, the dominant contribution to the denominator come from the vicinity of the well bottom, x_A^{N+1}. We can therefore replace $V(x^{N+1})$ in the denominator by

$$V(x^{N+1}) = \frac{1}{2} \sum_{i=0}^{N} \omega_{Ai}^2 x_{Ai}^2 \tag{14.27}$$

where x_{Ai} are the modes which diagonalize the Hessian of the potential at x_A^{N+1}. Similarly, the dominant contribution to the numerator comes from the neighborhood of the saddle point, x_{S}^{N+1}, where the potential may be expanded in the form

$$V(x^{N+1}) = E_{\mathrm{B}} - \frac{1}{2}\omega_{\mathrm{B}}^2 x_{\mathrm{B}}^2 + \frac{1}{2} \sum_{i=1}^{N} \omega_{\mathrm{S}i}^2 x_{\mathrm{S}i}^2 \tag{14.28}$$

Here $x_{\mathrm{S}i}$ are the modes which diagonalize the Hessian at the saddle point x_{S}^{N+1} and we have denoted the saddle point mode of imaginary frequency by the subscript B. Using (14.27) and (14.28) in (14.26) finally yields

$$k_{\mathrm{TST}} = \frac{1}{2\pi} \frac{\prod_{i=0}^{N} \omega_{Ai}}{\prod_{i=1}^{N} \omega_{\mathrm{Si}}} e^{-\beta E_{\mathrm{B}}} \tag{14.29}$$

Note that the product of frequencies in the denominator goes over the stable modes associated with the saddle point S, and exclude the imaginary frequency ω_{B} associated with the reaction coordinate. If the mode associated with this coordinate could also be identified in the well region, say the mode with frequency ω_{A0}, then Eq. (14.29) can be rewritten in the form

$$k_{\mathrm{TST}} = \frac{\omega_{A0}}{2\pi} e^{-\beta F_{\mathrm{B}}}; \qquad \beta = (k_B T)^{-1} \tag{14.30a}$$

where $F_B = W(x_{\mathrm{S}0})$ is the activation free energy, given by

$$F_{\mathrm{B}} = E_{\mathrm{B}} - TS_{\mathrm{B}} \quad \text{and} \quad S_{\mathrm{B}} = k_B \ln\left(\frac{\prod_{i=1}^{N} \omega_{Ai}}{\prod_{i=1}^{N} \omega_{\mathrm{Si}}}\right) \tag{14.30b}$$

In the one-dimensional form (14.30) F_{B} and S_{B} can be identified as the activation free energy and its entropic component, respectively, thus making explicit the observations already made in Eqs (14.23)–(14.25) and problem 14-1.

14.3.4 Some observations

In view of the simplifying assumptions which form the basis for TST, its success in many practical situations may come as a surprise. Bear in mind, however, that transition state theory accounts quantitatively for the most important factor affecting the rate—the activation energy. Dynamical theories which account for deviations from TST often deal with effects which are orders of magnitude smaller than that determined by the activation barrier. Environmental effects on the *dynamics* of chemical reactions in solution are therefore often masked by solvent effect on the activation free energy, $W(x_{S0})$, with $W(x)$ given by Eq. (14.25).

Transition state theory is important in one additional respect: It is clear from the formulation above that the rate (14.16), (14.26), or (14.29) constitutes an upper bound to the exact rate. The reason for this is that the correction factor discussed above, essentially the probability that an escaping equilibrium trajectory is indeed a reactive trajectory, is smaller than unity. This observation forms the basis to the so-called *variational TST*, [5] which exploits the freedom of choosing the dividing surface between reactants and products: Since any dividing surface will yield an upper bound to the exact rate, the best choice is that which minimizes the TST rate.

Corrections to TST arise from *dynamical effects* on the rate and may become significant when the coupling to the thermal environment is either too large or too small. In the first case the total outgoing flux out of the reactant region is not a good representative of the reactive flux because most of the trajectories cross the dividing surface many times—a general characteristic of a diffusive process. In the extreme strong coupling case the system cannot execute any large amplitude motion, and the actual rate vanishes even though the transition state rate is still given by the expressions derived above. In the opposite limit of a very small coupling between the system and its thermal environment it is the assumption that thermal equilibrium is maintained in the reactant region that breaks down.[6] In the extreme limit of this situation the rate is controlled not by the time it takes a thermal particle to traverse the barrier, but by the time it takes the reactant particle to accumulate enough energy to reach the barrier. This transition from *barrier dynamics to well dynamics* is the basis for the *Lindemann mechanism* discussed in Section 14.2. Indeed, we can rewrite Eq. (14.4) using $\bar{k}_{21} = k_{21}M = k_{12}e^{-\beta E_{21}}M$, where M is

[5] E. Pollak, Variational Transition State Theory for dissipative systems, in *Activated Barrier Crossing*, edited by Fleming, G. R. and Hänggi, P. (World Scientific, 1993), pp. 5–41.

[6] If the product space is bound as in the case of an isomerization rather than dissociation reaction, another source of error is the breakdown of the assumption of fast equilibration in the product region. Unrelaxed trajectories may find their way back into the reactant subspace.

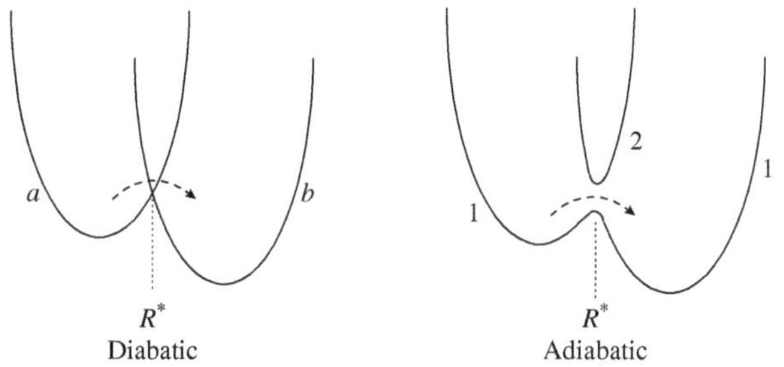

FIG. 14.4 A schematic display of a curve crossing reaction in the diabatic and adiabatic representations. See Section 2.4 for further details. (Note: This is same as Fig. 2.3).

the density of molecules in the bathing gas, to get

$$k = \frac{e^{-\beta E_{21}} M k_{12} k_2}{M k_{12} + k_2} \tag{14.31}$$

k_{12}, the downhill rate, may be assumed to be temperature-independent. We see that depending on the magnitude of M the nature of k changes: For large M it becomes independent of M, in analogy to the TST rate which does not depend on the system–bath coupling. For $M \to 0$ it is dominated by the first of the reactions in (14.1), accumulation of energy in A, and is proportional to M. Obviously, the extreme limit of very large system-environment coupling (e.g. the stoppage of all motion in the large M limit) is not described by this model.

The above discussion, as most of this chapter, is based on a classical picture of the chemical reactions. Quantum mechanical transition state theory is far less obvious; even a transition state cannot be well defined. A particularly simple case that can be formulated within TST is described next.

14.3.5 TST for nonadiabatic transitions

Consider again the TST rate expression (14.12) for a reaction $a \to b$, and assume that the assumptions underlying TST hold except that for a particle that crosses the transition point x_B with forward speed v (direction $a \to b$) there is a finite probability $P_{b \leftarrow a}(v)$ to end up in state b. In this case expression (14.12) should be modified to read

$$k_{TST} = \int_0^\infty dv v P(x_B, v) P_{b \leftarrow a}(v) \tag{14.32}$$

Consider now the reaction depicted in Fig. 14.4 that was discussed in Section 2.4. The classical picture treated in Section 14.3.4 may be thought as the extreme adiabatic limit where the splitting between the adiabatic states 1 and 2 (right panel of Fig. 14.4) is large relative to the thermal energy k_BT so that excitation to the upper potential surface 2 can be disregarded, and in fact this surface may be disregarded altogether. In the nonadiabatic limit the diabatic representation (left panel of Fig. 14.4) is more useful. The reaction $a \to b$ may be described by the Landau–Zener theory (Section 2.4) where the transition probability, Eq. (2.47), is dominated by the dynamics at the surface crossing point R^*:

$$P_{b\leftarrow a} = 1 - \exp\left\{-\frac{2\pi|V_{ab}|^2}{\hbar|\dot{R}\Delta F|}\right\}_{R=R^*} \tag{14.33}$$

Here V_{ab} is the coupling between the two states and $\Delta F \equiv F_b - F_a$ with $F_i = -\partial E_i/\partial R$ being the force that the system experiences while moving on the surface E_i. Using this in Eq. (14.32) where the crossing point R^* replaces the barrier point x_B the transition rate takes the form

$$k = \int_0^\infty d\dot{R}\dot{R}P(R^*, \dot{R})P_{b\leftarrow a}(\dot{R}) \tag{14.34}$$

In the adiabatic limit $P_{b\leftarrow a} \to 1$ and k becomes the standard TST rate that can be evaluated as above to yield Eq. (14.16) with ω_a, the frequency associated with the motion at the bottom of well a, replacing ω_0, and with E_B taken as the barrier height on the adiabatic surface 1 of Fig. 14.4 (right panel). In the non-adiabatic limit $P_{b\leftarrow a} = (2\pi|V_{ab}|^2/\hbar|\dot{R}\Delta F|)_{R=R^*}$. The rate (14.34) can be evaluated in this limit by using

$$\begin{aligned} P(R^*, \dot{R}) &= Z^{-1}\exp\left(-\beta E_a(R^*)\right)\exp\left(-(1/2)\beta m\dot{R}^2\right) \\ Z &= \int_{-\infty}^{\infty} dR \int_{-\infty}^{\infty} d\dot{R}e^{-\beta E_a(R)}e^{-(1/2)\beta m\dot{R}^2} \end{aligned} \tag{14.35}$$

The partition function Z may be evaluated as in the derivation of (14.15), by taking $E_a(R) \simeq (1/2)m\omega_a^2R^2$ to give $Z = 2\pi/\beta m\omega_a$. Using (14.35) in (14.34) then leads to

$$k = |V_{ab}|^2\sqrt{\frac{\pi\beta m}{2}}\frac{\omega_a}{\hbar|\Delta F|_{R=R^*}}e^{-\beta E_a(R^*)} \tag{14.36}$$

Note that $E_a(R^*)$ is the height of the curve-crossing energy above the bottom of the reactant well a. It is also noteworthy that in the non-adiabatic limit the rate depends explicitly on the interstate coupling V_{ab} (in additional to its dependence on the characteristic frequency ω_a and the slopes at the crossing point via ΔF). In the adiabatic limit dependence on V_{ab} enters only indirectly, through its effect on the adiabatic barrier height.

Problem 14.2. Show that the barrier height E_B in the adiabatic limit is related to that of the nonadiabatic limit by $E_B \cong E_a(R^*) - 2|V_{a,b}|$.

14.3.6 TST with tunneling

One can continue the same line of thought that leads to Eq. (14.32) to include tunneling transitions. For definiteness, consider a thermal population of free electrons on the left side of the barrier and let the tunneling transmission coefficient from left to right at energy E be $\mathcal{T}(E)$ (see Section 2.10). Standard TST assumptions: thermal equilibrium in the reactant well and no reflection following transition across the barrier, lead to the following expression for the number of electrons crossing per unit time in one dimension

$$\frac{dN}{dt} = \frac{1}{2}\int_0^\infty dEv(E)n(E)\mathcal{T}(E) \tag{14.37}$$

where $v(E)$ and $n(E)$ are the speed and density of electrons of energy E. The factor half comes from the fact that only half the electrons move in the barrier direction. The electron density is $n(E) = \rho(E)f(E)/L$, where $\rho(E)$ is the density of states, L is the the normalization length, and $f(E)$ is the Fermi–Dirac distribution function. Using Eq. (2.96) for the one-dimensional density of states and $v(E) = \sqrt{2E/m}$ we find $\rho v/L = (\pi\hbar)^{-1}$ so that

$$\frac{dN}{dt} = \frac{1}{\pi\hbar}\int_0^\infty dEf(E)\mathcal{T}(E) \tag{14.38}$$

where we have multiplied by another factor of 2 to account for the electron spin multiplicity. The escape rate (14.38) can be multiplied by the electron charge e to give the electric current going out of the well. When applied to metals under the effect of an electric field this gives and expression for the *field emission current.*

14.4 Dynamical effects in barrier crossing—The Kramers model

The Arrenius expression for the rate of a unimolecular reaction, $k = \kappa\exp(-E_B/k_BT)$, expresses the rate in terms of the activation energy E_B and a pre-exponential coefficient κ. The activation energy reflects the height of the barrier that separates the reactant and product configurations. The equilibrium-based analysis of Section 14.3.3 emphasizes the fact that the potential experienced by the reaction coordinate is a free energy surface associated with a free energy barrier. We have already made the observation that much of the solvent influence on the rate of chemical reactions stems from its effect on this free energy barrier. We have also indicated that dynamical solvent effect should show up in the limits where the system is coupled either very strongly or very weakly to its thermal environment. We already have some intuitive expectation about these effects. For weak system–bath coupling the rate determining step will be the accumulation of energy inside the well so the rate should increase with increasing system–bath coupling. In the fast thermal relaxation limit thermal equilibrium will prevail in the well, and the process will be dominated by the escape dynamics near the barrier. These modes of behavior were already seen in the two-level model of Section 14.2. We also expect a mode of behavior not described by that model: cessation of reaction when the system–bath coupling is so strong so that escape across the barrier is hindered by what is essentially solvent induced friction. For definiteness we will use the terms *strong and weak coupling*, or *high and low friction* to refer to situations in which TST fails for the reasons just indicated, and refer by *intermediate coupling* or *intermediate friction* to cases where the fundamental assumptions of TST approximately hold. The Kramers theory[7] described below provides a framework for analyzing these modes of behavior and the transitions between them.

14.4.1 Escape from a one-dimensional well

The starting point of the Kramers theory of activated rate processes is the one-dimensional Markovian Langevin equation, Eq. (8.13)

$$\begin{aligned} \dot{x} &= v \\ \dot{v} &= -\frac{1}{m}\frac{dV(x)}{dx} - \gamma v + \frac{1}{m}R(t) \end{aligned} \tag{14.39}$$

which is a Newton equation supplemented by random noise and damping which represent the effect of the thermal environment. They are related by the fluctuation dissipation theorem, Eq. (8.20),

$$\langle R(0)R(t)\rangle = 2\gamma m k_B T\delta(t); \qquad \langle R\rangle = 0 \tag{14.40}$$

Equations (14.39) and (14.40) describe a one-dimensional Brownian particle moving under the influence of a systematic force associated with the potential $V(x)$ and a random force and the associated damping that mimic the influence of a thermal environment. In applying it as a model for chemical reactions it is assumed that this

[7] H. A. Kramers, Physica (Utrecht) **7**, 284 (1940)

particle represents the reaction coordinate, and that all other molecular and environmental coordinates can be represented by the stochastic input ingrained in $R(t)$ and γ. This is obviously a highly simplified model in several respects: it assumes that the essential dynamics is one-dimensional, it makes the Markovian approximation with respect to the thermal environment, it characterizes this environment with a single parameter—the temperature T, and it describes the system–bath coupling with a single parameter, the friction γ. No realistic model of chemical reaction in condensed phases can be achieved without improving on these drastic simplifications. Still, we will see that even this simple model can give important insight on the way by which solvent dynamics affects chemical reactions.

We have already seen that Eqs (14.39) and (14.40) are equivalent to the Fokker–Planck equation, Eq. (8.144),

$$\frac{\partial P(x,v;t)}{\partial t} = \frac{1}{m}\frac{dV}{dx}\frac{\partial P}{\partial v} - v\frac{\partial P}{\partial x} + \gamma\left[\frac{\partial}{\partial v}(vP) + \frac{k_B T}{m}\frac{\partial^2 P}{\partial v^2}\right] \tag{14.41}$$

In the present context, Eq. (14.41) is sometimes referred to as the Kramers equation. We have also found that the Boltzmann distribution

$$P_{\text{eq}} = \mathcal{N}\exp\left[-\beta\left((1/2)mv^2 + V(x)\right)\right] \tag{14.42}$$

where $\mathcal{N}$ is a normalization constant, is a stationary zero current solution to this equation. It is convenient to make the substitution

$$P(x,v;t) = P_{\text{eq}}(x,v)f(x,v;t) \tag{14.43}$$

so that the function $f(x,v,t)$ represents the deviation from equilibrium. This leads to

$$\frac{\partial f}{\partial t} = \frac{1}{m}\frac{dV}{dx}\frac{\partial f}{\partial v} - v\frac{\partial f}{\partial x} + \gamma\left(\frac{k_B T}{m}\frac{\partial^2 f}{\partial v^2} - v\frac{\partial f}{\partial v}\right) \tag{14.44}$$

As before, it is assumed that the potential V is characterized by a reactant region, a potential well, separated from the product region by a high potential barrier, see Fig. 14.2. We want to calculate the reaction rate, defined within this model as the rate at which the particle escapes from this well.

The probability that the Brownian particle remains in the initial well at time t is given by $A(t) \equiv \int_{\text{well}} dxP(x,t)$. Following the general discussion in Section 14.2, the rate coefficient for escape from the initial well is given by $k = A^{-1}(-dA/dt)$ *provided that this quantity is time-independent.* For this to happen in an experimentally meaningful time, the timescale associated with the escape out of the well should be much longer than the time it takes to reach the quasi-steady-state characterized by a constant distribution (apart from overall normalization) in the well subspace. In analogy to the model of Section 14.2, we expect that the first time is dominated by the activation factor $\exp(\beta E_B)$, while the second does not depend on this factor. We may conclude that the inequality $\beta E_B \gg 1$, the high barrier limit, is sufficient to insure the conditions needed for a simple time evolution characterized by a constant rate. We therefore limit ourselves to this high barrier limit.

With this realization we may evaluate the rate by considering an artificial situation in which A is maintained strictly constant (so the quasi-steady-state is replaced by a true one) by imposing a source inside the well and a sink outside it. This source does not have to be described in detail: We simply impose the condition that the population (or probability) inside the well, far from the barrier region, is fixed, while outside the well we impose the condition that it is zero. Under such conditions the system will approach, at long time, a steady state in which $\partial P/\partial t = 0$ but $J \neq 0$. The desired rate k is then given by J/A. We will use this strategy to calculate the escape rate associated with Eq. (14.44).

Once the rate is found, the different modes of behavior expected for weak and strong system–bath coupling should reveal themselves through its dependence on the friction γ. As indicated above we expect the rate to increase with γ in the low friction limit, to decrease with increasing γ in the high friction regime and, perhaps, to approach the prediction of TST in intermediate cases. In what follows we start with the high friction limit which is mathematically the simplest, and will later work our way to lower friction regimes.

14.4.2 The overdamped case

The physical manifestation of friction is the relaxation of velocity. In the high friction limit velocity relaxes on a timescale much faster than any relevant observation time, and can therefore be removed from the dynamical equation, leading to a solvable equation in the position variable only, as discussed in Section 8.4.4. The Fokker–Planck or Kramers equation (14.41) then takes its simpler, Smoluchowski form, Eq. (8.132)

$$\frac{\partial P(x,t)}{\partial t} = -\frac{\partial}{\partial x}J(x,t) \tag{14.45a}$$

$$J(x,t) = -D\left(\frac{\partial}{\partial x} + \beta\frac{dV}{dx}\right)P(x,t) \tag{14.45b}$$

where D is the diffusion constant, related to the friction γ by

$$D = \frac{k_B T}{m\gamma} \tag{14.46}$$

and where J is identified as the probability flux. At steady state $J(x,t) = J$ is a constant which can be viewed as one of the integration constants to be determined by the boundary conditions. We then need to solve the equation

$$D\left(\beta\frac{dV}{dx}+\frac{d}{dx}\right)P_{\mathrm{ss}}(x)=-J \tag{14.47}$$

where P_{ss} denotes the steady-state probability distribution. We may also consider the more general case of a position-dependent diffusion coefficient, $D = D(x)$. The equation for $P_{\mathrm{ss}}(x)$ is then

$$\left(\beta\frac{dV}{dx}+\frac{d}{dx}\right)P_{\mathrm{ss}}(x)=-\frac{J}{D(x)} \tag{14.48}$$

As discussed above, we expect that a solution characterized by a constant nonzero flux out of the well will exist for imposed source and sink boundary conditions. In the model of Fig. 14.2 the source should be imposed in the well while the sink is imposed by requesting a solution with the property $P_{\mathrm{ss}}(x)\xrightarrow{x\to\infty}0$. Looking for a solution of the form

$$P_{\mathrm{ss}}(x)=f(x)e^{-\beta V(x)} \tag{14.49}$$

we find

$$\frac{df}{dx}=-\frac{J}{D(x)}e^{\beta V(x)} \tag{14.50}$$

which integrates to give the particular solution

$$f(x)=-J\int_{\infty}^{x}dx'\frac{e^{\beta V(x')}}{D(x')}=J\int_{x}^{\infty}dx'\frac{e^{\beta V(x')}}{D(x')} \tag{14.51}$$

The choice of ∞ as the upper integration limit corresponds to the needed sink boundary condition, $f(x\to\infty)=0$, while assuming a time-independent solution in the presence of such sink is equivalent to imposing a source. Equations (14.49) and (14.51) lead to

$$P_{\mathrm{ss}}(x)=Je^{-\beta V(x)}\int_{x}^{\infty}dx'\frac{e^{\beta V(x')}}{D(x')} \tag{14.52}$$

and integrating both sides from $x=-\infty$ to $x=x_{\mathrm{B}}$ finally yields

$$k=\frac{J}{\int_{-\infty}^{x_{\mathrm{B}}}dxP_{\mathrm{ss}}(x)}=\left[\int_{-\infty}^{x_{\mathrm{B}}}dxe^{-\beta V(x)}\int_{x}^{\infty}dx'\frac{e^{\beta V(x')}}{D(x')}\right]^{-1} \tag{14.53}$$

This result can be further simplified by using the high barrier assumption, $\beta(V(x_{\mathrm{B}})-V(0))\gg 1$, that was already recognized as a condition for meaningful unimolecular behavior with a time-independent rate constant. In this case the largest contribution to the inner integral in Eq. (14.53) comes from the neighborhood of the barrier, $x = x_B$, so $\exp[\beta V(x)]$ can be replaced by $\exp[\beta(E_B-(1/2)m\omega_B^2(x-x_B)^2)]$, while the main contribution to the outer integral comes from the bottom of the well at $x = 0$, so the $\exp[-\beta V(x)]$ can be replaced by $\exp(-(1/2)\beta m\omega_0^2 x^2)$. This then gives

$$k = \left[\int_{-\infty}^{\infty} dx e^{-(1/2)\beta m\omega_0^2 x^2} \int_{-\infty}^{\infty} dx \frac{e^{\beta[E_B-(1/2)m\omega_B^2(x-x_B)^2]}}{D(x_B)} \right]^{-1} \tag{14.54}$$

The integrals are now straightforward, and the result is (using $D = (\beta m\gamma)^{-1}$)

$$k = \frac{\omega_0\omega_B}{2\pi\gamma} e^{-\beta E_B} = \frac{\omega_B}{\gamma} k_{TST} \tag{14.55}$$

The resulting rate is expressed as a corrected TST rate. Recall that we have considered a situation where the damping γ is faster than any other characteristic rate in the system. Therefore, the correction term is smaller than unity, as expected.

Problem 14.3. Use the discussion in Section 8.5 to show that the rate, Eq. (14.53) or (14.55) can be obtained as the inverse mean first passage time to arrive at some point x_1, well to the right of the barrier (the well is on the left as in Fig. 14.2), starting from position x_0, well to the left of it. Quantify the meaning of the phrases "well to the right," and "well to the left."

Solution. The Smoluchowski equation (14.45) is of the form

$$\frac{\partial P(x,t;x_0)}{\partial t} = -\frac{d}{dx}\left[a(x) - b(x)\frac{d}{dx}\right] P(x,t;x_0); \quad P(x,t=0;x_0) = \delta(x-x_0) \tag{14.56}$$

where $b(x) = D(x)$ and $a(x) = \beta D(x)[-dV(x)/dx]$. Its equilibrium solution is, up to a normalization constant

$$P_{eq}(x) = \exp\left(\int^{x} dx' \frac{a(x')}{b(x')} \right) \tag{14.57}$$

The mean first passage time $\tau(x_1, x_0)$ to reach x_1 starting from x_0 is given by Eq. (8.159):

$$\tau(x_1, x_0) = \int_{x_0}^{x_1} dx' \left[D(x')P_{eq}(x')\right]^{-1} \int_{x_c}^{x'} dx'' P_{eq}(x'') \tag{14.58}$$

The constant x_c is the point at which $(d\tau(x_1, x_0)/dx_1)_{x_1=x_c} = 0$. We apply this result to a problem in which x_0 and x_1 on opposite sides of a high barrier so that both $|V(x_B) - V(x_0)|$ and $|V(x_B) - V(x_1)|$ are much larger than k_BT. This implies that in the outer integral in (14.58) the dominant contribution comes from $x' = x_B$. The inner integral is then $\int_{x_c}^{x_B} dx'' P_{eq}(x'')$, and will not depend on x_c if the latter is placed anywhere on the other side of the well opposite to x_B and far enough from the well bottom. This is so because in that case the inner integral is dominated by the neighborhood of the well bottom where $P_{eq}(x)$ has a sharp maximum. For the potential of Fig. 14.2 we can take $x_c = -\infty$ so that

$$\tau(x_1; x_0) = \int_{x_0}^{x_1} dx' \left[D(x')P_{eq}(x')\right]^{-1} \int_{-\infty}^{x_B} dx'' P_{eq}(x'') \tag{14.59}$$

where the integral between x_0 and x_1 is dominated by the barrier and that from $-\infty$ to x_B is dominated by the well. This result is essentially the inverse of (14.53).

We do not, however, need to solve Eq. (14.45) to reach the most important conclusion about this high friction limit. The structure of Eq. (14.45) implies at the outset that the time can be scaled by the diffusion coefficient, so any calculated rate should be proportional to D, and by (14.46) inversely proportional to the friction γ. *The rate is predicted to vanish like* γ^{-1} *as* $\gamma \to \infty$. Recalling that the TST rate does not depend on γ it is of interest to ask how the transition between these different modes of behavior takes place. We address this issue next.

14.4.3 Moderate-to-large damping

When the relaxation is not overdamped we need to consider the full Kramers equation (14.41) or, using Eqs (14.42) and (14.43), Eq. (14.44) for f. In contrast to Eq. (14.45) that describes the overdamped limit in terms of the stochastic position variable x, we now need to consider two stochastic variables, x and v, and their probability distribution. The solution of this more difficult problem is facilitated by invoking another simplification procedure, based on the observation that if the damping is not too low, then deep in the well thermal equilibrium prevails. Equilibrium is disturbed, and dynamical effects need to be considered, only near the barrier where we now place the origin, i.e., $x_B = 0$. We may therefore attempt to solve the dynamical problem by considering, near the barrier, a steady-state distribution that satisfies the boundary conditions:

$$\begin{aligned} &P_{\text{ss}}(x,v) \to P_{\text{eq}}(x,v) \qquad \text{for } x \to -\infty \quad \text{(reactant region)} \\ &P_{\text{ss}}(x,v) \to 0 \qquad \text{for } x \to \infty \quad \text{(product region)} \end{aligned} \tag{14.60}$$

Furthermore, for high barrier, these boundary conditions are satisfied already quite close to the barrier on both sides. In the relevant close neighborhood of the barrier we expand the potential up to quadratic terms and neglect higher-order terms,

$$V(x) = E_{\text{B}} - \frac{1}{2} m\omega_{\text{B}}^2 x^2 \tag{14.61}$$

Using this together with a steady-state condition ($\partial P/\partial t = \partial f/\partial t = 0$) in Eq. (14.44) leads to

$$v\frac{\partial f}{\partial x} + \omega_{\text{B}}^2 x\frac{\partial f}{\partial v} = \gamma\frac{k_B T}{m}\frac{\partial^2 f}{\partial v^2} - \gamma v\frac{\partial f}{\partial v} \tag{14.62}$$

with f satisfying the boundary conditions derived from (14.43) and (14.60)

$$f(x \to \infty) = 0 \quad \text{and} \quad f(x \to -\infty) = 1 \tag{14.63}$$

Note that with the simplified potential (14.61) our problem becomes mathematically similar to that of a harmonic oscillator, albeit with a negative force constant. Because of its linear character we may anticipate that a linear transformation on the variables x and v can lead to a separation of variables. With this in mind we follow Kramers by making the ansatz that Eq. (14.62) may be satisfied by a function f of one linear combination of x and v, that is, we seek a solution of the form

$$f(x,v) = f(v + \Gamma x) \equiv f(u) \tag{14.64}$$

Such solution may indeed be found; see Appendix 14A for technical details. The function $f(v,x)$ is found in the form

$$f(x,v) = \sqrt{\frac{\alpha m}{2\pi k_B T}} \int_{-\infty}^{v+\Gamma x} dz e^{(-\alpha m z^2/2k_B T)} \tag{14.65}$$

where

$$\alpha = -\frac{\Gamma + \gamma}{\gamma} \tag{14.66a}$$

and where the constant Γ is determined to be

$$\Gamma = -\left\{\frac{\gamma}{2} + \left[\left(\frac{\gamma}{2}\right)^2 + \omega_{\mathrm{B}}^2\right]^{1/2}\right\} \tag{14.66b}$$

Thus, finally,

$$P_{\mathrm{ss}}^{(B)}(x,v) = \sqrt{\frac{\alpha m}{2\pi k_B T}} P_{\mathrm{eq}}(x,v) \int_{-\infty}^{v-|\Gamma|x} dz e^{(-\alpha m z^2/2k_B T)} \tag{14.67}$$

where the superscript (B) indicates that this result is valid near the barrier top. Also in the barrier neighborhood, $P_{\mathrm{eq}}(x,v)$ takes the (unnormalized) form

$$P_{\mathrm{eq}}^{(B)}(x,v) = e^{-\beta(E_{\mathrm{B}}-(1/2)m\omega_{\mathrm{B}}^2 x^2)-(1/2)\beta m v^2} \tag{14.68}$$

Equations (14.67) and (14.68) constitutes a full steady-state solution to Eqs (14.41) and (14.60) *near the top of the barrier*. They can be used to compute the steady-state current

$$J = \int_{-\infty}^{\infty} dv v P_{\mathrm{ss}}^{(B)}(x,v) = \sqrt{\frac{\alpha m}{2\pi k_B T}} \left(\frac{k_B T}{m}\right)^{3/2} \left(\frac{2\pi}{\alpha+1}\right)^{1/2} e^{-\beta E_{\mathrm{B}}} \tag{14.69}$$

This result is obtained by integration by parts, using $vP_{\mathrm{eq}}(x,v) = -(\beta m)^{-1}\partial P_{\mathrm{eq}}(x,v)/\partial v$. Note that the constant α is given explicitly, in terms of γ and ω_B, by Eqs (14.66). Also note that J is independent of the position x, as expected from the current at steady state. Indeed, because the steady-state current is position independent, the result (14.69) is valid everywhere even though it was obtained from the distribution (14.67) that is valid only near the barrier top.

We are now in a position to calculate the steady-state escape rate, given by

$$k = \mathcal{N}^{-1} J \tag{14.70}$$

where

$$\mathcal{N} = \int_{\mathrm{well}} dx \int_{-\infty}^{\infty} dv P_{\mathrm{ss}}(x,v) \tag{14.71}$$

is the integrated probability to be in the reactant subspace. In the present one-dimensional problem $\int_{\mathrm{well}} dx = \int_{-\infty}^{x_{\mathrm{B}}} dx$. This integral is dominated by the bottom of the well, were the Maxwell Boltzmann form is a good approximation for the steady state-distribution and where the potential can be expanded to quadratic order near the well bottom (now taken as the origin)

$$P_{ss}^{(W)}(x,v) = P_{eq}(x,v) = e^{-\beta((1/2)mv^2+(1/2)m\omega_0^2x^2)} \tag{14.72}$$

Using (14.72) in (14.71) yields $\mathcal{N} \cong 2\pi/\beta m\omega_0$, and using this in (14.70) together with Eq. (14.69) for J and (14.66) for α finally leads to

$$k = \frac{\omega_r}{\omega_B}\frac{\omega_0}{2\pi}e^{-\beta E_B} = k_{TST}\frac{\omega_r}{\omega_B} \tag{14.73}$$

where

$$\omega_r = \left(\omega_B^2 + \frac{\gamma^2}{4}\right)^{1/2} - \frac{\gamma}{2} \tag{14.74}$$

Again we got the rate as a corrected TST expression. Examination of (14.74) shows that the correction factor ω_r/ω_B becomes ω_B/γ when $\gamma \to \infty$, yielding the large friction limit result (14.55). When $\gamma \to 0$, $\omega_r \to \omega_B$, and $k \to k_{TST}$.

The results (14.73) and (14.74) are seen to bridge between the TST solution and the overdamped solution, but cannot describe the expected vanishing of the rate when $\gamma \to 0$. The reason for this failure is that in using the boundary condition $P_{ss}(x,v) \overset{x\to-\infty}{\longrightarrow} P_{eq}(x,v)$ we have assumed that thermal equilibrium always prevails in the well. This can be true only if thermal relaxation in the well is fast enough to maintain this equilibrium in spite of the escape of the more energetic particles. It is this assumption that breaks down in the small friction limit. This limit therefore requires special handling to which we now turn.

14.4.4 The low damping limit

We start again from the one-dimensional Langevin equation (14.39), (14.40), but focus on the low friction, $\gamma \to 0$, limit. To understand the nature of this limit lets go back to the simple model, Figure 14.1, discussed in Section 14.2. We may use this model as a particularly simple analog to our problem by taking state 1 to represent the bottom of the well, therefore putting $k_1 = 0$, and state 2 as the barrier top. The rates $k_{21} = k_{12}e^{-\beta E_{21}}$ measure the system–thermal bath coupling (also their ratio conveys the information that the bath is in thermal equilibrium) and as such are equivalent to the friction γ. k_2 is the rate of escape once the barrier top is reached.

In the simple two-level model we have distinguished between two limits: The case $k_2 \ll k_{12}$ where thermal equilibrium prevails in the well and the reaction rate is dominated by k_2, and the opposite limit $k_2 \gg k_{12}$ where the rate was found to be controlled by the thermal relaxation between the states 1 and 2. The first of these is the analog of the relatively strong friction cases considered above, where thermal equilibrium was assumed to prevail in the well. The second corresponds to the $\gamma \to 0$ where the rate determining process is the energy accumulation and relaxation inside the reactant well. Note that unlike k_{12} and k_{21} that express rates of energy change in the system, γ expresses just the system–bath coupling strength and the rate of energy exchange is yet to be calculated.

This rate of energy exchange between an oscillator and the thermal environment was the focus of Chapter 13, where we have used a quantum harmonic oscillator model for the well motion. In the $\gamma \to 0$ limit of the Kramers model we are dealing with energy relaxation of a classical anharmonic oscillator. One may justifiably question the use of Markovian classical dynamics in this part of the problem, and we will come to this issue later. For now we focus on the solution of the mathematical problem posed by the low friction limit of the Kramers problem.

Our starting point is again Eqs (14.39)–(14.40) or (14.41), however we already know that our focus should be the energy variable. Recall that in the overdamped limit (Section 14.4.2) we have used the fact that the particle's velocity relaxes fast in order to derive a Fokker–Planck type equation (the Smoluchowski equation) for the position coordinate alone, and have argued that this equation is sufficient to describe the escape process on the relevant timescale. Here we have a similar situation with different variables: the oscillations inside the reactant well are characterized by a rapidly changing phase and a slowly changing (for $\gamma \to 0$) energy. More rigorously, this limit is characterized by the following inequalities between the fundamental rates in the system

$$k \ll \gamma \ll \omega \tag{14.75}$$

where k is the escape rate, the friction γ determines the energy relaxation, and ω is the well frequency—the rate at which phase changes. Again we consider the high barrier case for which k is the smallest rate in the system. The inequality $\gamma \ll \omega$, which characterizes the underdamped nature of the system in this limit, implies that the particle oscillates in the well many times during the characteristic time for energy loss. Therefore, if we use the energy E and phase ϕ as dynamical variables, we should be able to average the corresponding dynamical equations over the fast phase oscillations and get in this way an equation for E.

To accomplish this we transform to action–angle variables ($x, v \to K, \phi$). Recall that the action K is related to the energy E by

$$\frac{dE}{dK} = \omega(K) \tag{14.76a}$$

that is,

$$E(K) = \int_0^K dK'\omega(K'); \quad K(E) = \int_0^E dE'\omega^{-1}(E') \tag{14.76b}$$

and that the dependence of ω on K (or on E) reflects the well anharmonicity. After the stochastic evolution is expressed in terms of these variables an average over the fast phase oscillations yields the coarse-grained evolution in terms of K or E. This procedure, described in Appendix 14B, yields the following Smoluchowski-like equation for the probability density $P(E)$ to find the particle moving in the well with energy E

$$\frac{\partial P(E)}{\partial t} = \frac{\partial}{\partial E}\left[D(E)\left(1 + k_B T\frac{\partial}{\partial E}\right)\omega(E)P(E)\right] \tag{14.77}$$

where the energy diffusion function $D(E)$ is

$$D(E) = \gamma K(E) \tag{14.78}$$

For a harmonic oscillator, where the frequency ω is constant, Eq. (14.76) implies that $E = K\omega$ so the action K is the classical equivalent to the number of oscillator quanta. The linear relationship between D and K is the classical analog of the fact that the rate of relaxation out of the nth level of a harmonic oscillator, Eq. (13.18), is proportional to n.

Problem 14.4.

1. Show that the function J_E defined by

$$J_E(E) = -D(E)\left(1 + k_B T\frac{\partial}{\partial E}\right)\omega(E)P(E) \tag{14.79}$$

can be identified as the energy flux.

2. Show that the equilibrium solution of (14.77) is

$$P(E) \sim \omega(E)^{-1}\exp(-\beta E) \tag{14.80}$$

Note that the pre-exponential factor ω^{-1} in (14.80) is proportional in this one-dimensional model to the density of states on the energy axis (the classical analog of the inverse of the quantum level spacing).

The escape rate in this low damping limit can be found if we assume that Eq. (14.77) remains valid up to the barrier energy E_B, and that this energy provides an absorbing boundary, that is, escape takes place whenever the energy E_{B} is reached. In analogy with the procedure used in Section 14.4.2, we consider the following steady state situation: Particles are injected near the bottom of the well (the exact injection energy is immaterial), and every particle which reaches the energy barrier E_{B} is absorbed, so that $P(E_{\mathrm{B}}) = 0$.

In this state the flux J_E does not depend on E.[8] Starting from

$$D(E)\left(1 + k_B T \frac{d}{dE}\right)\omega(E)P_{\mathrm{ss}}(E) = -J_E \tag{14.81}$$

denote $\omega(E)P_{\mathrm{ss}}(E) = y(E)$ to get

$$y(E) = -\beta J_E e^{-\beta E}\int_{E_{\mathrm{B}}}^{E} dE' e^{\beta E'}\frac{1}{D(E')} \tag{14.82}$$

The choice of the lower limit in this integration is determined by the $P(E_{\mathrm{B}}) = 0$ boundary condition. Equation (14.82) leads to

$$P_{\mathrm{ss}}(E) = \frac{\beta J_E e^{-\beta E}}{\omega(E)}\int_{E}^{E_{\mathrm{B}}} dE' e^{\beta E'}\frac{1}{D(E')} \tag{14.83}$$

The escape rate is given by

$$k = \frac{J_E}{\int_0^{E_{\mathrm{B}}} dE' P_{\mathrm{ss}}(E')} = \frac{1}{\beta}\left[\int_0^{E_{\mathrm{B}}}\frac{dE}{\omega(E)}e^{-\beta E}\int_E^{E_{\mathrm{B}}}\frac{dE'}{D(E')}e^{\beta E'}\right]^{-1} \tag{14.84}$$

For $\beta E_{\mathrm{B}} \gg 1$ we can simplify this expression by noting that the integrands are heavily biased by the exponential factors: the outer integral is dominated by the well bottom, $E \approx 0$, where the frequency is ω_0, while the inner integral is dominated by the neighborhood of $E' = E_{\mathrm{B}}$. Therefore

$$k \simeq \frac{\omega_0 D(E_{\mathrm{B}})}{\beta}\left[\int_0^{E_{\mathrm{B}}} dE e^{-\beta E}\int_0^{E_{\mathrm{B}}} dE' e^{\beta E'}\right]^{-1} \simeq \frac{\omega_0 D(E_{\mathrm{B}})}{\beta}\beta^2 e^{-\beta E_{\mathrm{B}}} \tag{14.85}$$

[8] This is equivalent to the independence of the flux J in Eqs (14.47) and (14.69) on the position x.

and, using also (14.78)

$$k = \beta\omega_0 K(E_B)\gamma e^{-\beta E_B} \tag{14.86}$$

This is the Kramers low friction result for the escape rate k. As anticipated, the rate in this limit is proportional to the friction γ which determines the efficiency of energy accumulation and loss in the well.

Problem 14.5. Show that the escape rate (14.86) is equal to the inverse mean first passage time to reach E_B after starting in the well with energy well below E_B.

To end this discussion, it should be emphasized that using the Langevin equations (14.39) and (14.40) or the equivalent Kramers equation (14.41) in the low friction limit has led to the energy diffusion equation (14.77). An analogous energy relaxation process in a quantum harmonic oscillator model is the master equation treatment of Section 8.3.3. Such models can be classified as "weak collision" processes because energy is assumed to change in a continuous manner in Eq. (14.77) and by jumps between nearest harmonic oscillator levels in the analogous quantum case. Other, "strong collision," models were discussed in the literature, in which a collision event is assumed to cause immediate thermalization of the molecule. For example, for the master equation (8.93), the rate expressions (8.96) represent the weak collision model. Taking the strong collision model for the same master equation would mean to assume $k_{m\leftarrow n} = k_{mn} = \alpha P_{eq}(m)$ where α is a constant. For more details on the use and implications of such models see the review paper by Hänggi, Talkner, and Borkovec cited at the end of this chapter.

14.5 Observations and extensions

The results (14.73)–(14.74) and (14.86) for the unimolecular reaction rate in the moderate-high and low friction limits, respectively, where obtained by Kramers in 1940 (see footnote 7). As a quantitative tool for evaluating such rates this theory is useless. Its great importance stems from its conceptual value. It has provided a framework within which dynamic medium effects on thermal chemical reactions (and other barrier controlled rate processes) may be analyzed and discussed and which can be generalized and extended to cover realistic situations. Moreover, it gives insight about the different ways thermal interactions may affect barrier crossing processes. Whatever its failing as a quantitative theory (which it was never meant to be), the insight obtained through these considerations, has remained a key component in later development of this subject. In what follows we discuss the implications of Kramers theory, and outline some of the extensions developed since its inception.

14.5.1 Implications and shortcomings of the Kramers theory

Figure 14.5 shows how dynamical solvent effects (corrections to the TST rate) behave as a function of molecule–solvent interactions. Several observations should be pointed out:

1. The rate of a barrier crossing reaction decreases like γ in the $\gamma \to 0$ limit and decreases like γ^{-1} when $\gamma \to \infty$, where the friction γ measures the strength of the molecule–solvent interaction, or rather the interaction between the reaction coordinate and its environment.
2. As a function of γ the rate goes through a maximum which is smaller than the TST rate.
3. While not seen explicitly in the figure we have learned that in the low friction limit where $k \sim \gamma$ the rate is controlled by energy diffusion, essentially vertical motion in the well of Fig. 14.2, while in the moderate-strong friction regime it stems from configuration change—motion along the reaction coordinate.
4. While the general behavior displayed in Fig. 14.5 always has regimes where dynamical effects are important and TST fails, we can expect conditions under which TST will work well. Let the full line in the figure express the behavior of the model as developed above. Suppose now that we have another control parameter by which we can enhance the energy relaxation in the well without affecting the solvent friction, that is, without changing the molecule–solvent coupling and through that the hindrance to motion along the reaction coordinate. In this case thermal equilibration in the well (the condition whose breach causes the low friction failure of TST) becomes more efficient, and the resulting behavior of rate versus friction will be represented by the dotted line. While we do not have a practical control parameter that can accomplish this, we do have a conceptual one. We have seen in Chapter 13 that vibrational energy relaxation in small molecules is relatively inefficient, and can be very slow (timescale in the range of 1–10^{-6} s) in clean atomic hosts at low temperatures. In contrast, large molecules relax very quickly, on the picoseconds timescale. This difference between small and large molecules stems not from the strength of their coupling to the environment but because of the availability in large molecules of relaxation routes that release energy in relatively small quanta. As a result, thermal equilibration in the well is much more efficient in large molecule, resulting in the behavior shown by the dotted line in Fig. 14.5.

The last observation concerning the difference between small and large molecules points out one direction in which the one-dimensional Kramers theory

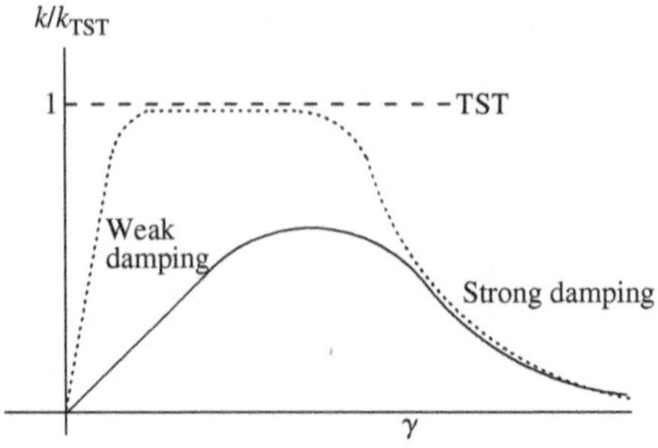

FIG. 14.5 The thermal rate, in units of k_{TST}, displayed as a function of the solvent friction γ. This dependence is characterized by a turnover from the low friction behavior $k \sim \gamma$ to the high friction behavior $k \sim \gamma^{-1}$. In the intermediate regime the rate approaches k_{TST} from below. The full line characterizes the behavior of a small molecule, where relaxation in the well is slow. The dotted line represents the expected behavior of a large molecule where, because of considerably faster energy relaxation in the well, the low friction regime turns over to a TST behavior at much smaller γ.

needs to be extended.[9] However, the most serious drawback of this theory, when considered against realistic situations involving molecular systems, is the use of Markovian stochastic models. The use of delta-correlated noise and the associated constant friction in Eq. (14.39) and (14.40) amount to the assumption that the timescale associated with the motion of the thermal environment is much shorter than the timescales associated with the molecular motion. In fact, the opposite is true. The shortest timescales in the problem are associated with the molecular frequencies which are typically in the range of $10^{14\pm1}$ s^{-1}, while the timescales of intermolecular motions associated with molecule–solvent interactions are in the range of $10^{12\pm1}$ s^{-1}. *A Markovian theory of vibrational energy relaxation can overestimate the rate of this process by many orders of magnitude*, rendering as meaningless the quantitative aspect of the Kramers rate in the low friction limit. The moderate-high friction result (14.73) and (14.74) may be more reliable: here the bath timescale should be compared to the barrier frequency ω_B that may be considerably lower than the well frequency ω_0. Still, non-Markovian effects should be considered also in this case. We expand on this issue below.

Problem 14.6. Examine the theory of vibrational energy relaxation of Chapter 13 in order to explain the above statement that a Markovian theory of vibrational energy relaxation can overestimate the rate of this process by many orders of magnitude.

[9] A. Nitzan and Z. Schuss, Multidimensional barrier crossing, in Fleming, G. R. and Hänggi, P. eds, *Activated Barrier Crossing* (World Scientific, London, 1993).

Another obvious shortcoming of the theory is its classical nature. The need for quantum mechanics can arise in two ways. First and obvious is the possibility that the transition of interest is affected by tunneling or by nonadiabatic curve-crossing transitions. We have discussed the TST aspects of these phenomena in Sections 14.3.5 and 14.3.6.[10] Less obvious is the fact that, as discussed in Sections 13.4.1 and 13.6, quantum mechanical effects in the vibrational energy relaxation of small molecules can be very large. Both these manifestations of quantum effects in barrier crossing become important, in particular, at low temperatures.

Finally, Kramers solution to the barrier crossing problem consists of two expressions: Eqs (14.73) and (14.74) correspond to moderate-to-high damping and span the range of behaviors between that corresponding to TST and the large damping regime, and Eq. (14.86) describes the low damping behavior. A practical bridging formula is

$$\frac{1}{k} = \frac{1}{k_{\text{low damping}}} + \frac{1}{k_{\text{moderate to large damping}}} \tag{14.87}$$

Given the qualitative character of the theory with regard to realistic situations, such ad hoc approach is both reasonable and practical. A rigorous theoretical treatment[11] is based on the normal mode approach to the barrier dynamics (see Section 14.5.3 below) supplemented by incorporating the rate at which the reactive barrier mode exchanges energy with other modes in the well. It yields the expression

$$k = k_{\text{TST}} \frac{\omega_r}{\omega_{\text{B}}} \exp\left\{ \frac{1}{\pi} \int_{-\infty}^{\infty} dy \frac{\ln\left[1 - \exp\left(-(\delta/4)\left(1 + y^2\right)\right)\right]}{1 + y^2} \right\} \tag{14.88}$$

where ω_{B} and ω_r are given by Eqs (14.61) and (14.74), respectively, and $\delta = D(E_{\text{B}})/k_B T$ with $D(E)$ given by (14.78).[12] Note that when δ increases beyond 1 the exponential correction term becomes unity and Eq. (14.73) is recovered.

[10] Dynamical corrections to rates associated with tunneling barrier crossing have been subjects of extensive theoretical studies. These are reviewed in the papers by Hänggi et al. and by Melnikov, cited at the end of this chapter.

[11] E. Pollak, H. Grabert, and P. Hänggi, J. Chem. Phys. **91**, 4073 (1989).

[12] The general result of this work[9] is of the same form (14.88) but the expression for δ, the dimensionless average energy loss by the reactive mode during its excursion into the well and back to the barrier top, is more general.

14.5.2 Non-Markovian effects

The starting point of a non-Markovian theory of barrier crossing is the generalized Langevin equation (cf. Eqs (8.61) and (8.62))

$$\ddot{x} = -\frac{1}{m}\frac{\partial V(x)}{\partial x} - \int_0^t d\tau Z(t-\tau)\dot{x}(\tau) + \frac{1}{m}R(t) \tag{14.89}$$

where the stationary stochastic force $R(t)$ and the friction kernel $Z(t)$ satisfy $\langle R \rangle = 0$ and $\langle R(0)R(t) \rangle = mk_BTZ(t)$. We have seen (Section 8.2.5) that this equation can be derived from the microscopic Hamiltonian, Eqs (8.47) and (8.48), in which a harmonic bath is assumed to couple linearly to the system of interest, in our case the reaction coordinate. In this case $R(t)$ and $Z(t)$ are found (Eqs (8.55) and (8.56)) in terms of the spectral properties of the system–bath coupling.

The dynamical contents of Eq. (14.89) is much more involved than its Markovian counterpart. Indeed, non-Markovian evolution is a manifestation of multidimensional dynamics, since the appearance of a memory kernel in an equation of motion signifies the existence of variables, not considered explicitly, that change on the same timescale. Still, the physical characteristics of the barrier crossing process remain the same, leading to similar modes of behavior:

1. In the barrier controlled regime, thermal relaxation in the well is assumed fast and the rate determining step is the barrier crossing itself. The problem can be solved using the simplified potential (14.61) and calculating the steady-state current for the same boundary conditions used in Section 14.4.3, that is, thermal distribution on the reactant side of the barrier and absorbing boundary (i.e. zero probability) on the product side. The result is of the same form as Eq. (14.73), except that the *reactive frequency* ω_{r} is different. It is given[13] as the largest (real and positive) root of the equation

$$\lambda^2 - \omega_{\mathrm{B}}^2 + \lambda\tilde{Z}(\lambda) = 0 \tag{14.90a}$$

   where $\tilde{Z}(\lambda)$ is the Laplace transform of the memory kernel

$$\tilde{Z}(\lambda) = \int_0^\infty dt e^{-\lambda t} Z(t) \tag{14.90b}$$

2. In the low damping limit the rate determining step is again the energy accumulation in the well. The idea that one can average over the fast phase oscillations

[13] R. F. Grote and J. T. Hynes, J. Chem. Phys. **73**, 2715 (1980); **74**, 4465 (1981).

in order to derive an energy diffusion equation for this process is still valid. Indeed, the same Eq. (14.77) is obtained, where the energy diffusion function $D(E)$ now given by[14]

$$D(E) = \frac{m}{\omega(E)} \int_0^{\infty} dt Z(t) \langle v(0)v(t) \rangle_E \tag{14.91}$$

where $\langle v(0)v(t) \rangle_E$ is the microcanonical average, over the initial phase at a given energy, of the velocity product $v(0)v(t)$ for the isolated particle. For a harmonic oscillator this microcanonical correlation function is just $(E/m)\cos(\omega t)$ and using $E/\omega = K(E)$ we get

$$D(E) = (1/2)K(E) \int_{-\infty}^{\infty} dt e^{i\omega t} Z(t) \tag{14.92}$$

This result is akin to Eq. (13.22), which relates the vibrational relaxation rate of a single harmonic oscillator of frequency ω to the Fourier transform of the force autocorrelation function at that frequency. In the Markovian limit, where (cf. Eq. (8.60)) $Z(t) = 2\gamma\delta(t)$, we recover Eq. (14.78).

The general behavior of the barrier crossing rate as a function of coupling strength to the surrounding environment, shown in Fig. 14.5, follows from general considerations that remain valid also in the non-Markovian case. There is however an interesting difference in the underlying physics that governs the high friction limit. To see this consider the model

$$Z(t) = \frac{\gamma}{\tau_c} e^{-|t|/\tau_c} \tag{14.93}$$

in which τ_c stands for the bath correlation time. From (14.92) we get for this model

$$D(E) = K(E) \frac{\gamma}{(\omega\tau_c)^2 + 1} \tag{14.94}$$

Now, if $\gamma \to \infty$ at constant τ_c, the energy diffusion becomes faster, the well distribution is rapidly thermalized and becomes irrelevant for the crossing dynamics. This is the moderate-high friction regime discussed above. However, increasing γ while maintaining a constant τ_c/γ ratio actually leads to decreasing D. Such a limit is potentially relevant: the experimental parameter pertaining to liquid friction is the liquid viscosity, and highly viscous, overdamped fluids are characterized by sluggish, large τ_c motions. When such a limit is approached, the assumption that

[14] B. Carmeli and A. Nitzan, Phys. Rev. Lett. **49**, 423 (1982); Chem. Phys. Lett., **102**, 517 (1983).

the well motion is quickly thermalized breaks down and the well dynamics may become rate determining again, as first discovered in numerical simulations.[15] The behavior in Fig. 14.5 is still expected, but the decreasing of the rate in the high friction limit may result from this different physical origin.

14.5.3 The normal mode representation

Many discussions in this text start by separating an overall system to a system of interest, referred to as "the system" and the "rest of the world," related to "the bath." The properties assigned to the bath, for example, assuming that it remains in thermal equilibrium throughout the process studied, constitute part of the model assumptions used to deal with a particular problem.

Obviously, the separation itself is not an approximation and can be done arbitrarily, in the same way that the Hamiltonian $\hat{H}$of a given system can be separated into components $\hat{H}_0$ and $\hat{V}$ in many ways. Different routes chosen in this way to treat the same problem should, and do, give the same result if treated rigorously. Sometimes, however identical results are conveyed in ways that show different physical aspects of the problem. Consider, for example, the decay of a discrete state, energetically embedded in, and coupled to, a continuous manifold of states. This model was discussed in Section 9.1. We have analyzed the problem by separating the Hamiltonian according to $\hat{H} = \hat{H}_0 + \hat{V}$, defining the discrete state $|1\rangle$ and the manifold $\{|l\rangle\}$ as eigenstates of $\hat{H}_0$ and have assigned some properties to the coupling elements $V_{1,l}$ and the density of states in the $\{l\}$ manifold. These model assumptions have led to exponential relaxation of the initially prepared state $|1\rangle$. The same problem can be done (see Problem 9.1) by working in the basis of exact eigenstates of $\hat{H}$. In the standard time evolution according to Eq. (2.6) the relaxation appears as destructive interference between the components of an initial wavepacket. A similar observation can be made in the simple model for vibrational relaxation advanced in Section 9.4. In the particular representation chosen, where the systems and bath oscillators are linearly coupled, the initially excited system oscillator is damped by this coupling to the bath. However, the Hamiltonian (9.44) is bilinear in the oscillator coordinates, implying that we could diagonalize it to find a set of exact normal modes. Again, what appears as damping in one representation is just a destructive interference within an initially prepared packet of such modes.[16]

---

[15] J. E. Straub, M. Berkovec, and B. J. Berne, J. Chem. Phys. **83**, 3172; **84**, 1788 (1986).

[16] This statement holds at zero temperature. At finite T we assign to the bath an additional property—being at thermal equilibrium at this temperature throughout the process. It is hard, though not impossible to use this attribute of the model in the exact normal mode basis. Also, if the coupling was nonlinear we could no longer cast the problem in terms of modes of an exactly separable Hamiltonian.

Consider now the barrier crossing problem in the barrier controlled regime discussed in Section 14.4.3. The result, the rate expressions (14.73) and (14.74), as well as its non-Markovian generalization in which ω_r is replaced by λ of Eq. (14.90), has the structure of a corrected TST rate. TST is exact, and the correction factor becomes 1, if all trajectories that traverse the barrier top along the "reaction coordinate" (x of Eq. (14.39)) proceed to a well-defined product state without recrossing back. Crossing back is easily visualized as caused by collisions with solvent atoms, for example, by solvent-induced friction.

This physical picture is perfectly reasonable, however the mathematical representation of friction in Section 8.2.5 as originated from linear coupling to a bath of harmonic oscillators makes it possible to construct an equivalent mathematical model.[17] In the barrier region the dynamics of the reaction coordinate is governed by the potential (14.61), a harmonic potential with an imaginary frequency. This is true for a general molecular system: the potential near the saddle point that marks the lowest energy path between the reactant and product configurations can be expanded about this point to quadratic order. If, in addition, solvent-induced friction is also represented by linear coupling to a harmonic bath as in Eqs (8.47) and (8.48), the Hamiltonian of this overall system is bilinear and can be diagonalized to yield true normal modes. In this new representation there is no coupling between degrees of freedom, therefore no friction. A reaction coordinate can still be identified as an imaginary frequency mode. Indeed, if the overall multidimensional potential surface (in the space of all system and bath degrees of freedom) supports two separate reactant and product configurations, there must be a saddle point that represents the minimum energy path between these configurations. The new set of independent modes are simply the normal modes obtained by diagonalizing this *overall multidimensional Hamiltonian* (as opposed to just the molecular Hamiltonian) after expanding the potential to quadratic order about this saddle point. Compared with the original reaction coordinate which resides in the molecular subspace, the new one will be rotated toward the bath subspace, that is, contain some components of the bath modes.

Consider now the motion along this reaction coordinate. This is a motion that (1) connects between the reactant and the product basins of attraction, and (2) proceeds at the top of the barrier, that is, through the saddle point, with no coupling to other modes therefore no interactions or "collisions" that may cause reflection. This implies, given the original assumption that thermal equilibrium prevails in the reactant well, that TST must hold exactly. In other words, by choosing the correct reaction coordinate, the Kramers model in the barrier-controlled regime can be cast in terms of TST.

---

[17] E. Pollak, J. Chem. Phys. **85**, 865 (1986).

This observation suggests a practical way for evaluating the barrier crossing rate in the moderate–large friction regime. Following the steps taken in Section 14.3.3, we need to diagonalize the harmonic Hamiltonians associated with the bottom of the reactant well and with the barrier, both in the overall space that includes the harmonic bath and the system–bath coupling Eqs (8.47) and (8.48). Following diagonalization, we should use the resulting frequencies in expression (14.29). It can be shown (footnote 16), that this procedure yields again the result (14.73) and (14.74) in the Markovian limit, and the same result with ω_r replaced by λ of Eq. (14.90) in the general case. The extension of this approach to the low friction regime requires additional considerations and leads to Eq. (14.88) as discussed above.

14.6 Some experimental observations

The qualitative aspects of barrier crossing dynamics, as expressed by Fig. 14.5, are model-independent. Once explained, it seems inevitable that the effect of solute–solvent interaction should behave in the way shown. Indeed, the high friction behavior is often observed as expected—rates do decrease with increasing solvent friction, as expressed, for example, by its viscosity. The search for the "Kramers turnover," that is, the transition from the low to high friction behavior has proven considerably more challenging. Figures 14.7 and 14.8 show experimental results for the rate of isomerization of excited trans-stilbene (see Fig. 14.6), identified as the nonradiative decay rate of this species. Figure 14.7 summarizes data obtained in various experiments in different gas and liquid phases, which show the rate as a function of the inverse self-diffusion coefficient of the solvent, taken as a measure of the friction (see Eq. (11.68)). The Kramers turnover is seen to be located at the borderline between gas and liquid so that solution phase reactions appear to belong to the overdamped Kramers regime. This explains why early attempts to find this turnover behavior in low viscosity solvents were unsuccessful, and is also compatible with the observation (Section 14.5.1) that vibrational energy relaxation in large molecules in solution is fast and can hardly expected to become rate-limiting.

Going into details, however, proves both difficult and illuminating. The strength of the Kramers theory originates from its generic character. This is also its weakness, as it cannot account for specific solvent effects that can mask generic trends. It was pointed out (see discussion below Eq. (14.25)) that the barrier experienced by the reaction coordinate has the character of a free energy barrier[18] and may reflect features that stem from the solute–solvent interaction. Figure 14.8 shows that the activation energy in the stilbene isomerization reaction does depend on

[18] This statement is rigorous only within TST.

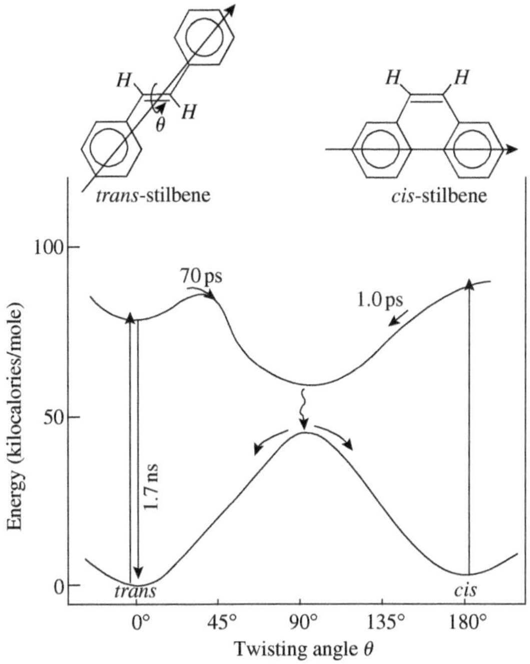

FIG. 14.6 Schematic one-dimensional potential surfaces showing the energy barriers for ground state stilbene (lower curve) and excited (S_1) state stilbene (upper curve). The arrow through the structure diagram of each isomer shows the direction of the transition dipole moment between the ground and excited state. The reaction coordinate involves twisting about the carbon–carbon double bond. A large barrier separates the *cis* and *trans* isomers on the ground state surface but not on the excited state curve. The *cis-trans* transition in the excited state is believed to proceed via an intermediate for which this twisting angle θ is 90°. The times indicated refer to measurements in Hexane. The experimental results reported below refer to the barrier crossing between the excited *trans*-stilbene and the intermediate 90° configuration on the excited state surface. (From G. R. Fleming and P. G. Wolynes, Phys. Tod. p. 36, May 1990).

solvent properties. This has to be taken into account in comparing predictions of the Kramers theory to experiments, and rationalizes the solvent-dependent results seen in Fig. 14.9. An interesting observation is that the deviations from the large friction Kramers behavior in Fig. 14.7, that could be attributed to non-Markovian effects, are absent in Fig. 14.9 that use solvent-specific fitting parameters.

We end this brief excursion into the experimental literature by noting that other issues should be of concern in addition to the uncertainties discussed above:

1. To what extent does "macroscopic friction", as indicated by solvent viscosity or by inverse self-diffusion coefficient, really reflects the microscopic friction experienced by the reaction coordinate?

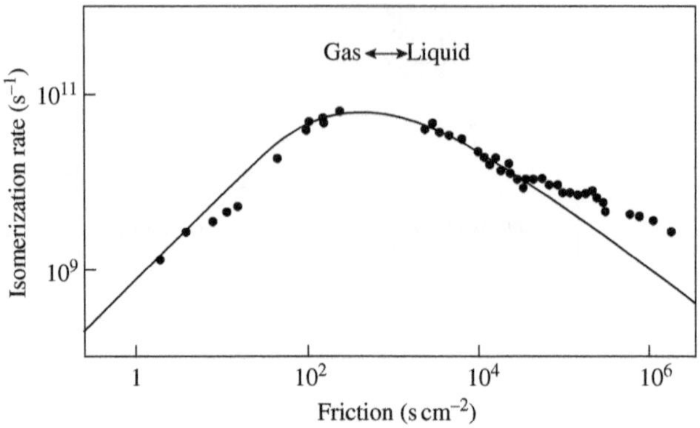

FIG. 14.7 A compilation of gas and liquid phase data showing the turnover of the photoisomerization rate of *trans*-stilbene as a function of the "friction" expressed as the inverse self-diffusion coefficient of the solvent, where the latter is varied over six orders of magnitude in systems ranging from supersonic expansion to low- and high-pressure gases and liquid solutions. The turnover occurs at the borderline between gas and liquid. (From G. R. Fleming and P. G. Wolynes, Phys. Tod. p. 36, May 1990. The solid line is a theoretical fit based on J. Schroeder and J. Troe, Ann. Rev. Phys. Chem. **38**, 163 (1987)).

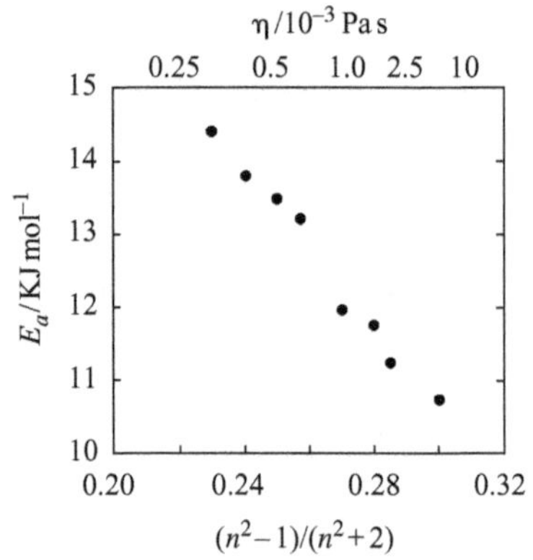

FIG. 14.8 The dependence of the activation energy of the *trans*-stilbene isomerization reaction on solvent (*n*-hexane) viscosity, η, and polarity expressed in terms of the refractive index n as shown. The input for this graph is obtained from rate, viscosity, and refraction index data over a wide range of temperature and pressure. (From J. Schroeder, J. Troe, and P. Vöhringer, Chem. Phys. Lett. **203**, 255 (1993)).

2. How do we know that the reaction under study proceeds adiabatically? An observed barrier can actually signify avoided crossing as in Fig. 14.4.
3. In applying ultrafast spectroscopy to kinetic measurements as done in the excited *trans*-stilbene isomerization, do we really look at thermal reaction rates? The alternative is that the reaction proceeds before complete vibrational

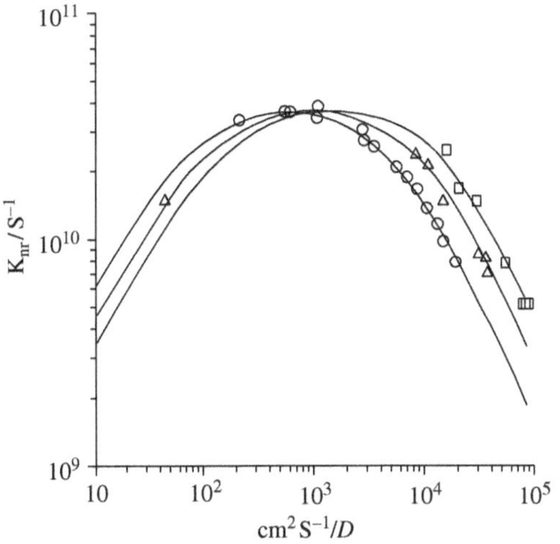

FIG. 14.9 The isomerization rate on *trans*-stilbene displayed as a function of the inverse self-diffusion coefficient of the solvent at $T = 298$ K. The measurements are done at different pressures in super-critical and liquid alkane solvents: Ethane (circles), propane (triangles), and *n*-butane (squares). The solid lines represent fit to the Markovian Kramers theory that use solvent modified barrier height (E_B) and barrier frequency (ω_B). From Jörg Schroeder, Ber. Bunsenges. Phys. Chem. **95**, 233 (1991).

relaxation in the excited state has taken place and therefore depends on the way the reactant is prepared, thus distorting the interpretation of the observed results.

Besides indicating areas for future concern, these questions exemplify the ever-present tension between our desire to explain observations by the most general and generic models, and between the ever-present system-specific and experiment-specific features.

14.7 Numerical simulation of barrier crossing

Why (said the queen), sometimes I've believed as many as six impossible things before breakfast. (Alice's Adventures in Wonderland—Lewis Carrol)

One way to bridge the gap between simple models used for insight, in the present case the Kramers model and its extensions, and realistic systems, is to use numerical simulations. Given a suitable force field for the molecule, the solvent, and their interaction we could run molecular dynamic simulations hoping to reproduce experimental results like those discussed in the previous section. Numerical simulations are also often used to test approximate solutions to model problems, for example, given the Langevin equations (14.39) and (14.40) for a particle moving in a potential V under the influence of random force and damping, and given that V is characterized by a high barrier separating two domains of attractions, we may try to examine numerically the dependence of the transition rate on the friction γ. In both cases the computer generates a trajectory, a series of positions and velocities of the entity that represents the reaction coordinate at successive time intervals.[19]

So, in principle we can start each trajectory by putting the particle (i.e. the reaction coordinate) in the well and generate the subsequent evolution. The particle will start near the bottom of the well and we can compute the average escape time $\langle \tau \rangle$, which is the mean first passage time (see Section 8.5) to reach the product side of the barrier. $\langle \tau \rangle$ is computed by (1) running a trajectory until the particle exits the well, (2) recording the time it took for this to happen, (3) repeating the calculation many times and averaging this exit time over all trajectories, that is, over all realizations of the stochastic force or over all molecular dynamic trajectories with initial conditions compatible with the solvent density and temperature. A criterion for "exit" has to be defined.[20] Once $\langle \tau \rangle$ is computed, the rate is given by $k = \langle \tau \rangle^{-1}$ (see Problem 14.3).

For high barriers, $E \gg k_B T$, this numerical approach is very difficult. The rate is very small in this case, being proportional to $\exp(-\beta E_B)$, so we have to integrate for a very long time before we see an exit event. And then we need to average over many such events. We face a problem analogous to that discussed in Section 13.6: We need to integrate using time steps short relative to $2\pi/\omega_0$ where ω_0 is the frequency of the oscillations in the well—a characteristic molecular frequency. At the same time we need to run trajectories as long as the exit time. The essence of

[19] To generate the trajectories that result from stochastic equations of motion (14.39) and (14.40) one needs to be able to properly address the stochastic input. For Eqs (14.39) and (14.40) we have to move the particle under the influence of the potential $V(x)$, the friction force—$\gamma v m$ and a time-dependent random force $R(t)$. The latter is obtained by generating a Gaussian random variable at each time step. Algorithms for generating realizations of such variables are available in the applied mathematics or numerical methods literature. The needed input for these algorithms are the two moments, $\langle R \rangle$ and $\langle R^2 \rangle$. In our case $\langle R \rangle = 0$, and (cf. Eq. (8.19)) $\langle R^2(t) \rangle = 2m\gamma k_B T/\Delta t$, where Δt is the time interval used by the integration routine, and $\langle R(t_1)R(t_2) \rangle = 0$ for t_1 and t_2 from different time intervals. Given these moments, the required sequence that represents $R(t)$ can be generated and Eq. (14.39) can be solved in standard ways. Obviously we need to generate many solutions with different realizations $R(t)$ and average every calculated result over this ensemble of solutions.

[20] As indicated in the solution to Problem 14.3, for high barrier, $E_B \gg k_B T$, the starting point x_0 and the end point x_1 (where "exit" is decided) can be taken anywhere well in the reactant region, and well in the product region, respectively. "Well" in these region imply a position x at which the potential is considerably lower (relative to $k_B T$) than its value at the barrier top. Variations in x_0 and x_1 that adhere to this condition affect the computed rate only marginally.

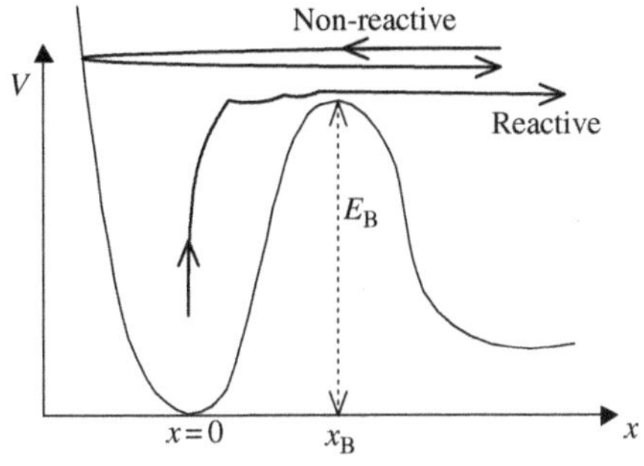

FIG. 14.10 A model for a barrier crossing reaction showing reactive and nonreactive trajectories.

our problem is that *we need to follow the system in great detail in order to extract information about very rare highly improbable events.*

A way to overcome this difficulty is to realize that the factor $e^{-\beta E_B}$ that makes exit events extremely rare has no *dynamical* origin. To put this observation to practice we reformulate our problem: Rather than attempting to calculate the rate k we look for the correction α to the TST rate

$$k = \alpha k_{TST} \tag{14.95}$$

where in the model (14.39) $k_{TST} = (\omega_0/2\pi)\exp(-\beta E_B)$, and in any case can be computed from equilibrium considerations. The correction α contains the friction-dependent dynamical information. It can be found from the following argument below.

Consider the system at equilibrium. The TST rate is obtained from the equilibrium flux in the outer direction, by dividing it by the well population. It was already argued that this is actually an upper bound to the true rate. One way to see this is to realize that only part of the trajectories that go out of the well are in fact reactive. This is seen in Fig. 14.10 which depicts a single well process as in dissociation or desorption.[21] Obviously, the correction factor α is the fraction of reactive trajectories relative to the total equilibrium flux.

An easy way to find this correction factor is to look at the history of an exit trajectory. This history is followed by starting at $x = x_B$ trajectories with velocity sampled from a Maxwell–Boltzmann distribution in the outward direction—these represent the outgoing equilibrium flux, then inverting the velocity ($v \to -v$) so that the particle is heading into the well, and integrating the equations of motion

[21] The argument is easily extended to a double well situation. In this case "reactive trajectories" should not only start deep enough inside the reactant well but should end deep enough in the product well.

with these new initial conditions. In other words, we send a thermal distribution of particles from x_B *into* the well. The resulting trajectories trace back the history of the exiting trajectories that constitute the equilibrium flux. Part of these trajectories will hit the inner wall, bounce back and eventually exit the well without being trapped (say, before loosing at least k_BT energy). The others will become trapped (by the same criterion) before exiting the well. For the single well problem α is simply the fraction of trapped trajectories.[22]

As a numerical simulation problem the computation of α is easy since it involves following short-time trajectories: these trajectories start at the barrier top going into the well, and end either by going through the barrier in the outward direction, or by loosing a predefined amount of energy, of the order k_BT that practically insures their trapping.

Problem 14.7. Describe how the correction factor α should be computed in calculating the reaction rate in a double well potential.

Another way to apply the same idea can be derived from formal considerations.[23] We now consider a double well system that correspond to the reaction $R \rightleftharpoons P$ and suppose that we can generate trajectories as before. Define the population function

$$h(x) = \begin{cases} 1; & x > x_B \quad (P \text{ region}) \\ 0; & x < x_B \quad (R \text{ region}) \end{cases} \tag{14.96}$$

so that $\langle h \rangle = f$ is the equilibrium probability that the system is in the P state. Define $q(t) = x(t) - x_B$ and let $q(0) = 0$. Now consider the function

$$C(t) = P_{\text{eq}}(x_B)\langle \dot{x}(0)h[q(t)] \rangle \tag{14.97}$$

For t infinitesimally larger than zero, $h[q(t = 0+)]$ is 1 if $\dot{x} > 0$ and is zero otherwise. Therefore,

$$C(0+) = P_{\text{eq}}(x_B)\langle \dot{x}\,\theta(\dot{x}) \rangle \tag{14.98}$$

This is the total equilibrium flux out of the R region, and if divided by the normalized equilibrium population of R, that is, $1 - f$, will give the transition state rate from

[22] In some systems α itself may be an observable: In surface physics it is known as the sticking coefficient, the *fraction* of particles that get stuck and adsorb on a surface upon hitting it. Note that trapping does not mean that the particle will never exit or desorb, only that it will equilibrate in the well before doing so.

[23] D. Chandler, A Story of rare events: From barriers to electrons to unknown pathways, in *Classical and Quantum Dynamics in Condensed Phase Simulations*, edited by B. J. Berne, G. Ciccotti, and D. F. Coker, *Proceedings of the International School of Physics on Classical and Quantum Condensed Phase Simulations*, 1997.

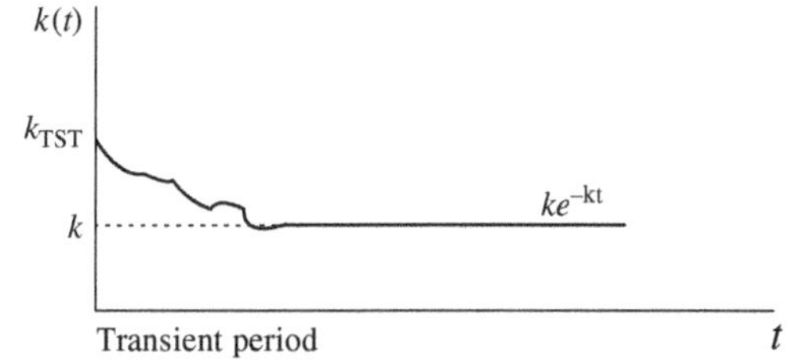

FIG. 14.11 The reactive flux correlation function, Eq. (14.99) plotted against time. After the initial transient period this function becomes essentially constant on the time scale shown.

R to P. Also, in the infinite time limit $C(t \to \infty) \to 0$ because at long time $\dot{x}(0)$ and $h[q(t)]$ are not correlated. However, this relaxation to zero is very slow, of the order of the reaction rate k, because losing this correlation implies that the trajectory has to go several times between the wells.

Next consider the function

$$k(t) = \frac{C(t)}{(1-f)} = \frac{P_{\mathrm{eq}}(x_{\mathrm{B}})}{1-f}\langle \dot{x}(0)h[q(t)]\rangle \tag{14.99}$$

This function is expected to behave as seen in Fig. 14.11. As just argued, at $t = 0+$ it includes all the equilibrium flux, therefore $k(t = 0) = k_{\mathrm{TST}}$. There follows a relative short time period during which the trajectory may be reflected before it relaxes. After this short transient time we reach a situation where the trajectory is well inside the R or the P regions. This is where we would have stopped the simulation for finding the sticking coefficient described above. The correlation function continues to relax to zero on the relatively long reaction timescale, however already at this "plateau" we have a good approximation (better for a higher barrier) for the reaction rate k.

14.8 Diffusion-controlled reactions

Our focus so far was on unimolecular reactions and on solvent effects on the dynamics of barrier crossing. Another important manifestation of the interaction between the reaction system and the surrounding condensed phase comes into play in bimolecular reactions where the process by which the reactants approach each other needs to be considered. We can focus on this aspect of the process by considering bimolecular reactions characterized by the absence of an activation barrier, or by a barrier small relative to $k_B T$. In this case the stage in which reactants approach each other becomes the rate determining step of the overall process.

In condensed phases the spatial motion of reactants takes place by diffusion, which is described by the Smoluchowski equation. To be specific we consider a particular reaction model: The reactants are two species, A and B, where the A molecules are assumed to be static while the B molecules undergo diffusion characterized by a diffusion coefficient D.[24] A chemical reaction in which B disappears occurs when B reaches a critical distance R^* from A. We will assume that A remains intact in this reaction. The macroscopic rate equation is

$$\frac{d[B]}{dt} = -k[B][A] \tag{14.100}$$

where $[A]$ and $[B]$ are molar concentrations. Our aim is to relate the rate coefficient k to the diffusion coefficient D.

It is convenient to define $A = \mathcal{A}[A]$ and $B = \mathcal{A}[B]$, where $\mathcal{A}$ is the Avogadro number and A and B are molecular number densities of the two species. In terms of these quantities Eq. (14.100) takes the form

$$\frac{dB}{dt} = -\frac{kBA}{\mathcal{A}} \tag{14.101}$$

Macroscopically the system is homogeneous. Microscopically however, as the reaction proceeds, the concentration of B near any A center becomes depleted and the rate becomes dominated by the diffusion process that brings fresh supply of B into the neighborhood of A. Focusing on one particular A molecule we consider the distribution of B molecules, $B(\mathbf{r}) = N_{\mathrm{B}}P(\mathbf{r})$ in its neighborhood. Here N_{B} is the total number of B molecules and $P(\mathbf{r})$ is the probability density for finding a B molecule at position $\mathbf{r}$ given that an A molecule resides at the origin. $P(\mathbf{r})$ and therefore $B(\mathbf{r})$ satisfy the Smoluchowski equation (cf. Eq. (8.137))

$$\frac{\partial B(\mathbf{r},t)}{\partial t} = -\nabla \cdot \mathbf{J} \qquad \mathbf{J} = -D(\beta\nabla V + \nabla)B(\mathbf{x},t) \tag{14.102}$$

where V is the A–B interaction potential.

In order to obtain an expression of the bimolecular rate coefficient associated with this process we follow a similar route as in the barrier crossing problem, by considering the flux associated with a steady state that is approached at long time by a system subjected to the following boundary conditions: (1) the bulk concentration of B remains constant and (2) B disappears when it reacts with A at a distance R^*

[24] It can be shown that if the molecules A diffuse as well, the same formalism applies, with D replaced by $D_A + D_{\mathrm{B}}$.

from A's center (placed at the origin).[25] For simplicity we assume that the A and B molecules are spherical, so that the interaction between them depends only on their relative distance r. Consequently the steady-state distribution is spherically symmetric. This implies that only the radial part of the flux **J** is nonzero

$$J(r) = -D\left[\beta\left(\frac{d}{dr}V(r)\right) + \frac{d}{dr}\right]B(r) \tag{14.103}$$

At steady state, the number of B molecules in each spherical shell surrounding A remains constant. This implies that the integral of $J(r)$ over any sphere centered about A is a constant independent of the sphere radius. Denoting this constant by $-J_0$ gives[26]

$$J(r) = -\frac{J_0}{4\pi r^2} \tag{14.104}$$

Using this in Eq. (14.103) leads to

$$J_0 = 4\pi Dr^2\left[\beta\left(\frac{d}{dr}V(r)\right) + \frac{d}{dr}\right]B(r) \tag{14.105}$$

It is convenient at this point to change variable, putting $B(r) = b(r)\exp(-\beta V(r))$. Equation (14.105) then becomes

$$\frac{db(r)}{dr} = \frac{J_0}{4\pi D}\frac{e^{\beta V(r)}}{r^2} \tag{14.106}$$

which may be integrated from R^* to ∞ to yield

$$b(\infty) - b(R^*) = \frac{J_0}{4\pi D\lambda}; \quad \text{with} \quad \lambda^{-1} \equiv \int_{R^*}^{\infty} dr\frac{e^{\beta V(r)}}{r^2} \tag{14.107}$$

λ is a parameter of dimension length. Note that in the absence of an A–B interaction, that is, when $V(r) = 0$, $\lambda = R^*$. In any case, the A–B interaction vanishes at large

[25] The rationale for this steady-state approach is the same as used in the barrier crossing problem: The assumption is that the system reaches a "quasi-steady-state" in which the overall number of B molecules reduces slowly, but the distribution of B in the space about A remains otherwise constant. There is another underlying assumption—that this approach to quasi-steady-state is fast relative to the timescale of the reaction itself. When this assumption does not hold, that is, if the reaction takes place in the transient period before steady state is established, the rate coefficient is not well defined (or may be defined as a function of time).

[26] This is the radial equivalent to the statement that in a one-dimensional flow the steady-state flux does not depend on position.

distances, that is, $V(r) \to 0$ as $r \to \infty$. This implies that $b(\infty) = B(\infty) = B$, where B the bulk number density of the B species. Denoting $B^* = B(R^*)$ and $V^* = V(R^*)$, Eq. (14.107) finally gives

$$B^* = \left(B - \frac{J_0}{4\pi D\lambda}\right) e^{-\beta V^*} \tag{14.108}$$

Consider now Eq. (14.101). The rate dB/dt at which B is consumed (per unit volume) is equal to the integrated B flux towards any A center, multiplied by the number of such centers per unit volume

$$-kB\frac{A}{\mathcal{A}} = -4\pi r^2 J(r)A = -J_0 A \tag{14.109}$$

whence

$$J_0 = \frac{kB}{\mathcal{A}} \tag{14.110}$$

Using this in Eq. (14.108) leads to

$$B^* = Be^{-\beta V^*}\left(1 - \frac{k}{4\pi D\lambda\mathcal{A}}\right) \tag{14.111}$$

If upon reactive contact, that is, when $r = R^*$, reaction occurs instantaneously with unit probability, then $B^* = 0$. The steady-state rate is then

$$k = 4\pi\mathcal{A}D\lambda \tag{14.112}$$

More generally, it is possible that B disappears at R^* with a rate that is proportional to B^*, that is,

$$\frac{dB}{dt} = -kB\frac{A}{\mathcal{A}} = -k^*B^*\frac{A}{\mathcal{A}}, \quad \text{that is,} \quad kB = k^*B^* \tag{14.113}$$

Using this in Eq. (14.111) leads to

$$k = \frac{4\pi D\lambda\mathcal{A}}{1 + (4\pi D\lambda\mathcal{A}/k^* e^{-\beta V^*})} \tag{14.114}$$

which yields the result (14.112) in the limit $k^* e^{-\beta V^*} \to \infty$. V^* is the interaction potential between the A and B species at the critical separation distance R^* (on a scale where $V(\infty) = 0$), and can be positive or negative. A strongly positive V^* amounts to a potential barrier to reaction. In the limit $k^* e^{-\beta V^*} \to 0$ we get $k = k^* e^{-\beta V^*}$.

The result (14.114) gives the bimolecular rate k in terms of the intermolecular diffusion constant D and the intermolecular potential $V(r)$. The rate coefficient k^* associated with the reaction between the A and B species after they are assembled at the critical separation R^* is a parameter of this theory. If we regard the assembled A–B complex as a single molecule we could in principle calculate k^* as a unimolecular rate involving this complex, using the methodologies discussed in Sections 14.4 and 14.5.

Finally, it should be kept in mind that we have treated diffusion-controlled reactions within a particular simple model. More complex situations arise when the diffusion itself is more complex, for example when it proceeds on restricted pathways[27] or when it is controlled by gating.[28] Also, the assumption that reaction occurs at one fixed distance does not always hold, as is the case when the species B are excited molecules that disappear by fluorescence quenching.

Appendix 14A: Solution of Eqs (14.62) and (14.63)

Here we seek a solution of the form $f(x, v) = f(u)$ with $u = v + \Gamma x$ to Eq. (14.62). Γ is an unknown constant at this stage. This form implies that $\partial f/\partial x = \partial f/\partial u \cdot \Gamma$ and $\partial f/\partial v = \partial f/\partial u$. Using these relationships in (14.62) leads to

$$\gamma^{-1}\left[(\Gamma + \gamma)v + \omega_B^2 x\right]\frac{df}{du} = \frac{k_B T}{m}\frac{d^2 f}{du^2} \tag{14.115}$$

To be consistent with our ansatz, that is, in order for f to be a function of the single variable u only, the coefficient of df/du on the left-hand side should be proportional to u, that is,

$$\gamma^{-1}[(\Gamma + \gamma)v + \omega_B^2 x] = -\alpha u = -\alpha v - \alpha\Gamma x \tag{14.116}$$

Equating the coefficients of x and v, that is, taking $\Gamma + \gamma = -\alpha\gamma$ and $\omega_B^2/\gamma = -\alpha\Gamma$ and eliminating α from these equations leads to an equation for Γ

$$\Gamma^2 + \gamma\Gamma - \omega_B^2 = 0 \tag{14.117}$$

which yields the solutions

$$\Gamma = \frac{-\gamma \pm \sqrt{\gamma^2 + 4\omega_B^2}}{2} \tag{14.118}$$

[27] D. Ben-Avraham and S. Havlin, *Diffusion and Reactions in Fractals and Disordered Systems*, (Cambridge University Press, Cambridge 2000).

[28] A. Szabo, D. Shoup, S. H. Northrup, and J. A. McCammon, Stochastically gated diffusion-influenced reactions, J. Chem. Phys. **77**, 4484 (1982).

The choice between these solutions will be made by the requirements imposed by the boundary conditions. Proceeding with Eq. (14.115) in the form

$$\frac{k_B T}{m}\frac{d^2 f}{du^2} + \alpha u \frac{df}{du} = 0; \qquad \alpha = -\frac{\Gamma + \gamma}{\gamma} \tag{14.119}$$

we can write its general solution

$$f(u) = F_1 + F_2 \int_0^u dz e^{(-\alpha m z^2/2k_B T)} \tag{14.120}$$

where F_1 and F_2 are to be determined by the boundary conditions and the choice of α, that is, of Γ made accordingly. First note that since $f(u)$ should not diverge for $|u| \to \infty$, α must be positive. Therefore, $\Gamma + \gamma$ should be negative, which implies that the physically acceptable solution of (14.117) is that with the minus sign in (14.118). Thus our final result for Γ is

$$\Gamma = -\left\{ \frac{\gamma}{2} + \left[\left(\frac{\gamma}{2} \right)^2 + \omega_B^2 \right]^{1/2} \right\} \tag{14.121}$$

Next, for any velocity v we require $f(x \to \infty, v) = 0$. This implies that $f(u \to -\infty) = 0$ (note that because Γ was determined to be negative, $u \to -\infty$ is equivalent to $x \to \infty$). Using this in (14.120) leads to

$$F_1 = -F_2 \int_0^{-\infty} dz e^{(-\alpha m z^2/2k_B T)} = \left[\frac{\pi k_B T}{2\alpha m} \right]^{1/2} F_2 \tag{14.122}$$

We actually use just the intermediate result in (14.122) to get

$$\begin{aligned} f(x, v) &= F_2 \left\{ \int_{-\infty}^{0} dz e^{(-\alpha m z^2/2k_B T)} + \int_0^u dz e^{(-\alpha m z^2/2k_B T)} \right\} \\ &= F_2 \int_{-\infty}^{v - |\Gamma| x} dz e^{(-\alpha m z^2/2k_B T)} \end{aligned} \tag{14.123}$$

(recall that Γ is negative so $u = v + \Gamma x = v - |\Gamma| x$). Now we can use the boundary condition $f(x \to -\infty) = 1$ to get $F_2 = \sqrt{\alpha m / 2\pi k_B T}$. So, finally,

$$f(x, v) = \sqrt{\frac{\alpha m}{2\pi k_B T}} \int_{-\infty}^{v - |\Gamma| x} dz e^{(-\alpha m z^2/2k_B T)} \tag{14.124}$$

Appendix 14B: Derivation of the energy Smoluchowski equation

It is convenient to use a Langevin starting point, so we begin with Eq. (14.39). We transform to action–angle variables (K, ϕ):

$$x(K,\phi) = \sum_{n=-\infty}^{\infty} x_n(K)e^{in\phi} \tag{14.125a}$$

$$v(K,\phi) = \sum_{n=-\infty}^{\infty} v_n(K)e^{in\phi} \tag{14.125b}$$

$$v_n(K) = in\omega(K)x_n(K); \quad \omega(K) = \dot{\phi} \tag{14.125c}$$

where, since x and v are real, $x_{-n} = x_n^*$ and $v_{-n} = v_n^*$. The action K is related to the energy E by

$$\frac{dE}{dK} = \omega(K) \tag{14.126}$$

and to the motion in phase space by

$$K = \frac{m}{2\pi}\oint v(x)dx = \frac{m}{2\pi}\int_0^{2\pi} v(K,\phi)\frac{\partial x(K,\phi)}{\partial \phi}d\phi \tag{14.127}$$

where $\oint$ denotes integration over an oscillation period. Inserting (14.125) into (14.127) and using $\int_0^{2\pi} d\phi e^{i(n+l)\phi} = 2\pi\delta_{n,-l}$ leads to an explicit expression for K in terms of x and v:

$$K = m\omega(K)\sum_n n^2|x_n|^2 = \frac{m}{\omega(K)}\sum_n |v_n|^2 \tag{14.128}$$

In order to derive a Langevin equation for E (or K) we start with Eq. (14.39),

$$m\ddot{x} = -\frac{\partial V}{\partial x} - m\gamma v + R \tag{14.129}$$

multiply both sides by $\dot{x} = v$ and use $E = (1/2)mv^2 + V(x)$ to get

$$\frac{dE}{dt} = \omega\frac{dK}{dt} = -m\gamma v^2 + vR \tag{14.130}$$

From (14.125b) we get, after averaging over phase oscillations and using (14.128)

$$v^2 = \left(\sum_{n=-\infty}^{\infty} v_n(K)e^{in\phi}\right)^2 \cong \sum_{n=-\infty}^{\infty} |v_n(K)|^2 = \frac{\omega K}{m} \tag{14.131}$$

which, with (14.130) gives

$$\frac{dK}{dt} = -\gamma K + S(t) \tag{14.132}$$

where

$$S(t) = \frac{R(t)}{\omega(K)} \sum_n v_n e^{in\phi} \tag{14.133}$$

note that all the fast variables, $R(t)$ and ϕ now appear only in $S(t)$, which may be regarded as a random noise source in (14.132). It satisfies

$$\langle S \rangle = 0$$

$$\langle S(t_1)S(t_2)\rangle = \frac{2\gamma m k_B T}{\omega^2(K)} \left(\sum_n v_n e^{in\phi}\right)^2 \delta(t_1 - t_2) = \frac{2\gamma k_B T K}{\omega(k)} \delta(t_1 - t_2) \tag{14.134}$$

Here the average was done both on $R(t)$ and on the phase oscillations, and we have used again Eq. (14.128). Equation (14.132) is a Langevin-type equation characterized by a random noise $S(t)$ whose statistical properties are given by (14.134). It is similar to others we had before, with one important difference: The random "force" $S(t)$ depends on the state K of the system. We can repeat the procedure we used in Section 8.4.4 to get a Fokker–Planck equation, but more caution has to be exercised in doing so.[29] The result for the probability $P(K,t)dK$ to find the action in the range $K \ldots K + dK$ is

$$\frac{\partial P(K,t)}{\partial t} = \gamma \frac{\partial}{\partial K}\left(K + \frac{K k_B T}{\omega(K)} \frac{\partial}{\partial K}\right) P(K,t) \tag{14.135}$$

Equation (14.135) has the form $\partial P/\partial t = -(\partial/\partial K)J_K$, where

$$J_K(K) \equiv -\gamma K \left(1 + \frac{k_B T}{\omega(K)} \frac{\partial}{\partial K}\right) P \tag{14.136}$$

[29] The term $\gamma\partial/\partial K(KP)$, that remains as $T \to 0$, is obtained as before. Extra care is needed in deriving the *second* term, in assessing the proper positions of the two K-derivatives with respect to the term $K k_B T/\omega(K)$.

is the *action flux*. At equilibrium this flux is zero, that is, $\partial P/\partial K = -(\omega(K)/k_B T)P$, leading to the Boltzmann distribution,

$$P(K) \propto \exp\left(-\beta \int^{K} dK' \omega(K')\right) = \exp\left(-\beta E(K)\right) \tag{14.137}$$

It is useful to recast Eq. (14.135) in terms of the more familiar energy variable E. To this end use $\partial/\partial K = \omega(\partial/\partial E)$ and $P(K,t) = \omega(E)P(E,t)$.[30] Denoting $\gamma K(E) \equiv D(E)$ this leads to the energy Smoluchowski equation,

$$\frac{\partial P(E)}{\partial t} = \frac{\partial}{\partial E}\left[D(E)\left(1 + k_B T \frac{\partial}{\partial E}\right)\omega(E)P(E)\right] \tag{14.138}$$

Further reading

Barrier crossing

P. Hänggi, P. Talkner, and M. Borkovec, Rev. Mod. Phys. **62**, 251 (1990).
G. R. Fleming and P. Hänggi, eds., *Activated Barrier Crossing* (World Scientific, 1993).
V. I. Melnikov, Phys. Rep. **209**, 1 (1991).
A. Nitzan, Adv. Chem. Phys. **70**, 489 (1988).

Diffusion controlled reactions

E. Kotomin and V. Kuzovkov, *Diffusion Controlled Reactions* (Elsevier, Amsterdam, 1996).
S. A. Rice, *Diffusion-Limited Reactions* (Elsevier, Amsterdam, 1985).
G. Wilemski and M. Fixman, J. Chem. Phys. **58**, 4009 (1973).

[30] For simplicity of the presentation we use the same notation for the probability densities $P(E)$ in energy space and $P(K)$ in action space, even though they have different functional forms. A more rigorous notation would be $P_E(E)$ for the former and $P_K(K)$ for the latter, which satisfy $(dE/dK)P_E(E) = P_K(K)$.

15

SOLVATION DYNAMICS

There are many things for which it's not enough
To specify one cause, although the fact
Is that there's only one. But just suppose
You saw a corpse somewhere, you'd better name
Every contingency—how could you say
Whether he died of cold, or of cold still,
Of poison, or disease? The one thing sure
Is that he's dead. It seems to work this way
In many instances . . .

Lucretius (c.99–c.55 BCE*) "The way things are"*
translated by Rolfe Humphries, Indiana University

Solvent dynamical effects on relaxation and reaction process were considered in Chapters 13 and 14. These effects are usually associated with small amplitude solvent motions that do not appreciably change its configuration. However, the most important solvent effect is often equilibrium in nature — modifying the free energies of the reactants, products, and transition states, thereby affecting the free energy of activation and sometime even the course of the chemical process. Solvation energies relevant to these modifications can be studied experimentally by calorimetric and spectroscopic methods, and theoretically by methods of equilibrium statistical mechanics.

With advances of experimental techniques that made it possible to observe timescales down to the femtosecond regime, the dynamics of solvation itself became accessible and therefore an interesting subject of study. Moreover, we are now able to probe molecular processes that occur on the same timescale as solvation, making it necessary to address solvation as dynamic in addition to energetic phenomenon. This chapter focuses on the important and most studied subclass of these phenomena—solvation dynamics involving charged and polar solutes in dielectric environments. In addition to their intrinsic importance, these phenomena play a central role in all processes involving molecular charge rearrangement, most profoundly in electron transfer processes that are discussed in the next chapter.

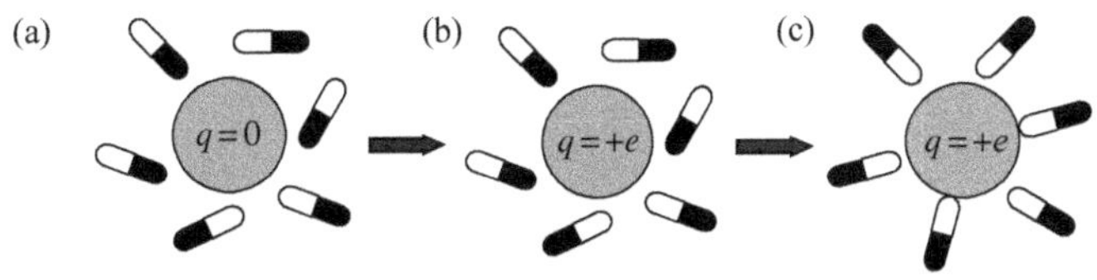

FIG. 15.1 A schematic view of instantaneous configurations of (a) solvent dipolar molecules about initial uncharged solute at equilibrium, (b) the same molecules following a sudden change of solute charge, and (c) solvent dipoles in the final equilibrium state. Solvent dipoles are represented by the small ellipses whose negative side is denoted by white color.

15.1 Dielectric solvation

Consider, as a particular example, a neutral ($q = 0$) atomic solute embedded in a dielectric solvent, that undergoes a sudden change of its charge to $q = e$, where e is the magnitude of the electron charge. This can be achieved, for example, by photoionization. The dipolar solvent molecules respond to this change in the local charge distribution by rotating in order to make their negative end point, on the average, to the just formed positive ion (see Fig. 15.1). Thus, the solvent configuration changes in response to the sudden change in a local charge distribution. The driving force for this change is the lowering of overall free energy that accompanies the buildup of solvent polarization.

Consider this process in more detail. Figure 15.1 describes it under the assumption that the ionization process is fast relative to the timescale of solvent motion. Shown are snapshots of the system configuration just before (a) and just after (b) the ionization event, as well as a snapshot from the final equilibrium state (c). Because the solvent is slow relative to the ionization process, its configurations in (a) and (b) are the same.[1] This is followed by the process (b)→(c) in which the newly formed ion is "solvated" by surrounding dipoles. Denoting by E_a, E_b, and E_c the energies of the system in these states, the difference $\langle E_b - E_a \rangle$ is referred to as the *vertical ionization energy* while $\langle E_c - E_a \rangle$ is the *adiabatic ionization energy*. Both are in principle experimentally observable. The former is obtained from the peak of the absorption lineshape associated with the photoionization process. The latter is essentially the free energy difference between two equilibrium configurations

[1] The actual situation is in fact more complicated, because solvent response about a newly formed charge distribution is characterized by more than one timescale. In particular, solvent polarization has a substantial electronic component whose characteristic timescale is fast or comparable to that of electronic transitions in the solute, and a nuclear component, here associated with the orientation of solvent dipoles, that is slow relative to that timescale. In the present introductory discussion we disregard the fast electronic component of the solvent response, but it is taken into account later.

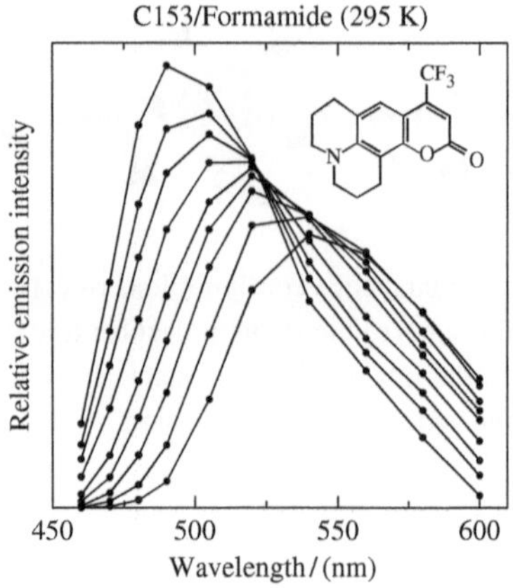

FIG. 15.2 Emission spectra of Coumarin 153 in formamide at different times. The times shown here are (in order of increasing peak-wavelength) 0, 0.05, 0.1, 0.2, 0.5, 1, 2, 5, and 50 ps (M. L. Horng, J. A. Gardecki, A. Papazyan, and M. Maroncelli, J. Phys. Chem. **99**, 17311 (1995)).

that can be estimates from experimental heats of solvation.[2] The difference between these energies $E_R \equiv \langle E_b - E_c \rangle$ is called the *solvent reorganization energy* (see also Chapter 16). Ultrafast spectroscopy has recently made the time evolution ("solvation dynamics") of the (b) $\to$ (c) process accessible to experimental observation. This process is the subject of this chapter.

Problem 15.1. (1) Suggest the reason why the average energy differences $\langle E_b - E_a \rangle$ and $\langle E_c - E_a \rangle$ are called vertical and adiabatic, respectively. (2) Why is $\langle E_b - E_a \rangle$ related to the peak energy of the corresponding absorption lineshape? (3) What is the distribution over which these averages are taken?

Solvation dynamics experiments probe the evolution of solvent structure following a sudden change in the solute. Most often the change is in the electronic charge density following an optical transition[3] and, as in the example discussed above, the subsequent rearrangement is most dramatic in polar solvents. An experimental manifestation of this evolution is shown in Fig. 15.2, where the solvation process is seen as an evolving red shift in the emission spectrum of the excited solute. Note that a shift to longer wavelengths indicates that the excited state of the fluorescing

---

[2] Solvation free energies are usually related to state in which the ion is out of, and far from, the solvent. To estimate the difference $\langle E_c - E_a \rangle$ one would need to construct a Born cycle that includes the process of vacuum ionization as well as the solvation of the neutral species.

[3] In most experimental studies this change is not an ionization process as in Fig. 15.1 but a change in the molecular dipole upon excitation to a different electronic state.

solute molecule is stabilized relative to the ground state, a likely scenario in a process in which the solvent responds to the excited state charge distribution. In the following sections we will develop a theoretical framework for discussing such phenomena.

15.2 Solvation in a continuum dielectric environment

15.2.1 General observations

We consider a polar solvent characterized by its dielectric response function $\varepsilon(\omega)$. Upon a sudden change in the charge distribution inside this solvent a relaxation process follows in which the solvent adjusts to the new charge distribution. We want to describe this relaxation process in terms of the (assumed known) dielectric response function.[4]

It should be emphasized that this description of solvation as a purely electrostatic process is greatly over-simplified. Short-range interactions exist as well, and the physical exclusion of the solvent from the space occupied by the solute must have its own dynamics. Still, for solvation of ions and dipolar molecules in polar solvents electrostatic solvent–solute and solvent–solvent interactions dominate, and disregarding short-range effects turns out to be a reasonable approximation.[5] Of main concern should be the use of continuum electrostatics to describe a local molecular process and the fact that the tool chosen is a linear response theory. We will come to these points later.

In a typical experiment the solute charge distribution is assumed to change abruptly, at $t = 0$, say, from $\rho_1(\mathbf{r})$ to $\rho_2(\mathbf{r})$, then stays constant. This means that the dielectric displacement, related to $\rho(\mathbf{r})$ by the Poisson equation $\nabla \cdot \mathcal{D} = 4\pi\rho$, is also switched from $\mathcal{D}_1$ to $\mathcal{D}_2$ at $t = 0$. In the process that follows the solvent structure adjusts itself to the new charge distribution. In our continuum model this appears as a local relaxation of the solvent polarization, which over time changes from $\mathbf{P}_1$ to $\mathbf{P}_2$.

This polarization is made of two contributions: electronic, $\mathbf{P}_e$, and nuclear, $\mathbf{P}_n$, characterized by relatively short and long response times. In what follows we assume that the response time of the electronic polarization is shorter than all other system timescales, including that on which $\rho(\mathbf{r})$ was changed. This implies that the

---

[4] Our discussion follows A. Mozumder in *Electron-solvent and anion-solvent interactions*, L. Kevan and B. Webster, Editors (Elsevier, Amsterdam, 1976).

[5] To get a feeling of the relevant orders of magnitude compare typical electrostatic Born solvation energies $(q^2/2a)[1 - (1/\varepsilon_s)] \approx$1–2 eV (where q is the ionic charge and ε_s is the static dielectric constant) to the pressure–volume work needed to form a cavity of radius a at atmospheric pressure, of order 10^{-6} eV for $a \cong 2$Å.

onset of electronic polarization in the solvent follows the change in ρ immediately: It changes instantly, at $t = 0$, from $\mathbf{P}_{1e}$ to $\mathbf{P}_{2e}$. The relaxation of the nuclear polarization follows more slowly. These induced polarizations affect the local electrostatic field at the solute, therefore its energy. We want to relate the time evolution of this local electrostatic field to the given $\varepsilon(\omega)$.

15.2.2 Dielectric relaxation and the Debye model

Assuming the validity of standard linear dielectric response theory, the electrostatic displacement $\boldsymbol{\mathcal{D}}$, and the electrostatic field $\boldsymbol{\mathcal{E}}$ in a dielectric medium are related to each other by

$$\mathcal{D}(\mathbf{r},t) = \int d\mathbf{r}' \int_{-\infty}^{t} dt' \boldsymbol{\varepsilon}(\mathbf{r}-\mathbf{r}',t-t')\mathcal{E}(\mathbf{r}',t') \tag{15.1a}$$

In what follows we assume that the response is local, that is, $\boldsymbol{\varepsilon}(\mathbf{r}-\mathbf{r}',t-t') = \boldsymbol{\varepsilon}(\mathbf{r},t-t')\delta(\mathbf{r}-\mathbf{r}')$. This assumption is not really valid for dielectric response on molecular lengthscales, but the errors that result from it appear to be small in many cases while the mathematical simplification is considerable. Also, while in general the dielectric response $\boldsymbol{\varepsilon}$ is a tensor, we take it for simplicity to be a scalar, that is, we consider only isotropic systems. In this case it is sufficient to consider the magnitudes $\mathcal{D}$ and $\mathcal{E}$ of $\boldsymbol{\mathcal{D}}$ and $\boldsymbol{\mathcal{E}}$. Thus, our starting point is the local scalar relationship

$$\mathcal{D}(\mathbf{r},t) = \int_{-\infty}^{t} dt' \varepsilon(t-t')\mathcal{E}(\mathbf{r},t') \tag{15.1b}$$

and its Fourier transform (defining, for example, $\mathcal{E}(\omega) = \int_{-\infty}^{\infty} dt e^{i\omega t}\mathcal{E}(t)$)

$$\mathcal{D}(\omega) = \varepsilon(\omega)\mathcal{E}(\omega) \tag{15.2}$$

where

$$\varepsilon(\omega) \equiv \int_{0}^{\infty} dt e^{i\omega t}\varepsilon(t) \tag{15.3}$$

To account for the fast and slow components of the dielectric response we take $\varepsilon(t)$ in the form

$$\varepsilon(t) = 2\varepsilon_e\delta(t) + \tilde{\varepsilon}(t) \tag{15.4}$$

to get

$$\mathcal{D}(t) = \varepsilon_e\mathcal{E}(t) + \int_{-\infty}^{t} dt'\tilde{\varepsilon}(t-t')\mathcal{E}(t') \tag{15.5}$$

$$\mathcal{D}(\omega) = \varepsilon_e\mathcal{E}(\omega) + \tilde{\varepsilon}(\omega)\mathcal{E}(\omega) \tag{15.6}$$

The *Debye model* takes for the slow part of the dielectric response the form

$$\tilde{\varepsilon}(t) = \frac{\varepsilon_s - \varepsilon_e}{\tau_D} e^{-t/\tau_D} \tag{15.7}$$

so that

$$\varepsilon(\omega) = \varepsilon_e + \int_0^\infty dt \frac{\varepsilon_s - \varepsilon_e}{\tau_D} e^{-t/\tau_D} e^{i\omega t} = \varepsilon_e + \frac{\varepsilon_s - \varepsilon_e}{1 - i\omega\tau_D} \tag{15.8}$$

The dielectric response in this model is thus characterized by three parameters: the electronic ε_e and static ε_s response constants, and the *Debye relaxation time* τ_D.

What are the experimental implications of this dielectric relaxation model? To answer this question let us start again from

$$\mathcal{D}(t) = \varepsilon_e \mathcal{E}(t) + \int_{-\infty}^{t} dt' \tilde{\varepsilon}(t - t')\mathcal{E}(t') \tag{15.9}$$

and take the time derivative of both sides with respect to t

$$\frac{d\mathcal{D}}{dt} = \varepsilon_e \frac{d\mathcal{E}}{dt} + E(t)\tilde{\varepsilon}(0) + \int_{-\infty}^{t} dt' \left(\frac{d\tilde{\varepsilon}}{dt}\right)_{t-t'} \mathcal{E}(t') \tag{15.10}$$

Next use the relations $\tilde{\varepsilon}(0) = (\varepsilon_s - \varepsilon_e)/\tau_D$ and

$$\int_{-\infty}^{t} dt' \left(\frac{d\tilde{\varepsilon}}{dt}\right)_{t-t'} \mathcal{E}(t') = -\frac{1}{\tau_D} \int_{-\infty}^{t} dt' \tilde{\varepsilon}(t - t')\mathcal{E}(t') = -\frac{1}{\tau_D}(\mathcal{D}(t) - \varepsilon_e \mathcal{E}(t)) \tag{15.11}$$

(cf. Eq. (15.7)), to get

$$\frac{d}{dt}(\mathcal{D} - \varepsilon_e \mathcal{E}) = -\frac{1}{\tau_D}(\mathcal{D} - \varepsilon_s \mathcal{E}) \tag{15.12}$$

An interesting outcome of (15.12) is that the implied relaxation depends on the way the experiment is conducted. Consider first a step function change in the electrostatic field:

$$\mathcal{E}(t) = \begin{cases} 0, & t < 0, \\ \mathcal{E}, & t \geq 0 \end{cases} \tag{15.13}$$

after which $\mathcal{E}$ remains constant so that $\mathcal{D}$ evolves in time under a constant $\mathcal{E}$. Equarton (15.12) then becomes

$$\frac{d\mathcal{D}}{dt} = -\frac{1}{\tau_D}\mathcal{D} + \frac{\varepsilon_s}{\tau_D}\mathcal{E} \tag{15.14}$$

whose solution is

$$\mathcal{D}(t) = \mathcal{D}(t=0)e^{-t/\tau_{\mathrm{D}}} + \varepsilon_{\mathrm{s}}\mathcal{E}(1 - e^{-t/\tau_{\mathrm{D}}}) \tag{15.15}$$

At long time, $t \to \infty$, $\mathcal{D}$ assumes its equilibrium value $\varepsilon_{\mathrm{s}}\mathcal{E}$. However, immediately after the field jump the solvent can respond only with the ε_{e} component of its dielectric function, so $\mathcal{D}(t=0) = \varepsilon_{\mathrm{e}}\mathcal{E}$. Equation (15.15) therefore becomes

$$\mathcal{D}(t) = [\varepsilon_{\mathrm{s}}(1 - e^{-t/\tau_{\mathrm{D}}}) + \varepsilon_{\mathrm{e}}e^{-t/\tau_{\mathrm{D}}}]\mathcal{E} \tag{15.16}$$

This result could also be obtained from (15.5) and (15.7). The displacement field $\mathcal{D}$ its seen to relax from its initial value $\varepsilon_{\mathrm{e}}\mathcal{E}$ to its final equilibrium value $\varepsilon_{\mathrm{s}}\mathcal{E}$ with the characteristic relaxation time τ_{D}. The experimental realization of this situation is, for example, a capacitor in which a dielectric solvent fills the space between two planar electrodes and a potential difference between the electrodes is suddenly switched on, then held constant while the solvent polarization relaxes. This relaxation proceeds at constant electric field (determined by the given potential difference divided by the distance between the electrodes). To keep the field constant as the solvent polarization changes the surface charge density on the electrodes must change—the needed charge is supplied by the voltage source. The Poisson equation, $\nabla \cdot \boldsymbol{\mathcal{D}} = 4\pi\rho$, then tells us that $\boldsymbol{\mathcal{D}}$ must change, as given explicitly by (15.16).

It was already stated in Section 15.2.1 that the experimental conditions pertaining to the observation of solvation dynamics are different. The jump is not in the voltage but in the charge distribution. This implies a jump in the dielectric displacement, so, Eq. (15.13) is replaced by

$$\mathcal{D}(t) = \begin{cases} 0, & t < 0, \\ \mathcal{D}, & t \geq 0 \end{cases} \tag{15.17}$$

In this case Eq. (15.12) describes the evolution of $\mathcal{E}$ under the constant displacement $\mathcal{D}$,

$$\frac{d}{dt}\mathcal{E} = -\frac{\varepsilon_{\mathrm{s}}}{\varepsilon_{\mathrm{e}}\tau_{\mathrm{D}}}\left(\mathcal{E} - \frac{1}{\varepsilon_{\mathrm{s}}}\mathcal{D}\right); \qquad t > 0 \tag{15.18}$$

which implies that at equilibrium ($d\mathcal{E}/dt = 0$), $\mathcal{E} = \varepsilon_{\mathrm{s}}^{-1}\mathcal{D}$. Immediately following the jump in $\mathcal{D}$, however, the electric field is $\mathcal{E}(t=0) = \varepsilon_{\mathrm{e}}^{-1}\mathcal{D}$. The corresponding solution of (15.18) is now

$$\mathcal{E}(t) = \frac{1}{\varepsilon_{\mathrm{s}}}\mathcal{D} + \left(\frac{1}{\varepsilon_{\mathrm{e}}} - \frac{1}{\varepsilon_{\mathrm{s}}}\right)\mathcal{D}e^{-t/\tau_{\mathrm{L}}} \tag{15.19}$$

where τ_L is the *longitudinal relaxation time*[6]

$$\tau_L = \frac{\varepsilon_e}{\varepsilon_s}\tau_D \tag{15.20}$$

We see that in this case the relaxation is characterized by the time τ_L which can be very different from τ_D: For example in water $\varepsilon_e/\varepsilon_s \cong 1/40$ and while $\tau_D \cong 8$ ps, τ_L is of the order of 0.2 ps!

15.3 Linear response theory of solvation

The continuum dielectric theory used above is a linear response theory,[7] as expressed by the linear relation between the perturbation $\mathcal{D}$ and the response $\mathcal{E}$, Eq. (15.1b). Thus, our treatment of solvation dynamics was done within a linear response framework. Linear response theory of solvation dynamics may be cast in a general form that does not depend on the model used for the dielectric environment and can therefore be applied also in molecular (as opposed to continuum) level theories. Here we derive this general formalism. For simplicity we disregard the fast electronic response of the solvent and focus on the observed nuclear dielectric relaxation.

Our starting point (see Eqs (11.1)–(11.3)) is the classical Hamiltonian for the atomic motions

$$H = H_0 + H_1 \tag{15.21}$$

where H_0 describes the unperturbed system that is characterized by a given potential surface on which the nuclei move, and where

$$H_1 = -\sum_j A_j F_j(t) \tag{15.22}$$

is some perturbation written as a sum of products of system variables A_j and external time dependent perturbations $F_j(t)$. The detailed structure of A and F depend on the particular experiment. If for example the perturbation is caused by a point charge $q(t)$ at position $\mathbf{r}_j$, $q(t)\delta(\mathbf{r} - \mathbf{r}_j)$, we may identify $F_j(t)$ with this charge

[6] The origin of the terms "transverse" and "longitudinal" dielectric relaxation times lies in the molecular theory of dielectric relaxation, where one finds that the decay of correlation functions involving transverse and longitudinal components of the induced polarization vector are characterized by different time constants. In a Debye fluid the relaxation times that characterize the transverse and longitudinal components of the polarization are τ_D and $\tau_L = (\varepsilon_e/\varepsilon_s)\tau_D$, respectively. See, for example, P. Madden and D. Kivelson, J. Phys. Chem. **86**, 4244 (1982).

[7] M. Maronelli and G. R. Fleming J. Chem. Phys. **89**, 5044 (1988); E. A. Carter and J. T. Hynes J. Chem. Phys. **94**, 2084 (1991).

and the corresponding A_j is minus the electrostatic potential at the charge position, $A_j = -\Phi(\mathbf{r}_j)$.[8] For a continuous distribution $\rho(\mathbf{r}, t)$ of such charge we may write $H_1 = \int d^3r\Phi(\mathbf{r})\rho(\mathbf{r}, t)$, and for $\rho(\mathbf{r}, t) = \sum_j q_j(t)\delta(\mathbf{r} - \mathbf{r}_j)$ this becomes $\sum_j \Phi(\mathbf{r}_j)q_j(t)$. In this case $\rho(\mathbf{r})$ is the "external force" and $\Phi(\mathbf{r})$ is the corresponding system response. Alternatively we may find it convenient to express the charge distribution in terms of point moments (dipoles, quadrupoles, etc.) coupled to the corresponding local potential gradient tensors, for example, H_1 will contain terms of the form $\boldsymbol{\mu} \cdot \nabla\Phi$ and $\boldsymbol{Q}: \nabla\nabla\Phi\cdot$, where $\boldsymbol{\mu}$ and $\boldsymbol{Q}$ are point dipoles and quadrupoles respectively.

In linear response theory the solvation energies are proportional to the corresponding products $q\langle\Phi\rangle$, $\boldsymbol{\mu} \cdot \langle\nabla\Phi\rangle$, and $\boldsymbol{Q}{:}\langle\nabla\nabla\Phi\rangle$ where $\langle\ \rangle$ denotes the usual average of the given observable in the presence of the perturbation. For example, the average potential $\langle\Phi\rangle$ formed in response to a charge q is proportional in linear response to this charge q, $\langle\Phi(q)\rangle = \alpha q$. The energy needed to create the charge q is therefore

$$\int_0^q dq'\langle\Phi(q')\rangle = \int_0^q dq'\alpha q' = \frac{1}{2}\alpha q^2 = \frac{1}{2}q\langle\Phi\rangle \tag{15.23}$$

We now apply linear response theory to the relaxation that follows a sudden change in the external force, see Section 11.1.2. Focusing on the simple case where $H_1 = -AF(t)$, we consider $F(t)$ of the following form

$$F(t) = \begin{cases} -q, & t < 0, \\ 0, & t \geq 0 \end{cases} \tag{15.24}$$

which amounts to a sudden increase of the charge at a given position, where $-A = \Phi$ is the potential in that position. We will use Eq. (11.15) with B replaced by Φ because we are interested in the response of the electrostatic potential at the position of the charge. We also insert a slight change in notation: $\langle B\rangle_0$, the average observable B under the Hamiltonian H_0, is also the value approached by $\langle B(t)\rangle$ at $t \to \infty$. We can therefore write $\langle B(\infty)\rangle$ instead of $\langle B\rangle_0$. Equation (11.15) now becomes, using $B = \Phi$, $A = -\Phi$, and $F = -q$

$$\langle\Phi(t)\rangle - \langle\Phi(\infty)\rangle = \beta q\langle\delta\Phi(0)\delta\Phi(t)\rangle_0 \tag{15.25}$$

[8] In Chapter 11 we discussed examples with essentially same perturbation H_1 but with a different assignment of terms: the electrostatic field was the external perturbation and the coordinate of a charged particle was the internal dynamic variable.

The subscript 0 on the right denotes that the average is taken with the unperturbed Hamiltonian, here taken as H_0 — the Hamiltonian at $t > 0$, however, within linear response theory we could equally take the Hamiltonian $H = H_0 + H_1$ for this purpose. Mathematically this is seen from the fact that $\beta q(\langle\delta\Phi(0)\delta\Phi(t)\rangle_{H_0} - \langle\delta\Phi(0)\delta\Phi(t)\rangle_H)$ must be of order q^2. Physically, we could repeat the calculation, going from $F(t) = 0$ for $t < 0$ to $F(t) = -q$ for $t > 0$, thereby interchanging the roles of H_0 and H without changing the linear response result.

We have found that upon a sudden change of the charge at some point in the solvent by q the potential at that point changes according to

$$\langle\Phi(t)\rangle - \langle\Phi(\infty)\rangle = \frac{q}{k_BT}(\langle\Phi(0)\Phi(t)\rangle - \langle\Phi\rangle^2) = \frac{q}{k_BT}\langle\delta\Phi(0)\delta\Phi(t)\rangle \quad (15.26)$$

The left-hand side of (15.26) is, by Eq. (15.23), a linear response approximation of the corresponding solvation energies difference. This makes it possible for us to write a linear response expression for the *solvation function* which is defined by

$$S(t) \equiv \frac{E_{\text{solv}}(t) - E_{\text{solv}}(\infty)}{E_{\text{solv}}(0) - E_{\text{solv}}(\infty)} \quad (15.27)$$

In linear response this becomes

$$S(t) \overset{\text{LR}}{=} \frac{\langle\Phi(t)\rangle - \langle\Phi(\infty)\rangle}{\langle\Phi(0)\rangle - \langle\Phi(\infty)\rangle} \quad (15.28)$$

and using (15.26) we find

$$S(t) \overset{\text{LR}}{=} C(t) \equiv \frac{\langle\delta\Phi(0)\delta\Phi(t)\rangle}{\langle\delta\Phi^2\rangle} \quad (15.29)$$

The nonequilibrium solvation function $S(t)$, which is directly observable (e.g. by monitoring dynamic line shifts as in Fig. 15.2), is seen to be equal in the linear response approximation to the time correlation function, $C(t)$, of equilibrium fluctuations in the solvent response potential at the position of the solute ion. This provides a route for generalizing the continuum dielectric response theory of Section 15.2 and also a convenient numerical tool that we discuss further in the next section.

The relationship (15.29) was found for the case of charge solvation. Solvation of higher moments of a given charge distribution can be treated in the same way. For dipole solvation we will find a similar relationship, except that the electrostatic potential in $C(t)$ is replaced by the electric field (because $H_1 = q\Phi$ is replaced by $H_1 = -\mu \cdot \mathcal{E}$). Similarly, higher gradients of the electrostatic potential will enter when solvation of higher moments of the charge distribution is considered.

15.4 More aspects of solvation dynamics

There are two ways in which polar solvation dynamics enters into our considerations. First, it must play an essential role in the dynamics of charge rearrangement processes, including and foremost the very important class of charge transfer reactions that are discussed in the next chapter. Second, we can use this process to learn about the short-time dynamics of the solvent themselves. With the second goal in mind we regard the solute molecule as a probe that is used, aided by a fast excitation and detection system, to study the solvent. An ideal probe would be an inert body of controllable size whose charge state can be changed at will. We could use it to study the limitations of the continuum dielectric picture of solvation dynamics, the adequacy or inadequacy of linear response theory and the nature of the solvent motions responsible for solvation dynamics on different time and length scales. Practically, the charge distribution of a molecular probe is changed by optical excitation. Realistic probes that absorb in convenient spectral ranges are however large molecules, forcing us to address other questions concerning the effect of the probe on the observed results. How does the probe size and the detailed charge distribution within it affect the results? How much of the observed signal results from the probe intramolecular vibrational relaxation[9] and what is the role played by specific probe–solvent interactions.

There are several ways by which these issues can be addressed:

(1) Continuum dielectric models of solvation can be generalized to include some aspects of the solvent molecularity. This has lead to the dynamic mean spherical approximation[10] which improves the agreement between these kind of theories and experimental observations.[11]
(2) The linear response formalism discussed in the previous section makes it possible to develop molecular level theoretical approaches to solvation

---

[9] In Chapter 11 we discussed examples with essentially same perturbation H_1 but with a different assignment of terms: the electrostatic field was the external perturbation and the coordinate of a charged particle was the internal dynamic variable.

[10] P. G. Wolynes, J. Chem. Phys. **86**, 5133 (1987); I. Rips, J. Klafter, and J. Jortner, J. Chem. Phys. **89**, 4288 (1988); A. L. Nichols and D. F. Calef, J. Chem. Phys. **89**, 3783 (1988).

[11] M. L. Horng, J. A. Gardecki, A. Papazyan, and M. Maroncelli, J. Phys. Chem. **99**, 17311 (1995).

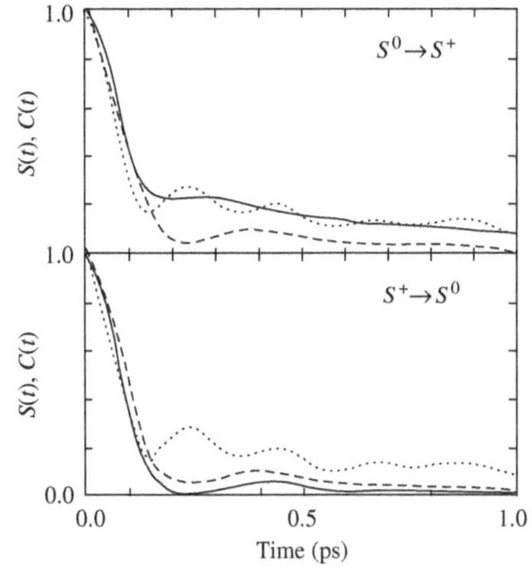

FIG. 15.3 The nonequilibrium solvation function $S(t)$ (full lines) and the solvation correlation functions $C(t)$ for a model solute ion of diameter 3.1 Å in acetonitrile computed with the positive solute (dotted line) and neutral solute (dashed line). (From M. Maroncelli, J. Chem. Phys. **94**, 2084 (1991).)

dynamics.[12] Such methods have the capacity to address specific aspects of the solvent molecular structure.

(3) Molecular dynamic simulations are very useful for solvation dynamic studies. In contrast to the difficulties described in applying numerical methods to the problems of vibrational relaxation (Section 13.6) and barrier crossing (Section 14.7), solvation dynamics is a short-time downhill process that takes place (in pure simple solvents) on timescales easily accessible to numerical work.

Such simulations can lead to new physical insight. Dielectric relaxation on timescales of ps and longer is a diffusive process. This implies that documented dielectric relaxation times inferred from relatively long-time measurements reflect solvent diffusive motion. The short timescales now accessible by ultrafast spectroscopy are shorter than characteristic times for solvent–solvent interactions, and dielectric response data may not contain the fast, and perhaps short lengthscale components relevant to this motion.

Figure 15.3 shows the results of computer simulations of solvation of a model ion in acetonitrile (CH_3CN). The simulations produce the solvation function $S(t)$ for

[12] See, for example, N. Nandi, S. Roy, and B. Bagchi, J. Chem. Phys. **102**, 1390 (1995); H. L. Friedman, F. O. Raineri, F. Hirata, and B. C. Perng, J. Stat. Phys. **78**, 239 (1955).

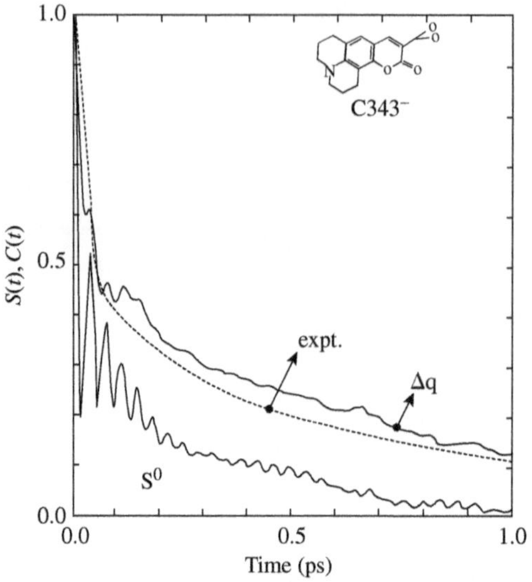

FIG. 15.4 The experimental solvation function for water using sodium salt of coumarin-343 as a probe. The line marked "expt." is the experimental solvation function $S(t)$ obtained from the shift in the fluorescence spectrum. The line marked "Δq" is a simulation result based on the linear response function $C(t)$. The line Marked S^0 is the linear response function for a neutral atomic solute with Lennard Jones parameters of the oxygen atom. (From R. Jimenez, G. R. Fleming, P. V. Kumar, and M. Maroncelli, Nature **369**, 471 (1994).)

the transitions $0 \rightarrow +e$ and $+e \rightarrow 0$, as well as the solvation correlation function $C(t)$ computed for a neutral and a charged solute. The differences between the curves show interesting deviations from linear response functions, but the most interesting observation is the prominent fast (50–100 fs) component that account for about 70% of the total solvation energy[13] and was not predicted at the time by dielectric solvation theory. Close examination of the simulated trajectory shows that this fast component results mainly from ballistic rotations of solvent molecules in the first solvation shell about the solute, on a timescale (following the charging of the probe ion) faster than intermolecular collisions. Figure 15.4 shows experimental data for Coumarin anion in water, showing that a fast (~50 fs) relaxation component indeed shows up in such processes.

[13] This number may be artificially enhanced because of the small sample, a few hundred solvent molecules, used in the simulations.

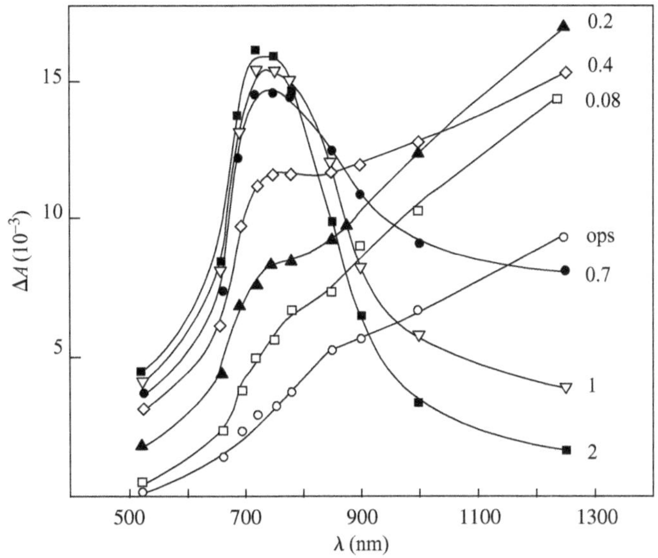

FIG. 15.5 The first observation of hydration dynamics of electron. Absorption profiles of the electron during its hydration are shown at 0, 0.08, 0.2, 0.4, 0.7, 1, and 2 ps. The absorption changes its character in a way that suggests that two species are involved, the one that absorbs in the infrared is generated immediately and converted in time to the fully solvated electron that absorbs near 700 nm. (From A. Migus, Y. Gauduel, J. L. Martin, and A. Antonetti, Phys. Rev Lett. **58**, 1559 (1987).) For later developments in this subject see, for example, K. Yokoyama, C. Silva, D. Hee Son, P. K. Walhout, and P. F. Barbara, J. Phys. Chem. A, **102**, 6957 (1998).)

15.5 Quantum solvation

How would solvation proceed if the solvated particle is an electron rather than essentially classical ion or dipole?

Experimentally, electrons can be injected into and dissolve in molecular liquids. In liquid ammonia solvated electrons form blue, relatively stable, solutions. In water, solvated electrons can be created by photoionizing solute anions or even neat water. These electrons eventually disappear by recombining with the parent species, but live long enough as distinct species with a typical absorption peak near 7000 Å. Provided that the injection and subsequent probing are done with ultrafast time resolution, it is possible to follow the solvation process of the electron via its evolving spectrum.

This fascinating subject has attracted much experimental and theoretical effort for the past two decades, but a conclusive word may still lie ahead. We will not discuss it in detail except to leave the reader with some observations on the difference between quantum and classical solvation. For definiteness, we will compare the solvation of an electron to that of an ion in, say, water.

(1) The ion size remains practically constant during the solvation. The electron size, measured for example by

$$\langle(\mathbf{r} - \langle\mathbf{r}\rangle)^2\rangle^{1/2} = \langle\Psi(\mathbf{r}, t)|(\mathbf{r} - \langle\mathbf{r}\rangle)^2|\Psi(\mathbf{r}, t)\rangle^{1/2} \tag{15.30}$$

where $\Psi(\mathbf{r}, t)$ is the electron wavefunction, decreases substantially during the solvation. Electron solvation is in fact a localization process, in this case by polaron formation.[14]

(2) Related to (1) is the fact that, while ion solvation is basically a relaxation of its potential energy, electron solvation involves both potential and kinetic energies. In fact, two forms of kinetic energy are involved at different stages of the process. An electron is usually injected into the liquid with excess kinetic energy associated with its center of mass motion, and proceeds to lose it at the initial stage of its accommodation in the liquid. As localization (or polaron formation) proceeds, the kinetic energy of the electron actually increases due to its localization.

(3) Ion solvation is essentially a classical process (though in water there must be some quantum effect associated with the motion of hydrogen atoms) that proceeds on a well-defined potential surface. Electron solvation involves several electronic states and proceeds by a combination of adiabatic relaxation on a single potential surface and nonadiabatic crossing between different surfaces. Indeed, theoretical analysis suggests that the two species whose spectroscopic signatures are seen in Fig. 15.5 are ground and excited states of the electron in its forming solvent cavity.

In addition to the intrinsic interest of quantum solvation phenomena, the process of electron solvation offers another example of a localized quantum process taking place in an otherwise essentially classical environment.[15] We have encountered a similar situation in the vibrational relaxation of high-frequency diatomic molecules

[14] A polaron is an electron attached to, and moving with, the polarization induced by it in a polar environment. This concept is used mostly in solid state physics; the liquid analog is the solvated electron.

[15] F. Webster, P. J. Rossky, and R. A. Friezner, Comp. Phys. Comm. **63**, 494 (1991); O. V. Prezhdo and P. J. Rossky, J. Chem. Phys. **107**, 5863 (1997); E. Neria and A. Nitzan, J. Chem. Phys. **99**, 1109 (1993).

(Chapter 13) and will see this again in electron transfer reactions discussed in the next chapter.

Further Reading

B. Ladanyi and M. S. Skaf, Annu. Rev. Phys. Chem. **44**, 335 (1993).
M. Maroncelli, J. Mol. Liq. **57**, 1 (1993).

16

ELECTRON TRANSFER PROCESSES

I might not know a thing about atoms,
But this much I can say, from what I see
Of heaven's way and many other features:
The nature of the world just could not be
A product of the god's devising; no,
There are too many things the matter with it...

Lucretius (c. 99–c. 55BCE) "The way things are"
translated by Rolfe Humphries, Indiana University Press, 1968

Electron transfer processes are at the core of all oxidation–reduction reactions, including those associated with electrochemistry and corrosion. Photoelectrochemistry and solar energy conversion, organic light emitting diodes, and molecular electronic devices, all dominated by electron transfer and electron transmission in molecular systems, are presently subjects of intensive research at the interface of science and technology. Similarly, electron transfer processes constitute fundamental steps in important biological phenomena such as photosynthesis and vision. This chapter is an introduction to the general phenomenology and theoretical concepts associated with these processes.

16.1 Introduction

Electron transfer is one of the most important, and most studied, elementary chemical processes. This most fundamental oxidation–reduction process lies at the core of many chemical phenomena ranging from photosynthesis to electrochemistry and from the essential steps governing vision to the chemical processes controlling corrosion. As other molecular phenomena that involve charges and charged particles, the natural environment for such processes is a polar solution; the solvation energy associated with the polarization of the environment is a major component in the energetics of such processes. Noting that in vacuum typical molecular ionization potentials are of the order of $(100–400)k_BT$ for $T = 300\,\text{K}$, it appears that the stabilization of ionic species by the solvent environment is the reason why electron transfer processes in solution can take place at room temperature.

When we try to go beyond this general statement, questions arise. Consider for example, the following *self-exchange electron transfer reaction*:

$$Fe^{+3} + Fe^{+2} \rightarrow Fe^{+2} + Fe^{+3} \tag{16.1}$$

or the *cross electron transfer reaction*

$$Fe^{+2} + Ce^{+4} \rightarrow Fe^{+3} + Ce^{+3} \tag{16.2}$$

both in, say, aqueous environment. In the first reaction the reactants and the products are the same (they can still be distinguished by using different isotopes), while in the second they are different. We will see later that the relative thermodynamic stability of reactants and products influences the reaction rate in an important way. Comparing these reactions to those discussed in Chapter 14, several other observations can be made:

1. No bonds are formed or broken in these reactions.
2. The reaction results in a substantial rearrangement of charge density.
3. As a consequence of (2), the reaction is expected to involve a substantial configurational change in the surrounding polar solvent.
4. Because electrons and nuclei move at very different speeds, their relative characteristic timescales may affect the reaction dynamics.

Let us consider the last point. The reader is already familiar with two important implications of the timescale separation between electronic and nuclear motions in molecular systems: One is the Born–Oppenheimer principle which provides the foundation for the concept of potential energy surfaces for the nuclear motion. The other is the prominent role played by the Franck–Condon principle and Franck–Condon factors (overlap of nuclear wavefunctions) in the vibrational structure of molecular electronic spectra. Indeed this principle, stating that electronic transitions occur at fixed nuclear positions, is a direct consequence of the observation that electronic motion takes place on a timescale short relative to that of the nuclei.

Accepting this as a guiding principle immediately points to an important energetic consequence: In the reactions (16.1) or (16.2) the structure of the polar solvent surrounding the reactants is very different from the corresponding structure about the products. This is most easily seen in the case where one side of the reaction involves neutral species, for example, $A \rightarrow A^+$, see Fig. 15.1. Consequently, if the electronic charge distribution changes on a timescale in which the solvent has not moved, the solvent configuration is no longer the most stable configuration. The free energy that would be released in the subsequent solvent relaxation to its stable configuration under the new charge distribution (the second stage in Fig. 15.1;

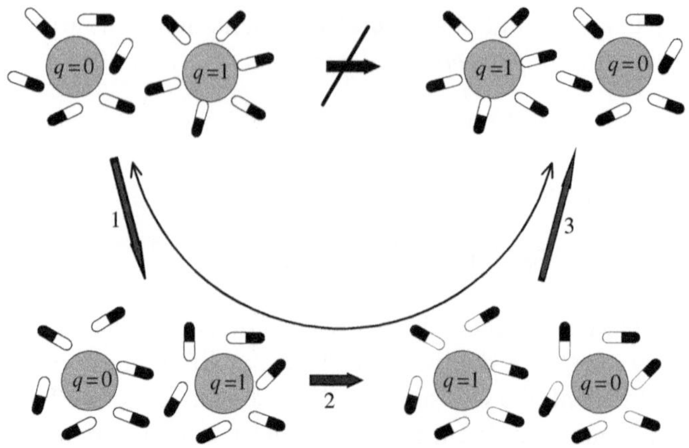

FIG. 16.1 The Marcus picture of electron transfer reaction. The initial and final states are described by the two upper diagrams, however the transition between them cannot take place directly because such a direct transition will involve simultaneous motions of electron and solvent dipoles. Instead the transitions proceeds through steps 1—preparation of a suitable solvation configuration, 2—electron transfer at fixed solvent polarization, and 3—relaxation of the solvent polarization in the final electronic state.

see also Fig. 16.1) is called the *solvent reorganization energy* of the given charge transfer process. This picture seems to suggest that the larger is the solvent reorganization energy the more uphill is the charge redistribution process, implying larger activation energy and a smaller rate. Indeed, in the early days of electron transfer studies such correlations were discussed. For example, electron transfer processes involving small ions are often slower than similar processes with bulky ions, perhaps because a smaller ion size, that is, a shorter distance between the centers of the ion and the nearest solvent molecules, implies a stronger Coulomb field and consequently larger reorganization energy.

There is however a fundamental flaw in this picture: Because the envisioned fast electronic transition, as described, would not conserve energy. In the reaction (16.1) it would generate a system with excess energy equal to the reorganization energy, of order of 0.1–1 eV in aqueous solutions, considerably larger than $k_B T$ at room temperature. The same situation is encountered in photoinduced electronic transitions, however there the needed energy is supplied by the absorbed photon. In the present case, where the only available energy is thermal, the mechanism has to be different.

This observation, first made by Marcus, has led to his Nobel Prize work on electron transfer. The basic idea is simple: In a closed system with two modes of motion, one fast and one slow, a change of state of the fast mode can take place only in such configuration(s) of the slow mode for which the fast transition is energy-conserving.[1] In our particular example, the transition between the two electronic states involved in the electron transfer reaction can take place only in nuclear configurations for which these states are degenerate. The subspace of these configurations therefore constitutes the transition state for the reaction and the dynamics that determines the rate includes the nuclear dynamics that brings the system to the transition state. The overall processes will include, in addition to the electron transfer—step 2 in Fig. 16.1—also the initial and final changes in the nuclear configuration—steps 1 and 3 in that figure.

Problem 16.1. Explain the following statement: In a symmetrical electron transfer process, where the donor and acceptor species comprise identical ionic centers, for example, Eq. (16.1), the "transition state" is given by all configurations that are in equilibrium with the donor and acceptor species when both are carrying a charge $q = 0.5\,(q_{\text{donor}} + q_{\text{acceptor}})$.

16.2 A primitive model

To see the consequence of Marcus' observation let us first consider a simple example in which the solvent configuration is represented by a single nuclear coordinate X. We consider two electronic states, a and b, that are assumed to be coupled to each other by the full electronic Hamiltonian of the system, that is, $H_{a,b} \equiv V_{a,b} \neq 0$.[2] Each electronic state corresponds to a stable solvent configuration—a point along the X coordinate at which the electronic energy, that is, the potential energy surface for the nuclear motion, is a minimum. We further assume that the electronic energy is a quadratic function of the deviation from this minimum. Thus, the two electronic states a and b correspond to two parabolic nuclear potential surfaces, $W_a(X)$ and $W_b(X)$, whose minima are at the stable configurations X_a and X_b. In classical mechanics these surfaces determine the probability for fluctuations of the solvent configuration from its most stable state, for example, $P_a(X) \sim \exp(-\beta(W_a(X) - W_a(X_a))$. These two surfaces can differ in their minimum energy and in their minimum energy configuration, however we assume for

[1] Obviously any process in a closed system must conserve the overall energy. The transition described here has to conserve the energy of the fast mode since, because of the timescale separation, energy cannot be exchanged between the two modes.

[2] This may result, for example, when considering electron transfer between two molecular species and working in a representation derived from the eigenstates of a system in which these species do not interact (e.g. are far from each other). The coupling results from the residual interaction that increases when the two species come closer to each other.

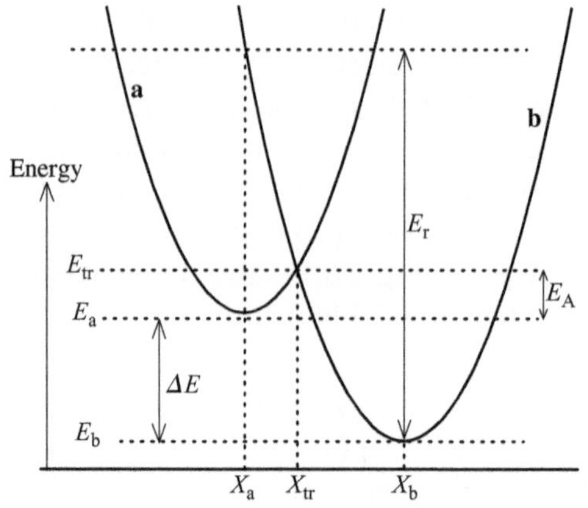

FIG. 16.2 A model for the energetics of electron transfer reactions. The two potential surfaces are associated with two electronic states that transform to each other when the electron is transferred. The coordinate X stands for the nuclear configuration. The model may be characterized by the curvature of these surfaces, by the energy gap ΔE between the two electronic origins and by the reorganization energy E_r. Other important parameters that can be expressed in terms of these are the equilibrium configuration shift $X_b - X_a$ and the activation energy E_A.

simplicity that their curvatures are the same. Thus (see Fig. 16.2)

$$W_a(X) = E_a + (1/2)K(X - X_a)^2; \quad W_b(X) = E_b + (1/2)K(X - X_b)^2 \quad (16.3)$$

As argued above, starting, say, in state a, the coupling $V_{a,b}$ cannot affect a transition to state b unless the electronic energy is conserved, that is, when the two potential surfaces cross at $X = X_{tr} \equiv (E_b - E_a + (1/2)K(X_b^2 - X_a^2))/K(X_b - X_a)$. In the neighborhood of this configuration the electronic states are nearly degenerate and the electron transfer process can take place. The configuration represented by X_{tr} is therefore the transition state for this reaction and the activation energy for the reaction $a \to b$ is $W_a(X_{tr}) - W_a(X_a)$.[3] This yields

$$E_A = \frac{[(E_a - E_b) - E_r]^2}{4E_r}; \qquad E_r = (1/2)K(X_a - X_b)^2 \qquad (16.4)$$

Note that E_r is the amount of nuclear energy which should be released following a *vertical* (i.e. without changing X) electronic transition from state a to b (in fact

[3] In fact, the degeneracy at X_{tr} is removed by the coupling $V_{a,b}$ (the noncrossing rule). This correction may be important in the adiabatic limit (see Section 2.4) where the activation energy along the adiabatic surface is (cf. Eq. (2.19)) $E_A - V_{a,b}$.

also from state b to a). It is therefore, by definition, the reorganization energy in this model.

As a model for a rate process, the surface crossing picture described above can be treated within the Landau–Zener theory (Section 2.4) that yields the probability that a transition between the two electronic states occurs during one dynamical crossing event. Here X_{tr} stands for R^* of Section 2.4. Using this theory to evaluate the rate of such electron transfer processes involves several assumptions:

1. The basic assumptions of the Landau–Zener theory need to be satisfied. These involve the applicability of classical mechanics (e.g. the neglect of tunneling) for the nuclear dynamics and the locality of the curve crossing event.
2. Thermal relaxation (solvent reorganization) is fast relative to the reaction rate, so that the distribution of nuclear configurations remains thermal throughout the reaction.
3. The dynamical interaction with the solvent can be disregarded on the timescale of a single surface crossing event. Furthermore, subsequent crossing events may be treated independently of each other. This makes it possible to use the Landau–Zener expression for the transition probability at each such event.

Note that assumptions (2) and (3) are about timescales. Denoting by τ_r, τ_s, and τ_{LZ} the characteristic times (inverse rates) of the electron transfer reaction, the solvent relaxation, and the Landau–Zener transition, respectively, (the latter is the duration of a single curve-crossing event) we are assuming that the inequalities $\tau_r \gg \tau_s \gg \tau_{LZ}$ hold. The validity of this assumption has to be addressed, but for now let us consider its consequences. When assumptions (1)–(3) are satisfied we can invoke the extended transition-state theory of Section 14.3.5 that leads to an expression for the electron transfer rate coefficient of the form (cf. Eq. 14.32)

$$k = \int_0^\infty d\dot{X}\, \dot{X}\, P(X_{tr}, \dot{X}) P_{b\leftarrow a}(\dot{X}) \tag{16.5}$$

where $P_{b\leftarrow a}(\dot{X})$ is the Landau–Zener transition probability, Eqs (2.47) and (2.48),

$$P_{b\leftarrow a}(\dot{X}) = 1 - \exp\left\{ -\frac{2\pi |V_{a,b}|^2}{\hbar |\dot{X}\Delta F|} \right\}_{X=X_{tr}} \tag{16.6}$$

and where (for the $a \to b$ transition)

$$P(X_{tr}, \dot{X}) = \left(\frac{\beta m}{2\pi}\right)^{1/2} e^{-\beta m\dot{X}^2/2} \frac{\exp(-\beta E_A)}{\int_{-\infty}^{X_{tr}} dX \exp(-\beta(W_a(X) - E_a))}$$

$$= \frac{\beta}{2\pi}(mK)^{1/2} \exp(-\beta E_A) e^{-\beta m\dot{X}^2/2} \tag{16.7}$$

is the Boltzmann probability density to reach the configuration X_{tr} with speed $\dot{X}$. Here m is the mass associated with the reaction coordinate X. The last equality in (16.7) is obtained by replacing X_{tr} by ∞ in the integral over X, an approximation that is valid if E_A is large relative to $k_B T$.

Problem 16.2. Show that for the present model $|\Delta F|$ in Eq. (16.6) is given by $|\Delta F| = K(X_b - X_a)$.

In the adiabatic limit (see Section 2.4) $P_{b\leftarrow a} \to 1$ and Eq. (16.5) yields the standard transition state theory result, Eq. (14.16)

$$k_{ad} = \frac{\omega_s}{2\pi} e^{-\beta \tilde{E}_A}; \qquad \omega_s = \left(\frac{K}{m}\right)^{1/2} \tag{16.8}$$

where $\tilde{E}_A = E_A - |V_{a,b}|$ is the barrier height on the adiabatic potential surface. In the nonadiabatic limit we get (cf. Eq. (14.36))

$$k_{na} = \sqrt{\frac{\pi\beta K}{2}} \frac{|V_{a,b}|^2}{\hbar|\Delta F|_{X=X_{tr}}} e^{-\beta E_A}; \qquad |\Delta F| = K(X_b - X_a) \tag{16.9}$$

Remarkably, this result depends on the force constant K, but not on the mass m, and its dependence on the square of the nonadiabatic coupling characterizes it as a perturbation theory expression.

In the continuation of our discussion we face two tasks. First, we need to replace the simple two-parabola model described above by a realistic model that uses input from the energetic and dynamic properties of the solvent. Second, we have to provide a reliable description of the process that takes place when the electronic states become nearly degenerate, that is, of the electronic transition itself, taking into account the quantum mechanical nature of motion on two coupled potential energy surfaces.

16.3 Continuum dielectric theory of electron transfer processes

16.3.1 The problem

An electron transfer process is essentially a change in the electronic charge distribution in the molecular system. For example, an electron transfer between species A and B is schematically described by the following transition from the electronic states 0 and 1

$$q_{\mathrm{A}}^{(0)} q_{\mathrm{B}}^{(0)} \quad \rightarrow \quad q_{\mathrm{A}}^{(1)} q_{\mathrm{B}}^{(1)} \tag{16.10}$$

Charge conservation implies that $q_{\mathrm{A}}^{(0)} + q_{\mathrm{B}}^{(0)} = q_{\mathrm{A}}^{(1)} + q_{\mathrm{B}}^{(1)}$.

As discussed above, this process involves a substantial reorganization of the surrounding dielectric environment. In the last section each electronic state was associated with a parabolic potential surface—a simple way to relate each such state to a stable configuration of the environment and to the energetic cost of deviating from it. However, the physical significance of such a surface is as yet unclear. In particular, the coordinate X used to express distance in Fig. 16.2 has not been specified. We have, it appears, a reasonable abstract picture that is still unconnected to the physical world. Marcus theory of electron transfer makes this connection using a continuum dielectric picture of the polar solvent. This macroscopic and classical approach to the energetics is then supplemented, when needed, by a dynamical theory that describes the electronic transition itself.

As in our simple treatment of solvation dynamics in Chapter 15, the solvent in Marcus theory is taken as a dielectric continuum characterized by a local dielectric function $\varepsilon(\omega)$. Thus, the relation between the *source*, $\mathcal{D}$ (electrostatic displacement) and the *response*, $\mathcal{E}$ (electric field) is (cf. Eqs (15.1) and (15.2))

$$\begin{aligned} \mathcal{D}(\omega) &= \varepsilon(\omega)\mathcal{E}(\omega) \\ \mathcal{D}(t) &= \int_{-\infty}^{t} d\tau \varepsilon(t-\tau)\mathcal{E}(\tau) \end{aligned} \tag{16.11}$$

Following Marcus, we simplify this picture by assuming that the solvent is characterized by only two timescales, fast and slow, associated, respectively with its electronic and the nuclear response. Correspondingly, the solvent dielectric response function is represented by the total, or *static*, dielectric coefficient ε_{s} and by its fast electronic component ε_{e} (sometimes called *optical response* and related to the refraction index n by $\varepsilon_{\mathrm{e}} = n^2$). ε_{s} includes, in addition to the fast electronic component, also contributions from solvent motions on slower nuclear timescales: Translational, rotational, and vibrational motions. The working assumption of the

Marcus theory is that *the actual change in the electronic charge distribution of the ET system is fast relative to the nuclear motions underlying the static response, but is slow relative to the electronic motions which determine* ε_e. In other words *the electron transfer occurs at constant nuclear polarization, or at fixed nuclear positions.* This is an expression of the Franck–Condon principle in this continuum dielectric theory of electronic transitions.

Problem 16.3. If the electronic degrees of freedom did not exist, the only contribution to ε_s would come from nuclear motions, so in this case ε_s can be identified as the nuclear contribution ε_n. However, in the real system we cannot identify $\varepsilon_s - \varepsilon_e$ as the nuclear contribution to the dielectric response, that is, $\varepsilon_n \neq \varepsilon_s - \varepsilon_e$. Explain why.

16.3.2 Equilibrium electrostatics

Let us examine the electrostatic consequence of this assumption. The Poisson equation, $\nabla \cdot \mathcal{D} = 4\pi\rho$ gives the electrostatic displacement $\mathcal{D}$ for a given electrostatic charge density ρ. This is the bare electrostatic field, a free space property, which is not related in any way to the presence of the solvent. The electrostatic field $\mathcal{E}$ is determined by the solvent response via the relation

$$\mathcal{E} = \mathcal{D} - 4\pi\mathbf{P}, \tag{16.12}$$

where $\mathbf{P}$ is the polarization (dipole density) induced in the solvent. For simplicity of notation we will assume that the dielectric response of the solvent is a scalar, so all relations between $\mathcal{D}$, $\mathcal{E}$, and P may be written for the corresponding magnitudes. *At equilibrium* P is related to $\mathcal{E}$ through the relation

$$P = \chi\mathcal{E} = \frac{\varepsilon_s - 1}{4\pi}\mathcal{E} \tag{16.13}$$

so that

$$\mathcal{D} = \varepsilon_s\mathcal{E} \tag{16.14}$$

The polarization P can be viewed as originating from two sources: Electronic (e) and nuclear (n): $P = P_n + P_e$. We may write

$$P_e = \frac{\varepsilon_e - 1}{4\pi}\mathcal{E}; \qquad \text{therefore } P_n = \frac{\varepsilon_s - \varepsilon_e}{4\pi}\mathcal{E} \tag{16.15}$$

16.3.3 Transition assisted by dielectric fluctuations

Let us reiterate the Marcus dilemma stated in Section 16.1. In vacuum, to remove a charge from the donor and put it on the acceptor is a process of very large activation energy. In solution, if we try to execute such a process without solvent motion the situation will be worse: The solvent is initially polarized in response to the original charge distribution of the donor/acceptor pair and an attempt to change this distribution in a frozen solvent will take us to a state of much higher energy. Since the electron transition itself occurs at fixed nuclear positions, it is not clear how this can possibly happen.

The answer to this dilemma is that the electronic transition can take place if the following conditions are satisfied: (1) the energies of the states immediately before and after the transition are equal, and (2) nuclear positions (therefore the nuclear polarization P_n) are fixed. Therefore, this transition can occur only after a fluctuation in the nuclear positions into a configuration in which condition (1) is satisfied. This fluctuation has to occur *before* the electronic transition took place, namely at *constant charge distribution.*

We are therefore interested in changes in solvent configuration that take place at constant solute charge distribution ρ that have the following characteristics:

1. P_n fluctuates because of thermal motion of solvent nuclei;
2. P_e, as a fast variable, satisfies the equilibrium relationship $P_e = ((\varepsilon_e - 1)/4\pi)\mathcal{E}$;
3. $\mathcal{D} =$ constant (depends on ρ only).

Note that the relations $\mathcal{E} = \mathcal{D} - 4\pi P$; $P = P_n + P_e$ are always satisfied per definition. However, in general $\mathcal{D} \neq \varepsilon_s\mathcal{E}$; equality between these quantities holds only at equilibrium.

16.3.4 Thermodynamics with restrictions

How do we calculate the probability of a fluctuation about an equilibrium state? Consider a system characterized by a classical Hamiltonian $H(\mathbf{r}^N, \mathbf{p}^N)$ where $\mathbf{p}^N$ and $\mathbf{r}^N$ denote the momenta and positions of all particles. The phase space probability distribution is $f(\mathbf{r}^N, \mathbf{p}^N) = Q^{-1}\exp(-\beta H(\mathbf{r}^N, \mathbf{p}^N))$, where Q is the canonical partition function,

$$Q = \iint d\mathbf{r}^N d\mathbf{p}^N e^{-\beta H(\mathbf{r}^N, \mathbf{p}^N)} = e^{-\beta F} \tag{16.16}$$

and F is the Helmholtz free energy. We may also write a partition function for a system in which a particular dynamical variable, $X(\mathbf{r}^N, \mathbf{p}^N)$ is restricted to have a fixed value $\bar{X}$

$$\bar{Q}(\bar{X}) = \iint d\mathbf{r}^N d\mathbf{p}^N e^{-\beta H(\mathbf{r}^N,\mathbf{p}^N)}\delta(\bar{X} - X(\mathbf{r}^N, \mathbf{p}^N)) \equiv e^{-\beta \bar{F}(\bar{X})} \tag{16.17}$$

Q and $\bar{Q}$ are obviously related by

$$Q = \int d\bar{X}\bar{Q}(\bar{X}) = \int d\bar{X} e^{-\beta \bar{F}(\bar{X})} \tag{16.18}$$

Now, by definition, the probability $P_{\bar{X}}(\bar{X})$ that the dynamical variable $X(\mathbf{r}^N, \mathbf{p}^N)$ will assume the value $\bar{X}$ is given by

$$P_{\bar{X}}(\bar{X}) = \iint d\mathbf{r}^N d\mathbf{p}^N f(\mathbf{r}^N, \mathbf{p}^N)\delta(\bar{X} - X(\mathbf{r}^N, \mathbf{p}^N)) = \bar{Q}(\bar{X})/Q \tag{16.19}$$

or

$$P_{\bar{X}}(\bar{X}) = e^{-\beta[\bar{F}(\bar{X}) - F]} \tag{16.20}$$

$\bar{F}(\bar{X}) - F$ is the difference between two equilibrium free energies: F is the regular equilibrium free energy of the system and $\bar{F}(\bar{X})$ is the free energy of a fictitious equilibrium state in which the dynamical variable X was restricted to have a particular value $\bar{X}$.

Note that the above formalism could be repeated for a system characterized not by given temperature and volume but by given temperature and pressure. This would lead to a similar result, except that the Helmholtz free energies F and $\bar{F}$ are replaced by Gibbs free energies G and $\bar{G}$ that are defined in an analogous way.

16.3.5 Dielectric fluctuations

We are interested in fluctuations of the nuclear polarization. As we have just seen, the probability for such a fluctuation is determined by the difference between the free energy of our equilibrium system and the free energy of a fictitious equilibrium state in which P_{n} was restricted by some means to a given value. In addition, we *assume* that the fluctuations relevant to our process are those for which the instantaneous value of P_{n} corresponds to some charge distribution ρ of the solute that will produce this value of P_{n} at equilibrium. We are particularly interested in such fluctuations of P_{n} because, as will be seen, these are the ones that lead to the charge rearrangement.

Specifically, let the initial state of our system ("state 0") be characterized by an electronic charge distribution ρ_0 and by the nuclear polarization $P_{\mathrm{n}0}$ that is in equilibrium with it. In the final state ("state 1") the electronic charge distribution is ρ_1 and the nuclear polarization is P_{n1}. Starting in state 0 we are interested in fluctuations in the nuclear polarization about P_{n0} in the "direction" from ρ_0 to ρ_1.

We will assign to such fluctuations a parameter θ that defines a fictitious charge distribution ρ_θ according to

$$\rho_\theta = \rho_0 + \theta(\rho_1 - \rho_0) \tag{16.21}$$

In turn ρ_θ defines a nuclear polarization $P_{n\theta}$ as that polarization obtained in an equilibrium system (state θ) in which the charge distribution is ρ_θ. Now, in state 0 (where $\rho = \rho_0$) this $P_{n\theta}$ is a fluctuation from equilibrium that is characterized by the parameters ρ_1 and θ.

To obtain the probability for such a fluctuation we introduce another state, state t, which is a fictitious restricted equilibrium system in which (1) the charge distribution is ρ_0 and (2) the *nuclear* polarization is $P_{n\theta}$, that is, same as in the equilibrium state θ in which the charge density is ρ_θ. We want to calculate the free energy difference, $\Delta G_{0\to t}$, between the restricted equilibrium state t and the fully equilibrated state 0. This is the reversible work needed to go, at constant temperature and pressure, from state 0 to state t.

The required difference is calculated in terms of two other free energy differences:

$$\Delta G_{0\to t} = \Delta G_{\theta\to t} - \Delta G_{\theta\to 0} \tag{16.22}$$

Three states are involved here: The two equilibrium states 0 and θ, and the nonequilibrium state t. Consider these states in more detail:

Equilibrium state 0.

$$\begin{aligned}
&\rho = \rho_0 \text{ implies } \mathcal{D} = \mathcal{D}_0 \text{ such that } \nabla\cdot\mathcal{D}_0 = 4\pi\rho_0\\
&\mathcal{E} = \mathcal{E}_0 \text{ satisfies } \mathcal{E}_0 = \varepsilon_s^{-1}\mathcal{D}_0 \text{ and } \mathcal{E}_0 = \mathcal{D}_0 - 4\pi\mathbf{P}_0\\
&\mathbf{P}_0 = \mathbf{P}_{0n} + \mathbf{P}_{0e} \text{ where}\\
&\mathbf{P}_{0e} = \frac{(\varepsilon_e - 1)}{4\pi}\mathcal{E}_0; \qquad \mathbf{P}_{0n} = \frac{(\varepsilon_s - \varepsilon_e)}{4\pi}\mathcal{E}_0; \qquad \mathbf{P}_0 = \frac{(\varepsilon_s - 1)}{4\pi}\mathcal{E}_0
\end{aligned} \tag{16.23}$$

Equilibrium state θ.

$$\begin{aligned}
&\rho = \rho_\theta = \rho_0 + \theta(\rho_1 - \rho_0) \text{ implies } \mathcal{D} = \mathcal{D}_\theta \text{ such that } \nabla\cdot\mathcal{D}_\theta = 4\pi\rho_\theta\\
&\mathcal{E} = \mathcal{E}_\theta \text{ satisfies } \mathcal{E}_\theta = \varepsilon_s^{-1}\mathcal{D}_\theta \text{ and } \mathcal{E}_\theta = \mathcal{D}_\theta - 4\pi\mathbf{P}_\theta\\
&\mathbf{P}_\theta = \mathbf{P}_{\theta n} + \mathbf{P}_{\theta e} \text{ where}\\
&\mathbf{P}_{\theta e} = \frac{(\varepsilon_e - 1)}{4\pi}\mathcal{E}_\theta; \quad \mathbf{P}_{\theta n} = \frac{(\varepsilon_s - \varepsilon_e)}{4\pi}\mathcal{E}_\theta; \quad \mathbf{P}_\theta = \frac{(\varepsilon_s - 1)}{4\pi}\mathcal{E}_\theta
\end{aligned} \tag{16.24}$$

These are the same relations as in Eq. (16.23), with θ replacing 0 everywhere.

Restricted equilibrium state t.

$$\rho = \rho_t = \rho_0 \text{ implies } \mathcal{D} = \mathcal{D}_t = \mathcal{D}_0 \text{ such that } \nabla \cdot \mathcal{D}_0 = 4\pi\rho_0$$

$$\mathcal{E} = \mathcal{E}_t \text{ satisfies } \mathcal{E}_t = \mathcal{D}_t - 4\pi \mathbf{P}_t \text{ (but not } \mathcal{E}_t = \varepsilon_s^{-1}\mathcal{D}_t)$$

$$\mathbf{P}_t = \mathbf{P}_{tn} + \mathbf{P}_{te} \text{ where} \tag{16.25}$$

$$\mathbf{P}_{te} = \frac{\varepsilon_e - 1}{4\pi}\mathcal{E}_t; \qquad \mathbf{P}_{tn} = \frac{\varepsilon_s - \varepsilon_e}{4\pi}\mathcal{E}_\theta$$

Note that ρ, $\mathcal{D}$, $\mathcal{E}$, and $\mathbf{P}$ are all functions of position. Two types of relationships appear in these equations: First (terms with white background) there are those that stem from electrostatic definitions. Another type (terms with light-grey background) are constitutive linear response relationships that are assumed valid at equilibrium. In the equilibrium states described by Eqs (16.23) and (16.24) both are satisfied. The restricted equilibrium state described by Eq. (16.25) is characterized by the fact that the nuclear polarization is not "allowed" to relax to its equilibrium value for the given electric field, but instead restricted to the same value it would have in the equilibrium state θ (last equation in (16.25)with dark-grey background).

To see the consequences of these relationships we will focus first on a particular simple example, where the system contains one spherical ion of charge q and radius a in equilibrium with an infinite dielectric solvent characterized by the dielectric constants ε_e and ε_s. The different states "0" and "θ" correspond to different values of the ion charge, q_0 and q_θ.

16.3.5.1 *Calculation of $\Delta G_{\theta \to t}$*

The free energy difference $\Delta G_{\theta \to t}$ is the reversible work associated with the transition from state θ to state t at constant temperature and pressure. In the initial state θ the potential on the surface of the ion is written as a sum of a bare (vacuum) term and a term derived from the solvent polarization

$$\Phi_\theta = \frac{q_\theta}{\varepsilon_s a} = \frac{q_\theta}{a} + \left(\frac{1}{\varepsilon_s} - 1\right)\frac{q_\theta}{a} \tag{16.26}$$

If we consider the associated radial field

$$\frac{q_\theta}{\varepsilon_s a^2} = \frac{q_\theta}{a^2} + \left(\frac{1}{\varepsilon_s} - 1\right)\frac{q_\theta}{a^2} \tag{16.27}$$

the first term on the right is the displacement field $\mathcal{D}$ and the second one is $4\pi P$. Now add to the ion a small amount, ξ, of charge, *under the condition that the nuclear polarization remains fixed.* Under this condition only the electronic polarization responds. The resulting potential on the ion surface is

$$\Phi(\xi) = \frac{q_\theta}{a} + \left(\frac{1}{\varepsilon_s} - 1\right)\frac{q_\theta}{a} + \frac{\xi}{a} + \left(\frac{1}{\varepsilon_e} - 1\right)\frac{\xi}{a} = \frac{q_\theta}{\varepsilon_s a} + \frac{\xi}{\varepsilon_e a} \tag{16.28}$$

When this charging process proceeds until $\xi = q_0 - q_\theta$ the state of the system will become t. To get the reversible work that affects this change we need to integrate $\Phi(\xi)d\xi$ from $\xi = 0$ to $\xi = q_0 - q_\theta$. This leads to

$$\Delta G_{\theta\to t} = \int_0^{q_0 - q_\theta} d\xi\,\Phi(\xi) = \frac{q_\theta(q_0 - q_\theta)}{\varepsilon_s a} + \frac{(q_0 - q_\theta)^2}{2\varepsilon_e a} \tag{16.29}$$

We may substitute here relation (16.21) in the form $q_\theta = q_0 + \theta(q_1 - q_0)$ to get

$$\Delta G_{\theta\to t} = \frac{q_0(q_0 - q_1)}{\varepsilon_s a}\theta + \frac{(q_1 - q_0)^2}{a}\left(\frac{1}{2\varepsilon_e} - \frac{1}{\varepsilon_s}\right)\theta^2 \tag{16.30}$$

16.3.5.2 Calculation of $\Delta G_{\theta\to 0}$

The process $\theta \to 0$ is a transition between two unrestricted equilibrium states, and the change in free energy is again calculated from the corresponding reversible work. Consider a general equilibrium state θ' with charge $q_{\theta'}$. The potential on the ion surface in this state is:

$$\Phi'_\theta = \frac{q_{\theta'}}{\varepsilon_s a} \tag{16.31}$$

and the charging work is $\int_{q_\theta}^{q_0} dq_{\theta'}\Phi_{\theta'}$. This leads to

$$\Delta G_{\theta\to 0} = \frac{(q_0 - q_1)q_0}{\varepsilon_s a}\theta - \frac{(q_0 - q_1)^2\theta^2}{2\varepsilon_s a} \tag{16.32}$$

16.3.5.3 Evaluation of $\Delta G_{0\to t}$ and its significance

Finally, using (16.22), (16.30), and (16.32), we get a "potential surface" for fluctuations of the nuclear polarization of the solvent about the equilibrium state 0 in which the charging state of the solute is q_0,

$$W_0(\theta) \equiv \Delta G_{0\to t} = \frac{(q_1 - q_0)^2}{2a}\left(\frac{1}{\varepsilon_e} - \frac{1}{\varepsilon_s}\right)\theta^2 \tag{16.33}$$

This is thus the physical picture that underlines the abstract potential surfaces of Section 16.2. It is important to emphasize that $W_0(\theta)$ is a *free energy surface*, useful for evaluating probabilities: The probability that in the equilibrium associated with the solute at charging state q_0, the nuclear polarization will assume a value associated with another equilibrium state in which the charging state is $q_0+\theta(q_1-q_0)$, satisfies $P_0(\theta)/P_0(0) = \exp(-\beta W_0(\theta))$, with $\beta = (k_BT)^{-1}$. This form is analogous to the corresponding result for a particle moving in a harmonic potential $W(X) = (1/2)KX^2$, where the thermal probability $P(X)$ to be at position X satisfies $P(X)/P(0) = \exp(-\beta W(X))$. The analogy is only partial however: In the latter case the potential surface has also dynamical implications while in our case it was derived solely from equilibrium considerations.

We end this discussion with several observations:

1. Continuum dielectric-linear response theory yields a free energy surface for dielectric fluctuations which is a parabola in the "reaction coordinate" θ. This "harmonic oscillator property" is quite nonobvious and very significant.
2. This theory gives us the "force constant" associated with this parabola in terms of physical parameters: The dielectric response parameters ε_e and ε_s, the initial and final charge distributions and a geometric factor, here the ionic radius a. Note that the dimensionality of this force constant is energy, in correspondence with the dimensionless nature of the reaction coordinate θ.
3. We could repeat the above calculation using state 1, where the ion charge is q_1, as the reference point. The corresponding free energy surface is

$$W_1(\theta) = \frac{(q_1-q_0)^2}{2a}\left(\frac{1}{\varepsilon_e}-\frac{1}{\varepsilon_s}\right)(1-\theta)^2 \tag{16.34}$$

so that the probability for a fluctuation in the dielectric environment into a polarization state that would have been in equilibrium with an ion in charging state $q_0+(1-\theta)(q_1-q_0) = q_1+\theta(q_0-q_1)$ is $P_1(\theta)/P_1(0) = \exp(-\beta W_1(\theta))$. Now θ designates the state of the dielectric environment using the equilibrium state 1 as a reference.

To complete the energetic picture it is natural to use a single energy reference, and to measure the free energies of both states 0 and 1 relative to this common reference. Equations (16.33) and (16.34) then become

$$W_0(\theta) = E_0 + \frac{(q_1-q_0)^2}{2a}\left(\frac{1}{\varepsilon_e}-\frac{1}{\varepsilon_s}\right)\theta^2 \tag{16.35a}$$

$$W_1(\theta) = E_1 + \frac{(q_1-q_0)^2}{2a}\left(\frac{1}{\varepsilon_e}-\frac{1}{\varepsilon_s}\right)(1-\theta)^2 \tag{16.35b}$$

The energies $E_0 = W_0(\theta=0)$ and $E_1 = W_1(\theta=1)$ are sometimes referred to as the *electronic origins* of the states 0 and 1, respectively. These energies are properties of the corresponding ions that do not result from the present theory but rather from a quantum mechanical calculation of the ground state energies of these species in the given dielectric environment. Note that the forms of the expressions for the fluctuations probabilities should be modified accordingly, for example, $P_0(\theta)/P_0(0) = \exp[-\beta(W_0(\theta) - E_0)]$.

4. The terms (16.33) and (16.34), that add to E_0 and E_1 in the right-hand sides of Eqs. (16.35), can be regarded as the nuclear energy components of the total state energies. Indeed, these are energies associated with changes in the nuclear polarization of the solvent from its equilibrium values in these states, and are therefore analogs of the nuclear energy as defined in molecular spectroscopy. Obviously, this is only part of the total nuclear energy, associated with the solvent environment in the continuum dielectric model.
5. In constructing the free energy surface we have specified the kind of dielectric fluctuations of interest by limiting considerations only to the subspace of nuclear polarization fields associated with the state of a single charged ion. This is why the resulting surface is specified in terms of one ionic radius. (Note that specifying q_0 and q_1 as the initial and final charges does not impose any limitations; using other values would just amount to rescaling the reaction coordinate θ.) A different example is discussed in the next section.

16.3.6 Energetics of electron transfer between two ionic centers

We shall now repeat the above calculation for the process (16.10) in which electronic charge is transferred between two atomic centers A and B. In this case

$$\rho_0 = (q_{\mathrm{A}}^{(0)}, q_{\mathrm{B}}^{(0)}) \quad \rightarrow \quad \rho_1 = (q_{\mathrm{A}}^{(1)}, q_{\mathrm{B}}^{(1)}) \tag{16.36}$$

and charge conservation implies

$$q_{\mathrm{A}}^{(0)} + q_{\mathrm{B}}^{(0)} = q_{\mathrm{A}}^{(1)} + q_{\mathrm{B}}^{(1)} \tag{16.37}$$

We consider two ions of radii R_A and R_B positioned at a distance R_{AB} between them. We assume that $R_{\mathrm{AB}} \gg R_A, R_B$ so that simple electrostatic relations can be used. The relevant fluctuations of the nuclear polarization about state 0 are those that take place in the "direction" of state 1. The reaction coordinate θ defines equilibrium states along this direction according to Eq. (16.21), that is,

$$\begin{aligned} q_{\theta A} &= q_{\mathrm{A}}^{(0)} + \theta\left[q_{\mathrm{A}}^{(1)} - q_{\mathrm{A}}^{(0)}\right] \\ q_{\theta B} &= q_{\mathrm{B}}^{(0)} + \theta\left[q_{\mathrm{B}}^{(1)} - q_{\mathrm{B}}^{(0)}\right] \end{aligned} \tag{16.38}$$

So that $q_{0A} = q_{\mathrm{A}}^{(0)}$; $q_{1A} = q_{\mathrm{A}}^{(1)}$ and same for B. The same calculation as in Section 16.3.5 is now repeated, however with two modifications. First, the potential on the surface of A has a contribution also from ion B, so an equation such as (16.31) is replaced by:

$$\Phi_{\theta A} = \frac{q_{\theta A}}{\varepsilon_s R_A} + \frac{q_{\theta B}}{\varepsilon_s R_{AB}} \tag{16.39}$$

Second, the required charging work is a sum of the works done to charge both A and B. Even though these two charging processes always occur simultaneously, we may calculate their contribution to the overall charging energy separately.

Consider the contribution to the free energy change from changing the charging state of A. As before, we first consider the reversible work needed to go from an unrestricted equilibrium state θ to a restricted state t. For the present process this restricted state is obtained, starting from the state θ, by moving an amount of charge ξ from B to A (or $-\xi$ from A to B) while keeping the nuclear polarization frozen. The process is complete when $\xi = \xi_{\text{final}} = q_{0A} - q_{\theta A} = q_{\theta B} - q_{0B}$ (so that the final charges on A and B are q_{0A} and q_{0B}). The equation analogous to (16.28) is now

$$\Phi_A(\xi) = \frac{q_{\theta A}}{\varepsilon_s R_A} + \frac{\xi}{\varepsilon_e R_A} + \frac{q_{\theta B}}{\varepsilon_s R_{AB}} - \frac{\xi}{\varepsilon_e R_{AB}} \tag{16.40}$$

and integrating over ξ from $\xi = 0$ to $\xi = \xi_{\text{final}}$ results in

$$\Delta G^A_{\theta\to t} = \left(\frac{q_{\theta A}}{\varepsilon_s R_A} + \frac{q_{\theta B}}{\varepsilon_s R_{AB}}\right)(q_{0A} - q_{\theta A}) + \left(\frac{1}{2\varepsilon_e R_A} - \frac{1}{2\varepsilon_e R_{AB}}\right)(q_{0A} - q_{\theta A})^2 \tag{16.41}$$

Next, for the transition $\theta \to 0$ between the two equilibrium states θ and 0, ξ varies in the same way but the potential on A is

$$\Phi_A(\xi) = \frac{q_{\theta A} + \xi}{\varepsilon_s R_A} + \frac{q_{\theta B} - \xi}{\varepsilon_s R_{AB}} \tag{16.42}$$

and the charging energy in the process where ξ goes from 0 to ξ_{final} is

$$\Delta G^A_{\theta\to 0} = \left(\frac{q_{\theta A}}{\varepsilon_s R_A} + \frac{q_{\theta B}}{\varepsilon_s R_{AB}}\right)(q_{0A} - q_{\theta A}) + \left(\frac{1}{2\varepsilon_s R_A} - \frac{1}{2\varepsilon_s R_{AB}}\right)(q_{0A} - q_{\theta A})^2 \tag{16.43}$$

Subtracting (16.43) from (16.41) and putting $(q_{0A} - q_{\theta A})^2 = (q_{0B} - q_{\theta B})^2 \equiv (q_0 - q_1)^2\theta^2$ leads to

$$\Delta G^A_{0\to t} = \left(\frac{1}{\varepsilon_e} - \frac{1}{\varepsilon_s}\right)\left(\frac{1}{2R_A} - \frac{1}{2R_{AB}}\right)\Delta q^2\theta^2 \tag{16.44}$$

$\Delta q = |q_0 - q_1|$ is the amount of charge transferred between A and B in the $0 \rightarrow 1$ process.

Equation (16.44) is the contribution to the free energy difference between states t and 0 that is associated with the work involved in the reversible change of the charging state of A from $q_A^{(0)} = q_{0A}$ to $q_{\theta A}$. The equivalent work involved in changing the charging state of B is obtained from (16.44) by interchanging A and B everywhere. Their sum, the reversible work needed to go from state 0 to state t gives the free energy for this process

$$\Delta G_{0\rightarrow t} = \left(\frac{1}{\varepsilon_e} - \frac{1}{\varepsilon_s}\right)\left(\frac{1}{2R_A} + \frac{1}{2R_B} - \frac{1}{R_{AB}}\right)\Delta q^2\theta^2 \tag{16.45}$$

Comparing this with the result (16.33) obtained for the single ion case, we see again a parabolic free energy surface with a different force constant. The difference is seen to arise from a different geometric factor.

The free energy surface (16.45) is associated with fluctuations about state $0 = (q_A^{(0)}, q_B^{(0)})$. Repeating the argument that leads to Eq. (16.35), we denote the minimum energy on this surface, that is, the energy at $\theta = 0$, by E_0.[4] The free energy surface associated with state 0 is therefore,

$$\begin{aligned} W_0(\theta) &= E_0 + (1/2)K\theta^2 \\ K &= 2\left(\frac{1}{\varepsilon_e} - \frac{1}{\varepsilon_s}\right)\left(\frac{1}{2R_A} + \frac{1}{2R_B} - \frac{1}{R_{AB}}\right)\Delta q^2 \end{aligned} \tag{16.46}$$

This free energy can be used to calculate the probability for fluctuations in the dielectric environment about state $0 = (q_A^{(0)}, q_B^{(0)})$ of the solute system. Note again that the variable θ specifies the state of the dielectric environment: It is a state with such polarization as will be in equilibrium with the solute system in charging state $(q_A^{(0)} + \theta(q_A^{(1)} - q_A^{(0)}), q_B^{(0)} + \theta(q_B^{(1)} - q_B^{(0)}))$.

Obviously, the same calculation could be done about state $1 = (q_A^{(1)}, q_B^{(1)})$. This state has a different electronic origin E_1, and the most stable state of the dielectric environment is $\theta = 1$. The corresponding free energy surface is therefore,

$$W_1(\theta) = E_1 + (1/2)K(1-\theta)^2 \tag{16.47}$$

with the same force constant K. Note again that K has the dimensionality [energy].

---

[4] In this chapter we use the notation E_j and $E_{el}^{(j)}$ interchangeably to denote the electronic origin of state j. Note that E_r is used exclusively to denote reorganization energy.

16.3.7 The electron transfer rate

Now compare the results (16.46) and (16.47) to the model considered in Section 16.2. There we have constructed an abstract model for an electron transfer reaction, where the electronic states a and b were associated with potential energy surfaces given as a functions of the solvent configuration, represented by a coordinate X. Here we identified these potential surfaces as free energy surfaces, $W_a(\theta)$ and $W_b(\theta)$, expressed in terms of a coordinate θ that characterizes the nuclear polarization of the solvent. Now "a" and "b" replace "0" and "1" as state designators. Furthermore, the assumption made in Section 16.2, that the two surfaces are equal curvature parabolas, has turned out to be a property of our electrostatic model! We can now identify the reorganization and activation energies (cf. Eq. (16.4), noting that the distance between the minima of the surfaces $W_a(\theta)$ and $W_b(\theta)$ is 1) for this electron transfer reaction

$$E_{\mathrm{r}} = \left(\frac{1}{\varepsilon_{\mathrm{e}}} - \frac{1}{\varepsilon_{\mathrm{s}}}\right)\left(\frac{1}{2R_{\mathrm{A}}} + \frac{1}{2R_{\mathrm{B}}} - \frac{1}{R_{\mathrm{AB}}}\right)\Delta q^2 \tag{16.48}$$

$$E_{\mathrm{A}} = \frac{[(E_a - E_b) - E_{\mathrm{r}}]^2}{4E_{\mathrm{r}}}; \qquad \text{for the } a \rightarrow b \text{ transition} \tag{16.49}$$

and we have already found the force constant K, Eq. (16.46), that characterizes the electron transfer rate. We have found two expressions for this rate. In the adiabatic limit we have (cf. Eq. (16.8))

$$k_{\mathrm{ad}} = \frac{\omega_{\mathrm{s}}}{2\pi} e^{-\beta \tilde{E}_{\mathrm{A}}}; \qquad \omega_{\mathrm{s}} = \left(\frac{K}{m}\right)^{1/2}; \qquad \tilde{E}_{\mathrm{A}} = E_{\mathrm{A}} - |V_{a,b}| \tag{16.50}$$

and the nonadiabatic rate, using Eq. (16.9), is

$$k_{\mathrm{na}} = \sqrt{\frac{\pi\beta K}{2}} \frac{|V_{a,b}|^2}{\hbar K} e^{-\beta E_{\mathrm{A}}} \tag{16.51}$$

At this stage, the theory did not yield the "mass" (a parameter of dimensionality [mass] $\times$ [length]^2) associated with the dielectric fluctuations, which is needed to determine ω_{s} that appears in the adiabatic rate expression. On the other hand, short of the nonadiabatic coupling itself, all parameters needed to calculate the nonadiabatic rate from Eq. (16.51) have been identified within this dielectric theory.

16.4 A molecular theory of the nonadiabatic electron transfer rate

Equation (16.51) was derived from the Landau–Zener theory for transitions between electronic states at the crossing of the corresponding potential surfaces, however the use of free energy rather than potential energy surfaces as input for a rate expression of the Landau–Zener type is far from obvious. We can derive an alternative expression for the nonadiabatic rate, using as input only the activation energy E_{A}. The starting point for this derivation is the Fermi golden-rule for the transition rate between two electronic states a and b, Eq. (12.34), written in the form

$$k_{b\leftarrow a}(E_{ab}) = \frac{2\pi}{\hbar}|V_{a,b}|^2 \frac{1}{Q_a}\sum_i e^{-\beta E_{a,i}} \sum_f |\langle \chi_{a,i}|\chi_{b,f}\rangle|^2 \delta(E_{ab} + E_{a,i} - E_{b,f}) \tag{16.52}$$

where $E_{a,i}$ are the energies of vibrational levels on the multidimensional nuclear potential energy surface of electronic state a measured from the bottom (referred to as the electronic origin) of that surface, and $\chi_{a,i}$ are the corresponding nuclear wavefunctions. A similar notation applies to $E_{b,f}$ and $\chi_{b,f}$. The terms $|\langle\chi_{a,i}|\chi_{b,f}\rangle|^2$ are the Franck–Condon (FC) factors associated with this electronic transition. Also, $E_{ab} = E_a - E_b \equiv \Delta E$ is the difference between the electronic origins of states a and b and $Q_a = \sum_i \exp(-\beta E_{a,i})$ is the nuclear partition function in electronic state a. In writing Eq. (16.52) we have invoked the Condon approximation, Eq. (12.26), by which the coupling matrix element between any two vibronic wavefunctions in the a and b states was written as a product of the electronic matrix element $V_{a,b}$ and a nuclear overlap function.

Now, Eq. (16.52) can be written in the form

$$k_{b\leftarrow a}(E_{ab}) = \frac{2\pi}{\hbar}|V_{a,b}|^2 F(E_{ab}) \tag{16.53}$$

where the function

$$F(\Delta E) = \frac{1}{Q_a}\sum_i e^{-\beta E_{a,i}} \sum_f |\langle\chi_{a,i}|\chi_{b,f}\rangle|^2 \delta(\Delta E + E_{a,i} - E_{b,f}) \tag{16.54}$$

is the *thermally averaged Franck–Condon factor*. This function satisfies the sum rule

$$\int_{-\infty}^{\infty} d(\Delta E) F(\Delta E) = 1 \tag{16.55}$$

and the detailed balance relation

$$F(\Delta E) = F(-\Delta E) \times e^{\beta\Delta E}; \qquad \beta = k_B T \tag{16.56}$$

In addition, in the absence of coupling between the electronic and vibrational motions it becomes

$$F(\Delta E) = \delta(\Delta E) \tag{16.57}$$

Problem 16.4. Prove Eqs (16.55)–(16.57).

It is important to realize that the only approximations that enter into this rate expression is the use of the Fermi golden-rule, which is compatible with the weak coupling nonadiabatic limit, and the Condon approximation, which is known to be successful in applications to electronic spectroscopy. The solvent effect on the electronic process, including the slow dielectric response, must arise from the FC factor that contains contributions from all the surrounding intermolecular and intramolecular nuclear degrees of freedom. In fact, if the nuclear component of the solvent polarization was the only important nuclear motion in the system, then on the classical level of treatment used by Marcus Eqs (16.53) and (16.51) with E_A given by (16.49) should be equivalent. This implies that in this case

$$F(\Delta E) = \frac{1}{\sqrt{4\pi E_{\mathrm{r}} k_B T}} \exp\left(-\frac{(\Delta E - E_{\mathrm{r}})^2}{4E_{\mathrm{r}} k_B T}\right) \tag{16.58}$$

where the pre-exponential term was taken according to the sum rule (16.55). Equations (16.53) and (16.58) constitute the required rate expression, that is,

$$k_{b\leftarrow a,na}(\Delta E) = \frac{1}{\hbar}\sqrt{\frac{\pi}{E_{\mathrm{r}} k_B T}} |V_{a,b}|^2 \exp\left(-\frac{(E_{ab} - E_{\mathrm{r}})^2}{4E_{\mathrm{r}} k_B T}\right) \tag{16.59}$$

where the subscript "*na*" again emphasizes the nonadiabatic character of the process considered. Remarkably, this result is identical to Eq. (12.69), obtained for the transition rate between two molecular electronic states in the spin–boson model in the thermal activation limit. The treatment of Chapter 12 was not based on a continuum dielectric model and the reorganization energy, cf. Eq. (12.22)

$$E_{\mathrm{r}} = \hbar \sum_{\alpha} \omega_{\alpha} \bar{\lambda}_{\alpha}^2, \tag{16.60}$$

expressed in terms of frequencies and coupling coefficients associated with the normal modes of the nuclear subsystem, did not necessarily stem from dielectric relaxation. Geometrically, however, as can be seen by comparing Figs 16.2 and 12.3, the reorganization energies that appear in Chapters 12 and 16 are similar.

Problem 16.5. Verify that $F(\Delta E)$, Eq. (16.58), satisfies the equality (16.56) and (16.57).

We have thus found that the reorganization of the nuclear environment that accompanies electron transfer involves the change in the molecular configuration, that is, shifts in the equilibrium positions of the nuclei between the two electronic states, as well as the dielectric response of the solvent. Equation (16.52) represents the effect of both, and in the thermal activation limit it leads to (16.59) in which both kinds of nuclear reorganization contribute additively to the reorganization energy.[5] These two contributions are sometimes referred to as *inner sphere* and *outer sphere* response. It should be emphasized that the separation between molecular and solvent nuclear degrees of freedom is to some extent a matter of choice. We may, for example, include solvent motions within the first solvation shell about the molecule with the molecular subsystem.

Equation (16.59) was obtained from (16.52) in the thermal activation limit described in Section 12.5.4, which is valid in high-temperature strong-coupling situations. It is expected to hold at room temperature for the solvent dielectric response that involve mostly intermolecular low-frequency motions that are strongly coupled to the electronic charge distribution. On the other hand, intramolecular motions involve high-frequency modes for which the high-temperature approximation is questionable.[6] In fact, the behavior of such modes may be better described by the nuclear tunneling limit of Section 12.5.3. With this in mind we can reformulate Eq. (16.52) so as to make it possible to use different limits for different groups of modes. To this end we assume that the nuclear wavefunctions can be written as products of high-frequency and low-frequency components, for example,

$$\chi_{a,i} = \chi^h_{a,i}\chi^l_{a,i}$$

and accordingly

$$E_{a,i} = E^h_{a,i} + E^l_{a,i} \qquad \text{and} \qquad Q_a = Q^h_a Q^l_a \tag{16.61}$$

Using $\delta(\Delta E + E^h_{a,i} + E^l_{a,i} - E^h_{b,i} - E^l_{b,i}) = \int du\delta(\Delta E - u + E^h_{a,i} - E^h_{b,i})\delta(u + E^l_{a,i} - E^l_{b,i})$ it is easily shown that

$$k_{b\leftarrow a,na}(\Delta E) = \frac{2\pi}{\hbar}|V_{a,b}|^2 \int du\, F^h(\Delta E - u)F^l(u) \tag{16.62}$$

where the functions $F(E)$ are defined as in Eq. (16.54), except that F^h and F^l contain modes from the high- and low-frequency groups, respectively. Equation (16.62) can

[5] This statement is obviously subject to the assumption that molecular nuclear reorganization can be described by the electron–phonon model used in Chapter 12. Readers accustomed to notations used in the electron transfer literature should note that λ, commonly used in that literature for the reorganization energy (here denoted E_r), is used here (and in Chapter 12) to denote nuclear shift.

[6] The terms *inner sphere* and *outer sphere* are sometimes used to distinguish between the contributions to the solvent response associated with motions close to the charge transfer centers (e.g. the first solvation shell) and the bulk of the solvent, respectively. Intramolecular motions are in this sense part of the inner sphere response.

be used as a starting point for treatments that use different approximation schemes for F^h and F^l.[7]

16.5 Comparison with experimental results

The assumptions underlying the theory of electron transfer, most strongly the dielectric continuum model of the solvent and the assumed linearity of dielectric response on the molecular scale, together with the fact that the electronic coupling matrix elements involved are usually not known, direct us to seek experimental verifications not in its absolute quantitative predictions but in relative quantities, qualitative aspects, and expected trends. Indeed, the theory has been very successful on this level of predictive power. An early demonstration of its success was its ability to predict the rates of cross exchange reactions, for example, $A+B \to (A+e)+(B-e)$ from measured rates of the corresponding self-exchange reactions $X + X \to (X + e) + (X - e)$ with $X = A$ or $X = B$. The argument is based on the fact that in rate expressions such as (16.53) the electronic coupling elements between different species of about the same size may be assumed to be approximately equal, and on the observation that the reorganization energies $E_{\mathrm{r}}^{(AB)}, E_{\mathrm{r}}^{(AA)}, E_{\mathrm{r}}^{(BB)}$ of the above reactions satisfy

$$E_{\mathrm{r}}^{(AB)} = \frac{E_{\mathrm{r}}^{(AA)} + E_{\mathrm{r}}^{(BB)}}{2} \tag{16.63}$$

This follows from expressions such as (16.60), and more generally from the harmonic medium representation of dielectric response (Section 16.9) in which E_{r} is made of additive contributions of independent normal modes including intramolecular modes.

Consider now the implications of (16.63) with regards to the four rates defined by the following rate diagrams

$$A + B \underset{k_{ab}}{\overset{k_{ba}}{\rightleftharpoons}} (A + e) + (B - e) \tag{16.64}$$

$$\begin{aligned} A + A &\xrightarrow{k_{aa}} (A + e) + (A - e) \\ B + B &\xrightarrow{k_{bb}} (B + e) + (B - e) \end{aligned} \tag{16.65}$$

The reorganization energy associated with either rate in (16.64) will be denoted $E_{\mathrm{r}}^{(ab)}$ $(= E_{\mathrm{r}}^{(ba)})$, and the corresponding energies associated with the processes in

[7] See, for example, J. Jortner, J. Chem. Phys. **64**, 4860 (1976).

(16.65) will be denoted $E_r^{(aa)}$ for the first, and $E_r^{(bb)}$ for the second. We will also denote the electronic energy gap $E_{A+B} - E_{(A+e)+(B-e)}$ by ΔE so that

$$K_e = \frac{k_{ba}}{k_{ab}} = \exp(\beta \Delta E); \qquad \Delta E = E_a - E_b \tag{16.66}$$

Following Eq. (16.53) and assuming that the electronic coupling is the same for all these reactions[8] we also have

$$k_{ij} = \frac{2\pi}{\hbar} |V|^2 F_{ij}(\Delta E) \tag{16.67}$$

where

$$F_{ba} = C_{ab} \exp\left(-\beta \frac{(\Delta E - E_r^{(ab)})^2}{4E_r^{(ab)}}\right); \qquad F_{ab} = C_{ab} \exp\left(-\beta \frac{(\Delta E + E_r^{(ab)})^2}{4E_r^{(ab)}}\right) \tag{16.68a}$$

$$F_{ii} = C_{ii} \exp\left(-\beta \frac{E_r^{(ii)}}{4}\right); \qquad i = a, b \tag{16.68b}$$

Equations (16.67) and (16.68) together with (16.63) written in the form $E_r^{(ab)} = (E_r^{(aa)} + E_r^{(bb)})/2$ can now be used to write the following equality:

$$\begin{aligned}
\frac{k_{ba}}{(k_{aa}k_{bb})^{1/2}} &= \frac{F_{ba}}{(F_{aa}F_{bb})^{1/2}} \\
&= \frac{C_{ab}}{\sqrt{C_{aa}C_{bb}}} \exp\left[-\beta\left(\frac{(\Delta E - (E_r^{(aa)} + E_r^{(bb)})/2)^2}{2(E_r^{(aa)} + E_r^{(bb)})} - \frac{E_r^{(aa)}}{8} - \frac{E_r^{(bb)}}{8}\right)\right] \\
&= \frac{C_{ab}}{\sqrt{C_{aa}C_{bb}}} \exp\left(-\beta \frac{\Delta E^2}{2(E_r^{(aa)} + E_r^{(bb)})}\right) \exp(+\beta \Delta E/2)
\end{aligned} \tag{16.69}$$

or

$$k_{ba} = \left(\frac{k_{aa}k_{bb}K_e}{C_{aa}C_{bb}}\right)^{1/2} C_{ab} \exp\left(-\beta \frac{\Delta E^2}{2(E_r^{(aa)} + E_r^{(bb)})}\right) \tag{16.70}$$

Therefore k_{ba} can be predicted from the equilibrium constant of the reaction (16.64) together with parameters that characterize the rates of the reactions (16.65). The

[8] This assumption can of course fail, in particular for nonadiabatic electron transfer.

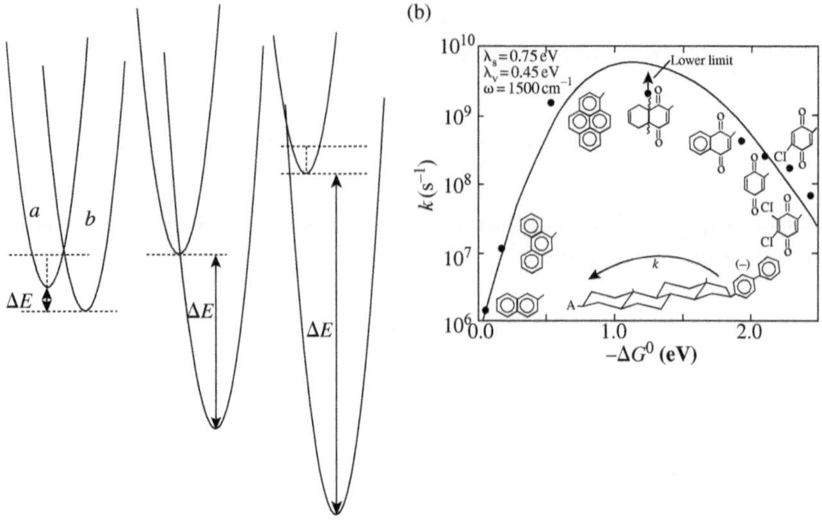

FIG. 16.3 (a) Change of activation energy (vertical dotted lines) from positive ('normal regime', left), to zero (middle) and back to positive ('inverted regime', right) as the driving free energy gap (ΔE; double arrowed vedrtical lines) is increased. Right: Electron transfer rate for a series of molecules characterized by different energy gaps, displayed against the energy gap (From J. R. Miller, L. Calcaterra and G. L. Closs, J. Am. Chem. Soc. **106**, 3047 (1984), where the (free) energy gap (ΔE in our notation) is denoted $-\Delta G^0$.)

success of such predictions (L. E. Bennett, Prog. Inorg. Chem. **18**, 1 (1973)) has provided a strong early support to the validity of this theory.

A more dramatic prediction is the existence of the "inverted regime" in the dependence of $k_{b\leftarrow a}$ on the "driving force" $E_{ab} = E_a - E_b$. Equation (16.59) shows that the rate increases as E_{ab} grows from zero, however beyond $E_{ab} = E_{\rm r}$ this dependence is inverted, and the rate decreases with further increase in this "force." Reflection on the geometrical origin of this observation shows it to be related to the way by which the crossing point between two shifted parabolas changes when one of these parabolas move vertically with respect to the other, see Fig. 16.3(a). The verification of this prediction, pictorially displayed in Fig. 16.3(b) has provided a dramatic evidence in support of this picture.[9]

It is of interest to note another characteristic of the inverted regime: The anti-correlation between the nuclear shift λ and the activation energy $E_{\rm A}$ as seen from Eq. (16.49) and from the fact that $E_{\rm r}$ scales like λ^2. In this limit of small nuclear

[9] The inclusion of high-frequency molecular modes in the spirit of Eq. (16.62) is critical for the quantitative aspects of this behavior, in particular in the inverted regime (see Fig. 13 in the Review by Bixon and Jortner cited at the end of this chapter).

shifts and large activation energy we may expect the Marcus theory to break down because of the onset of nuclear tunneling. In this limit and at low temperatures the weak coupling formalism of Section 12.5.3 may provide a better description of the electronic transition rate.

16.6 Solvent-controlled electron transfer dynamics

The Marcus theory, as described above, is a transition state theory (TST, see Section 14.3) by which the rate of an electron transfer process (in both the adiabatic and nonadiabatic limits) is assumed to be determined by the probability to reach a subset of solvent configurations defined by a certain value of the reaction coordinate. The rate expressions (16.50) for adiabatic, and (16.59) or (16.51) for nonadiabatic electron transfer were obtained by making the TST assumptions that (1) the probability to reach transition state configuration(s) is thermal, and (2) once the reaction coordinate reaches its transition state value, the electron transfer reaction proceeds to completion. Both assumptions rely on the supposition that the overall reaction is slow relative to the thermal relaxation of the nuclear environment. We have seen in Sections 14.4.2 and 14.4.4 that the breakdown of this picture leads to dynamic solvent effects, that in the Markovian limit can be characterized by a friction coefficient γ: The rate is proportional to γ in the low friction, $\gamma \to 0$, limit where assumption (1) breaks down, and varies like γ^{-1} when $\gamma \to \infty$ and assumption (2) does. What stands in common to these situations is that in these opposing limits the solvent affects *dynamically* the reaction rate. Solvent effects in TST appear only through its effect on the free energy surface of the reactant subspace.

How will similar considerations manifest themselves in electron transfer reactions? It should be obvious that the above limits of solvent controlled rates should still exist. This follows from the fact that any activated molecular reaction can be reduced to a succession of processes (that in reality are not necessarily temporally distinguishable): (1) preparation of the transition configuration in the reactant subspace, (2) evolution of the transition configuration from reactants to products, and (3) relaxation of the product to final equilibrium. Irrespective of details this remains true also here, as shown in Fig. 16.1. When solvent-induced relaxation (namely steps 1 and 3 above and in Fig. 16.1) is fast enough to keep the reactant(s) in thermal equilibrium and to affect efficient relaxation in the product subspace, TST holds and the rate expressions (16.50) and (16.59), (16.51) are valid in their appropriate limits. When the solvent relaxation cannot "keep up" with the intrinsic charge transfer process it becomes the rate determining step. In this limit the observed rate will reflect the relaxation properties of the nuclear environment, whose manifestation in the context of solvation dynamics was discussed in Chapter 15. Figure 16.4 shows an example of this behavior, where the fluorescence that follows a charge transfer transitions in two molecules is used to monitor the rate of that transition.

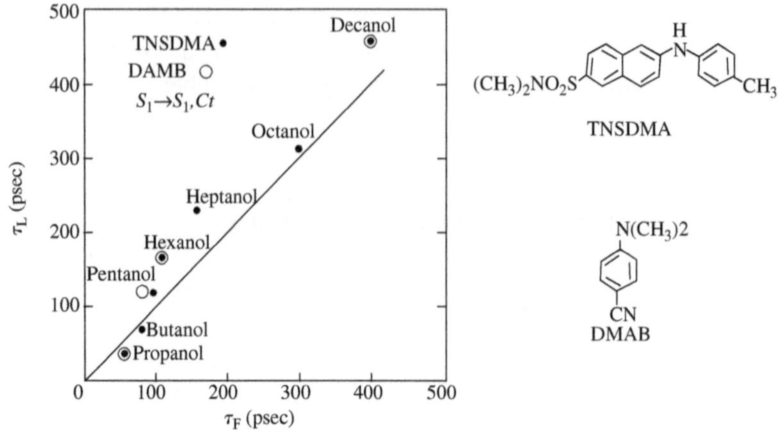

FIG. 16.4 Correlation between the fluorescence lifetime τ_F and the longitudinal dielectric relaxation time, τ_L (Eq. (15.19)) of 6-N-(4-methylphenylamino-2-naphthalene-sulfon-N,N-dimethylamide) (TNSDMA) and 4-N,N-dimethylaminobenzonitrile (DMAB) in linear alcohol solvents. The fluorescence signal is used to monitor an electron transfer process that precedes it. The line is drawn with a slope of 1. (From E. M. Kosower and D. Huppert, Ann. Rev. Phys. Chem. **37**, 127 (1986); see there for original publications.)

If electron transfer was indeed a combination of three consecutive rate processes, the overall rate should satisfy a relationship of the kind $k^{-1} = k_1^{-1} + k_2^{-1} + k_3^{-1}$. As in Eq. (14.87) this is an oversimplification which still conveys the qualitative nature of the process. It also emphasizes an important observation, that the rate of the overall process cannot be faster than any of its individual components. In particular, given that at least one of the processes involved is the solvent reorganization, we might have expected that the inverse dielectric relaxation time $(\tau_L)^{-1}$ will constitute an upper limit to the observed rate.[10] Two counter arguments can be raised, however. First, as indicated by the discussion of Section 15.4, dielectric relaxation theory, at least in its usually practiced long wavelength limit, does not describe correctly the fast, inertial component of solvent relaxation associated with a sudden change of a solute charge distribution. Second, as was indicated in Section 16.3.7, in addition to the solvent dielectric response, a substantial contribution to the nuclear reorganization arises from intramolecular nuclear motion that

[10] In the Markovian Kramers model discussed in Section 14.4, the friction coefficient γ describes the coupling of the reaction coordinate to the thermal environment. In the low friction (underdamped) limit it is equal to the thermal relaxation rate in the reactant well, which is equivalent in the present case to the solvation well of the initial charge distribution. More generally, this rate should depend also on the frequency ω_S of this well. The theory of solvation dynamics, Chapter 15, does not use a Langevin equation such as (14.39) as a starting point, however it still yields an equivalent relaxation rate, the inverse solvation time $(\tau_L)^{-1}$, which is used in the present discussion.

involves high-frequency molecular modes. The existence of these fast relaxation modes of the nuclear environment of the elecronic system may be the origin of several recent observations of electron transfer rates that are considerably faster than the dielectric relaxation time of the host solvent. Such behavior is particularly likely to be encountered in the inverted regime, where electron tunneling may be assisted by nuclear tunneling. An example is shown in the following table.

Electron transfer rates in betaine-30 correlated with solvent dielectric relaxation
E. Åkesson et al., J. Chem. Phys. **95**, 4188 (1991)

Solvent	k_{et} $(10^{12}\,\mathrm{s}^{-1})$	$1/\tau_L$ $(10^{12}\,\mathrm{s}^{-1})$
Propylene carbonate	0.91	0.29
Acetonitrile	2.0	2.0
Acetone	1.4	1.2
Benzonitrile	0.27	0.21
Triacetin 263 K	0.22	10^{-4}
Triacetin 293 K	0.29	0.01

In this system electron transfer is believed to be in the inverted regime. The transfer rate is seen to correlate with the solvent dielectric relaxation when this relaxation is fast enough, but decouples from it in a slow solvent, in this case triacetine.

16.7 A general expression for the dielectric reorganization energy

The reorganization energy was seen to be a characteristic attribute of molecular electron transfer processes. Equation (16.48) gives an expression for the dielectric solvent contribution to this energy for electron transfer between two atomic centers embedded in a dielectric environment. For completeness we give below the general result for the reorganization energy associated with a general change in the charge distribution from $\rho_0(\mathbf{r})$ to $\rho_1(\mathbf{r})$. The corresponding electric displacement vectors are the solutions of the Poisson equation $\nabla \cdot \mathcal{D}_i = 4\pi\rho_i$ and the reorganization energy is obtained in the form[11]

$$E_r = \frac{1}{8\pi}\left(\frac{1}{\varepsilon_e} - \frac{1}{\varepsilon_s}\right)\int d\mathbf{r}\,(\mathcal{D}_0(\mathbf{r}) - \mathcal{D}_1(\mathbf{r}))^2 \tag{16.71}$$

[11] This result is obtained by using the general electrostatic theory reviewed in Section 1.6.2 along lines similar to that used to obtained Eq. (16.48).

Problem 16.6. Show that for the charge transfer (16.36) between two spherical centers A and B, Eq. (16.71) yields the result (16.48).

Solution: For the reaction $\rho_0 = (q_A^{(0)}, q_B^{(0)}) \rightarrow \rho_1 = (q_A^{(1)}, q_B^{(1)})$ we have

$$\mathcal{D}_0 = \frac{q_A^{(0)}(\mathbf{r}-\mathbf{r}_A)}{|\mathbf{r}-\mathbf{r}_A|^3} + \frac{q_B^{(0)}(\mathbf{r}-\mathbf{r}_B)}{|\mathbf{r}-\mathbf{r}_B|^3} \tag{16.72a}$$

$$\mathcal{D}_1 = \frac{q_A^{(1)}(\mathbf{r}-\mathbf{r}_A)}{|\mathbf{r}-\mathbf{r}_A|^3} + \frac{q_B^{(1)}(\mathbf{r}-\mathbf{r}_B)}{|\mathbf{r}-\mathbf{r}_B|^3} \tag{16.72b}$$

$$\begin{aligned}\int d\mathbf{r}(\mathcal{D}_0(\mathbf{r}) - \mathcal{D}_1(\mathbf{r}))^2 &= 4\pi(q_A^{(0)} - q_A^{(1)})^2 \int_{a_A}^{\infty} dr \frac{1}{r^2} + 4\pi(q_B^{(0)} - q_B^{(1)})^2 \int_{a_B}^{\infty} dr \frac{1}{r^2} \\ &- 2(q_A^{(0)} - q_A^{(1)})(q_B^{(0)} - q_B^{(1)}) \int d\mathbf{r} \left(\nabla \cdot \frac{1}{|\mathbf{r}-\mathbf{r}_A|}\right)\left(\nabla \cdot \frac{1}{|\mathbf{r}-\mathbf{r}_B|}\right)\end{aligned} \tag{16.73}$$

Using $\Delta q = q_A^{(0)} - q_A^{(1)} = q_B^{(1)} - q_B^{(0)}$ we get

$$\begin{aligned}\int d\mathbf{r}(\mathcal{D}_0(\mathbf{r}) - \mathcal{D}_1(\mathbf{r}))^2 &= 4\pi(\Delta q)^2 \left(\frac{1}{a_A} + \frac{1}{a_B}\right) \\ &+ 2(\Delta q)^2 \int d\mathbf{r} \left(\nabla \cdot \frac{1}{|\mathbf{r}-\mathbf{r}_A|}\right)\left(\nabla \cdot \frac{1}{|\mathbf{r}-\mathbf{r}_B|}\right)\end{aligned} \tag{16.74}$$

The integral in the last term can be done by using $\nabla\phi \cdot \nabla\psi = \nabla \cdot (\phi\nabla\psi) - \phi\nabla^2\psi$, where ϕ and ψ are scalar functions. Using the divergence theorem the integral $\int d\mathbf{r}\nabla \cdot (\phi\nabla\psi)$ is shown to vanish, so

$$\begin{aligned}&\int d\mathbf{r} \left(\nabla \cdot \frac{1}{|\mathbf{r}-\mathbf{r}_A|}\right)\left(\nabla \cdot \frac{1}{|\mathbf{r}-\mathbf{r}_B|}\right) \\ &= -\int d\mathbf{r} \left(\frac{1}{|\mathbf{r}-\mathbf{r}_A|}\right)\left(\nabla^2 \frac{1}{|\mathbf{r}-\mathbf{r}_B|}\right) = \frac{4\pi}{|\mathbf{r}_A - \mathbf{r}_B|}\end{aligned} \tag{16.75}$$

The last equality was obtained by using $\nabla^2(|\mathbf{r}-\mathbf{r}_B|^{-1}) = 4\pi\delta(\mathbf{r}-\mathbf{r}_B)$. Equations (16.71), (16.74), and (16.75) indeed lead to (16.48).

16.8 The Marcus parabolas

The result (16.46) is remarkable in its apparent simplicity. Not only do we get an expression for the free energy needed to distort the nuclear configuration about the reactant charge distribution, but we can also use the resulting expression to find a complete expression for the nonadiabatic transfer rate. This simplicity is however not without important caveats. We have already identified one difficulty—the use of free energy rather than potential energy surfaces. Also, the reaction coordinate θ that was used to characterize solvent configurations has not been defined in terms of the microscopic solvent structure. Finally, the parabolic form of the free energy was obtained from a continuum model using linear dielectric response, and it is not clear that these assumptions hold on the molecular scale.

It is possible to define the reaction coordinate as an explicit function of the solvent structure in the following way. Let $V_\theta(\mathbf{R}^N; R_{\text{AB}})$ be the potential energy surface of the system, a function of the nuclear configuration $\mathbf{R}^N = (\mathbf{R}_1, \ldots, \mathbf{R}_N)$ (N is the number of all nuclear centers, including those associated with the donor and acceptor species), when A and B are held at a fixed distance R_{AB} and when the electronic charge distribution is $\rho_\theta(\mathbf{r})$, Eq. (16.21).[12] Also define a new variable, the difference in potential energies,

$$X(\mathbf{R}^N) = V_1(\mathbf{R}^N) - V_0(\mathbf{R}^N) \tag{16.76}$$

and the probability to observe a particular value X of $X(\mathbf{R}^N)$ in state θ,

$$P_\theta(X; R_{\text{AB}}) = \frac{\int d\mathbf{R}^N e^{-\beta V_\theta(\mathbf{R}^N)} \delta(R_{\text{AB}} - |\mathbf{R}_{\text{A}} - \mathbf{R}_{\text{B}}|)\delta(X - X(\mathbf{R}^N))}{\int d\mathbf{R}^N e^{-\beta V_\theta(\mathbf{R}^N)} \delta(R_{\text{AB}} - |\mathbf{R}_{\text{A}} - \mathbf{R}_{\text{B}}|)} \tag{16.77}$$

We now define X as the reaction coordinate for the electron transfer reaction for any given R_{AB}. In what follows we suppress the label R_{AB} but the presence of this parameter as an important factor in the transfer reaction should be kept in mind. Note that a given fixed R_{AB} is also implicit in the calculations of Section 16.3.6 and 16.4. By the discussion of Section 16.3.4, the functions

$$W_0(X) = E_0 - k_B T \ln P_0(X) \tag{16.78a}$$

$$W_1(X) = E_1 - k_B T \ln P_1(X) \tag{16.78b}$$

are the corresponding free energy surfaces analogous to (16.46) and (16.47), with X replacing θ as the reaction coordinate. The constants E_0 and E_1 are chosen so that

---

[12] For $\theta = 0$ or $\theta = 1$ this is the Born–Oppenheimer nuclear potential surface of the corresponding electronic state. We assume that we can define an analogous potential surface for any θ. In fact, the treatment here makes use only of the standard surfaces V_0 and V_1.

the maximum values of $\ln P_0(X)$ and $\ln P_1(X)$ are zeros. The point $X = 0$ along this reaction coordinate is the transition point for the electron transfer reaction because at this point the two electronic states are degenerate by definition.

As defined, the new reaction coordinate X is different from the reaction coordinate θ defined and used in Sections 16.3.5 and 16.3.6. However, if the potential surfaces $W_0(X)$ and $W_1(X)$ also turned out to be shifted parabolas as the corresponding functions of θ were in Eqs (16.46) and (16.47) it has to follow that X and θ are proportional to each other, so they can be used equivalently as reaction coordinates. This argument can be reversed: The validity of the Marcus theory, which relies heavily on the parabolic form of Marcus' free energy surfaces, implies that $W_0(X)$ and $W_1(X)$ should be quadratic functions of X.

The parabolic form of the Marcus surfaces was obtained from a linear response theory applied to a dielectric continuum model, and we are now in a position to verify this form by using the microscopic definition (16.76) of the reaction coordinate, that is, by verifying that $\ln(P(X))$, where $P(X)$ is defined by (16.77), is quadratic in X. Evaluating $P(X)$ is relatively simple in systems where the initial and final charge distributions ρ_0 and ρ_1 are well localized at the donor and acceptor sites so that $\rho_0(\mathbf{r}) = q_{\mathrm{A}}^{(0)}\delta(\mathbf{r} - \mathbf{r}_{\mathrm{A}}) + q_{\mathrm{B}}^{(0)}\delta(\mathbf{r} - \mathbf{r}_{\mathrm{B}})$ and $\rho_1(\mathbf{r}) = q_{\mathrm{A}}^{(1)}\delta(\mathbf{r} - \mathbf{r}_{\mathrm{A}}) + q_{\mathrm{B}}^{(1)}\delta(\mathbf{r} - \mathbf{r}_{\mathrm{B}})$. In this case, and for transfer of one electron, $X(\mathbf{R}^N) = \Delta\Phi(\mathbf{R}^N)$ is the difference between the electrostatic potentials at the A and B centers that is easily evaluated in numerical simulations.[13,14] An example of such result, the free energy surfaces for electron transfer within the Fe^{+2}/Fe^{+3} redox pair, is shown in Fig. 16.5. The resulting curves are fitted very well by identical shifted parabolas. Results of such numerical simulations indicate that the origin of the parabolic form of these free energy curves is more fundamental than what is implied by continuum linear dielectric theory.

16.9 Harmonic field representation of dielectric response

In Sections (16.3) and (16.4) we have seen that the description of motion along the reaction coordinate in the two electronic states involved contains elements of displaced harmonic potential surfaces. Here[15] we develop this picture in greater

[13] J.-K. Hwang and A. Warshel, Microscopic examination of free-energy relationships for electron transfer in polar solvents, J . Am. Chem. Soc. **109**, 715 (1987).

[14] R. A. Kuharski, J. S. Bader, D. Chandler, M. Sprik, M. L. Klein, and R. W. Impey, Molecular model for aqueous ferrous–ferric electron transfer, J. Chem. Phys. **89**, 3248 (1988).

[15] V. G. Levich and R. R. Dogonadze, Doklady. Akad. Nauk. USSR **124**, 123 (1954); Coll. Czech. Chem. Comm. **26**, 293 (1961).

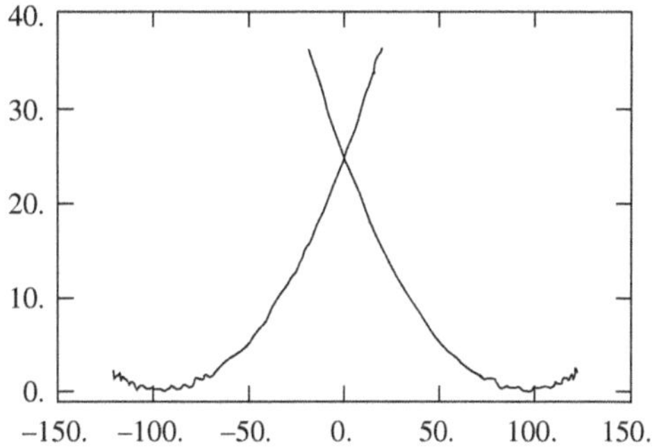

FIG. 16.5 The diabatic free energy curves ($W(X)$) plotted against X, Eq. (16.76) (both energies in Kcal mol^{-1}) obtained from a numerical simulation of the $Fe^{+2}/Fe^{+3} \rightleftharpoons Fe^{+3}/Fe^{+2}$ electron transfer reaction in water. The distance between the iron centers is $R_{AB} = 6.5$ Å, and the temperature is $T = 298$ K. The simulation (Kuharski et al.[14]) was done with the SPC water force-field and an umbrella technique was used to sample nonequilibrium configurations.

detail with the purpose of identifying the origin and the physical significance of this behavior.

Our starting point is the expression for the reorganization energy E_r, Eq. (16.71). We use it in a more general form that expresses the energy that will be released when we start from a fluctuation in the nuclear polarization about a given charge distribution and let the system relax to equilibrium:

$$E_r^{(l)} = \frac{1}{8\pi}\left(\frac{1}{\varepsilon_e} - \frac{1}{\varepsilon_s}\right)\int d\mathbf{r}\,(\mathcal{D}(\mathbf{r}) - \mathcal{D}_l(\mathbf{r}))^2 \tag{16.79}$$

Here the index l is used to denote a particular charge distribution (i.e. a particular electronic state of the system). The displacement field $\mathcal{D}_l(\mathbf{r})$ represents a charge distribution $\rho_l(\mathbf{r})$ according to the Poisson equation $\nabla \cdot \mathcal{D}_l = 4\pi\rho_l$. In (16.79) $\mathcal{D}(r)$ and the associated $\rho(\mathbf{r})$ represent a fluctuation in the nuclear polarization, defined by the *equilibrium relationship* between the nuclear polarization and the displacement vector (cf. Eqs (16.14) and (16.15))

$$\mathbf{P}_n(\mathbf{r}) = \frac{\varepsilon_e}{4\pi}\left(\frac{1}{\varepsilon_e} - \frac{1}{\varepsilon_s}\right)\mathcal{D}(\mathbf{r}) \tag{16.80}$$

Thus, in terms of the fluctuation $\mathbf{P}_n(\mathbf{r}) - \mathbf{P}_n^{(l)}(\mathbf{r})$ away from the equilibrium nuclear polarization $\mathbf{P}_n^{(l)}(\mathbf{r})$, Eq. (16.79) reads

$$E_r^{(l)} = \frac{2\pi\varepsilon_e^{-2}}{(1/\varepsilon_e) - (1/\varepsilon_s)}\int d^3r(\mathbf{P}_n(\mathbf{r}) - \mathbf{P}_n^{(l)}(\mathbf{r}))^2 \tag{16.81}$$

As before, we limit ourselves only to a subspace of such fluctuations that can be characterized as equilibrium values under some other charge distribution $\rho(\mathbf{r})$ (i.e. some other $\mathcal{D}(\mathbf{r})$).

Written in this way, $E_r^{(l)}$ can be interpreted as the potential energy associated with the corresponding fluctuation. We rewrite Eq. (16.81) in the form

$$W_l[\mathbf{X}_l(\mathbf{r})] = \frac{2\pi C}{\varepsilon_e^2} \int d^3r \mathbf{X}_l(\mathbf{r})^2; \qquad \mathbf{X}_l = \mathbf{P}_n - \mathbf{P}_n^{(l)} \tag{16.82}$$

$$\frac{1}{C} = \frac{1}{\varepsilon_e} - \frac{1}{\varepsilon_s} \tag{16.83}$$

$\mathbf{X}_l(\mathbf{r})$ is the coordinate that measures polarization fluctuation at position $\mathbf{r}$ when the system is in electronic state l, and $W_l[\mathbf{X}_l(\mathbf{r})]$ which is a functional of $\mathbf{X}_l(\mathbf{r})$ is the corresponding energy. The integral over $\mathbf{r}$ is a sum over many such coordinates and the potential energy W_l is seen to be a quadratic function of these coordinates, that is, a harmonic field. Note that if we assume that all coordinates $\mathbf{P}_n(\mathbf{r})$ change together, say according to $\mathbf{P}_n = \mathbf{P}_n^{(l)} + \theta(\mathbf{P}_n^{(1)} - \mathbf{P}_n^{(l)})$, then Eq. (16.82) yields $W(\theta) = \lambda\theta^2$ with $\lambda = 2\pi C\varepsilon_e^{-2} \int d^3r(P_n^{(1)} - P_n^{(l)})^2$, that is, a potential surface defined in terms of a single variable θ, as in Sections 16.3.5 and 16.3.6. Equation (16.82) defines this potential surface in terms of the infinitely many local variables $\mathbf{X}_l(\mathbf{r})$.

Consider now one of these variable and its contribution to the potential energy, $W_l(\mathbf{r}) = 2\pi\varepsilon_e^{-2} C\mathbf{X}_l(\mathbf{r})^2$. This is the potential energy of a three-dimensional isotropic harmonic oscillator. The total potential energy, Eq. (16.82) is essentially a sum over such contributions. This additive form indicates that these oscillators are independent of each other. Furthermore, all oscillators are characterized by the same force constant. We now also assume that all masses associated with these oscillators are the same, namely we postulate the existence of a single frequency ω_s, sometimes referred to as the "Einstein frequency" of the solvent polarization fluctuations. m_s and ω_s are related as usual by the force constant

$$m_s = \frac{4\pi C}{\varepsilon_e^2 \omega_s^2} \tag{16.84}$$

so that

$$W_l(\mathbf{r}) = \frac{1}{2} m_s \omega_s^2 \mathbf{X}_l(\mathbf{r})^2 \tag{16.85}$$

We can now define a Hamiltonian $h_l(\mathbf{r})$ for the local polarization coordinate $\mathbf{X}_l$

$$\begin{aligned} h_l(\mathbf{r}) &\equiv \frac{1}{2} m_s \dot{\mathbf{X}}_l(\mathbf{r})^2 + W_l(\mathbf{r}) \\ &= \frac{2\pi C}{\varepsilon_e^2} (\mathbf{X}_l(\mathbf{r})^2 + \omega_s^{-2} \dot{\mathbf{X}}_l(\mathbf{r})^2) \end{aligned} \tag{16.86}$$

which implies that the entire dielectric environment can be associated with the total Hamiltonian

$$H_l = \frac{2\pi C}{\varepsilon_e^2} \int d^3r(\mathbf{X}_l(\mathbf{r})^2 + \omega_s^{-2}\dot{\mathbf{X}}_l(\mathbf{r})^2)$$
$$\mathbf{X}_l(\mathbf{r}) = \mathbf{P}_n(\mathbf{r}) - \mathbf{P}_n^{(l)}(\mathbf{r}) \tag{16.87}$$

This is a Hamiltonian for independent harmonic oscillators specified by the continuous index $\mathbf{r}$. The equilibrium nuclear polarization $\mathbf{P}_n^{(l)}(\mathbf{r})$ (which is determined by the given charge distribution through the Poisson equation together with Eq. (16.80)) defines the equilibrium position of the local oscillator at $\mathbf{r}$. Hence, changing the charge distribution corresponds to changing these equilibrium positions. Taking into account that a change in the charge distribution in a molecular system is also usually associated with a change in the energy origin, we can write the nuclear polarization Hamiltonian that corresponds to a charge distribution associated with a given electronic state l in the form[4]

$$H^{(l)} = E_{el}^{(l)} + \frac{2\pi C}{\varepsilon_e^2} \int d^3r[(\mathbf{P}_n(\mathbf{r}) - \mathbf{P}_n^{(l)}(\mathbf{r}))^2 + \omega_s^{-2}\dot{\mathbf{P}}_n(\mathbf{r})^2] \tag{16.88}$$

Problem 16.7. Show that the reorganization energy associated with the transition between two electronic states l and l' described in this way is

$$E_r^{(l,l')} = \frac{2\pi C}{\varepsilon_e^2} \int d^3r(\mathbf{P}_n^{(l)}(\mathbf{r}) - \mathbf{P}_n^{(l')}(\mathbf{r}))^2 \tag{16.89}$$

To summarize, we have found that the dielectric response of a polar medium can be described in terms of the Hamiltonian (16.88) that corresponds to a system of independent harmonic oscillators indexed by the spatial poison $\mathbf{r}$.[16] These oscillators are characterized by given equilibrium "positions" $\mathbf{P}_n^{(l)}(\mathbf{r})$ that depend on the electronic state l. Therefore a change in electronic state corresponds to shifts in these equilibrium positions.

The Hamiltonian (16.88) describes a system of classical harmonic oscillators. It is reasonable to assume that it represents a classical approximation to a more fundamental quantum theory. Such a theory may be obtained in analogy to the quantization of the electromagnetic field carried in Section 3.2.2, by "quantizing"

[16] Advanced readers will notice that in case of nonlocal response where, for example, $P(\mathbf{r}) = \int d^3r'\chi(\mathbf{r}-\mathbf{r}')E(\mathbf{r})$, a similar harmonic field Hamiltonian is obtained in $\mathbf{k}$-space.

this harmonic field, that is, replacing the Hamiltonian (16.88) by its quantum analog. The result is a system of independent quantum harmonic oscillators whose interaction with the electronic system is expressed by the fact that a change in the electronic charge distribution affects a parallel[17] shift of the harmonic potential surfaces. This is exactly the model used in Chapter 12, see in particular Section 12.3.

Electron transfer processes, more generally transitions that involve charge reorganization in dielectric solvents, are thus shown to fall within the general category of shifted harmonic oscillator models for the thermal environment that were discussed at length in Chapter 12. This is a result of linear dielectric response theory, which moreover implies that the dielectric response frequency ω_S does not depend on the electronic charge distribution, namely on the electronic state. This rationalizes the result (16.59) of the dielectric theory of electron transfer, which is identical to the rate (12.69) obtained from what we now find to be an equivalent spin–boson model.

We end our discussion of this quasi-microscopic electron transfer theory with two important comments: First, once electron transfer was identified as a charge reorganization process accompanied by shifting equilibrium positions of an underlying harmonic oscillator field, it can be generalized in an obvious way to take into account also the role played by the intramolecular nuclear motions of the molecular species involved. The dielectric solvent dynamics ("outer sphere" dynamics) and the intramolecular nuclear motions ("inner sphere" dynamics) correspond in such a unified theory to two groups of harmonic modes whose coupling to the electronic transition is characterized by shifting their equilibrium positions. Rate expressions such as (12.34) apply, with both groups of modes included. In evaluating these rates we may use representations such as Eq. (16.62) to apply different approximation schemes to different modes, based, for example, on the recognition that intramolecular motions are characterized by much higher frequencies than intermolecular dynamics.

Second, we note that the *dynamical* aspect of the dielectric response is still incomplete in the above treatment, since (1) no information was provided about the dielectric response frequency ω_S[18] and (2) a harmonic oscillator model for the local dielectric response is oversimplified and a damped oscillator may provide a more complete description. These dynamical aspects are not important in equilibrium considerations such as our transition-state-theory level treatment, but become so in other limits such as solvent-control electron-transfer reactions discussed in Section 16.6.

[17] As in Chapter 12 the word "parallel" implies that the harmonic frequencies remain unchanged.

[18] Note that this frequency appears in Eqs (16.8) and (16.50), where it was expressed in terms of some unknown "dielectric response mass." It is possible to address the magnitude of ω_S numerically, see, for example, E. A. Carter and J. T. Hynes, J. Phys. Chem. **93**, 2184 (1989).

Problem 16.8. Let us use the superscript (0) for the state of the dielectric system where the charge distribution is zero. This implies that $\mathbf{P}_{\mathrm{n}}^{(0)}(\mathbf{r}) = 0$ so that the reorganization energy of any other state relative to this one is $E_{\mathrm{r}}^{(l,0)} = (2\pi C/\varepsilon_{\mathrm{e}}^2)\int d^3r(\mathbf{P}_{\mathrm{n}}^{(l)}(\mathbf{r}))^2$. This reorganization energy can be then identified as the contribution of the nuclear dielectric response to the solvation energy associated with the corresponding charge distribution

$$W_{\mathrm{n}}^{(l)} = -\frac{2\pi C}{\varepsilon_{\mathrm{e}}^2}\int d^3r(\mathbf{P}_{\mathrm{n}}^{(l)}(\mathbf{r}))^2 \tag{16.90}$$

Show that this energy can be written in the form

$$W_{\mathrm{n}}^{(l)} = -\frac{1}{8\pi}\left(\frac{1}{\varepsilon_{\mathrm{e}}} - \frac{1}{\varepsilon_{\mathrm{s}}}\right)\int d^3r\mathcal{D}_l^2(r) \tag{16.91}$$

and that the full solvation energy of the charge distribution associated with $\mathcal{D}_l$ ($\rho_l = (4\pi)^{-1}\nabla\cdot\mathcal{D}_l$) is

$$W_{\mathrm{sol}}^{(l)} = -\frac{1}{8\pi}\left(1 - \frac{1}{\varepsilon_{\mathrm{s}}}\right)\int d^3r\mathcal{D}_l^2(r) \tag{16.92}$$

what is the significance of the difference

$$W_{\mathrm{sol}}^{(l)} - W_{\mathrm{n}}^{(l)} = -\frac{1}{8\pi}\left(1 - \frac{1}{\varepsilon_{\mathrm{e}}}\right)\int d^3r\mathcal{D}_l^2(r) \tag{16.93}$$

Solution: Using Eq. (16.80)

$$\mathbf{P}_{\mathrm{n}}^{(l)}(\mathbf{r}) = \frac{\varepsilon_{\mathrm{e}}}{4\pi C}\mathcal{D}_l(\mathbf{r}) \tag{16.94}$$

in (16.90) leads directly to (16.91). The total energy in an electrostatic field inside a dielectric medium is

$$W = \frac{1}{8\pi}\int d^3r\mathcal{E}\cdot\mathcal{D} = \frac{1}{8\pi\varepsilon_{\mathrm{s}}}\int d^3r\mathcal{D}^2 \tag{16.95}$$

So the solvation energy, the energy difference between assembling a charge distribution in vacuum and inside a dielectric is given by (16.92). Since (16.91) is the contribution of the nuclear dielectric response to this solvation energy, the contribution (16.93) is the contribution from the solvent electronic polarization.

16.10 The nonadiabatic coupling

In the expression for the nonadiabatic electron transfer rate (cf. Eqs (16.53) and (16.58))

$$k_{b\leftarrow a} = \frac{2\pi}{\hbar}|V_{a,b}|^2 \frac{1}{\sqrt{4\pi E_r k_B T}} \exp\left(-\frac{(\Delta E - E_r)^2}{4E_r k_B T}\right) \tag{16.96}$$

ΔE can be determined spectroscopically and the reorganization energy E_r is evaluated from, for example, (16.48) or more generally (16.71). Thus the only unknown parameter is the nonadiabatic coupling $V_{a,b}$. In fact this parameter can sometimes also be evaluated from spectroscopical data. This is the case where the initial and final states of the electron transfer process are two diabatic states (see Sections 2.4 and 2.5) that can be expressed as linear combinations of two adiabatic states that constitute the initial and final states of optical absorption. The corresponding absorption lineshape, a charge transfer optical transition, can be analyzed to yield the needed electronic coupling. This is shown in Appendix 16A where we derive an expression relating the coupling between two nonadiabatic electronic states a and b to the optical transition dipole between the corresponding adiabatic states 1 and 2

$$|\mu_{1,2}| = \frac{|V_{a,b}|}{(E_2 - E_1)}|\mu_b - \mu_a| = \frac{e|V_{a,b}|r_{ab}}{\hbar\omega_{\max}} \tag{16.97}$$

where μ_a and μ_b are the dipole moments of the localized diabatic states

$$\mu_a = -e\langle\psi_a|\sum_i r_i|\psi_a\rangle; \qquad \mu_b = -e\langle\psi_b|\sum_i r_i|\psi_b\rangle \tag{16.98}$$

and where in the second equality of (16.97) we represented $E_2 - E_1$ by the frequency $\omega_{\max}$ of the maximum absorption in the optical transition from 1 to 2, and have defined

$$r_{ab} \equiv \left|\frac{\mu_b - \mu_a}{e}\right| \tag{16.99}$$

If the only change in the molecular charge distribution between the states a and b is the position of the transferred electron (i.e. if we assume that the other electrons are not affected) then r_{ab} is the transfer distance, that is, the separation between the donor and acceptor centers. Eq. (16.97) is known as the Mulliken–Hush formula.

Equation (16.97) is useful for estimating the nonadiabatic coupling associated with electron transfer reactions mirrored by an equivalent optical transition.[19] An

[19] The derivation of this formula, outlined in Appendix 16A, is limited by the assumption that the electronic coupling can be described in a two-state framework. For a discussion of this point and extension to more general situations see the paper by Bixon and Jortner cited at the end of this chapter.

important application is also found in bridge-mediated electron transfer transitions (see next section). In such processes the transition between a donor site D and an acceptor site A is promoted by their mutual coupling with a connecting ligand bridge B (see Fig. 16.7). In this case and under conditions discussed below, the effective DA coupling is obtained from Eq. (16.113). For a single bridge site it takes the form

$$V_{\mathrm{AD}}^{\mathrm{eff}} = \frac{V_{\mathrm{AB}}V_{\mathrm{BD}}}{\Delta E} \tag{16.100}$$

In this case the donor acceptor coupling $V_{\mathrm{AD}}^{\mathrm{eff}}$ may be obtained by using (16.97)–(16.99) to estimate the couplings V_{AB} and V_{BD} from spectroscopic parameters obtained from charge transfer donor-to-ligand and acceptor-to-ligand transitions.

To end this section it should be mentioned that another route to the nonadiabatic electronic coupling can be found in theory. Ab initio quantum chemical calculations of this coupling in medium-size systems are now feasible,[20] and semi-empirical calculations are also becoming increasingly reliable.[21]

16.11 The distance dependence of electron transfer rates

How do electron transfer rates depend on the donor acceptor distance R_{DA}? It will be useful to consider this question from the point of view of the transferred electron. In reality we are obviously not concerned with a single electron but with electronic states comprising many electrons, however to obtain a simple picture it helps to think of the transfer process as a single electron event. Particularly simple descriptions are obtained for the case where the donor and acceptor centers are far from each other. Figure 16.6 depicts these centers as two potential wells for the transferring electron. The lower and upper diagrams correspond to different nuclear configurations: In the lower diagram the nuclear configuration is in equilibrium with the electronic charge distribution, while the upper diagram corresponds to the transition state where the energies of electronic states localized on the donor and the acceptor are equal, that is, to the crossing point $(X_{\mathrm{tr}}, E_{\mathrm{tr}})$ in the Marcus picture, Fig. 16.2. It is important to realize that these diagrams, showing the electron potential energy

[20] M. D. Newton, Quantum chemical probes of electron-transfer kinetics—the nature of donor–acceptor interactions, Chem. Rev. **91**, 767 (1991).

[21] S. S. Skourtis and D. Beratan, Theories for structure-function relationships for bridge-mediated electron transfer reactions, in *Electron Transfer—From Isolated Molecules to Biomolecules*, edited by M. Bixon and J. Jortner, *Advances in Chemical Physics*, Vol. 106 (Wiley, New York, 1999), Part I, p. 377.

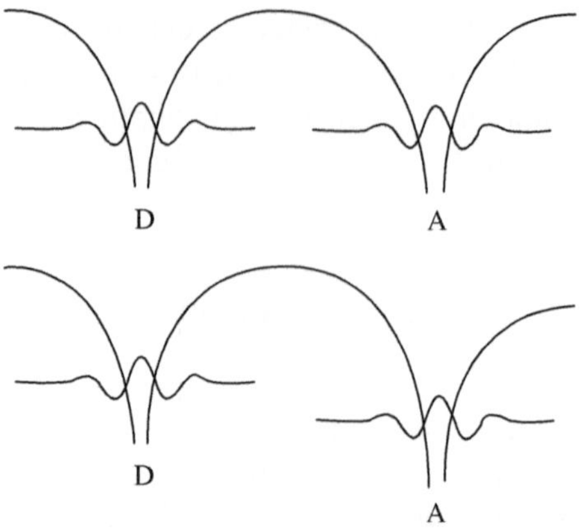

FIG. 16.6 A schematic representation of the potential surface for the electron between two centers (D—donor, A—acceptor). Shown also are the relevant diabatic electronic wavefunctions, localized on each center. The lower diagram corresponds to a stable nuclear configuration and the upper one—to a nuclear fluctuation that brings the system into the transition state where the diabatic electronic energies are equal. The electronic transition probability depends on the overlap between the electronic wavefunctions ψ_D and ψ_A in this transition state.

as a function of its position, have nothing to do with the Marcus parabolas that are nuclear potential surfaces. The latter show the nuclear potential energy displayed as function of nuclear configuration expressed in terms of some reaction coordinate.

In the diagrams of Fig. 16.6, the diabatic states shown are the electronic states obtained as eigenstates of one electronic well, in the absence of the other. The non-adiabatic coupling is the matrix element of the full electronic Hamiltonian (kinetic energy plus potential energy of the two wells) between such states localized in different wells. As discussed in Section 16.10 we can estimate the magnitude of this coupling from spectroscopic data, however let us now consider this problem theoretically. The desired coupling determines the tunneling rate between the two centers, which was seen in Section 2.10 to be proportional to the factor $\exp(-2\hbar^{-1}\int dx\sqrt{2m_e(U(x)-E)})$. Here m_e is the electron mass, U is the potential, and the integral is calculated over the barrier separating the wells between the two turning points at energy E. This calculation should be done in the transition state configuration (upper diagram of Fig. (16.6)). If the two centers are far enough from each other it is reasonable to approximate the barrier by a rectangular shape of width R_{AD} and height above the ground state of the donor (say) given by the donor ionization potential I_D. For the D $\rightarrow$ A transfer one might think that $U-E$ is equivalent $I_D - E_{tr}(D)$ where both the donor ionization potential I_D and the transition energy are measured from the ground state donor energy. However, recalling the fundamental assumption that the electronic transition takes place at a fixed nuclear configuration, the solvent configuration about the donor immediately before and immediately after the electron tunneling is the same; in other words, the energy E_{tr} cannot be "used" to overcome the ionization barrier.[22] Therefore, $\int dx\sqrt{2m_e(U-E)}$ should be evaluated for $U-E=I_D$, which yields $R_{DA}\sqrt{2m_eI_D}$ and the relevant tunneling factor then becomes $\exp(-\beta' R_{AD})$ where[23]

$$\beta' \sim 2\hbar^{-1}\sqrt{2m_eI_D}. \tag{16.101}$$

Note that I_D is the donor ionization potential *in the polar solvent* that may be different from the vacuum value.

Note that this tunneling factor is not the only term in the nonadiabatic electron transfer rate expression (16.96) that depends on the donor–acceptor distance. The reorganization energy E_r also depends on this distance as seen, for example, in Eq. (16.48). Still, the exponential form of the tunneling factor dominates the distance dependence at large separations. We finally get rate $\sim \exp(-\beta' R_{DA})$, or

$$|V_{a,b}| = V_0 e^{-(1/2)\beta' R_{DA}} \tag{16.102}$$

16.12 Bridge-mediated long-range electron transfer

The appearance of the ionization potential in (16.101) reflects the understanding that for transfer to take place over large distances the electron has to traverse the space between the donor/acceptor centers, that is, to move in free space outside these centers. If we use a vacuum-based estimate $I_D \approx 5$ eV we find $\beta' \approx (2.3\ \text{Å})^{-1}$. This implies that the coupling effectively vanishes for distances larger than just a few angstroms. Experiments show, however, that in suitable environments electron transfer can take place over considerably longer distances.

It is not difficult to point out a plausible reason for this observation when we realize that in order to move an electron out of any of the wells in Fig. 16.6 it is not necessary to move it energetically up to the vacuum levels. In fact any unoccupied molecular levels of the solvent can supply the needed transient residence at a lower cost of energy. An example is shown in Fig. 16.7(a). D and A represent the electronic potential wells on the donor and acceptor centers, while B_i, $i = 1, 2, \ldots$ represent

[22] The term *vertical transition* is often used to describe an electronic transition that takes place at fixed nuclear configuration.

[23] In much of the electron transfer literature the inverse range parameter that characterizes bridge mediated electron transfer reactions is denoted β. In the present text we use the notation β' to distinguish it from the usual notation for the inverse temperature $\beta = (k_BT)^{-1}$.

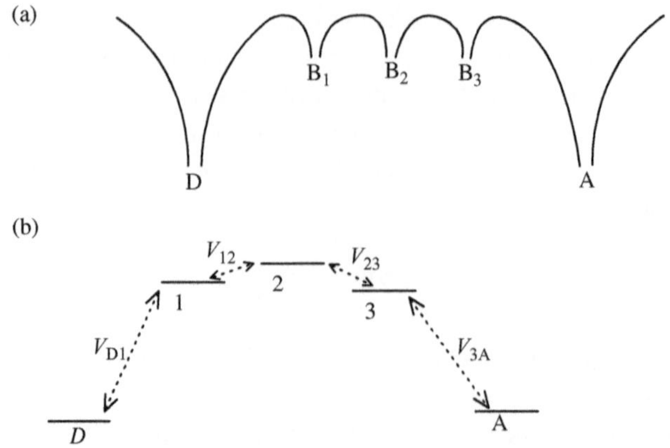

FIG. 16.7 Bridge-assisted electron transfer. (a) The existence of intermediate electron binding sites between the donor and acceptor changes the effective tunneling barrier and can promote the transfer rate. (b) In a simple model of bridge-assisted transfer each of the donor, acceptor, and intermediate sites are represented by a single electronic level.

similar wells on atomic or molecular centers along the path connecting these centers. A schematic representation of a corresponding electronic Hamiltonian is shown in Fig. 16.7(b), where, as before, we have represented A and D by a single electronic state on each center (e.g. the ground electronic state in the corresponding well), and similarly represented each bridge site by its lowest electronic state (ground state of the isolated bridge unit). The coupling matrix elements between these states are the non-diagonal elements of the total electronic Hamiltonian in the chosen basis set of one level per well.

Some clarification concerning the nature of these states is needed here. As already emphasized, the initial and final states involved in the electron transfer process are many electron states characterized by an excess electron localized initially on the donor and finally on the acceptor. The intermediate states $1, 2, \ldots$ are similarly many-electron states that can be described in different representations. In the local site representation we designate consecutive molecular segments along the bridge as "bridge sites" and the bridge states $1, 2, \ldots$ are the lowest energy site states—each a ground state of a site with one excess electron.[24] The relative energies displayed in Fig. 16.7(a) and (b) express the actual state of affairs in many real situations, where the donor and acceptor electronic states are lower in energy

[24] The number of sites on the bridge is a matter of choice, although physically motivated choices often lead to simpler analysis. It is also possible to take the whole bridge as a single site, which amounts to using the many-electron basis of the bridge Hamiltonian.

than local bridge states with an excess electron.[25] Still these states often provide a lower energy path, hence a lower tunneling barrier, than vacuum. For historical reasons this mechanism is often referred to as "*superexchange*." Particularly simple pictures, convenient for qualitative discussions, are obtained by disregarding electron–electron interaction except for Pauli spin exclusion. In this case the transfer is described in a one-electron language by which an electron goes from the highest occupied molecular orbital (HOMO) on the donor site to the lowest unoccupied molecular orbital (LUMO) on the acceptor side through a sequence of LUMO orbitals of the bridge sites.[25]

Let us consider a general N-level bridge and assume that a model with nearest-level interactions is sufficient. The model Hamiltonian is

$$\begin{aligned}\hat{H} &= E_{\mathrm{D}}|D\rangle\langle D| + \sum_{j=1}^{N} E_j|j\rangle\langle j| + E_{\mathrm{A}}|A\rangle\langle A| \\ &+ V_{\mathrm{D}1}|D\rangle\langle 1| + V_{1\mathrm{D}}|1\rangle\langle D| + V_{\mathrm{A}N}|A\rangle\langle N| + V_{N\mathrm{A}}|N\rangle\langle A| \\ &+ \sum_{j=1}^{N-1} (V_{j,j+1}|j\rangle\langle j+1| + V_{j+1,j}|j+1\rangle\langle j|)\end{aligned} \tag{16.103}$$

We will focus on the case where the group of bridge levels $\{j\}$ is energetically distinct from the donor and acceptor levels, so that, denoting by E_{B} the mean energy of a bridge level we have

$$|E_j - E_{\mathrm{B}}|, |V_{j,j+1}| \ll |E_{\mathrm{B}} - E_{\mathrm{D/A}}| \tag{16.104}$$

(note that $E_{\mathrm{D}} = E_{\mathrm{A}} \equiv E_{\mathrm{D/A}}$ at the transition state). In this case the electron transfer still takes place between the D and A states because the bridge states cannot be populated, being so different in energy. We therefore expect that the process may still be described as a two-states problem, with an effective coupling between the A and D states.

To find this effective coupling we look for an eigenstates of the Hamiltonian (16.103) in the form $\psi = c_{\mathrm{D}}\psi_{\mathrm{D}} + c_{\mathrm{A}}\psi_{\mathrm{A}} + \sum_{j=1}^{N} c_j\psi_j$. The Schrödinger equation

[25] The actual situation may be more involved than what is shown in Fig. 16.7. First the coupling may be longer range than shown, for example, one may have to consider 1–3 coupling, etc. Neglecting such coupling is usually not a bad approximation since $\langle\phi_i|\hat{H}|\phi_j\rangle$ quickly vanishes with increasing distance between the centers on which the orbitals i and j are located. More important is to realize that a route for electron transfer is provided not only by bridge states with an excess electron (or bridge LUMOs), but also by bridge states that miss an electron (or bridge HOMOs). In the language of solid-state physics (Chapter 4), one can have a transition dominated by electrons, holes, or combination of both.

$\hat{H}\psi = E\psi$ takes the form

$$\begin{pmatrix} E_{\mathrm{D}}-E & V_{D1} & 0 & 0 & \cdots & 0 \\ V_{1\mathrm{D}} & E_1-E & V_{12} & 0 & \cdots & 0 \\ 0 & V_{21} & E_2-E & V_{23} & & \vdots \\ 0 & 0 & V_{32} & \ddots & \ddots & 0 \\ \vdots & \vdots & & \ddots & E_N-E & V_{N\mathrm{A}} \\ 0 & 0 & \cdots & 0 & V_{\mathrm{A}N} & E_{\mathrm{A}}-E \end{pmatrix} \begin{pmatrix} c_{\mathrm{D}} \\ c_1 \\ c_2 \\ \vdots \\ c_N \\ c_{\mathrm{A}} \end{pmatrix} = 0 \tag{16.105}$$

or, alternatively

$$\begin{pmatrix} E_1-E & V_{12} & 0 & \cdots & 0 \\ V_{21} & E_2-E & V_{23} & & \vdots \\ 0 & V_{32} & \ddots & & 0 \\ \vdots & & & E_{N-1}-E & V_{N-1,N} \\ 0 & \cdots & 0 & V_{N,N-1} & E_N-E \end{pmatrix} \begin{pmatrix} c_1 \\ c_2 \\ \vdots \\ \vdots \\ c_N \end{pmatrix} = - \begin{pmatrix} V_{1\mathrm{D}}c_{\mathrm{D}} \\ 0 \\ \vdots \\ 0 \\ V_{N\mathrm{A}}c_{\mathrm{A}} \end{pmatrix} \tag{16.106}$$

$$\begin{aligned} (E_{\mathrm{D}}-E)c_{\mathrm{D}} + V_{D1}c_1 &= 0 \\ (E_{\mathrm{A}}-E)c_{\mathrm{A}} + V_{\mathrm{A}N}c_N &= 0 \end{aligned} \tag{16.107}$$

Equation (16.106) is a nonhomogeneous equation for the coefficients of the bridge states. We can write it in the compact form

$$(\hat{H}_{\mathrm{B}} - E\mathbf{I}_{\mathrm{B}})\mathbf{c}_{\mathrm{B}} = \mathbf{u} \tag{16.108}$$

where $\hat{H}_{\mathrm{B}}$ is the bridge Hamiltonian (in the absence of coupling to the D/A system), $\mathbf{I}_{\mathrm{B}}$ is a unit operator in the bridge subspace, $\mathbf{c}_{\mathrm{B}}$ is the vector of bridge coefficients, and $\mathbf{u}$ is the vector on the right side of (16.106). If E is outside the range of eigenvalues of $\hat{H}_{\mathrm{B}}$ the solution of (16.108) is $\mathbf{c}_{\mathrm{B}} = -\hat{G}^{(B)}\mathbf{u}$ where

$$\hat{G}^{(B)} = (E\mathbf{I}_{\mathrm{B}} - \hat{H}_{\mathrm{B}})^{-1} \tag{16.109}$$

In fact, in (16.107) we need only c_1 and c_N, which are given by

$$\begin{aligned} c_1 &= G_{11}^{(B)}V_{1\mathrm{D}}c_{\mathrm{D}} + G_{1N}^{(B)}V_{N\mathrm{A}}c_{\mathrm{A}} \\ c_N &= G_{N1}^{(B)}V_{1\mathrm{D}}c_{\mathrm{D}} + G_{NN}^{(B)}V_{N\mathrm{A}}c_{\mathrm{A}} \end{aligned} \tag{16.110}$$

Using these in (16.107) leads to

$$\begin{aligned} (\tilde{E}_{\mathrm{D}}-E)c_{\mathrm{D}} + \tilde{V}_{\mathrm{DA}}c_{\mathrm{A}} &= 0 \\ (\tilde{E}_{\mathrm{A}}-E)c_{\mathrm{A}} + \tilde{V}_{\mathrm{AD}}c_{\mathrm{D}} &= 0 \end{aligned} \tag{16.111}$$

where

$$\tilde{E}_{\mathrm{D}} = E_{\mathrm{D}} + V_{\mathrm{D}1} G_{11}^{(B)} V_{1\mathrm{D}}; \qquad \tilde{E}_{\mathrm{A}} = E_{\mathrm{A}} + V_{\mathrm{A}N} G_{NN}^{(B)} V_{N\mathrm{A}} \tag{16.112}$$

are donor and acceptor energies, slightly shifted because of their couplings to the bridge, and where

$$\tilde{V}_{\mathrm{DA}} = V_{\mathrm{D}1} G_{1N}^{(B)} V_{N\mathrm{A}}; \qquad \tilde{V}_{\mathrm{AD}} = V_{\mathrm{A}N} G_{N1}^{(B)} V_{1\mathrm{D}} = \tilde{V}_{\mathrm{DA}}^{*} \tag{16.113}$$

is the desired effective coupling that should be used in the rate expression (cf. (16.53) and (16.58))

$$k_{\mathrm{A}\leftarrow\mathrm{D}} = \frac{2\pi}{\hbar} |\tilde{V}_{\mathrm{DA}}|^2 F(\Delta E); \qquad F(\Delta E) = \frac{1}{\sqrt{4\pi E_r k_B T}} \exp\left(-\frac{(\Delta E - E_{\mathrm{r}})^2}{4E_r k_B T}\right) \tag{16.114}$$

Unlike a "regular" coupling, the effective coupling $\tilde{V}_{\mathrm{DA}}$ depends on the energy E through the Green's function (16.109), however since we are interested in the D/A subspace and because the two eigenvalues in this subspace are expected, under the inequalities (16.104), to remain close to $E_{\mathrm{D/A}}$ we can substitute E by $E_{\mathrm{D/A}}$ in (16.113). Furthermore, these inequalities imply that we can retain only the lowest-order term in the Dyson expansion, Eq. (9.11),[26] applied to $\hat{G}_{1N}^{(B)}$

$$G_{1N}^{(B)} = \frac{1}{(E_{\mathrm{D/A}} - E_1)} V_{12} \frac{1}{(E_{\mathrm{D/A}} - E_2)} V_{23} \cdots \frac{1}{(E_{\mathrm{D/A}} - E_{N-1})} V_{N-1,N}$$
$$\times \frac{1}{(E_{\mathrm{D/A}} - E_N)} \tag{16.115}$$

where, again, $E_{\mathrm{D/A}}$ stands for the equal donor and acceptor energies at the transition configuration. In many applications the bridge is made of connected identical units. Within our model this will be represented by taking all coupling elements the same, denoted V_{B}, and all energy denominators equal to $E_{\mathrm{D/A}} - E_{\mathrm{B}}$. This leads to $G_{1N}^{(B)} = (V_{\mathrm{B}})^{N-1}(E_{\mathrm{D/A}} - E_{\mathrm{B}})^{-N}$ and the effective coupling (16.113) then takes the form

$$\tilde{V}_{\mathrm{DA}} = \frac{V_{\mathrm{D}1} V_{N\mathrm{A}}}{V_{\mathrm{B}}} \left(\frac{V_{\mathrm{B}}}{E_{\mathrm{D/A}} - E_{\mathrm{B}}}\right)^N \tag{16.116}$$

[26] The Dyson expansion starts form the identity (9.11), for example, $\hat{G}(z) = \hat{G}_0(z) + \hat{G}_0(z)\hat{V}\hat{G}(z)$ and uses repeated substitutions to get the perturbation series $\hat{G}(z) = \hat{G}_0(z) + \hat{G}_0(z)\hat{V}\hat{G}_0(z) + \hat{G}_0(z)\hat{V}\hat{G}_0(z)\hat{V}\hat{G}_0(z) + \ldots$. Equation (16.115) is the lowest-order term in this series that can contribute to the Green function matrix element $G_{1N}^{(B)}$.

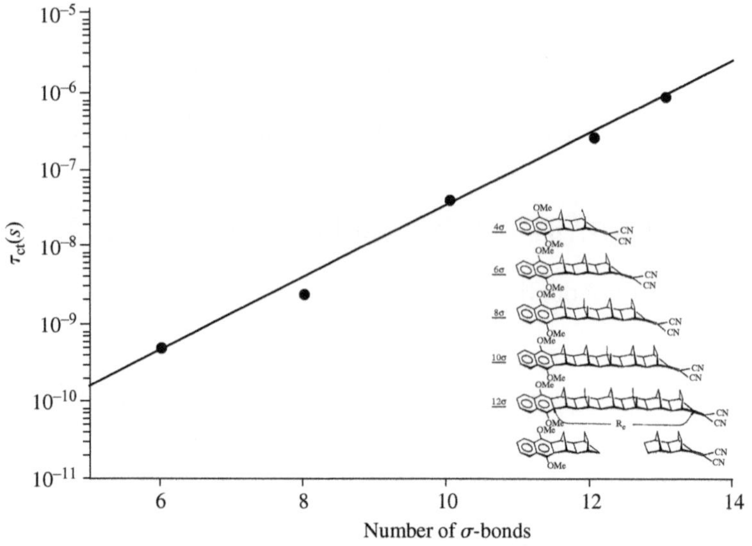

FIG. 16.8 Charge recombination lifetimes in the compounds shown in the inset in dioxane solvent. (J. M. Warman, M. P. de Haas, J. W. Verhoeven, and M. N. Paddon-Row, Adv. Chem. Phys. **106**, Electron transfer—from isolated molecules to bio-molecules, Part I, edited by J. Jortner and M. Bixon (Wiley, New York, 1999). The technique used is time-resolved microwave conductivity (TRMC), in which the change in dielectric response of a solution is monitored following photoinduced electron transfer—a charge separation process that changes the solute molecular dipole. The lifetimes shown as a function of bridge length (number of σ-bonds separating the donor and acceptor sites in the compounds shown in the inset) are for the back electron transfer (charge recombination) process.

Comparing to Eq. (16.102) and using (16.114) we see that, in this model,

$$k_{\mathrm{et}} \sim \exp(-\beta' L); \qquad \beta' = -\frac{2}{b} \ln \left| \frac{V_{\mathrm{B}}}{E_{\mathrm{D/A}} - E_{\mathrm{B}}} \right| \tag{16.117}$$

where b is the length, in the tunneling direction, of a single bridge unit.

The prediction of exponential dependence of k_{et} on the bridge length L has been repeatedly verified in many systems and values of β' have been tabulated for different bridge types. An example is shown in Fig. 16.8, where a value $\beta' = 1\ \text{Å}^{-1}$ was inferred from the distance-dependent lifetimes.

16.13 Electron tranport by hopping

The bridge-assisted electron transfer discussed above is a coherent quantum-mechanical process. Its tunneling nature is manifested by the exponentially decreasing rate with increasing donor acceptor distance, and limits observable transfer phenomena of this kind to relatively short bridges. Another mode of transfer, more akin to electron transport in macroscopic systems, may become important for longer bridges and at higher temperatures. In this mode the electron, if thermally activated onto the bridge, can migrate along the bridge by successive transfer steps. Thus, transitions between intermediate centers such as shown in Fig. 16.7 can be viewed as individual electron transfer steps characterized by rates that can be obtained from the Marcus theory. The overall D→A electron transfer then becomes a successive hopping process that may be described by the coupled kinetic equations (a master equation, see Section 8.3)

$$\begin{aligned} \dot{P}_0 &= -k_{1,0}P_0 + k_{0,1}P_1 \\ \dot{P}_1 &= -(k_{0,1} + k_{2,1})P_1 + k_{1,0}P_0 + k_{1,2}P_2 \\ &\vdots \\ \dot{P}_N &= -(k_{N-1,N} + k_{N+1,N})P_N + k_{N,N-1}P_{N-1} + k_{N,N+1}P_{N+1} \\ \dot{P}_{N+1} &= -k_{N,N+1}P_{N+1} + k_{N+1,N}P_N \end{aligned} \tag{16.118}$$

or

$$\dot{\mathbf{P}} = \mathbf{KP} \tag{16.119}$$

where $\mathbf{P}$ is the vector with elements P_j and $\mathbf{K}$ is the matrix of kinetic coefficients. Here P_j is the probability to be in the center j, where $j = 0$ and $j = N + 1$ denote the donor and acceptor centers, respectively, and $k_{i,j} = k_{i \leftarrow j}$ is the hopping rate coefficient from center j to center i. These rate coefficients are not all independent; detailed balance (see Section 8.3.3) requires that ratios between forward and backward rates should be compatible with Boltzmann statistics, that is, $k_{i,j}/k_{j,i} = \exp((E_j - E_i)/k_BT)$ where E_j and E_i are energies of the corresponding states.

Problem 16.9. Show that the matrix $\mathbf{K}$ has one zero eigenvalue, and that the corresponding eigenvector is the equilibrium state that satisfies $P_i/P_j = \exp((E_j - E_i)/k_BT)$.

In general, the time evolution inferred from such a model is characterized by $N + 1$ characteristic times, or rate coefficients, associated with the nonvanishing eigenvalues of the kinetic matrix $\mathbf{K}$. In our application, however, the energies of the donor and acceptor states $j = 0, N + 1$ are considerably lower than those of the bridge states. Specifically, we assume that the inequalities

$$E_j - E_0,\ E_j - E_{N+1} \gg k_BT \tag{16.120}$$

hold for all $j \neq 0, N + 1$. Under such conditions the transition between states 0 and N+1 is a discrete version of the barrier crossing problem discussed in Chapter 14, and the transition D→A will be dominated by a single rate coefficient $k_{A\leftarrow D}$. Starting from $P_0 = 1$ at $t = 0$ we will observe (after a short transient period) an exponential decay, $P_0(t) = \exp(-k_{A\leftarrow D}t)$, and an equivalent increase in the acceptor population $P_{N+1}(t)$. There is one caveat in this description: It is based on the assumption that to a good approximation $P_0(t) + P_{N+1}(t) = 1$ at all times, that is, that the population in the intermediate bridge states is very small. This is usually justified by the inequality (16.120), however may not hold if the bridge is too long. In the latter case the process may become dominated by diffusion on the bridge and simple first-order kinetics will not apply.

Barring this possibility, we proceed to evaluate the effective transfer rate $k_{A\leftarrow D}$. We will use the steady-state flux method, now a classical version of the scheme developed in Section 9.5. In this approach we consider the steady-state obtained at long time under the restriction that state 0 is a constant source ($P_0 =$ constant) while state $N+1$ is a drain ($P_{N+1} = 0$). We also limit ourselves to the case where all rate coefficients that are not associated with species 0 and N+1 are the same, denoted k. Equations (16.118), without the first and last equations, then lead to the N steady-state equations

$$\begin{aligned}
0 &= -(k_{0,1} + k)P_1 + k_{1,0}P_0 + kP_2 \\
0 &= -2kP_2 + k(P_1 + P_3) \\
&\vdots \\
0 &= -2kP_{N-1} + k(P_{N-2} + P_N) \\
0 &= -(k + k_{N+1,N})P_N + kP_{N-1}
\end{aligned} \tag{16.121}$$

while the last equation in (16.118) is rewritten in the form

$$\dot{P}_{N+1} = 0 = k_{N+1,N}P_N - J \tag{16.122}$$

Here J is the flux through the system that drains population out of state N+1. At steady state this flux also satisfies

$$J = k_{A\leftarrow D}P_0 \tag{16.123}$$

that defines the effective transfer rate $k_{A\leftarrow D}$. To solve the system of N equations (16.121) we first create another set of N equations as the sums, for $n = N, N-1, \ldots, 1$, of the last n equations in the set (16.121). This yields

$$\begin{aligned}
0 &= -k_{01}P_1 + k_{10}P_0 - k_{N+1,N}P_N \\
0 &= -kP_2 + kP_1 - k_{N+1,N}P_N \\
0 &= -kP_3 + kP_2 - k_{N+1,N}P_N \\
&\vdots \\
0 &= -kP_{N-1} + kP_{N-2} - k_{N+1,N}P_N \\
0 &= -kP_N + kP_{N-1} - k_{N+1,N}P_N
\end{aligned} \tag{16.124}$$

(the last equations in the sets (16.121) and (16.124) are identical). The first equation in (16.124) and the sum of the last $N-1$ equations in this set yield the set of two equations

$$\begin{aligned} 0 &= -k_{01}P_1 + k_{10}P_0 - k_{N+1,N}P_N \\ 0 &= kP_1 - [k + (N-1)k_{N+1,N}]P_N \end{aligned} \qquad (16.125)$$

that can be solved to yield the steady-state relationship between P_N and P_0

$$P_N = \frac{(k/k_{N+1,N})(k_{1,0}/k_{0,1})P_0}{(k/k_{N+1,N}) + (k/k_{0,1}) + N - 1} \qquad (16.126)$$

whence, using (16.122) and (16.123),

$$k_{A\leftarrow D} = \frac{e^{-(E_{BD}/k_BT)}}{1/(k_{A\leftarrow N}) + 1/(k_{D\leftarrow 1}) + (N-1)/k} \qquad (16.127)$$

In obtaining the form (16.127) we have used the detailed balance relation $k_{1,0}/k_{0,1} = \exp(-E_{BD}/k_BT)$ where $E_{BD} = E_1 - E_0$ is the "barrier height" above the donor energy, and have identified $k_{0,1}$ and $k_{N+1,N}$ as the rate coefficients $k_{D\leftarrow 1}$ and $k_{A\leftarrow N}$ to go from the first and last bridge sites to the donor and acceptor, respectively.

The rate coefficient $k_{A\leftarrow D}$, Eq. (16.127), is seen to have an Arrhenius form, with an activation energy associated with the thermal population of the bridge and a pre-exponential coefficient which is the inverse of a sum of three times. Two of them, $(k_{D\leftarrow 1})^{-1}$ and $(k_{A\leftarrow N})^{-1}$ are, respectively the lifetimes of the electron on the edge bridge states for going into the donor and acceptor levels. Recalling that D and A denote stabilized donor and acceptor states (the bottoms of the corresponding Marcus parabolas) we may be tempted to identify states 1 and N of the bridge as nonequilibrium donor and acceptor states raised by a solvent fluctuation to a height for which transfer to the bridge is possible. In this case $k_{D\leftarrow 1}$ and $k_{A\leftarrow N}$ are the corresponding solvent relaxation (solvation) rates. The third time, $(N-1)/k$, a product of the hopping time k^{-1} and the number of hops $N-1$ in the bridge may be interpreted as the time spent on the bridge itself.[27]

Consider now the bridge-length dependence of such electron transfer processes. For short bridges a tunneling-dominated transfer will show an exponential length dependence. However, for long bridges we expect the behavior shown in (16.127): A weak N dependence that becomes $1/N$ for long bridges. Figure 16.9 shows an experimental example in which the donor and acceptor species are placed at

[27] When the process becomes dominated by diffusion on the bridge the actual time spent on the bridge should be derived from the diffusion law, and is proportional to N^{-2}. As discussed above, when we apply the steady-state formalism to calculate a rate, we in fact assume that the bridge is not long enough to yield this limit.

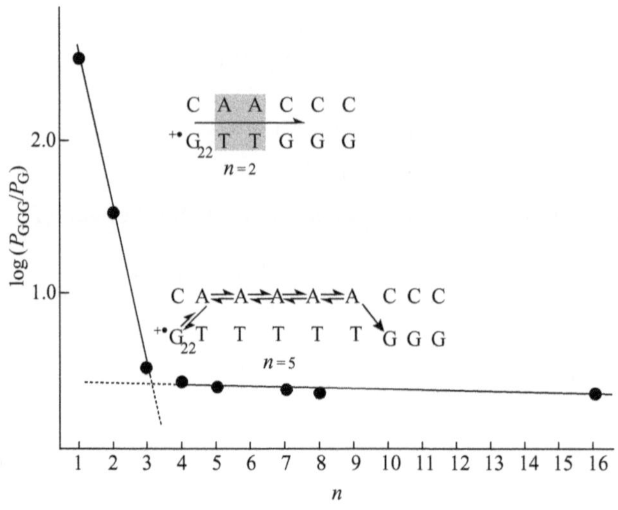

FIG. 16.9 Electron transfer rates in DNA: Shown are the yields for transfer between guanine (G) and GGG groups separated by Adenine–Thymine (A) bridges for different bridge lengths. The transition from exponential dependence on bridge length to practically bridge independent rates (on this scale) marks the transition from tunneling to hopping transfer. (From B. Giese, J. Amaudrut, A.-K. Koehler, M. Spormann, and S. Wessely, Nature **412**, 318 (2001).

different distances on the backbone of a DNA molecule. The electron transfer yield displayed as a function of distance (expressed in terms of amino acid segments) shows what appears to be a crossover from tunneling to hopping behavior with increasing distance.

The transition from tunneling to activated transport can be manifested also by the temperature dependence of the process. One expects that at low temperature such processes will be dominated by tunneling, therefore show weak or no dependence on temperature. As the temperature increases we expect a crossover to the activated mode of transport, characterized by an Arrhenius behavior. Fig. 16.10 shows an example of such crossover phenomenon.

16.14 Proton transfer

Proton transfer could certainly be another full chapter in this book. With applications ranging from photosynthesis to fuel cells this is one of the most important elementary reactions and as such was and is intensively investigated. This section does not pretend to provide any coverage of this process, and is included here mainly as a reminder that this important reaction should be on the mind of a researcher in condensed phase chemical dynamics. It is also of interest to point out an interesting

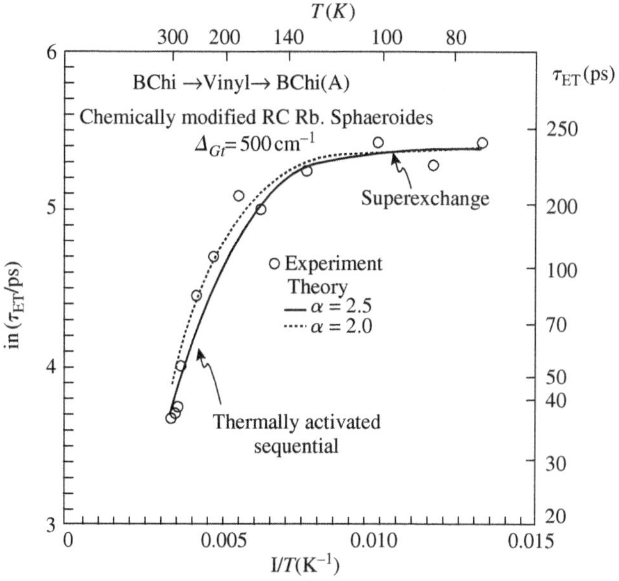

FIG. 16.10 The electron transfer time (inverse rate) in a chemically modified photoreaction center of bacteriochlorophyll, showing a crossover from thermally activated sequential hopping behavior at high temperature to a "superexchange" tunneling behavior at low temperature. (Open circles are experimental data from M. E. Michel-Beyerle et al., unpublished; full and dashed lines are theoretical fits from the articles by M. Bixon and J. Jortner cited at the end of this chapter.)

conceptual dilemma related to the placement of this reaction relative to two other processes studied in this text: electron transfer in this chapter and barrier crossing in Chapter 14.

For specificity let us focus on a reaction of the form

$$AH \ldots B \rightarrow A^{-} \ldots HB^{+} \tag{16.128}$$

The question is: Should we better regard proton transfer in the same framework as electron transfer, namely solvent rearrangement as a precursor to tunneling transition, or is it better to use a description more akin to the barrier crossing reactions discussed in Chapter 14? In the first case the reaction coordinate is associated with the solvent rearrangement as was the case for electron transfer. In the second—it is the position of the proton on its way across the barrier separating its two binding sites on A or on B. In either case the motion across this barrier can be classical-like or tunneling-like depending on the barrier height relative to the zero point energy of the proton in its local well.

What makes this reaction conceptually special is that there is no simple answer to this question. Rather, proton transfer should probably be described with respect to two coordinates: The solvent reorganization energy that constituted the reaction coordinate of electron transfer reactions *and* the proton position between its two sites. On this two-dimensional free energy surface one coordinate (proton position) is quantum. The other (solvent reorganization) is essentially classical. This combination of higher dimensionality and mixed quantum and classical dynamics, together with the availability of an additional observable—the kinetic isotope effect associated with the reaction, make proton transfer a unique process.[28]

Appendix 16A: Derivation of the Mulliken–Hush formula

Here we present the derivation[29] of the expression (16.97) that relates the coupling between two nonadiabatic electronic states a and b to the optical transition dipole between the corresponding adiabatic states 1 and 2, as described in Section 16.10.

Our discussion refers to a given fixed nuclear configuration. The electron transfer reaction is assumed to take place between two states, a state a localized on the center A and a state b localized on the center B. ψ_a and ψ_b are the corresponding wavefunctions with the energies E_a and E_b, respectively. We assume that $S_{ab} = \langle\psi_a|\psi_b\rangle = 0$, an assumption valid when these centers are far enough from each other. These states diagonalize that part of the system's Hamiltonian from which the interaction V that leads to the electron transfer is excluded. In the literature one often refers to these zero-order states as "diabatic" states, and to the representation defined by this basis as the diabatic representation (see Section 2.5). We further assume that the coupling V that leads to transition between these states has no diagonal elements (i.e. does not modify the zero-order energies). The full Hamiltonian in the diabatic representation is then

$$\mathbf{H} = \begin{pmatrix} E_a & V_{ab} \\ V_{ba} & E_b \end{pmatrix} \tag{16.129}$$

The eigenstates ψ_1 and ψ_2 of this Hamiltonian are, by definition, the adiabatic states, which are exact states in the Born–Oppenheimer approximation. They are

[28] Further reading: K. D. Kreuer, Proton conductivity: Materials and applications, Chem. Mater. **8**, 610 (1996); K. Ando and J. T. Hynes, Adv. Chem. Phys. **110**, 381 (1999); Philip M. Kiefer and J. T. Hynes, Sol. St. Ionics, **168**, 219 (2004).

[29] R. S. Mulliken, J. Am. Chem. Soc. **64**, 811–824 (1952); R. S. Mulliken and W. B. Persson, *Molecular Complexes* (Wiley, New York, 1969); N. S. Hush, Prog. Inorg. Chem. **8**, 391–444 (1967); N. S. Hush, Electrochim. Acta, **13**, 1005–1023 (1968); C. Creutz, M. D. Newton, and N. Sutin, Photochem. Photobiol. A: Chem., **82**, 47–59 (1994).

written as

$$\begin{aligned}\psi_1 &= c_a\psi_a + c_b^*\psi_b \\ \psi_2 &= -c_b\psi_a + c_a^*\psi_b\end{aligned} \tag{16.130}$$

normalized so that $|c_a|^2 + |c_b|^2 = 1$. The corresponding energies E_1 and E_2 are solutions of the secular equation

$$\begin{vmatrix} E_a - E & V_{ab} \\ V_{ba} & E_b - E \end{vmatrix} = 0 \tag{16.131}$$

Denote

$$E_{ab} = E_b - E_a \geq 0 \tag{16.132}$$

Then

$$E_1 = \frac{(E_a + E_b)}{2} - \frac{(E_{ab}^2 + 4V_{ab}^2)^{1/2}}{2} \tag{16.133}$$

$$E_2 = \frac{(E_a + E_b)}{2} + \frac{(E_{ab}^2 + 4V_{ab}^2)^{1/2}}{2} \tag{16.134}$$

$$E_2 - E_1 = +\sqrt{E_{ab}^2 + 4V_{ab}^2} \tag{16.135}$$

The critical assumption in the following derivation is that an optical transition, that is, absorption or emission of a photon, takes place between the exact eigenstates 1 and 2.

This statement is not obvious. We consider two transitions: Electron transfer and photon absorption, and state that the former is a transition between states a and b while the latter takes place between states 1 and 2. Why should these transitions viewed as processes that transfer populations between different states?

The answer lies in the realization that light absorption is a process that starts with a system in equilibrium, disturbed only by the external radiation field. At low temperature the initial molecular state is the electronic ground state, an eigenstate of the full molecular Hamiltonian. On the other hand, in many experimental situations electron transfer takes place in a system that was brought into a nonequilibrium state by some preparation event (e.g. a would-be donor was suddenly brought into the neighborhood of a potential acceptor or, more easily, a donor state was prepared optically). There are no external perturbations; the only reason for the transition that follows is that the system was not prepared in an eigenstate. This nonequilibrium initial state may be taken as an eigenstate of some zero-order Hamiltonian—that Hamiltonian in which the terms responsible for the electron transfer process are not included.

In most situations we do not find this Hamiltonian by an analytical process. Rather, we first identify the donor state a and the acceptor state b using chemical intuition. Their assumed (approximate) mutual orthogonality is based on their physical character: They are localized on donor and acceptor sites that are relatively far from each other. The zero-order Hamiltonian may be then formally written as $E_a|a\rangle\langle a| + E_b|b\rangle\langle b|$. *The full system Hamiltonian can be formally written in the representation defined by these states, in the form* $E_a|a\rangle\langle a| + E_b|b\rangle\langle b| + H_{ab}|a\rangle\langle b| + H_{ba}|b\rangle\langle a|$ *(equivalent to Eq. (16.129) with* $V_{ab} = H_{ab}$*). The adiabatic states 1 and 2 are obviously those that diagonalize the full Hamiltonian, leading to the final conclusions that these are the states between which optical transitions take place.*

The equations for the coefficients c_a and c_b of Eq. (16.130) are

$$\begin{pmatrix} E_a - E_1 & V_{ab} \\ V_{ba} & E_b - E_1 \end{pmatrix} \begin{pmatrix} c_a \\ c_b^* \end{pmatrix} = 0$$
$$\begin{pmatrix} E_a - E_2 & V_{ab} \\ V_{ba} & E_b - E_2 \end{pmatrix} \begin{pmatrix} -c_b \\ c_a^* \end{pmatrix} = 0 \tag{16.136}$$

that yield

$$E_1 = E_a + \frac{c_b^*}{c_a} V_{ab} \tag{16.137a}$$
$$= E_b + \frac{c_a}{c_b^*} V_{ba} \tag{16.137b}$$

$$E_2 = E_a - \frac{c_a^*}{c_b} V_{ab} \tag{16.138a}$$
$$= E_b - \frac{c_b}{c_a^*} V_{ba} \tag{16.138b}$$

(Note that the corrections to the zero-order energies must be real numbers and their signs are determined by our choice $E_1 < E_a < E_b$ and $E_2 > E_b > E_a$.) From (16.137a) and (16.138a) or from (16.137b) and (16.138b) we get that

$$E_2 - E_1 = -\frac{V_{ab}}{c_a c_b} = -\frac{V_{ba}}{c_a^* c_b^*} \quad \text{(real)} \tag{16.139}$$

or

$$|c_a c_b| = \frac{|V_{ab}|}{(E_2 - E_1)} = \frac{1}{2}\left[1 - \left(\frac{E_{ab}}{E_2 - E_1}\right)^2\right]^{1/2} \tag{16.140}$$

Consider now the absorption lineshape, which, as discussed above, corresponds to an optical transition between states 1 and 2. What is measured is the extinction coefficient, $\varepsilon(\nu)$, that determines the reduction in light intensity as a beam of frequency ν travels a distance d through a sample of concentration C according to $I = I_0 e^{-\varepsilon(\nu)Cd}$. The *oscillator strength*, essentially a measure for the integrated lineshape is defined by

$$f_{\text{osc}} = \frac{2.303 \times 10^3 m_e c^2}{\pi \mathcal{A} e^2} \int d\nu \varepsilon(\nu) \tag{16.141}$$

where m_e and e are the electron mass and charge, c is the speed of light, and $\mathcal{A}$ is the Avogadro number. (The numerical factors corresponds to the integral $\int d\nu\varepsilon(\nu)$ evaluated in cgs length unit: ε in cm^2 and ν in cm^{-1}. f_{osc} itself is dimensionless, and if ε and ν are expressed in these units we get $f_{\text{osc}} = 4.33 \times 10^{-9} \int d\nu\varepsilon(\nu)$). For dipole-allowed absorption the oscillator strength is related to the absolute-squared transition dipole matrix element, $|\mu_{12}|^2$, between states 1 and 2 according to

$$f_{\text{osc}} = \frac{4\pi m_e \nu_{\max}}{3\hbar e^2} |\mu_{12}|^2 \qquad (\nu_{\max} \text{ is in inverse time units}) \tag{16.142}$$

$$= 1.08 \times 10^{-5} \nu_{\max} |\mu_{12}|^2 \qquad (\nu_{\max} \text{ is in cm}^{-1}) \tag{16.143}$$

Thus, a measurement of f_{osc} yields $|\mu_{12}|$. What we need is a relationship between $|\mu_{12}|$ and H_{ab}. To this end we start with the expression for the electronic-dipole matrix element,

$$\mu_{12} = -e \langle \psi_1 | \sum_i r_i | \psi_2 \rangle \tag{16.144}$$

where the sum is over all electrons, and use Eq. (16.130) together with the assumption made above that $S_{ab} = \langle \psi_a | \psi_b \rangle = 0$ and the equivalent assumption that $\mu_{ab} = -e\langle \psi_a | \sum_i r_i | \psi_b \rangle = 0$ (both assumptions rely on the locality of the a and b states on their corresponding centers, which are assumed to be far enough from each other). We get

$$\mu_{12} = c_a^* c_b (\mu_b - \mu_a) \tag{16.145a}$$

where

$$\mu_a = -e\langle \psi_a | \sum_i r_i | \psi_a \rangle; \qquad \mu_b = -e\langle \psi_b | \sum_i r_i | \psi_b \rangle \tag{16.145b}$$

are the dipole moments of the localized zero-order states. Using Eq. (16.140) we find

$$|\mu_{12}| = \frac{|V_{ab}|}{(E_2 - E_1)} |\mu_b - \mu_a| = \frac{e|V_{ab}| r_{ab}}{\hbar \omega_{\max}} \tag{16.146}$$

where in the second equality we represented $E_2 - E_1$ by the frequency $\omega_{\max} = 2\pi\nu_{\max}$ of maximum absorption and have defined

$$r_{ab} \equiv \left| \frac{\mu_b - \mu_a}{e} \right| \tag{16.147}$$

If the only change in the molecular charge distribution between the states a and b is the position of the transferred electron (i.e. if we assume that the other electrons are not affected) then r_{ab} is the transfer distance, that is, the separation between the donor and acceptor centers.

Further reading

M. Bixon and J. Jortner, editors, *Electron transfer–from isolated molecules to biomolecules*, Advances in Chemical Physics, Vol. **106** (Wiley, New York, 1999) Parts I and II.

M. Bixon and J. Jortner in *Electron transfer – from isolated molecules to biomolecules*, edited by M. Bixon and J. Jortner, Advances in Chemical Physics, Vol. **106** (Wiley, New York, 1999) Parts I, pp. 35–202.

A. M. Kuznetsov, *Charge Transfer in Physics, Chemistry and Biology: Physical Mechanisms of Elementary Processes and an Introduction to the Theory* (Gordon & Breach, New York, 1995).

A. M. Kuznetsov and J. Ulstrup, *Electron Transfer in Chemistry and Biology: An Introduction to the Theory* (Wiley, New York, 1999).

17

ELECTRON TRANSFER AND TRANSMISSION AT MOLECULE–METAL AND MOLECULE–SEMICONDUCTOR INTERFACES

Our world I think is very young,
Has hardly more than started; some of our arts
Are in the polishing stage, and some are still
In the early phases of their growth; we see
Novel equipments on our ships, we hear
New sound in our music, new philosophies ...

Lucretius (c.99–c.55 BCE*) "The way things are"*
translated by Rolfe Humphries, Indiana University Press, 1968

This chapter continues our discussion of electron transfer processes, now focusing on the interface between molecular systems and solid conductors. Interest in such processes has recently surged within the emerging field of *molecular electronics*, itself part of a general multidisciplinary effort on *nanotechnology*. Notwithstanding new concepts, new experimental and theoretical methods, and new terminology, the start of this interest dates back to the early days of electrochemistry, marked by the famous experiments of Galvani and Volta in the late eighteenth century. The first part of this chapter discusses electron transfer in what might now be called "traditional" electrochemistry where the fundamental process is electron transfer between a molecule or a molecular ion and a metal electrode. The second part constitutes an introduction to molecular electronics, focusing on the problem of *molecular conduction*, which is essentially electron transfer (in this context better termed electron transmission) between two metal electrodes through a molecular layer or sometimes even a single molecule.

17.1 Electrochemical electron transfer

17.1.1 Introduction

In Chapter 16 we have focused on electron transfer processes of the following characteristics: (1) Two electronic states, one associated with the donor species, the other with the acceptor, are involved. (2) Energetics is determined by the electronic energies of the donor and acceptor states and by the electrostatic solvation of the initial and final charge distributions in their electronic and nuclear environments. (3) The energy barrier to the transfer process originates from the fact that electronic and nuclear motions occur on vastly different timescales. (4) Irreversibility is driven by nuclear relaxation about the initial and final electronic charge distributions.

How will this change if one of the two electronic species is replaced by a metal? We can imagine an electron transfer process between a metal substrate and a molecule adsorbed on its surface, however the most common process of this kind takes place at the interface between a metal electrode and an electrolyte solution, where the molecular species is an ion residing in the electrolyte, near the metal surface. Electron transfer in this configuration is the fundamental process of electrochemistry. Knowledge of the atomic and electrostatic structure of metal–electrolyte interfaces (more generally interfaces between an ionic conductor such as an electrolyte or a molten salt and an electronic conductor such as a metal or a semiconductor) is a prerequisite to understanding electron transfer at such interfaces. In the present discussion we assume that this knowledge is available and focus on the electron transfer itself. At issue is the comparison between a process such as (16.1)

$$\mathrm{Fe}^{+3}\,(\text{solution}) + \mathrm{Fe}^{+2}\,(\text{solution}) \rightleftharpoons \mathrm{Fe}^{+2}\,(\text{solution}) + \mathrm{Fe}^{+3}\,(\text{solution}) \qquad (17.1)$$

in which the electron is transferred between solvated molecular species and the analogous process

$$\mathrm{Fe}^{+3}\,(\text{solution}) + \text{e-(metal)} \rightleftharpoons \mathrm{Fe}^{+2}\,(\text{solution}) \qquad (17.2)$$

where the electron is transferred between a solvated molecular species and the metal. Note that we could write (17.2) in analogy to (17.1) in the form

$$\mathrm{Fe}^{+3}\,(\text{solution}) + \text{metal}\,(N) \rightleftharpoons \mathrm{Fe}^{+2}\,(\text{solution}) + \text{metal}\,(N-1) \qquad (17.3)$$

where N is the number of electrons on the metal. The form (17.3) is similar to (17.1) except that the metal replaces one of the reactants. There are, however, three important differences between these two processes:

1. While the process (17.1) involves two electronic states, one on each reactant, a macroscopic metal electrode is characterized by a continuum of electronic states with average occupation given by the Fermi function in terms of the

electronic chemical potential μ,

$$f(E) = \frac{1}{e^{\beta(E-\mu)} + 1} \tag{17.4}$$

2. In a typical electrochemical setup the potential difference between the interiors of the metal and the solution is controlled, so that the direction and rate of electron transfer can be monitored as functions of this voltage.
3. In addition to providing this new control parameter, this setup also provides a new observable: the electronic current flowing under a given voltage. It is this element of electrochemical electron transfer that makes Eq. (17.2) a better representation of this process: First, the electron current is the direct observable and second, the state of the metal remains unchanged during the process. This results from the fact that at a given voltage the electron density in the metal is constantly readjusted by the source of this voltage.

The current flowing under a given applied voltage is the most important observable of an electrochemical system, and understanding the factors that determine and control it is the central issue of electrochemistry. It is not always the case that the observed current is directly related to the electron transfer rate; for example, the rate determining process may be diffusion of redox components toward the working electrode. As stated above, in the following discussion we limit ourselves to a simpler question, just focusing on the rate of electron transfer. Given a molecule or an ionic species in solution at a given distance from the metal surface, our aim is to evaluate the electron transfer rate, the equivalent of the rate of electron transfer between two species in solution given by Eqs (16.50–16.59).

17.1.2 The electrochemical measurement

Before addressing the rate issue we need to understand the way in which such rates are measured, keeping in mind that the observable in a typical experiment is electrical current measured as a function of voltage.

Let us consider the voltage first. When a metal electrode M (— the electrode whose interface with the solution we investigate; henceforth referred to the *working electrode*) is dipped into an electrolyte solution and equilibrium is established, an electrostatic potential is established between the two phases. What is usually measured (see Fig. 17.1) is the potential difference Φ between this electrode and a reference half cell, R—say a platinum electrode in contact with some fixed redox solution which in turn is connected by a capillary to the close neighborhood of

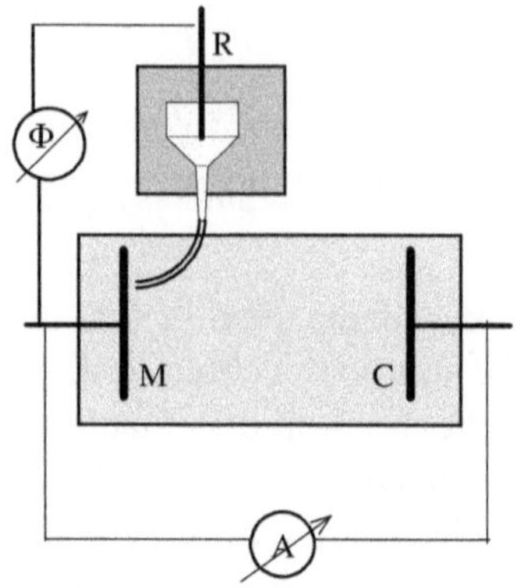

FIG. 17.1 An electrochemical measurement with a three-electrode configuration: M is the working electrode, C is the counter electrode, and R is the reference electrode. In a typical measurement the current between M and C is measured against the voltage between M and R.

the working electrode.[1] At equilibrium $\Phi = \Phi_0$. In a typical electrochemical experiment the current flows between the working electrode M and a counter electrode C (the current between M and R is negligible in comparison) and is measured as a function of Φ. When we change Φ we affect the potential difference between M and the molecular species in its solution neighborhood and consequently affect the rate of electron exchange between them as discussed below. The difference $\eta = \Phi - \Phi_0$ between the potential Φ and the equilibrium potential Φ_0 is referred to as the *overpotential*. In this chapter a positive η corresponds to the electrode biased positively relative to the solution.

It should be intuitively obvious (and is further clarified below) that the effect of applied potential on the electron transfer rate between the electrode M and a molecular species S in its solution neighborhood reflects the way by which this potential translates into a potential drop between M and S. This follows from the fact that the rate depends on the relative positions of electronic levels in the electrode and the molecule, which in turn depend on this drop. In much of the electrochemical literature it is assumed that when the electrode potential changes by $\delta\Phi$ so does this potential drop. This amounts to the assumption that the species S does not feel the potential change on M, that is, that the electrolyte solution effectively screens the electrode potential at the relevant S–M distance. Such an assumption holds at high supporting electrolyte concentration[2] (order of 1 mole per liter). However, even

[1] In this arrangement variations in Φ imply similar variations in the potential between the working electrode and the solution, irrespective of what happens at the counter electrode.

[2] The term "supporting electrolyte" refers to an electrolyte species in the solution that is inert to the electrode process under consideration.

at lower electrolyte concentrations the corrections to this behavior are small, and will be disregarded below. Indeed, assuming that the molecular species is separated from the metal surface by two hydration shells (one surrounding the ion and one on the metal) a typical distance between S and M is 5–6 Å, (the so-called *outer Helmholtz plane*) and most of the potential falls within this range.

17.1.3 The electron transfer process

Consider now the electron transfer process. In contrast to the problem discussed in Sections 16.3 and 16.4, of electron transfer between molecular donor and acceptor states, where the role of nuclear motion was critical for converting a two-state dynamics into a rate process, in the present situation a rate exists even in the absence of nuclear relaxation because of the presence of a continuum of metal levels. We will start by considering this problem, disregarding nuclear motion.

17.1.3.1 Electron transfer to/from a metal electrode without nuclear relaxation

The theoretical treatment in this case is similar to that of a level interacting with a continuum (Section 9.1) but with some new twists. The result will have a golden-rule form, that is, contains a product $V^2\rho$, but we need to identify the coupling and the density of states involved.

We use a simple picture in which the molecular species S is a two-state system, where the oxidized state $|a\rangle$ has one electron more than the reduced state $|b\rangle$. The corresponding energies are E_a and E_b, and their difference is denoted $E_{ab} = E_a - E_b$. For the metal electrode we use a free electron model, so that a state of the metal is specified by a set of occupation numbers $\mathbf{m} = (m_1, m_2, \ldots)$ of the single electron levels. For the single electron level j of energy E_j, $m_j = 1$ with probability $f(E_j)$ and $m_j = 0$ with probability $1 - f(E_j)$, where $f(E)$ is given by (17.4). A basis of states for the overall SM system is written $|s, \mathbf{m}\rangle = |s\rangle\, |\mathbf{m}\rangle$ where $s = a, b$ and $|\mathbf{m}\rangle$ is an antisymmetrized product of single electron metal states.

Consider the coupling that gives rise to electron tunneling between molecule and metal. We assume that it is independent of the spin state of the electron, so consideration of spin and spin multiplicity appears only in the electron density of states. We will not go into details of this coupling (suitable models can be constructed using pictures like Fig. 12.2 and using a basis of electronic states localized on the molecules or in the metal); except for *assuming* that it can be written as a sum of one-electron coupling terms

$$\hat{V} = \sum_{s=a,b} \sum_{\mathbf{m}} \sum_{j} V^{\mathbf{m}}_{a0_j,b1_j} |a, (\mathbf{m}, 0_j)\rangle\langle b, (\mathbf{m}, 1_j)| \tag{17.5}$$

where $|a, (\mathbf{m}, 0_j)\rangle = |a\rangle|\mathbf{m}, 0_j\rangle$ describes the molecule in state a and the metal in state $\mathbf{m}$, with 0_j emphasizing that no electron occupies the single electron level j in the metal, while $|b, (\mathbf{m}, 1_j)\rangle = |b\rangle|\mathbf{m}, 1_j\rangle$ is the state obtained from it by transferring an electron from the molecule to the metallic level j.

Next consider the rate of such electron transfer process. The golden-rule expression for this rate is

$$k_{b\leftarrow a} = \frac{2\pi}{\hbar}\sum_{\mathbf{m}}\sum_{j} P(\mathbf{m}, 0_j)|V^{\mathbf{m}}_{a0_j,b1_j}|^2\delta(E_{ab} - E_j) \tag{17.6a}$$

$$k_{a\leftarrow b} = \frac{2\pi}{\hbar}\sum_{\mathbf{m}}\sum_{j} P(\mathbf{m}, 1_j)|V^{\mathbf{m}}_{a0_j,b1_j}|^2\delta(E_{ba} + E_j) \tag{17.6b}$$

where $P(\mathbf{m}, n_j)$ is the thermal probability to find the metal in the electronic state $|\mathbf{m}, n_j\rangle$ ($n_j = 0$ or 1). In what follows we assume that the coupling elements $V^{\mathbf{m}}_{a0_j,b1_j}$ do not depend on $\mathbf{m}$. This amounts to the assumption that the electron tunneling between the molecule and the metal is a purely one-electron process that does not depend on the state of the other electrons. In this case we can use the identities

$$\sum_{\mathbf{m}} P(\mathbf{m}, 0_j) = 1 - f(E_j); \qquad \sum_{\mathbf{m}} P(\mathbf{m}, 1_j) = f(E_j) \tag{17.7}$$

so that Eqs (17.6) may be written in the forms

$$k_{b\leftarrow a} = \frac{2\pi}{\hbar}\sum_{j}(1 - f(E_j))|V_{a,b}|^2\delta(E_{ab} - E_j) \tag{17.8a}$$

$$k_{a\leftarrow b} = \frac{2\pi}{\hbar}\sum_{j} f(E_j)|V_{a,b}|^2\delta(E_{ba} + E_j) \tag{17.8b}$$

Note that the δ functions that appear in Eqs. (17.6) and (17.8) are all identical. The different forms become meaningful when nuclear relaxation is taken into account (see below). In writing Eqs (17.8) in these forms we have assumed that the coupling element $V_{a0_j,b1_j}$ depends on j only through the energy E_j, which is already specified by the a and b indices. We can convert the sums in these equations into integrals, using the density $\rho_{\mathrm{M}}(E)$ of single electron states in the metal

$$k_{b\leftarrow a} = \int dE\Gamma(E)(1 - f(E))\delta(E - E_{ab}) = \Gamma(E_{ab})(1 - f(E_{ab})) \tag{17.9a}$$

and similarly

$$k_{a\leftarrow b} = \Gamma(E_{ab})f(E_{ab}) \tag{17.9b}$$

where

$$\Gamma(E_{ab}) = \frac{2\pi}{\hbar}|V_{a,b}|^2\rho_M(E_{ab})$$

is the rate of electron transfer from S to M for the case where all single electron levels on M are unoccupied (or from M to S if all these levels were occupied) in the absence of nuclear relaxation effects.

17.1.3.2 The effect of nuclear relaxation

Equations (17.8) and (17.9) were obtained under the assumption that electron transfer takes place in the absence of nuclear motions. How do such motions, that were found to play a central role in molecular electron transfer, affect the dynamics in the present case? In analogy to Eq. (16.52) we can now write for electron transfer to the metal

$$
\begin{aligned}
k_{b\leftarrow a} &= \frac{2\pi}{\hbar}\sum_j (1-f(E_j))\frac{1}{Q_a} \\
&\quad \times \sum_i e^{-\beta E_{a,i}} \sum_f |V_{ai,bf}|^2 \delta(E_{a,b} - E_j + E_{a,i} - E_{b,f}) \\
&= \frac{2\pi}{\hbar}|V_{a,b}|^2 \sum_j (1-f(E_j))\frac{1}{Q_a}\sum_i e^{-\beta E_{a,i}} \\
&\quad \times \sum_f |\langle \chi_i^{(a)} | \chi_f^{(b)}\rangle|^2 \delta(E_{a,b} - E_j + E_{E_{a,i}} - E_{E_{b,f}}) \\
&= \frac{2\pi}{\hbar}|V_{a,b}|^2 \sum_j (1-f(E_j))F(E_{a,b} - E_j)
\end{aligned}
\tag{17.10}
$$

where the sums over i and f are for the nuclear states associated with the molecular electronic states a and b, respectively, and where the function $F(E)$ was defined by (16.54) with a high-temperature/strong electron–phonon coupling limit given by (16.58). Converting again the sum over single electron metal levels to an integral we now get

$$
\begin{aligned}
k_{b\leftarrow a} &= \frac{2\pi}{\hbar}|V_{a,b}|^2 \int dE\rho_M(E)(1-f(E))F(E_{ab} - E) \\
&= \int dE\,\Gamma(E)(1-f(E))F(E_{ab} - E)
\end{aligned}
\tag{17.11}
$$

The equivalent expression for electron transfer from the metal is obtained by repeating the same procedure starting from Eq. (17.8b). This leads to

$$
\begin{aligned}
k_{a\leftarrow b} &= \frac{2\pi}{\hbar}|V_{a,b}|^2 \int dE\rho_M(E)f(E)F(E_{ba} + E) \\
&= \int dE\Gamma(E)f(E)F(E_{ba} + E)
\end{aligned}
\tag{17.12}
$$

Note that while the δ-functions in the two equations (17.8) are equivalent, that is, $\delta(E_{ab} - E_j) = \delta(E_{ba} + E_j)$, including nuclear transitions leads to the F functions for which $F(E_{ab} - E) \neq F(E_{ba} + E)$. Also note that in the absence of nuclear relaxation $F(E) = \delta(E)$ (see Eq. (16.57)) and Eqs (17.11) and (17.12) lead back to (17.9).

17.1.4 The nuclear reorganization

In the high-temperature/strong electron–phonon coupling limit the functions $F(E)$ take the form (16.58).

$$F(x) = \frac{1}{\sqrt{4\pi E_r k_B T}} \exp\left(-\frac{(x - E_r)^2}{4E_r k_B T}\right) \tag{17.13}$$

where E_r is the reorganization energy. It is important to note that while E_r is defined exactly as before, for example, Eq. (16.60), its magnitude for electrode reactions is smaller than in molecular electron transfer in solution. The reason is that when the electron transfer takes place between two molecular species in solution more nuclear modes are involved in the ensuing reorganization. A rough estimate, valid when the transfer takes place between two species that are far enough from each other so that their contributions to the solvent reorganization are additive, is that when one of them is replaced by an electrode, its contribution to the reorganization energy is eliminated. By this argument, the solvent reorganization energy for the process (17.2) should be roughly half of that for the process (17.1).

This difference stems from an important difference in the nature of electron transfer in the two cases. In (17.1), as in any molecular electron transfer, the electron moves from one *localized* state to another. The reorganization energy is in fact the energy that is released when the solvent starts in a configuration that was in equilibrium with the electron on one localization center, and relaxes to a configuration equilibrated with the electron on another center. In contrast, an electron on the metal is not localized, and after electron transfer to or from the metal, the excess charge created spread in the metal on a timescale fast relative to solvent motion. The solvent therefore responds to the electronic charge only when it localizes on the molecule, and only this localization contributes to the reorganization energy.

17.1.5 Dependence on the electrode potential: Tafel plots

In deriving Eqs (17.11) and (17.12) we did not address the imposed electrode potential but the information must implicitly exist in the electronic energies that appear in these equations. To make this information explicit we redefine the electronic energies E_a, E_b, E_j, as the values at equilibrium relative to some specified fixed reference (e.g. ground state of electron in vacuum), keeping in mind that already at equilibrium there is usually some electrostatic potential difference between the metal and the bulk electrolyte. We are concerned with the additional overpotential η. As discussed at the end of Section 17.1.2, we assume that the full bias η falls in the electrolyte side of the interface, between the metal M and the molecular species S, and take $\eta > 0$ to mean that the metal is biased positively relative to the solution. Since in state a the molecular species S has one electron more than in state b this implies that with the new definition of the energy scale $E_{ab} \to E_{ab} + e\eta$ where e is the magnitude of the electron charge. We then have

$$k_{b\leftarrow a} = \int_{-\infty}^{\infty} dE\, \Gamma(E)\, (1 - f(E))\, F\, (E_{ab} + e\eta - E) \tag{17.14a}$$

$$k_{a\leftarrow b} = \int_{-\infty}^{\infty} dE \Gamma(E) f(E) F\, (E_{ba} - e\eta + E) \tag{17.14b}$$

Taking the integration limits to infinity rests on the assumption that the integrand is well contained within the metallic band.

Eqs. (17.14) can be used together with expressions (17.4) for the Fermi function and (17.13) for the function F to evaluate the rates. These expressions can be cast in alternative forms that bring out the dependence on physical parameters. First note again that as the metal electrode comes to contact with the redox solution some charge is transferred until the system comes to equilibrium in which some potential bias has been established between the metal and the molecular species. This potential bias has to be of just such magnitude that stops further charge transfer between the molecular and metallic phases. This implies that the free energy for removing an electron from a molecule in state a and putting it on the metal, $\mu - E_{ab}$ has to vanish, which implies that, $E_{ab} = \mu$.[3] Taking this as one approximation and assuming also that the energy dependence of $\Gamma(E)$ can be disregarded, we can change the energy variable in (17.14) so as to measure all energies from the metal chemical potential μ, for example,

$$k_{b\leftarrow a} \approx \frac{\Gamma}{\sqrt{4\pi E_{\mathrm{r}} k_B T}} \int_{-\infty}^{\infty} dE \frac{\exp(E/(k_B T))}{\exp(E/(k_B T)) + 1} \exp\left[-\frac{(E_{\mathrm{r}} + E - e\eta)^2}{4E_{\mathrm{r}} k_B T}\right] \tag{17.15}$$

[3] This argument is used in the electrochemistry literature, but it is only qualitative since it disregards the role of the reorganization energy in determining the free energy. Indeed, if we use the zero-temperature approximation for the Fermi functions in (17.14) we find that the equality $k_{b\leftarrow a} = k_{a\leftarrow b}$, which must be satisfied at equilibrium, leads to $E_{ab} = \mu$ only for $E_{\mathrm{r}} \to 0$.

If the reorganization energy E_r is large relative to k_BT while $-e\eta$ is small,[4] we can use the fact that the integrand vanishes quickly when $E - e\eta$ increases beyond k_BT to make the expansion $(E_r + E - e\eta)^2 = E_r^2 + 2E_r(E - e\eta)$, which leads to

$$\begin{aligned} k_{b\leftarrow a} &\approx \frac{\Gamma}{\sqrt{4\pi E_r k_B T}} \exp\left(-\frac{E_r - 2e\eta}{4k_BT}\right) \\ &\times \int_{-\infty}^{\infty} dE \frac{1}{\exp(E/(2k_BT)) + \exp(-E/(2k_BT))} \\ &= \Gamma\sqrt{\frac{\pi k_BT}{4E_r}} \exp\left(-\frac{E_r - 2e\eta}{4k_BT}\right) \end{aligned} \tag{17.16}$$

The current at the electrode where oxidation $a \to b$ takes place is referred to as the *anodic current*. If the density c_a of the reduced species a is kept constant near the electrode, the current is $I_A = ek_{b\leftarrow a}c_a$. The result (17.16) predicts that under the specified conditions (large E_r, small η) a logarithmic plot of the current with respect to $e\eta/(k_BT)$ increases linearly with the overpotential η, with a slope 1/2. This behavior is known in electrochemistry as Tafel's law, and the corresponding slope is related to the so called Tafel's slope.[5] An example where this "law" is quantitatively observed is shown in Fig. 17.2. In fact, a linear dependence of log(I) on the overpotential is often seen, however the observed slope can considerably deviate from $\frac{1}{2}$ and is sometimes temperature-dependent. Observed deviations are usually associated with the approximations made in deriving (17.16) from (17.15) and may be also related to the assumption made above that all the overpotential is realized as a potential drop between the molecule and the metal, an assumption that is better satisfied when the ionic strength of the solution increases.

17.1.6 Electron transfer at the semiconductor–electrolyte interface

With respect to electron transfer processes, semiconductor electrodes are different from their metal counterparts in two ways. First, the band structure, characterized by a band gap separating the conduction and valence band, will express

[4] Note that $|e\eta|$ should not be too small—the Tafel law holds only beyond a bias that satisfies $|e\eta| > k_BT$. When $\eta \to 0$ the net current which results from the balance between the direct and reverse reactions, must vanish like η. This implies that the Tafel behavior is always preceded by a low bias Ohmic regime.

[5] The Tafel slope is defined in the electrochemistry literature as $b \equiv (\partial\eta/\partial \log I)_{c,T} = 2.3k_BT/(\alpha e)$ where α is the slope defined above (sometimes referred to as the *transfer coefficient*), which this theory predicts to take the value 0.5.

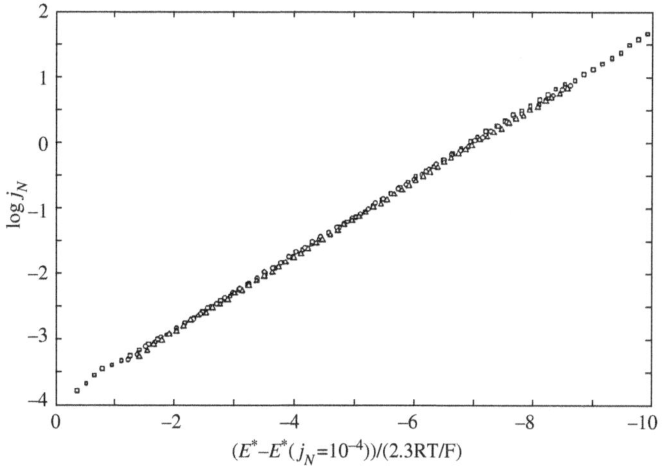

FIG. 17.2 Tafel plots for the (normalized, dimensionless) current, j_N, that accompanies hydrogen evolution in a solution containing 3.4 mM HCl + 1.0 M KCl, corrected for diffuse-double-layer effects, mass transport controlled kinetics and ohmic potential drop, measured at three temperatures (5, 45, 75°C; all results fall on the same line of this reduced plot) at a dropping mercury electrode. The slope obtained from this plot is 0.52, independent of temperature. (Based on data from E. Kirowa-Eisner, M. Schwarz, M. Rosenblum, and E. Gileadi, J. Electroanal. Chem. **381**, 29 (1995) and reproduced by the authors.)

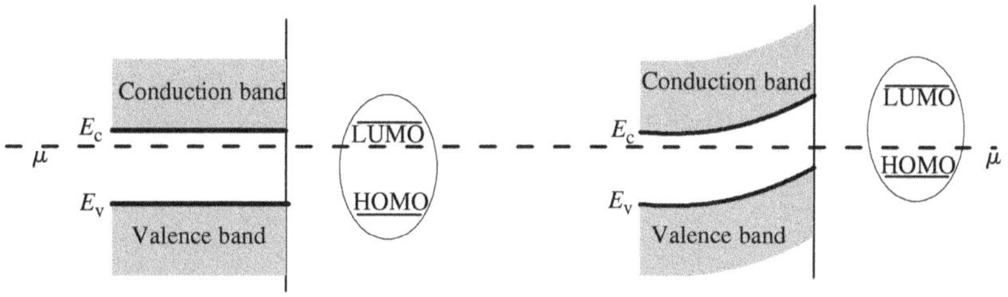

FIG. 17.3 An unbiased (left) and biased (right) semiconductor—electrolyte interface, showing the semiconductor to the left of the electrolyte, and a molecule represented by its HOMO and LUMO, respectively, in the solution. The potential bias (here the semiconductor is biased positively relative to the electrolyte) is assumed to fall on the semiconductor side of the interface. The dashed line represents the electronic chemical potential μ of the semiconductor.

itself in applying Eqs (17.11) and (17.12) to this interface. Second, in most cases the semiconductor screening length (see Section 4.5) is much larger than that of the electrolyte solution, implying that a given voltage between the bulks of these phases falls mostly on the semiconductor side of their interface (see Eq. (4.157)).

Let us assume for simplicity that all the potential falls on the semiconductor. In this case (Fig. 17.3) the energy levels on the electrolyte side remain unchanged with respect to the semiconductor valence and conduction band edges at the interface, which means that their alignment relative to the semiconductor chemical potential changes according to the given bias. If we assume that the electron transfer takes place between the molecule and the semiconductor surface, the rates equivalent to (17.14) are now

$$k_{b\leftarrow a} = \int_{-\infty}^{E_v+e\eta} dE\,\Gamma(E)\,(1-f(E))\,F\,(E_{ab}+e\eta-E) + \int_{E_c+e\eta}^{\infty} dE\,\Gamma(E)\,(1-f(E))\,F\,(E_{ab}+e\eta-E) \tag{17.17a}$$

and

$$k_{a\leftarrow b} = \int_{-\infty}^{E_v+e\eta} dE\Gamma(E)f(E)F\,(E_{ba}-e\eta+E) + \int_{E_c+e\eta}^{\infty} dE\Gamma(E)f(E)F\,(E_{ba}-e\eta+E) \tag{17.17b}$$

Again, in taking infinities as limits we assume that the energy range between the bottom of the valence band and the top of the conduction band fully encompasses the range in which the integrands in (17.17) are nonzero.

17.2 Molecular conduction

The last decade of the twentieth century has been revolutionary in the study of molecular electron transfer processes. For the preceding century scientists have investigated three types of such processes: transfer between a donor and an acceptor species, transfer between two sites on the same molecule and transfer between a molecular species in solution and a metal or a semiconductor electrode. The main observable in such studies is the electron transfer rate, though in studies of photoinduced electron transfer processes the quantum yield, defined as the number of electrons transferred per photon absorbed, is also a useful observable. The invention of the tunneling microscope and later experimental developments have now made it possible to investigate another manifestation of electron transfer: electronic conduction by a molecule connecting two bulk metal or semiconductor electrodes. In this section we outline the theoretical description of this phenomenon, its relationship to "traditional" electron transfer processes and its experimental realization.

17.2.1 Electronic structure models of molecular conduction

Figure 17.4 shows a cartoon of an experimental setup of a junction in which a molecule connects between two metal leads, themselves extensions of two bulk electrodes (sometimes referred to as "source" and "drain" leads). An ideal experimental setup will have an additional electrode, a "gate," which does not carry current but can affect the junction energetics by imposing an external potential. Such "three-terminal molecular devices" are now starting to be realized. Figure 17.5 shows a theoretical model for treating such systems. These are energy level diagrams that depict the right and left electrodes with their corresponding Fermi energies, and some of the molecular levels. Let us consider the different scenarios shown here. For simplicity of the following discussion we take the metals on the two sides to be the same and the temperature to be zero, and consider an independent electron model where electronic correlations are disregarded.

In Fig. 17.5(a) the junction is at equilibrium: the Fermi energies on the two sides are equal, $E_{FL} = E_{FR} = E_F$, and the molecular levels are arranged so that the energy of the highest occupied molecular orbital (HOMO) is below, while that of the lowest unoccupied molecular orbital (LUMO) is above E_F. Figures 17.5(b) and (c) show biased junctions. The potential bias Φ appears as the difference between the local Fermi-energies or the electrochemical potentials on the left and the right electrodes, $e\Phi = e(\Phi_L - \Phi_R) = E_{FR} - E_{FL}$, where e is the magnitude of the electron charge. When the bias is large enough so that either the HOMO or the LUMO enters into the energy window between the right and left Fermi energies, current can flow as indicated by the arrows. In Fig. 17.5(b) the LUMO level is first filled by the right electrode, then can transfer its electron to the left. In Fig. 17.5(c) an electron has to move first from the HOMO level to the right electrode, creating an electron vacancy that can be filled by an electron from the left. We sometimes use the names "electron conduction" and "hole conduction" for the processes depicted in Figs 17.5(b) and (c), respectively. As long as the potential bias between the two sides is maintained, current will flow through the system as the molecule tries (and fails) to reach equilibrium with both electrodes.

Several comments should be made at this point. First, the simple independent electron picture discussed above is only useful for qualitative descriptions of electron transport in such systems; electron–electron interactions and electronic correlations should be taken into account in realistic treatments. Second, the discussion above is appropriate at zero temperature. For finite temperature the Fermi energies should be replaced by the corresponding electron chemical potentials on the two sides, and the energy thresholds will be broadened by the

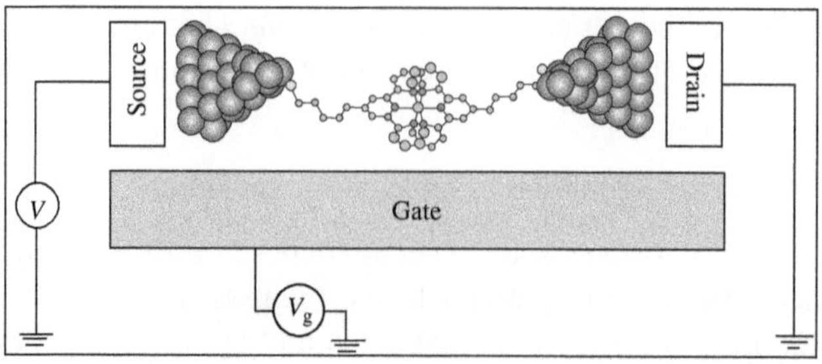

FIG. 17.4 A schematic three-terminal molecular junction, made of a molecule that connects between two (source and drain) electrodes, with a third electrode that functions as a gate. (From J. Park, A. N. Pasupathy, J. I. Goldsmith, C. Chang, Y. Yaish, J. R. Petta, M. Rinkoski, J. P. Sethna, H. D. Abruna, P. L. McEuen, and D. C. Ralph, Nature **417**, 722 (2002).)

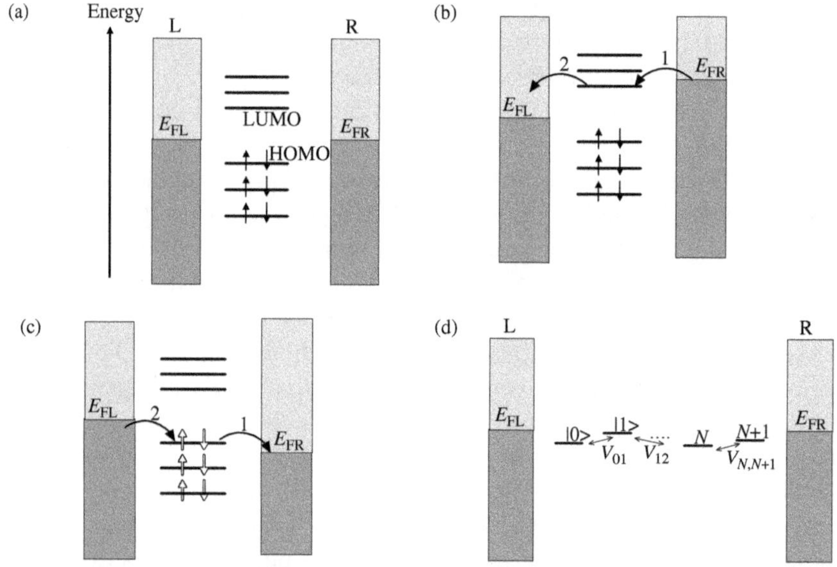

FIG. 17.5 Electron transmission through a molecular bridge connecting two metal leads. (a) An unbiased junction, (b) electron conduction, (c) hole conduction, (d) a local representation/nearest neighbor coupling model of a molecular bridge.

thermal distributions. Next, note that current can flow in the system also for potential bias smaller than the threshold needed to position the HOMO or LUMO in the window between the Fermi levels. This will be a tunneling current, assisted by the molecular levels but not occurring through them—an analog of the bridge assisted electron transfer in the super-exchange model discussed in Section 16.12.

Finally, we note that in Figs 17.5(a)–(c) the molecule is represented by the set of its eigenstates. It is often instructive to use an alternative picture, of local states (e.g. atomic orbitals) that are coupled to each other and to the electrodes, as seen in Fig. 17.5(d). This representation is akin to our description of bridge-assisted electron transfer in Section 16.12. In the simplest model of this kind it is assumed that only nearest neighbor states are coupled to each other, as indicated in Fig. 17.5(d). This is the nearest-neighbor tight binding approximation discussed in Section 4.3.4, also known in the chemistry literature as the Huckel model. The theoretical analysis described below can be carried using any of these representations.

17.2.2 Conduction of a molecular junction

In most models of molecular conduction the system is divided into three subregions associated with the two leads and the molecule(s) that bridge them. This division is not unique and different choices can be made according to the chosen calculation strategy. In particular, it is often advantageous to combine the molecular bridge, the molecule–metal bond, and small sections of the leads themselves into a "super-molecule" that connects between the remaining (still infinite) parts of the leads. In accordance, the Hamiltonian of the overall junction is written as

$$\hat{H} = \hat{H}_{\mathrm{S}} + \hat{H}_{\mathrm{L}} + \hat{H}_{\mathrm{R}} + \hat{H}_{\mathrm{SL}} + \hat{H}_{\mathrm{SR}} \tag{17.18}$$

where $\hat{H}_{\mathrm{S}}, \hat{H}_{\mathrm{L}}$, and $\hat{H}_{\mathrm{R}}$ are the Hamiltonians of the molecular bridge, the left, and the right leads, respectively, while $\hat{H}_{\mathrm{SL}}$ and $\hat{H}_{\mathrm{SR}}$ are the interactions between these subsystems. Direct interactions between the left and right leads are disregarded. These operators can be expressed in terms of any suitable basis sets that span the corresponding subsystems; most simply the eigenstates of $\hat{H}_{\mathrm{S}}, \hat{H}_{\mathrm{L}}$, and $\hat{H}_{\mathrm{R}}$ that are localized in the corresponding regions[6]

$$\hat{H}_{\mathrm{S}} = \sum_n E_n|n\rangle\langle n|; \qquad \hat{H}_{\mathrm{L}} = \sum_l E_l|l\rangle\langle l|; \qquad \hat{H}_{\mathrm{R}} = \sum_r E_{\mathrm{r}}|r\rangle\langle r| \tag{17.19a}$$

$$\hat{H}_{\mathrm{SL}} = \sum_{n,l} (H_{n,l}|n\rangle\langle l| + H_{l,n}|l\rangle\langle n|); \qquad \hat{H}_{\mathrm{SR}} = \sum_{n,r} (H_{n,r}|n\rangle\langle r| + H_{r,n}|r\rangle\langle n|) \tag{17.19b}$$

Alternatively, one often uses for the bridge a local representation, where the basis $\{|n\rangle\}$ comprises functions that are localized on different bridge segments.

---

6 In most practical applications different basis sets are used, for example, atomic orbitals associated or other wavefunctions localized in the different subsystems.

The model (17.19) remains the same, except that $\hat{H}_S$ is written in this non-diagonal representation as

$$\hat{H}_S = \sum_n \sum_{n'} H_{n,n'} |n\rangle\langle n'| \tag{17.19c}$$

By the nature of our problem, the molecular subsystem S is a finite system, and we will assume that it can be adequately described by a finite basis $\{|n\rangle\}, n = 1, 2, \ldots, N$. The leads are obviously infinite, at least in the direction of current flow, and consequently the eigenvalue spectra $\{E_l\}$ and $\{E_r\}$ constitute continuous sets that are characterized by density of states functions $\rho_L(E)$ and $\rho_R(E)$, respectively. Below we also use the index k to denote states belonging to either the L or the R leads.

From the results of Section 9.5.3 and Appendix 9C we can obtain an expression for the conduction of this model system. Indeed, using Eq. (9.139) and noting that the electron flux acquires an additional factor of 2 because of contributions from the two spin populations, we get the unidirectional transmitted flux per unit energy in the form

$$\left(\frac{dJ_{L\to R}(E)}{dE}\right)_{E=E_0} = \frac{1}{\pi\hbar}\mathcal{T}(E_0) f_L(E_0) \tag{17.20}$$

where we have identified the population $|c_0|^2$ with the Fermi–Dirac distribution

$$|c_0|^2 = f_L(E_0) = \frac{1}{e^{(E_0-\mu_L)/k_BT} + 1} \tag{17.21a}$$

$$\mu_L = \mu - e\phi_L \tag{17.21b}$$

(μ_L is the electrochemical potential of electrons in the left lead in the presence of a bias potential ϕ_L) and where $\mathcal{T}(E_0)$ is the "all-to-all" transmission coefficient at energy E_0

$$\mathcal{T}(E) = \sum_{\alpha,\alpha'} \mathcal{T}_{\alpha,\alpha'}(E) = \mathrm{Tr}_S\left[\hat{\Gamma}^{(L)}(E)\hat{G}^{(B)\dagger}(E)\hat{\Gamma}^{(R)}(E)\hat{G}^{(B)}(E)\right] \tag{17.22}$$

This transmission coefficient, a scattering property, contains all the dynamical information relevant to the conduction process under discussion. In (17.22) it is given in two forms:

1. As explained in Appendix 9C, the double-sum form of $\mathcal{T}(E)$ expresses its "all-to-all" nature. The initial and final states of the transmission process are enumerated by the indices α and α', respectively, that characterize the electronic states in directions perpendicular to the transmission process, and $\mathcal{T}_{\alpha,\alpha'}(E)$ is the coefficient of transmission between these states.

This notation is useful in particular when the perpendicular dimensions of the leads are finite so these states constitute discrete sets. When these dimensions become infinite these indices become continuous parameters that may denote the incident and outgoing directions of the transmission process.

2. The other form of $\mathcal{T}(E)$ in (17.22) is remarkable in that it is expressed as a trace of some operator over states defined on the scattering zone, that is, the molecular species S, only. The operator $\hat{\Gamma}(E)\hat{G}^{(B)\dagger}(E)\hat{\Gamma}(E)\hat{G}^{(B)}(E)$ is expressed in terms of the bridge's Green operator

$$G^{(B)}(E) = (E\hat{I}^{(B)} - \hat{H}^{(B)})^{-1}; \qquad \hat{H}^{(B)} = \hat{H}_S + \hat{B}; \qquad \hat{B} = \hat{B}^{(L)} + \hat{B}^{(R)} \tag{17.23}$$

$\hat{B}$ and $\hat{\Gamma}$ are the self energy operator and minus twice its imaginary part, defined by (cf. Eqs (9.133))

$$B_{n,n'}(E) = B^{(L)}_{n,n'}(E) + B^{(R)}_{n,n'}(E)$$

$$B^{(K)}_{n,n'}(E) \equiv \sum_{k\in K} \frac{H_{n,k}H_{k,n'}}{E - H_{k,k} + i\eta/2} = \Lambda^{(K)}_{n,n'}(E) - \frac{1}{2}i\Gamma^{(K)}_{n,n'}(E); \qquad K = L, R$$

$$\Gamma^{(K)}_{n,n'}(E) = 2\pi(H_{n,k}H_{k,n'}\rho_K(E_k))_{E_k=E}; \qquad k \in K, \quad K = L, R$$

$$\Lambda^{(K)}_{n,n'}(E) = PP\int_{-\infty}^{\infty} dE_k \frac{H_{n,k}H_{k,n'}\rho_K(E_k)}{E - E_k}; \qquad k \in K, \quad K = L, R \tag{17.24}$$

$\hat{\Gamma}$ is associated with the imaginary part of $\hat{B}$ via $\hat{\Gamma} = -2\mathrm{Im}\hat{B}$. Note that $\hat{B}$ and $\hat{\Gamma}$ are operators in the molecular bridge subspace. If this subspace is spanned by a finite basis of N states, then these operators are represented as $N \times N$ matrices. Examples of specific models are described below.

The net electronic current in the junction is now obtained by (1) multiplying the particle flux (17.20) by the electron charge $-e$, (2) taking the difference between the leftward and rightward fluxes, and (3) integrating over all energies. This yields

$$I = \frac{e}{\pi\hbar}\int_{-\infty}^{\infty} dE\mathcal{T}(E)\,(f_R(E) - f_L(E)) \tag{17.25}$$

Note that, as defined, the current will be positive (i.e. going from left to right with electrons flowing leftwards) when $\Phi = \Phi_L - \Phi_R$ is positive, that is, when the Fermi energy on the left electrode is lower then on the right. In metals $\mu \gg k_B T$, so the Fermi functions are nearly step functions.

Consider now the case where the potential bias Φ is small, $|e\Phi| \ll k_B T$. We can then use the expansion

$$f_{\rm R}(E) - f_{\rm L}(E) = \frac{1}{e^{\beta(E-\mu+e\Phi_R)}+1} - \frac{1}{e^{\beta(E-\mu+e\Phi_{\rm L})}+1}$$
$$\approx \delta(E-\mu)e\Phi \tag{17.26}$$

to get the final low bias expression, the *Landauer formula*[7]

$$I = \frac{e^2}{\pi\hbar}\mathcal{T}(E=\mu)\Phi \tag{17.27}$$

or

$$g \equiv \frac{I}{\Phi} = \frac{e^2}{\pi\hbar}\mathcal{T}(E=\mu) \tag{17.28}$$

Equation (17.27) implies that at low bias the junction response is linear: the current is proportional to the bias voltage and the proportionality coefficient is given by the conductance g, Eq. (17.28). It is given as the product of a universal constant

$$g_0 = \frac{e^2}{\pi\hbar} = (1.290 \times 10^4\ \Omega)^{-1} \tag{17.29}$$

and the all-to-all transmission coefficient evaluated at the electrode's chemical potential (or, at $T = 0$, at the electrode's Fermi energy). At finite bias one may define the voltage dependent *differential conductance*

$$g(\Phi) = \frac{dI}{d\Phi} \tag{17.30}$$

with I given by (17.25).

Equation (17.25), together with (17.22)–(17.24) and the definition of the Fermi functions

$$f_K(E) = \frac{1}{e^{(E-\mu+e\Phi_K/k_BT)}+1}; \qquad K = L, R \tag{17.31}$$

provide a theoretical framework for evaluating the conduction properties of molecular junctions. Molecular information (geometry and electronic structure) enters in these expressions through the elements of the Green operators while metal properties as well as the molecule–metal interaction enter via the self energy terms, all

[7] R. Landauer, IBM J. Res. Dev. **1**, 223 (1957); Phil. Mag. **21**, 863–867 (1970).

defined by Eqs (17.23) and (17.24). The actual calculations needed in order to evaluate these terms are however not simple. A reader familiar with standard molecular electronic structure calculations may appreciate the nature of the problem. Such calculations are usually done for a closed molecular system, where the number of electrons is a given constant. Here we require electronic structure information for a molecular bridge that is, (1) open to its electronic environment so that the electronic chemical potential, not the number of electrons, is given, and (2) in contact with two or more "electron baths" that may be out of equilibrium with each other, that is, characterized by different chemical potentials. A full solution of this problem should yield the total charge, the charge density distribution as well as the electrostatic potential distribution on the molecular bridge under the applied bias, and in addition to the matrix elements required for the evaluation of the transmission function $\mathcal{T}$ and the current I. An immediate result from such calculations is that the molecular electronic structure and consequently the function $\mathcal{T}$ depends, sometime critically, on the applied voltage.

This is a formidable problem, a subject of current research, which will not be discussed further here. Instead, we limit ourselves in what follows to some simple examples and qualitative observations.

17.2.3 The bias potential

As indicated above, when the junction is biased by a finite potential Φ Eq. (17.25) applies, however the transmission function $\mathcal{T}$ depends on the bridge's electronic structure which in turn depends on the bias Φ. To make this explicit we rewrite Eq. (17.25) in the form

$$I(\Phi) = \frac{e}{\pi\hbar}\int_{-\infty}^{\infty} dE\mathcal{T}(E,\Phi)\,(f_R(E,\Phi) - f_L(E,\Phi)) \tag{17.32}$$

At $T = 0$ this becomes

$$I(\Phi) = \frac{e}{\pi\hbar}\int_{\mu-e\Phi_L}^{\mu-e\Phi_R} dE\mathcal{T}(E,\Phi) = \frac{e}{\pi\hbar}\int_{e\Phi_R}^{e(\Phi_R+\Phi)} dE\mathcal{T}(\mu - E,\Phi) \tag{17.33}$$

with $\Phi_L - \Phi_R = \Phi$.

The actual way by which an imposed potential bias distributes itself on the molecular bridge depends on the molecular response to this bias, and constitutes part of the electronic structure problem. Starting from the unbiased junction in Fig. 17.6(a) (shown in the local representation of a tight binding model similar to

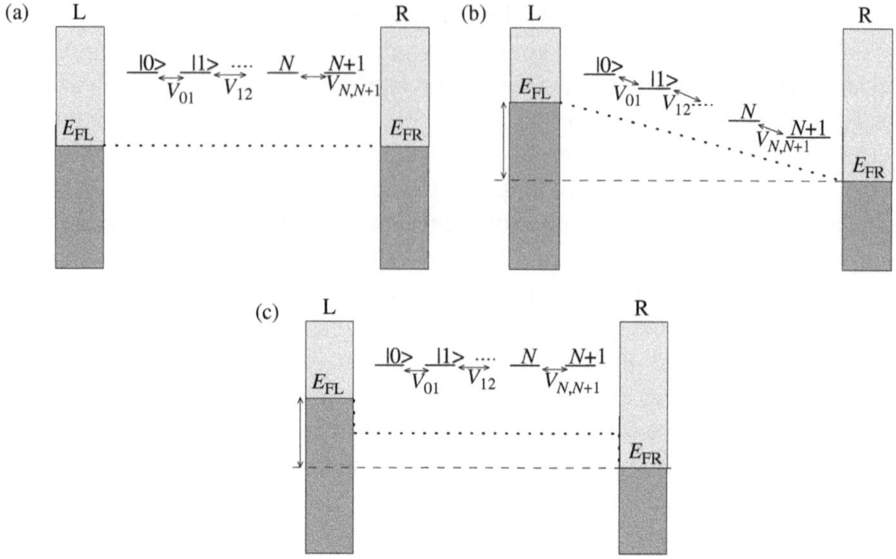

FIG. 17.6 Different scenarios for the way the potential bias is distributed (dotted line) along a molecular conductor. (a) An unbiased junction; (b) the potential drops linearly along the molecular bridge; (c) the potential drops only at the molecule–metal contacts.

Fig. 17.5(d)), two extreme limits can be distinguished. In Fig. 17.6(b) the potential drops linearly along the molecular bridge, as it would be if there was vacuum (or any other unpolarizable medium) between the electrodes. In Fig. 17.6(c) the potential drops only at the electrode–molecule interface, while remaining constant along the molecule. This behavior characterizes an easily polarizable object, for example, a metal rod, weakly bonded to the electrodes. These potential distributions are reflected in the energies associated with local electronic bridge orbitals as shown in Figs 17.6(b) and (c) that enter in the calculation of the transmission function via Eqs (17.22)–(17.24). A particular example is discussed in the following section.

17.2.4 The one-level bridge model

Further insight can be gained by considering specific cases of the transmission function $\mathcal{T}(E)$, Eq. (17.22). Here we consider the case where the bridge is adequately represented by a single level, $|1\rangle$, with energy E_1 (see Fig. 17.7). In this case the matrices $\hat{G}^{(B)}$ and $\hat{\Gamma}^{(K)}; K = L, R$ are scalar functions. From Eqs (17.23) and (17.24) we get

$$G^{(B)}(E) = \frac{1}{E - \tilde{E}_1 + (1/2)i\Gamma_1(E)} \tag{17.34}$$

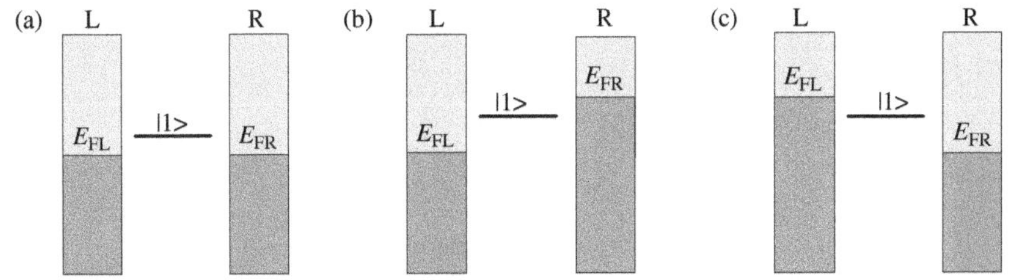

FIG. 17.7 A single-level bridge between two leads at different bias potentials: (a) An unbiased junction, (b) the right electrode is negatively biased, and (c) the left electrode is negatively biased.

$$\Gamma_1(E) = \Gamma_1^{(L)}(E) + \Gamma_1^{(R)}(E) \tag{17.35a}$$

$$\Gamma_1^{(K)}(E) = 2\pi \sum_{k\in K} |H_{1,k}|^2 \delta(E - E_k) = 2\pi \left(\overline{|H_{1,k}|^2 \rho_K(E_k)}\right)_{E_k=E}; \qquad K = L, R \tag{17.35b}$$

$$\tilde{E}_1 = E_1 + \Lambda_1(E) = E_1 + \Lambda_1^{(L)}(E) + \Lambda_1^{(R)}(E) \tag{17.35c}$$

$$\Lambda_1^{(K)} = \frac{1}{2\pi} PP \int_{-\infty}^{\infty} dE_k \frac{\Gamma_1(E_k)}{E - E_k}; \qquad k \in K, \quad K = L, R \tag{17.35d}$$

Equation (17.34) implies that $\mathcal{T}(E)$ is small except for E in the neighborhood (within a distance of order Γ_1) of $\tilde{E}_1$. If the function $\Gamma_1(E)$ does not change appreciably with E in this neighborhood it is reasonable to replace it by the constant $\Gamma_1 \equiv \Gamma_1(E_1)$. Consequently, in Eq. (17.35c) Λ_1 can be disregarded so that $\tilde{E}_1 = E_1$. We refer to this procedure as the *wide band approximation*, valid when the continua L and R are wide and the coupling $H_{1,k}$ ($k \in L, R$) is weakly dependent on k. From Eqs (17.22) and (17.34) it then follows that

$$\mathcal{T}(E) = \Gamma_1^{(L)}\Gamma_1^{(R)} G^{(B)\dagger}(E) G^{(B)}(E) = \frac{\Gamma_1^{(L)}\Gamma_1^{(R)}}{(E - E_1)^2 + [(1/2)\Gamma_1]^2} \tag{17.36}$$

in agreement with Eq. (9.91). The following observations can now be made:

1. The low bias conductance, Eq. (17.28) is given by

$$g = \frac{e^2}{\pi\hbar} \frac{\Gamma_1^{(L)}\Gamma_1^{(R)}}{(\mu - E_1)^2 + [(1/2)\Gamma_1]^2} \tag{17.37}$$

   where μ is the chemical potential of the electrons in the electrodes.
2. On resonance, where $E_1 = \mu$, and in the symmetric case, $\Gamma_1^{(L)} = \Gamma_1^{(R)} = (1/2)\Gamma_1$ the transmission coefficient is 1 irrespective of the coupling strength.

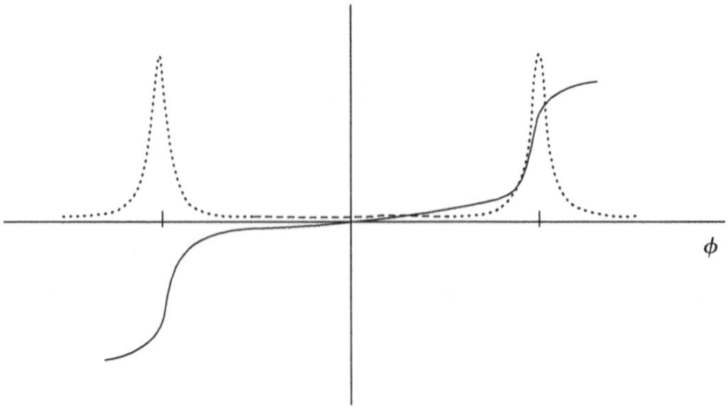

FIG. 17.8 Current I (full line) and differential conduction $g(\Phi) = dI/d\Phi$ (dotted line) displayed as a function of voltage for a junction characterized by a single resonance state.

The zero-bias conduction in this case is given by the "quantum conductance unit" $e^2/(\pi\hbar)$.

3. Viewed as a function of the potential bias Φ (the source–drain potential), the molecular conduction also shows resonance behavior. This can be realized by examining the integral in Eq. (17.33). The transmission function $\mathcal{T}(E,\Phi)$ is given by Eq. (17.36), perhaps with a bias-dependent resonance energy, $E_1 = E_1(\Phi)$, as explained in Section 17.2.3. At any bias Φ the magnitude of the integral in (17.33) depends on whether the resonance position $E_1(\Phi)$ is located in the integration window. When Φ is small, the resonance is outside this window and the resulting current is small. As Φ increases beyond some threshold value the resonance enters the integration window as seen in Fig. 17.7, so as a function of Φ we will see a step increase in the current. The current-voltage characteristic of this one level bridge model therefore assumes the form seen in Fig. 17.8. Obviously, the differential conductance $g(\Phi)$, Eq. (17.30), will show a peak at the same position where the $I(\Phi)$ steps up.
4. Viewed as a function of E_1 (that in principle can be changed with a gate potential), the conductance again shows an upward step whenever E_1 enters into the window between the two Fermi energies, and a downward step when it exits this range. For small bias ($e\Phi \ll \Gamma_1$) these positive and negative steps coalesce into a peak (in the conduction (or current) displayed against the gate voltage) whose width is determined by Γ_1 at low T.
5. The position of the resonance seen in Fig. 17.8 marks the onset potential beyond which the resonant level enters into the window between the left and right Fermi energies as seen in Fig. 17.7. The width of this conducting feature is determined at low temperature by the inverse lifetime, Γ_1, that an electron would remain on the resonant level if placed there at $t = 0$. An additional

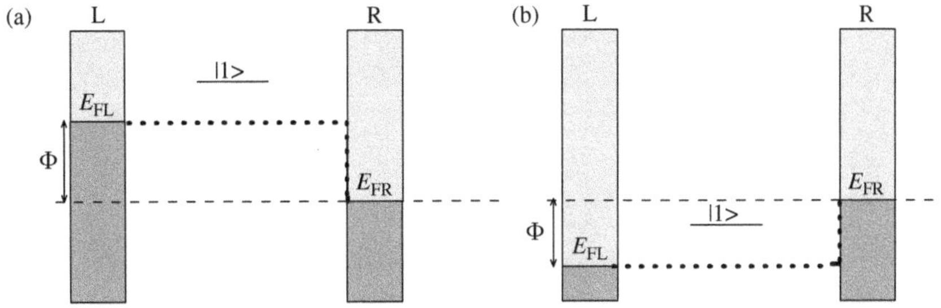

FIG. 17.9 The bias-dependent energy level structure in a one-level bridge model in which the bridge level is assumed pinned to the left electrode. The potential distribution in the junction is represented by the thick dotted line.

contribution to the width may arise from the bias dependence of E_1, and, at higher temperatures, from the increasing width of the Fermi step.

6. The entrance of the resonant level into the window between the left and right Fermi energies generally takes place under both directions of the bias potential as seen in Figs 17.7(b) and (c). This leads to the appearance of two resonance peaks in the conductance as shown in Fig. 17.8. However, depending on the way in which the bias potential is distributed along the junction, the $I(\Phi)$ behavior does not have to be symmetric under bias reversal. As an extreme example consider the situation depicted in Fig. 17.9. Here the potential bias drops exclusively at the interface between the molecule and the right lead and the position of E_1 relative to E_{FL} does not change. In this case we say that the molecular level is *pinned* to the left electrode, that is, the potential bias moves the energies E_1 and E_{FL} together. Alternatively the applied bias can be thought of as a change in E_{FR} while E_{FL} and E_1 remain fixed. Now, Fig. 17.9(a) describes the result of putting a positive bias on the right electrode. This cannot cause level 1 to enter the window between E_{FL} and E_{FR}. On the other hand, putting a negative bias on that electrode (Fig. 17.9(b)) does lead to such a situation. This then is a model for a current rectifier, however it should be kept in mind that reality is more complex since many more molecular levels, both occupied and empty, can contribute to the conduction.

17.2.5 A bridge with several independent levels

Consider next a model with several molecular levels bridging between the two electrodes, and assume that these levels contribute independently to the conduction. These independent levels can be identified as the molecular eigenstates, provided that their mixing by the coupling to the electrodes can be disregarded, that is, that non-diagonal elements in the molecular eigenstates representation of the self-energy

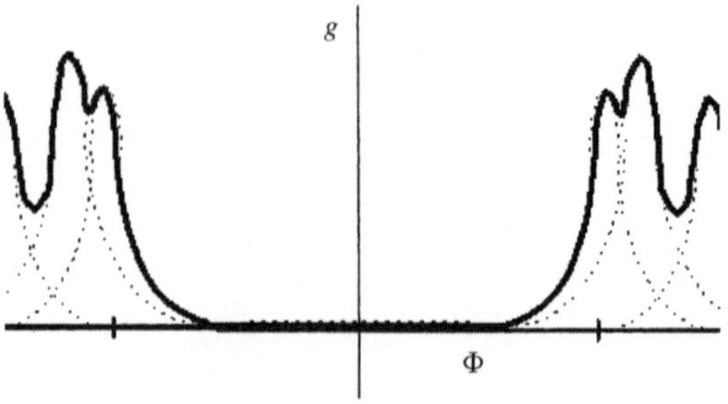

FIG. 17.10 A schematic display of molecular conduction. The full line represents the overall conductance of a molecular junction, as expected from Eqs (17.38), (17.25), and (17.30). The thin dotted lines trace individual resonances.

matrix, Eq. (17.24), are negligible. In this case the generalization of Eq. (17.36) is simply

$$\mathcal{T}(E) = \sum_j \frac{\Gamma_j^{(L)}\Gamma_j^{(R)}}{(E - E_j)^2 + [(1/2)\Gamma_j]^2} \tag{17.38}$$

where the sum is over all relevant bridge levels. Using this expression in (17.33) yields again the current voltage characteristic, with the same qualitative features of the conduction process that we found for a single resonant level, except that both occupied (HOMO and below) and unoccupied (LUMO and above) levels may contribute.

At small bias the Fermi levels on of the metal electrodes are positioned in the HOMO-LUMO gap of the bridge, as shown in Fig. 17.5(a). As the bias grows the current increases in steps as additional levels enter into the window between the electrodes' Fermi energies. The resulting conduction spectrum may look as in Fig. 17.10 (though the symmetry with respect to voltage inversion is a special case, expected only for ideally symmetric junction structures). The most prominent feature is the low voltage conduction gap that characterizes the molecular behavior as semiconducting in nature. As indicated by the discussion of Fig. 17.5, this gap may reflect the HOMO-LUMO gap of the molecule. Quantitative relationship is not to be expected both because the energy level structure of an isolated molecule is quite different from that of a molecule between metal surfaces and because the HOMO-LUMO gap of an isolated molecule reflects the electronic structure of the neutral molecule while the conduction gap is associated with the energy to put an excess electron or an excess hole on the molecule. It is also important to realize that the observed conduction gap may be unrelated to the distance between the molecular HOMO and LUMO. Figures 17.5 and 17.9 display a particular case (that characterizes scanning tunneling microscope geometries) where a bias potential moves the Fermi energy of one electrode relative to the molecular levels that remain "pinned" to the other electrode. In another common situation, which characterizes symmetric

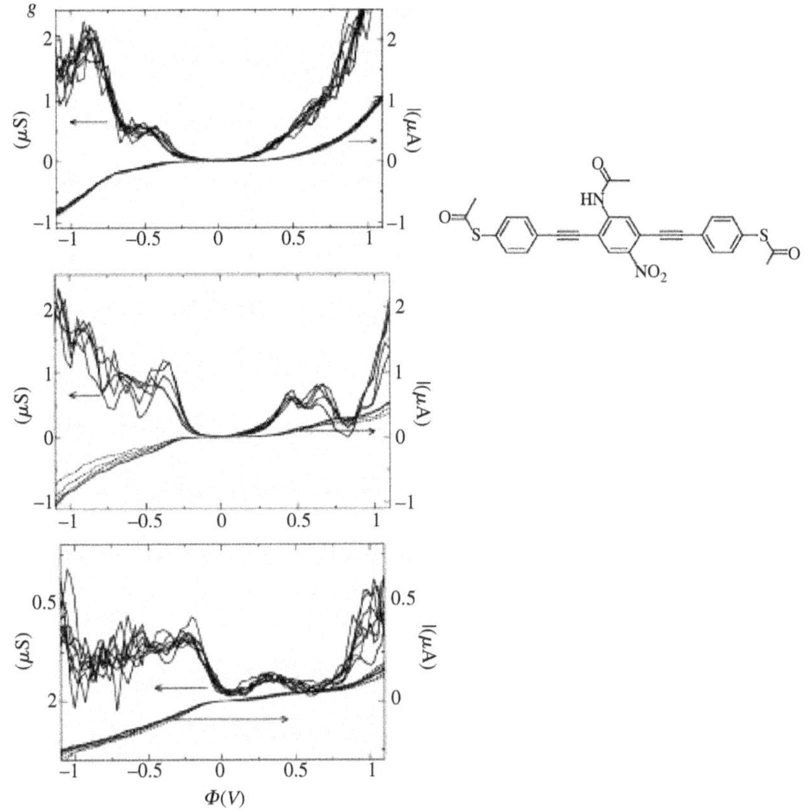

FIG. 17.11 The current (right axis) and conductance (left axis) of a molecular junction plotted against the applied voltage. Each plot shows several sweeps of the potential. The different plots correspond to different junctions prepared by the mechanically controlled break junction technique using gold contacts with the molecule shown. (From H. B. Weber, J. Reichert, F. Weigend, R. Ochs, D. Beckmann, M. Mayor, R. Ahlrichs, and H.v. Löhneysen, Chem. Phys. **281**, 113 (2002).)

junctions, the Fermi energies of the two electrodes are displaced symmetrically with respect to the molecular levels. It is easy to realize that in this case, exemplified by Fig. 17.8, the conduction gap is expected to be twice the smaller of the distances between the molecular HOMO and LUMO energies and the Fermi energy of the unbiased junction.

Beyond the onset of conduction, its voltage dependence is associated with the molecular level structure, though overlap between levels and experimental noise wash out much of this structure as shown schematically in figure 17.10.

As may be expected, experimental reality is not as neat as the results of theoretical toy models. Figure 17.11 shows such results obtained using a mechanically controlled break junction technique with gold contacts and the molecule shown.

The semiconducting character of molecular junctions, expressed by the low conduction regime at low bias followed by a conduction threshold, and the subsequent conduction spectrum beyond this threshold, are manifestations of the discrete quantum level structure of small molecular systems. However, the small system nature of molecular junctions inevitably expresses itself in the appearance of noise and relatively poor reproducibility.

17.2.6 Experimental statistics

Noise and poor reproducibility are the main drawbacks of using single molecule junctions as components of electronic devices. Their existence also suggests that useful analysis of experimental results must rely on the statistics of many experiments. Figure 17.12 shows an example. The conducting molecule is 1,8-octane-dithiol. Such molecules are inserted into a monolayer of octanethiol on a gold substrate. The other metallic conduct is a gold nanoparticle, which connects to the rest of the circuit via the tip of a conducting atomic force microscope. The resulting current–voltage signals from many such junctions fall into groups that can scale into a single line by dividing by different integers (see inset). This suggests that one main origin of irreproducibility in these junctions is the different numbers of dithiol molecules that are found between the gold substrate and the gold nanodot.

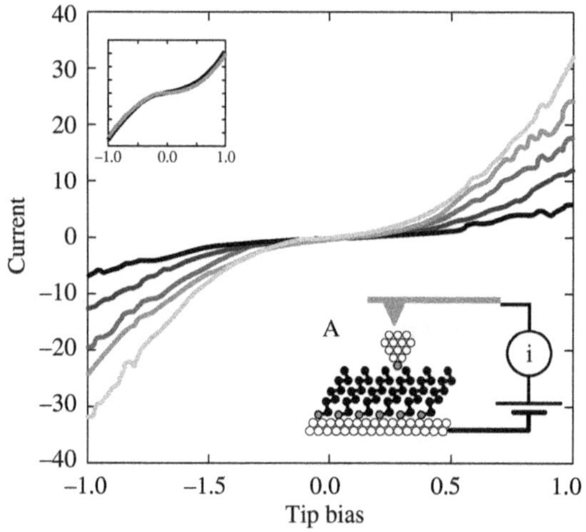

FIG. 17.12 Current–voltage characteristics of different gold–octanedithiol–gold junctions. The results fall into distinct groups that can be scaled into a single line (inset) by dividing by different integers. (From X. D. Cui, A. Primak, X. Zarate, J. Tomfohr, O. F. Sankey, A. L. Moore, T. A. Moore, D. Gust, G. Harris, and S. M. Lindsay, Science **294**, 571 (2001).)

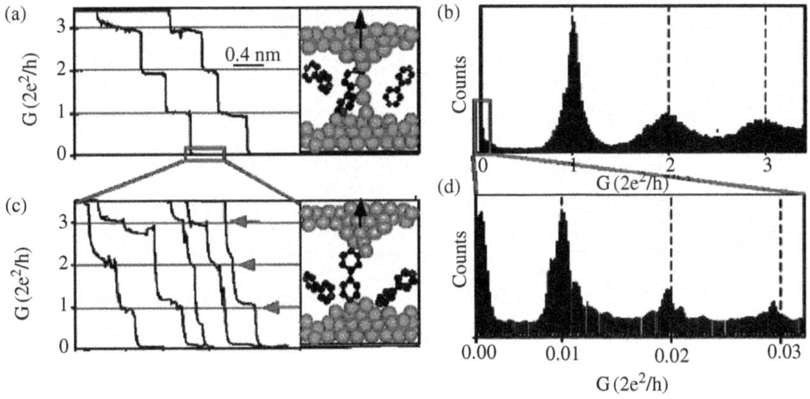

FIG. 17.13 Conductance of gold–bipyridine–gold junctions. (a) The decrease of conductance during pullout of a gold tip from a gold substrate in solution. (b) Conductance histogram for gold–gold contacts. (c) After the gold chain breaks, a much lower conductance that decreases in successive steps with pullout is observed, suggesting the presence of molecular bridges between the gold leads (d) 'Conductance histogram in the molecular regime, suggesting a transmission coefficient ~0.01 for the molecular bridge. (From B. Xu and N. J. Tao, Science **301**, 1221 (2003).)

Another example is shown in Fig. 17.13. Here individual molecular junctions are formed by repeatedly moving a gold scanning tunneling microscope into and out of contact with a gold substrate in a solution containing 4, 4′ bipyridine. Both in the absence and presence of this molecule one observes a stepwise decrease of the conductance when the tip is pulled away from contact, with successive steps occurring preferentially at an integer multiples of the conductance quantum $g_0 = e^2/\pi\hbar = 2e^2/h$ (Fig. 17.13(a)). A histogram constructed from many such results shows pronounced peaks at $1g_0$, $2g_0$, and $3g_0$ (Fig. 17.13(b)) suggesting that repeated loss of gold atoms, each contributing a conduction channel with near unity transmission, results in these steps. After the last chain of gold atoms is broken the signal appears to go to zero, but in fact the same phenomenon seems to repeat itself on a different scale, now with conductance units of $\sim 0.01\mathrm{G}_0$ (Figs 17.13(c) and (d)). The last phenomenon (which is not observed in the absence of bipyridine) suggests that after the breakup of the gold contacts, different numbers of bipyridine molecules may still connect between the separated gold leads; the number decreases in successive order as pulling continues. The conduction histogram in this regime shows pronounced peaks at $1\times$, $2\times$, and $3 \times 0.01\mathrm{G}_0$.

17.2.7 The tight-binding bridge model

As a last example consider a model of molecular bridge similar to that used in the electron transfer problem in Section 16.12. We consider the model of Fig. 17.6(a) where a tight-binding model for the molecular bridge in the local representation comprises $N + 2$ levels with nearest-neighbor coupling. Furthermore, we focus on the low bias case and assume that the bridge levels are energetically distinct from the leads' Fermi energy. The local character of these bridge levels makes it reasonable to disregard intersite coupling beyond nearest neighbor. In particular, we assume that only level 0 couples to the left lead and only level $N + 1$ couples to the right lead. The Hamiltonian (17.18) and (17.19) now takes the form

$$\hat{H} = \hat{H}_{\mathrm{L}} + \hat{H}_{\mathrm{R}} + \hat{H}_{\mathrm{S}} + \hat{H}_{\mathrm{SL}} + \hat{H}_{\mathrm{SR}} \tag{17.39}$$

where

$$\begin{aligned}\hat{H}_{\mathrm{S}} =& E_0|0\rangle\langle 0| + \sum_{j=1}^{N} E_j|j\rangle\langle j| + E_{N+1}|N+1\rangle\langle N+1| \\ &+ V_{01}|0\rangle\langle 1| + V_{10}|1\rangle\langle 0| + V_{N+1,N}|N+1\rangle\langle N| + V_{N,N+1}|N\rangle\langle N+1| \\ &+ \sum_{j=1}^{N-1}(V_{j,j+1}|j\rangle\langle j+1| + V_{j+1,j}|j+1\rangle\langle j|)\end{aligned} \tag{17.40}$$

is the bridge (molecular) Hamiltonian,

$$\hat{H}_{\mathrm{L}} = \sum_l E_l|l\rangle\langle l|; \qquad \hat{H}_{\mathrm{R}} = \sum_r E_r|r\rangle\langle r| \tag{17.41}$$

are Hamiltonians of the free (or Bloch) electron states on the left and right electrodes, and

$$\begin{aligned}\hat{H}_{\mathrm{SL}} &= \sum_l (H_{0,l}|0\rangle\langle l| + H_{l,0}|l\rangle\langle 0|); \\ \hat{H}_{\mathrm{SR}} &= \sum_r (H_{N+1,r}|N+1\rangle\langle r| + H_{r,N+1}|r\rangle\langle N+1|)\end{aligned} \tag{17.42}$$

are the operators that couple the bridge with the left and right leads. We will use E_{S}, V_{S}, and E_{F} to denote the order of magnitudes of the bridge energies (E_0, E_1, ..., E_{N+1}), the bridge couplings ($V_{0,1}$, ..., $V_{N,N+1}$) and Fermi energies (E_{FL}, E_{FR}), respectively, so the model assumption concerning the energetic separation of the bridge from the Fermi energies can be expressed by the inequality (analog of (16.104)

$$|E_{\mathrm{S}} - E_{\mathrm{F}}| \gg V_{\mathrm{S}} \tag{17.43}$$

In the local representation the self-energy matrix $\hat{B}$, Eq. (17.24), is an $(N + 2) \times (N + 2)$ matrix with only two nonzero terms, $B_{0,0} = -(1/2)i\Gamma_0^{(L)}$ and $B_{N+1,N+1} =$

$-(1/2)i\Gamma_{N+1}^{(R)}$, where

$$\begin{aligned}\Gamma_0^{(L)} &\equiv \Gamma_{0,0}^{(L)} = 2\pi(|H_{0,l}|^2\rho_L(E_l))_{E_l=E_S} \\ \Gamma_{N+1}^{(R)} &\equiv \Gamma_{N+1,N+1}^{(R)} = 2\pi(|H_{N+1,r}|^2\rho_R(E_r))_{E_r=E_S}\end{aligned} \tag{17.44}$$

Note that in writing Eqs (17.44) we have invoked the wide band approximation (Section 9.1). The bridge Green's function, Eq. (17.23), then satisfies

$$(\hat{G}^{(B)}(E))^{-1} = \begin{pmatrix} E-E_0+(1/2)i\Gamma_0^{(L)} & V_{01} & 0 & 0 & \cdots & 0 \\ V_{10} & E-E_1 & V_{12} & 0 & \cdots & 0 \\ 0 & V_{21} & E-E_2 & V_{23} & & \vdots \\ 0 & 0 & V_{32} & \ddots & \ddots & 0 \\ \vdots & \vdots & & \ddots & E-E_N & V_{N,N+1} \\ 0 & 0 & \cdots & 0 & V_{N+1,N} & E-E_{N+1}+(1/2)i\Gamma_{N+1}^{(R)} \end{pmatrix} \tag{17.45}$$

and the transmission coefficient, Eq. (17.22), evaluated at the electrodes' electrochemical potential μ (or Fermi energy E_F) takes the form

$$\mathcal{T}(E_F) = Tr_B[\hat{\Gamma}^{(L)}\hat{G}^{(B)\dagger}(E_F)\hat{\Gamma}^{(R)}\hat{G}^{(B)}(E_F)] = \Gamma_0^{(L)}\Gamma_{N+1}^{R}|G_{0,N+1}^{(B)}(E_F)|^2 \tag{17.46}$$

The inequality (17.43) suggests that the needed Green function matrix element can be evaluated to lowest order in $|V_S/(E_S - E_F)|$ using the Dyson expansion (compare Eq. (16.115))

$$\begin{aligned}G_{0,N+1}^{(B)}(E_F) =& \frac{1}{(E_F - E_0 + (1/2)i\Gamma_0^{(L)})}V_{01} \\ &\times \frac{1}{(E_F - E_1)}V_{12}\frac{1}{(E_F - E_2)}V_{23}\cdots\frac{1}{(E_F - E_{N-1})}V_{N-1,N}\frac{1}{(E_F - E_N)} \\ &\times V_{N,N+1}\frac{1}{(E_F - E_{N+1} + (1/2)i\Gamma_{N+1}^{(R)})}\end{aligned} \tag{17.47}$$

This can be rewritten in the form

$$G^{(B)}_{0,N+1}(E_{\mathrm{F}}) = \frac{1}{(E_{\mathrm{F}} - E_0 + (1/2)i\Gamma_0^{(L)})} \times V_{01}G^{(B)}_{1,N}(E_{\mathrm{F}})V_{N,N+1}\frac{1}{(E_{\mathrm{F}} - E_{N+1} + (1/2)i\Gamma^{(R)}_{N+1})} \tag{17.48}$$

leading to the final expression for the low bias conduction

$$g = \frac{e^2}{\pi\hbar}|V_{01}V_{N,N+1}|^2|G^{(B)}_{1,N}(E_{\mathrm{F}})|^2\chi$$

$$\chi = \frac{\Gamma_0^{(L)}\Gamma^{(R)}_{N+1}}{((E_{\mathrm{F}} - E_0)^2 + (\Gamma_0^{(L)}/2)^2)((E_{\mathrm{F}} - E_{N+1})^2 + (\Gamma^{(R)}_{N+1}/2)^2)}$$

$$\approx \frac{\Gamma_0^{(L)}\Gamma^{(R)}_{N+1}}{((E_{\mathrm{F}} - E_S)^2 + (\Gamma_0^{(L)}/2)^2)((E_{\mathrm{F}} - E_S)^2 + (\Gamma^{(R)}_{N+1}/2)^2)} \tag{17.49}$$

It is instructive to compare this result for the conduction through an $N + 2$ level bridge to the parallel expression for the bridge-assisted electron transfer rate (cf. Eqs (16.114) and (16.113)

$$k_{\mathrm{A}\leftarrow\mathrm{D}} = \frac{2\pi}{\hbar}|V_{\mathrm{D}1}V_{N\mathrm{A}}|^2|G^{(B)}_{1N}(E_{\mathrm{A/D}})|^2F(\Delta E)$$

$$F(\Delta E) = \frac{1}{\sqrt{4\pi E_r k_B T}}\exp\left(-\frac{(\Delta E - E_r)^2}{4E_r k_B T}\right) \tag{17.50}$$

Both were obtained in the "super-exchange" limit, where inequalities (16.104) and (17.43) are satisfied. Both involve factors that convey information about the electronic structure of the $(N + 2)$-level molecular system, where the donor and acceptor levels D and A in the electron transfer case are replaced by the levels 0 and $N + 1$ that couple directly to the electrodes. These factors are essentially the same: a product of coupling elements from the edge levels onto the center bridge, and a Green function element connecting sites 1 and N of this center bridge. In the electron transfer case this Green function element is calculated at the energy $E_{\mathrm{A/D}}$ at which the Marcus parabolas cross, while in the low bias conduction case it is evaluated at the electronic chemical potential of the electrodes. In both cases these are the energies at which the electron transfer process actually takes place.

The main difference between these expressions lies in the factors χ of Eq. (17.49) and F of Eq. (17.50), that express the different ways by which the processes are terminated. The word "termination" is used here to express the relaxation mechanism that makes the transition a $\rightleftharpoons$ b between two stable species a and b different from the two-state dynamics of Section 2.2. In the latter case, starting from state a, the system oscillates coherently and forever between states a and b. In contrast, the present case corresponds to a process where the a $\rightarrow$ b transition is followed by a fast stabilization (or "termination") that establishes b as a distinct species and makes the reverse transition b $\rightarrow$ a an independent rate process.

In the electron transfer case, the termination of the D$\rightarrow$A (or A$\rightarrow$D) electron transfer process is caused by the solvent reorganization about the newly formed charge distribution. This reorganization dissipates electronic energy into a continuum of nuclear modes and establishes both sides of the redox reaction as distinct stable species, so that the direct and reverse processes have no memory of each other and proceed independently. This stabilization also implies that the transition must be thermally activated. In the rate expression (17.50) this process expresses itself via the factor F that depends on the reorganization energy and on the temperature through a distinct activation term.

In the conduction process the termination has a completely different origin. Instead of a continuum of nuclear modes, an electron on the edge levels 0 and $N+1$ sees the continuum of quasi-free electronic states of the metal electrodes. An electron reaching one of these edge states can proceed into the electrodes and is lost as far as the transport process is concerned. The process is thus irreversible in the sense that the reverse process again takes place completely independently. In expression (17.49) for the conduction this termination expresses itself via the factor χ that depends on the inverse lifetimes, $\Gamma_0^{(L)}$ and $\Gamma_{N+1}^{(R)}$, for electrons on the edge states to decay into the metals.

The appearance of the factor

$$G_{1,N}^{(B)}(E_F) = \frac{1}{E_F - E_1} V_{12} \frac{1}{E_F - E_2} V_{23} \cdots \frac{1}{E_F - E_{N-1}} V_{N-1,N} \frac{1}{E_F - E_N}$$
$$\approx \frac{1}{V_S} \left(\frac{V_S}{E_F - E_S} \right)^N \tag{17.51}$$

in Eq. (17.49) has the same important experimental implication for the bridge-length dependence of the low bias conductance as we found before (see Eq. (16.117)) for the bridge-assisted electron transfer rate, namely both depend exponentially on this

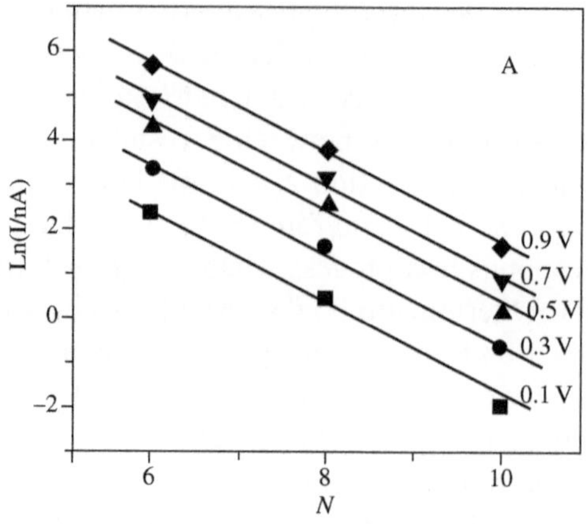

FIG. 17.14 Same experiment as described in Fig. 17.12 using as molecular species three alkanes: hexanedithiol ($N = 6$), octanedithiol ($N = 8$) and decanedithiol ($N = 10$). The lowest peaks in conductance histograms are $0.0012g_0$, $0.00025g_0$, and $0.00002g_0$, which satisfy Eq. (17.52) with $\beta' b = 1.00 + 0.05$. Note the weak dependence of β' on the applied voltage.

length. In analogy to Eq. (16.117) we find

$$g \sim e^{-\beta' L} \tag{17.52a}$$

$$\beta' = -\frac{2}{b} \ln \left| \frac{V_S}{E_F - E_S} \right| \tag{17.52b}$$

where b is the length of a bridge unit and $L = bN$ is the bridge length. Figure 17.14 shows an experimental example of this behavior.

Finally we note that our treatment of conduction was based on viewing the electron transport as a coherent quantum-mechanical transition that acquires an irreversible character only because of the dissipative nature of the macroscopic electrodes. As in the electron transfer case, another mode of conduction exists at higher temperatures and longer chain lengths, namely thermally activated hopping. In this case, the exponential dependence on bridge length, Eq. (17.52), is replaced by the algebraic dependence

$$g \sim \frac{1}{\alpha_1 + \alpha_2 L} \tag{17.53}$$

that originates from solving hopping equations similar to (16.118) (compare Eq. (16.127)). We see that g becomes inversely proportional to the length L for long bridges, establishing a connection to the macroscopic Ohm's law.

Further reading

Electrode processes

R. J. D. Miller, G. L. McLendon, A. J. Nozik, W. Schmickler, and F. Willig, *Surface Electron Transfer Processes* (VCH, New York, 1995).

W. Schmickler, *Interfacial Electrochemistry* (Oxford University Press, Oxford, 1996).

Molecular Conduction

G. Cuniberti, G. Fagas, and K. Richter, eds, *Introducing Molecular Electronics* (Springer, Berlin, 2005).

A. Aviram and M. A. Ratner, Chem. Phys. Lett. **29**, 277 (1974).

S. Datta, *Electric Transport in Mesoscopic Systems* (Cambridge University Press, Cambridge, 1995).

S. Datta, Quantum Transport: Atom to Transistor, (Cambridge University Press, Cambridge, 2005).

A. Nitzan, Electron transmission through molecules and molecular interfaces, Ann. Rev. Phys. Chem. **52**, 681 (2001).

A. Salomon, D. Cahen, S. Lindsay, J. Tomfohr, V. B. Engelkes, and C. D. Frisbie, Comparison of Electronic Transport Measurements on Organic Molecules, Adv. Mater. **15**, 1881 (2003)

18

SPECTROSCOPY

Since, without light, color cannot exist,
And since the atoms never reach the light,
They must be colorless. In the blind dark
What color could they have? Even in a bright day
Hues change as light-fall comes direct or slanting.
The plumage of a dove, at nape or throat,
Seems in the sunlight sometimes ruby-red
And sometimes emerald-green suffused with coral.
A peacock's tail, in the full blaze of light,
Changes in color as he moves and turns.
Since the light's impact causes this, we know
Color depends on light...

Lucretius (c.99–c.55 BCE*) "The way things are"*
translated by Rolfe Humphries, Indiana University Press, 1968

The interaction of light with matter provides some of the most important tools for studying structure and dynamics on the microscopic scale. Atomic and molecular spectroscopy in the low pressure gas phase probes this interaction essentially on the single particle level and yields information about energy levels, state symmetries, and intramolecular potential surfaces. Understanding environmental effects in spectroscopy is important both as a fundamental problem in quantum statistical mechanics and as a prerequisite to the intelligent use of spectroscopic tools to probe and analyze molecular interactions and processes in condensed phases.

Spectroscopic observables can be categorized in several ways. We can follow a temporal profile or a frequency resolved spectrum; we may distinguish between observables that reflect linear or nonlinear response to the probe beam; we can study different energy domains and different timescales and we can look at resonant and nonresonant response. This chapter discusses some concepts, issues, and methodologies that pertain to the effect of a condensed phase environment on these observables. For an in-depth look at these issues the reader may consult many texts that focus on particular spectroscopies.[1]

---

[1] A unified theoretical approach to many of these phenomena is provided in the text by Mukamel cited at the end of this chapter.

18.1 Introduction

With focus on the optical response of molecular systems, effects of condensed phase environments can be broadly discussed within four categories:

1. Several important effects are equilibrium in nature, for example spectral shifts associated with solvent induced changes in solute energy levels are equilibrium properties of the solvent–solute system. Obviously, such observables may themselves be associated with dynamical phenomena, in the example of solvent shifts it is the dynamics of solvation that affects their dynamical evolution (see Chapter 15). Another class of equilibrium effects on radiation–matter interaction includes properties derived from symmetry rules. A solvent can affect a change in the equilibrium configuration of a chromophore solute and consequently the associated selection rules for a given optical transition. Some optical phenomena are sensitive to the symmetry of the environment, for example, surface versus bulk geometry.
2. The environment affects the properties of the radiation field; the simplest example is the appearance of the dielectric coefficient ε in the theory of radiation–matter interaction (e.g. see Eq. (3.27)). Particularly interesting are effects derived from the behavior of the local electromagnetic field in inhomogeneous environments. For example, Raman scattering by molecules adsorbed on rough surfaces of certain metals is strongly affected by the fact that the local field can be strongly enhanced at certain wavelengths for distances comparable to the characteristic lengthscale of roughness features.
3. The environment induces relaxation processes of many kinds and at many timescales. Spectroscopy, a sensitive probe of molecular energy levels, their populations, and sometimes their relative phases, is also a sensitive probe of population and phase relaxation.
4. Spectroscopy is a sensitive probe of the interactions between chromophores that can be very important at distances characteristic of condensed phases. Excitation transfer between chromophores is a simple example. In pure phases of excitable molecules we observe coherent many-body effects. The interesting excitation transport, localization and dephasing phenomena that take place in such systems are largely beyond the scope of this text but they are obviously important manifestations of optical response of condensed phase molecular systems.

In addition, two other aspects of spectroscopy in condensed phases should be mentioned. First, the density of chromophore molecules is usually high enough that many such molecules may exist within the coherence length of the incident radiation field and consequently respond coherently as a group. This gives rise to optical response phenomena associated with the formation and subsequent destruction of a relative phase between chromophore molecules. On the other hand, in low temperature condensed phases molecules are localized within their small neighborhood making it possible, using advanced techniques, to monitor the optical response of single molecules.

Some fundamental concepts pertaining to our subject were discussed in earlier chapters. The necessary concepts from electromagnetic theory and radiation–matter interaction were discussed in Chapter 3. A simple framework suitable for treating linear spectroscopy phenomena was described in Sections 9.2 and 9.3. A prototype model for many problems in optical spectroscopy involves two electronic states, ground and excited, and at least two continuous manifolds of states associated with the radiative and nonradiative environments. Such models were discussed in Sections 9.3 and 10.5.2.

In these discussions we have used two representations for the radiation field and its interaction with the molecular system: (1) In the derivation of the Bloch equations in Section 10.5.2 we described the radiation field as an external oscillating field $\mathcal{E}(t) = \mathcal{E}_0 \cos \omega t$ that couples to the system dipole operator, resulting in the term $\bar{V} \to -\mathcal{E}(t)\hat{\mu}$ in the Hamiltonian. (2) In Section 9.3 we have used a picture based on a truncated *dressed states* basis $|m, \mathbf{p}\rangle$ of zero-order states, where m stands for the molecular state and $\mathbf{p}$ denotes the state of the radiation field expressed in terms of the number of photons $p_{\mathbf{k}}$ in each mode $\mathbf{k}$.[2] The usefulness of the latter approach stems from the fact that the corresponding equations of motion are derived from time-independent Hamiltonians and from the ability to use truncated basis representations tailored to given physical conditions. For example, in linear spectroscopy problems and near resonance between the incident radiation and a particular molecular transition, we can use a model that involves only molecular states that are coupled by the radiation field (and possible other states that provide relaxation channels) and apply arguments based on the rotating wave approximation (see Section 10.5.2) to consider only transitions that (nearly) conserve energy. The study of molecular absorption then involves the interaction between a state $|g, 1_{\mathbf{k}}\rangle$—a lower energy molecular state dressed by one photon in mode $\mathbf{k}$ of frequency $\omega = kc$ (c is the speed of light), and the state $|s, 0\rangle$—an higher-energy molecular state with no photons. The energies of the bare molecular states $|g\rangle$ and $|s\rangle$ satisfy $E_s - E_g \simeq \hbar\omega$.[3]

[2] To simplify notation we will specify the photon state by the wavevector $\mathbf{k}$, and suppress, unless otherwise needed, the polarization vector $\boldsymbol{\sigma}$.

[3] Note that in Section 9.3 we have used the notations g and s for the ground and excited states. Here we use 1 and 2 for the ground and excited electronic states, keeping labels such as s, l, and g for individual vibronic levels.

It should be noted that instead of dressing the molecular states g and s by 1 and 0 photons, respectively, we could use any photon numbers p and $p-1$. The corresponding matrix elements are than proportional to p. In processes pertaining to linear spectroscopy it is convenient to stick with photon populations 1 and 0, keeping in mind that all observed fluxes should be proportional to the incident photon number p or, more physically, to the incident field intensity $|\mathcal{E}_0|^2$. With this in mind we will henceforth use the notation $|g, \mathbf{k}\rangle$ (or $|g, \omega\rangle$ if the incident direction is not important for the discussion) as a substitute for $|g, 1_{\mathbf{k}}\rangle$.

In Section 9.3 we have used this truncated dressed state picture to discuss photo-absorption and subsequent relaxation in a model described by a zero-order basis that includes the following states: a molecular ground state with one photon of frequency ω, $|0\rangle = |g, \omega\rangle$, an excited "doorway" state with no-photons, $|s, 0\rangle$, and a continuous manifold of states $\{|l\rangle\}$ that drives the relaxation. This model is useful for atomic spectroscopy, however, in molecular spectroscopy applications it has to be generalized in an essential way—by accounting also for molecular nuclear motions. In the following section we make this generalization before turning to consider effects due to interaction with the thermal environment.

18.2 Molecular spectroscopy in the dressed-state picture

Because processes of interest to us take place in condensed phases, we can usually exclude rotational levels from our discussion: gas phase rotational motions become in the condensed phase librations and intermolecular vibrations associated with the molecular motion in its solvent cage.[4]

We therefore envision the molecule as an entity characterized by its vibronic spectrum, interacting with a dissipative environment. As indicated above, a useful characteristic of the truncated dressed state approach is the simplification provided by considering only states that pertain to the process considered.[5] In experiments involving a weak incident field of frequency ω these states are found in the energy range about $\hbar\omega$ above the molecular ground state. For simplicity we assume that in this range there is only one excited electronic state. We therefore focus in what follows on a model characterized by two electronic states (See Fig. 18.1) and include also their associated vibrational manifolds. The lower molecular state $|g\rangle$

[4] In some cases involving diatomic hydrides HX embedded in solid low temperature matrices rotational motion is only slightly perturbed and explicit consideration of this motion is useful, see Chapter 13.

[5] One should keep in mind the dangerous aspect of this practice: In selecting the model we already make an assumption about the nature of the process under discussion. Our results are then relevant to the chosen model, and their relevance to the physical system under discussion is assumed.

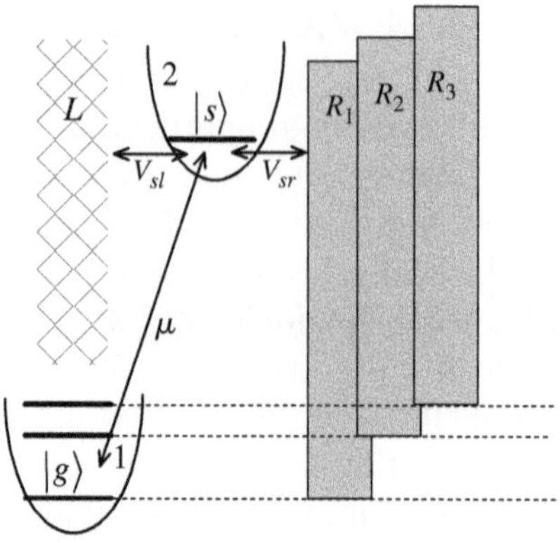

FIG. 18.1 A dressed-state model that is used in the text to describe absorption, emission, and elastic (Rayleigh) and inelastic (Raman) light scattering. $|g\rangle$ and $|s\rangle$ represent particular vibronic levels associated with the lower (1) and upper (2) electronic states, respectively. These are levels associated with the nuclear potential surfaces of electronic states 1 and 2 (schematically represented by the parabolas). R_j are radiative continua—1-photon-dressed vibronic levels of the lower electronic states. The quasi-continuum L represents a nonradiative channel—the high-energy regime of the vibronic manifold of electronic state 1. Note that the molecular dipole operator $\hat{\mu}$ couples ground (g) and excited (s) molecular states, but the ensuing process occurs between quasi-degenerate dressed states $|g, k\rangle$ and $|s, 0\rangle$.

now becomes a manifold of vibronic levels $|1, \mathbf{v}_1\rangle$ associated with the ground electronic state $|1\rangle$ and shown on the left of the figure. For a large molecule[6] and high energy this manifold becomes dense and turns into the quasi-continuum[7] of levels $L = \{|l\rangle\}$, also shown in the figure. Each of these vibronic levels provides a baseline for a radiative continuum $|1, \mathbf{v}_1, \mathbf{k}\rangle$ (see footnote 2). The first few radiative continua, denoted R_1, R_2, and R_3, are shown in the right of the figure. In a similar way, the higher-energy molecular level $|s\rangle$ has now become a manifold of vibronic levels $|2, \mathbf{v}_2\rangle$ associated with the excited electronic state $|2\rangle$.

Before continuing, we pause for a remark on notation. Our discussion uses a basis of zero-order states that are eigenfunctions of a Hamiltonian given by a sum of the molecular Hamiltonian in the Born–Oppenheimer approximation and the Hamiltonian of the radiation field. As noted above, for linear spectroscopy problems we can limit ourselves to the 0- and 1-photon states of the latter. We will use

[6] Depending on the particular process under observation, what we call a "molecule" here can be a "supermolecule" comprised of the entity of interest together with its immediate environment.

[7] The term "quasi-continuum" is used to describe a dense discrete manifold of energy levels under circumstances where the effective broadening of these level (from physical origins or because of poor experimental resolution) is larger than their spacing.

$|1, \mathbf{v}_1\rangle$ and $|2, \mathbf{v}_2\rangle$ interchangeably with $|1, \mathbf{v}_1, 0\rangle$ and $|2, \mathbf{v}_2, 0\rangle$ to denote zero-photon vibronic states, while vibronic levels of the ground electronic state dressed by one photon **k** will be written $|1, \mathbf{v}_1, \mathbf{k}\rangle$. When convenient, we will also continue to use $|g\rangle$ and $|s\rangle$ to denote particular vibronic levels $|1, \mathbf{v}_1\rangle$ and $|2, \mathbf{v}_2\rangle$, respectively, in the lower and upper molecular manifolds. Also, we will use $|r\rangle$ to denote individual states in the radiative channels R_i, and $|l\rangle$ to denote individual states in the nonradiative channel L. Thus $|r\rangle$ is a state of the type $|1, \mathbf{v}_1, \mathbf{k}\rangle$ while $|l\rangle$ belongs to the $|1, \mathbf{v}_1\rangle = |1, \mathbf{v}_1, 0\rangle$ group. Note that with this notation, $|g\rangle$ and $|l\rangle$ are respectively low and high-energy vibronic levels of the electronic state 1. The energies of these states are sums of contributions from the different degrees of freedom involved. For example, the energy of state $|r\rangle = |1, \mathbf{v}_1, \mathbf{k}\rangle$ is $E_r = E_{1,\mathbf{v}_1,\mathbf{k}} = E_{\mathrm{el}}^{(1)} + E_{\mathbf{v}_1}^{(1)} + \hbar\omega(\mathbf{k})$ ($\omega(\mathbf{k}) = kc$ with c being the speed of light) and the energy of $|s\rangle = |2, \mathbf{v}_2\rangle$ is $E_s = E_{2,\mathbf{v}_2} = E_{\mathrm{el}}^{(2)} + E_{\mathbf{v}_2}^{(2)}$, where $E_{\mathrm{el}}^{(n)}$ stands for the electronic origin of state n (in our two electronic states model $n = 1, 2$) and where, for harmonic molecules, $E_{\mathbf{v}_n}^{(n)} = \sum_\alpha \hbar\omega_\alpha^{(n)}(v_\alpha^{(n)} + 1/2)$ with the sum going over all normal modes α. Here $\mathbf{v}_n$ stands for the set of occupation numbers $\{v_\alpha^{(n)}\}$ and $\omega_\alpha^{(n)}$ is the frequency of normal mode α in the electronic state n.

Consider now a particular vibronic level $|s\rangle = |2, \mathbf{v}_2\rangle$ in the excited electronic state, $|2\rangle$. As seen in Fig. 18.1 this state is energetically embedded within continuous manifolds of states associated with the lower electronic state $|1\rangle$. These include the *radiative* continua R_i and the *nonradiative* manifold L of higher vibrational levels $|1, \mathbf{v}_1\rangle$ of the lower electronic state that, at this high energy, constitute an effective continuum.[8] As discussed in Section 9.1 (see also below), the coupling of $|s\rangle$ to these continua implies that a system prepared in this state will relax—radiatively (with the emission of a photon) into the R channels and nonradiatively into the L channel. Concerning the latter, it should be noted that, in addition to highly lying bound vibrational levels of the lower electronic state, other nonradiative relaxation pathways may exist, for example, dissociation if E_s is larger than the molecular dissociation threshold, or ionization if it exceeds the molecular ionization potential. While below we refer to the L manifold as a dense vibrational manifold, the general formalism apply also to these other cases.

We can use this model to describe several fundamental optical processes. For example, absorption is the transition from the 1-photon-dressed lower vibrational levels of the ground electronic state $|1, \mathbf{v}_1, \mathbf{k}\rangle$ to the no-photon state $|s\rangle = |2, \mathbf{v}_2\rangle$, which can also be written as $|s, 0\rangle$ or $|2, \mathbf{v}_2, 0\rangle$ to emphasize that this is a zero-photon

[8] A dense manifold of states is effectively continuous in the relevant energy regime if the level spacing (inverse density of states) is smaller than the inverse experimental timescale. The latter is determined in the present case by the decay rate Γ_s of the level s, implying the condition, for example, for the manifold L, $\Gamma_s \rho_L(E_s) \gg 1$.

state. This state is broadened by its interaction with the radiative continua R_{l} and the nonradiative manifold L. Spontaneous emission (i.e. fluorescence) is the process in which the state $|s\rangle$ decays into the radiative continua R_{l}, that is, to vibronic levels of the ground electronic states plus a photon. A competitive decay channel is the nonradiative decay of $|s\rangle$ into the nonradiative manifold(s) L. Light scattering is a process that starts with a 1-photon level of the ground electronic state and ends in another such state, that is, $|1, \mathbf{v}_1, \mathbf{k}\rangle \rightarrow |1, \mathbf{v}_1', \mathbf{k}'\rangle$. The elastic process $\mathbf{v}_1 = \mathbf{v}_1'$ and $|\mathbf{k}| = |\mathbf{k}'|$ is called *Rayleigh scattering*. The inelastic process where these equalities are not satisfied is *Raman scattering*.

These processes are associated with corresponding coupling terms in the Hamiltonian. The radiative coupling V_{sr} involves the operator $\hat{H}_{\mathrm{MR}}$, Eq. (3.27), that couples the two electronic states via the dipole operator $\hat{\mu}$ and changes the number of photons by one. Nonradiative interactions between molecular electronic states, which arise from the effect of intramolecular or intermolecular nuclear motions on the molecular electronic structure, give rise to the V_{sl} coupling elements. In the intramolecular context these interactions are associated with corrections to the Born–Oppenheimer approximation or, for states of different spin multiplicity, with spin–orbit coupling. In the common case where the L channel involves vibrational states, a popular model for this coupling is the electron–phonon interaction given by Eqs (12.16) or (12.29) that represent horizontally shifted nuclear potential surfaces between the two electronic states.

It is important to note that the model of Fig. 18.1 cannot account for *thermal interactions* between the molecule and its environment. As discussed above, it contains elements of relaxation, and if the continuum $\{l\}$ represents excited states of the environment the transition $s \rightarrow \{l\}$ describes relaxation by energy transfer from the molecule to environmental motions. However, the opposite process of heating the molecule by energy transfer from a hot environment cannot be described in this way. The R and L continua can represent only zero-temperature radiative and nonradiative baths. The theory is therefore applicable to zero temperature situations only.

Deferring to later the consideration of finite temperature effects, several experimental observables can be described by this picture. Some of these observables were already discussed in preliminary ways in Chapters 9 and 12:

1. Following an initial preparation of state $|s\rangle$, it will subsequently decay into the radiative and the non radiative channels. Under specified conditions discussed in Section 9.1 this relaxation is exponential, that is, the probability to stay in state $|s\rangle$ at time t is

$$P_s(t) = e^{-k_s t}; \qquad k_s = \Gamma_s/\hbar \tag{18.1}$$

where Γ_s is a sum,

$$\Gamma_s = \Gamma_{L\leftarrow s} + \Gamma_{R\leftarrow s} \tag{18.2}$$

with

$$\Gamma_{J\leftarrow s} = 2\pi \sum_j |V_{sj}|^2 \delta(E_s - E_j) = 2\pi(|V_{sj}|^2 \rho_J(E_j))_{E_j=E_s}; \qquad j = l, r \quad J = L, R \tag{18.3}$$

and where ρ_L and ρ_R are the corresponding densities of states. Note that $\Gamma_{R\leftarrow s}$ is a sum over the different radiative channels, $\Gamma_{R\leftarrow s} = \sum_j \Gamma_{R_j\leftarrow s}$, each associated with a different vibronic origin $\mathbf{v}_1$, with states $|r\rangle = |1, \mathbf{v}_1, \mathbf{k}\rangle$. In each such channel the sum over r is a sum over photon wavevectors $\mathbf{k}$ and the density of states is given by Eq. (3.20). In the nonradiative channel the states are $|l\rangle = |1, \mathbf{v}_1\rangle$ and the density ρ_L is a density of vibrational levels.

Note this difference between the physical origins of L and R continua. Each R_i continuum is characterized by a single vibronic origin and its continuous character stems from the photon states. On the other hand, the effectively continuous nature of the L continuum originates from the fact that, with no photon emitted, the vibrational energy involved is high and, for large molecules, the corresponding density of vibrational states is huge.

2. The state $|s\rangle$ is a particular vibronic level $|2, \mathbf{v}_2\rangle$ in the upper electronic state. If thermal relaxation within this electronic manifold is fast relative to the timescale Γ_s^{-1}, then the overall population in this electronic manifold decays into channel J ($J = L, R_\mathrm{i}$) at a rate given by the thermal average of (18.3), that is, (compare Eqs (12.34) and (12.35))

$$\Gamma_{J\leftarrow 2} = \sum_s P_s \Gamma_{J\leftarrow s}; \qquad P_s = \exp(-\beta E_s) \Big/ \sum_s \exp(-\beta E_s) \tag{18.4}$$

For example, the thermally averaged nonradiative transition rate from electronic state 2 to state 1 is

$$k_{L\leftarrow 2} = \frac{\Gamma_{L\leftarrow 2}}{\hbar} = \frac{2\pi}{\hbar} \sum_{\mathbf{v}_2} P_{\mathbf{v}_2} \sum_{\mathbf{v}_1} |\langle 2, \mathbf{v}_2|\hat{V}|1, \mathbf{v}_1\rangle|^2 \delta(E_{2,\mathbf{v}_2} - E_{1,\mathbf{v}_1}) \tag{18.5}$$

while the equivalent radiative emission (fluorescence) rate is

$$\begin{aligned} k_{R\leftarrow 2} &= \frac{\Gamma_{R\leftarrow 2}}{\hbar} \\ &= \frac{2\pi}{\hbar} \sum_{\mathbf{v}_2} P_{\mathbf{v}_2} \sum_{\mathbf{v}_1} |\langle 2, \mathbf{v}_2, 0|\hat{H}_{\mathrm{MR}}|1, \mathbf{v}_1, \omega\rangle|^2 \rho_R(\omega = (E_{2,\mathbf{v}_2} - E_{1,\mathbf{v}_1})/\hbar) \end{aligned} \tag{18.6}$$

where $\rho_R(\omega)$ is the density of radiation field modes at frequency ω. We have seen (Sections 6.2.2 and 10.5.2) that these rates can be cast in forms of Fourier transforms of appropriate correlation functions.

3. Experimentally, the total decay rate is obtained by following the time evolution of the fluorescence, that is counting the total number of photons emitted per unit time. Let $k_{\mathrm{R}\leftarrow s} = \sum_i k_{\mathrm{R_i}\leftarrow s}$ be the total radiative decay rate out of the state $|s\rangle$, so that the flux of emitted light at time t per excited molecule following its initial preparation at $t = 0$ is

$$I_{\mathrm{R}}(t) = k_{\mathrm{R}\leftarrow s}P_s(t) \tag{18.7}$$

Using (18.1) and integrating from $t = 0$ to infinity yields the *quantum yield* for photon emission

$$Y_{\mathrm{R}\leftarrow s} = \frac{k_{\mathrm{R}\leftarrow s}}{k_s} = \frac{\Gamma_{\mathrm{R}\leftarrow s}}{\Gamma_{\mathrm{R}\leftarrow s} + \Gamma_{L\leftarrow s}} \tag{18.8}$$

which measures the fraction of the absorbed radiation that is re-emitted as fluorescence. Measuring both $Y_{\mathrm{R}\leftarrow s}$ and Γ_s yield both the radiative and the nonradiative decay rates of state s. It is easily seen that a similar ratio between thermally averaged rates of the form (18.5) results in the quantum yield for emission from the excited electronic state 2 in the limit where thermal relaxation within a given electronic manifold is fast relative to the emission process.

4. Starting from a particular vibronic level of the ground electronic state, the absorption lineshape is obtained by monitoring the attenuation of transmission as a function of incident photon wavelength and using the Beer–Lambert law. For the transition between two individual vibronic levels $|g\rangle$ and $|s\rangle$ the lineshape was predicted (see Section 9.3) to be Lorentzian with a width Γ_s that reflects the inverse lifetime $\hbar/\Gamma_s$ of level $|s\rangle$, peaked about a slightly shifted energy $\tilde{E}_s$.

$$L_{s\leftarrow g}(\omega) \propto \frac{|\mu_{gs}|^2\Gamma_s}{(E_g + \hbar\omega - \tilde{E}_s)^2 + (\Gamma_s/2)^2} \tag{18.9}$$

The shift is given by contributions of the form (9.29) (with 1 replaced by s) for each continuous manifold that interacts with level s.

5. An interesting variation on the result (18.9) is that if the initial level is also characterized by a Lorentzian width, Γ_g, then the total width associated with the transition is the sum $\Gamma_s + \Gamma_g$. One way to see this is to realize that in this case the

observed lineshape is a convolution

$$\begin{aligned} L_{s\leftarrow g}(\omega) &\propto |\mu_{g,s}|^2 \int_{-\infty}^{\infty} dx \frac{\Gamma_g/2\pi}{(x - E_g)^2 + (\Gamma_g/2)^2} \frac{\Gamma_s}{(x + \hbar\omega - \tilde{E}_s)^2 + (\Gamma_s/2)^2} \\ &= |\mu_{g,s}|^2 \frac{\Gamma_g + \Gamma_s}{(E_g + \hbar\omega - \tilde{E}_s)^2 + ((\Gamma_g + \Gamma_s)/2)^2} \end{aligned} \tag{18.10}$$

6. Actual molecular spectra usually involve superpositions of such terms. In large molecules and in molecules embedded in condensed host environments individual vibronic transitions cannot usually be resolved (unless the host is a low temperature solid matrix). The relevant spectrum is then the full vibronic lineshape for the given electronic transition.

$$\begin{aligned} L_{2\leftarrow 1}(\omega) &\propto \frac{1}{\hbar} \sum_{\mathbf{v}_1} P_{\mathbf{v}_1} \sum_{\mathbf{v}_2} |\langle 2, \mathbf{v}_2|\hat{\mu}|1, \mathbf{v}_1\rangle|^2 \\ &\quad \times \frac{\Gamma_{\mathbf{v}_2}^{(2)} + \Gamma_{\mathbf{v}_1}^{(1)}}{(E_{2,1} + E_{\mathbf{v}_2}^{(2)} - E_{\mathbf{v}_1}^{(1)} - \hbar\omega)^2 + ((\Gamma_{\mathbf{v}_2}^{(2)} + \Gamma_{\mathbf{v}_1}^{(1)})/2)^2} \\ &\xrightarrow{\text{Condon approx}} \frac{|\mu_{12}|^2}{\hbar} \sum_{\mathbf{v}_1} P_{\mathbf{v}_1} \sum_{\mathbf{v}_2} |\langle \mathbf{v}_2|\mathbf{v}_1\rangle|^2 \\ &\quad \times \frac{\Gamma_{\mathbf{v}_2}^{(2)} + \Gamma_{\mathbf{v}_1}^{(1)}}{(E_{2,1} + E_{\mathbf{v}_2}^{(2)} - E_{\mathbf{v}_1}^{(1)} - \hbar\omega)^2 + ((\Gamma_{\mathbf{v}_2}^{(2)} + \Gamma_{\mathbf{v}_1}^{(1)})/2)^2} \end{aligned} \tag{18.11}$$

where $E_{2,1} = E_{\text{el}}^{(2)} - E_{\text{el}}^{(1)}$. This expression is similar to Eq. (12.60),[9] except that it is written for finite temperature and that the δ functions associated with individual vibronic transitions are replaced by Lorentzian profiles with widths that depend on the excited vibronic levels. As already said, in most spectra involving large molecules in condensed phases these individual transitions cannot be resolved and therefore the exact Lorentzian forms in (18.11) are not important. For practical purposes expressions like (18.11) are often replaced by similar expressions in which $\Gamma_{\mathbf{v}_2}^{(2)} + \Gamma_{\mathbf{v}_1}^{(1)}$ are substituted by a constant Γ independent of the particular vibronic level, or even disregarded altogether. The expression obtained in the

[9] Expressions of the forms $|\langle \mathbf{v}_2|e^{i(\hat{\Omega}_x - \hat{\Omega}_g)}|\mathbf{v}_1\rangle|^2$ used in (12.60) and $|\langle \mathbf{v}_2|\mathbf{v}_1\rangle|^2$ that appears in (18.11) are different representations of the same quantity.

limit $\Gamma \to 0$

$$L_{2\leftarrow 1}(\omega) \propto \frac{2\pi}{\hbar} \sum_{\mathbf{v}_1} P_{\mathbf{v}_1} \sum_{\mathbf{v}_2} |\langle 2, \mathbf{v}_2|\hat{\mu}|1, \mathbf{v}_1\rangle|^2 \delta(E_{2,1} + E_{\mathbf{v}_2}^{(2)} - E_{\mathbf{v}_1}^{(1)} - \hbar\omega) \tag{18.12}$$

is mathematically meaningful only when we deal with real continua, but we often use such expressions even for dense discrete spectra that were referred to as "quasi-continua." In practical evaluation each δ function is replaced by a Lorentzian whose width is inconsequential as long as it is larger than the average level spacing.

Problem 18.1. Show that if the δ functions in golden-rule rate expressions like (18.5) are replaced by Lorentzians with constant width Γ independent of the individual transitions, the corresponding correlation function expressions for these rates, for example, Eqs (12.41) and (12.44) become

$$k_{1\leftarrow 2} = \int_{-\infty}^{\infty} dt e^{iE_{2,1}t/\hbar} e^{-(\Gamma/2)|t|/\hbar} C_{21}(t);$$
$$k_{2\leftarrow 1} = \frac{1}{\hbar^2} \int_{-\infty}^{\infty} dt e^{-iE_{2,1}t/\hbar} e^{-(\Gamma/2)|t|/\hbar} C_{12}(t) \tag{18.13}$$

Problem 18.2. A well-known result from the theory of optical absorption lineshapes is that the integrated lineshape associated with the transition between two quantum levels is equal, up to known numerical factors, to the squared radiative coupling element between these levels. For example, using Eq. (18.9) or (18.10) yields $\int d\omega L(\omega) \propto |\mu_{1,2}|^2$. Show that, under the Condon approximation, the integrated absorption lineshape of an overall transition between two vibronic manifolds of two electronic states 1 and 2 is also proportional to the squared radiative electronic coupling $|\mu_{1,2}|^2$.

Solution The absorption lineshape is given by Eq. (18.11). Without invoking yet the Condon approximation it reads

$$L_{2\leftarrow 1}(\omega) \propto \sum_{\mathbf{v}_1} P_{\mathbf{v}_1} \sum_{\mathbf{v}_2} |\langle 2, \mathbf{v}_2|\hat{\mu}|1, \mathbf{v}_1\rangle|^2 \times \frac{\Gamma_{\mathbf{v}_2}^{(2)} + \Gamma_{\mathbf{v}_1}^{(1)}}{(E_{2,1} + E_{\mathbf{v}_2}^{(2)} - E_{\mathbf{v}_1}^{(1)} - \hbar\omega)^2 + ((\Gamma_{\mathbf{v}_2}^{(2)} + \Gamma_{\mathbf{v}_1}^{(1)})/2)^2} \tag{18.14}$$

Integrating over ω we get

$$\int d\omega L_{2\leftarrow 1}(\omega) \propto \sum_{\mathbf{v}_1} P_{\mathbf{v}_1} \sum_{\mathbf{v}_2} |\langle 2, \mathbf{v}_2|\hat{\mu}|1, \mathbf{v}_1\rangle|^2$$

$$= \sum_{\mathbf{v}_1} P_{\mathbf{v}_1} \sum_{\mathbf{v}_2} |\langle \mathbf{v}_2|\hat{\mu}_{1,2}(\mathbf{R})|\mathbf{v}_1\rangle|^2 \qquad (18.15)$$

where $\hat{\mu}_{1,2}(\mathbf{R}) = \langle 1|\hat{\mu}|2\rangle$ is a matrix element in the electronic subspace and is an operator in the nuclear configuration $\mathbf{R}$. Summing over $\mathbf{v}_2$, that is, over all the vibronic levels of the excited electronic manifold and using $\sum_{\mathbf{v}_2} |\mathbf{v}_2\rangle\langle\mathbf{v}_2| = \hat{I}_N$ (unit operator in nuclear space) leads to

$$\int d\omega L_{2\leftarrow 1}(\omega) \propto \sum_{\mathbf{v}_1} P_{\mathbf{v}_1} \langle \mathbf{v}_1||\mu_{1,2}(\mathbf{R})|^2|\mathbf{v}_1\rangle \qquad (18.16)$$

which is the thermal average of $|\mu_{1,2}(\mathbf{R})|^2$ over the nuclear configurations of electronic state 1. In the Condon approximation the $\mathbf{R}$ dependence of the latter is disregarded, leading to

$$\int d\omega L_{2\leftarrow 1}(\omega) \propto |\mu_{1,2}|^2 \qquad (18.17)$$

Note that this result is obtained irrespective of the distribution $P_{\mathbf{v}_1}$.

18.3 Resonance Raman scattering

Here we use the level structure of Fig. 18.1 as a simplified model for light scattering. Within this model such a process can be described as the transition between two 1-photon-dressed vibronic states that belong to the ground electronic state, $|1, \mathbf{v}_1, \mathbf{k}_1\rangle \to |1, \mathbf{v}, \mathbf{k}\rangle$. In this process the molecule scatters a photon from state $\mathbf{k}_1$ to state $\mathbf{k}$ while possibly changing its own vibrational state from $\mathbf{v}_1$ to $\mathbf{v}$. In Rayleigh (elastic) scattering $\mathbf{v} = \mathbf{v}_1$ and the initial and final states of the scattering process belong to the same radiative continuum in Fig. 18.1. In Raman scattering $\mathbf{v} \neq \mathbf{v}_1$ and these states belong to different radiative continua.

Such processes can be realized experimentally both in the time domain and in the energy domain. We may send a pulse of light of finite width, that is a wavepacket in momentum and energy spaces, onto the molecular system and monitor the scattered light as a function of frequency, direction, and time. Alternatively we may use a continuous wave (CW) field, a wave of infinite duration (relative to relevant experimental timescales) with a well-defined frequency and propagation direction and monitor the light scattered at different frequencies and directions. These choices represent a general dichotomy in spectroscopic probing of molecular systems: Both time domain and frequency domain spectroscopies are important tools that often yield complementary information.

In what follows we focus on long time, frequency-domain Raman scattering, which is easier to analyze. To simplify notation we denote the initial state, $|1, \mathbf{v}_1, \mathbf{k}_1\rangle$ by $|in\rangle$ and the final state $|1, \mathbf{v}, \mathbf{k}\rangle$ by $|out\rangle$. We also assume that a single zero-photon excited state $|s\rangle = |2, \mathbf{v}_2, 0\rangle$ is close to resonance with the incident radiation, that is, $E_{\text{in}}(\mathbf{v}_1, \omega_1) = E_{1,\mathbf{v}_1} + \hbar\omega_1 \simeq E_s = E_{2,\mathbf{v}_2}$, where $\omega_1 = |\mathbf{k}_1|c$ and c is the speed of light. This state therefore dominates the scattering and other states of the electronic manifold 2 will be disregarded. The mathematical problem then is to describe the transition between states $|in\rangle$ and $|out\rangle$ due to their mutual couplings to state $|s\rangle$ with the corresponding coupling elements $V_{\text{in},s}$ and $V_{\text{out},s}$. In addition $|s\rangle$ is coupled to radiative and nonradiative continua, as discussed above. The Hamiltonian in the truncated dressed-state-basis is

$$\begin{aligned}\hat{H} =& E_{\text{in}}|in\rangle\langle in| + E_{\text{out}}|\text{out}\rangle\langle\text{out}| + E_s|s\rangle\langle s| + V_{\text{in},s}|in\rangle\langle s| + V_{s,\text{in}}|s\rangle\langle in| \\ &+ V_{\text{out},s}|\text{out}\rangle\langle s| + V_{s,\text{out}}|s\rangle\langle\text{out}| + \sum_{j\neq\text{out}}(V_{j,s}|j\rangle\langle s| + V_{s,j}|s\rangle\langle j|) \qquad (18.18)\end{aligned}$$

In (18.18) we have lumped together all the relevant continuous state manifolds that overlap with E_s into the group $\{j\}$. In fact, the state $|out\rangle$ formally belongs to this group as a member of the radiative continua, however, it has special status as the outgoing state of the process under discussion.

Before proceeding, let us consider the expected dependence on the intensity of the incident field. The scattering process is obviously not linear in the molecule–field interaction, however, it is intuitively expected that for weak incident radiation the scattering signal will be linear in the incident intensity. To see this note that as in the previous section we simplify the theory by considering the scattering process $|1, \mathbf{v}_1, \mathbf{k}_1\rangle \to (|s\rangle = |2, \mathbf{v}_2, 0\rangle) \to |1, \mathbf{v}, \mathbf{k}\rangle$, while in reality the process is $|1, \mathbf{v}_1, n_{\mathbf{k}_1}\rangle \to (|s\rangle = |2, \mathbf{v}_2, n_{\mathbf{k}_1} - 1\rangle) \to |1, \mathbf{v}, (n_{\mathbf{k}_1} - 1), 1_{\mathbf{k}}\rangle$ with $n_{\mathbf{k}_1}$ photons in the incident mode $\mathbf{k}_1$. (In our notation $|1, \mathbf{v}, \mathbf{k}\rangle$ and $|1, \mathbf{v}, 1_{\mathbf{k}}\rangle$ are equivalent descriptions of a 1-photon state). Therefore the matrix element of the molecule–radiation field coupling operator (3.27) between $|in\rangle$ and $|s\rangle$ is proportional to $\langle n_{\mathbf{k}_1}|\hat{a}^{\dagger}_{\mathbf{k}_1}|n_{\mathbf{k}_1} - 1\rangle \sim (n_{\mathbf{k}_1})^{1/2}$, which is essentially the incident field amplitude, while the corresponding matrix element between states $|out\rangle$ and $|s\rangle$ is $\sim \langle 1_{\mathbf{k}}|\hat{a}^{\dagger}_{\mathbf{k}}|0\rangle = 1$. We will see below (Eq. (18.23)) that the scattering intensity is proportional to the absolute square of the product of these elements, and will be therefore linear in $n_{\mathbf{k}_1}$, that is in the incident intensity.

Consider now the time evolution under the Hamiltonian (18.18). Writing a general solution of the time-dependent Schrödinger equation as a linear combination, $\Psi(t) = \sum_k C_k(t)|k\rangle$; $k = in, out, s, \{j\}$, we get the following equations for the time evolution of the coefficients $C_k(t)$

$$\hbar\dot{C}_{\text{in}} = -iE_{\text{in}}C_{\text{in}} - iV_{\text{in},s}C_s \tag{18.19a}$$

$$\hbar\dot{C}_s = -iE_sC_s - iV_{s,\text{in}}C_{\text{in}} - iV_{s,\text{out}}C_{\text{out}} - i\sum_{j\neq\text{out}} V_{sj}C_j \tag{18.19b}$$

$$\hbar\dot{C}_{\text{out}} = -iE_{\text{out}}C_{\text{out}} - iV_{\text{out},s}C_s - \frac{\eta_{\text{out}}}{2}C_{\text{out}} \tag{18.19c}$$

$$\hbar\dot{C}_j = -iE_jC_j - iV_{js}C_s - \frac{\eta_j}{2}C_j \tag{18.19d}$$

where, as in Section 9.5, we have added damping terms with rates $(1/2)\eta_{\text{out}}$ and $(1/2)\eta_j$ that force outgoing boundary conditions in the corresponding channels. These damping terms will be taken to zero at the end of the calculation. We are interested in the long-time behavior of this system when driven by a weak incident field. This driving can be accounted for by solving Eqs (18.19b)–(18.19c) under the "driving boundary condition" $c_{\text{in}}(t) = \exp(-iE_{\text{in}}t/\hbar)c_{\text{in}}(0)$ that expresses the fact that state *in*, the molecular initial state dressed by the incident field, drives the system dynamics. The observable of interest is then the outgoing steady state flux, given by

$$J_{\text{out}} = \lim_{\eta_{\text{out}}\to 0} \frac{\eta_{\text{out}}}{\hbar}|C_{\text{out}}|^2 \tag{18.20}$$

We have solved a steady state problem of this kind in Section 9.5. Indeed, Eqs (18.19) are identical to Eqs (9.77) in which level 1 plays the same role as level s here and where the continuous manifold of states $L = \{l\}$ and $R = \{r\}$ that may represent nonradiative and radiative relaxation channels, are now lumped together in $\{j\}$. There is a difference in the question asked. In solving Eqs. (9.77) we were interested in the total flux into manifolds L and R. Here we are interested in the flux into one state, $|out\rangle$, of the radiative manifold—the state selected by the detector which is positioned to detect scattered photons of particular frequency and propagation direction. We follow the steps taken in Eqs (9.78)–(9.88) and use Eqs (9.80) and (9.83) with $c_{\text{in}}, c_{\text{out}}$ and c_s replacing c_0, c_r and c_1, respectively ($c_k = C_k(t)\exp(-iE_{\text{in}}t/\hbar)$ for all k), to get

$$c_{\text{out}} = \frac{V_{\text{out},s}}{E_{\text{in}} - E_{\text{out}} + i\eta_{\text{out}}/2}\frac{V_{s,\text{in}}c_{\text{in}}}{E_{\text{in}} - \tilde{E}_s + (i/2)\Gamma_s(E_{\text{in}})} \tag{18.21}$$

or (an equation equivalent to (9.86))

$$|C_{\text{out}}|^2 = \frac{|V_{\text{out},s}|^2}{(E_{\text{in}} - E_{\text{out}})^2 + (\eta_{\text{out}}/2)^2} \; \frac{|V_{s,\text{in}}|^2 |C_{\text{in}}|^2}{(E_{\text{in}} - \tilde{E}_s)^2 + (\Gamma_s(E_{\text{in}})/2)^2} \tag{18.22}$$

Here, in analogy to Eqs (9.84)–(9.85), $\Gamma_s(E)$ is the total width of state s which is an additive combination of contributions from all relaxation channels (i.e. continuous state-manifolds), and $\tilde{E}_s = \tilde{E}_s(E_{\text{in}}) = E_s + \Lambda_1(E_{\text{in}})$ where $\Lambda_1(E)$ is the level shift function, also expressed by additive contributions from the different relaxation channels. Inserting (18.22) into (18.20) and taking the limit $\eta_{\text{out}}/((E_{\text{in}} - E_{\text{out}})^2 + (\eta_{\text{out}}/2)^2) \overset{\eta_{\text{out}}\to 0}{\longrightarrow} 2\pi\delta(E_{\text{in}} - E_{\text{out}})$ yields the scattered flux per molecule

$$J_{\text{out}} = \frac{2\pi}{\hbar} |C_{\text{in}}|^2 \frac{|V_{\text{out},s}|^2 |V_{s,\text{in}}|^2}{(E_{\text{in}} - \tilde{E}_s)^2 + (\Gamma_s(E_{\text{in}})/2)^2} \delta(E_{\text{in}} - E_{\text{out}}) \tag{18.23}$$

Apart from numerical constants this result is a product of three terms. The delta-function conveys the expected conservation of the total energy, $E_{\text{in}} = E_{1,\mathbf{v}_1} + \hbar\omega_1 = E_{\text{out}} = E_{1,\mathbf{v}} + \hbar\omega$, where $\omega_1 = |\mathbf{k}_1|c, \quad \omega = |\mathbf{k}|c$. The expression preceding it tells us that in the present case, where light scattering is dominated by a single excited level s that is close to resonance with the incident radiation, the scattering probability is proportional to the absorption lineshape into this resonance level (compare, e.g. Eq. (18.9)) and to the absolute square of the product of coupling elements between the intermediate level s and the *in* and *out* states. Finally $|C_{\text{in}}|^2$ is the probability that the *in* state is populated. If the molecule is thermally equilibrated in the initial (ground) electronic manifold Eq. (18.23) is replaced by

$$J_{\text{out}} = \frac{2\pi}{\hbar} \sum_{\mathbf{v}_1} P^{(1)}_{\mathbf{v}_1} \frac{|V_{\text{out},s}|^2 |V_{s,\text{in}(\mathbf{v}_1)}|^2}{(E_{\text{in}}(\mathbf{v}_1) - \tilde{E}_s)^2 + (\Gamma_s(E_{\text{in}}(\mathbf{v}_1))/2)^2} \delta(E_{\text{in}}(\mathbf{v}_1) - E_{\text{out}}) \tag{18.24}$$

where we have emphasized the dependence of different incident states on $\mathbf{v}_1$, and where $P_{\mathbf{v}} = \exp(-\beta E^{(1)}_{\mathbf{v}}) / \sum_{\mathbf{v}} \exp(-\beta E^{(1)}_{\mathbf{v}})$.

The form of the result (18.23) suggests an interpretation of resonance light scattering as a 2-stage process—preparation of the intermediate level s followed by emission out of this level. We will see, however, that such a picture is too simplistic. For example, if several intermediate levels s contribute to the light scattering an approximate generalization of (18.21) is[10]

$$c_{\text{out}} = \frac{c_{\text{in}}}{E_{\text{in}} - E_{\text{out}} + i\eta_{\text{out}}/2} \sum_s \frac{V_{\text{out},s} V_{s,\text{in}}}{E_{\text{in}} - \tilde{E}_s + (i/2)\Gamma_s(E_{\text{in}})} \tag{18.25}$$

which leads to

$$J_{\text{out}} = \frac{2\pi}{\hbar} |C_{\text{in}}|^2 \left| \sum_s \frac{V_{\text{out},s} V_{s,\text{in}}}{E_{\text{in}} - \tilde{E}_s + (i/2)\Gamma_s(E_{\text{in}})} \right|^2 \delta(E_{\text{in}} - E_{\text{out}}) \tag{18.26}$$

Here the scattering amplitudes associated with the different levels s add coherently to form the total scattering flux, and a picture of additive two-step processes $in \rightarrow s \rightarrow out$ clearly does not hold.

In fact, two scenarios can appear as two limiting cases of the resonance light scattering phenomenon. In one, described by (18.26), the scattering process is coherent. This coherence is expressed by the fact that the scattering amplitude, hence the scattering intensity, depends on the way by which the intermediate electronic state 2 interacts with the photon-dressed ground state. In Eq. (18.25) this determines the way by which different amplitudes superimpose to yield the total scattering amplitude. In the other extreme case, coherence is destroyed by interaction with the thermal environment and the process is a truly two-stage process in the sense that state 2 relaxes by photon emission in a way (Eq. (18.6)) that does not depend on how it was prepared. Measurement-wise, these two modes of molecular response to an incident radiation field are observed in the same way—by sending a photon onto the molecular system, and monitoring the light that comes out. We refer to the inelastic ($\mathbf{v}_1 \neq \mathbf{v}$) signal observed in the coherent case as resonance Raman scattering, and to the light seen in the incoherent limit where the light emission is decoupled from the excitation process as resonance fluorescence. *Resonance Raman scattering results from the system response to the incident radiation in the absence of thermal interactions, while resonance fluorescence is the light emission process that takes place after the molecular excited state defines itself as an independent species by thermal interaction with the environment.*

The reader should ponder about this last phrase: "*defines itself as an independent species by thermal interaction with the environment.*" As long as the process is

[10] This can be obtained by extending the set of equations (18.19) to include several levels $\{s\}$ that couple to states *in* and *out* with an additional approximation of disregarding cross coupling between levels s due to their mutual interaction with the continuous manifold $\{j\}$.

fully coherent, that is, in the absence of phase destroying thermal interactions, the intermediate electronic states 2 cannot be regarded as an independent excited species because it carries phase relationships with the ground state that depend on details of the excitation process. It is only after these phases are destroyed that state 2 is established as an independent species that does not depend on how it was prepared. It is the emission from this independent species that we refer to as fluorescence.

Two additional points should be made. First, as was already pointed out, the coherence of a process does not require the absence of a thermal environment, only that the timescale for decoherence is longer than the time, τ, that characterizes the optical response of the molecular system. For a process involving a photon of frequency ω interacting with a molecule characterized by ground and excited levels 1 and 2 respectively, an energy gap E_{21} and a lifetime $\hbar/\Gamma$ of level 2, we can qualitatively estimate this time from the Heisenberg uncertainty principle in the form $\tau \approx \hbar/|E_{21} - \hbar\omega + i\Gamma|$.[11] This implies that for off-resonance processes where $E_{21} - \hbar\omega \gg \Gamma$ the timescale for optical response is determined by the inverse off-resonance energy gap $(E_{21} - \hbar\omega)^{-1}$ while for resonance processes where $E_{21} - \hbar\omega \ll \Gamma$ it is determined by Γ^{-1}. Therefore off resonance or "regular" Raman scattering, where $E_{21} - \hbar\omega$ is of the order $\sim$1 eV (while Γ is typically of order 10^{-3}–10^{-2} eV) and τ is of order $\sim$1 fs, is practically always a coherent scattering process while resonance scattering may be either coherent or incoherent.

Second, we should keep in mind that between the two extreme limits discussed above there exists a regime of intermediate behavior, where dephasing/decoherence and molecular response occur on comparable timescales. In this case the scattering process may exhibit partial coherence. Detailed description of such situations requires treatment of optical response within a formalism that explicitly includes thermal interactions between the system and its environment. In Section 18.5 we will address these issues using the Bloch–Redfield theory of Section 10.5.2.

18.4 Resonance energy transfer

When two molecular species reside in proximity to each other and one is excited, the excitation energy can be transferred to the other. In the simplest experimental observation of this kind the emission from the originally excited species (donor) decreases, and emission from the other species (acceptor) increases with increasing acceptor concentration. This phenomenon, first analyzed theoretically by Förster,[12] plays a central role in several fundamental processes such as sensitization and

[11] This statement is not obvious, but can be shown to be correct.

[12] Th Förster, Ann. Phys. **2**, 55 (1948).

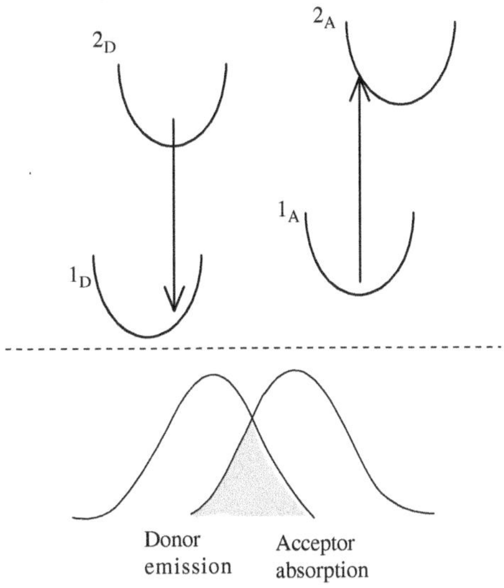

FIG. 18.2 A schematic description of energy transfer according to Förster: De-excitation of the donor molecule is accompanied by excitation of an acceptor molecule. The rate is shown (See Eq. (18.33)) to be proportional to the overlap integral between the emission profile of the donor molecule and the absorption profile of the other.

photosynthesis that are initiated by light absorption by a chromophore site followed by energy transfer to the reaction center. It has found widespread applications, many based on *fluorescence resonance energy transfer* (FRET), in which the detection of fluorescence from the acceptor molecule is used to measure distances and distance distributions between fluorescent tags in proteins and other polymers. Time resolved FRET is similarly used to observe the kinetics of conformational changes in such systems, for example to follow the folding of proteins. Natural and artificial light harvesting systems is an arena of current research and development work that relies principally on this phenomenon (Fig. 18.2).

In what follows we derive the Förster expression for the rate of electronic energy transfer between two chromophore molecules. We consider two such molecules, donor D and acceptor A, each represented by its ground and excited electronic states and the associated vibrational manifolds: $\{|1_D, \chi_d^{(1D)}\rangle\}$, $\{|2_D, \chi_{d'}^{(2D)}\rangle\}$ for the ground and excited state manifolds of molecule D and $\{|1_A, \chi_a^{(1A)}\rangle\}$, $\{|2_A, \chi_{a'}^{(2A)}\rangle\}$ for the corresponding states of molecule A, where d, d', a, a' are used as vibronic state indices (equivalent to the vector indices $\mathbf{v}$ used above). As in similar treatments above, these vibrational manifolds correspond to both intramolecular and environmental modes. Starting from the electronic state $|2_D 1_A\rangle$ in which the donor is in the excited state and the acceptor in the ground state we seek an expression for the rate at which the system converts into the electronic state $|1_D 2_A\rangle$.

Förster theory relies on the following assumptions:

1. The distance between the two molecules D and A is large enough so that the relevant intermolecular coupling is electrostatic. Moreover, the intermolecular distance is assumed large relative to the spatial extent of the individual molecular charge distributions. Under these circumstances the dominant electrostatic interaction is dipole–dipole coupling

$$\hat{V} = \frac{\hat{\boldsymbol{\mu}}_D \cdot \hat{\boldsymbol{\mu}}_A - 3(\hat{\boldsymbol{\mu}}_D \cdot \mathbf{u}_R)(\hat{\boldsymbol{\mu}}_A \cdot \mathbf{u}_R)}{R^3} \tag{18.27}$$

   where $\hat{\boldsymbol{\mu}}_D$ and $\hat{\boldsymbol{\mu}}_A$ are the dipole operators associated with the donor and acceptor molecules and $\mathbf{u}_R$ is a unit vector in the direction from donor to acceptor.
2. In the absence of this coupling the two-molecule Hamiltonian is the sum of Hamiltonians of the individual chromophores. Consequently, the zero-order wavefunctions are products of terms associated with the individual molecules, for example,

$$|1_D \chi_d^{(1D)}, 2_A \chi_a^{(2A)}\rangle = |1_D, \chi_d^{(1D)}\rangle |2_A, \chi_a^{(2A)}\rangle \tag{18.28a}$$

   Furthermore the Born-Oppenheimer approximation is used to describe the wavefunctions of the individual molecules, that is,

$$|1_D, \chi_d^{(1D)}\rangle = |1_D\rangle |\chi_d^{(1D)}\rangle; \qquad |2_A, \chi_a^{(2A)}\rangle = |2_A\rangle |\chi_a^{(2A)}\rangle \tag{18.28b}$$

3. The rate can be calculated by the Golden-rule formula. As discussed at length in Chapter 9, this assumes that the quasi-continuum of final states is broad and relatively unstructured.
4. The energy transfer process within a pair of donor acceptor molecules D and A does not depends on the existence of other donor and acceptor molecules in the system.

Evaluation of the golden-rule rate involves the absolute square of matrix elements of the form $\langle 2_D \chi_{d'}^{(2D)}, 1_A \chi_a^{(1A)} | \hat{V} | 1_D \chi_d^{(1D)}, 2_A \chi_{a'}^{(2A)} \rangle$. Equations (18.27) and (18.28) imply that such matrix elements can be constructed from dipole matrix elements of the individual molecules that take forms like $\langle 2_D, \chi_{d'} | \hat{\mu}_D | 1_D, \chi_d \rangle \mathbf{u}_D$, $\langle 1_A, \chi_a | \hat{\mu}_A | 2_A, \chi_{a'} \rangle \mathbf{u}_A$, etc., where the matrix elements of the vector dipole operators were written as the corresponding magnitudes multiplied by unit vectors in the corresponding directions. This in turn implies that

$$|\langle 2_{\rm D}\chi_{d'}^{(2{\rm D})}, 1_{\rm A}\chi_a^{(1{\rm A})}|\hat{V}|1_{\rm D}\chi_d^{(1{\rm D})}, 2_{\rm A}\chi_{a'}^{(2{\rm A})}\rangle|^2$$
$$= |\langle 2_{\rm D}, \chi_{d'}|\hat{\mu}_{\rm D}|1_{\rm D}, \chi_d\rangle|^2 |\langle 1_{\rm A}, \chi_a|\hat{\mu}_{\rm A}|2_{\rm A}, \chi_{a'}\rangle|^2 \times \frac{\kappa^2(\Theta_{\rm D}, \Theta_{\rm A})}{R^6} \tag{18.29}$$

where

$$\kappa(\Theta_{\rm D}, \Theta_{\rm A}) = \mathbf{u}_{\rm D} \cdot \mathbf{u}_{\rm A} - 3(\mathbf{u}_{\rm D} \cdot \mathbf{u}_R)(\mathbf{u}_{\rm A} \cdot \mathbf{u}_R) \tag{18.30}$$

depends on the spatial orientation of the two molecules (denoted $\Theta_{\rm D}$, $\Theta_{\rm A}$); in fact only on their relative orientations. We thus find that the golden-rule expression for the energy transfer rate can be written in the form

$$k_{\rm ET} = \frac{2\pi}{\hbar}\frac{\kappa^2(\Theta_{\rm D}, \Theta_{\rm A})}{R^6} \sum_{a,a',d,d'} P_{d'}^{(2{\rm D})} P_a^{(1{\rm A})} |\langle 2_{\rm D}, \chi_{d'}|\hat{\mu}_{\rm D}|1_{\rm D}, \chi_d\rangle|^2$$
$$\times |\langle 1_{\rm A}, \chi_a|\hat{\mu}_{\rm A}|2_{\rm A}, \chi_{a'}\rangle|^2 \delta(E_d^{(1{\rm D})} + E_{a'}^{(2A)} - E_{d'}^{(2{\rm D})} - E_a^{(1{\rm A})}) \tag{18.31}$$

where the product $P_{d'}^{(2{\rm D})} P_a^{(1{\rm A})}$ is the thermal probability that the donor molecule is in the vibrational state d' of its excited electronic state while the acceptor molecule is in the vibrational state a of its ground electronic state. Next use

$$\delta\left(E_d^{(1D)} + E_{a'}^{(2{\rm A})} - E_{d'}^{(2{\rm D})} - E_a^{(1{\rm A})}\right)$$
$$= \int_{-\infty}^{\infty} dE \delta\left(E + E_a^{(1{\rm A})} - E_{a'}^{(2{\rm A})}\right)\delta\left(E + E_d^{(1{\rm D})} - E_{d'}^{(2{\rm D})}\right) \tag{18.32}$$

in (18.31), to get

$$k_{\rm ET} = \frac{\kappa^2(\Theta_{\rm D}, \Theta_{\rm A})}{R^6}\frac{\hbar}{2\pi} \int_{-\infty}^{\infty} dE\, k_{2\leftarrow 1}^{\rm A}(E) k_{1\leftarrow 2}^{\rm D}(E) \tag{18.33}$$

where

$$k_{1\leftarrow 2}^{\rm D}(E) = \frac{2\pi}{\hbar} \sum_{d,d'} P_{d'}^{(2{\rm D})} |\langle 2_{\rm D}, \chi_{d'}|\hat{\mu}_{\rm D}|1_{\rm D}, \chi_d\rangle|^2 \delta\left(E + E_d^{(1{\rm D})} - E_{d'}^{(2{\rm D})}\right) \tag{18.34a}$$

is the emission lineshape of the donor, while

$$k^{\mathrm{A}}_{2\leftarrow 1}(E) = \frac{2\pi}{\hbar}\sum_{a,a'} P^{(1\mathrm{A})}_{a}|\langle 2_{\mathrm{A}}, \chi_{a'}|\hat{\mu}_{\mathrm{A}}|1_{\mathrm{A}}, \chi_{a}\rangle|^{2}\delta\big(E + E^{(1\mathrm{A})}_{a} - E^{(2\mathrm{A})}_{a'}\big) \tag{18.34b}$$

is the absorption lineshape for the acceptor. The energy transfer rate is seen in (18.33) to be proportional to the overlap between these lineshapes.

It is convenient to recast Eq. (18.33) in terms of the distance R_0 at which k_{ET} is equal to the decay rate $k^{(0)}_{\mathrm{D}}$ of the donor molecule in the absence of acceptor molecules,

$$k_{\mathrm{ET}} = k^{(0)}_{\mathrm{D}}\left(\frac{R_0}{R}\right)^{6} \tag{18.35a}$$

$$(R_0)^{6} = \left(k^{(0)}_{\mathrm{D}}\right)^{-1}\kappa^{2}(\Theta_{\mathrm{D}}, \Theta_{\mathrm{A}})\frac{\hbar}{2\pi}\int_{-\infty}^{\infty} dE\, k^{\mathrm{A}}_{2\leftarrow 1}(E)k^{\mathrm{D}}_{1\leftarrow 2}(E) \tag{18.35b}$$

R_0 is referred to as the *Förster radius*.[13]

The observables of an energy transfer measurement are rates and yields of the different relaxation processes that follow the donor excitation. In the absence of acceptor molecules the overall relaxation rate of the donor is

$$k^{(0)}_{\mathrm{D}} = k_{\mathrm{D},r} + k_{\mathrm{D},nr} \tag{18.36}$$

where the subscripts r and nr denote radiative and nonradiative processes, respectively. Here and above the superscript (0) marks the absence of acceptor molecules, and we assume that $k_{\mathrm{D},r} = k^{(0)}_{\mathrm{D},r}$ and $k_{\mathrm{D},nr} = k^{(0)}_{\mathrm{D},nr}$, that is, the presence of the acceptor does not affect the relaxation channels that exist in its absence. The donor emission yield, that is, the fraction of absorbed energy that is reemitted by the donor, is in this case

$$Y^{(0)}_{\mathrm{D},r} = \frac{k_{\mathrm{D},r}}{k^{(0)}_{\mathrm{D}}} = \frac{k_{\mathrm{D},r}}{k_{\mathrm{D},r} + k_{\mathrm{D},nr}} \tag{18.37}$$

[13] In some texts R_0 is defined by $(R_0)^6 = (k^{(0)}_{\mathrm{D},r})^{-1}\kappa^2(\Theta_{\mathrm{D}}, \Theta_{\mathrm{A}})(\hbar/2\pi)\int_{-\infty}^{\infty} dE\, k^{\mathrm{A}}_{2\leftarrow 1}(E)k^{\mathrm{D}}_{1\leftarrow 2}(E)$ where $k^{0}_{\mathrm{D},r} = Y^{(0)}_{\mathrm{D},r}/\tau^{(0)}_{\mathrm{D}}$ is the *radiative* decay rate of the donor in the absence of acceptor molecules ($Y^{(0)}_{\mathrm{D},r}$ is the emission yield and $\tau^{(0)}_{\mathrm{D}}$ is the lifetime of the excited donor in the absence of acceptor). With this definition, $k_{\mathrm{ET}} = k^{(0)}_{\mathrm{D},r}(R_0/R)^6$. Note that elsewhere in our treatment we assume that $k^{(0)}_{\mathrm{D},r} = k_{\mathrm{D},r}$, that is, the radiative relaxation rate of the donor is not affected by the acceptor.

When acceptor molecules are present the overall relaxation rate becomes

$$k_D = k_D^{(0)} + k_{ET} \tag{18.38}$$

and the yields, Y_{ET} of the energy transfer process and $Y_{D,r}$ of donor radiative emission, are given by

$$Y_{ET} = \frac{k_{ET}}{k_D} = \frac{1}{(k_D^{(0)}/k_{ET}) + 1} \tag{18.39}$$

$$Y_{D,r} = \frac{k_{D,r}}{k_D} = \frac{k_{D,r}}{k_D^{(0)} + k_{ET}} \tag{18.40}$$

Problem 18.3.

(1) Show that R_0 can be defined as the donor–acceptor distance for which half the energy absorbed by the donor is transferred to the acceptor molecule.
(2) Show that the quantum yield of the energy transfer, Y_{ET}, is given by

$$Y_{ET} = 1 - \frac{Y_{D,r}}{Y_{D,r}^{(0)}} = \frac{R_0^6}{R^6 + R_0^6} \tag{18.41}$$

Solution The first equality in (18.41) is obtained by using $Y_{D,r} = k_{D,r}/k_D$ and $Y_{D,r}^{(0)} = k_{D,r}/k_D^{(0)}$ to find $1 - Y_{D,r}/Y_{D,r}^{(0)} = 1 - k_D^{(0)}/k_D = k_{ET}/k_D = Y_{ET}$. The second equality is obtained from

$$Y_{ET} = \frac{k_{ET}}{k_D} = \frac{1}{(k_D^{(0)}/k_{ET}) + 1} = \frac{1}{(R/R_0)^6 + 1} = \frac{R_0^6}{R^6 + R_0^6}$$

where we have used (18.35a).

Figure 18.3 is an example of an experimental verification of the distance dependence predicted by the Förster theory. Some more points are notable:

1. In (18.27) we have used the vacuum expression for the dipole–dipole interaction. If the process takes place in a medium of dielectric constant ε, then a factor ε^{-1} enters in this expression. Consequently, the factor κ should be redefined to

$$\kappa(\Theta_D, \Theta_A) = \frac{\mathbf{u}_D \cdot \mathbf{u}_A - 3(\mathbf{u}_D \cdot \mathbf{u}_R)(\mathbf{u}_A \cdot \mathbf{u}_R)}{\varepsilon} \tag{18.42}$$

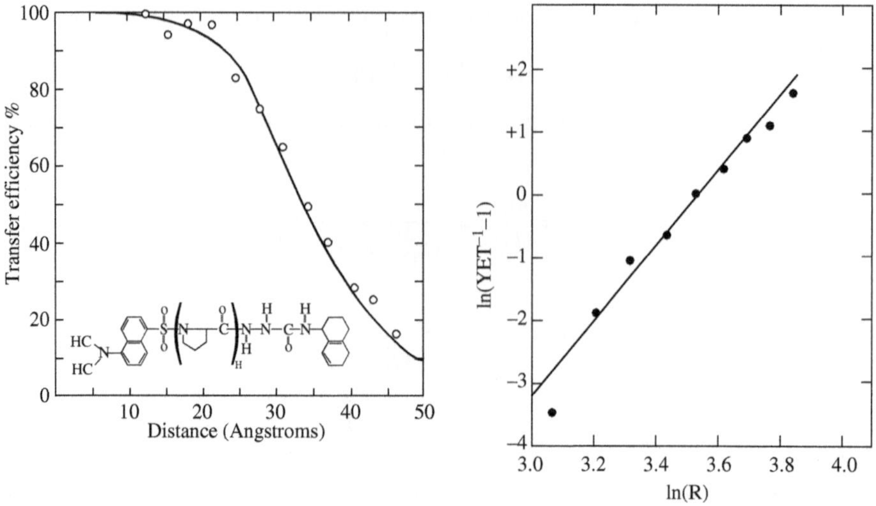

FIG. 18.3 An early experimental verification of Eq. (18.41), showing (circles) the energy transfer yield between two chromophores (1-dimethylaminonaphtalene-5-sulfonyl group and a α-naphthyl group) connected by a molecular bridge (olygomers of poly-L-proline) of variable length (inset) as a function of distance, and a fit to (18.41) with $R_0 \approx 32$ Å. The straight line in the right panel has a slope of 5.9. (From L. Stryer and R. P. Haugland, PNAS, **58**, 719 (1967).)

For the optical frequencies associated with electronic transitions, $\varepsilon = n^2$ where n is the refractive index.

2. The energy transfer process occurs also in the gas phase. However in this environment a primary source for energy transport are molecules that carry their excitation energy as they move. Intermolecular transfer is usually dominated by molecular collisions, that is, at such small distances for which the dipole–dipole interaction is not necessarily the dominant coupling (see below).
3. Continuing on a related theme, the practical importance of observations such as FRET is their ability to convey information about the spatial distributions of chromophores and consequently structural information about the condensed phase under study. It is important to keep in mind that diffusion of the donor and acceptor species, if taking place on the fluorescence timescale, will influence the observed kinetics.
4. The dipole–dipole coupling, an essential ingredient in the Förster theory, is a convenient approximation, valid when the donor and acceptor molecules are far enough (relative to their molecular sizes) from each other. The obvious correction to take, in situations where the donor–acceptor distance becomes comparable to molecular size, is the inclusion of higher-multipole interactions. Without going into details, one can readily infer that such corrections

to the rate (18.33) will fall off as R^{-2n}, $n = 4, 5, \ldots$. At even smaller distances, when the overlap between the electronic densities of the two molecules becomes significant, we expect that this overlap contributes an important interaction term that will fall exponentially with intermolecular distance, $V \sim \exp(-\alpha R)$. This short-range mechanism for energy exchange is known as the Dexter version of the theory.[14] A related mechanism of energy transfer which also stems from orbital overlap is the 2-electron exchange mechanism that dominates triplet–triplet energy transfer.[15]

5. The dipole–dipole interaction (18.27) is the electrostatic approximation to a more general result of electromagnetic theory, which includes retardation effects associated with the finite speed of light. The electrostatic approximation is valid for donor–acceptor distances that are much smaller than the radiation wavelength λ at the frequency of the transition under discussion. In the opposite limit, where the distance greatly exceeds λ, one should use the general electrodynamics formulation to describe the energy transfer. Not surprisingly, energy transfer in this limit becomes simply a photon emission by the donor followed by its absorption by the acceptor. The qualitative difference between the two limits becomes evident by noting that the rate of photon emission followed by absorption should fall with distance like the intensity of the emitted radiation, that is, R^{-2}, while the electrostatic treatment yields the R^{-6} falloff seen above.

   Yet other mechanisms for long-range energy exchange are the analog of the superexchange mechanism of bridge mediated electron transfer discussed in Section 16.12 (it should be noted though that Fig. 18.3 represents a case in which the Förster mechanism dominates), and the analog of the electron hopping mechanism described in Section 16.13.

6. Another analogy to electron transfer can be drawn by considering the extension of the scenario in which an electron is exchanged between two centers to the case of an electron moving in an infinite chain (or a three-dimensional array) of repeated identical centers, for example, a crystal. The electronic problem becomes in this limit the electronic band structure theory of solid crystals (Section 4.3.3). In analogy, the model of electronic energy transfer yields, in the same limit, a picture of energy bands associated with delocalized excitation modes called *excitons*. In both cases we find that in this

---

[14] D. L. Dexter, J. Chem. Phys. **21**, 836 (1953).

[15] Triplet-triplet, or T-T energy transfer is a transition of the type $^3(^3\mathrm{D}^*\,^1\mathrm{A}) \to {}^3(^1\mathrm{D}\,^3\mathrm{A}^*)$. It is overall spin allowed, however a coupling such as $\langle ^3\mathrm{D}^* | \mu_\mathrm{D} | ^1\mathrm{D} \rangle$ is zero because of the orthogonality of the spin states. The coupling that promotes such a transition amounts to simultaneous exchange of two electrons in opposite directions between the donor and the acceptor molecules, and depends strongly on orbital overlap between the two molecules.

infinite crystal limit the physics is better described in terms of delocalized states (Bloch states in one case, excitons in the other) rather than local ones.

7. The energy transfer rate, Eq. (18.33) or (18.35) pertains to a donor interacting with a single acceptor. The contribution of this process to the overall relaxation of the excited donor is obtained by summing this rates over all relevant acceptor molecules. For high homogeneous density ρ of acceptors this amounts to replacing the factor R^{-6} by $\rho \cdot 4\pi \int_a^\infty dRR^{-4} = (4/3)\pi\rho/a^3$ where a is some characteristic distance of nearest approach (assuming that the theory is valid at this distance). More interesting is the opposite limit of low acceptor density. In this case the relaxation due to energy transfer of any donor molecule is dominated by one or just a few acceptor molecules and we expect a large distribution of lifetimes that reflect particular realizations of acceptor positions about individual donor molecules. This is the lifetime equivalent of inhomogeneous line broadening (discussed in Section 18.5.5), and we can take advantage of single molecule spectroscopy techniques (Section 18.6.3) that are suitable for studying such distributed phenomena to probe the distribution of donor–acceptor pairs in our sample.

Further reading

Th. Förster, *Intermolecular Energy Migration and Fluorescence*, Ann. Phys. **2**, 55 (1948).

J. R. Lakowicz, Energy transfer, In *Principles of Fluorescence Spectroscopy*, 2nd ed (Plenum, New York, 1999, p. 367).

G. D. Scholes, *Long Range Resonance Energy Transfer in Molecular Systems*, Ann. Rev. Phys. Chem. **54**, 57 (2003).

T. Ha, *Single-Molecule Fluorescence Resonance Energy Transfer*, Methods **25**, 78 (2001).

18.5 Thermal relaxation and dephasing

In the previous sections we have considered basic processes: Absorption, relaxation of excited states, fluorescence, light scattering and energy transfer. We have taken into account the fact that highly excited molecular states are embedded in, and interact with, continuous manifolds of states that induce relaxation processes. Such processes affect the width of excitation spectra, the lifetimes of excited states and the yield of re-emission in the forms of fluorescence and light scattering. We have argued that modeling relaxation channels in this way amounts to assuming that the system interacts with the corresponding baths at $T = 0$. We have also noted that a clear distinction between a coherent light scattering process and the two-step process of absorption of radiation followed by emission can be made only for systems that undergo dephasing interactions with their thermal environment. In this section we treat this thermal interaction explicitly by considering the optical response of a molecule that interacts both with the radiation field and with its thermal environment.

18.5.1 The Bloch equations

As before we limit ourselves to near resonance processes involving weak radiation fields and model the molecule as a two-state system: the ground state and an excited state selected by the frequency of the incident radiation. Hence, our starting point are the Bloch equations, Eqs. (10.181), for the reduced density matrix σ_{ij} $(i,j = 1,2)$ of such systems

$$\frac{d\sigma_{11}}{dt} = -\frac{d\sigma_{22}}{dt} = -\frac{i}{2}\Omega(\tilde{\sigma}_{21} - \tilde{\sigma}_{12}) - k_{2\leftarrow1}\sigma_{11} + k_{1\leftarrow2}\sigma_{22} \tag{18.43a}$$

$$\frac{d\tilde{\sigma}_{12}}{dt} = -i\eta\tilde{\sigma}_{12} - \frac{i}{2}\Omega(\sigma_{22} - \sigma_{11}) - k_{\mathrm{d}}\tilde{\sigma}_{12} \tag{18.43b}$$

$$\frac{d\tilde{\sigma}_{21}}{dt} = i\eta\tilde{\sigma}_{21} + \frac{i}{2}\Omega(\sigma_{22} - \sigma_{11}) - k_{\mathrm{d}}\tilde{\sigma}_{21} \tag{18.43c}$$

where $\Omega = \mathcal{E}_0\mu/\hbar$ denotes the radiative coupling. Here $\tilde{\sigma}_{12}(t) = e^{-i\omega t}\sigma_{12}(t), \tilde{\sigma}_{21}(t) = e^{i\omega t}\sigma_{21}(t), \tilde{\sigma}_{ii}(t) = \sigma_{ii}(t); (i = 1,2)$, and $\eta = \omega - (E_2 - E_1)/\hbar = \omega - \omega_{21}$ correspond to an implementation of the dressed-state picture: given the molecular states $|1\rangle$ and $|2\rangle$ with energy spacing $E_{21} = E_2 - E_1$, Eqs (18.43) are written for the density matrix elements in the representation of the dressed states $|1,\omega\rangle$ (the molecular state 1 plus a photon of frequency ω) and $|2,0\rangle$ (molecular state 2 with no photons) whose spacing is η.

It should be kept in mind that, as already noted, the 2 states system is greatly oversimplified as a molecular model. We will nevertheless see that it provides important insight that remains useful also for realistic molecular applications. In what follows we consider the implications of Eqs (18.43) on the optical observables that were discussed in Sections 18.2 and 18.3.

18.5.2 Relaxation of a prepared state

Suppose that following the excitation pulse the system was somehow prepared in the excited state $|2\rangle$, so that at $t = 0$, the starting time of our observation, $\Omega = 0$ and the only non-vanishing element of $\hat{\boldsymbol{\sigma}}$ is $\sigma_{22} = 1$. Keeping in mind that $\sigma_{11} + \sigma_{22} = 1$, the time evolution implied by Eq. (18.43a) is

$$\sigma_{22} = \exp[-(k_{1\leftarrow 2} + k_{2\leftarrow 1})t] \tag{18.44}$$

The rates $k_{1\leftarrow 2}$ and $k_{2\leftarrow 1}$ satisfy the detailed balance condition $k_{1\leftarrow 2}/k_{2\leftarrow 1} = \exp(\beta(E_2 - E_1))$ and both may depend on the temperature T. The limit of zero temperature in which $k_{2\leftarrow 1} = 0$ is often relevant in optical spectroscopy, even at room temperature if $E_2 - E_1 \gg k_B T$. In this case, $k_{1\leftarrow 2}$ is the total decay rate of level 2—sum of rates associated with different relaxation channels represented by the continuous state manifolds of the model of Fig. 18.1. However, as discussed below, pure dephasing does not exist in this limit.

In some cases the model of Eqs (18.43) may be modified by replacing Eq. (18.43a) by

$$\frac{d\sigma_{11}}{dt} = -\frac{i}{2}\Omega(\tilde{\sigma}_{21} - \tilde{\sigma}_{12}) - k_{2\leftarrow 1}\sigma_{11} + k_{1\leftarrow 2}\sigma_{22} \tag{18.45a}$$

$$\frac{d\sigma_{22}}{dt} = \frac{i}{2}\Omega(\tilde{\sigma}_{21} - \tilde{\sigma}_{12}) + k_{2\leftarrow 1}\sigma_{11} - k_{1\leftarrow 2}\sigma_{22} - K\sigma_{22} \tag{18.45b}$$

where the second equation contains an additional damping term with damping rate K. In correspondence, Eqs (18.43b) and (18.43c) are augmented by adding the corresponding damping terms $-(1/2)K\sigma_{12}$ and $-(1/2)K\sigma_{21}$, respectively, to their right-hand sides. In this way we account for processes that destroy the system (note that $\sigma_{11} + \sigma_{22}$ is no longer conserved) in the upper state, for example, by ionization or dissociation. This is a trivial modification of the treatment and will not be henceforth considered.

18.5.3 Dephasing (decoherence)

If at time zero, after the field has been switched off, the system is found in a state with nonvanishing coherences, σ_{ij} $(i \neq j)$, Eqs (18.43b,c) tell us that these coherences decay with the dephasing rate constant k_{d}. k_{d} was shown in turn to consist of two parts (cf. Eq. (10.176): The lifetime contribution to the decay rate of σ_{ij} is the sum of half the population relaxation rates out of states i and j, in the present case for σ_{12} and σ_{21} this is $(1/2)(k_{2\leftarrow 1} + k_{1\leftarrow 2})$. Another contribution that we called "pure dephasing" is of the form (again from (10.176)) $\hbar^{-2}\tilde{C}(0)(V_{11}^S - V_{22}^S)^2$. $\hat{V}^S$ is the system operator that couples to the thermal bath so that $V_{11}^S - V_{22}^S$ represents modulation in the system energy spacing E_{21} due to this coupling. $\tilde{C}(0)$ is the zero frequency Fourier transform of the time correlation function of the bath operator that couples to the system, as determined by the dynamics in the unperturbed bath subspace. We have argued (see last paragraph of Sect. 10.4.9) that pure dephasing vanishes at zero temperature. Consequently, only zero temperature population relaxation rates, that can be obtained by modeling relaxation channels as continua of states, are relevant in this limit.

18.5.4 The absorption lineshape

We next consider the effect of thermal relaxation on the absorption lineshape. We start with the Bloch equations in the form (10.184),

$$\frac{d\sigma_z}{dt} = \frac{\mathcal{E}_0\mu}{\hbar}\tilde{\sigma}_y - k_r(\sigma_z - \sigma_{z,\mathrm{eq}}) \tag{18.46a}$$

$$\frac{d\tilde{\sigma}_x}{dt} = -\eta\tilde{\sigma}_y - k_d\tilde{\sigma}_x \tag{18.46b}$$

$$\frac{d\tilde{\sigma}_y}{dt} = \eta\tilde{\sigma}_x - \frac{\mathcal{E}_0\mu}{\hbar}\sigma_z - k_d\tilde{\sigma}_y \tag{18.46c}$$

where $\sigma_{z,\mathrm{eq}}$ and k_r are given by Eqs. (10.185) and (10.186), respectively. Let us assume that the system has reached a steady state under a constant field $\mathcal{E}_0$, and consider the rate at which it absorbs energy from the field. We can identify this rate by observing that in (18.46a) there are two kinds of terms that cancel each other when $d\sigma_z/dt = 0$,

$$\frac{\mathcal{E}_0\mu}{\hbar}\tilde{\sigma}_{y,\mathrm{ss}} = k_r(\sigma_{z,\mathrm{ss}} - \sigma_{z,\mathrm{eq}}) \tag{18.47}$$

We use the subscript 'ss' to denote steady state. The rate of energy change in the system is $-\hbar\omega_{21}(d\sigma_z/dt)$. $\hbar\omega_{21}k_r(\sigma_{z,\mathrm{ss}} - \sigma_{z,\mathrm{eq}})$ must therefore be the rate at which energy is being dissipated, while $-\omega_{21}\mathcal{E}_0\mu\tilde{\sigma}_{y,\mathrm{ss}}$ is the rate at which it is absorbed from the radiation field. To find an explicit expression we solve (18.46b,c) for steady state, that is, with the time derivatives put to zero, and find

$$\tilde{\sigma}_{y,\mathrm{ss}} = -\frac{\mathcal{E}_0\mu}{\hbar}\frac{k_d}{\eta^2 + k_d^2}\sigma_{z,\mathrm{ss}} \tag{18.48}$$

so that

$$\left(\frac{d\sigma_z}{dt}\right)_{\mathrm{absorption}} = \omega_{21}(\mathcal{E}_0\mu)^2\frac{\hbar k_d}{[\hbar(\omega - \omega_{21})]^2 + (\hbar k_d)^2}\sigma_{z,\mathrm{ss}} \tag{18.49}$$

The resulting absorption rate is proportional to the population difference $\sigma_{z,\mathrm{ss}} = \sigma_{11,\mathrm{ss}} - \sigma_{22,\mathrm{ss}}$, as may have been expected. We have obtained a Lorentzian lineshape whose width is determined by the total phase relaxation rate k_d, Eq. (10.176).

18.5.5 Homogeneous and inhomogeneous broadening

The fact that the lineshape (18.49) is Lorentzian is a direct consequence of the fact that our starting point, the Redfield equations (10.174) correspond to the limit were the thermal bath is fast relative to the system dynamics. A similar result was obtained in this limit from the stochastic approach that uses Eq. (10.171) as a starting point for the classical treatment of Section 7.5.4. In the latter case we were also able to consider the opposite limit of slow bath that was shown to yield, in the model considered, a Gaussian lineshape.

To understand the physical difference between these limits we have to realize that interaction with the environment can affect the spectral behavior of a molecular system in two ways, static and dynamic, both derived from the random character of this interaction:

1. In the static limit the medium is much slower than all relevant molecular processes, and can be regarded frozen on the experimental timescale. Medium induced population relaxation cannot take place in this limit because a static medium cannot exchange energy with the molecule. Medium induced line broadening still takes place *in an ensemble of molecules* because each molecule sees a slightly different local configuration of the medium surrounding it. It therefore experiences a slightly different interaction with its local environment and consequently a slightly different frequency shift. If we could perform an experiment on a single molecule, we would observe a narrow absorption lineshape (upper panel of Fig. 18.4) whose width is determined by processes unrelated to the thermal environment (radiative decay, intramolecular relaxation, etc), with peak positions different for different molecules (Fig. 18.4, middle panel). These lines superimpose in the observed many-molecule spectrum to yield the broad absorption profile seen in the lower panel of Fig. 18.4. The lineshape in this case is called *inhomogeneous* and the broadening is referred to as *inhomogeneous broadening.*

It is important to understand that the origin of the different frequency shifts experienced by different molecules is the same stochastic frequency modulation $\delta\omega(t)$ of Eq. (10.171), only that in the limit considered each molecule encounters a different instantaneous realization of this stochastic variable, which persists on the timescale of the measurement. In this limit the observed lineshape is determined not by the dynamics of $\omega(t)$ but by the probability $P(\omega')$ that at any time the molecule is characterized by the instantaneous transition frequency ω'. If the normalized absorption profile of an individual molecule is given by $a(\omega - \omega')$, where $a(\omega)$ peaks at $\omega = 0$ and $\int_{-\infty}^{\infty} d\omega a(\omega) = 1$, the observed lineshape is

$$L(\omega) = \int_{-\infty}^{\infty} d\omega P(\omega')a(\omega - \omega') \simeq P(\omega) \tag{18.50}$$

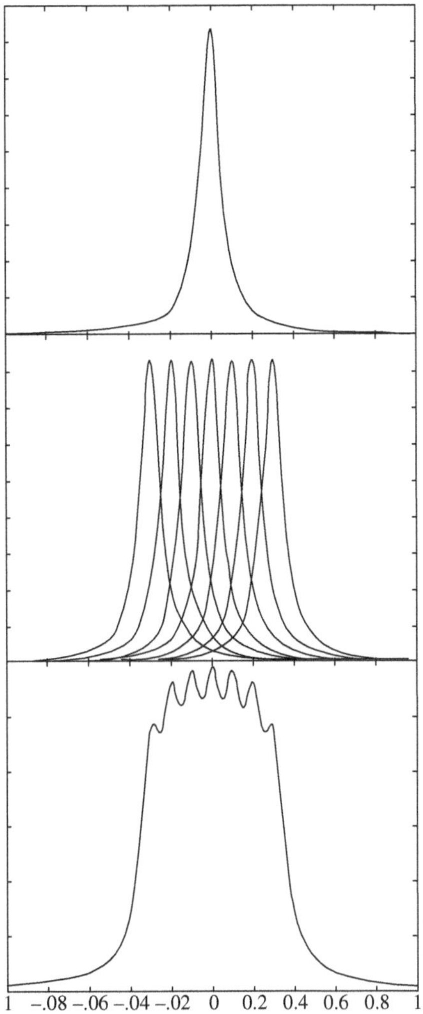

FIG. 18.4 The spectrum of a single molecule (upper panel) is superimposed on similar spectra of other molecules (middle panel) to yield the inhomogeneously broadened line shape (lower panel).

where the second equality was written under the assumption that the $a(\omega)$ profile is much narrower than the distribution $P(\omega)$. The Gaussian lineshape, Eq. (7.107) that results in the static limit of the stochastic theory of lineshape (Section 7.5.4) reflects the assumption made there that $\delta\omega$ is a Gaussian stochastic variable so that $P(\omega)$ is a Gaussian function, in accord with Eq. (18.50).

2. Now consider the opposite limit where the thermal motion in the environment is fast relative to the molecular processes under discussion, in particular relative to the timescale of the molecule–photon interaction that leads to absorption. Now each

single molecule experiences, on the relevant experimental timescale, all the possible configurations of its local neighborhood. Consequently, each molecule provides a full representation of the molecular ensemble. Therefore the ensemble-averaged lineshape is the same as what would be observed by monitoring a single molecule. We refer to the lineshape and the broadening in this limit as *homogeneous*.

The Redfield equations that lead to Eqs (18.43) or (18.46) were obtained under the assumption that thermal environment is fast relative to the system and therefore correspond to this homogeneous limit. Consequently the absorption spectrum (18.49) obtained from these equations corresponds to a homogeneous lineshape. In contrast, the classical stochastic theory of lineshape, Section 7.5.4, can account for both limits and the transition between them. We will see in the next section that an equivalent theory can be also constructed as an extension of the Bloch equations (18.43).

18.5.6 Motional narrowing

An important consequence of the lineshape theory discussed above concerns the effect of the bath dynamics on the linewidths of spectral lines. We have already seen this in the discussion of Section 7.5.4, where a Gaussian power spectrum has evolved into a Lorentzian when the timescale associated with random frequency modulations became fast. Let us see how this effect appears in the context of our present discussion based on the Bloch–Redfield theory.

As noted above, in their straightforward implementation, the Bloch equations correspond to the homogeneous broadening limit. To account for contributions to the linewidth that do not originate from the thermal environment, for example, radiative and intramolecular relaxation, we may replace k_d in Eq. (18.49) by $\bar{k}_\mathrm{d} = k_\mathrm{d} + k_\mathrm{i}$ where k_i is the combined rate of these other relaxation processes that will be taken constant in the following discussion. In what follows we focus on the width k_d which is associated with the thermal environment.

This width is affected by the dynamics of the thermal bath through the bath correlation function $C(t)$, Eq. (10.121). According to Eq. (10.176) two components of the transform $\tilde{C}(\omega) = \int_0^\infty d\tau e^{i\omega\tau} C(\tau)$ are involved: $\tilde{C}(0) = \int_0^\infty d\tau C(\tau)$ determines the "pure dephasing" $k_2^{(12)}$ according to Eq. (10.168), while $\mathrm{Re}\tilde{C}(\pm\omega_{12})$ determines the lifetime contribution to the dephasing, $k_1^{(12)}$, as implied by Eqs (10.167) and (10.160).

An important characteristic of the bath dynamics is its correlation time, essentially the lifetime of the correlation function $C(t)$. Using the simple model $C(t) = C(0)e^{-t/\tau_\mathrm{c}}$ we find $\tilde{C}(\omega) = -C(0)(i\omega - \tau_\mathrm{c}^{-1})^{-1}$,[16] so that $k_2^{(12)} \sim \tau_\mathrm{c}$

[16] Note that this result is a high-temperature approximation, since $\tilde{C}(\omega)$ has to satisfy Eq. (6.72), that is, $\mathrm{Re}\tilde{C}(\omega)/\mathrm{Re}\tilde{C}(-\omega) = \exp(\beta\hbar\omega)$. This, however, does not affect the present discussion.

and $k_1^{(12)} \sim \tau_c/((\omega\tau_c)^2 + 1)$. Both vanish when $\tau_c \to 0$, implying that the absorption profile (18.49) narrows when the bath moves faster and its width approaches $\bar{k}_d \to k_i$ in this limit. This is a manifestation of a phenomenon known as *motional narrowing*. In reality however, for optical transitions, the inequality $\omega\tau_c \gg 1$ holds: Thermal motions in condensed phase molecular systems are characterized by $\tau_c \sim 10^{-10} - 10^{-12}$ s at room temperature, while frequencies associated with vibrational and electronic transitions satisfy $\omega > 10^{13}$ s. Therefore we are usually in a situation where $k_2^{(12)} \sim \tau_c$ while $k_1^{(12)} \sim \tau_c^{-1}$, that is, the width associated with pure dephasing decreases, while lifetime broadening increases, when the bath becomes faster. Still in the common situation where pure dephasing dominates an observed linewidth, a faster bath dynamics may lead to a narrower line. It should be kept in mind, however, that because the only parameter at our disposal for controlling the dynamics of a thermal bath is the temperature, the effect described above may be obscured by other temperature-dependent phenomena. For example the effective strength (as opposed to the timescale) of the system–bath interaction can be stronger at higher temperatures because of larger amplitudes of motions and larger collision velocities. For this reason motional narrowing is usually observed in a different dynamical regime—during the transition from inhomogeneous to homogeneous line broadening, as discussed below.

Before turning to this other interesting case we briefly dwell on a potential source of confusion. We have seen above that in the extreme (and unphysical) $\tau_c \to 0$ limit all spectral broadening associated with the system–bath interaction vanishes. This seems to contradict a popular stochastic model, the Langevin equation discussed in Section 8.2, where the random force associated with the thermal bath was taken to be δ-correlated in time (Eq. (8.20)). We can resolve this apparent contradiction by noting that a model with a δ-correlated random force is just a convenient mathematical framework. We get this description of the physical system in the *mathematical* limit where the force correlation time is infinitely short but its amplitude is infinitely large (as again seen in Eq. (8.20)). This yields a mathematically simple stochastic equation that describes the system dynamics on timescale long relative to τ_c. Again, this mathematical model compensates for the vanishing correlation time by taking an appropriately diverging force. In contrast, in the above analysis we have considered the *physical* limit in which $\tau_c \to 0$ with the forces remaining within their fixed physical range. We find that in this case the dynamic system–bath coupling effectively vanishes. Indeed, in this extremely fast bath limit the system cannot follow the bath motion and the latter appears to it as a static environment characterized by its averaged configuration.

Our discussion of motional narrowing has focused so far on the homogeneous spectrum described by the Bloch–Redfield theory, which is valid only when the bath is fast relative to the system timescale. In this case we could investigate the $\tau_c \to 0$ limit, but the opposite case, $\tau_c \to \infty$, cannot be taken. In contrast, the classical stochastic theory of lineshapes (Section 7.5.4) holds in both limits and can describe the transition between them. In what follows we develop an equivalent picture from a generalization of the Bloch equations, and use it to demonstrate motional narrowing that accompanies a change in the character of the bath from "slow" to "fast."

Let the two-level molecule of interest be embedded in a solvent that itself can be in either of two states, a and b,[17] and let the corresponding molecular transition frequency be ω_{21}^a or ω_{21}^b, respectively. Assuming that the molecular transition dipole is the same in these two states and denoting $\Omega = \mathcal{E}_0\mu/\hbar$, Eq. (18.43b) is

$$\frac{d\tilde{\sigma}_{12}^a}{dt} = -i\eta^a\tilde{\sigma}_{12}^a - \frac{i}{2}\Omega(\sigma_{22}^a - \sigma_{11}^a) - k_d\tilde{\sigma}_{12}^a \tag{18.51}$$

when the solvent (or the molecule) is in configuration a, and a similar equation with b replacing a when it is in configuration b. We assume that thermal fluctuations cause transitions $a \rightleftarrows b$ between these configurations, characterized by a rate k. We further assume that this dynamics is not fast enough to affect transitions between the molecular states 1 and 2, so that $\sigma_{22}^a - \sigma_{11}^a = \sigma_{22}^b - \sigma_{11}^b = -\sigma_z$ is constant.[18] This dynamics is then expressed by the two coupled equations

$$\frac{d\tilde{\sigma}_{12}^a}{dt} = -i\eta^a\tilde{\sigma}_{12}^a + \frac{i}{2}\Omega\sigma_z - k_d\tilde{\sigma}_{12}^a - k(\tilde{\sigma}_{12}^a - \tilde{\sigma}_{12}^b) \tag{18.52a}$$

$$\frac{d\tilde{\sigma}_{12}^b}{dt} = -i\eta^b\tilde{\sigma}_{12}^b + \frac{i}{2}\Omega\sigma_z - k_d\tilde{\sigma}_{12}^b - k\left(\tilde{\sigma}_{12}^b - \tilde{\sigma}_{12}^a\right) \tag{18.52b}$$

The steady state ($d\tilde{\sigma}_{12}^a/dt = d\tilde{\sigma}_{12}^b/dt = 0$) solution of these equations is

$$\begin{aligned} \sigma_{12}^a &= \frac{(i/2)\Omega(i\eta^b + k_d + 2k)}{(i\eta^a + k_d + k)(i\eta^b + k_d + k) - k^2}\sigma_z \\ \sigma_{12}^b &= \frac{(i/2)\Omega(i\eta^a + k_d + 2k)}{(i\eta^a + k_d + k)(i\eta^b + k_d + k) - k^2}\sigma_z \end{aligned} \tag{18.53}$$

---

[17] This may happen when the close environment of the molecule, for example, the cage in which it is trapped, can have two conformations.

[18] It is important to realize that we are not discussing a transition between two molecular conformations. The molecule remains the same; only the way in which the non-diagonal elements of its density matrix evolve is changing because of the change in its environment. $\sigma_{12}^{(a)}$ and $\sigma_{12}^{(b)}$ do not represent elements of different density matrices, but of the same density matrix that evolve differently in time because of the change in frequency.

The rate at which energy is absorbed from the radiation field is, in analogy to Eq. (18.49)

$$\left(\frac{dE}{dt}\right)_{\text{abs}} = \Omega(\hbar\omega_{21}^a \text{Im}\sigma_{12}^a + \hbar\omega_{21}^b \text{Im}\sigma_{12}^b)\sigma_z \approx \hbar\Omega\omega_{21}(\text{Im}\sigma_{12}^a + \text{Im}\sigma_{12}^b)\sigma_z \tag{18.54}$$

In the inhomogeneous limit $k \to 0$ this yields

$$\left(\frac{dE}{dt}\right)_{\text{abs}} = \frac{1}{2}\Omega^2\hbar\omega_{21}\left(\frac{k_{\text{d}}}{(\eta^a)^2 + k_{\text{d}}^2} + \frac{k_{\text{d}}}{(\eta^b)^2 + k_{\text{d}}^2}\right)\sigma_z \tag{18.55}$$

that is, a sum of the two Lorentzian lineshapes. In the opposite, homogeneous limit where $k \to \infty$ we get

$$\left(\frac{dE}{dt}\right)_{\text{abs}} = \Omega^2\hbar\omega_{21}\frac{k_{\text{d}}}{[(\eta^a + \eta^b)/2]^2 + k_{\text{d}}^2}\sigma_z \tag{18.56}$$

that is, a single Lorentzian lineshape peaked about the average frequency. The absorption lineshape obtained from this model for different values of the exchange rate k is displayed in Fig. 18.5.

In general, many more than two configurations will be involved in inhomogeneous broadening, but the physics that leads to the collapse of a broad envelope of

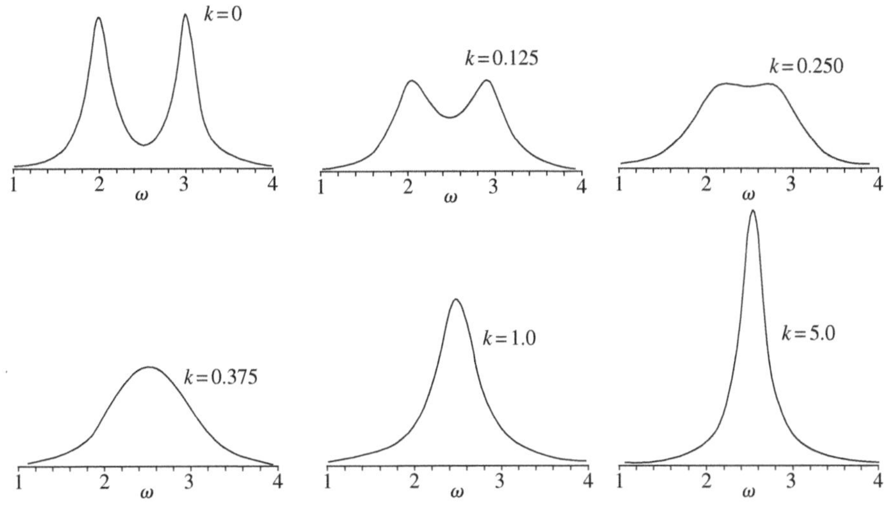

FIG. 18.5 Absorption lineshape (arbitrary units) from Eq. (18.54): Demonstration of motional narrowing using $\omega_{21}^a = \omega_{21} - 0.5$, $\omega_{21}^b = \omega_{21} + 0.5$ ($\omega_{21} \gg 1$), and $k_{\text{d}} = 0.15$.

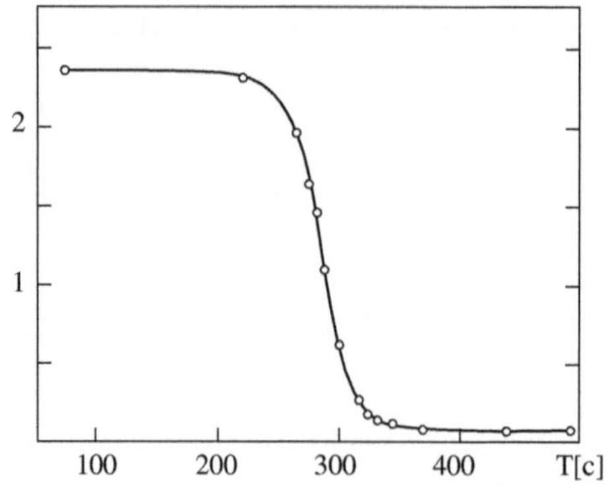

FIG. 18.6 The NMR linewidth (in gauss) of sodium in sodium metal (from H. S. Gutowsky, Phys. Rev. **83**, 1073 (1951)). The thermal motion that causes the observed narrowing at elevated temperature is due to self diffusion of sodium atoms that create local fluctuations in the magnetic field, and consequently in the spin splitting.

lines to a single relatively narrow feature remains the same. Experimental manifestations of this effect have long been known in the NMR literature. Figure 18.6 shows an example in which the NMR linewidth of sodium atom in sodium metal narrows at higher temperatures. The original inhomogeneous broadening arises from small random variations in the local magnetic field, and the narrowing reflects the increasing rate of diffusion (hence more rapid changes in the local configuration and the local magnetic field) of sodium atoms at higher temperatures. k in Eqs (18.52) is related in this case to the hopping rate between lattice sites. Furthermore, according to (18.53) the transition out of the inhomogeneous limit occurs when k becomes of order η_a, η_b in the relevant frequency range, namely of the order of the inhomogeneous width.[19] We can therefore use results such as Fig. 18.6 to estimate hopping rates. In particular, the temperature dependence of k can provide an estimate for the activation energy of hopping, that is, for diffusion.

18.5.7 Thermal effects in resonance Raman scattering

We have already argued that the phenomena of Raman scattering and fluorescence cannot be distinguished from each other unless the system interacts with its thermal environment. Next we extend the model discussed in Section 18.3 to explicitly include thermal relaxation effects. Our model now consists of four levels: The incoming state $|in\rangle = |1, \mathbf{v}_1, \mathbf{k}_1\rangle$ with energy $E_{\text{in}} = E_{\text{el}}^{(1)} + E_{\mathbf{v}_1}^{(1)} + \hbar\omega_1$ $(\omega_1 = c|\mathbf{k}_1|$

[19] In the two-configuration model (Eqs (18.52)) of inhomogeneous broadening a natural choice is to take $\eta_a = -\eta_b$ where $|\eta_a|$ is of the order of the inhomogeneous width.

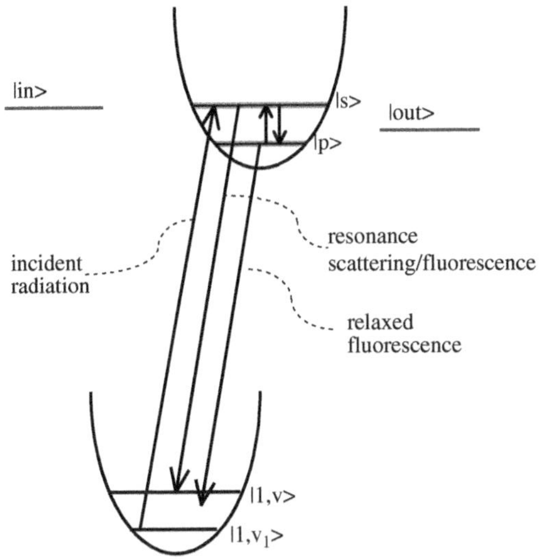

FIG. 18.7 A schematic display of light scattering/excitation-fluorescence process. Shown are the relevant molecular states and the dressed states $|in\rangle$ and $|out\rangle$ used in the calculation. The arrows denote thermal population transfer within the intermediate state manifold. The shading on levels p and s corresponds to energy level fluctuations that leads to pure dephasing.

where c is the speed of light) is a vibronic state of the ground electronic manifold dressed by a photon whose quantum state is defined by the wavevector $\mathbf{k}_1$. $|out\rangle = |1, \mathbf{v}, \mathbf{k}\rangle$ with energy $E_{\text{out}} = E_{\text{el}}^{(1)} + E_{\mathbf{v}}^{(1)} + \hbar\omega$ ($\omega = c|\mathbf{k}|$) is another vibronic state of the same electronic manifold dressed by another photon. $|p\rangle = |2, \mathbf{v}_p, 0\rangle$ and $|s\rangle = |2, \mathbf{v}_s, 0\rangle$ with energies $E_p = E_{\text{el}}^{(2)} + E_{\mathbf{v}_p}^{(2)}$ and $E_s = E_{\text{el}}^{(2)} + E_{\mathbf{v}_s}^{(2)}$, respectively are vibronic levels of the excited electronic state. For specificity we will think of states $|p\rangle$ and $|s\rangle$ as the lowest and a higher-vibrational levels of the excited electronic manifold. The incoming photon is assumed to be in resonance with the state $|s\rangle$, that is, $E_{\text{in}} \simeq E_s$, so we will disregard the interaction between states $|in\rangle$ and $|p\rangle$. In the presence of such incident radiation we expect that outgoing radiation originates by three routes (see Fig. 18.7): First, scattering of light by the molecule, second, absorption into state $|s\rangle$ followed by emission from the same state (resonance fluorescence) and, third, absorption into state $|s\rangle$ followed by thermal relaxation and emission from state $|p\rangle$ (relaxed or thermalized fluorescence). We want to see how these processes appear in a quantum mechanical treatment of the relevant dynamics. We emphasize that the model is far too primitive to describe realistic systems that have far more than two ground and two excited state levels, however, the principles involved remain the same in more complex situations, as will be demonstrated below.

The Hamiltonian of the truncated state model described above is

$$\begin{aligned}\hat{H} &= E_{\text{in}}|in\rangle\langle in| + E_{\text{out}}|out\rangle\langle out| + E_p|p\rangle\langle p| + E_s|s\rangle\langle s| \\ &\quad + V_{\text{in},s}|in\rangle\langle s| + V_{s,\text{in}}|s\rangle\langle in| + V_{s,\text{out}}|s\rangle\langle out| + V_{\text{out},s}|out\rangle\langle s| \\ &\quad + V_{p,\text{out}}|p\rangle\langle out| + V_{\text{out},p}|out\rangle\langle p|\end{aligned} \tag{18.57}$$

The coupling elements $V_{i,j} = V_{j,i}^*$ $(i,j = in, out, p, s)$ are products of the electric field of the incident radiation and the dipole coupling elements between the corresponding molecular states. The steady-state light scattering process is described by letting state $|in\rangle$ drive the system, and finding the flux carried by state $|out\rangle$ under steady-state conditions. This again constitutes a quantum mechanical steady-state problem, however, unlike in Sections 9.5 and 18.3 where such problems were handled using the Schrödinger equations for amplitudes, now in the presence of thermal interactions we have to formulate such problem within the Liouville equation for the density matrix.

A proper way to proceed would be to add to the Hamiltonian (18.57) terms that describe the thermal environment and its interaction with the system, then to derive kinetic equations akin to the Redfield approximation (Section 10.4.8) using the rotating wave approximation to get the Bloch form of the Redfield equations in the presence of the driving field (as in Section 10.5.2). The desired flux would then be obtained by solving the resulting generalized Bloch equations at steady state to yield an analog of the steady-state dynamics scheme used in Section 18.3 as described by Eqs (18.19). Instead of following such rigorous but tedious procedure we will use a phenomenological shortcut. We start from the general form of the Liouville equation $\dot{\hat{\sigma}} = -i\mathcal{L}\hat{\sigma} + \mathcal{R}\hat{\sigma}$, where $\mathcal{L} \equiv \hbar^{-1}[\hat{H},]$ is the Liouville operator associated with the Hamiltonian (18.57). Rather than deriving the relaxation terms that constitute $\mathcal{R}\hat{\sigma}$ from, for example, the Redfield procedure, we postulate a reasonable form for these terms based on the experience gained from deriving the Bloch equations before (Section 10.5.2). The resulting set of Liouville equations is

$$\hbar\frac{d\sigma_{\text{in,in}}}{dt} = -2\text{Im}(V_{s,\text{in}}\sigma_{\text{in},s}) \tag{18.58a}$$

$$\hbar\frac{d\sigma_{s,s}}{dt} = 2\text{Im}(V_{s,\text{in}}\sigma_{\text{in},s}) + 2\text{Im}(V_{s,\text{out}}\sigma_{\text{out},s}) - k_{ps}\sigma_{s,s} + k_{sp}\sigma_{p,p} - \Gamma_s\sigma_{s,s} \tag{18.58b}$$

$$\hbar\frac{d\sigma_{p,p}}{dt} = 2\text{Im}(V_{p,\text{out}}\sigma_{\text{out},p}) + k_{ps}\sigma_{s,s} - k_{sp}\sigma_{p,p} - \Gamma_p\sigma_{p,p} \tag{18.58c}$$

$$\hbar\frac{d\sigma_{\text{out,out}}}{dt} = -2\text{Im}(V_{p,\text{out}}\sigma_{\text{out},p}) - 2\text{Im}(V_{s,\text{out}}\sigma_{\text{out},s}) - \eta\sigma_{\text{out,out}} \tag{18.58d}$$

$$\hbar\frac{d\sigma_{\text{in,out}}}{dt} = -iE_{\text{in,out}}\sigma_{\text{in,out}} - iV_{\text{in},s}\sigma_{s,\text{out}} + iV_{s,\text{out}}\sigma_{\text{in},s} + iV_{p,\text{out}}\sigma_{\text{in},p} - (1/2)\eta\sigma_{\text{in,out}} \tag{18.58e}$$

$$\hbar\frac{d\sigma_{\text{in},s}}{dt} = -iE_{\text{in},s}\sigma_{\text{in},s} + iV_{\text{in},s}(\sigma_{\text{in,in}} - \sigma_{s,s}) + iV_{\text{out},s}\sigma_{\text{in,out}} - (1/2)\gamma_s\sigma_{\text{in},s} \tag{18.58f}$$

$$\hbar\frac{d\sigma_{\text{out},s}}{dt} = -iE_{\text{out},s}\sigma_{\text{out},s} + iV_{\text{in},s}\sigma_{\text{out,in}} - iV_{\text{out},s}(\sigma_{s,s} - \sigma_{\text{out,out}}) - iV_{\text{out},p}\sigma_{p,s} - (1/2)\gamma_s\sigma_{\text{out},s} \tag{18.58g}$$

$$\hbar\frac{d\sigma_{\text{out},p}}{dt} = -iE_{\text{out},p}\sigma_{\text{out},p} - iV_{\text{out},p}(\sigma_{p,p} - \sigma_{\text{out,out}}) - iV_{\text{out},s}\sigma_{s,p} - (1/2)\gamma_p\sigma_{\text{out},p} \tag{18.58h}$$

In these equations $E_{i,j} = E_i - E_j$. The terms with white as well as light-grey backgrounds arise from $-i[\hat{H}, \hat{\sigma}]$ with $\hat{H}$ given by (18.57). The terms with dark-grey backgrounds that describe relaxation processes were added phenomenologically as follows:

1. Thermal transitions (population relaxation) between levels s and p is accounted for by the rates $k_{sp} = k_{s\leftarrow p}$ and $k_{ps} = k_{p\leftarrow s}$ that connect between $\sigma_{s,s}$ and $\sigma_{p,p}$. These rates should satisfy the detailed balance condition $k_{sp}/k_{ps} = \exp(-\beta E_{sp})$.
2. The molecule in the excited electronic state (levels p and s) can undergo nonthermal relaxation processes, for example, dissociation, ionization, and radiative damping. These processes are irreversible because their products are removed from the system, and they are accounted for by the damping rates Γ_s and Γ_p in Eqs (18.58b) and (18.58c), respectively.
3. Non-diagonal elements of σ that involve levels s and p relax with rates derived from the population relaxation processes described above. In addition, pure dephasing associated with thermal fluctuations of molecular energy spacing is assigned for simplicity to the upper levels s and p (i.e. we picture these levels as fluctuating against a static ground state). Correspondingly, the relaxation rates γ_s and γ_p that appear in Eqs (18.58f–h) are given by

$$\gamma_p = \kappa_p + k_{sp} + \Gamma_p; \qquad \gamma_s = \kappa_s + k_{ps} + \Gamma_s \tag{18.59}$$

where κ_p and κ_s are the pure dephasing rates assigned to levels p and s, respectively.

4. As in Eqs (18.19) we impose outgoing boundary conditions on the $|out\rangle$ state, by assigning a small damping rate η to this channel. The corresponding terms containing η that appear in (18.58d,e) insure that a steady state is achieved if we also impose a constant $\sigma_{\text{in,in}}$ on the dynamics (equivalent to the driving boundary condition $c_{\text{in}}(t) = \exp(-iE_{\text{in}}t/\hbar)c_{\text{in}}(0)$ imposed on Eqs (18.19)).

The solution of Eqs (18.58) for this steady state will be obtained in the lowest order of the interaction between the molecule and the driving radiation field. Since the resonant light scattering process involves a photon coming in then going out, the lowest order for the scattering amplitude is 2, and therefore the lowest order for the observed scattering flux is 4. In Appendix 18A we show that the terms marked by light-grey backgrounds in Eqs (18.58) can then be disregarded since they contribute only in higher order. The steady-state solution of (18.58) is obtained by disregarding Eq. (18.58a), imposing a constant $\sigma_{\text{in,in}}$ on the other equations and putting all time derivatives on the left to zero.

Once the steady-state solution is obtained, we seek an expression for the outgoing flux F_{out} under the constant driving imposed by $\sigma_{\text{in,in}}$. This flux contributes to $d\sigma_{\text{out,out}}/dt$ in Eq. (18.58d), and can be evaluated (see Appendix 18A) as the steady state value of the term $-2\text{Im}(V_{p,\text{out}}\sigma_{\text{out},p}) - 2\text{Im}(V_{s,\text{out}}\sigma_{\text{out},s})$ in that equation. Further technical details are given in Appendix 18A. The result is

$$\frac{F_{\text{out}}}{\sigma_{\text{in,in}}} = \frac{2\pi}{\hbar}\frac{|V_{\text{in},s}|^2|V_{\text{out},s}|^2}{(E_{\text{in},s}^2 + ((1/2)\gamma_s)^2)}\left(\delta(E_{\text{in}} - E_{\text{out}}) + \frac{\tilde{\kappa}_s}{\tilde{\Gamma}_s}\frac{\gamma_s/2\pi}{(E_{\text{out},s}^2 + ((1/2)\gamma_s)^2)}\right.$$
$$\left. + \frac{|V_{\text{out},p}|^2}{|V_{\text{out},s}|^2}\frac{\gamma_s}{\tilde{\Gamma}_s}\frac{k_{ps}}{k_{sp} + \Gamma_p}\frac{\gamma_p/2\pi}{(E_{\text{out},p}^2 + ((1/2)\gamma_p)^2)}\right) \tag{18.60}$$

where $\tilde{\Gamma}_s$ and $\tilde{\kappa}_s$ are constants defined by Eqs. (18.128) and (18.138), respectively. In the absence of thermal relaxation from s to p, that is, if k_{ps} vanishes, the scattered flux contains only the first two terms. The first of these, which remains when also the pure dephasing rate of level s, κ_s, vanishes, strictly conserves energy as implied by the $\delta(E_{\text{in}} - E_{\text{out}})$ term. This contribution can be interpreted as resonance Raman scattering. (Note that our result does not contain off-resonance scattering from level p because we have disregarded the corresponding radiative coupling $V_{\text{in},p}$). The other term, proportional to κ_s in which the δ-function is replaced by a Lorentzian of width γ_s, can be identified as resonance fluorescence—emission of light after absorption into $|s\rangle$.

The last term, proportional to the population relaxation rate k_{ps} and broadened by γ_p rather than γ_s is obviously relaxed fluorescence. Note that all terms are proportional to the absorption lineshape (a Lorentzian of width γ_s), as expected. In fact, the ability to monitor resonance light scattering/resonance fluorescence as a function of the incident frequency is an important technique for measuring absorption lineshapes.

18.5.8 A case study: resonance Raman scattering and fluorescence from Azulene in a Naphthalene matrix

In matrix isolation spectroscopy chromophore molecules are embedded in a low temperature inert host matrix at low enough density so that intermolecular interactions between them can be disregarded. Because rotational motions are inhibited and since at the low temperatures studied the molecule occupies its lowest vibronic level, the absorption and fluorescence/Raman scattering spectra are relatively "clean" and simple. The electronic transition associated with the 0–0 line (i.e. transition between the ground vibrational levels of the molecule in the two electronic states) is characterized by a zero-phonon peak and a "phonon sideband" on its high energy side. This sideband is equivalent to excitation into higher vibrational states of the molecule, except that these vibrations are phonons of the embedding host matrix whose frequencies are considerably lower than those of most molecular vibrational modes.

Figure 18.8 shows the emission spectrum observed following excitation of Azulene embedded in a Naphthalene matrix at 2 K. Note that E on the horizontal axis corresponds to E_{out} in Eq. (18.60) while the "excitation energy" is our E_{in}. The upper panel shows the spectrum obtained following excitation into the zero-phonon line while the lower panel shows the spectrum obtained after excitation with energy higher by 30 cm^{-1}, that is, into the phonon sideband. In the latter spectrum lines appear in pairs separated by 30 cm^{-1} with the lower energy line having the same energy as the corresponding emission following the zero-phonon excitation (upper panel). What we are seeing in the lower panel spectrum are emission peaks that originate at the excited region (those are marked "R" and assigned by the authors to resonant Raman scattering) and those that correspond to fluorescence following relaxation into the zero-phonon level ("0"). Note that some peaks, marked "NR," are assigned to the Naphthalene host and are not relevant to our discussion. In the spirit of Eq. (18.60) the lower peaks in each pair correspond to the third term, relaxed fluorescence. The upper peak must be a combination of the first two contributions, Raman and resonant fluorescence, in Eq. (18.60).

The resonance Raman and resonance fluorescence contributions are seen separated in the closer look seen in Fig. 18.9. The observed spectrum is assigned to emission from the origin of the electronic excited state onto a particular vibronic level in the ground electronic state. For reasons irrelevant to our discussion the Raman line appears as a doublet in this high-resolution spectrum. The emission

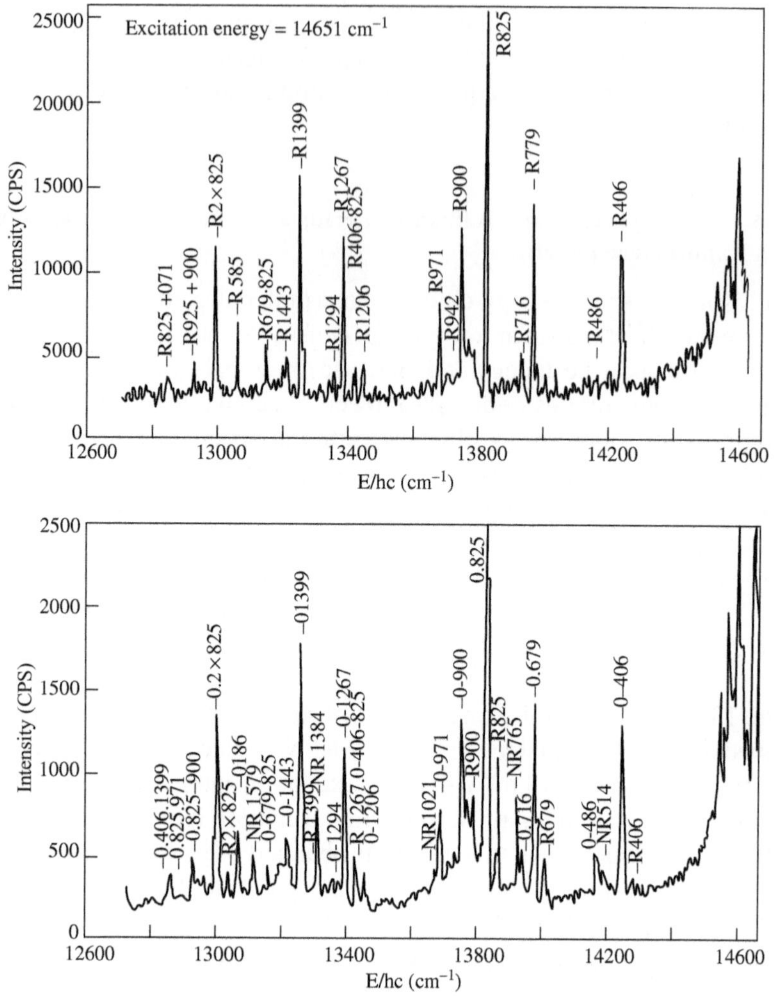

FIG. 18.8 The emission spectrum observed following excitation of Azulene in Naphthalene matrix at $T = 2$ K. Upper panel: emission following excitation into the zero phonon line. Lower panel—emission following excitation into the phonon sideband, 30 cm^{-1} above the zero-phonon line. (Fig. 1 of R. M. Hochstrasser and C. A. Nyi, J. Chem. Phys. **70**, 1112 (1979).)

spectrum is shown for different incident frequencies and we see that the Raman doublet shifts in correspondence with this frequency, as expected for the first term of (18.60) that peaks when $E_{\text{out}} = E_{\text{in}}$. We also see a broad low peak that does not follow the excitation frequency (the shaded rectangle is added to guide the eye onto this peak). This appears to be the resonance fluorescence peak, associated with the second term of (18.60), with expected maximum at $E_{\text{out}} = E_s$ independent of the excitation frequency.

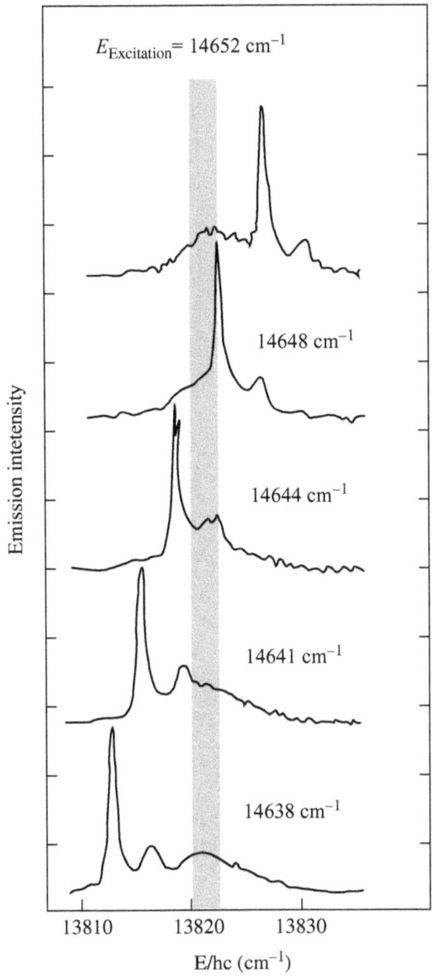

FIG. 18.9 An enlarged view of a narrow spectral region out of the spectrum of Fig. 18.8, assigned to emission from the vibrational origin of the excited electronic state onto the fundamental of the 825 cm^{-1} mode in the ground electronic state at $T = 30$ K. In this high-resolution spectrum the Raman line appears as a doublet that shifts with the excitation wavelength. In addition, a broader emission that does not shift with the excitation frequency is seen (its location on the energy axis is emphasized by the shaded rectangle). (Fig. 7 of R. M. Hochstrasser and C. A. Nyi, J. Chem. Phys. **70**, 1112 (1979).)

Further support for this interpretation is given by the temperature dependence of the same emission spectrum shown in Fig. 18.10. At $T = 4$ K we see only the doublet that corresponds to the Raman scattering signal. As T increases this coherent peak is reduced, and intensity is growing in the broad resonance fluorescence peak

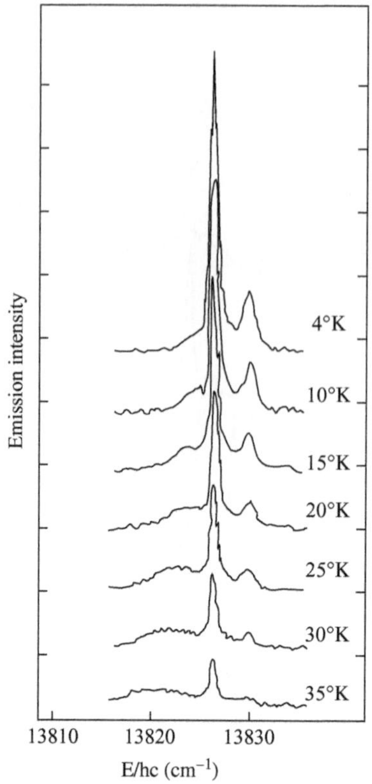

FIG. 18.10 The temperature dependence of the emission seen in Fig. 18.9. (Fig. 8 of R. M. Hochstrasser and C. A. Nyi, J. Chem. Phys. **70**, 1112 (1979).)

seen on its low energy (left) side. Both the reduction in the Raman signal and the increase in the resonance fluorescence are compatible with the expectation that the pure dephasing rate κ_s in Eq. (18.60) increases for increasing temperature. This is obvious for the resonance fluorescence which vanishes when $\kappa_s = 0$. The reduction in the Raman intensity with increasing γ_s and therefore with increasing κ_s (which adds additively to γ_s) also follows from (18.60) if $E_{\text{in}} = E_s$.

18.6 Probing inhomogeneous bands

In Section 9.3 we have seen that there is in principle a close relationship between an absorption lineshape and the underlying dynamics of a molecule excited to the corresponding spectral region. The discussion in the previous section however has taught us that life is less simple: In many systems the absorption lineshape is an average over many individual molecules that experience different local environments, and is not directly associated with any molecular relaxation process. Still, we have seen that the homogeneous and inhomogeneous contributions to line broadening can be told apart, for example, by their dependence on the temperature. The question that we pose now is, can we go further in uncovering the dynamical information that lies underneath inhomogeneous spectra? We note in passing that in gas phase spectroscopy a similar problem exists. Doppler broadening, resulting from the fact that molecules moving at different velocities absorb at slightly different frequencies, is an inhomogeneous contribution to gas phase spectral linewidth. However, Doppler free spectroscopy can be achieved. It is based on saturation of the optical transition by zero-speed molecules interacting with two laser beams traveling in opposite directions, and demonstrates that one can indeed use physical principles to probe underneath inhomogeneous broadening effects. In this section we briefly discuss three different techniques that can achieve this goal in condensed phase spectroscopy.

18.6.1 Hole burning spectroscopy

An inhomogeneously broadened spectral band is obviously just a superposition of individual contributions from different molecules in our sample (Fig. 18.11, upper panel). The width of each individual lineshape is the homogeneous linewidth. Suppose now that we have at our disposal a light source whose spectral width is narrower than this homogeneous width, and suppose this light source is intense enough to destroy or transform those molecules that absorbed it. Then the effect of illuminating our sample with such light is to effectively remove from the sample a subset of molecules whose interaction with their immediate neighborhood has put their transition frequency within that of the light source. If, following this illumination, we interrogate our system with a second tunable narrow-band light we should see a hole in the inhomogeneous envelope—hence the name hole burning (Fig. 18.11, bottom panel).[20] Ideally the spectral width of this hole is the homogeneous linewidth.

Note that the same concept could be used differently: If molecules that are excited by the first light beam fluoresce, we would expect the fluorescence spectrum observed following such excitation to be considerably narrower than what would be normally observed after exciting the full inhomogeneous band.

Both this 'fluorescence line narrowing' and hole burning spectroscopy are conceptually trivial. Still, they can provide very useful information on the homogeneously broadened lines as illustrated in Fig. 18.11. What makes life less simple and more interesting is that other dynamical effects can express themselves in this

[20] For further reading see S. Völker, Ann. Rev Phys. Chem. **40**, 499–530 (1989).

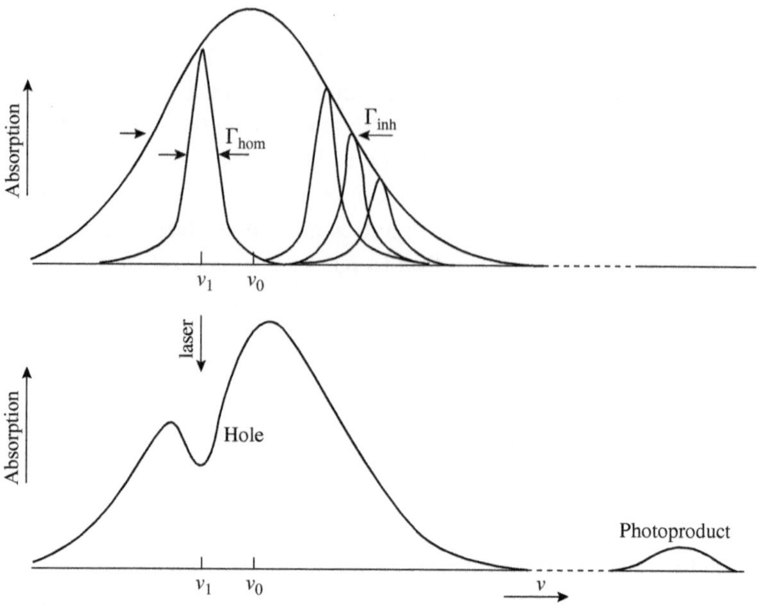

FIG. 18.11 Upper panel: Diagram of an inhomogeneously broadened absorption band of width Γ_{inh}, consisting of a superposition of individual transitions of homogeneous width Γ_{hom}. Bottom: Laser induced hole burnt at low temperature and absorption of the photoproduct. (Fig. 1 from S. Völker, Ann. Rev Phys. Chem. **40**, 499–530 (1989).)

kind of spectroscopy. The extent and importance of these effects depend on the lifetime of the hole vis-à-vis the timescale of the measurement. For example, if the hole formation results from transferring population from the ground state, through the excited state of interest, to a metastable state (e.g. a long living triplet state), the lifetime of the hole corresponds to the lifetime of the metastable state.

Another reason for a finite hole lifetime is the phenomenon of *spectral diffusion*. We have already noted that the distinction between homogeneous broadening and inhomogeneous broadening is to some extent the artificial division of dynamical phenomena to those that are faster than the timescale of our measurement, and those that are slower. "Spectral diffusion" is a phrase used to describe spectral changes resulting from motions that cannot be put into these categories because they take place on a timescale comparable to that of our experiment. Thus a hole formed within an inhomogeneous spectral band may eventually fill up due to environmental motions that change the chromophore neighborhood. Monitoring this fill-up time as a function of temperature is another potentially valuable source of information on the nature of the interaction of a chromophore molecule with its environment.

Finally, even if small amplitude configuration changes about the chromophore molecules are not sufficient for filling up the hole on the experimental lifetime,

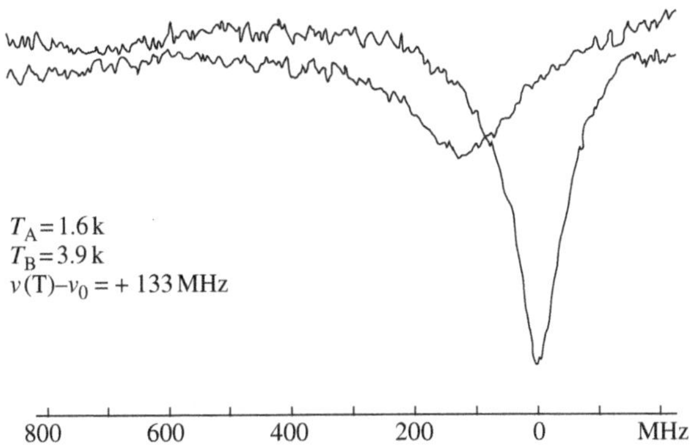

FIG. 18.12 Frequency shift and broadening of the hole burnt in the B_1 site of the 0–0 transition of free-base porphin in *n*-octane. Shown are excitation spectra of hole burnt at 1.6 K (the deeper hole) and 3.9 K. (Fig. 4 from S. Völker, Ann. Rev Phys. Chem. **40**, 499–530 (1989).)

they can induce temperature dependent shift and broadening of the hole, as seen in Fig. 18.12.

18.6.2 Photon echoes

The photon echo phenomenon is perhaps the most dramatic manifestation of using physical principles to overcome what appears to be the erasure of molecular information caused by inhomogeneous broadening. The effect was first demonstrated by Hahn in NMR spectroscopy (Phys. Rev. **80**, 580 (1950)) and extended to optical spectroscopy by Kernit, Abella and Hartmann (Phys. Rev. Letters, **13**, 567 (1964)).[21] Here we only outline its basic physical principle.

Consider a system of N two-level molecules, characterized by level spacing $E_2 - E_1 = \hbar\omega_{21}$ and transition dipole moment $\mu = \mu_{12} = \mu_{21}$, and subjected to a light source represented by $\mathcal{E}(t) = \mathcal{E}_0 \cos(\omega t)$. We assume that the dynamics of this system can be described by the Bloch equations (18.46),

$$\frac{d\sigma_z}{dt} = \frac{\mathcal{E}_0\mu}{\hbar}\tilde{\sigma}_y - k_r(\sigma_z - \sigma_{z,\text{eq}}) \tag{18.61a}$$

$$\frac{d\tilde{\sigma}_x}{dt} = -\eta\tilde{\sigma}_y - k_d\tilde{\sigma}_x \tag{18.61b}$$

$$\frac{d\tilde{\sigma}_y}{dt} = \eta\tilde{\sigma}_x - \frac{\mathcal{E}_0\mu}{\hbar}\sigma_z - k_d\tilde{\sigma}_y \tag{18.61c}$$

[21] See also U. Kh. Kopvillem and V. R. Nagibarov, Fiz. Metal i Matlloved. **15**, 313 (1963).

where

$$\sigma_z \equiv \sigma_{11} - \sigma_{22}$$
$$\tilde{\sigma}_x = \sigma_{12}e^{-i\omega t} + \sigma_{21}e^{i\omega t} \qquad (18.62)$$
$$\tilde{\sigma}_y = i(\sigma_{12}e^{-i\omega t} - \sigma_{21}e^{i\omega t})$$

are elements of the molecular density matrix and $\eta \equiv \omega - \omega_{21} \ll \omega$ is the detuning. In Eqs (18.61) $k_\mathrm{d} = (1/2)k_\mathrm{r}$ + (pure dephasing rate) is the total dephasing rate so that $\hbar/k_\mathrm{d}$ is the corresponding homogeneous broadening.

The Bloch equations by themselves cannot describe spontaneous emission, because they contain the effect of the electromagnetic field on the molecule but not vice versa. To include the effect of the molecules on the radiation field within the semiclassical formalism that led to these equations we should supplement them by a description of the radiation field using the Maxwell equations in the presence of the molecular sources, as described in Appendix 3A (see Eq. (3.75). For our present purpose we can however make a shortcut. We know that one result of Eq. (3.75) is that an oscillating dipole emits radiation, so we can obtain the intensity of emitted radiation by calculating the expectation value $P(t)$ of the oscillating dipole induced in the system and evaluate the emission intensity (energy per unit time) from the classical formula

$$I = |P|^2\omega_{21}^4/(3c^3). \qquad (18.63)$$

Consider first the case $N = 1$. Using $\hat{\mu} = \mu(|1\rangle\langle 2| + |2\rangle\langle 1|)$ and $\hat{\sigma}(t) = \sum_{i,j=1}^{2} \sigma_{i,j}(t)|i\rangle\langle j|$ we find that the dipole induced in the system is given by

$$\langle\hat{\mu}\rangle(t) = \mathrm{Tr}(\hat{\mu}\hat{\sigma}(t)) = \mu(\sigma_{12}(t) + \sigma_{21}(t)) \equiv P_1(t) \qquad (18.64)$$

To find the needed density matrix element we start from Eqs (18.61) and disregard for now the relaxation terms involving k_r and k_d. Let the molecule start at $t = 0$ in state 1, so $\sigma_{i,j}(t = 0) = \delta_{1,i}\delta_{1,j}$, that is, $\sigma_z(t = 0) = 1$ and $\sigma_x(t = 0) = \sigma_y(t = 0) = 0$. Now apply a light pulse of frequency ω, whose duration τ satisfies $\eta\tau \ll 1$. On this timescale, terms in (18.61) that involve η can be disregarded. The dynamics during the pulse is then described by

$$\frac{d\sigma_z}{dt} = \frac{\mathcal{E}_0\mu}{\hbar}\tilde{\sigma}_y; \qquad \frac{d\tilde{\sigma}_y}{dt} = -\frac{\mathcal{E}_0\mu}{\hbar}\sigma_z \qquad (18.65)$$

which yields

$$\sigma_z(t) = \cos\left(\frac{\mathcal{E}_0\mu}{\hbar}t\right); \qquad \tilde{\sigma}_y(t) = -\sin\left(\frac{\mathcal{E}_0\mu}{\hbar}t\right) \qquad (18.66)$$

For definiteness, let us take the pulse duration τ to be such that $(\mathcal{E}_0\mu/\hbar)\tau = \pi/2$. At the end of such pulse we have $\sigma_z = 0$ and $\tilde{\sigma}_y = -1$. Note that the condition $\eta\tau \ll 1$ implies $\eta \ll 2\mathcal{E}_0\mu/(\pi\hbar)$ that can always be achieved with a suitable choice of light intensity.

The two-level system then evolves freely in time, starting from $\tilde{\sigma}_y(t = 0) = -1$ and $\tilde{\sigma}_x(t = 0) = \sigma_z(t = 0) = 0$, where $t = 0$ now refers to the time when the pulse stopped. The time evolution is now given by

$$\frac{d\tilde{\sigma}_x}{dt} = -\eta\tilde{\sigma}_y; \qquad \frac{d\tilde{\sigma}_y}{dt} = \eta\tilde{\sigma}_x \tag{18.67}$$

that is, $\sigma_y(t) = -\cos\eta t$ and $\sigma_x(t) = \sin\eta t$. From (18.62) we get

$$\sigma_{1,2}(t) = \sigma_{2,1}^*(t) = \frac{1}{2}e^{i\omega t}(\sigma_x(t) - i\sigma_y(t)) = \frac{1}{2}ie^{i\omega t}e^{-i\eta t} = \frac{1}{2}ie^{i\omega_{21}t} \tag{18.68}$$

Now consider the N molecules case. For simplicity we assume that the spatial extent of this N molecule system is much smaller than the radiation wavelength. In this case all molecules are subject to the same incident beam and respond coherently. Following the short pulse of duration $\tau = \pi\hbar/(2\mathcal{E}_0\mu)$ the density matrix of each molecule j evolves as before. The density matrix of the whole system is a direct product of these molecular contributions. The expectation value of the total system dipole operator $\hat{P}_N = \sum_{j=1}^{N}\hat{\mu}_j$ is

$$P_N(t) \equiv \langle\hat{P}_N\rangle(t) = \sum_{j=1}^{N}\langle\hat{\mu}_j\rangle(t) = 2\mu\text{Re}\sum_{j=1}^{N}\sigma_{12}^{(j)}(t) = -\mu\text{Im}\left(e^{i\omega t}\sum_{j=1}^{N}e^{-i\eta_j t}\right) \tag{18.69}$$

where t again measures the time following the termination of the light pulse. A distribution of detuning frequencies η_j is a manifestation of inhomogeneous broadening where each molecule is characterized by a slightly different $\omega_{21}^{(j)} = \omega - \eta_j$.

Equation (18.69) tells us that following the termination of the exciting pulse the system is found in a state characterized by a macroscopic dipole, $P_N(t = 0) = NP_1(t = 0)$. Such state is called *superradiant*. In the absence of inhomogeneous broadening all $\omega_{21}^{(j)}$ are the same so that $P_N(t) = NP_1(t)$ and the emitted intensity (18.63) is seen to be proportional to N^2. This stands in contrast to regular fluorescence where no phase relationships exist between individual molecules, so each molecule emits individually implying that the signal is simply proportional to the number of molecules N.

Now, in the presence of inhomogeneous broadening this coherent signal disappears quickly because terms oscillating at different frequencies go out of phase. It

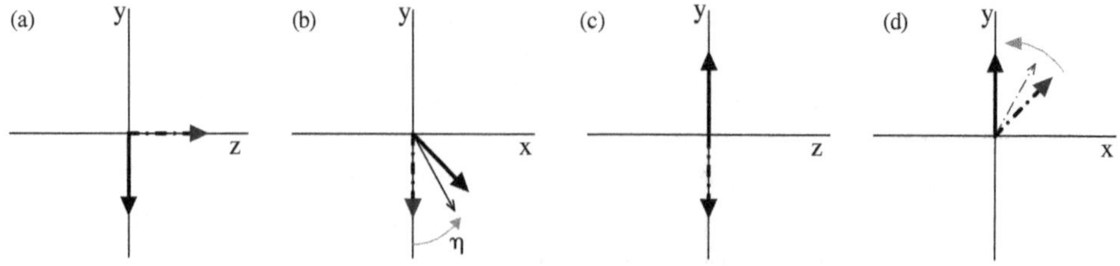

FIG. 18.13 Dephasing and rephasing of superradiance emission (see text).

is easier to analyze this behavior using the complex analog of (18.69), $P_N(t) = \mu e^{i\omega t}\sum_{j=1}^{N} e^{-i\eta_j t}$, which yields

$$|P_N(t)|^2 = \mu^2\left(N + \sum_j \sum_{j'\neq j} e^{i(\eta_j - \eta_{j'})t}\right) \tag{18.70}$$

The second contribution on the right vanishes on a timescale of the order of the inverse inhomogeneous width, leaving the first term, of order N, that corresponds to regular fluorescence.

This disappearance of the superradiance emission is due to dephasing of the coherently excited macroscopic dipole initially induced in the system. The mathematical origin of this dephasing was seen in Eqs (18.67)–(18.68) and an intuitive picture is displayed in Fig. 18.13. Each of the four panels in this figure describes a process that starts in the state described by the thick-dashed arrow and ends up in the state characterized by the thick-full arrow. Panel (a) describes the operation of the pulse on the system, Eq. (18.65)–(18.66) that changes the state of the system from $\sigma_z = 1, \sigma_x = \sigma_y = 0$ to $\sigma_y = -1, \sigma_x = \sigma_z = 0$. The vector $(\sigma_x, \sigma_y, \sigma_z)$ has simply rotated in the (σ_y, σ_z) plane from its original (dashed) orientation to its final (full) one. This final configuration appears as an initial (dashed) one on panel (b). The process shown in panel (b) is the dephasing process described by Eqs (18.67)–(18.68), that corresponds for positive η to a rotation in the counterclockwise direction by the angle ηt (grey arrow in panel (b)), leading to the orientation assumed by the full-line vector. The dephasing process is expressed by the fact that different molecules are associated with different angular speeds η—the two thick and thin full-line vectors in panel (b) represent the states of two such molecules.

Suppose now that after such time t during which the system evolves freely and this dephasing of the macroscopic polarization takes place, the molecule experiences interaction with another light pulse, and suppose that the duration of that light is short enough so that the dynamics associated with the terms involving η in Eqs (18.61) can again be disregarded. The process undergone by the molecule is then described again by Eqs (18.65)–(18.66), that is, a rotation in the (σ_y, σ_z)

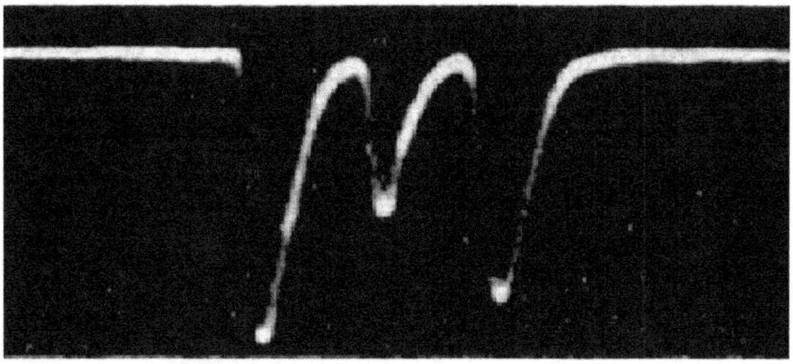

FIG. 18.14 One of the first photon echo experiments, showing the effect in Ruby (N. A. Kernit, I. D. Abella and S. R. Hartmann, Phys. Rev. **141**, 391 (1966). Time increases to the right. The first two peaks shown are the excitation pulses. The third one is the echo.

plane. Assume furthermore that the duration τ' of the second pulse is chosen so that $(\mathcal{E}_0\mu/\hbar)\tau = \pi$. This amounts to sign reversal shown in panel (c)—a process in which $(\sigma_x, \sigma_y, \sigma_z) \rightarrow (\sigma_x, -\sigma_y, \sigma_z)$. The result of this sign reversal in the (σ_x, σ_y) plane is shown in panel (d): The dashed-line arrows result from flipping the sign of the y-component of the full-line arrows in panel (b).

Following the second pulse the systems proceed with its free evolution marked by the grey arrow in panel (d). However, this free evolution now causes *rephasing* of the dephased molecular dipoles! After time t, which is equal to the time elapsed between the two light pulses, the system will be completely rephased as indicated by the final (full-line) vector in panel (d). This analysis then predicts that at that point in time the superradiance emission by the molecular system will resume. In other words, following a second light-pulse at time t after the first pulse, the system will respond with an "echo" at time $2t$. An experimental example is shown in Fig. 18.14.

Note that the above analysis was simplified by disregarding the relaxation terms in Eq. (18.61). Indeed, it should be expected that the amplitude of the echo signal will be lower than that of the initial emission burst by an amount that will reflect the relaxation that take places during the time $2t$.

18.6.3 Single molecule spectroscopy

If the photon echo effect is a most dramatic way to probe underneath inhomogeneous spectral bands, single molecule spectroscopy is undoubtedly the most direct approach. Clearly, if we could observe in each measurement just a single molecule from the given sample, the issue of inhomogeneous broadening would be reflected very differently: Rather than seeing the average response of many molecules, we would, by repeated measurements, obtain a full distribution of responses from

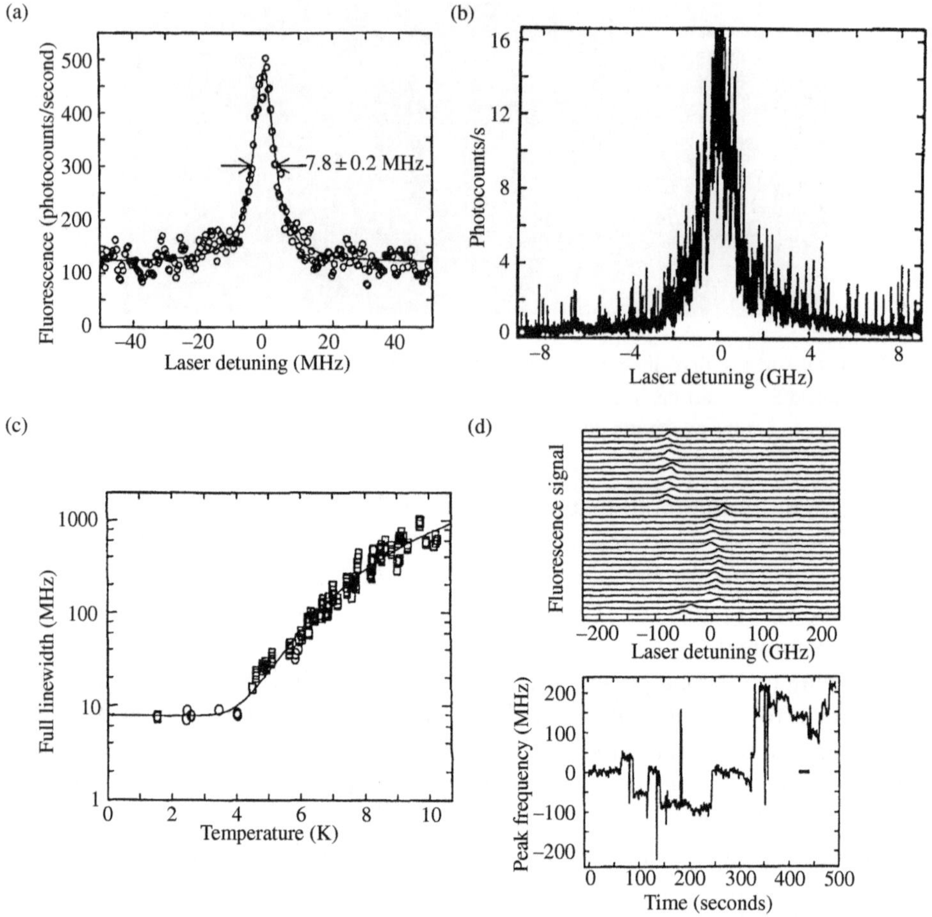

FIG. 18.15 Spectroscopy of single pentacene molecules in p-terphenyl crystal (W. P. Ambrose, Th. Basche and W. E. Moerner, J. Chem. Phys. **95**, 7150 (1991). (a) Fluorescence excitation spectrum of a single molecule at 1.5 K (0 MHz detuning = 592.407 nm, at the wing of the inhomogeneous lineshape) (b) Fluorescence excitation spectrum of the full inhomogeneous line at 1.5 K. (c) The dependence of the single molecule homogeneous linewidth on temperature (the solid line is a fit to the data). (d) Two views of spectral diffusion: The upper panel shows a time sequence of excitation spectra (each taken over a period of 1s). The lower panel shows the jumps in the peak frequency as a function of time.

different molecules. We could still take an average, but we could also look at new aspects of the distribution. For example, the standard way of thinking about inhomogeneous lineshapes is to assume that molecules at different sites of the host environment absorb at slightly different frequencies, however the average lineshape cannot distinguish between contributions by many sites absorbing weakly or a few sites with strong absorption, while a full distribution of single molecule responses will easily tell the difference. Perhaps more significant is the fact that by observing single molecules we could also follow their local dynamics, albeit within restrictions imposed by the experimental time resolution.

Single molecule spectroscopy is the branch of spectroscopy that has developed from the realization that such single molecule detection is in fact possible. This combines two factors: insuring that the density of absorbing molecules is low enough so that only a single molecule is bound within the illuminating region, and improving the detection techniques while eliminating as much as possible the noise associated with the optical response of the far greater number of the nominally transparent host molecules.

Single molecules spectra are usually studied in the wing of the inhomogeneous absorption line; one way to address a smaller density of molecules in resonance with the incident laser beam. An example is shown in Fig. 18.15, which displays the fluorescence excitation spectrum (i.e. fluorescence intensity plotted against the exciting laser frequency) for single pentacene molecules embedded in p-terphenyl crystal at low (1–10 K) temperature. The single molecule spectrum (18.15(a)) should be compared to the much broader full inhomogeneous absorption (18.15(b)). The ability to observe a single homogeneous line makes it possible to study, for example, its dependence on temperature (18.15(c)). Spectral diffusion is clearly observed (18.15(d)) as "jumps" in the peak-position of the fluorescence excitation spectrum. Obviously using the word "jumps" with respect to the observed time evolution merely expresses the time resolution (1 s) of this experiment. In any case, the observed "telegraphic noise" reflects the existence of several considerably different timescales in this evolution.

18.7 Optical response functions

Our discussion in the previous section was based on the conceptual framework of the Bloch–Redfield theory using the dressed-state approach within the rotating wave approximation (RWA). This limits our considerations to weak radiation fields, consistent with the application to processes linear in the incident field. Also, apart from our discussions of inhomogeneous broadening, we have focused on single molecule models that are often not sufficient for discussing the dynamical response of the system as a whole. This section provides an introduction to a more general treatment of optical response of molecular systems in the framework of optical response functions (ORFs). Our aim is not to obtain new results pertaining to nonlinear optical phenomena, only to introduce a formalism that can be used to reproduce the results of linear spectroscopy and can be generalized to the much richer realm of nonlinear optical response.

18.7.1 The Hamiltonian

Our starting point is again the Hamiltonian for the molecular system, M, the radiation field, R, and their mutual interaction,

$$\hat{H} = \hat{H}_{\mathrm{M}} + \hat{H}_{\mathrm{R}} + \hat{H}_{\mathrm{MR}} \tag{18.71}$$

where the molecular Hamiltonian $\hat{H}_{\mathrm{M}}$ now refers to the full molecular system, that is, a collection of molecules. Again we could attempt to treat the problem fully quantum mechanically, with $\hat{H}_{\mathrm{R}}$ describing a collection of bosons (photons), Eq. (3.64), and $\hat{H}$ given by Eq. (3.72). However, since we no longer limit ourselves to weak radiation fields[22] and to the RWA, we adopt the semiclassical level of description, treating the radiation field as a classical object. In this case our starting point is the Hamiltonian for the material system under the influence of an external oscillating field (cf. Eq. 3.73)

$$\begin{aligned} \hat{H} &= \hat{H}_{\mathrm{M}} + \hat{V}(t) \\ \hat{V}(t) &= -\int d\mathbf{r}\hat{P}(\mathbf{r}) \cdot \mathcal{E}^{\perp}(\mathbf{r}, t) \end{aligned} \tag{18.72}$$

in which $\mathcal{E}^{\perp}(\mathbf{r}, t)$ is the transverse part of the classical electric field and $\hat{P}(\mathbf{r})$ is the polarization density operator. The latter is given, in the point dipole approximation (cf. Eq. (3.74)), by

$$\hat{\mathbf{P}}(\mathbf{r}) = \sum_{m} \hat{\mu}_m \delta(\mathbf{r} - \mathbf{r}_m) \tag{18.73}$$

where $\hat{\mu}_m$ is the dipole operator of the mth molecule. We should keep in mind that $\hat{H}_{\mathrm{M}}$, the Hamiltonian of the material system, corresponds in principle not only to the molecular system of interest but also to the thermal environment in which this molecular system is embedded.

The Hamiltonian (18.72) has the form (11.3), generalized to the continuous case, $\hat{H}_1 = -\sum_j \hat{A}_j F_j \to -\int d\mathbf{r}\hat{A}(\mathbf{r})F(\mathbf{r})$, that was the starting point of our discussion of linear response theory. Linear spectroscopy processes (e.g. absorption, but not light scattering) can be treated within this framework, however many important spectroscopical methods are derived from the nonlinear optical response of the material system and their description makes it necessary to go beyond linear

[22] In fact, we continue to assume that the field is weak enough to allow the use of a perturbation series in the field–molecule interaction to any desired order.

response. We therefore present below the general formulation before specifying again to the linear response level.

18.7.2 Response functions at the single molecule level

In what follows we will simplify notation in several ways, one of which is to drop the $\perp$ superscript on the transverse electric field. Let us assume for the moment that only a single chromophore molecule exists, so that

$$\hat{V} = -\hat{\boldsymbol{\mu}}_m \cdot \boldsymbol{\mathcal{E}}(\mathbf{r}_m) \tag{18.74}$$

Both $\hat{\boldsymbol{\mu}}$ and $\boldsymbol{\mathcal{E}}$ are generally vectors and $\hat{V}$ is their scalar product, however again to simplify notation we will suppress this aspect in the formulation below.

Our starting point is the equation of motion (10.21) for the density operator $\hat{\rho}$ of the overall system in the interaction representation

$$\frac{d\hat{\rho}_I}{dt} = -\frac{i}{\hbar}[\hat{V}_I(t), \hat{\rho}_I(t)] \tag{18.75}$$

where

$$\hat{\rho}_I(t) \equiv e^{i\hat{H}_{\mathrm{M}}t/\hbar}\hat{\rho}(t)e^{-i\hat{H}_{\mathrm{M}}t/\hbar} \tag{18.76}$$

and

$$\hat{V}_I(t) \equiv e^{i\hat{H}_{\mathrm{M}}t/\hbar}\hat{V}e^{-i\hat{H}_{\mathrm{M}}t/\hbar} \tag{18.77}$$

are the interaction representation forms of $\hat{\rho}$ and $\hat{V}$. Note that, as defined, $\hat{H}_{\mathrm{M}}$ contains the molecule, its thermal environment and the interaction between them. To further simplify notation we will drop henceforth the subscript I from all operators except $\hat{\rho}$, and keep in mind that all the time-dependent operators encountered in this section are defined as in Eqs (18.76) and (18.77).[23] The formal solution of (18.75) is

$$\hat{\rho}_I(t_0) = \hat{\rho}_I(t_p) - \frac{i}{\hbar}\int_{t_p}^{t_0} dt_1[\hat{V}(t_1), \hat{\rho}_I(t_1)] \tag{18.78}$$

[23] The notation is kept for $\hat{\rho}$ in order to distinguish between $\hat{\rho}(t)$ in the Schrödinger representation and $\hat{\rho}_I(t)$ in the interaction representation. Any other operator $\hat{A}$ will appear as such in the Schrödinger representation, and as $\hat{A}(t)$ in the interaction representation.

This equation can be iterated (similar to the solution (2.76) of Eq. (2.73)) to give the interaction representation of the density operator at time t_0 starting from some time t_p in the past

$$\hat{\rho}_I(t_0) = \hat{\rho}_I(t_p) + \sum_{n=1}^{\infty}\left(-\frac{i}{\hbar}\right)^n \int_{t_p}^{t_0} dt_1 \int_{t_p}^{t_1} dt_2 \cdots \int_{t_p}^{t_{n-1}} dt_n [\hat{V}(t_1), [\hat{V}(t_2), \ldots,$$
$$\times [\hat{V}(t_n), \hat{\rho}_I(t_p)] \cdots]] \tag{18.79}$$

We will consider a process in which the external field is switched on at the distant past, when the system was at thermal equilibrium, $\hat{\rho}(t_p \to -\infty) = \hat{\rho}_{\text{eq}} = \exp(-\beta\hat{H}_{\text{M}})/\text{Tr}\exp(-\beta\hat{H}_{\text{M}})$. Referring to this initial time, Eq. (18.79) represents the deviation from thermal equilibrium caused by the external field, written as a sum of terms of increasing orders in this field

$$\hat{\rho}_I(t) = \sum_{n=0}^{\infty} \hat{\rho}_I^{(n)}(t) = \hat{\rho}_{\text{eq}} + \sum_{n=1}^{\infty} \hat{\rho}_I^{(n)}(t) \tag{18.80}$$

where

$$\hat{\rho}_I^{(n)}(t) = \left(-\frac{i}{\hbar}\right)^n \int_{-\infty}^{t} dt_n \int_{-\infty}^{t_n} dt_{n-1} \cdots$$
$$\times \int_{-\infty}^{t_2} dt_1 [\hat{V}(t_n), [\hat{V}(t_{n-1}), \ldots, [\hat{V}(t_1), \hat{\rho}_{\text{eq}}] \cdots]]; \quad n = 1, 2 \ldots \tag{18.81}$$

is the nth order response to the external field. In writing this expression we have renamed the time variables according to $t_0 \to t$ and $t_k \to t_{n-(k-1)}$; $(k = 1, \ldots, n)$. Using (18.74) this can be recast in the form

$$\hat{\rho}_I^{(n)}(t) = \left(\frac{i}{\hbar}\right)^n \int_{-\infty}^{t} dt_n \int_{-\infty}^{t_n} dt_{n-1} \cdots \int_{-\infty}^{t_2} dt_1 \mathcal{E}(t_n)\mathcal{E}(t_{n-1}) \cdots \mathcal{E}(t_1)$$
$$\times [\hat{\mu}(t_n), [\hat{\mu}(t_{n-1}), \ldots, [\mu(\hat{t}_1), \hat{\rho}_{\text{eq}}] \cdots]] \tag{18.82}$$

where $\hat{\mu}(t) \equiv e^{i\hat{H}_{\text{M}}t/\hbar}\hat{\mu}e^{-i\hat{H}_{\text{M}}t/\hbar}$. We have suppressed here the molecular index m, but will keep in mind that the dipole operator above is $\hat{\mu}_m$ and $\mathcal{E}(t) = \mathcal{E}(\mathbf{r}_m, t)$ is the electric field at the position of the molecule m. Now go back to the Schrödinger representation

$$\hat{\rho}^{(n)}(t) = \left(\frac{i}{\hbar}\right)^n \int_{t_p}^{t} dt_n \int_{t_p}^{t_n} dt_{n-1} \ldots \int_{t_p}^{t_2} dt_1 \mathcal{E}(t_n)\mathcal{E}(t_{n-1}) \cdots \mathcal{E}(t_1)$$
$$\times e^{-i\hat{H}_{\mathrm{M}}t/\hbar}[\hat{\mu}(t_n), [\hat{\mu}(t_{n-1}), \ldots, [\mu(\hat{t}_1), \hat{\rho}_{\mathrm{eq}}] \cdots]]e^{i\hat{H}_{\mathrm{M}}t/\hbar}$$
$$= \left(\frac{i}{\hbar}\right)^n \int_{t_p}^{t} dt_n \int_{t_p}^{t_n} dt_{n-1} \ldots \int_{t_p}^{t_2} dt_1 \mathcal{E}(t_n)\mathcal{E}(t_{n-1}), \ldots, \mathcal{E}(t_1)$$
$$\times [\hat{\mu}(t_n - t), [\hat{\mu}(t_{n-1} - t), \ldots, [\mu(\hat{t}_1 - t), \hat{\rho}_{\mathrm{eq}}] \cdots]] \tag{18.83}$$

and change variables $\tau_n = t - t_n$ and $\tau_k = t_{k+1} - t_k; k = 1, 2, \ldots, n-1$ to get

$$\hat{\rho}^{(n)}(t) = \left(\frac{i}{\hbar}\right)^n \int_0^{\infty} d\tau_n \int_0^{\infty} d\tau_{n-1} \cdots$$
$$\times \int_0^{\infty} d\tau_1 \mathcal{E}(t - \tau_n)\mathcal{E}(t - \tau_n - \tau_{n-1}) \ldots \mathcal{E}(t - \tau_n - \cdots - \tau_1)$$
$$\times [\hat{\mu}(-\tau_n), [\hat{\mu}(-\tau_n - \tau_{n-1}), \ldots, [\mu(-\tau_n - \tau_{n-1} \cdots - \tau_1), \hat{\rho}_{\mathrm{eq}}] \cdots]] \tag{18.84}$$

The molecular response pertaining to its optical properties is the dipole induced in the molecule, that can be calculated from

$$\langle \mu \rangle = \mathrm{Tr}(\hat{\mu}\hat{\rho}) = \sum_{n=1}^{\infty} \langle \mu \rangle^{(n)} \tag{18.85}$$

where

$$\langle \mu \rangle^{(n)} = \mathrm{Tr}(\hat{\mu}\hat{\rho}^{(n)}) \tag{18.86}$$

In writing Eqs (18.85)–(18.86) we have assumed that the molecule has no permanent dipole moment, that is $\mathrm{Tr}(\hat{\mu}\hat{\rho}_{\mathrm{eq}}) = 0$. From (18.84) we then find

$$\langle \mu \rangle^{(n)}(t) = \int_{-\infty}^{\infty} d\tau_n \int_{-\infty}^{\infty} d\tau_{n-1} \cdots \int_{-\infty}^{\infty} d\tau_1 \mathcal{E}(t - \tau_n)\mathcal{E}(t - \tau_n - \tau_{n-1}) \cdots$$
$$\times \mathcal{E}(t - \tau_n - \cdots - \tau_1)\alpha^{(n)}(\tau_1, \ldots, \tau_n) \tag{18.87}$$

where the nth order single molecule response functions are

$$\begin{aligned}
\alpha^{(n)}(\tau_1,\ldots,\tau_n) &= \left(\frac{i}{\hbar}\right)^n \theta(\tau_1)\theta(\tau_2)\cdots\theta(\tau_n) \\
&\quad \times \mathrm{Tr}\{\hat{\mu}(0)[\hat{\mu}(-\tau_n),[\hat{\mu}(-\tau_n-\tau_{n-1}),\ldots, \\
&\quad \times [\hat{\mu}(-\tau_n-\tau_{n-1}\cdots-\tau_1),\hat{\rho}_{\mathrm{eq}}]\cdots]]\} \\
&= \left(\frac{i}{\hbar}\right)^n \theta(\tau_1)\theta(\tau_2)\cdots\theta(\tau_n) \\
&\quad \times \mathrm{Tr}\{\hat{\mu}(\tau_n+\tau_{n-1}+\ldots+\tau_1)[\hat{\mu}(\tau_{n-1}+\cdots+\tau_1), \\
&\quad \times [\hat{\mu}(\tau_{n-2}+\cdots+\tau_1),\ldots,[\hat{\mu}(\tau_1)[\mu(0),\hat{\rho}_{\mathrm{eq}}]]\cdots]]\} \\
&= \left(\frac{i}{\hbar}\right)^n \theta(\tau_1)\theta(\tau_2)\cdots\theta(\tau_n) \\
&\quad \times \mathrm{Tr}\{[[\cdots[[\hat{\mu}(\tau_n+\tau_{n-1}+\ldots+\tau_1),\hat{\mu}(\tau_{n-1}\cdots+\tau_1)], \\
&\quad \times \hat{\mu}(\tau_{n-2}+\cdots+\tau_1)],\ldots,\hat{\mu}(\tau_1)],\mu(0)]\hat{\rho}_{\mathrm{eq}}\}
\end{aligned} \tag{18.88}$$

Problem 18.4. Show that $\langle \hat{A}(t_{\mathrm{A}})\hat{B}(t_B)\hat{C}(t_C)\cdots\rangle = \mathrm{Tr}\{\hat{A}(t_{\mathrm{A}})\hat{B}(t_B)\hat{C}(t_C)\cdots\hat{\rho}_{\mathrm{eq}}\}$ is invariant to a uniform time shift, that is, equal to $\mathrm{Tr}\{\hat{A}(t_{\mathrm{A}}+t)\hat{B}(t_B+t)\hat{C}(t_C+t)\cdots\hat{\rho}_{\mathrm{eq}}\}$. The second equality in (18.88) is based on this identity.

Problem 18.5. Show that $\mathrm{Tr}(\hat{A}[\hat{B},[\hat{C},\hat{D}]]) = Tr([[\hat{A},\hat{B}],\hat{C}]D)$. The third equality in (18.88) is based on a generalization of this identity.

18.7.3 Many body response theory

How should Eqs (18.87) and (18.88) be modified when the system contains many molecules $\{m\}$ at various position $\mathbf{r}_m$? In this case $\hat{V}(t) = -\int d\mathbf{r}\hat{P}(\mathbf{r})\cdot\mathcal{E}(\mathbf{r},t) = -\sum_m \hat{\mu}_m\cdot\mathcal{E}(\mathbf{r}_m,t)$ replaces $\hat{V} = -\hat{\mu}_m\cdot\mathcal{E}(\mathbf{r}_m,t)$ in the derivation above, but the procedure can proceed in the same way. Using the integral expression for $\hat{V}(t)$,

Eq. (18.82) takes the form

$$\hat{\rho}_I^{(n)}(\mathbf{r},t) = \left(\frac{i}{\hbar}\right)^n \int d\mathbf{r}_n \cdots \int d\mathbf{r}_1 \int_{t_p}^{t} dt_n \int_{t_p}^{t_n} dt_{n-1} \cdots$$
$$\times \int_{t_p}^{t_2} dt_1 \mathcal{E}(\mathbf{r}_n, t_n)\mathcal{E}(\mathbf{r}_{n-1}, t_{n-1}) \cdots \mathcal{E}(\mathbf{r}_1, t_1)$$
$$\times [\hat{P}(\mathbf{r}_n, t_n), [\hat{P}(\mathbf{r}_{n-1}, t_{n-1}), \ldots, [\hat{P}(\mathbf{r}_1, t_1), \hat{\rho}_{\text{eq}}] \cdots]] \tag{18.89}$$

Following the steps that lead to (18.85) and (18.87) now yields

$$\langle P(\mathbf{r})\rangle = \sum_{n=1}^{\infty} \langle P(\mathbf{r})\rangle^{(n)} \tag{18.90}$$

$$\langle P(\mathbf{r})\rangle^{(n)}(t) = \int d\mathbf{r}_n \int d\mathbf{r}_{n-1} \cdots \int d\mathbf{r}_1 \int_{-\infty}^{\infty} d\tau_n \int_{-\infty}^{\infty} d\tau_{n-1} \cdots \int_{-\infty}^{\infty} d\tau_1$$
$$\times \mathcal{E}(\mathbf{r}_n, t-\tau_n)\mathcal{E}(\mathbf{r}_{n-1}, t-\tau_n-\tau_{n-1}) \ldots \mathcal{E}(\mathbf{r}_1, t-\tau_n-\cdots-\tau_1)$$
$$\chi^{(n)}(\mathbf{r}; \mathbf{r}_1, \ldots, \mathbf{r}_n, \tau_1, \ldots, \tau_n) \tag{18.91}$$

with the many-body response function

$$\chi^{(n)}(\mathbf{r}; \mathbf{r}_1, \ldots, \mathbf{r}_n, \tau_1, \ldots, \tau_n)$$
$$= \left(\frac{i}{\hbar}\right)^n \theta(\tau_1)\theta(\tau_2) \cdots \theta(\tau_n)$$
$$\times \text{Tr}\{[[\cdots [[\hat{P}(\mathbf{r}, \tau_n + \tau_{n-1} \cdots + \tau_1), \hat{P}(\mathbf{r}_n, \tau_{n-1} \cdots + \tau_1)]$$
$$\times \hat{P}(\mathbf{r}_{n-1}, \tau_{n-2} + \cdots + \tau_1)], \ldots, \hat{P}(\mathbf{r}_2, \tau_1)], \hat{P}(\mathbf{r}_1, 0)]\hat{\rho}_{\text{eq}}\} \tag{18.92}$$

Note that the many-body response functions are in general non-local, implying that the response (i.e. the polarization) at point $\mathbf{r}$ depends on the electric field at other locations. This makes sense: In a system of interacting molecules the response of molecules at location $\mathbf{r}$ arises not only from the field at that location, but also from molecules located elsewhere that were polarized by the field, then affected other molecules by their mutual interactions. Also note that by not stressing the vector forms of $\boldsymbol{\mathcal{E}}$ and $\hat{\mathbf{P}}$ we have sacrificed notational rigor for relative notational simplicity. In reality the response function is a tensor whose components are derived from the components of the polarization vector, and the tensor product $\boldsymbol{\mathcal{E}}\boldsymbol{\mathcal{E}} \ldots \boldsymbol{\mathcal{E}}\chi$ is the corresponding sum over vector components of $\boldsymbol{\mathcal{E}}$ and tensor components of $\boldsymbol{\chi}$.

Equations (18.90)–(18.92) (and their Liouville space equivalents) can be used as the starting point for a general treatment of nonlinear spectroscopy phenomena. On the linear response level this will yield absorption, dielectric response, and propagation of electromagnetic waves. In second order we can use this approach to describe, for example, second harmonic generation (more generally sum-frequency generation) and in third order this will yield a variety of four wave mixing phenomena. A detailed discussion of these phenomena together with practical methods for calculating the corresponding optical response functions is given in the book by Mukamel cited at the end of this chapter. Here we will only consider the relationship of this approach to our earlier discussion of the linear optical properties of noninteracting molecular species.

18.7.4 Independent particles

First, consider a homogeneous system in which the molecular dipoles do not interact with each other, either directly or through their interaction with the thermal environment. This approximation becomes better for lower molecular density. In this case $\hat{H}_\text{M}$ is a sum over individual terms, $\hat{H}_\text{M} = \sum_m \hat{h}_m$, each associated with a different molecular dipole and its thermal environment. This implies that in $\hat{P}(\mathbf{r}, t) = \sum_m \hat{\mu}_m(t)\delta(\mathbf{r}-\mathbf{r}_m)$, the different operators $\hat{\mu}_m(t) = \exp(i\hat{h}_m t/\hbar)\hat{\mu}_m \exp(-i\hat{h}_m t/\hbar)$ commute with each other. In this case

$$
\begin{aligned}
&[\cdots[[\hat{P}(\mathbf{r}, \tau_n + \tau_{n-1} \cdots + \tau_1), \hat{P}(\mathbf{r}_n, \tau_{n-1} \cdots + \tau_1)],\\
&\quad \hat{P}(\mathbf{r}_{n-1}, \tau_{n-2} + \cdots + \tau_1)], \ldots, \hat{P}(\mathbf{r}_2, \tau_1)], \hat{P}(\mathbf{r}_1, 0)]\\
&\quad = \sum_m \delta(\mathbf{r} - \mathbf{r}_m)\delta(\mathbf{r}_n - \mathbf{r}_m)\delta(\mathbf{r}_{n-1} - \mathbf{r}_m)\cdots\delta(\mathbf{r}_2 - \mathbf{r}_m)\delta(\mathbf{r}_1 - \mathbf{r}_m)\\
&\qquad \times [[\cdots[[\hat{\mu}_m(\tau_n + \tau_{n-1} \cdots + \tau_1), \hat{\mu}_m(\tau_{n-1} \cdots + \tau_1)], \hat{\mu}_m(\tau_{n-2} + \cdots + \tau_1)]\\
&\qquad , \cdots, \hat{\mu}_m(\tau_1)], \hat{\mu}_m(0)]
\end{aligned}
\tag{18.93}
$$

Next we use (18.93) and (18.88) together with

$$
\begin{aligned}
&\sum_m \delta(\mathbf{r} - \mathbf{r}_m)\delta(\mathbf{r}_n - \mathbf{r}_m)\delta(\mathbf{r}_{n-1} - \mathbf{r}_m)\cdots\delta(\mathbf{r}_2 - \mathbf{r}_m)\delta(\mathbf{r}_1 - \mathbf{r}_m)\\
&\quad = \delta(\mathbf{r}_n - \mathbf{r})\delta(\mathbf{r}_{n-1} - \mathbf{r})\cdots\delta(\mathbf{r}_2 - \mathbf{r})\delta(\mathbf{r}_1 - \mathbf{r})\sum_m \delta(\mathbf{r} - \mathbf{r}_m)\\
&\quad = \delta(\mathbf{r}_n - \mathbf{r})\cdots\delta(\mathbf{r}_1 - \mathbf{r})\rho(\mathbf{r})
\end{aligned}
\tag{18.94}
$$

to find

$$
\chi^{(n)}(\mathbf{r}; \mathbf{r}_1, \ldots, \mathbf{r}_n, \tau_1, \ldots, \tau_n) = \delta(\mathbf{r}_n - \mathbf{r})\cdots\delta(\mathbf{r}_1 - \mathbf{r})\chi^{(n)}(\mathbf{r}; \tau_1, \ldots, \tau_n)
\tag{18.95}
$$

with

$$\chi^{(n)}(\mathbf{r};\tau_1,\ldots,\tau_n) = \rho(\mathbf{r})\alpha^{(n)}(\tau_1,\ldots,\tau_n) \tag{18.96}$$

where $\alpha^{(n)}(\tau_1,\ldots,\tau_n)$ is the single molecule response function of Eq. (18.88). Here and in (18.94) $\rho(\mathbf{r}) = \sum_m \delta(\mathbf{r}-\mathbf{r}_m)$ is the density of diploes at position $\mathbf{r}$. Furthermore, Eqs (18.91), (18.95), and (18.96) imply that

$$\begin{aligned}\langle P(\mathbf{r})\rangle^{(n)}(t) = \rho(\mathbf{r}) \int_{-\infty}^{\infty} d\tau_n \int_{-\infty}^{\infty} d\tau_{n-1}\cdots \int_{-\infty}^{\infty} d\tau_1 \\ \times\, \mathcal{E}(\mathbf{r},t-\tau_n)\mathcal{E}(\mathbf{r},t-\tau_n-\tau_{n-1})\cdots\mathcal{E}(\mathbf{r},t-\tau_n-\cdots-\tau_1) \\ \times\, \chi^{(n)}(\tau_1,\ldots,\tau_n) = \rho(\mathbf{r})\langle\mu\rangle^{(n)}(t)\end{aligned} \tag{18.97}$$

We have thus found that in a system of noninteracting particles the response is local and of an intuitively obvious form: The polarization at position $\mathbf{r}$ (in any order of the calculation) is the induced dipole on a single molecule at that position, multiplied by the local density.

18.7.5 Linear response

Next consider the lowest order response function. Using the last form of (18.88) we find that

$$\begin{aligned}\alpha^{(1)}(t) &= \left(\frac{i}{\hbar}\right)\theta(t)\mathrm{Tr}\{[\hat{\mu}(t),\hat{\mu}(0)]\hat{\rho}_{\mathrm{eq}}\} \\ &= \left(\frac{i}{\hbar}\right)\theta(t)(\langle\hat{\mu}(t)\hat{\mu}(0)\rangle - \langle\hat{\mu}(0)\hat{\mu}(t)\rangle) = \left(\frac{i}{\hbar}\right)\theta(t)(J(t)-J^*(t))\end{aligned} \tag{18.98}$$

where

$$J(t) = \langle\hat{\mu}(t)\hat{\mu}(0)\rangle = \langle\hat{\mu}(0)\hat{\mu}(t)\rangle^* \tag{18.99}$$

Problem 18.6. Show that in the basis of eigenstates of $\hat{H}_{\mathrm{M}}$ Eq. (18.98) takes the form

$$\alpha^{(1)}(t) = 2\frac{\theta(t)}{\hbar}\sum_a P_a \sum_b |\mu_{ab}|^2 \sin(\omega_{ba}t) \tag{18.100}$$

where $P_a = e^{-\beta E_a}/\sum_{a'} e^{-\beta E_{a'}}$, $H_{\mathrm{M}}|l\rangle = E_l|l\rangle (l = a, b, \ldots)$, $\omega_{ba} = (E_b - E_a)/\hbar$ and $\mu_{ab} = \langle a|\hat{\mu}|b\rangle$

In this order Eq. (18.97) yields

$$\langle P(\mathbf{r})\rangle^{(1)}(t) = \int_{-\infty}^{\infty} d\tau_1 \chi^{(1)}(\tau_1)\mathcal{E}(\mathbf{r}, t-\tau_1); \qquad \chi^{(1)}(t) = \rho\alpha^{(1)}(t) \quad (18.101)$$

and, by taking Fourier transform of all time-dependent functions, for example, $\mathcal{E}(\omega) = \int_{-\infty}^{\infty} dt e^{i\omega t}\mathcal{E}(t)$,

$$\langle P(\mathbf{r})\rangle^{(1)}(\omega) = \chi^{(1)}(\omega)\mathcal{E}(\mathbf{r}, \omega) \quad (18.102)$$

Using Eqs (18.100) together with the identity[24]

$$\int_{-\infty}^{\infty} dt e^{ixt}(\theta(t)e^{ix't}) = \lim_{\eta\to 0+} \frac{i}{x + x' + i\eta} \quad (18.103)$$

leads to

$$\alpha^{(1)}(\omega) = \frac{1}{\hbar}\sum_a P_a \sum_b |\mu_{ab}|^2 \left(\frac{1}{\omega + \omega_{ba} + i\eta} - \frac{1}{\omega - \omega_{ba} + i\eta} \right) \quad (18.104)$$

In continuum electrostatic theory the response function $\chi^{(1)}$ is known as the susceptibility. A more common linear response function, the dielectric response, is defined by the linear expression

$$\mathcal{D}(t) = \int_{-\infty}^{t} dt' \varepsilon(t - t')\mathcal{E}(t') \quad (18.105)$$

or

$$\mathcal{D}(\omega) = \varepsilon(\omega)\mathcal{E}(\omega) \quad (18.106)$$

with

$$\varepsilon(\omega) = \int_{0}^{\infty} dt e^{i\omega t}\varepsilon(t), \quad (18.107)$$

[24] Note that the existence of the θ function is important in the identity (18.103). The inverse Fourier transform is

$$\lim_{\eta\to 0+} \frac{1}{2\pi}\int_{-\infty}^{\infty} dx e^{-ixt} \frac{i}{x + x' + i\eta} = \theta(t)e^{ix't}$$

as is easily shown by contour integration.

relating the electric and displacement fields. Together with the relationship

$$\mathcal{E} = \mathcal{D} - 4\pi P \tag{18.108}$$

this leads to

$$\varepsilon(\omega) = 1 + 4\pi\chi^{(1)}(\omega) \tag{18.109}$$

Thus, Eqs (18.98)–(18.109) provide a microscopic expression for the dielectric response function in a system of noninteracting particles

$$\varepsilon(\omega) = 1 + 4\pi\rho\frac{i}{\hbar}\int_0^\infty dt e^{i\omega t}(J(t) - J^*(t)) \tag{18.110}$$

Problem 18.7. Repeat the derivation of these linear response equations taking the vector nature of $\boldsymbol{\mathcal{E}}$ and $\hat{\boldsymbol{\mu}}$ into account, so that Eq. (18.74) is $\hat{V} = -\hat{\boldsymbol{\mu}}_m \cdot \boldsymbol{\mathcal{E}}(\mathbf{r}_m) = -\sum_{j=x,y,z}\mu_{mj}\mathcal{E}_j(\mathbf{r}_m)$ and show that Eq. (18.102) becomes $\langle\mathbf{P}(\mathbf{r})\rangle^{(1)}(\omega) = \boldsymbol{\chi}^{(1)}(\omega)\cdot\boldsymbol{\mathcal{E}}(\mathbf{r},\omega)$ where $\boldsymbol{\chi}$ is the matrix $\boldsymbol{\chi} = \rho\boldsymbol{\alpha}$ with the matrix $\boldsymbol{\alpha}$ defined as in (18.98) except that the matrix $\mathbf{J}$ replaces J, with $J_{j,j'}(t) = \langle\hat{\mu}_j(t)\hat{\mu}_{j'}(0)\rangle$; $(j,j' = x,y,z)$, and $\mathbf{J}^*$ replaced by $\mathbf{J}^\dagger$.

18.7.6 Linear response theory of propagation and absorption

Having obtained expressions for the dielectric susceptibility and the dielectric response functions in terms of microscopic variables, we may proceed to express other observables in microscopic terms. Consider an electromagnetic mode whose electric component is described by a plane wave propagating in the x direction in an isotropic medium, and assume that the field is weak enough to make linear response theory valid. The field is given by

$$\mathcal{E}(x,t) = \mathcal{E}_0 e^{ikx - i\omega t} \tag{18.111}$$

and should be a solution of the Maxwell equation. Start with Eq. (3.75)

$$\nabla\times\nabla\times\mathcal{E}(\mathbf{r},t) + \frac{1}{c^2}\frac{\partial^2\mathcal{E}(\mathbf{r},t)}{\partial t^2} = -\frac{4\pi}{c^2}\frac{\partial^2\langle\mathbf{P}(\mathbf{r},t)\rangle}{\partial t^2} \tag{18.112}$$

and recall that the electric field above is transverse, $\mathcal{E}_r = \mathcal{E}^\perp$, so that (from (1.29) and (1.34a)) $\nabla\times\nabla\times\mathcal{E}^\perp(\mathbf{r},t) = -\nabla^2\mathcal{E}^\perp(\mathbf{r},t)$. Using (18.108) and (18.105), and keeping in mind that χ and ε are scalars in an isotropic medium, then leads to

$$-\frac{\partial^2}{\partial x^2}\mathcal{E}(x,t) + \frac{1}{c^2}\frac{\partial^2}{\partial t^2}\left(\int_{-\infty}^t \varepsilon(t-t')\mathcal{E}(x,t')\right) = 0 \tag{18.113}$$

which indeed yields (18.111) as a solution, provided that the following *dispersion relationship* (dependence of ω on k)

$$k^2 - \frac{\omega^2}{c^2}\varepsilon(\omega) = 0, \text{ or } \frac{kc}{\omega} = \sqrt{\varepsilon(\omega)} \tag{18.114}$$

is satisfied.

Next, the complex character of ε is taken into account by writing (18.114) in the form

$$\frac{kc}{\omega} = n(\omega) + i\kappa(\omega) \tag{18.115}$$

where n is called *index of refraction* and κ is the *extinction coefficient*. Rewriting (18.115) as $k = \omega n(\omega)/c + i\omega\kappa(\omega)/c$ we find that the position dependence in (18.111) translates into the position dependence of the intensity $I = |\mathcal{E}|^2$ in the form

$$I(x) = I_0 e^{-a(\omega)\rho x} \tag{18.116}$$

where the absorption coefficient is

$$a(\omega) \equiv \frac{2\omega\kappa(\omega)}{c\rho} \tag{18.117}$$

Here ρ is the number density of the absorbing species. We have selected this form with the expectation that $a(\omega)$ as defined by (18.117) does not depend on ρ, so that absorption, as defined by (18.116), depends linearly on the density of absorbers.

Next we find a relationship between $a(\omega)$ and the susceptibility. To this end we write the susceptibility χ in terms of its real and imaginary parts, $\chi^{(1)} = \chi' + i\chi''$ and use (18.109) to rewrite Eq. (18.115) in the form

$$\sqrt{\varepsilon} = \sqrt{1 + 4\pi(\chi' + i\chi'')} = n + i\kappa \tag{18.118}$$

or $1 + 4\pi\chi' + 4\pi i\chi'' = n^2 - \kappa^2 + 2in\kappa$ whence

$$\kappa = \frac{2\pi\chi''}{n} \quad \text{or} \quad a(\omega) \equiv \frac{4\pi\omega}{n(\omega)c\rho}\chi''(\omega) \tag{18.119}$$

namely, the absorption coefficient is proportional to the imaginary part of the susceptibility. Also, in the common case where $n^2 \gg \kappa^2$, we find $n(\omega) = \sqrt{1 + 4\pi\chi'(\omega)}$. Note that from Eq. (18.101) and its frequency space equivalent,

$\chi^{(1)}(\omega) = \rho\alpha^{(1)}(\omega)$, and defining $\alpha^{(1)} = \alpha' + i\alpha''$, we can rewrite (18.119) in the form

$$a(\omega) \equiv \frac{4\pi\omega}{n(\omega)c}\alpha''(\omega) \tag{18.120}$$

An explicit expression for the absorption coefficient can be obtained using the imaginary part of Eq. (18.104)

$$\begin{aligned}\alpha''(\omega) = \mathrm{Im}(\alpha^{(1)}(\omega)) &= \frac{\pi}{\hbar}\sum_a P_a \sum_b |\mu_{ab}|^2(\delta(\omega - \omega_{ba}) - \delta(\omega + \omega_{ba}))\\ &= \frac{\pi}{\hbar}\sum_a\sum_b (P_a - P_b)|\mu_{ab}|^2\delta(\omega - \omega_{ba})\end{aligned} \tag{18.121}$$

We have obtained what is essentially the golden rule, except that both stimulated absorption and emission processes contribute to give the *net* absorption rate.

Appendix 18A. Steady-state solution of Eqs (18.58): the Raman scattering flux

Here we outline the procedure by which the steady-state solution of Eqs. (18.58) is obtained, en route to evaluate the steady-state scattering flux. The calculation is facilitated by making an approximation—disregarding all the terms with light-grey background in these equations. The rationale for this approximation is that we want to evaluate the scattered flux in the lowest order of interaction between the molecule and the driving field. Obviously, the lowest order in which the scattering amplitude can be obtained is 2, therefore the observed flux is of order 4 in this interaction. This can be pictorially seen in Fig. 18.16 which outlines the propagation from the *in* state ($\sigma_{\mathrm{in,in}}$) to the *out* state ($\sigma_{\mathrm{out,out}}$). In this diagram each junction represents a particular matrix element of $\hat{\sigma}$ and each line joining these junctions represents an operation by the coupling $\hat{V}$. En route from $\sigma_{\mathrm{in,in}}$ to $\sigma_{\mathrm{out,out}}$ we have to go through other matrix elements of $\hat{\sigma}$: Each operation by the coupling $\hat{V}$ can change only one of the two indices in $\sigma_{i,j}$. Lowest order transitions are obtained by following routes in which all transitions take place in the direction of the arrows. Going against the arrow implies the need to go back, increasing the order of the calculated amplitude. It is easily seen that the terms with light-grey background in Eqs (18.58) express such backward transitions, for example the corresponding term in (18.58c) affects the transition $\sigma_{\mathrm{out},p} \to \sigma_{p,p}$. (To see that the terms involving σ_{sp} and σ_{ps} are of this type write down their equations of motion). This is the rationale for disregarding them in what follows.

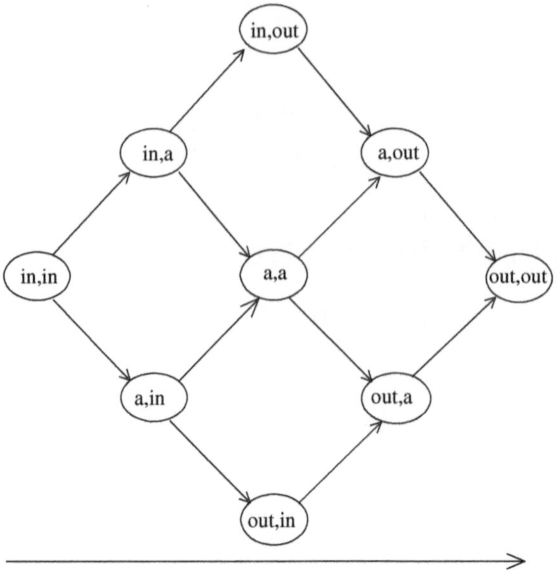

FIG. 18.16 A Liouville space pathway diagram describing the $\sigma_{\text{in,in}} \rightarrow \sigma_{\text{out,out}}$ transition. The state a represents either p or s (one can think of two diagrams like this, one for p the other for s, which are connected only at the $(\sigma_{p,p} \leftrightarrow \sigma_{s,s})$ junction.

Under this approximation, imposing a constant $\sigma_{\text{in,in}}$ (thus disregarding Eq. (18.58a) and putting all time derivatives on the left to zero, we obtain the following set of steady-state equations in Liouville space

$$0 = 2\text{Im}(V_{s,\text{in}}\sigma_{\text{in},s}) - k_{ps}\sigma_{s,s} + k_{sp}\sigma_{p,p} - \Gamma_s\sigma_{s,s} \tag{18.122a}$$

$$0 = k_{ps}\sigma_{s,s} - k_{sp}\sigma_{p,p} - \Gamma_p\sigma_{p,p} \tag{18.122b}$$

$$0 = -2\text{Im}(V_{p,\text{out}}\sigma_{\text{out},p}) - 2\text{Im}(V_{s,\text{out}}\sigma_{\text{out},s}) - \eta\sigma_{\text{out,out}} \tag{18.122c}$$

$$0 = -iE_{\text{in,out}}\sigma_{\text{in,out}} + iV_{s,\text{out}}\sigma_{\text{in},s} - (1/2)\eta\sigma_{\text{in,out}} \tag{18.122d}$$

$$0 = -iE_{\text{in},s}\sigma_{\text{in},s} + iV_{\text{in},s}\sigma_{\text{in,in}} - (1/2)\gamma_s\sigma_{\text{in},s} \tag{18.122e}$$

$$0 = -iE_{\text{out},s}\sigma_{\text{out},s} + iV_{\text{in},s}\sigma_{\text{out,in}} - iV_{\text{out},s}\sigma_{s,s} - (1/2)\gamma_s\sigma_{\text{out},s} \tag{18.122f}$$

$$0 = -iE_{\text{out},p}\sigma_{\text{out},p} - iV_{\text{out},p}\sigma_{p,p} - (1/2)\gamma_p\sigma_{\text{out},p} \tag{18.122g}$$

Equation (18.122c) give $d\sigma_{\text{out,out}}/dt = 0$ as a balance of two fluxes,

$$-\frac{\eta}{\hbar}\sigma_{\text{out,out}} = \frac{1}{\hbar}(2\text{Im}(V_{p,\text{out}}\sigma_{\text{out},p}) + 2\text{Im}(V_{s,\text{out}}\sigma_{\text{out},s})) \tag{18.123}$$

each of which can be used as the steady-state scattering flux. The required flux is then

$$F_{\text{out}} = F^{p}_{\text{out}} + F^{s}_{\text{out}} = -\frac{1}{\hbar}(2\text{Im}(V_{p,\text{out}}\sigma_{\text{out},p}) + 2\text{Im}(V_{s,\text{out}}\sigma_{\text{out},s})) \qquad (18.124)$$

This is a sum of terms that can be interpreted as fluxes through the intermediate states p and s.

Consider first the flux through the intermediate state p

$$F^{p}_{\text{out}} = \frac{2}{\hbar}\text{Im}(V_{p,\text{out}}\sigma_{\text{out},p}) \qquad (18.125)$$

$\sigma_{\text{out},p}$ is obtained from (18.122g) in terms of $\sigma_{p,p}$

$$\sigma_{\text{out},p} = \frac{V_{\text{out},p}\sigma_{p,p}}{E_{p,\text{out}} + (1/2)i\gamma_p} \qquad (18.126)$$

while $\sigma_{p,p}$ and $\sigma_{s,s}$ are obtained from solving (18.122a,b) in terms of $X \equiv 2\text{Im}(V_{s,\text{in}}\sigma_{\text{in},s})$

$$\sigma_{p,p} = \frac{k_{ps}}{k_{sp} + \Gamma_p}\frac{X}{\tilde{\Gamma}_s}; \qquad \sigma_{ss} = \frac{X}{\tilde{\Gamma}_s} \qquad (18.127)$$

where

$$\tilde{\Gamma}_s = \Gamma_s + k_{ps} - \frac{k_{sp}k_{ps}}{k_{sp} + \Gamma_p} \qquad (18.128)$$

X can be obtained from (18.122e) in terms of $\sigma_{\text{in,in}}$

$$X = \frac{\gamma_s|V_{s,\text{in}}|^2}{E^2_{s,\text{in}} + ((1/2)\gamma_s)^2}\sigma_{\text{in,in}} \qquad (18.129)$$

Combining Eqs (18.125)–(18.129) finally yields

$$\frac{F^{p}_{\text{out}}}{\sigma_{\text{in,in}}} = \frac{1}{\hbar}\frac{\gamma_p|V_{p,\text{out}}|^2}{E^2_{p,\text{out}} + ((1/2)\gamma_p)^2}\frac{\gamma_s|V_{s,\text{in}}|^2}{E^2_{s,\text{in}} + ((1/2)\gamma_s)^2}\frac{k_{ps}}{(k_{sp} + \Gamma_p)\tilde{\Gamma}_s} \qquad (18.130)$$

Next, consider the contribution F^{s}_{out} to the total scattered flux

$$F^{s}_{\text{out}} = \frac{2}{\hbar}\text{Im}(V_{s,\text{out}}\sigma_{\text{out},s}) \qquad (18.131)$$

$\sigma_{\text{out},s}$ is obtained from (18.122f) in terms of $\sigma_{\text{out,in}}$ and $\sigma_{s,s}$

$$\sigma_{\text{out},s} = \frac{V_{\text{in},s}\sigma_{\text{out,in}} - V_{\text{out},s}\sigma_{s,s}}{E_{\text{out},s} - (1/2)i\gamma_s} \tag{18.132}$$

$\sigma_{s,s}$ is already known in terms of $\sigma_{\text{in,in}}$ via Eqs (18.127) and (18.129), so let us focus attention on the term that contains $\sigma_{\text{out,in}}$. The latter element of $\hat{\sigma}$ is obtained from the complex conjugate of (18.122d) in terms of $\sigma_{s,\text{in}}$

$$\sigma_{\text{out,in}} = \frac{V_{\text{out},s}\sigma_{s,\text{in}}}{E_{\text{in,out}} + (1/2)i\eta} \tag{18.133}$$

and $\sigma_{s,\text{in}}$ is obtained from the complex conjugate of (18.122e)

$$\sigma_{s,\text{in}} = \frac{V_{s,\text{in}}\sigma_{\text{in,in}}}{E_{\text{in},s} + (1/2)i\gamma_s} \tag{18.134}$$

Using Eqs (18.127), (18.129), and (18.132)–(18.134) in (18.131) leads to

$$\begin{aligned}\frac{F^s_{\text{out}}}{\sigma_{\text{in,in}}} = {} & -\frac{2}{\hbar}|V_{\text{in},s}|^2|V_{\text{out},s}|^2\text{Im} \\ & \times \left(\frac{1}{(E_{\text{in},s} + (1/2)i\gamma_s)(E_{\text{out},s} - (1/2)i\gamma_s)(E_{\text{in,out}} + (1/2)i\eta)}\right) \\ & + \frac{1}{\hbar}\frac{\gamma_s|V_{s,\text{out}}|^2}{E^2_{s,\text{out}} + (\gamma_s/2)^2}\frac{|V_{s,\text{in}}|}{E^2_{s,\text{in}} + (\gamma_s/2)^2}\frac{\gamma_s}{\tilde{\Gamma}_s}\end{aligned} \tag{18.135}$$

Consider now the term

$$\begin{aligned}&\text{Im}\left(\frac{1}{(E_{\text{in},s} + (1/2)i\gamma_s)(E_{\text{out},s} - (1/2)i\gamma_s)(E_{\text{in,out}} + (1/2)i\eta)}\right) \\ &= \frac{\text{Im}[(E_{\text{in},s} - (1/2)i\gamma_s)(E_{\text{out},s} + (1/2)i\gamma_s)(E_{\text{in,out}} - (1/2)i\eta)]}{(E^2_{\text{in},s} + (\gamma_s/2)^2)(E^2_{\text{out},s} + (\gamma_s/2)^2)(E^2_{\text{in,out}} + (\eta/2)^2)}\end{aligned} \tag{18.136}$$

The term linear in η yields, in the limit $\eta \to 0$, $-\pi\delta(E_{\text{in}} - E_{\text{out}})(E^2_{\text{in},s} + (\gamma_s/2)^2)^{-1}$. The other term is easily simplified to

$$\begin{aligned}&\lim_{\eta\to 0}\frac{\text{Im}[(E_{\text{in},s} - (1/2)i\gamma_s)(E_{\text{out},s} + (1/2)i\gamma_s)E_{\text{in,out}}]}{(E^2_{\text{in},s} + (\gamma_s/2)^2)(E^2_{\text{out},s} + (\gamma_s/2)^2)(E^2_{\text{in,out}} + (\eta/2)^2)} \\ &= \frac{(1/2)\gamma_s}{(E^2_{\text{in},s} + (\gamma_s/2)^2)(E^2_{\text{out},s} + (\gamma_s/2)^2)}\end{aligned}$$

Using these in (18.135) we get after some algebra

$$\frac{F^s_{\text{out}}}{\sigma_{\text{in,in}}} = \frac{2\pi}{\hbar}\delta(E_{\text{in}} - E_{\text{out}})\frac{|V_{\text{in},s}|^2|V_{\text{out},s}|^2}{(E^2_{\text{in},s} + ((1/2)\gamma_s)^2)}$$
$$+ \frac{1}{\hbar}\frac{\tilde{\kappa}_s}{\tilde{\Gamma}_s}\frac{|V_{\text{in},s}|^2|V_{\text{out},s}|^2\gamma_s}{(E^2_{\text{in},s} + ((1/2)\gamma_s)^2)(E^2_{\text{out},s} + ((1/2)\gamma_s)^2)} \quad (18.137)$$

where

$$\tilde{\kappa}_s = \kappa_s + \frac{\kappa_{sp}\kappa_{ps}}{\kappa_{sp} + \Gamma_p} \quad (18.138)$$

and where Eq. (18.59) has been used. Combining (18.130) and (18.137) finally gives Eq. (18.60).

Further reading

L. Allen and J. H. Eberly, *Optical Resonance and Two-Level Atoms*. (Wiley, New York, 1975)

C. Cohen-Tannoudji, J. Dupont-Roc, and G. Grynberg, *Atom-Photon Interactions: Basic Processes and Applications* (Wiley, New York, 1998).

W. H. Louisell, *Quantum Statistical Properties of Radiation* (Wiley, New York, 1973).

S. Mukamel, *Nonlinear Optical Spectroscopy* (Oxford University Press, Oxford, 1995).

M. Orszag, *A quantum statistical model of interacting two level systems and radiation* (Worcester Polytechnic Institute, 1973).

INDEX

Page numbers in italic, e.g. *319*, refer to figures. Page numbers in bold, e.g. **579**, signify entries in tables.